MATHEMATICAL PAPERS.

London: C. J. CLAY & SONS,
CAMBRIDGE UNIVERSITY PRESS WAREHOUSE,
AVE MARIA LANE.

Cambridge: DEIGHTON, BELL AND CO.
Leipzig: F. A. BROCKHAUS.

THE COLLECTED

MATHEMATICAL PAPERS

OF

ARTHUR CAYLEY, Sc.D., F.R.S.,

SADLERIAN PROFESSOR OF PURE MATHEMATICS IN THE UNIVERSITY OF CAMBRIDGE.

VOL. II.

CAMBRIDGE:
AT THE UNIVERSITY PRESS.
1889

CAMBRIDGE:
PRINTED BY C. J. CLAY, M.A. AND SONS,
AT THE UNIVERSITY PRESS.

ADVERTISEMENT.

THE present volume contains fifty-eight papers (numbered 101, 102,..., 158) originally published, all but two of them, in the years 1851 to 1860: they are here reproduced nearly but not exactly in chronological order. The two excepted papers are 142, Numerical Tables Supplementary to Second Memoir on Quantics, now first published (1889); and, 143, Tables of the Covariants M to W of the Binary Quintic: from the Second, Third, Fourth, Fifth, Eighth, Ninth and Tenth Memoirs on Quantics (arranged in the present form, 1889): the determination of the finite number, 23, of the covariants of the quintic was made by Gordan in the year 1869, and the calculation of them having been completed in my Ninth and Tenth Memoirs, it appeared to me convenient in the present republication to unite together the values of all the covariants: viz. those of A to L are given in the Second Memoir 141, and the remainder M to W in the paper 143. I have added to the Third Memoir 144, in the notation thereof, some formulæ which on account of a difference of notation were omitted from a former paper, 35.

I remark that the present volume comprises the first six of the ten Memoirs on Quantics, viz. these are 139, 141, 144, 155, 156 and 158. I have, in the Notes and References, inserted a discussion of some length in reference to the paper 121, Note on a Question in the Theory of Probabilities: and also some remarks in reference to the theory of Distance developed in the Sixth Memoir on Quantics, 158.

CONTENTS.

CLASSIFICATION.

101.

NOTES ON LAGRANGE'S THEOREM.

[From the *Cambridge and Dublin Mathematical Journal,* vol. VI. (1851), pp. 37—45.]

I.

IF in the ordinary form of Lagrange's theorem we write $(x + a)$ for x, it becomes

$$x = hf(a+x),$$

$$F(a+x) = Fa + \frac{h}{1} F'afa + \&c. \quad \ldots\ldots\ldots\ldots\ldots\ldots (1)$$

It follows that the equation

$$F(a+x) = Fa + \frac{1}{1}\frac{x}{f(a+x)}(F'afa) + , \quad \ldots\ldots\ldots\ldots\ldots\ldots(2)$$

must reduce itself to an identity when the two sides are expanded in powers of x;. or writing for shortness F, f instead of Fa, fa, and δ for $\frac{d}{da}$, we must have

$$\frac{1}{[r]^r}\delta^r F = S\left\{\frac{1}{[p]^p}\delta^{p-1}(\delta F \,.\, f^p)\frac{1}{[r-p]^{r-1}}\delta^{r-p}f^{-p}\right\}, \quad \ldots\ldots\ldots\ldots(3)$$

(where p extends from 0 to r). Or what comes to the same,

$$\frac{1}{[r]^r}\delta^r F = S\left\{\frac{1}{p\,[p-s]^{p-s}\,[r-p]^{r-p}\,[s-1]^{s-1}}\delta^{p-s}f^p \,.\, \delta^{r-p}f^{-p} \,.\, \delta^s F\right\}, \quad \ldots\ldots(4)$$

where s extends from 0 to $(r-p)$. The terms on the two sides which involve $\delta^r F$ are immediately seen to be equal; the coefficients of the remaining terms $\delta^s F$ on the second side must vanish, or we must have

$$S\left\{\frac{1}{p\,[p-s]^{p-s}\,[r-p]^{r-p}}(\delta^{p-s}f^p)(\delta^{r-p}f^{-p})\right\} = 0, \quad \ldots\ldots\ldots\ldots (5)$$

(s being less than r). Or in a somewhat more convenient form, writing p, q and k for $p-s$, $r-p$ and $r-s$,

$$S\left\{\frac{1}{(p+s)\,[p]^p\,[s]^s}(\delta^p f^{p+s})\,(\delta^q f^{-p-s})\right\}=0, \quad \ldots\ldots\ldots\ldots\ldots\ldots\ldots\ldots (6)$$

where s is constant and p and q vary subject to $p+q=k$, k being a given constant different from zero (in the case where $k=0$, the series reduces itself to the single term $\frac{1}{s}$). The direct proof of this theorem will be given presently.

II.

The following symbolical form of Lagrange's theorem was given by me in the *Mathematical Journal*, vol. III. [1843], pp. 283—286, [8].

If

$$x=a+hfx, \quad \ldots\ldots\ldots\ldots\ldots\ldots\ldots\ldots\ldots\ldots\ldots\ldots (7)$$

then

$$Fx=\left(\frac{d}{da}\right)^{h\frac{d}{dh}-1}F'a\,e^{hfa}.$$

Suppose $fx=\phi\,(b+k\psi x)$, or $x=a+h\phi\,(b+k\psi x)$, then

$$Fx=\left(\frac{d}{da}\right)^{h\frac{d}{dh}-1}F'a\,e^{h\phi\,(b+k\psi a)}.$$

But

$$e^{h\phi\,(b+k\psi a)}=\left(\frac{d}{db}\right)^{k\frac{d}{dk}}e^{h\phi b+k\psi a}.$$

(In fact the two general terms

$$\{\phi\,(b+k\psi a)\}^m \text{ and } \left(\frac{d}{db}\right)^{k\frac{k}{dk}}e^{k\psi a}\,(\phi b)^m,$$

of which the former reduces itself to $e^{k\psi a\frac{d}{db}}(\phi b)^m$, are equal on account of the equivalence of the symbols

$$e^{k\psi a\frac{d}{db}} \text{ and } \left(\frac{d}{db}\right)^{k\frac{d}{dk}}e^{k\psi a}).$$

Hence

$$x=a+h\phi\,(b+k\psi x), \quad \ldots\ldots\ldots\ldots\ldots\ldots\ldots\ldots\ldots\ldots\ldots (8)$$

$$Fx=\left(\frac{d}{da}\right)^{h\frac{d}{dh}-1}\left(\frac{d}{db}\right)^{k\frac{d}{dk}}F'ae^h\phi^{b+k\psi a};$$

and the coefficient of h^mk^n is

$$\frac{1}{[m]^m\,[n]^n}\left(\frac{d}{da}\right)^{m-1}F'a\,(\psi a)^n.\left(\frac{d}{db}\right)^n(\phi b)^n.$$

A similar formula evidently applies to the case of any finite number of functions ϕ, ψ, &c.: in the case of an infinite number we have

$$F(a+h\phi(b+k\psi(c+l\chi(d+\ldots) = \left(\frac{d}{da}\right)^{h\frac{d}{dh}-1}\left(\frac{d}{db}\right)^{k\frac{d}{dk}}\left(\frac{d}{dc}\right)^{l\frac{d}{dl}}\ldots F'a\, e^{h\phi b+k\psi c+l\chi d+\ldots}; \ldots(9)$$

or the coefficient of $h^m k^n l^p \ldots$ is

$$\frac{1}{[m]^m [n]^n [p]^p \ldots}\left(\frac{d}{da}\right)^m Fa \,.\left(\frac{d}{db}\right)^n (\phi b)^m \,.\left(\frac{d}{dc}\right)^p (\psi c)^r \ldots$$

the last of the series $m, n, p \ldots\ldots$ being always zero; e.g. in the coefficient of $h^m k^n$, account must be had of the factor $\left(\frac{d}{dc}\right)^p (\psi c)^n$ or $(\psi c)^n$. The above form is readily proved independently by Taylor's theorem, without the assistance of Lagrange's. If in it we write $h=k$, &c., $a=b=$ &c., and $\phi=\psi=$ &c. $=f$, we have $F(a+hf(a+hf(a+\ldots)=Fx$, where $x=a+hfx$. Hence, comparing the coefficient of h^s with that given by Lagrange's theorem,

$$\frac{1}{[s]^s}\,\delta^{s-1}(\delta F\,.f^s) = S\left\{\frac{1}{[m]^m [n]^n [p]^p \ldots}\,\delta^m F\,.\,\delta^n f^m\,.\,\delta^p f^r \ldots\right\}, \;\ldots\ldots\; (10)$$

where $m+n+$ &c. $=s$, and as before Fa, fa, $\frac{d}{da}$ have been replaced by F, f, δ. By comparing the coefficients of $\delta^m F$,

$$\frac{1}{[t]^t}\,\frac{s-t}{s}\,\delta^t f^s = \Sigma\left\{\frac{1}{[n]^n [p]^p \ldots}(\delta^n f^{s-t})(\delta^p f^n)\ldots\right\}, \;\ldots\ldots\ldots\ldots\ldots(11)$$

where $n+p+\ldots=t$, the last of the series $n, p\ldots$ always vanishing. The formula (10) deduced, as above mentioned, from Taylor's theorem, and the subsequent formula (11) with an independent demonstration of it, not I believe materially different from that which will presently be given, are to be found in a memoir by M. Collins (volume II. (1833) of the Memoirs of the Academy of St Petersburg), who appears to have made very extensive researches in the theory of developments as connected with the combinatorial analysis.

III.

To demonstrate the formula (6), consider, in the first place, the expression

$$S\,\frac{\phi p}{[p]^p [q]^q}\{(\delta^p f^{p+s})(\delta^q f^{-p-s-\theta})\},$$

where $p+q=k$. Since

$$\frac{1}{[p]^p [q]^q} = \frac{1}{k}\left(\frac{1}{[p-1]^{p-1}[q]^q}+\frac{1}{[p]^p [q-1]^{q-1}}\right),$$

this is immediately transformed into

$$\frac{1}{k} S\phi p \left\{\frac{p+s}{[p-1]^{p-1}[q]^q}(\delta^{p-1} f^{p+s-1}\,\delta f)(\delta^q f^{-p-s-\theta}) - \frac{(p+s+\theta)}{[p]^p[q-1]^{q-1}}(\delta^p f^{p+s})(\delta^{q-1}.f^{-p-s-\theta-1}\,\delta f)\right.$$

$$= \frac{1}{k} S \frac{1}{[p]^p[q]^q}\{\phi(p+1)(p+s+1)(\delta^p.f^{p+s}\delta f)(\delta^q f^{-p-s-\theta-1})$$

$$-\phi p(p+s+\theta)(\delta^p f^{p+s})(\delta^q.f^{-p-s-\theta-1}\delta f)\},$$

in which last expression $p+q=(\rho-1)$. Of this, after separating the factor δf, the general term is

$$\frac{1}{k}\frac{1}{[\alpha]^\alpha}\delta^{\alpha+1}f.\,S\left\{\frac{1}{[p-\alpha]^{p-\alpha}[q]^q}\phi(p+1)(p+s+1)(\delta^{p-\alpha}f^{p+s})(\delta^q f^{-p-s-\theta-1})\right.$$

$$\left.-\frac{1}{[p]^p[q-\alpha]^{q-\alpha}}\phi p(p+s+\theta)(\delta^p f^{p+s})(\delta^{q-\alpha}f^{-p-s-\theta-1})\right\},$$

equivalent to

$$\frac{1}{k}\frac{1}{[\alpha]^\alpha}\delta^{\alpha+1}f.\,S\frac{1}{[p]^p[q]^q}\{\phi(p+\alpha+1)(p+s+\alpha+1)(\delta^p f^{p+s+\alpha})(\delta^q f^{-p-s-\alpha-\theta-1})$$

$$-\phi p(p+s+\theta)(\delta^p f^{p+s})(\delta^q f^{-p-s-\theta-1})\},$$

in which last expression $p+q=k-\alpha-1$. By repeating the reduction j times, the general term becomes

$$\frac{1}{k(k-\alpha-1)(k-\alpha-\beta-2)\ldots}\,\frac{1}{[\alpha]^\alpha[\beta]^\beta\ldots}\,\delta^{\alpha+1}f.\,\delta^{\beta+1}f\ldots$$

$$\times S\frac{1}{[p^p][q]^q}\,\Sigma\{(-)^{j-j'}\phi(p+\alpha+\beta\ldots+j')[p+s+\alpha+\beta\ldots+j']^{j'}$$

$$\times[p+s+\theta+\alpha+\beta\ldots+j-1]^{j-j'}(\delta^p f^{p+s+\alpha+\beta\ldots})(\delta^q f^{-p-s-\theta-j-\alpha-\beta\ldots})\},$$

where the sums $\alpha+\beta\ldots$ contain j' terms, j' being less than j or equal to it, and Σ extends to all combinations of the quantities α, $\beta\ldots$ taken j' and j' together (so that the summation contains 2^j terms). Also $p+q=k-\alpha-\beta\ldots$ (j terms) $-j$, and the products $k(k-\alpha-1)(k-\alpha-\beta-2)\ldots$ and $[\alpha]^\alpha[\beta]^\beta,\ldots\delta^{\alpha+1}f.\,\delta^{\beta+1}f\ldots$ contain each of them j terms. Suppose the reduction continued until $k-\alpha-\beta\ldots$ (j terms) $-j=0$, then the only values of p, q are $p=0$, $q=0$; and the general term of

$$S\frac{\phi p}{[p]^p[q]^q}\{(\delta^p f^{p+s})(\delta^q f^{-p-s-\theta})\}$$

becomes

$$\frac{1}{k(k-\alpha-1)(k-\alpha-\beta-2)\ldots}\;\frac{1}{[\alpha]^\alpha[\beta]^\beta\ldots}\,\delta^{\alpha+1}f.\,\delta^{\beta+1}f\ldots f^{-j-\theta}$$

$$\times\Sigma\{(-)^{j-j'}\phi(\alpha+\beta\ldots+j')[s+\alpha+\beta\ldots+j']^{j'}[s+\theta+\alpha+\beta\ldots+j-1]^{j-j'}\}.$$

If $\theta = 0$, the general term reduces itself to

$$\frac{1}{k(k-\alpha-1)(k-\alpha-\beta-2)\ldots}\;\frac{1}{[\alpha]^\alpha[\beta]^\beta\ldots}\;\delta^{\alpha+1}f.\,\delta^{\beta+1}f\ldots f^{-j}.$$

$$\Sigma\{(-)^{j-j'}(s+\alpha+\beta\ldots+j')\,\phi(\alpha+\beta\ldots+j')\,[s+\alpha+\beta\ldots+j-1]^{j-1}\};$$

whence finally, if $\phi p = \dfrac{1}{p+s}$, the general term of

$$S\,\frac{1}{(p+s)\,[p]^p\,[q]^q}\,\{(\delta^p f^{p+s})\,(\delta^q f^{-p-s})\}$$

becomes

$$\frac{1}{k(k-\alpha-1)(k-\alpha-\beta-2)\ldots}\cdot\frac{1}{[\alpha]^\alpha[\beta]^\beta\ldots}\,\delta^{\alpha+1}f.\,\delta^{\beta+1}f\ldots f^{-j}\,\Sigma\,\{(-)^{j-j'}\,[s+\alpha+\beta\ldots+j-1]^{j-1}\};$$

and it is readily shown that the sum contained in this formula vanishes, which proves the equation in question.

IV.

The demonstration of the equation (11) is much simpler. We have

$$\delta^{t-1}(f^{s-1}\delta f) = \Sigma\left\{\frac{[t-1]^{n-1}}{[n-1]^{n-1}}\,\delta^{n-1}(f^{s-t-1}\delta f)\,.\,(\delta^{t-n}f^t)\right\},$$

that is,

$$\delta^t f^s = \frac{s}{s-t}\,\Sigma\left\{\frac{[t-1]^{n-1}}{[n-1]^{n-1}}\,(\delta^n f^{s-t})\,(\delta^{t-n}f^t)\right\},$$

where n extends from $n=1$ to $n=t$. Similarly

$$\delta^{t-n}f^t = \frac{t}{n}\,\Sigma\left\{\frac{[t-n-1]^{p-1}}{[p-1]^{p-1}}\,(\delta^p f^n)\,(\delta^{t-n-p}f^{t-n})\right\},$$

$$\delta^{t-n-p}f^{t-n} = \frac{t-n}{p}\,\Sigma\left\{\frac{[t-n-p-1]^{q-1}}{[q-1]^{q-1}}\,(\delta^q f^p)\,(\delta^{t-n-p-q}f^{t-n-p})\right\},$$

&c.

Hence, substituting successively, and putting $t-n-p-q=r$, &c.,

$$\delta^t f^s = \frac{s}{s-t}\,\Sigma\,\frac{[t]^{t-r-1}}{[n]^n\,[p]^p\,(q+r)\,[q-1]^{q-1}}\,(\delta^n f^{s-t})\,(\delta^q f^p)\,(\delta^r f^{q+r})\Big\},$$

&c.; and the last of these corresponding to a zero value of the last of the quantities $n, p, q\ldots$ is evidently the required equation (11).

V.

The formula (18) in my paper on Lagrange's theorem (before referred to) is incorrect. I propose at present, after giving the proper form of the formula in question, to develope the result of the substitution indicated at the conclusion of the paper. It will be convenient to call to mind the general theorem, that when any number of variables $x, y, z\ldots$ are connected with as many other variables $u, v, w\ldots$ by the same number of equations (so that the variables of each set may be considered as functions of those of the other set) the quotient of the expressions $dxdy\ldots$ and $dudv\ldots$ is equal to the quotient of two determinants formed with the functions which equated to zero express the relations between the two sets of variables; the former with the differential coefficients of these functions with respect to $u, v\ldots$, the latter with the differential coefficients with respect to $x, y\ldots$. Consequently the notation $\frac{dxdy\ldots}{dudv\ldots}$ may be considered as representing the quotient of these determinants. This being premised, if we write

$$x - u - h\theta(x, y\ldots) = 0,$$
$$y - v - k\phi(x, y\ldots) = 0,$$

then the formula in question is

$$F(x, y\ldots)\frac{dxdy\ldots}{dudv\ldots} = \delta_u{}^{h\delta_h}\delta_v{}^{k\delta_k}\ldots e^{h\theta+k\phi\ldots} F,$$

if for shortness the letters $\theta, \phi, \ldots, F$ denote what the corresponding functions become when $u, v, \ldots$ are substituted for $x, y, \ldots$. Let $\frac{1}{\Delta}$ denote the value which $\frac{dxdy\ldots}{dudv\ldots}$, considered as a function of $x, y\ldots$, assumes when these variables are changed into $u, v, \ldots$, we have

$$\nabla = \begin{vmatrix} 1 - h\delta_u\theta, & -h\delta_v\theta\ldots \\ -k\delta_u\phi, & 1 - k\delta_v\phi\ldots \\ \vdots & \end{vmatrix}.$$

By changing the function F, we obtain

$$F(x, y\ldots) = \delta_u{}^{h\delta_h}\delta_v{}^{kd_k}\ldots e^{h\theta+k\phi\ldots} F\nabla;$$

where, however, it must be remembered that the $h, k, \ldots$, in so far as they enter into the function ∇, are not affected by the symbols $h\delta_h, k\delta_k, \ldots$ In order that we may consider them to be so affected, it is necessary in the function ∇ to replace h, k, &c. by $\frac{h}{\delta_u}, \frac{k}{\delta_v}$, &c. Also, after this is done, observing that the symbols $h\delta_u\theta, h\delta_v\theta\ldots$ affect a function $e^{h\theta+k\phi+\ldots}F$, the symbols $h\delta_u\theta, h\delta_v\theta, \ldots$ may be replaced by $\delta_u{}^\theta, \delta_v{}^\theta, \ldots$, where the θ is not an index, but an affix denoting that the differentiation is only to be performed with respect to $u, v\ldots$ so far as these variables respectively enter into the function θ. Transforming the other lines of the determinant in the same manner, and taking out from $\delta_u{}^{h\delta_h}\delta_v{}^{k\delta_k}\ldots$ the factor $\delta_u\delta_v\ldots$ in order to multiply this last factor into the determinant, we obtain

$$F(x, y\ldots) = \delta_u{}^{h\delta_h-1}\delta_v{}^{k\delta_k-1}\ldots e^{h\theta+k\phi\ldots} F\square;$$

where

$$\square = \begin{vmatrix} \delta_u - \delta_u{}^\theta, & -\delta_u{}^\phi, \ldots \\ -\delta_v{}^\theta, & \delta_v - \delta_v{}^\phi, \\ \vdots & \end{vmatrix}$$

in which expression $\delta_u, \delta_v \dots$ are to be replaced by

$$\delta_u{}^F + \delta_u{}^\theta + \delta_u{}^\phi \dots \delta_v{}^F + \delta_v{}^\theta + \delta_v{}^\phi \dots .$$

The complete expansion is easily arrived at by induction, and the form is somewhat singular. In the case of a single variable u we have $\square = \delta_u{}^F$, in the case of two variables, $\square = \delta_u{}^F \delta_v{}^F + \delta_u{}^F \delta_v{}^\theta + \delta_u{}^\phi \delta_v{}^F$. Or writing down only the affixes, in the case of a single variable we have F; in the case of two variables FF, $F\theta$, ϕF; and in the case of three variables FFF, ϕFF, χFF, $F\chi F$, $F\theta F$, $FF\theta$, $FF\phi$, $F\theta\theta$, $F\theta\phi$, $F\chi\theta$, $\phi F\phi$, $\chi F\phi$, $\phi F\theta$, $\chi\chi F$, $\phi\chi F$, $\chi\theta F$; where it will be observed that θ never occurs in the first place, nor ϕ in the second place, nor θ, ϕ (in any order) in the first and second places, &c., nor θ, ϕ, χ (in any order) in the first, second, and third places. And the same property holds in the general case for each letter and binary, ternary, &c. combination, and for the entire system of letters, and the system of affixes contains every possible combination of letters not excluded by the rule just given. Thus in the case of two letters, forming the system of affixes FF, $F\theta$, ϕF, θF, $F\phi$, $\theta\phi$, $\phi\theta$, the last four are excluded, the first three of them by containing θ in the first place or ϕ in the second place, the last by containing ϕ, θ in the first and second places: and there remains only the terms FF, $F\theta$, ϕF forming the system given above. Substituting the expanded value of $\square$ in the expression for $F(x, y\dots)$, the equation may either be permitted to remain in the form which it thus assumes, or we may, in order to obtain the finally reduced form, after expanding the powers of h, $k\dots$, connect the symbols $\delta_u{}^\theta$, $\delta_u{}^\phi \dots \delta_u{}^F$, &c. with the corresponding functions θ, $\phi \dots F$, and then omit the affixes; thus, in particular, in the case of a single variable the general term of Fx is

$$\frac{h^p}{[p]^p} \delta_u{}^{p-1} (\theta^p \delta_u F),$$

(the ordinary form of Lagrange's theorem). In the case of two letters the general term of $F(x, y)$ is

$$\frac{h^p k^q}{[p]^p [q]^q} \delta_u{}^{p-1} \delta_v{}^{q-1} \{\theta^p \phi^q \delta_u \delta_v F + \phi^q \delta_v \theta^p \delta_u F + \theta^p \delta_u \phi^q \delta^v F\}$$

(see the *Mécanique Céleste*, [Ed. 1, 1798] t. I. p. 176). In the case of three variables, the general term is

$$\frac{h^p k^q l^r}{[p]^p [q]^q [r]^r} \delta_u{}^{p-1} \delta_v{}^{q-1} \delta_w{}^{r-1} \{\theta^p \phi^q \chi^r \delta_u \delta_v \delta_w F + \dots\},$$

the sixteen terms within the { } being found by comparing the product $\delta_u \delta_v \delta_w$ with the system FFF, ϕFF, &c., given above, and then connecting each symbol of differentiation with the function corresponding to the affix. Thus in the first term the δ_u, δ_v, δ_w, each affect the F, in the second term the δ_u affects ϕ^q, and the δ_v and δ_w each affect the F, and so on for the remaining terms. The form is of course deducible from Laplace's general theorem, and the actual development of it is given in Laplace's Memoir in the *Hist. de l'Acad.* 1777. I quote from a memoir by Jacobi which I take this opportunity of referring to, "De resolutione equationum per series infinitas," *Crelle*, t. VI. [1830], pp. 257—286, founded on a preceding memoir, "Exercitatio algebraica circa discerptionem singularem fractionum quæ plures variabiles involvunt," t. V. [1830], pp. 344—364.

Stone Buildings, April 6, 1850.

102.

ON A DOUBLE INFINITE SERIES.

[From the *Cambridge and Dublin Mathematical Journal*, vol. VI. (1851), pp. 45—47.]

THE following completely paradoxical investigation of the properties of the function Γ (which I have been in possession of for some years) may perhaps be found interesting from its connexion with the theories of expansion and divergent series.

Let $\Sigma_r \phi r$ denote the sum of the values of ϕr for all integer values of r from $-\infty$ to ∞. Then writing

$$u = \Sigma_r [n-1]^r x^{n-1-r}, \qquad (1)$$

(where n is any number whatever), we have immediately

$$\frac{du}{dx} = \Sigma_r [n-1]^{r+1} x^{n-2-r} = \Sigma_r [n-1]^r x^{n-1-r} = u;$$

that is,

$$\frac{du}{dx} = u, \text{ or } u = C_n e^x,$$

(the constant of integration being of course in general a function of n). Hence

$$C_n e^x = \Sigma_r [n-1]^r x^{n-1-r}; \qquad (2)$$

or e^x is expanded in general in a *doubly infinite necessarily divergent series of fractional powers of* x, (which resolves itself however in the case of n a positive or negative integer, into the ordinary singly infinite series, the value of C_n in this case being immediately seen to be Γn).

The equation (2) in its general form is to be considered as a definition of the function C_n. We deduce from it

$$\Sigma_r [n-1]^r (ax)^{n-1-r} = C_n e^{ax},$$
$$\Sigma_{r'} [n'-1]^{r'} (ax')^{n-1-r'} = C_{n'} e^{ax'};$$
$$\vdots$$

and also

$$\Sigma_k\,[n+n'\ldots-1]^k\,\{a\,(x+x'\ldots)\}^{n+n'\ldots-1-k}=C_{n+n'\ldots}\,e^{a\,(x+x'\ldots)}.$$

Multiplying the first set of series, and comparing with this last,

$$C_{n+n'\ldots}\Sigma_{r,\,r'\ldots}\,[n-1]^r\,[n'-1]^{r'}\ldots x^{n-1-r}\,x'^{\,n'-1-r'}\ldots$$
$$=C_nC_{n'}\ldots[n+n'\ldots-1]^k\,(x+x'\ldots)^{n+n'\ldots-1-k},\;\ldots\ldots\ldots\ldots(3)$$

(where r, r' denote any positive or negative integer numbers satisfying $r+r'+\ldots=k+1-p$, p being the number of terms in the series n, n', ...). This equation constitutes a multinomial theorem of a class analogous to that of the exponential theorem contained in the equation (2).

In particular

$$C_{n+n'\ldots}\,\Sigma_{r,\,r'\ldots}\,[n-1]^r\,[n'-1]^{r'}\ldots=C_nC_{n'}\ldots[n+n'\ldots-1]^k\,p^{n+n'\ldots-1-k},\;\ldots\ldots\ldots(4)$$

and if $p=2$, writing also m, n for n, n', and $k-1-r$ for r',

$$C_{m+n}\,\Sigma_r\,[m-1]^r\,[n-1]^{k-1-r}=C_mC_n\,[m+n-1]^k\,2^{m+n-1-k},\;\ldots\ldots\ldots\ldots\ldots\ldots(5)$$

or putting $k=0$ and dividing,

$$C_mC_n\div C_{m+n}=\frac{1}{2^{m+n-1}}\,\Sigma_r\,[m-1]^r\,[n-1]^{-1-r}.\;\ldots\ldots\ldots\ldots\ldots(6)$$

Now the series on the second side of this equation is easily seen to be convergent (at least for positive values of m, n). To determine its value write

$$\mathsf{F}\,(m,\;n)=\int_0^1 x^{m-1}\,(1-x)^{n-1}\,dx\,;$$

then

$$\mathsf{F}\,(m,\;n)=\int_0^{\frac{1}{2}} x^{m-1}\,(1-x)^{n-1}\,dx+\int_0^{\frac{1}{2}} x^{n-1}\,(1-x)^{m-1}\,dx\,;$$

and by successive integrations by parts, the first of these integrals is reducible to $\frac{1}{2^{m+n-1}}\,\Sigma_r\,[m-1]^r\,[n-1]^{-1-r}$, r extending from -1 to $-\infty$ inclusively, and the second to $\frac{1}{2^{m+n-1}}\,\Sigma_r\,[m-1]^r\,[n-1]^{-1-r}$, r extending from 0 to ∞; hence

$$\mathsf{F}\,(m,\;n)=\frac{1}{2^{m+n-1}}\,\Sigma_r\,[m-1]^r\,[n-1]^{-1-r},$$

or

$$C_mC_n\div C_{m+n}=\mathsf{F}\,(m,\;n),\;\ldots\ldots\ldots\ldots\ldots\ldots\ldots\ldots\ldots\ldots\ldots\ldots(7)$$

which proves the identity of C_m with the function $\Gamma(m)$. {Substituting in two of the preceding equations, we have

$$\Gamma n \Gamma n' \ldots \div \Gamma(n+n'\ldots) = \frac{1}{[n+n'\ldots-1]^k p^{n+n'\ldots-1-k}} \Sigma_{r,r'\ldots} [n-1]^r [n'-1]^{r'} \ldots, \quad \ldots (8)$$

(where, as before, p denotes the number of terms in the series $n, n', \ldots$ and $r+r'+\ldots=k+1-p$), the first side of which equation is, it is well known, reducible to a multiple definite integral by means of a theorem of M. Dirichlet's. And

$$\mathsf{F}(m, n) = \frac{1}{[m+n-1]^k 2^{m+n-1-k}} \Sigma_r [m-1]^r [n-1]^{k-1-r}, \quad \ldots\ldots\ldots\ldots (9)$$

where r extends from $-\infty$ to $+\infty$, and k is arbitrary. By giving large negative values to this quantity, very convergent series may be obtained for the calculation of $\mathsf{F}(m, n)$}.

103.

ON CERTAIN DEFINITE INTEGRALS.

[From the *Cambridge and Dublin Mathematical Journal*, vol. VI. (1851), pp. 136—140.]

SUPPOSE that for any positive or negative integral value of r, we have $\psi(x+ra) = U_r\psi x$, U_r being in general a function of x, and consider the definite integral

$$I = \int_{-\infty}^{\infty} \psi x \Psi x\, dx;$$

Ψx being any other function of x. In case of either of the functions ψx, Ψx becoming infinite for any real value a of x, the principal value of the integral is to be taken, that is, I is to be considered as the limit of

$$\left(\int_{a+\epsilon}^{\infty} + \int_{-\infty}^{a-\epsilon}\right) \psi x\, \Psi x\, dx, \quad (\epsilon = 0),$$

and similarly, when one of the functions becomes infinite for several of such values of x.

We have

$$I = \left(\ldots \int_{ra}^{(r+1)a} + \ldots\right) \psi x \Psi x\, dx;$$

or changing the variables in the different integrals so as to make the limits of each a, 0, we have

$$I = \int_0^a [\Sigma\, \psi(x+ra)\, \Psi(x+ra)]\, dx,$$

Σ extending to all positive or negative integer values of r, that is,

$$I = \int_0^a \psi x\, [\Sigma\, U_r \Psi(x+ra)]\, dx, \quad \text{.............................(A)}$$

which is true, even when the quantity under the integral sign becomes infinite for particular values of x, provided the integral be replaced by its principal value, that is, provided it be considered as the limit of

$$\left(\int_{\alpha+\epsilon}^{a} + \int_{0}^{\alpha-\epsilon}\right) \psi x \left[\Sigma U_r \Psi (x + ra)\right] dx,$$

or

$$\int_{\epsilon}^{a-\epsilon} \psi x \left[\Sigma U_r \Psi (x + ra)\right] dx;$$

where α, or one of the limiting values a, 0, is the value of x, for which the quantity under the integral sign becomes infinite, and ϵ is ultimately evanescent.

In particular, taking for simplicity $a = \pi$, suppose

$$\psi (x + \pi) = \pm \psi x, \quad \text{or} \quad \psi (x + r\pi) = (\pm)^r \psi x;$$

then observing the equation

$$\Sigma \frac{(\pm)^r 1}{x + r\pi} = \cot x, \quad \text{or} = \operatorname{cosec} x,$$

according as the upper or under sign is taken, and assuming $\Psi x = x^{-\mu}$, we have finally

$$\int_{-\infty}^{\infty} \frac{\psi x \, dx}{x^\mu} = \frac{(-)^{\mu-1}}{\Gamma\mu} \int_0^\pi \psi x \left[\left(\frac{d}{dx}\right)^{\mu-1} \cot x\right] dx,$$

$$\int_{-\infty}^{\infty} \frac{\psi x \, dx}{x^\mu} = \frac{(-)^{\mu-1}}{\Gamma\mu} \int_0^\pi \psi x \left[\left(\frac{d}{dx}\right)^{\mu-1} \operatorname{cosec} x\right] dx,$$

the former equation corresponding to the case of $\psi (x + \pi) = \psi x$, the latter to that of $\psi (x + \pi) = - \psi x$.

Suppose $\psi_{,} x = \psi g x$, g being a positive integer. Then

$$\int_{-\infty}^{\infty} \frac{\psi_{,} x \, dx}{x^\mu} = g^{\mu-1} \int_{-\infty}^{\infty} \frac{\psi x \, dx}{x^\mu};$$

also if $\psi (x + \pi) = \psi x$, then $\psi_{,} (x + \pi) = \psi_{,} x$; but if $\psi (x + \pi) = - \psi x$, then $\psi_{,} (x + \pi) = \pm \psi_{,} x$, the upper or under sign according as g is even or odd. Combining these equations, we have

$\psi (x + \pi) = \psi x$, g even or odd,

$$\int_{-\infty}^{\infty} \frac{\psi g x \, dx}{x^\mu} = \frac{(-)^{\mu-1}}{\Gamma(\mu)} g^{\mu-1} \int_0^\pi \psi x \left[\left(\frac{d}{dx}\right)^{\mu-1} \cot x\right] dx = \frac{(-)^{\mu-1}}{\Gamma\mu} \int_0^\pi \psi g x \left[\left(\frac{d}{dx}\right)^{\mu-1} \cot x\right] dx;$$

$\psi (x + \pi) = - \psi x$, g even,

$$\int_{-\infty}^{\infty} \frac{\psi g x \, dx}{x^\mu} = \frac{(-)^{\mu-1}}{\Gamma\mu} g^{\mu-1} \int_0^\pi \psi x \left[\left(\frac{d}{dx}\right)^{\mu-1} \operatorname{cosec} x\right] dx = \frac{(-)^{\mu-1}}{\Gamma\mu} \int_0^\pi \psi g x \left[\left(\frac{d}{dx}\right)^{\mu-1} \operatorname{cosec} x\right] dx;$$

$\psi (x + \pi) = - \psi x$, g odd,

$$\int_{-\infty}^{\infty} \frac{\psi g x \, dx}{x^\mu} = \frac{(-)^{\mu-1}}{\Gamma\mu} g^{\mu-1} \int_0^\pi \psi x \left[\left(\frac{d}{dx}\right)^{\mu-1} \operatorname{cosec} x\right] dx = \frac{(-)^{\mu-1}}{\Gamma\mu} \int_0^\pi \psi g x \left[\left(\frac{d}{dx}\right)^{\mu-1} \operatorname{cosec} x\right] dx.$$

In particular

$$\int_{-\infty}^{\infty} \frac{\sin x\,dx}{x} = \pi,$$

$$\int_{-\infty}^{\infty} dx\,\frac{\sin x}{x^\mu} = \frac{(-)^{\mu-1}}{\Gamma\mu}\int_0^\pi \sin x\left[\left(\frac{d}{dx}\right)^{\mu-1}\operatorname{cosec} x\right]dx,$$

$$\int_0^\pi \sin gx\left[\left(\frac{d}{dx}\right)^{\mu-1}\cot x\right]dx = \quad g^{\mu-1}\int_0^\pi \sin x\left[\left(\frac{d}{dx}\right)^{\mu-1}\operatorname{cosec} x\right]dx, \quad g \text{ even},$$

$$\int_0^\pi \sin gx\left[\left(\frac{d}{dx}\right)^{\mu-1}\operatorname{cosec} x\right]dx = \quad g^{\mu-1}\int_0^\pi \sin x\left[\left(\frac{d}{dx}\right)^{\mu-1}\operatorname{cosec} x\right]dx, \quad g \text{ odd},$$

$$\int_0^\pi \sin gx \cot x\,dx = \pi, \quad g \text{ even},$$

$$\int_0^\pi \sin gx \operatorname{cosec} x\,dx = \pi, \quad g \text{ odd},$$

$$\int_0^\pi \frac{\tan x\,dx}{x} = 0, \text{ \&c.},$$

the number of which might be indefinitely extended.

The same principle applies to multiple integrals of any order: thus for double integrals, if $\psi(x+ra,\ y+rb) = U_{r.s}\psi(x,\ y)$, then

$$\int_{-\infty}^{\infty}\int_{-\infty}^{\infty} \psi(x,\ y)\,\Psi(x,\ y)\,dx\,dy = \int_0^a\int_0^b \psi(x,\ y)\,\Sigma U_{r,s}\,\Psi(x+ra,\ y+sb). \;\ldots\; \text{(B)}$$

In particular, writing w, v for a, b, and assuming $\psi(x+rw,\ y+sv) = (\pm)^r(\pm)^s\psi(x,\ y)$; also $\Psi(x,\ y) = (x+iy)^{-\mu}$, where as usual $i = \sqrt{-1}$,

$$\int_{-\infty}^{\infty}\int_{-\infty}^{\infty} \frac{\psi(x,\ y)\,dx\,dy}{(x+iy)^\mu} = \frac{(-)^{\mu-1}}{\Gamma\mu}\int_0^w\int_0^v \psi(x,\ y)\left[\left(\frac{d}{dx}\right)^{\mu-1}\Theta(x+iy)\right]dx\,dy, \ldots\text{(B}')$$

where

$$\Theta(x+iy) = \Sigma\,\frac{(\pm)^r(\pm)^s\,1}{(x+iy+rw+svi)},$$

Σ extending to all positive or negative integer values of r and s. Employing the notation of a paper in the *Cambridge Mathematical Journal*, "On the Inverse Elliptic Functions," t. IV. [1845], pp. 257—277, [24], we have for the different combinations of the ambiguous sign,

$$1. \quad -, \quad -, \quad \Theta(x+iy) = \frac{\mathfrak{G}(x+iy)}{\gamma(x+iy)} = \frac{1}{\phi(x+iy)},$$

$$2. \quad -, \quad +, \quad \Theta(x+iy) = \frac{G(x+iy)}{\gamma(x+iy} = \frac{F(x+iy)}{\phi(x+iy)},$$

$$3. \quad +, \quad -, \quad \Theta(x+iy) = \frac{g(x+iy)}{\gamma(x+iy)} = \frac{f(x+iy)}{\phi(x+iy)},$$

$$4. \quad +, \quad +, \quad \Theta(x+iy) = \frac{\gamma'(x+iy)}{\gamma(x+iy)};$$

where ϕ, f, F are in fact the symbols of the inverse elliptic functions (Abel's notation) corresponding very nearly to sin am, cos am, Δ am. It is remarkable that the last value of Θ cannot be thus expressed, but only by means of the more complicated transcendant γx, corresponding to the $H(x)$ of M. Jacobi. The four cases correspond obviously to

$$\begin{aligned}
&1. \quad \psi(x+rw,\ y+sv) = (-)^{r+s}\,\psi(x,\ y),\\
&2. \quad \psi(x+rw,\ y+sv) = (-)^{r}\ \psi(x,\ y),\\
&3. \quad \psi(x+rw,\ y+sv) = (-)^{s}\ \psi(x,\ y),\\
&4. \quad \psi(x+rw,\ y+sv) = \quad\ \psi(x,\ y).
\end{aligned}$$

The above formulæ may be all of them modified, as in the case of single integrals, by means of the obvious equation

$$\iint \frac{\psi(gx,\ gy)\,dx\,dy}{(x+iy)^{\mu}} = g^{\mu-2} \iint \frac{\psi(x,\ y)\,dx\,dy}{(x+iy)^{\mu}}, \text{ [limits } \infty,\ -\infty].$$

The most important particular case is

$$\int_{-\infty}^{\infty}\int_{-\infty}^{\infty} \frac{\phi(x+iy)\,dx\,dy}{(x+iy)} = wv,$$

for in almost all the others, for example in

$$\int_{-\infty}^{\infty}\int_{-\infty}^{\infty} \frac{\phi(x+iy)\,dx\,dy}{(x+iy)^{\mu}} = \frac{(-)^{\mu-1}}{\Gamma\mu}\int_0^w\int_0^v \phi(x+iy)\left[\left(\frac{d}{dx}\right)^{\mu-1}\frac{1}{\phi(x+iy)}\right]dx\,dy,$$

the second integration cannot be effected.

Suppose next $\psi(x,\ y)$ is one of the functions $\gamma(x+iy)$, $g(x+iy)$, $G(x+iy)$, $\mathfrak{G}(x+iy)$, so that

$$\psi(x+rw,\ y+sv) = (\pm)^{r}(\pm)^{s}\,U_{r,s}\,\psi(x,\ y),$$

where

$$U_{r,s} = (-)^{rs}\,\epsilon^{\beta x(rw-svi)}\,q_{,}^{-\frac{1}{2}r^2}\,q^{-\frac{1}{2}s^2},$$

(see memoir quoted). Then, retaining the same value as before of $\Psi(x,\ y)$, we have still the formula (B), in which

$$\Theta(x+iy) = \Sigma\,\frac{(\pm)^{r}(\pm)^{s}\,U_{r,s}}{x+iy+rw+svi}.$$

But this summation has not yet been effected; the difficulty consists in the variable factor $\epsilon^{\beta x(rw-svi)}$ in the numerator, nothing being known I believe of the decomposition of functions into series of this form.

On the subject of the preceding paper may be consulted the following memoirs by Raabe, "Ueber die Summation periodischer Reihen," *Crelle*, t. XV. [1836], pp. 355—364, and "Ueber die Summation harmonisch periodischer Reihen," t. XXIII. [1842], pp. 105—125, and t. XXV. [1843], pp. 160—168. The integrals he considers, are taken between the limits 0, ∞ (instead of $-\infty$, ∞). His results are consequently more general than those given above, but they might be obtained by an analogous method, instead of the much more complicated one adopted by him: thus if $\phi(x+2\pi)=\phi x$, the integral $\int_0^\infty \phi x \frac{dx}{x}$ reduces itself to

$$\Sigma_0^\infty \int_0^{2\pi} \phi x \frac{dx}{x+2r\pi} = \int_0^{2\pi} dx\, \phi x \left[\frac{1}{x} + \Sigma_1{}^\infty \left(\frac{1}{x+2r\pi} - \frac{1}{2r\pi}\right)\right],$$

provided $\int_0^{2\pi} dx\phi x = 0$. The summation in this formula may be effected by means of the function Γ and its differential coefficient, and we have

$$\int_0^\infty \phi x \frac{dx}{x} = -\frac{1}{2\pi}\int_0^{2\pi} dx \frac{\Gamma'\left(\dfrac{x}{2\pi}\right)}{\Gamma\left(\dfrac{x}{2\pi}\right)} dx,$$

which is in effect Raabe's formula (10), *Crelle*, t. XXV. p. 166.

By dividing the integral on the right-hand side of the equation into two others whose limits are 0, π, and π, 2π respectively, and writing in the second of these $2\pi - x$ instead of x, then

$$\int_0^\infty \phi x \frac{dx}{x} = -\frac{1}{2\pi}\int_0^\pi \left(\phi x \frac{\Gamma'\left(\dfrac{x}{2\pi}\right)}{\Gamma\left(\dfrac{x}{2\pi}\right)} + \phi(2\pi - x) \frac{\Gamma'\left(1-\dfrac{x}{2\pi}\right)}{\Gamma\left(1-\dfrac{x}{2\pi}\right)}\right) dx;$$

or reducing by

$$\frac{\Gamma'\left(\dfrac{x}{2\pi}\right)}{\Gamma\left(\dfrac{x}{2\pi}\right)} = \frac{\Gamma'\left(1-\dfrac{x}{2\pi}\right)}{\Gamma\left(1-\dfrac{x}{2\pi}\right)} - \pi \cot \tfrac{1}{2}x,$$

we have

$$\int_0^\infty \phi x \frac{dx}{x} = \tfrac{1}{2}\int_0^\pi \phi x \cot \tfrac{1}{2}x\, dx - \frac{1}{2\pi}\int_0^\pi [\phi x + \phi(2\pi - x)] \frac{\Gamma'\left(1-\dfrac{x}{2\pi}\right)}{\Gamma\left(1-\dfrac{x}{2\pi}\right)} dx,$$

which corresponds to Raabe's formula (10′). If $\phi(-x) = -\phi x$, so that $\phi x + \phi(2\pi - x) = 0$, the last formula is simplified; but then the integral on the first side may be replaced by $\tfrac{1}{2}\int_{-\infty}^{\infty} \phi x \frac{dx}{x}$, so that this belongs to the preceding class of formulæ.

104.

ON THE THEORY OF PERMUTANTS.

[From the *Cambridge and Dublin Mathematical Journal*, vol. VII. (1852), pp. 40—51.]

A FORM may by considered as composed of blanks which are to be filled up by inserting in them specializing characters, and a form the blanks of which are so filled up becomes a symbol. We may for brevity speak of the blanks of a symbol in the sense of the blanks of the form from which such symbol is derived. Suppose the characters are 1, 2, 3, 4, ..., the symbol may always be represented in the first instance and without reference to the nature of the form, by $V_{1234}\ldots$ And it will be proper to consider the blanks as having an invariable order to which reference will implicitly be made; thus, in speaking of the characters 2, 1, 3, 4, ... instead of as before 1, 2, 4, ... the symbol will be $V_{2134}\ldots$ instead of $V_{1234}\ldots$. When the form is given we shall have an equation such as

$$V_{1234} = P_{12}\, Q_3\, R_4 \ldots \quad \text{or} \quad = P_{12}\, P_{34} \ldots \quad \&\text{c.},$$

according to the particular nature of the form.

Consider now the characters 1, 2, 3, 4, ..., and let the primitive arrangement and every arrangement derivable from it by means of an even number of inversions or interchanges of two characters be considered as positive, and the arrangements derived from the primitive arrangement by an odd number of inversions or interchanges of two characters be considered as negative; a rule which may be termed "the rule of signs." The aggregate of the symbols which correspond to every possible arrangement of the characters, giving to each symbol the sign of the arrangement, may be termed a "Permutant;" or, in distinction from the more general functions which will presently be considered, a simple permutant, and may be represented by enclosing the symbol in brackets, thus $(V_{1234}\ldots)$. And by using an expression still more elliptical than the blanks of a symbol, we may speak of the blanks of a permutant, or the characters of a permutant.

As an instance of a simple permutant, we may take

$$(V_{123}) = V_{123} + V_{231} + V_{312} - V_{132} - V_{213} - V_{321};$$

and if in particular $V_{123} = a_1 b_2 c_3$, then

$$(V_{123}) = a_1 b_2 c_3 + a_2 b_3 c_1 + a_3 b_1 c_2 - a_1 b_3 c_2 - a_2 b_1 c_3 - a_3 b_2 c_1.$$

It follows at once that a simple permutant remains unaltered, to the sign *près* according to the rule of signs, by any permutations of the characters entering into the permutant. For instance,

$$(V_{123}) = (V_{231}) = (V_{312}) = -(V_{132}) = -(V_{213}) = -(V_{321}).$$

Consequently also when two or more of the characters are identical, the permutant vanishes, thus

$$V_{113} = 0.$$

The form of the symbol may be such that the symbol remains unaltered, to the sign *près* according to the rule of signs, for any permutations of the characters in certain particular blanks. Such a system of blanks may be termed a quote. Thus, if the first and second blanks are a quote,

$$V_{123} = -V_{213}, \quad V_{132} = -V_{312}, \quad V_{231} = -V_{321},$$

and consequently

$$(V_{123}) = 2(V_{123} + V_{231} + V_{312});$$

and if the blanks constitute one single quote,

$$(V_{123} \ldots) = N V_{123} \ldots,$$

where $N = 1.2.3 \ldots n$, n being the number of characters. An important case, which will be noticed in the sequel, is that in which the whole series of blanks divide themselves into quotes, each of them containing the same number of blanks. Thus, if the first and second blanks, and the third and fourth blanks, form quotes respectively,

$$\tfrac{1}{6}(V_{1234}) = V_{1234} + V_{1342} + V_{1423} + V_{3412} + V_{4213} + V_{2314}.$$

It is easy now to pass to the general definition of a "Permutant." We have only to consider the blanks as forming, not as heretofore a single set, but any number of distinct sets, and to consider the characters in each set of blanks as permutable *inter se* and not otherwise, giving to the symbol the sign compounded of the signs corresponding to the arrangements of the characters in the different sets of blanks. Thus, if the first and second blanks form a set, and the third and fourth blanks form a set,

$$((V_{1234})) = V_{1234} - V_{2134} - V_{1243} + V_{2143}.$$

The word 'set' will be used throughout in the above technical sense. The particular mode in which the blanks are divided into sets may be indicated either in words or by some superadded notation. It is clear that the theory of permutants depends ultimately on that of simple permutants; for if in a compound permutant we first write down all the terms which can be obtained, leaving unpermuted the characters of a particular set, and replace each of the terms so obtained by a simple permutant having for its characters the characters of the previously unpermuted set, the result is obviously the original compound permutant. Thus, in the above-mentioned case, where the first and second blanks form a set and the third and fourth blanks form a set,

$$((V_{1234})) = (V_{1234}) - (V_{1243}),$$

or

$$((V_{1234})) = (V_{1234}) - (V_{2134}),$$

in the former of which equations the first and second blanks in each of the permutants on the second side form a set, and in the latter the third and fourth blanks in each of the permutants on the second side form a set, the remaining blanks being simply supernumerary and the characters in them unpermutable. It should be noted that the term quote, as previously defined, is only applicable to a system of blanks belonging to the same set, and it does not appear that anything would be gained by removing this restriction.

The following rule for the expansion of a simple permutant (and which may be at once extended to compound permutants) is obvious. Write down all the distinct terms that can be obtained, on the supposition that the blanks group themselves in any manner into quotes, and replace each of the terms so obtained by a compound permutant having for a distinct set the blanks of each assumed quote; the result is the original simple permutant. Thus in the simple permutant (V_{1234}), supposing for the moment that the first and second blanks form a quote, and that the third and fourth blanks form a quote, this leads to the equation

$$(V_{1234}) = +((V_{1234})) + ((V_{1342})) + ((V_{1423})) + ((V_{3412})) + ((V_{4213})) + ((V_{2314})),$$

where in each of the permutants on the second side the first and second blanks form a set, and also the third and fourth blanks.

The blanks of a simple or compound permutant may of course, without either gain or loss of generality, be considered as having any particular arrangement in space, for instance, in the form of a rectangle: thus $V_{\substack{12\\34}}$ is neither more nor less general than V_{1234}. The idea of some such arrangement naturally presents itself as affording a means of showing in what manner the blanks are grouped into sets. But, considering the blanks as so arranged in a rectangular form, or in lines and columns, suppose in the first instance that this arrangement is independent of the grouping of the blanks into sets, or that the blanks of each set or of any of them are distributed at random in the different lines and columns. Assume that the form is such that a symbol

$$V_{\substack{\alpha\beta\gamma\ldots\\ \alpha'\beta'\gamma'\ldots\\ \vdots}}$$

is a function of symbols $V_{\alpha\beta\gamma\ldots}$, $V_{\alpha'\beta'\gamma'\ldots}$, &c. Or, passing over this general case, and the case (of intermediate generality) of the function being a symmetrical function, assume that

$$V_{\substack{\alpha\beta\gamma\ldots\\ \alpha'\beta'\gamma'\ldots\\ \vdots}}$$

is the product of symbols $V_{\alpha\beta\gamma\ldots}$, $V_{\alpha'\beta'\gamma'\ldots}$, &c. Upon this assumption it becomes important to distinguish the different ways in which the blanks of a set are distributed in the different lines and columns. The cases to be considered are: (A). The blanks of a single set or of single sets are situated in more than one column. (B). The blanks of each single set are situated in the same column. (C). The blanks of each single set form a separate column. The case (B) (which includes the case (C)) and the case (C) merit particular consideration. In fact the case (B) is that of the functions which I have, in my memoir on Linear Transformations in the *Journal*, [13, 14] called hyperdeterminants, and the case (C) is that of the particular class of hyperdeterminants previously treated of by me in the *Cambridge Philosophical Transactions*, [12] and also particularly noticed in the memoir on Linear Transformations. The functions of the case (B) I now propose to call "Intermutants," and those in the case (C) "Commutants." Commutants include as a particular case "Determinants," which term will be used in its ordinary signification. The case (A) I shall not at present discuss in its generality, but only with the further assumption that the blanks form a single set (this, if nothing further were added, would render the arrangement of the blanks into lines and columns valueless), and moreover that the blanks of each line form a quote: the permutants of this class (from their connexion with the researches of Pfaff on differential equations) I shall term "Pfaffians." And first of commutants, which, as before remarked, include determinants.

The general expression of a commutant is

$$\left(V_{\substack{11\ldots\\22\\\vdots\\nn}}\right); \qquad \text{or} \qquad \begin{pmatrix}11\ldots\\22\\\vdots\\nn\end{pmatrix}$$

and (stating again for this particular case the general rule for the formation of a permutant) if, permuting the characters in the same column in every possible way, considering these permutations as positive or negative according to the rule of signs, one system be represented by

$$\begin{matrix}r_1 s_1 \ldots\\ r_2 s_2\\ \vdots\\ r_n s_n\end{matrix}$$

the commutant is the sum of all the different terms

$$\pm\pm\ldots V_{r_1 s_1 \ldots} V_{r_2 s_2 \ldots} V_{r_n s_n \ldots}$$

The different permutations may be formed as follows: first permute the characters in all the columns except a single column, and in each of the arrangements so obtained permute entire lines of characters. It is obvious that, considering any one of the arrangements obtained by permutations of the characters in all the columns but one, the permutations of entire lines and the addition of the proper sign will only reproduce the same symbol—in the case of an even number of columns constantly with the positive sign, but in the case of an odd number of columns with the positive or negative sign, according to the rule of signs. For the inversion or interchange of two entire lines is equivalent to as many inversions or interchanges of two characters as there are characters in a line, that is, as there are columns, and consequently introduces a sign compounded of as many negative signs as there are columns. Hence

THEOREM. A commutant of an even number of columns may be calculated by considering the characters of any one column (no matter which) as supernumerary unpermutable characters, and multiplying the result by the number of permutations of as many things as there are lines in the commutant.

The mark † added to a commutant of an even number of columns will be employed to show that the numerical multiplier is to be omitted. The same mark placed over any one of the columns of the commutant will show that the characters of that particular column are considered as non-permutable.

A determinant is consequently represented indifferently by the notations

$$\begin{pmatrix} 11 \\ 22 \\ \vdots \\ nn \end{pmatrix}^{\dagger} \quad \begin{pmatrix} \overset{\dagger}{1}1 \\ 22 \\ \vdots \\ nn \end{pmatrix}, \quad \begin{pmatrix} 1\overset{\dagger}{1} \\ 22 \\ \vdots \\ nn \end{pmatrix};$$

and a commutant of an odd number of columns vanishes identically.

By considering, however, a commutant of an odd number of columns, having the characters of some one column non-permutable, we obtain what will in the sequel be spoken of as commutants of an odd number of columns. This non-permutability will be denoted, as before, by means of the mark † placed over the column in question, and it is to be noticed that it is not, as in the case of a commutant of an even number of columns, indifferent over which of the columns the mark in question is placed; and consequently there would be no meaning in simply adding the mark † to a commutant of an odd number of columns.

A commutant is said to be symmetrical when the symbols $V_{\alpha\beta\gamma\ldots}$ are such as to remain unaltered by any permutations *inter se* of the characters α, β, $\gamma\ldots$ A commutant is said to be skew when each symbol $V_{\alpha\beta\gamma\ldots}$ is such as to be altered in sign only according to the rule of signs for any permutations *inter se* of the characters α, β, $\gamma\ldots$, this of course implies that the symbol $V_{\alpha\beta\gamma\ldots}$ vanishes when any two of the characters α, β, $\gamma\ldots$ are identical. The commutant is said to be demi-skew when $V_{\alpha,\beta,\gamma\ldots}$ is altered in sign only, according to the rule of signs for any permutation *inter se* of non-identical characters α, β, $\gamma,\ldots$

An intermutant is represented by a notation similar to that of a commutant. The sets are to be distinguished, whenever it is possible to do so, by placing in contiguity the symbols of the same set, and separating them by a stroke or bar from the symbols of the adjacent sets. If, however, the symbols of the same set cannot be placed contiguously, we may distinguish the symbols of a set by annexing to them some auxiliary character by way of suffix or otherwise, these auxiliary symbols being omitted in the final result. Thus

$$\begin{pmatrix} 1 & 1 & 1a \\ 2 & 2 & 2b \\ \overline{3} & 3 & 5a \\ 4 & \overline{3} & 6b \end{pmatrix}$$

would show that 1, 2 of the first column and the 3, 4 of the same column, the 1, 2 and the upper 3 of the second column, and the lower 3 of the same column, the 1, 5 of the third column, and the 2, 6 of the same column, form so many distinct sets,—the intermutant containing therefore

$$(2 \,.\, 2 \,.\, 6 \,.\, 1 \,.\, 2 \,.\, 2 =)\ 96 \text{ terms.}$$

A commutant of an even number of columns may be considered as an intermutant such that the characters of some one (no matter which) of its columns form each of them by itself a distinct set, and in like manner a commutant of an odd number of columns may be considered as an intermutant such that the characters of some one determinate column form each of them by itself a distinct set

The distinction of symmetrical, skew and demi-skew applies obviously as well to intermutants as to commutants. The theory of skew intermutants and skew commutants has a connexion with that of Pfaffians.

Suppose $V_{\alpha\beta\gamma\ldots} = V_{\alpha+\beta+\gamma\ldots}$ (which implies the symmetry of the intermutant or commutant) and write for shortness $V_0 = a$, $V_1 = b$, $V_2 = c$, &c. Then

$$\begin{bmatrix} 0 & 0 \\ 1 & 1 \end{bmatrix} = 2(ac - b^2),$$

$$\begin{bmatrix} 0 & 0 & 0 & 0 \\ 1 & 1 & 1 & 1 \end{bmatrix} = 2(ae - 4bd + 3c^2),$$

$$\begin{bmatrix} 0 & \overset{\dagger}{0} \\ 1 & 1 \end{bmatrix} = (ac - b^2), \text{ \&c.}$$

The functions on the second side are evidently hyperdeterminants such as are discussed in my memoir on Linear Transformations, and there is no difficulty in forming directly from the intermutant or commutant on the first side of the equation the symbol of derivation (in the sense of the memoir on Linear Transformations) from which the hyperdeterminant is obtained. Thus

$$\begin{bmatrix} 0 & 0 \\ 1 & 1 \end{bmatrix} \text{ is } \overline{12}^2 . UU, \qquad \begin{bmatrix} 0 & 0 & 0 & 0 \\ 1 & 1 & 1 & 1 \end{bmatrix} \text{ is } \overline{12}^4 . UU,$$

$$\begin{bmatrix} 0 & \overset{\dagger}{0} \\ 1 & 1 \end{bmatrix} \text{ is } \overline{12}\, U^{,0} U^{,1}, \qquad \begin{bmatrix} 0 & 0 & 0 & \overset{\dagger}{0} \\ 1 & 1 & 1 & 1 \end{bmatrix} \text{ is } \overline{12}^3 U^{,0} U^{,1},$$

each permutable column $\begin{matrix}0\\1\end{matrix}$ corresponding to a $\overline{12}$ [1] and a non-permutable column $\begin{matrix}\dagger\\0\\1\end{matrix}$ changing UU into $U^{,0}U^{,1}$. Similarly

$$\begin{pmatrix}0&0\\1&1\\2&2\end{pmatrix} \text{ becomes } (\overline{12}\,.\,\overline{13}\,.\,\overline{23})^2\,.\,UUU,$$

$$\begin{pmatrix}0&\overset{\dagger}{0}\\1&1\\2&2\end{pmatrix} \text{ becomes } \overline{12}\,.\,\overline{13}\,.\,\overline{23}\;U^{,0}U^{,1}U^{,2},$$

$$\begin{pmatrix}0&0\\1&1\\2&2\\3&3\end{pmatrix} \text{ becomes } (\overline{12}\,.\,\overline{13}\,.\,\overline{14}\,.\,\overline{23}\,.\,\overline{24}\,.\,\overline{34})^2\;UUUU, \text{ \&c.}$$

The analogy would be closer if in the memoir on Linear Transformations, just as $\overline{12}$ is used to signify $\begin{vmatrix}\xi_1, & \eta_1\\ \xi_2, & \eta_2\end{vmatrix}$, $\overline{123}$ had been used to signify $\begin{vmatrix}\xi_1^2, & \xi_1\eta_1, & \eta_1^2\\ \xi_2^2, & \xi_2\eta_2, & \eta_2^2\\ \xi_3^2, & \xi_3\eta_3, & \eta_3^2\end{vmatrix}$ &c., for then $\begin{pmatrix}0&0\\1&1\\2&2\end{pmatrix}$ would have corresponded to $\overline{123}^2\,.\,UUU$, $\begin{pmatrix}0&\overset{\dagger}{0}\\1&1\\2&2\end{pmatrix}$ to $\overline{123}\;U^{,0}U^{,1}U^{,2}$; and this would not only have been an addition of some importance to the theory, but would in some instances have facilitated the calculation of hyperdeterminants. The preceding remarks show that the intermutant

$$\begin{pmatrix}0&0&0\\1&1&\overline{1}\\\overline{0}&\overline{0}&0\\1&1&\overline{1}\end{pmatrix}$$

(where the first and fourth blanks in the last column are to be considered as belonging to the same set) is in the hyperdeterminant notation $(12\,.\,34)^2\,.\,(14\,.\,23)\;UUUU$.

[1] Viz. $\begin{matrix}0\\1\end{matrix}$ corresponds to $\overline{12}$ because 0 and 1 are the characters occupying the *first* and *second* blanks of a column. If 0 and 1 had been the characters occupying the *second* and *third* blanks in a column, the symbol would have been $\overline{23}$ and so on. It will be remembered, that the symbolic numbers 1, 2...... in the hyperdeterminant notation are merely introduced to distinguish from each other functions which are made identical after certain differentiations are performed.

It will, I think, illustrate the general theory to perform the development of the last-mentioned intermutant. We have

$$\begin{pmatrix} 0 & 0 & 0 \\ 1 & 1 & \overline{1} \\ \overline{0} & \overline{0} & 0 \\ 1 & 1 & \overline{1} \end{pmatrix} = \begin{pmatrix} 0 & 0 & \overset{\dagger}{0} \\ 1 & 1 & 1 \\ \overline{0} & \overline{0} & 0 \\ 1 & 1 & 1 \end{pmatrix} - \begin{pmatrix} 0 & 0 & \overset{\dagger}{0} \\ 1 & 1 & 0 \\ \overline{0} & \overline{0} & 1 \\ 1 & 1 & 1 \end{pmatrix} - \begin{pmatrix} 0 & 0 & \overset{\dagger}{1} \\ 1 & 1 & 1 \\ \overline{0} & \overline{0} & 1 \\ 1 & 1 & 0 \end{pmatrix} + \begin{pmatrix} 0 & 0 & \overset{\dagger}{1} \\ 1 & 1 & 0 \\ \overline{0} & \overline{0} & 1 \\ 1 & 1 & 0 \end{pmatrix}$$

$$= 2\left\{\begin{bmatrix} 0 & 0 & \overset{\dagger}{0} \\ 1 & 1 & 1 \end{bmatrix}\begin{bmatrix} 0 & 0 & \overset{\dagger}{0} \\ 1 & 1 & 1 \end{bmatrix} - \begin{bmatrix} 0 & 0 & \overset{\dagger}{0} \\ 1 & 1 & 0 \end{bmatrix}\begin{bmatrix} 0 & 0 & \overset{\dagger}{1} \\ 1 & 1 & 1 \end{bmatrix}\right\}$$

$$= 2\,\{(ad - bc)^2 - 4\,(ac - b^2)\,(bd - c^2)\},$$

$$= 2\,(a^2d^2 + 4ac^3 + 4b^3d - 3b^2c^2 - 6abcd),$$

the different steps of which may be easily verified.

The following important theorem (which is, I believe, the same as a theorem of Mr Sylvester's, published in the *Philosophical Magazine*) is perhaps best exhibited by means of a simple example. Consider the intermutant

$$\begin{pmatrix} x & 1 \\ y & 4 \\ \overline{x} & 3 \\ y & 2 \end{pmatrix}$$

where in the first column the sets are distinguished as before by the horizontal bar, but in the second column the 1, 2 are to be considered as forming a set, and the 3, 4 as forming a second set. Then, partially expanding, the intermutant is

$$\begin{pmatrix} \overset{\dagger}{x} & 1 \\ y & 4 \\ x & 3 \\ y & 2 \end{pmatrix} - \begin{pmatrix} \overset{\dagger}{y} & 1 \\ x & 4 \\ x & 3 \\ y & 2 \end{pmatrix} - \begin{pmatrix} \overset{\dagger}{x} & 1 \\ y & 4 \\ y & 3 \\ x & 2 \end{pmatrix} + \begin{pmatrix} \overset{\dagger}{y} & 1 \\ x & 4 \\ y & 3 \\ x & 2 \end{pmatrix},$$

or, since entire horizontal lines may obviously be permuted,

$$\begin{pmatrix} \overset{\dagger}{x} & 1 \\ y & 2 \\ x & 3 \\ y & 4 \end{pmatrix} - \begin{pmatrix} \overset{\dagger}{y} & 1 \\ y & 2 \\ x & 3 \\ x & 4 \end{pmatrix} - \begin{pmatrix} \overset{\dagger}{x} & 1 \\ x & 2 \\ y & 3 \\ y & 4 \end{pmatrix} + \begin{pmatrix} \overset{\dagger}{y} & 1 \\ x & 2 \\ y & 3 \\ x & 4 \end{pmatrix};$$

and, observing that the 1, 2 form a permutable system as do also the 3, 4, the second and third terms vanish, while the first and fourth terms are equivalent to each other; we may therefore write

$$2\left(\begin{array}{cc} \overset{\dagger}{x} & 1 \\ y & 2 \\ \hline x & 3 \\ y & 4 \end{array}\right) = \left(\begin{array}{cc} x & 1 \\ y & 4 \\ \hline x & 3 \\ y & 2 \end{array}\right)$$

where on the first side of the equation the bar has been introduced into the second column, in order to show that *throughout* the equation the 1, 2 and the 3, 4 are to be considered as forming distinct sets.

Consider in like manner the expression

$$\left(\begin{array}{cc} x & 1 \\ y & 7 \\ \underline{z} & 6 \\ x & 8 \\ y & 2 \\ \underline{z} & 9 \\ x & 4 \\ y & 5 \\ z & 3 \end{array}\right)$$

where in the first column the sets are distinguished by the horizontal bars and in the second column the characters 1, 2, 3 and 4, 5, 6 and 7, 8, 9 are to be considered as belonging to distinct sets. The same reasoning as in the former case will show that this is a multiple of

$$\left(\begin{array}{cc} x & 1 \\ y & 2 \\ z & 3 \\ \hline x & 4 \\ y & 5 \\ z & 6 \\ \hline x & 7 \\ y & 8 \\ z & 9 \end{array}\right)$$

and to find the numerical multiplier it is only necessary to inquire in how many ways, in the expression first written down, the characters of the first column can be permuted so that x, y, z may go with 1, 2, 3 and with 4, 5, 6 and with 7, 8, 9. The order of the x, y, z in the second triad may be considered as arbitrary; but once assumed, it determines the place of one of the letters in the first triad; for instance, $x8$ and $z9$ determine $y7$. The first triad must therefore contain $x1$ and $z6$ or $x6$ and $z1$. Suppose the former, then the third triad must contain $z3$, but the remaining two combinations may be either $x4$, $y5$, or $x5$, $y4$. Similarly, if the first triad contained $x6$, $z1$, there would be two forms of the third triad, or a given form of the second triad gives four different forms. There are therefore in all 24 forms, or

$$24\overset{\dagger}{\begin{pmatrix} x & 1 \\ y & 2 \\ z & 3 \\ \overline{x} & \overline{4} \\ y & 5 \\ z & 6 \\ \overline{x} & \overline{7} \\ y & 8 \\ z & 9 \end{pmatrix}} = \begin{pmatrix} x & 1 \\ y & 7 \\ z & 6 \\ \overline{x} & 8 \\ y & 2 \\ z & 9 \\ \overline{x} & 4 \\ y & 5 \\ z & 3 \end{pmatrix}$$

where the bars in the second column on the first side show that *throughout* the equation 1, 2, 3 and 4, 5, 6 and 7, 8, 9 are to be considered as forming distinct sets. The above proof is in reality perfectly general, and it seems hardly necessary to render it so in terms.

To perceive the significance of the above equation it should be noticed that the first side is a product of determinants, viz.

$$24\begin{pmatrix} x & 1 \\ y & 2 \\ z & 3 \end{pmatrix}^{\dagger} \begin{pmatrix} x & 5 \\ y & 6 \\ z & 7 \end{pmatrix}^{\dagger} \begin{pmatrix} x & 7 \\ y & 8 \\ z & 9 \end{pmatrix}^{\dagger};$$

and if the second side be partially expanded by permuting the characters of the second column, each of the terms so obtained is in like manner a product of determinants, so that

$$24\begin{pmatrix} x & 1 \\ y & 2 \\ z & 3 \end{pmatrix}^{\dagger} \begin{pmatrix} x & 4 \\ y & 5 \\ z & 6 \end{pmatrix}^{\dagger} \begin{pmatrix} x & 7 \\ y & 8 \\ z & 9 \end{pmatrix}^{\dagger} = \begin{pmatrix} x & 1 \\ y & 7 \\ z & 6 \end{pmatrix}^{\dagger} \begin{pmatrix} x & 8 \\ y & 2 \\ z & 9 \end{pmatrix}^{\dagger} \begin{pmatrix} x & 4 \\ y & 5 \\ z & 3 \end{pmatrix}^{\dagger} \pm \&c.,$$

the permutations on the second side being the permutations *inter se* of 1, 2, 3, of 4, 5, 6, and of 7, 8, 9.

It is obvious that the preceding theorem is not confined to intermutants of two columns.

POSTSCRIPT.

I wish to explain as accurately as I am able, the extent of my obligations to Mr Sylvester in respect of the subject of the present memoir. The term permutant is due to him—intermutant and commutant are merely terms framed between us in analogy with permutant, and the names date from the present year (1851). The theory of commutants is given in my memoir in the *Cambridge Philosophical Transactions*, [12], and is presupposed in the memoir on Linear Transformations, [13, 14]. It will appear by the last-mentioned memoir that it was by representing the coefficients of a biquadratic function by $a=1111$, $b=1112=1121=$&c., $c=1122=$&c., $d=1222=$&c., $e=2222$, and forming the commutant $\begin{pmatrix} 1111 \\ 2222 \end{pmatrix}$ that I was led to the function $ae-4bd+3c^2$. The function $ace+2bcd-ad^2-b^2e-c^3$ or $\begin{vmatrix} a, & b, & c \\ b, & c, & d \\ c, & d, & e \end{vmatrix}$ is mentioned in the memoir on Linear Transformations, as brought into notice by Mr Boole. From the particular mode in which the coefficients a, b,... were represented by symbols such as 1111, &c., I did not perceive that the last-mentioned function could be expressed in the commutant notation. The notion of a permutant, in its most general sense, is explained by me in my memoir, "Sur les déterminants gauches," *Crelle*, t. XXXVII. pp. 93—96, [69]; see the paragraph (p. 94) commencing "On obtient ces fonctions, &c." and which should run as follows: "On obtient ces fonctions (dont je reprends ici la théorie) par les propriétés génerales d'un determinant défini comme suit. En exprimant &c.;" the sentence as printed being "......défini. Car en exprimant &c.," which confuses the sense. [The paragraph is printed correctly 69, p. 411.] Some time in the present year (1851) Mr Sylvester, in conversation, made to me the very important remark, that as one of a class the above-mentioned function,

$$ace+2bcd-ad^2-b^2e-c^3,$$

could be expressed in the commutant notation $\begin{pmatrix} 0 & 0 \\ 1 & 1 \\ 2 & 2 \end{pmatrix}$, viz. by considering $00=a$, $01=10=b$, $02=11=20=c$, $12=21=d$, $22=e$; and the subject being thereby recalled to my notice, I found shortly afterwards the expression for the function

$$a^2d^2+4ac^3+4b^3d-3b^2c^2-6abcd$$

(which cannot be expressed as a commutant) in the form of an intermutant, and I was thence led to see the identity, so to say, of the theory of hyperdeterminants, as given in the memoir on Linear Transformations, with the present theory of intermutants. It is understood between Mr Sylvester and myself, that the publication of the present memoir is not to affect Mr Sylvester's right to claim the origination, and to be considered as the first publisher of such part as may belong to him of the theory here sketched out.

105.

CORRECTION OF THE POSTSCRIPT TO THE PAPER ON PERMUTANTS.

[From the *Cambridge and Dublin Mathematical Journal*, vol. VII. (1852), pp. 97—98.]

MR SYLVESTER has represented to me that the paragraph relating to his communications conveys an erroneous idea of the nature, purport, and extent of such communications; I have, in fact, in the paragraph in question, singled out what immediately suggested to me the expression of the function $6abcd + 3b^2c^2 - 4ac^3 - 4b^3d - a^2d^2$ as a partial commutant or intermutant, but I agree that a fuller reference ought to have been made to Mr Sylvester's results, and that the paragraph in question would more properly have stood as follows:

"Under these circumstances Mr Sylvester communicated to me a series of formal statements, not only oral but in writing, to the effect that he had discovered a permutation method of obtaining as many invariants—viz. commutantive invariants—by direct inspection from a function of any degree of any number of letters as the index of the degree contains even factors; one of these commutantive invariants being in fact the function $ace + 2bcd - ae^2 - bd^2 - c^3$, expressible, according to Mr Sylvester's notation, by $\begin{pmatrix} a^2, & ab, & b^2 \\ a^2, & ab, & b^2 \end{pmatrix}$; and, according to the notation of my memoir in the *Camb. Phil. Trans.*, supposing $00 = a$, $01 = 10 = b$, $02 = 11 = 20 = c$, &c. by $\begin{vmatrix} 00 \\ 11 \\ 22 \end{vmatrix}$."

Mr Sylvester and I shall, I have no doubt, be able to agree to a joint statement of any further correction or explanation which may be required.

106.

ON THE SINGULARITIES OF SURFACES.

[From the *Cambridge and Dublin Mathematical Journal*, vol. VII. (1852), pp. 166—171.]

In the following paper, for symmetry of nomenclature and in order to avoid ambiguities, I shall, with reference to plane curves and in various phrases and compound words, use the term "node" as synonymous with double point, and the term "spinode" as synonymous with cusp. I shall, besides, have occasion to consider the several singularities which I call the "flecnode," the "oscnode," the "fleflecnode," and the "tacnode:" the flecnode is a double point which is a point of inflexion on one of the branches through it; the oscnode is a double point which is a point of osculation on one of the branches through it; the fleflecnode is a double point which is a point of inflexion on each of the branches through it; and the tacnode is a double point where two branches touch. And it may be proper to remark here, that a tacnode may be considered as a point resulting from the coincidence and amalgamation of two double points (and therefore equivalent to twelve points of inflexion); or, in a different point of view, [?] as a point uniting the characters of a spinode and a flecnode. I wish to call to mind here the definition of conjugate tangent lines of a surface, viz. that a tangent to the curve of contact of the surface with any circumscribed developable and the corresponding generating line of the developable, are conjugate tangents of the surface.

Suppose, now, that an absolutely arbitrary surface of any order be intersected by a plane: the curve of intersection has not in general any singularities other than such as occur in a perfectly arbitrary curve of the same order; but as a plane can be made to satisfy one, two, or three conditions, the curve may be made to acquire singularities which do not occur in such absolutely arbitrary curve.

Let a single condition only be imposed on the plane. We may suppose that the curve of intersection has a node; the plane is then a tangent plane and the node is the point of contact—of course any point on the surface may be taken for the node. We may if we please use the term "nodes of a surface," "node-planes of a surface," as synonymous with the points and tangent planes of a surface. And it will be convenient also to use the word node-tangents to denote the tangents to the curve of intersection at the node; it may be noticed here that the node-tangents are conjugate tangents of the surface.

Next let two conditions be imposed upon the plane: there are three distinct cases to be considered.

First, the curve of intersection may have a flecnode. The plane (which is of course still a tangent plane at the flecnode) may be termed a flecnode-plane; the flecnodes are singular points on the surface lying on a curve which may be termed the "flecnode-curve[1]," and the flecnode-planes give rise to a developable which may be termed the flecnode-develope. The "flecnode-tangents" are the tangents to the curve of intersection at the flecnode; the tangent to the inflected branch may be termed the "singular flecnode-tangent," and the tangent to the other branch the "ordinary flecnode-tangent."

Secondly, the curve of intersection may have a spinode. The plane (which is of course still a tangent plane at the spinode) may be termed a spinode-plane; the spinodes are singular points on the surface lying on a curve which may be termed the "spinode-curve[2]." And the spinode-planes give rise to a developable which may be termed the "spinode-develope." Also the "spinode-tangent" is the tangent to the curve of intersection at the spinode.

Thirdly, the curve of intersection may have two nodes, or what may be termed a "node-couple." The plane (which is a tangent plane at each of the nodes and therefore a double tangent plane) may be also termed a "node-couple-plane." The node-couples are pairs of singular points on the surface lying in a curve which may be termed the "node-couple-curve," and the node-couple-planes give rise to a developable which may be termed the "node-couple-develope." The tangents to the curve of intersection at the two nodes of a node-couple might, if the term were required, be termed the "node-couple-tangents." Also one of the nodes of a node-couple may be termed a "node-with-node," and the tangents to the curve of intersection at such point will be the "node-with-node-tangents."

[1] The flecnode-curve, defined as the locus of the points through which can be drawn a line meeting the surface in four consecutive points, was, so far as I am aware, first noticed in Mr Salmon's paper "On the Triple Tangent Planes of a Surface of the Third Order" (*Journal*, t. IV. [1849], pp. 252—260), where Mr Salmon, among other things, shows that the order of the surface being n, the curve in question is the intersection of the surface with a surface of the order $11n-24$.

[2] The notion of a spinode, considered as the point where the indicatrix is a parabola (on which account the spinode has been termed a parabolic point) may be found in Dupin's *Développements de Géométrie*: the most important step in the theory of these points is contained in Hesse's memoir "Ueber die Wendepuncte der Curven dritter Ordnung" (*Crelle*, t. XXVIII. [1848], pp. 97—107), where it is shown that the spinode-curve is the curve of intersection of the surface supposed as before of the order n, with a certain surface of the order $4(n-2)$. See also Mr Salmon's memoir "On the Condition that a Plane should touch a surface along a Curve Line" (*Journal*, t. III. [1848], pp. 44—46).

It is hardly necessary to remark that the flecnode-curve is *not* the edge of regression of the flecnode-develope, and the like remark applies *m.m.* to the spinode-curve and the node-couple curve.

Finally, let three conditions be imposed upon the plane: there are six distinct cases to be considered, in each of which we have no longer curves and developes, but only singular points and singular tangent planes determinate in number.

First, the curve of intersection may have an oscnode. The plane (which continues a tangent plane at the oscnode) is an "oscnode-plane." The "oscnode-tangents" are the tangents to the curve of intersection at the oscnode; the tangent to the osculating branch is the "singular oscnode-tangent;" and the tangent to the other branch the "ordinary oscnode-tangent."

Secondly, the curve of intersection may have a fleflecnode. The plane (which continues a tangent plane at the fleflecnode) is a "fleflecnode-plane." The "fleflecnode-tangents" are the tangents to the curve of intersection at the fleflecnode.

Thirdly, the curve of intersection may have a tacnode. The plane (which continues a tangent plane at the tacnode) is a "tacnode-plane." The "tacnode-tangent" is the tangent to the curve of intersection at the tacnode.

Fourthly, the curve of intersection may have a node and a flecnode, or what may be termed a node-and-flecnode. The plane (which is a tangent plane at the node and also at the flecnode, where it is obviously a flecnode-plane) is a "node-and-flecnode-plane." The "node-and-flecnode-tangents," if the term were required, would be the tangents to the curve of intersection at the node and at the flecnode of the node-and-flecnode. The node of the node-and-flecnode may be distinguished as the node-with-flecnode, and the flecnode as the flecnode-with-node, and we have thus the terms "node-with-flecnode-tangents," "flecnode-with-node-tangents," "singular flecnode-with-node-tangent," and "ordinary flecnode-with-node-tangent."

Fifthly, the curve of intersection may have a node and also a spinode, or what may be termed a "node-and-spinode." The plane (which is a tangent plane at the node, and is also a tangent plane at the spinode, where it is obviously a spinode-plane) is a "node-and-spinode-plane." The node-and-spinode-tangents, if the term were required, would be the tangents at the node and the tangent at the spinode of the node-and-spinode to the curve of intersection. The node of the node-and-spinode may be distinguished as the "node-with-spinode," and the spinode as the "spinode-with-node," and we have thus the terms "node-with-spinode-tangent," "spinode-with-node-tangent."

Sixthly, the curve of intersection may have three nodes, or what may be termed a "node-triplet." The plane (which is a triple tangent plane touching the surface at each of the nodes) is a "node-triplet-plane." The "node-triplet-tangents," if the term were required, would be the tangents to the curve of intersection at the nodes of the node-triplet. Each node of the node-triplet may be distinguished as a "node-with-node-couple," and the tangents to the curve of intersection at such nodes are "node-with-node-couple-tangents." The terms "node-couple-with-node," "node-couple-with-node-tangent," might be made use of if necessary.

It should be remarked that the oscnodes lie on the flecnode-curve, as do also the fleflecnodes; these latter points are real double points of the flecnode-curve. The tacnodes are points of intersection and (what will appear in the sequel) points of contact of the flecnode-curve, the spinode-curve, and the node-couple-curve. The spinode-with-nodes are points of intersection of the spinode-curve and node-couple-curve, and the flecnode-with-nodes are points of intersection of the flecnode-curve and node-couple-curve; the node-with-node-couples are real double points (entering in triplets) of the node-couple-curve.

Consider for a moment an arbitrary curve on the surface; the locus of the node-tangents at the different points of this curve is in general a skew surface, which may however, in cases to be presently considered, degenerate in different ways.

Reverting now to the flecnode-curve, it may be shown that the singular flecnode-tangent coincides with the tangent of the flecnode-curve. For consider on a surface two consecutive points such that the line joining them meets the surface in two points consecutive to the first-mentioned two points. The line meets the surface in four consecutive points, it is therefore a singular flecnode-tangent; *each* of the first-mentioned two points must be on the flecnode-curve, or the singular flecnode-tangent touches the flecnode-curve. The two flecnode-tangents are by a preceding observation conjugate tangents. It follows that the skew surface, locus of the flecnode-tangents, breaks up into two surfaces, each of which is a developable, viz. the locus of the singular flecnode-tangents is the developable having the flecnode-curve for its edge of regression, and the locus of the ordinary flecnode-tangents is the flecnode-develope. Of course at the tacnode, the tacnode-tangent touches the flecnode-curve.

Passing next to the spinode-curve, the spinode-plane and the tangent-plane at a consecutive point along the spinode-tangent are identical[1], or their line of intersection is indeterminate. The spinode-tangent is therefore the conjugate tangent to *any* other tangent line at the spinode, and therefore to the tangent to the spinode-curve. It follows that the surface locus of the spinode-tangents degenerates into a developable surface twice repeated, viz. the spinode-develope. Consider the tacnode as two coincident nodes; each of these nodes, by virtue of its constituting, in conjunction with the other, a tacnode, is on the spinode-curve; or, in other words, the tacnode-tangent touches the spinode-curve, and the same reasoning proves that it touches the node-couple-curve. It has already been seen that the tacnode-tangent touches the flecnode-curve; consequently the tacnode is a point, not of simple intersection only, but of contact, of the flecnode-curve, the spinode-curve, and the node-couple-curve.

In virtue of the principle of the spinode-plane being identical with the tangent plane at a consecutive point along the spinode tangent, it appears that the tacnode-

[1] It must not be inferred that the tangent plane at such consecutive point is a spinode-plane; this is obviously not the case.

plane is a stationary plane, as well of the flecnode-develope as of the spinode-develope, and it would at first sight appear that it must be also a stationary tangent plane of the node-couple-develope. But this is not so; the node-with-node-planes envelope, not the node-couple-develope, but the node-couple-develope twice repeated: the tacnode-plane is in a sense a stationary plane on such duplicate developable, but not in any manner on the single developable. The tacnode-plane is an ordinary tangent plane of the node-couple-develope.

Consider now a spinode-with-node, which we have seen is a point of intersection of the spinode-curve and node-couple-curve. The tangent plane at a consecutive point along the spinode-with-node-tangent, is *identical* with the spinode-with-node-plane; the curve of intersection of the tangent plane at such consecutive point has therefore a node at the node-with-spinode, or the tangent plane in question is a node-couple-plane, and the point of contact is a point on the node-couple-curve. Consequently the spinode-with-node-tangent touches the node-couple-curve, and thence also the spinode-with-node-plane is a stationary tangent plane of the node-couple-develope.

It should be remarked that no circumscribed developable can have a stationary tangent plane except the tangent planes at the points where the curve of contact meets the spinode-curve, and any one of these planes is only a stationary plane when the curve of contact touches the spinode-tangent; and that the node-couple-curve and the flecnode-curve do not intersect the spinode-curve except in the points which have been discussed.

Recapitulating, the node-couple-curve and the spinode-curve touch at the tacnodes, and intersect at the spinode-with-nodes: moreover, the tacnode-planes are stationary planes of the spinode-develope, and the spinode-with-node-planes are stationary planes of the node-couple-develope. Besides this, the two curves are touched at the tacnodes by the flecnode-curve, and the tacnode-planes are stationary planes of the flecnode-develope.

107.

ON THE THEORY OF SKEW SURFACES.

[From the *Cambridge and Dublin Mathematical Journal*, vol. VII. (1852), pp. 171—173.]

A SURFACE of the nth order is a surface which is met by an indeterminate line in n points. It follows immediately that a surface of the nth order is met by an indeterminate plane in a curve of the nth order.

Consider a skew surface or the surface generated by a singly infinite series of lines, and let the surface be of the nth order. Any plane through a generating line meets the surface in the line itself and in a curve of the $(n-1)^{\text{th}}$ order. The generating line meets this curve in $(n-1)$ points. Of these points one, viz. that adjacent to the intersection of the plane with the consecutive generating line, is a unique point; the other $(n-2)$ points form a system. Each of the $(n-1)$ points are *sub modo* points of contact of the plane with the surface, but the proper point of contact is the unique point adjacent to the intersection of the plane with the consecutive generating line. Thus every plane through a generating line is an ordinary tangent plane, the point of contact being a point on the generating line. It is not necessary for the present purpose, but I may stop for a moment to refer to the known theorems that the anharmonic ratio of any four tangent planes through the same generating line is equal to the anharmonic ratio of their points of contact, and that the locus of the normals to the surface along a generating line is a hyperbolic paraboloid. Returning to the $(n-2)$ points in which, together with the point of contact, a generating line meets the curve of intersection of the surface and a plane through the generating line, these are fixed points independent of the particular plane, and are the points in which the generating line is intersected by other generating lines. There is therefore on the surface a double curve intersected in $(n-2)$ points by each generating line of the surface—a property which, though insufficient to determine the order of this double curve, shows that the order cannot be less than $(n-2)$. (Thus for $n=4$, the above reasoning shows that the double-curve must be at least of the second order: assuming for a moment that it is in any case precisely of this order, it obviously cannot be a plane curve, and must therefore be two non-intersecting lines. This suggests at any rate the existence of a class of skew surfaces of the fourth order generated by a line which always passes through two fixed lines and by some other condition not yet ascertained; and it would appear that surfaces of the second order constitute a degenerate species belonging to the class in question.)

In particular cases a generating line will be intersected by the consecutive generating line. Such a generating line touches the double curve.

Consider now a point not on the surface; the planes determined by this point and the generating lines of the surface are the tangent planes through the point; the intersections of consecutive tangent planes are the tangent lines through the point; and the cone generated by these tangent lines or enveloped by the tangent planes is the tangent cone corresponding to the point. This cone is of the nth class. For considering a line through the point, this line meets the surface in n points, i.e. it meets n generating lines of the surface; and the planes through the line and these n generating lines, are of course tangent planes to the cone: that is, n tangent planes can be drawn to the cone through a given line passing through the vertex. The cone has not in general any lines of inflexion, or, what is the same thing, stationary tangent planes. For a stationary tangent plane would imply the intersection of two consecutive generating lines of the surface. And since the number of generating lines intersected by a consecutive generating line, and therefore the number of planes through two consecutive generating lines, is finite, no such plane passes through an indeterminate point. The tangent cone will have in general a certain number of double tangent planes; let this number be x. We have therefore a cone of the class n, number of double tangent planes x, number of stationary tangent planes 0. Hence, if m be the order of the cone, α the number of its double lines, and β the number of its cuspidal or stationary lines,

$$m = n(n-1) - 2x,$$

$$\beta = 3n(n-2) - 6x,$$

$$\alpha = \tfrac{1}{2}n(n-2)(n^2-9) - 2x(n^2-n-6) + 2x(x-1).$$

This is the proper tangent cone, but the cone through the double curve is *sub modo* a tangent cone, and enters as a square factor into the equation of the general tangent cone of the order $n(n-1)$. Hence, if X be the order of the double curve, and therefore of the cone through this curve,

$$m + 2X = n(n-1), \quad \text{and therefore } X = x;$$

that is, the number of double tangent planes to the tangent cone is equal to the order of the double curve. It does not appear that there is anything to determine x; and if this is so, skew surfaces of the nth order may be considered as forming different families according to the order of the double curve upon them.

To complete the theory, it should be added that a plane intersects the surface in a curve of the nth order having x double points but no cusps.

108.

ON CERTAIN MULTIPLE INTEGRALS CONNECTED WITH THE THEORY OF ATTRACTIONS.

[From the *Cambridge and Dublin Mathematical Journal*, vol. VII. (1852), pp. 174—178.]

It is easy to deduce from Mr Boole's formula, given in my paper "On a Multiple Integral connected with the theory of Attractions," *Journal*, t. II. [1847], pp. 219—223, [44], the equation

$$\int \frac{d\xi\, d\eta \ldots}{[(\xi-\alpha)^2+(\eta-\beta)^2+\ldots v^2]^{\frac{1}{2}n-q}} = \frac{fg\ldots \pi^{\frac{1}{2}n}}{\theta_1{}^n\Gamma(\frac{1}{2}n-q)\,\Gamma(q+1)}\int_\epsilon^\infty \frac{s^{q-1}(\theta_1{}^2-\sigma)^q\, ds}{\sqrt{\left\{\left(s+\frac{f^2}{\theta_1{}^2}\right)\left(s+\frac{g^2}{\theta_1{}^2}\right)\ldots\right\}}}$$

where n is the number of variables of the multiple integral, and the condition of the integration is

$$\frac{(\xi-\alpha_1)^2}{f^2}+\frac{(\eta-\beta_1)^2}{g^2}+\ldots \overset{=}{<} 1;$$

also where

$$\sigma = \frac{(\alpha-\alpha_1)^2}{s+\frac{f^2}{\theta_1{}^2}}+\frac{(\beta-\beta_1)^2}{s+\frac{g^2}{\theta_1{}^2}}\ldots+\frac{v^2}{s},$$

and ϵ is the positive root of

$$\theta_1{}^2 = \frac{(\alpha-\alpha_1)^2}{\epsilon+\frac{f^2}{\theta_1{}^2}}+\frac{(\beta-\beta_1)^2}{\epsilon+\frac{g^2}{\theta_1{}^2}}\ldots+\frac{v^2}{\epsilon}.$$

Suppose $f=g\ldots=\theta_1$, and write $(\alpha-\alpha_1)^2+\ldots=k^2$, we obtain

$$\int \frac{d\xi \ldots}{[(\xi-\alpha)^2+\ldots v^2]^{\frac{1}{2}n-q}} = \frac{\pi^{\frac{1}{2}n}}{\Gamma(\frac{1}{2}n-q)\,\Gamma(q+1)}\int_\epsilon^\infty \frac{s^{q-1}(\theta_1{}^2-\sigma)^q\, ds}{(1+s)^{\frac{1}{2}n}},$$

the limiting condition for the multiple integral being

$$(\xi-\alpha_1)^2+\ldots \gtreqless \theta_1^2,$$

and the function σ, and limit ϵ, being given by

$$\sigma=\frac{k^2}{1+s}+\frac{v^2}{s}, \quad \theta_1^2=\frac{k^2}{1+\epsilon}+\frac{v^2}{\epsilon},$$

ϵ denoting, as before, the positive root. Observing that the quantity under the integral sign on the second side vanishes for $s=\epsilon$, there is no difficulty in deducing, by a differentiation with respect to θ_1, the formula

$$\int\frac{d\Sigma}{[(\xi-\alpha)^2\ldots+v^2]^{\frac{1}{2}n-q}}=\frac{2\theta_1\pi^{\frac{1}{2}n}}{\Gamma(\frac{1}{2}n-q)\,\Gamma(q)}\int_\epsilon^\infty\frac{s^{q-1}(\theta_1^2-\sigma)^{q-1}\,ds}{(1+s)^{\frac{1}{2}n}},$$

where $d\Sigma$ is the element of the surface $(\xi-\alpha_1)^2+\ldots=\theta_1^2$, and the integration is extended over the entire surface.

A slight change of form is convenient. We have

$$\theta_1^2-\sigma=\theta_1^2-\frac{k^2}{1+s}-\frac{v^2}{s}=\frac{1}{s(1+s)}(\theta_1^2s^2+\chi s-v^2),$$

if we suppose

$$\chi=\theta_1^2-k^2-v^2.$$

The formulæ then become

$$\int\frac{d\xi\ldots}{[(\xi-\alpha)^2\ldots+v^2]^{\frac{1}{2}n-q}}=\frac{\pi^{\frac{1}{2}n}}{\Gamma(\frac{1}{2}n-q)\,\Gamma(q+1)}\int_\epsilon^\infty\frac{(\theta_1^2s^2+\chi s-v^2)^q\,ds}{s(1+s)^{\frac{1}{2}n+q}},$$

$$\int\frac{d\Sigma}{[(\xi-\alpha)^2\ldots+v^2]^{\frac{1}{2}n-q}}=\frac{2\pi^{\frac{1}{2}n}\theta_1}{\Gamma(\frac{1}{2}n-q)\,\Gamma q}\int_\epsilon^\infty\frac{(\theta_1^2s^2+\chi s-v^2)^{q-1}\,ds}{(1+s)^{\frac{1}{2}n+q-1}},$$

in which ϵ is the positive root of the equation

$$\theta_1^2\epsilon^2+\chi\epsilon-v^2=0.$$

I propose to transform these formulæ by means of the theory of images; it will be convenient to investigate some preliminary formulæ. Suppose $\lambda^2=\alpha^2+\beta^2\ldots$, $\lambda_1^2=\alpha_1^2+\beta_1^2\ldots$; also consider the new constants $a, b, \ldots, a_1, b_1, \ldots, u, f_1$, determined by the equations

$$\frac{\delta^2\alpha}{\lambda^2+v^2}=a, \quad \frac{\delta^2\alpha_1}{\lambda_1^2-\theta_1^2}=a_1,$$
$$\vdots \qquad\qquad \vdots$$
$$\frac{\delta^2 v}{\lambda^2+v^2}=u, \quad \frac{\delta^2\theta_1}{\lambda_1^2\sim\theta_1^2}=f_1,$$

where δ is arbitrary. Then, putting

$$l^2=a^2+b^2\ldots, \quad l_1^2=a_1^2+b_1^2\ldots,$$

it is easy to see that

$$(\lambda^2+v^2)(l^2+u^2)=\delta^4,\quad (\lambda_1^2-\theta_1^2)(l_1^2-f_1^2)=\delta^4,$$

and

$$\frac{\delta^2 a}{l^2+u^2}=\alpha,\qquad \frac{\delta^2 a_1}{l_1^2-f_1^2}=\alpha_1,$$
$$\vdots\qquad\qquad \vdots$$
$$\frac{\delta^2 u}{l^2+u^2}=v,\qquad \frac{\delta^2 f_1}{l_1^2-f_1^2}=\theta_1.$$

Proceeding to express the single integrals in terms of the new constants, we have in the first place $k^2=\delta^4 k^2$, where

$$k^2=\left(\frac{a}{l^2+u^2}-\frac{a_1}{l_1^2-f_1^2}\right)^2+\ldots;$$

or if we write

$$aa_1+bb_1\ldots=ll_1\cos\omega,$$

we have

$$k^2=\frac{l^2}{(l^2+u^2)^2}+\frac{l_1^2}{(l_1^2-f_1^2)^2}-\frac{2ll_1\cos\omega}{(l^2+u^2)(l_1^2-f_1^2)}.$$

Hence also $\chi=\delta^4 j$, where

$$j=\frac{f_1^2}{(l_1^2-f_1^2)^2}-k^2-\frac{u^2}{(l^2+u^2)^2},$$

whence

$$-j=\frac{1}{l^2+u^2}+\frac{1}{l_1^2-f_1^2}+\frac{2ll_1\cos\omega}{(l^2+u^2)(l_1^2-f_1^2)},$$
$$=\frac{1}{(l^2+u^2)(l_1^2-f_1^2)}\{p^2+u^2-f_1^2\},$$

where $p^2=l^2+l_1^2-2ll_1\cos\omega$, that is

$$p^2=(a-a_1)^2+(b-b_1)^2+\ldots;$$

consequently $\theta_1^2 s^2+\chi s-v^2=\delta^4\Pi$, where Π is given by

$$\Pi=\frac{f_1^2}{(l_1^2-f_1^2)^2}s^2-\frac{(p^2+u^2-f_1^2)}{(l^2+u^2)(l_1^2-f_1^2)}s-\frac{u^2}{(l^2+u^2)^2};$$

and it is clear that ϵ will be the positive root of

$$0=\frac{f_1^2}{(l_1^2-f_1^2)^2}\epsilon^2-\frac{(p^2+u^2-f_1^2)}{(l^2+u^2)(l_1^2-f_1^2)}\epsilon-\frac{u^2}{(l^2+u^2)^2}.$$

It may be noticed that, in the particular case of $u=0$, the roots of this equation are 0, and $\dfrac{(p^2-f_1^2)(l_1^2-f_1^2)}{l^2f_1^2}$. Consequently if $p^2-f_1^2$ and $l_1^2-f_1^2$ are of opposite signs, we have $\epsilon=0$; but if $p^2-f_1^2$ and $l_1^2-f_1^2$ are of the same sign, $\epsilon=\dfrac{(p^2-f_1^2)(l_1^2-f_1^2)}{l^2f_1^2}$.

In order to transform the double integrals, considering the new variables $x, y, \ldots$, I write $x^2+y^2\ldots=r^2$ and

$$\xi=\frac{\delta^2 x}{r^2}, \ldots$$

whence also, if $\xi^2+\eta^2+\ldots=\rho^2$ (which gives $r\rho=\delta^2$), we have

$$x=\frac{d^2\xi}{\rho^2}, \ldots;$$

also it is immediately seen that

$$(\xi-\alpha)^2+\ldots\,\upsilon^2=\frac{\delta^4}{(l^2+u^2)\,r^2}\{(x^2-a)^2+\ldots+u^2\},$$

$$(\xi-\alpha_1)^2\ldots-\theta_1{}^2=\frac{\delta^4}{(l_1{}^2-f_1{}^2)\,r^2}\{(x-a_1)^2+\ldots-f_1{}^2\};$$

and from the latter equation it follows that the limiting condition for the first integral is $(x-a_1)^2+\ldots \mathrel{\overline{\gtrless}} f_1{}^2$ (there is no difficulty in seeing that the sign $<$ in the former limiting condition gives rise here to the sign $>$), and that the second integral has to be extended over the surface $(x-a_1)^2+\ldots=f_1{}^2$. Also if dS represent the element of this surface, we may obtain

$$d\xi\,d\eta\ldots=\frac{\delta^{2n}}{r^{2n}}\,dx\,dy\ldots,\quad d\Sigma=\frac{\delta^{2n-2}}{r^{2n-2}}\,dS;$$

and, combining the above formulæ, we obtain

$$\int\frac{dx\,dy\ldots}{(x^2+y^2\ldots)^{\frac{1}{2}n+q}\{(x-a)^2+(y-b)^2\ldots+u^2\}^{\frac{1}{2}n-q}}$$
$$=\frac{\pi^{\frac{1}{2}n}}{\Gamma(\frac{1}{2}n-q)\,\Gamma(q+1)\,(l^2+u^2)^{\frac{1}{2}n-q}}\int_\epsilon^\infty\frac{\Pi^q ds}{s\,(1+s)^{\frac{1}{2}n+q}},$$

the limiting condition of the multiple integral being

$$(x-a_1)^2+(y-b_1)^2\ldots \mathrel{\overline{\gtrless}} f_1{}^2;$$

and

$$\int\frac{dS}{(x^2+y^2\ldots)^{\frac{1}{2}n+q-1}\{(x-a)^2+(y-b)^2+u^2\}^{\frac{1}{2}n-q}}$$
$$=\frac{2\pi^{\frac{1}{2}n}f_1}{\Gamma(\frac{1}{2}n-q)\,\Gamma q\,(l^2+u^2)^{\frac{1}{2}n-q}\,(l_1{}^2\sim f_1{}^2)}\int_\epsilon^\infty\frac{\Pi^{q-1}ds}{(1+s)^{\frac{1}{2}n+q-1}},$$

where dS is the element of the surface $(x-a_1)^2+(y-b_1)^2\ldots=f_1{}^2$, and the integration extends over the entire surface. In these formulæ, l, l_1, p, Π denote as follows:

$$l^2=a^2+b^2+\ldots,\quad l_1{}^2=a_1{}^2+b_1{}^2+\ldots,\quad p^2=(a-a_1)^2+(b-b_1)^2+\ldots,$$

$$\Pi=\frac{f_1{}^2}{(l_1{}^2-f_1{}^2)^2}s^2-\frac{(p^2+u^2-f_1{}^2)}{(l_1{}^2-f_1{}^2)(l^2+u^2)}s-\frac{u^2}{(l^2+u^2)^2}:$$

and ϵ is the positive root of the equation $\Pi=0$.

The only obviously integrable case is that for which in the second formula $q=1$; this gives

$$\int \frac{dS}{(x^2+y^2 \ldots)^{\frac{1}{2}n} \{(x-a)^2+(y-b)^2+u^2\}^{\frac{1}{2}n-1}} = \frac{2\pi^{\frac{1}{2}n} f_1}{\Gamma(\frac{1}{2}n)\,(l^2+u^2)^{\frac{1}{2}n-1}\,(l_1^2 \sim f_1^2)\,(1+\epsilon)^{\frac{1}{2}n-1}}.$$

In the case of $u=0$, we have, as before, when $p^2-f_1^2$ and $l_1^2-f_1^2$ are of opposite signs, $\epsilon=0$, and therefore $1+\epsilon=1$; but when $p^2-f_1^2$ and $l_1^2-f_1^2$ are of the same sign, the value before found for ϵ gives

$$1+\epsilon=\frac{1}{l^2 f_1^2}\{l^2 f_1^2+(p^2-f_1^2)(l_1^2-f_1^2)\}.$$

Consider the image of the origin with respect to the sphere $(x-a_1)^2+(y-b_1)^2 \ldots = f_1^2$, the coordinates of this image are

$$\frac{a_1}{l_1^2}(l_1^2-f_1^2), \quad \frac{b_1}{l_1^2}(l_1^2-f_1^2), \ldots,$$

and consequently, if μ be the distance of this image from the point $(a, b \ldots)$, we have

$$\mu^2=\{a-\frac{a}{l_1^2}(l_1^2-f_1^2)\}^2+\ldots$$

$$=\frac{1}{l_1^2}\{l^2 f_1^2+(p^2-f_1^2)(l_1^2-f_1^2)\};$$

whence, by a simple reduction,

$$1+\epsilon=\frac{l_1^2\mu^2}{l^2 f_1^2},$$

or the values of the integral are

$p^2-f_1^2$ and $l_1^2-f_1^2$ opposite signs, $I=\dfrac{2\pi^{\frac{1}{2}n}}{\Gamma(\frac{1}{2}n)}\,\dfrac{f_1}{l^{n-2}(l_1^2 \sim f_1^2)}$,

$p^2-f_1^2$ and $l_1^2-f_1^2$ the same sign, $I=\dfrac{2\pi^{\frac{1}{2}n}}{\Gamma(\frac{1}{2}n)}\,\dfrac{f_1^{n-1}}{l_1^{n-2}\mu^{n-2}(l_1^2 \sim f_1^2)}$,

where μ is the distance from the point $(a, b \ldots)$ of the image of the origin with respect to the sphere $(x-a_1)^2+\ldots-f_1^2=0$.

Stone Buildings, August 6, 1850.

109.

ON THE RATIONALISATION OF CERTAIN ALGEBRAICAL EQUATIONS.

[From the *Cambridge and Dublin Mathematical Journal*, vol. VIII. (1853), pp. 97—101.]

SUPPOSE

$$x + y = 0, \quad x^2 = a, \quad y^2 = b;$$

then if we multiply the first equation by 1, xy, and reduce by the two others, we have

$$x + \ \ y = 0,$$
$$bx + ay = 0,$$

from which, eliminating x, y,

$$\begin{vmatrix} 1, & 1 \\ b, & a \end{vmatrix} = 0;$$

which is the equation between a and b; or, considering x, y as quadratic radicals, the rational equation between x, y. So if the original equation be multiplied by x, y, we have

$$a + xy = 0,$$
$$b + xy = 0;$$

or, eliminating 1, xy,

$$\begin{vmatrix} a, & 1 \\ b, & 1 \end{vmatrix} = 0,$$

which may be in like manner considered as the rational equation between x, y.

The preceding results are of course self-evident, but by applying the same process to the equations

$$x + y + z = 0, \quad x^2 = a, \quad y^2 = b, \quad z^2 = c,$$

we have results of some elegance. Multiply the equation first by 1, yz, zx, xy, reduce and eliminate the quantities x, y, z, xyz, we have the rational equation

$$\left|\begin{array}{cccc} & 1 & 1 & 1 \\ 1 & . & c & b \\ 1 & c & . & a \\ 1 & b & a & . \end{array}\right| = 0;$$

and again, multiply the equation by x, y, z, xyz, reduce and eliminate the quantities 1, yz, zx, xy, the result is

$$\left|\begin{array}{cccc} & a & b & c \\ a & . & 1 & 1 \\ b & 1 & . & 1 \\ c & 1 & 1 & . \end{array}\right| = 0,$$

which is of course equivalent to the preceding one (the two determinants are in fact identical in value), but the form is essentially different. The former of the two forms is that given in my paper "On a theorem in the Geometry of Position" (*Journal*, vol. II. [1841] p. 270 [1]): it was only very recently that I perceived that a similar process led to the latter of the two forms.

Similarly, if we have the equations

$$x + y + z + w = 0, \quad x^2 = a, \quad y^2 = b, \quad z^2 = c, \quad w^2 = d,$$

then multiplying by 1, yz, zx, xy, xw, yw, zw, $xyzw$, reducing and eliminating the quantities in the outside row,

we have the result

$$\begin{array}{c} \begin{array}{cccccccc} x, & y, & z, & w, & yzw, & zwx, & wxy, & xyz \end{array} \\ \left|\begin{array}{ccc|c|ccc|c} 1 & 1 & 1 & 1 & . & . & . & . \\ \hline . & c & b & . & 1 & . & . & 1 \\ c & . & a & . & . & 1 & . & 1 \\ b & a & . & . & . & . & 1 & 1 \\ \hline d & . & . & a & . & 1 & 1 & . \\ . & d & . & b & 1 & . & 1 & . \\ . & . & d & c & 1 & 1 & . & . \\ \hline . & . & . & . & a & b & c & d \end{array}\right| = 0; \end{array}$$

so if we multiply the equations by x, y, z, w, yzw, zwx, wxy, and xyz, reduce and eliminate the quantities in the outside row, we have the result

$$\begin{array}{c|ccc|ccc|c|l}
1, & yz, & zx, & xy, & xw, & yw, & zw, & xyzw & \\
\hline
a & . & 1 & 1 & 1 & . & . & . & = 0, \\
b & 1 & . & 1 & . & 1 & . & . & \\
c & 1 & 1 & . & . & . & 1 & . & \\
\hline
d & . & . & . & 1 & 1 & 1 & . & \\
\hline
. & d & . & . & . & c & b & 1 & \\
. & . & d & . & c & . & a & 1 & \\
. & . & . & d & b & a & . & 1 & \\
\hline
. & a & b & c & . & . & . & 1 &
\end{array}$$

which however is not essentially distinct from the form before obtained, but may be derived from it by an interchange of lines and columns.

And in general for any *even* number of quadratic radicals the two forms are not essentially distinct, but may be derived from each other by interchanging lines and columns, while for an *odd* number of quadratic radicals the two forms cannot be so derived from each other, but are essentially distinct.

I was indebted to Mr Sylvester for the remark that the above process applies to radicals of a higher order than the second. To take the simplest case, suppose

$$x + y = 0, \quad x^3 = a, \quad y^3 = b;$$

and multiply first by 1, x^2y, xy^2; this gives

$$\begin{array}{l}
x + y \quad . \quad = 0 \\
. \quad ay + x^2y^2 = 0 \\
bx \quad . \quad + x^2y^2 = 0\,;
\end{array}$$

or, eliminating,

$$\begin{vmatrix} 1 & 1 & . \\ . & a & 1 \\ b & . & 1 \end{vmatrix} = 0\,;$$

next multiply by x, y, x^2y^2; this gives

$$\begin{array}{l}
x^2 \quad . \quad + xy = 0 \\
. \quad y^2 + xy = 0 \\
bx^2 + ay^2 \quad . \quad = 0\,;
\end{array}$$

or, eliminating,

$$\begin{vmatrix} 1 & . & 1 \\ . & 1 & 1 \\ b & a & . \end{vmatrix} = 0\,;$$

and lastly, multiply by x^2, y^2, xy; this gives

$$\begin{array}{l}
a + x^2y \quad . \quad = 0 \\
b \quad . \quad + xy^2 = 0 \\
. \quad x^2y + xy^2 = 0\,;
\end{array}$$

or, eliminating,

$$\left|\begin{array}{ccc} a & 1 & . \\ b & . & 1 \\ . & 1 & 1 \end{array}\right| = 0;$$

where it is to be remarked that the second and third forms are not essentially distinct, since the one may be derived from the other by the interchange of lines and columns.

Applying the preceding process to the system

$$x + y + z = 0, \quad x^3 = a, \quad y^3 = b, \quad z^3 = c;$$

multiply first by 1, xyz, $x^2y^2z^2$, x^2z, y^2x, z^2y, x^2y, y^2z, z^2x, reduce and eliminate the quantities in the outside row,

the result is

$$\begin{array}{ccc|ccc|ccc} x, & y, & z, & y^2z^2, & x^2yz, & y^2zx, & z^2xy, & z^2x^2, & x^2y^2 \\ \hline 1 & 1 & 1 & . & . & . & . & . & . \\ \hline . & . & . & 1 & 1 & 1 & . & . & . \\ \hline . & . & . & . & . & . & a & b & c \\ \hline . & . & a & 1 & . & . & . & 1 & . \\ b & . & . & . & 1 & . & . & . & 1 \\ . & c & . & . & . & 1 & 1 & . & . \\ \hline . & a & . & 1 & . & . & . & . & 1 \\ . & . & b & . & 1 & . & 1 & . & . \\ c & . & . & . & . & 1 & . & 1 & . \end{array} = 0;$$

next multiply by x, y, z, y^2z^2, z^2x^2, x^2y^2, x^2yz, y^2zx, z^2xy, reduce and eliminate the quantities in the outside row,

the result is

$$\begin{array}{ccc|ccc|ccc} x^2, & y^2, & z^2, & yz, & zx, & xy, & xy^2z^2, & yz^2x^2, & zx^2y^2 \\ \hline 1 & . & . & . & 1 & 1 & . & . & . \\ . & 1 & . & 1 & . & 1 & . & . & . \\ . & . & 1 & 1 & 1 & . & . & . & . \\ \hline . & c & b & . & . & . & 1 & . & . \\ c & . & a & . & . & . & . & 1 & . \\ b & a & . & . & . & . & . & . & 1 \\ \hline . & . & . & a & . & . & . & 1 & 1 \\ . & . & . & . & b & . & 1 & . & 1 \\ . & . & . & . & . & c & 1 & 1 & . \end{array} = 0;$$

lastly, multiply by x^2, y^2, z^2, yz, zx, xy, xy^2z^2, yz^2x^2, xy^2z^2, reduce and eliminate the quantities in the outside row,

the result is

$$\begin{array}{c|c|c|ccc|ccc|}
1 & xyz, & x^2y^2z^2, & yz^2, & zx^2, & xy^2, & y^2z, & z^2x, & x^2y \\
\hline
a & . & . & . & 1 & . & . & . & 1 \\
b & . & . & . & . & 1 & 1 & . & . \\
c & . & . & 1 & . & . & . & 1 & . \\
\hline
. & 1 & . & 1 & . & . & 1 & . & . \\
. & 1 & . & . & 1 & . & . & 1 & . \\
. & 1 & . & . & . & 1 & . & . & 1 \\
\hline
. & . & 1 & . & . & c & . & b & . \\
. & . & 1 & a & . & . & . & . & c \\
. & . & 1 & . & b & . & a & . & . \\
\hline
\end{array} = 0;$$

where, as in the case of two cubic radicals, two forms, viz. the first and third forms of the rational equation, are not essentially distinct, but may be derived from each other by interchanging lines and columns.

And in general, whatever be the number of cubic radicals, two of the three forms are not essentially distinct, but may be derived from each other by interchanging lines and columns.

110.

NOTE ON THE TRANSFORMATION OF A TRIGONOMETRICAL EXPRESSION.

[From the *Cambridge and Dublin Mathematical Journal*, vol. IX. (1854), pp. 61—62.]

THE differential equation

$$\frac{dx}{(a+x)\sqrt{(c+x)}} + \frac{dy}{(a+y)\sqrt{(c+y)}} + \frac{dz}{(a+z)\sqrt{(c+z)}} = 0,$$

integrated so as to be satisfied when the variables are simultaneously infinite, gives by direct integration

$$\tan^{-1}\sqrt{\left(\frac{a-c}{c+x}\right)} + \tan^{-1}\sqrt{\left(\frac{a-c}{c+y}\right)} + \tan^{-1}\sqrt{\left(\frac{a-c}{c+z}\right)} = 0;$$

and, by Abel's theorem,

$$\begin{vmatrix} 1, & x, & (a+x)\sqrt{(c+x)} \\ 1, & y, & (a+y)\sqrt{(c+y)} \\ 1, & z, & (a+z)\sqrt{(c+z)} \end{vmatrix} = 0.$$

To show *à posteriori* the equivalence of these two equations, I represent the determinant by the symbol $\square$, and expressing it in the form

$$\square = \begin{vmatrix} 1, & a+x, & (a+x)\sqrt{(c+x)} \\ \vdots & & \end{vmatrix},$$

I write for the moment $\xi = \sqrt{\left(\frac{a-c}{c+x}\right)}$ &c.; this gives

$$\square = \left| \begin{array}{lll} 1, & (a-c)\left(1+\dfrac{1}{\xi^2}\right), & (a-c)^{\frac{3}{2}}\left(\dfrac{1}{\xi}+\dfrac{1}{\xi^3}\right) \\ \vdots & & \end{array} \right|$$

$$= \frac{(a-c)^{\frac{5}{2}}}{\xi^3\eta^3\zeta^3} \left| \begin{array}{lll} \xi^3, & \xi^3+\xi, & \xi^2+1 \\ \vdots & & \end{array} \right|$$

$$= \frac{(a-c)^{\frac{5}{2}}}{\xi^3\eta^3\zeta^3} \left| \begin{array}{lll} \xi^3, & \xi, & \xi^2+1 \\ \vdots & & \end{array} \right|$$

$$= -\frac{(a-c)^{\frac{5}{2}}}{\xi^3\eta^3\zeta^3} \left\{ \left| \begin{array}{lll} 1, & \xi, & \xi^3 \\ \vdots & & \end{array} \right| - \xi\eta\zeta \left| \begin{array}{lll} 1, & \xi, & \xi^2 \\ \vdots & & \end{array} \right| \right\}$$

$$= -\frac{(a-c)^{\frac{5}{2}}(\xi+\eta+\zeta-\xi\eta\zeta)}{\xi^3\eta^3\zeta^3} \left| \begin{array}{lll} 1, & \xi, & \xi^2 \\ \vdots & & \end{array} \right|;$$

or, replacing ξ, η, ζ by their values, we have identically

$$\left| \begin{array}{lll} 1, & x, & (a+x)\sqrt{(c+x)} \\ 1, & y, & (a+y)\sqrt{(c+y)} \\ 1, & z, & (a+z)\sqrt{(c+z)} \end{array} \right| =$$

$$-\frac{(c+x)^{\frac{3}{2}}(c+y)^{\frac{3}{2}}(c+z)^{\frac{3}{2}}}{(a-c)^2} \left\{ \sqrt{\frac{a-c}{c+x}} + \sqrt{\frac{a-c}{c+y}} + \sqrt{\frac{a-c}{c+z}} - \sqrt{\frac{a-c}{c+x}}\sqrt{\frac{a-c}{c+y}}\sqrt{\frac{a-c}{c+z}} \right\} \left| \begin{array}{lll} 1, & \sqrt{\dfrac{a-c}{c+x}}, & \dfrac{a-c}{c+x} \\ 1, & \sqrt{\dfrac{a-c}{c+y}}, & \dfrac{a-c}{c+y} \\ 1, & \sqrt{\dfrac{a-c}{c+z}}, & \dfrac{a-c}{c+z} \end{array} \right|,$$

and the equation

$$\sqrt{\frac{a-c}{c+x}} + \sqrt{\frac{a-c}{c+y}} + \sqrt{\frac{a-c}{c+z}} - \sqrt{\frac{a-c}{c+x}}\sqrt{\frac{a-c}{c+y}}\sqrt{\frac{a-c}{c+z}} = 0$$

is of course equivalent to the trigonometrical equation

$$\tan^{-1}\sqrt{\frac{a-c}{c+x}} + \tan^{-1}\sqrt{\frac{a-c}{c+y}} + \tan^{-1}\sqrt{\frac{a-c}{c+z}} = 0,$$

which shows the equivalence of the two equations in question.

111.

ON A THEOREM OF M. LEJEUNE-DIRICHLET'S.

[From the *Cambridge and Dublin Mathematical Journal*, vol. IX. (1854), pp. 163—165.]

THE following formula,

$$\Sigma q^{ax^2+2bxy+cy^2}+\Sigma q^{a'x^2+2b'xy+c'y^2}+\ldots=2\Sigma\delta^{\frac{1}{2}(n-1)}\,\epsilon^{\frac{1}{8}(n^2-1)}\left(\frac{n}{P}\right)q^{nn'},\ldots\ldots\ldots\ldots\ldots\ldots(3)$$

is given in Lejeune-Dirichlet's well-known memoir "Recherches sur diverses applications &c." (*Crelle*, t. XXI. [1840] p. 8). The notation is as follows:—On the left-hand side (a, b, c), (a', b', c'), ... are a system of properly primitive forms to the *negative* determinant D (i.e. a system of positive forms); x, y are positive or negative integers including zero, such that in the sum $\Sigma q^{ax^2+2bxy+cy^2}$, $ax^2+2bxy+cy^2$ is prime to $2D$, and similarly in the other sums; q is indeterminate and the summations extend to the values first mentioned, of x and y. On the right-hand side we have to consider the form of D, viz. we have $D=PS^2$ or else $D=2PS^2$, where S^2 is the greatest square factor in D and where P is odd: this obviously defines P, and the values of δ, ϵ, which are always ± 1 (or, as I prefer to express it, are always $\pm$) are given as follows, viz.

$$\begin{array}{llll} D=PS^2\,, & P\equiv 1 \pmod 4, & \delta, & \epsilon=++, \\ D=PS^2\,, & P\equiv 3 \pmod 4, & \delta, & \epsilon=-+, \\ D=2PS^2, & P\equiv 1 \pmod 4, & \delta, & \epsilon=+-, \\ D=2PS^2, & P\equiv 3 \pmod 4, & \delta, & \epsilon=--, \end{array}$$

n, n' are any positive numbers prime to $2D$, $\left(\frac{n}{P}\right)$ is Legendre's symbol as generalized by Jacobi, viz. in general if p be a positive or negative prime not a factor of n, then $\left(\frac{n}{p}\right) = +$ or $-$ according as n is or is not a quadratic residue of p (or, what is the same thing, p being positive, $\left(\frac{n}{\pm p}\right) \equiv n^{\frac{1}{2}(n-1)} \pmod{p}$), and for $P = pp'p'' \ldots$,

$$\left(\frac{n}{P}\right) = \left(\frac{n}{p}\right)\left(\frac{n}{p'}\right)\left(\frac{n}{p''}\right)\ldots,$$

and the summation extends to all the values of n, n' of the form above mentioned. In the particular case $D = -1$, it is necessary that the second side should be doubled. The method of reducing the equation is indicated in the memoir. The following are a few particular cases.

$D = -1$, $$\Sigma q^{x^2+y^2} = 4\Sigma\,(-)^{\frac{1}{2}(n-1)}\,q^{nn'},$$

or $$(1 + 2q^4 + 2q^{16} + 2q^{36} + \ldots)(q + q^9 + q^{25} + q^{49} + \ldots)$$

$$= \frac{q}{1-q^2} - \frac{q^3}{1-q^6} + \frac{q^5}{1-q^{10}} - \frac{q^7}{1-q^{14}} + \ldots$$

$D = -2$, $$\Sigma q^{x^2+2y^2} = 2\Sigma\,(-)^{\frac{1}{2}(n-1)+\frac{1}{8}(n^2-1)}\,q^{nn'},$$

or $$(1 + 2q^2 + 2q^8 + 2q^{18}\ldots)(q + q^9 + q^{25} + q^{49} + \ldots)$$

$$= \frac{q}{1-q^2} + \frac{q^3}{1-q^6} - \frac{q^5}{1-q^{10}} - \frac{q^7}{1-q^{14}} + \&c.$$

an example given in the memoir.

$D = -3$, $$\Sigma q^{x^2+3y^2} = 2\Sigma\left(\frac{n}{3}\right)q^{nn'},$$

or $$(q^1 + q^{25} + q^{49} + q^{121} + q^{169}\ldots)(1 + 2q^{12} + 2q^{48} + 2q^{108}\ldots)$$
$$+ 2(q^3 + q^{27} + q^{75} + q^{147} + \ldots)(q^4 + q^{16} + q^{64} + q^{100}\ldots)$$

$$= \frac{q+q^5}{1-q^6} - \frac{q^5+q^{25}}{1-q^{30}} + \frac{q^7+q^{35}}{1-q^{42}} - \frac{q^{11}+q^{55}}{1-q^{66}} + \ldots$$

I am not aware that the above theorem is quoted or referred to in any subsequent memoir on Elliptic Functions, or on the class of series to which it relates; and the theorem is so distinct in its origin and form from all other theorems relating to the same class of series, and, independently of the researches in which it originates, so remarkable as a result, that I have thought it desirable to give a detached statement of it in this paper.

112.

DEMONSTRATION OF A THEOREM RELATING TO THE PRODUCTS OF SUMS OF SQUARES.

[From the *Philosophical Magazine*, vol. IV. (1852), pp. 515—519.]

MR KIRKMAN, in his paper "On Pluquaternions and Homoid Products of Sums of n Squares" (*Phil. Mag.* vol. XXXIII. [1848] pp. 447—459 and 494—509), quotes from a note of mine the following passage:—"The complete test of the possibility of the product of 2^n squares by 2^n squares reducing itself to a sum of 2^n squares is the following: forming the complete systems of triplets for (2^n-1) things, if eab, ecd, fac, fdb be any four of them, we must have, paying attention to the signs alone,

$$(\pm\, eab)\,(\pm\, ecd) = (\pm\, fac)\,(\pm\, fdb)\,;$$

i. e. if the first two are of the same sign, the last two must be so also, and *vice versâ;* I believe that, for a system of seven, two conditions of this kind being satisfied would imply the satisfaction of all the others: it remains to be shown that the complete system of conditions cannot be satisfied for fifteen things." I propose to explain the meaning of the theorem, and to establish the truth of it, without in any way assuming the existence of imaginary units.

The identity to be established is

$$(w^2 + a^2 + b^2 + \ldots)\,(w_{\prime}^2 + a_{\prime}^2 + b_{\prime}^2 \ldots) = w_{\prime\prime}^2 + a_{\prime\prime}^2 + b_{\prime\prime}^2 + \ldots$$

where the 2^n quantities $w, a, b, c, \ldots$ and the 2^n quantities $w_{\prime}, a_{\prime}, b_{\prime}, c_{\prime}, \ldots$ are given quantities in terms of which the 2^n quantities $w_{\prime\prime}, a_{\prime\prime}, b_{\prime\prime}, c_{\prime\prime}, \ldots$ have to be determined.

Without attaching any meaning whatever to the symbols $a_{\circ}, b_{\circ}, c_{\circ} \ldots$ I write down the expressions

$$w + aa_{\circ} + bb_{\circ} + cc_{\circ} \ldots, \quad w_{\prime} + a_{\prime}a_{\circ} + b_{\prime}b_{\circ} + c_{\prime}c_{\circ} \ldots,$$

and I multiply as if a_0, b_0, c_0 ... really existed, taking care to multiply without making any transposition in the order *inter se* of two symbols a_0, b_0 combined in the way of multiplication. This gives a quasi-product

$$\begin{aligned} & ww_{,} + (aw_{,} + a_{,}w)\, a_0 + (bw_{,} + b_{,}w)\, b_0 + \ldots \\ & + aa_{,}a_0^2 \ + bb_{,}b_0^2 \ + \ldots \\ & + ab_{,}a_0 b_0 + a_{,}bb_0 a_0 + \ldots. \end{aligned}$$

Suppose, now, that a quasi-equation, such as

$$a_0 b_0 c_0 = +,$$

means that in the expression of the quasi-product

$$b_0 c_0, \quad c_0 a_0, \quad a_0 b_0, \quad c_0 b_0, \quad a_0 c_0, \quad b_0 a_0$$

are to be replaced by

$$a_0, \quad b_0, \quad c_0, \quad -a_0, \quad -b_0, \quad -c_0;$$

and that a quasi-equation, such as $a_0 b_0 c_0 = -$, means that in the expression of the quasi-product

$$b_0 c_0, \quad c_0 a_0, \quad a_0 b_0, \quad c_0 b_0, \quad a_0 c_0, \quad b_0 a_0$$

are to be replaced by

$$-a_0, \quad -b_0, \quad -c_0, \quad a_0, \quad b_0, \quad c_0.$$

It is in the first place clear that the quasi-equation, $a_0 b_0 c_0 = +$, may be written in any one of the six forms

$$\begin{aligned} & a_0 b_0 c_0 = +, \quad b_0 c_0 a_0 = +, \quad c_0 a_0 b_0 = +, \\ & a_0 c_0 b_0 = -, \quad c_0 b_0 a_0 = -, \quad b_0 a_0 c_0 = -; \end{aligned}$$

and so for the quasi-equation $a_0 b_0 c_0 = -$. This being premised, if we form a system of quasi-equations, such as

$$a_0 b_0 c_0 = \pm, \quad a_0 d_0 e_0 = \pm, \text{ \&c.}$$

where the system of triplets contains each duad once, and once only, and the arbitrary signs are chosen at pleasure; if, moreover, in the expression of the quasi-product we replace a_0^2, b_0^2, ... each by -1, it is clear that the quasi-product will assume the form

$$w_{,,} + a_{,,}a_0 + b_{,,}b_0 + c_{,,}c_0 + \ldots,$$

$w_{,,}$, $a_{,,}$, $b_{,,}$, $c_{,,}$ ... being determinate functions of w, a, b, c, ...; $w_{,}$, $a_{,}$, $b_{,}$, $c_{,}$ ..., homogeneous of the first order in the quantities of each set; the value of $w_{,,}$ being obviously in every case

$$w_{,,} = ww_{,} - aa_{,} - bb_{,} - cc_{,} \ldots,$$

and $a_{,,}$, $b_{,,}$, $c_{,,}$, ... containing in every case the terms $aw_{,} + a_{,}w$, $bw_{,} + b_{,}w$, $cw_{,} + c_{,}w$, ... but the form of the remaining terms depending as well on the triplets entering into the system of quasi-equations as on the values given to the signs $\pm$; *the quasi-equations serving, in fact, to prescribe a rule for the formation of certain functions* $w_{\prime\prime}, a_{\prime\prime}, b_{\prime\prime}, c_{\prime\prime}, \ldots$, the properties of which functions may afterwards be investigated.

Suppose, now, that the system of quasi-equations is such that

$$e_\circ a_\circ b_\circ,\quad e_\circ c_\circ d_\circ$$

being any two of its triplets, with a common symbol $e_\circ$, there occur also in the system the triplets

$$f_\circ a_\circ c_\circ,\quad f_\circ d_\circ b_\circ,\quad g_\circ a_\circ d_\circ,\quad g_\circ b_\circ c_\circ;$$

and suppose that the corresponding portion of the system is

$$\begin{aligned} e_\circ a_\circ b_\circ &= \epsilon, \quad e_\circ c_\circ d_\circ = \epsilon', \\ f_\circ a_\circ c_\circ &= \zeta, \quad f_\circ d_\circ b_\circ = \zeta', \\ g_\circ a_\circ d_\circ &= \iota, \quad g_\circ b_\circ c_\circ = \iota', \end{aligned}$$

where $\epsilon, \zeta, \iota, \epsilon', \zeta', \iota'$ each of them denote one of the signs $+$ or $-$; then $e_{\prime\prime}, f_{\prime\prime}, g_{\prime\prime}$ will contain respectively the terms

$$\begin{aligned} &\epsilon(ab_\prime - a_\prime b) + \epsilon'(cd_\prime - c_\prime d), \\ &\zeta(ac_\prime - a_\prime c) + \zeta'(db_\prime - d_\prime b), \\ &\iota(ad_\prime - a_\prime d) + \iota'(bc_\prime - b_\prime c); \end{aligned}$$

and $e_{\prime\prime}^2 + f_{\prime\prime}^2 + g_{\prime\prime}^2$ contains the terms

$$\begin{aligned} (a^2 + b^2 + c^2 + d^2)(a_\prime^2 + b_\prime^2 + c_\prime^2 + d_\prime^2) - a^2a_\prime^2 - b^2b_\prime^2 - c^2c_\prime^2 - d^2d_\prime^2 \\ + 2\,[\epsilon\epsilon'(ab_\prime - a_\prime b)(cd_\prime - c_\prime d) \\ + \zeta\zeta'(ac_\prime - a_\prime c)(db_\prime - d_\prime b) \\ + \iota\iota'(ad_\prime - a_\prime d)(bc_\prime - b_\prime c)]; \end{aligned}$$

and by taking account of the terms $ew_\prime + e_\prime w$, $fw_\prime + f_\prime w$, $gw_\prime + g_\prime w$ in $e_{\prime\prime}, f_{\prime\prime}, g_{\prime\prime}$ respectively, we should have had besides in $e_{\prime\prime}^2 + f_{\prime\prime}^2 + g_{\prime\prime}^2$ the terms

$$\begin{aligned} (e^2 + f^2 + g^2)\, w_\prime^2 + (e_\prime^2 + f_\prime^2 + g_\prime^2)\, w^2 \\ + 2\,(ee_\prime + ff_\prime + gg_\prime)\, ww_\prime. \end{aligned}$$

Also $w_{\prime\prime}^2$ contains the terms

$$\begin{aligned} w^2w_\prime^2 + a^2a_\prime^2 + b^2b_\prime^2 + c^2c_\prime^2 + d^2d_\prime^2 \\ - 2\,(ee_\prime + ff_\prime + gg_\prime)\, ww_\prime; \end{aligned}$$

whence it is easy to see that

$$\begin{aligned} &w_{\prime\prime}^2 + a_{\prime\prime}^2 + b_{\prime\prime}^2 + c_{\prime\prime}^2 + \ldots = \\ &(w^2 + a^2 + b^2 + c^2 + \ldots)(w_\prime^2 + a_\prime^2 + b_\prime^2 + c_\prime^2 + \ldots) \\ &\quad + 2\Sigma\,[\epsilon\epsilon'(ab_\prime - a_\prime b)(cd_\prime - c_\prime d) \\ &\qquad + \zeta\zeta'(ac_\prime - a_\prime c)(db_\prime - d_\prime b) \\ &\qquad + \iota\iota'(ad_\prime - a_\prime d)(bc_\prime - b_\prime c)]. \end{aligned}$$

where the summation extends to all the quadruplets formed each by the combination of two duads such as ab and cd, or ac and db, or ad and bc, i.e. two duads, which, combined with the same common letter (in the instances just mentioned e, or f, or g), enter as triplets into the system of quasi-equations—so that if $\nu = 2^n - 1$, the number of quadruplets is

$$\tfrac{1}{2}\{\tfrac{1}{2}(\nu-1).\tfrac{1}{2}(\nu-3)\}\,\nu.\tfrac{1}{3}, = \tfrac{1}{24}\nu(\nu-1)(\nu-3),$$

and the terms under the sign Σ will vanish identically if only

$$\epsilon\epsilon' = \zeta\zeta' = \iota\iota';$$

but the relation $\epsilon\epsilon' = \iota\iota'$ is of the same form as the equation $\epsilon\epsilon' = \zeta\zeta'$; hence if all the relations

$$\epsilon\epsilon' = \zeta\zeta'$$

are satisfied, the terms under the sign Σ vanish, and we have

$$(w_{\prime\prime}^2 + a_{\prime\prime}^2 + b_{\prime\prime}^2 + c_{\prime\prime}^2 + \ldots) = (w^2 + a^2 + b^2 + c^2 + \ldots)(w_\prime^2 + a_\prime^2 + b_\prime^2 + c_\prime^2 + \ldots)$$

which is thus shown to be true, upon the suppositions—

1. That the system of quasi-equations is such that

$$e_\circ a_\circ b_\circ,\quad e_\circ c_\circ d_\circ$$

being any two of its triplets with a common symbol $e_\circ$, there occur also in the system the triplets

$$f_\circ a_\circ c_\circ,\quad f_\circ d_\circ b_\circ,$$
$$g_\circ a_\circ d_\circ,\quad g_\circ b_\circ c_\circ.$$

2. That for any two pairs of triplets, such as

$$e_\circ a_\circ b_\circ,\quad e_\circ c_\circ d_\circ \text{ and } f_\circ a_\circ c_\circ,\quad f_\circ d_\circ b_\circ,$$

the product of the signs of the triplets of the first pair is equal to the product of the signs of the triplets of the second pair.

In the case of fifteen things $a, b, c, \ldots$ the triplets may, as appears from Mr Kirkman's paper, be chosen so as to satisfy the first condition; but the second condition involves, as Mr Kirkman has shown, a contradiction; and therefore the product of two sums, each of them of sixteen squares, is *not* a sum of sixteen squares. It is proper to remark, that this demonstration, although I think rendered clearer by the introduction of the idea of the system of triplets furnishing the rule for the formation of the expressions $w_{\prime\prime}, a_{\prime\prime}, b_{\prime\prime}, c_{\prime\prime}$, &c., is not in principle different from that contained in Prof. Young's paper "On an Extension of a Theorem of Euler, &c.", *Irish Transactions*, vol. XXI. [1848 pp. 311—341].

113.

NOTE ON THE GEOMETRICAL REPRESENTATION OF THE INTEGRAL $\int dx \div \sqrt{(x+a)(x+b)(x+c)}$.

[From the *Philosophical Magazine*, vol. v. (1853), pp. 281—284.]

The equation of a conic passing through the points of intersection of the conics

$$x^2 + y^2 + z^2 = 0,$$
$$ax^2 + by^2 + cz^2 = 0,$$

is of the form

$$w(x^2 + y^2 + z^2) + ax^2 + by^2 + cz^2 = 0,$$

where w is an arbitrary parameter. Suppose that the conic touches a given line, we have for the determination of w a quadratic equation, the roots of which may be considered as parameters for determining the line in question. Let one of the values of w be considered as equal to a constant quantity k, the line is always a tangent to the conic

$$k(x^2 + y^2 + z^2) + ax^2 + by^2 + cz^2 = 0;$$

and taking $w=p$ for the other value of w, p is a parameter determining the particular tangent, or, what is the same thing, determining the point of contact of this tangent.

The equation of the tangent is easily seen to be

$$x\sqrt{b-c}\sqrt{a+k}\sqrt{a+p} + y\sqrt{c-a}\sqrt{b+k}\sqrt{b+p} + z\sqrt{a-b}\sqrt{c+k}\sqrt{c+p} = 0;$$

suppose that the tangent meets the conic $x^2+y^2+z^2=0$ (which is of course the conic corresponding to $w=\infty$) in the points P, P', and let θ, ∞ be the parameters of the point P, and θ', ∞ the parameters of the point P', i.e. (repeating the definition of the terms) let the tangent at P of the conic $x^2+y^2+z^2=0$ be also touched by the conic $\theta(x^2+y^2+z^2)+ax^2+by^2+cz^2=0$, and similarly for θ'. The coordinates of the point P are given by the equations

$$x:y:z=\sqrt{b-c}\sqrt{a+\theta}:\sqrt{c-a}\sqrt{b+\theta}:\sqrt{a-b}\sqrt{c+\theta};$$

and substituting these values in the equation of the line PP', we have

$$(b-c)\sqrt{a+k}\sqrt{a+p}\sqrt{a+\theta}+(c-a)\sqrt{b+k}\sqrt{b+p}\sqrt{b+\theta}+(a-b)\sqrt{c+k}\sqrt{c+p}\sqrt{c+\theta}$$
$$=0\ldots(*),$$

an equation connecting the quantities p, θ. To rationalize this equation, write

$$\sqrt{(a+k)(a+p)(a+\theta)}=\lambda+\mu a,$$
$$\sqrt{(b+k)(b+p)(b+\theta)}=\lambda+\mu b,$$
$$\sqrt{(c+k)(c+p)(c+\theta)}=\lambda+\mu c,$$

values which evidently satisfy the equation in question. Squaring these equations, we have equations from which λ^2, $\lambda\mu$, μ^2 may be linearly determined; and making the necessary reductions, we find

$$\lambda^2=abc+kp\theta,$$
$$-2\lambda\mu=bc+ca+ab-(p\theta+kp+k\theta),$$
$$\mu^2=a+b+c+k+p+\theta;$$

or, eliminating λ, μ,

$$\{bc+ca+ab-(p\theta+kp+k\theta)\}^2-4(a+b+c+k+p+\theta)(abc+kp\theta)=0,\ \ldots\ldots(*),$$

which is the rational form of the former equation marked (*). It is clear from the symmetry of the formula, that the same equation would have been obtained by the elimination of L, M from the equations

$$\sqrt{(k+a)(k+b)(k+c)}=L+Mk,$$
$$\sqrt{(p+a)(p+b)(p+c)}=L+Mp,$$
$$\sqrt{(\theta+a)(\theta+b)(\theta+c)}=L+M\theta;$$

and it follows from Abel's theorem (but the result may be verified by means of Euler's fundamental integral in the theory of elliptic functions), that if

$$\Pi x=\int_\infty \frac{dx}{\sqrt{(x+a)(x+b)(x+c)}},$$

then the algebraical equations (*) are equivalent to the transcendental equation

$$\pm\Pi k\pm\Pi p\pm\Pi\theta=0;$$

the arbitrary constant which should have formed the second side of the equation having been determined by observing that the algebraical equation gives for $p=\theta$, $k=\infty$, a system of values, which, when the signs are properly chosen, satisfy the transcendental equation. In fact, arranging the rational algebraical equation according to the powers of k, it becomes

$$k^2(p-\theta)^2 - 2k\{p\theta(p+\theta) + 2(a+b+c)p\theta + (bc+ca+ab)(p+\theta) + 2abc\}$$
$$+ p^2\theta^2 - 2(bc+ca+ab)p\theta - 4abc(p+\theta) + b^2c^2 + c^2a^2 + a^2b^2 - 2a^2bc - 2b^2ca - 2c^2ab = 0\,;\,(*)$$

which proves the property in question, and is besides a very convenient form of the algebraical integral. The ambiguous signs in the transcendental integral are not of course arbitrary (indeed it has just been assumed that for $p=\theta$, Πp and $\Pi\theta$ are to be taken with opposite signs), but the discussion of the proper values to be given to the ambiguous signs would be at all events tedious, and must be passed over for the present.

It is proper to remark, that $\theta=p$ gives not only, as above supposed, $k=\infty$, but another value of k, which, however, corresponds to the transcendental equation

$$\pm\Pi k \pm 2\Pi p = 0\,;$$

the value in question is obviously

$$4k = \frac{p^4 - 2(bc+ca+ab)p^2 - 8abcp + b^2c^2 + c^2a^2 + a^2b^2 - 2a^2bc - 2b^2ca - 2c^2ab}{(p+a)(p+b)(p+c)}.$$

Consider, in general, a cubic function $\mathrm{a}x^3 + 3\mathrm{b}x^2y + 3\mathrm{c}xy^2 + \mathrm{d}y^3$, or, as I now write it in the theory of invariants, $(\mathrm{a}, \mathrm{b}, \mathrm{c}, \mathrm{d})(x, y)^3$, the Hessian of this function is

$$\left(\mathrm{ac}-\mathrm{b}^2,\ \tfrac{1}{2}(\mathrm{ad}-\mathrm{bc}),\ \mathrm{bd}-\mathrm{c}^2\right)(x, y)^2,$$

and applying this formula to the function $(p+a)(p+b)(p+c)$, it is easy to write the equation last preceding in the form

$$4k = p - (a+b+c) - \frac{9\ \text{Hessian}\ \{(p+a)(p+b)(p+c)\}}{(p+a)(p+b)(p+c)},$$

which is a formula for the duplication of the transcendent Πx.

Reverting now to the general transcendental equation

$$\pm\Pi k \pm \Pi p \pm \Pi\theta = 0,$$

we have in like manner

$$\pm\Pi k \pm \Pi p \pm \Pi\theta' = 0\,;$$

and assuming a proper correspondence of the signs, the elimination of Πp gives

$$\Pi\theta' - \Pi\theta = 2\Pi k\,;$$

i.e. if the points P, P' upon the conic $x^2+y^2+z^2=0$ are such that their parameters θ, θ' satisfy this equation, the line PP' will be constantly a tangent to the conic

$$k(x^2+y^2+z^2)+(ax^2+by^2+cz^2)=0.$$

Hence also, if the parameters k, k', k'' of the conics

$$k\ (x^2+y^2+z^2)+ax^2+by^2+cz^2=0,$$
$$k'\ (x^2+y^2+z^2)+ax^2+by^2+cz^2=0,$$
$$k''(x^2+y^2+z^2)+ax^2+by^2+cz^2=0,$$

satisfy the equation

$$\Pi k+\Pi k'+\Pi k''=0,$$

there are an infinity of triangles inscribed in the conic $x^2+y^2+z^2=0$, and the sides of which touch the last-mentioned three conics respectively.

Suppose $2\Pi k=\Pi\kappa$ (an equation the algebraic form of which has already been discussed), then

$$\Pi\theta'-\Pi\theta=\Pi\kappa,$$

$\theta=\infty$ gives $\theta'=\kappa$; or, observing that $\theta=\infty$ corresponds to a point of intersection of the conics $x^2+y^2+z^2=0$, $ax^2+by^2+cz^2=0$, κ is the parameter of the point in which a tangent to the conic $k(x^2+y^2+z^2)+ax^2+by^2+cz^2=0$ at any one of its intersections with the conic $x^2+y^2+z^2=0$ meets the last-mentioned conic. Moreover, the algebraical relation between θ, θ' and κ (where, as before remarked, κ is a given function of k) is given by a preceding formula, and is simpler than that between θ, θ' and k.

The preceding investigations were, it is hardly necessary to remark, suggested by a well-known memoir of the late illustrious Jacobi, and contain, I think, the extension which he remarks it would be interesting to make of the principles in such memoir to a system of two conics. I propose reverting to the subject in a memoir to be entitled "Researches on the Porism of the in- and circumscribed triangle." [This was, I think, never written.]

114.

ANALYTICAL RESEARCHES CONNECTED WITH STEINER'S EXTENSION OF MALFATTI'S PROBLEM.

[From the *Philosophical Transactions of the Royal Society of London*, vol. CXLII. *for the year* 1852, pp. 253—278: Received April 12,—Read May 27, 1852]

THE problem, in a triangle to describe three circles each of them touching the two others and also two sides of the triangle, has been termed after the Italian geometer by whom it was proposed and solved, Malfatti's problem. The problem which I refer to as Steiner's extension of Malfatti's problem is as follows:—"To determine three sections of a surface of the second order, each of them touching the two others, and also two of three given sections of the surface of the second order," a problem proposed in Steiner's memoir, "Einige geometrische Betrachtungen," *Crelle*, t. I. [1826 pp. 161—184]. The geometrical construction of the problem in question is readily deduced from that given in the memoir just mentioned for a somewhat less general problem, viz. that in which the surface of the second order is replaced by a sphere; it is for the sake of the analytical developments to which the problem gives rise, that I propose to resume here the discussion of the problem. The following is an analysis of the present memoir:—

§ 1. Contains a lemma which appears to me to constitute the foundation of the analytical theory of the sections of a surface of the second order.

§ 2. Contains a statement of the geometrical construction of Steiner's extension of Malfatti's problem.

§ 3. Is a verification, founded on a particular choice of coordinates, of the construction in question.

§ 4. In this section, referring the surface of the second order to absolutely general coordinates, and after an incidental solution of the problem to determine a section touching three given sections, I obtain the equations for the solution of Steiner's extension of Malfatti's problem.

§ 5. Contains a separate discussion of a system of equations, including as a particular case the equations obtained in the preceding section.

§§ 6 and 7. Contain the application of the formulæ for the general system to the equations in § 4, and the development and completion of the solution.

§ 8. Is an extension of some preceding formulæ to quadratic functions of any number of variables.

§ 1. *Lemma relating to the sections of a surface of the second order.*

If

$$ax^2 + by^2 + cz^2 + dw^2 + 2fyz + 2gzx + 2hxy + 2lxw + 2myw + 2nzw = 0$$

be the equation of a surface of the second order, and

$$\mathfrak{A}x^2 + \mathfrak{B}y^2 + \mathfrak{C}z^2 + \mathfrak{D}w^2 + 2\mathfrak{F}yz + 2\mathfrak{G}zx + 2\mathfrak{H}xy + 2\mathfrak{L}xw + 2\mathfrak{M}yw + 2\mathfrak{N}zw = 0$$

the reciprocal equation, the condition that the two sections

$$\lambda x + \mu y + \nu z + \rho w = 0,$$
$$\lambda' x + \mu' y + \nu' z + \rho' w = 0,$$

may touch, is

$$\begin{aligned}
&(\mathfrak{A}\lambda^2 + \mathfrak{B}\mu^2 + \mathfrak{C}\nu^2 + \mathfrak{D}\rho^2 + 2\mathfrak{F}\mu\nu + 2\mathfrak{G}\nu\lambda + 2\mathfrak{H}\lambda\mu + 2\mathfrak{L}\lambda\rho + 2\mathfrak{M}\mu\rho + 2\mathfrak{N}\nu\rho)^{\frac{1}{2}}\\
&\times(\mathfrak{A}\lambda'^2 + \mathfrak{B}\mu'^2 + \mathfrak{C}\nu'^2 + \mathfrak{D}\rho'^2 + 2\mathfrak{F}\mu'\nu' + 2\mathfrak{G}\nu'\lambda' + 2\mathfrak{H}\lambda'\mu' + 2\mathfrak{L}\lambda'\rho' + 2\mathfrak{M}\mu'\rho' + 2\mathfrak{N}\nu'\rho')^{\frac{1}{2}}\\
&= (\mathfrak{A}\lambda\lambda' + \mathfrak{B}\mu\mu' + \mathfrak{C}\nu\nu' + \mathfrak{D}\rho\rho' + \mathfrak{F}(\mu\nu' + \mu'\nu) + \mathfrak{G}(\nu\lambda' + \nu'\lambda) + \mathfrak{H}(\lambda\mu' + \lambda'\mu)\\
&\qquad + \mathfrak{L}(\lambda\rho' + \lambda'\rho) + \mathfrak{M}(\mu\rho' + \mu'\rho) + \mathfrak{N}(\nu\rho' + \nu'\rho));
\end{aligned}$$

and in particular if the equation of the surface be

$$ax^2 + by^2 + cz^2 + 2fyz + 2gzx + 2hxy + pw^2 = 0,$$

the condition of contact is

$$\begin{aligned}
&\left(\mathfrak{A}\lambda^2 + \mathfrak{B}\mu^2 + \mathfrak{C}\nu^2 + 2\mathfrak{F}\mu\nu + 2\mathfrak{G}\nu\lambda + 2\mathfrak{H}\lambda\mu + \frac{K}{p}\rho^2\right)^{\frac{1}{2}}\\
&\times\left(\mathfrak{A}\lambda'^2 + \mathfrak{B}\mu'^2 + \mathfrak{C}\nu'^2 + 2\mathfrak{F}\mu'\nu' + 2\mathfrak{G}\nu'\lambda' + 2\mathfrak{H}\lambda'\mu' + \frac{K}{p}\rho'^2\right)^{\frac{1}{2}}\\
&= \left(\mathfrak{A}\lambda\lambda' + \mathfrak{B}\mu\mu' + \mathfrak{C}\nu\nu' + \mathfrak{F}(\mu\nu' + \mu'\nu) + \mathfrak{G}(\nu\lambda' + \nu'\lambda) + \mathfrak{H}(\lambda\mu' + \lambda'\mu) + \frac{K}{p}\rho\rho'\right),
\end{aligned}$$

in which last formula

$$\begin{aligned}
&\mathfrak{A} = bc - f^2, \quad \mathfrak{B} = ca - g^2, \quad \mathfrak{C} = ab - h^2,\\
&\mathfrak{F} = gh - af, \quad \mathfrak{G} = hf - bg, \quad \mathfrak{H} = fg - ch,\\
&K = abc - af^2 - bg^2 - ch^2 + 2fgh.
\end{aligned}$$

§ 2.

In order to state in the most simple form the geometrical construction for the solution of Steiner's extension of Malfatti's problem, let the given sections be called for conciseness the determinators[1]; any two of these sections lie in two different cones, the vertices of which determine with the line of intersection of the planes of the determinators, two planes which may be termed bisectors; the six bisectors pass three and three through four straight lines; and it will be convenient to use the term bisectors to denote, not the entire system, but any three bisectors passing through the same line. Consider three sections, which may be termed tactors, each of them touching a determinator and two bisectors, and three other sections (which may be termed separators) each of them passing through the point of contact of a determinator and tactor and touching the other two tactors; the separators will intersect in a line which passes through the point of intersection of the determinators. The three required sections, or as I shall term them the resultors, are determined by the conditions that each resultor touches two determinators and two separators, the possibility of the construction being implied as a theorem. The *à posteriori* verification may be obtained as follows:—

§ 3.

Let $x=0$, $y=0$, $z=0$ be the equations of the resultors, $w=0$ the equation of the polar of the point of intersection of the resultors. Since the resultors touch two and two, the equation of the surface is easily seen to be of the form

$$2yz + 2zx + 2xy + w^2 = 0. \,(^2)$$

The determinators are sections each of them touching two resultors, but otherwise arbitrary; their equations are

$$-\alpha x + \frac{1}{2\alpha} y + \frac{1}{2\alpha} z + w = 0,$$

$$\frac{1}{2\beta} x - \beta y + \frac{1}{2\beta} z + w = 0,$$

$$\frac{1}{2\gamma} x + \frac{1}{2\gamma} y - \gamma z + w = 0.$$

The separators are sections each of them touching two resultors at their point of contact (or what is the same thing, passing through the line of intersection of two resultors), and all of them having a line in common. Their equations may be taken to be

$$cy - bz = 0, \quad az - cx = 0, \quad bx - ay = 0,$$

[1] I use the words "determinators," &c. to denote indifferently the sections or the planes of the sections; the context is always sufficient to prevent ambiguity.

[2] The reciprocal form is, it should be noted,

$$x^2 + y^2 + z^2 - 2yz - 2zx - 2xy - 2w^2 = 0.$$

8—2

the values of a, b, c remaining to be determined. Now before fixing the values of these quantities, we may find three sections each of them touching a determinator at a point of intersection with the section which corresponds to it of the sections $cy - bz = 0$, $az - cx = 0$, $bx - ay = 0$, and touching the other two of the last-mentioned sections; and when a, b, c have their proper values the sections so found are the tactors. For, let $\lambda x + \mu y + \nu z + \rho w = 0$ be the equation of a section touching the determinator $-\alpha x + \frac{1}{2\alpha} y + \frac{1}{2\alpha} z + w = 0$, and the two sections $bx - ay = 0$, $az - cx = 0$; and suppose

$$\Delta^2 = \lambda^2 + \mu^2 + \nu^2 - 2\mu\nu - 2\nu\lambda - 2\lambda\mu - 2\rho^2;$$

the conditions of contact with the sections $bx - ay = 0$, $az - cx = 0$ are found to be

$$(b + a)\,\Delta = (b + a)\,\lambda - (b + a)\,\mu - (b - a)\,\nu,$$

$$(c + a)\,\Delta = (c + a)\,\lambda - (c - a)\,\mu - (c + a)\,\nu,$$

values, however, which suppose a correspondence in the signs of the radicals. Thence $(b + a)\,\mu = (c + a)\,\nu$; or since the ratios only of the quantities λ, μ, ν, ρ are material, $\mu = c + a$, $\nu = b + a$, and therefore

$$\Delta^2 = \lambda^2 - 2\,(2a + b + c)\,\lambda + (b - c)^2 - 2\rho^2, = (\lambda - b - c)^2,$$

or $$\rho^2 = -\,2\,(a\lambda + bc).$$

Hence the equation to a section touching $bx - ay = 0$, $az - cx = 0$ is

$$\lambda x + (c + a)\,y + (b + a)\,z + \sqrt{-2\,(a\lambda + bc)}\,w = 0;$$

and to express that this touches the determinator in question, we have

$$\pm\,\alpha\,(\lambda - b - c) = \left(\alpha + \frac{1}{\alpha}\right)\lambda - \alpha\,(2a + b + c) + 2\sqrt{-2\,(a\lambda + bc)};$$

and selecting the upper sign,

$$\frac{1}{\alpha}\lambda - 2a\alpha = -\,2\sqrt{-2\,(a\lambda + bc)};$$

whence

$$\lambda = -\,2\alpha\,(a\alpha - \sqrt{-2bc}), \quad \sqrt{-2\,(a\lambda + bc)} = (2a\alpha - \sqrt{-2bc});$$

or the section touching the determinator and the sections $bx - ay = 0$, $az - cx = 0$ is

$$-\,2\alpha\,(a\alpha - \sqrt{-2bc}\,)\,x + (c + a)\,y + (b + a)\,z + (2a\alpha - \sqrt{-2bc}\,)\,w = 0;$$

and at the point of contact with the determinator

$$-\,\alpha x + \frac{1}{2\alpha} y + \frac{1}{2\alpha} z + w = 0,$$

$$2yz + 2zx + 2xy + w^2 = 0.$$

Eliminating w between the first and second equations and between the second and third equations,

$$\sqrt{-2bc}\left(ax+\frac{1}{2\alpha}y+\frac{1}{2\alpha}z\right)+cy+bz=0,$$

$$\left(ax+\frac{1}{2\alpha}y+\frac{1}{2\alpha}z\right)^2+2yz \quad =0;$$

and from these equations $(cy-bz)^2=0$, or the point of contact lies in the section $cy-bz=0$. It follows that the equations of the tactors are

$$-2\alpha(a\alpha-\sqrt{-2bc})\,x+(c+a)\,y+(b+a)\,z+(2a\alpha-\sqrt{-2bc})\,w=0,$$

$$(c+b)\,x-2\beta(b\beta-\sqrt{-2ca})\,y+(a+b)\,z+(2b\beta-\sqrt{-2ca})\,w=0,$$

$$(b+c)\,x+(a+c)\,y-2\gamma(c\gamma-\sqrt{-2ab})\,z+(2c\gamma-\sqrt{-2ab})\,w=0,$$

where a, b, c still remain to be determined.

Now the separators pass through the point of intersection of the determinators; the equations of these give for the point in question,

$$\begin{aligned} x:y:z:w = &\ (2\beta\gamma+1)(-\alpha+\beta+\gamma+2\alpha\beta\gamma)\\ &:(2\gamma\alpha+1)(\ \ \alpha-\beta+\gamma+2\alpha\beta\gamma)\\ &:(2\alpha\beta+1)(\ \ \alpha+\beta-\gamma+2\alpha\beta\gamma)\\ &:\ \ 4\alpha^2\beta^2\gamma^2-1+\alpha^2+\beta^2+\gamma^2;\end{aligned}$$

and the values of a, b, c are therefore

$$\begin{aligned} a:b:c = &\ (2\beta\gamma+1)(-\alpha+\beta+\gamma+2\alpha\beta\gamma)\\ &:(2\gamma\alpha+1)(\ \ \alpha-\beta+\gamma+2\alpha\beta\gamma)\\ &:(2\alpha\beta+1)(\ \ \alpha+\beta-\gamma+2\alpha\beta\gamma),\end{aligned}$$

which are to be substituted for a, b, c in the equations of the separators and tactors respectively.

Now proceeding to find the bisectors, let $\lambda x+\mu y+\nu z+\rho w=0$ be the equation of a section touching the determinators,

$$\frac{1}{2\beta}x-\beta y+\frac{1}{2\beta}z+w=0,\quad \frac{1}{2\gamma}x+\frac{1}{2\gamma}y-\gamma z+w=0;$$

and suppose, as before, $\Delta^2=\lambda^2+\mu^2+\nu^2-2\mu\nu-2\nu\lambda-2\lambda\mu-2\rho^2$; the conditions of contact are

$$\pm\beta\Delta=\beta\lambda-\left(\beta+\frac{1}{\beta}\right)\mu+\beta\nu-2\rho,$$

$$\mp\gamma\Delta=\gamma\lambda+\gamma\mu-\left(\gamma+\frac{1}{\gamma}\right)\nu-2\rho,$$

where it is necessary, for the present purpose, to give opposite signs to the radicals. For if the radicals had the same sign, it would follow that

$$\frac{1}{\beta}\left[\beta\lambda-\left(\beta+\frac{1}{\beta}\right)\mu+\beta\nu-2\rho\right]-\frac{1}{\gamma}\left[\gamma\lambda+\gamma\mu-\left(\gamma+\frac{1}{\gamma}\right)\nu-2\rho\right]=0\,;$$

hence the section $\lambda x+\mu y+\nu z+\rho w=0$ would pass through the point

$$x:y:z:w=0:\frac{1}{\beta^2}:\frac{1}{\gamma^2}:-\frac{2}{\beta}+\frac{2}{\gamma}\,;$$

or the section would be a tangent section of the two determinators of the same class with the resultor $x=0$, which ought not to be the case. The proper formula is

$$\frac{1}{\beta}\left[\beta\lambda-\left(\beta+\frac{1}{\beta}\right)\mu+\beta\nu-2\rho\right]+\frac{1}{\gamma}\left[\gamma\lambda+\gamma\mu-\left(\gamma+\frac{1}{\gamma}\right)\nu-2\rho\right]=0;$$

and this equation being satisfied, the section

$$\lambda x+\mu y+\nu z+\rho w=0$$

passes through a point

$$x:y:z:w=2:-\frac{1}{\beta^2}:-\frac{1}{\gamma^2}:-\frac{2}{\beta}-\frac{2}{\gamma}.$$

The bisector passes through this point and the line of intersection of the determinators; its equation is

$$\frac{1}{\beta}\left(\frac{1}{2\beta}x-\beta y+\frac{1}{2\beta}z+w\right)-\frac{1}{\gamma}\left(\frac{1}{2\gamma}x+\frac{1}{2\gamma}y-\gamma z+w\right)=0\,;$$

or reducing and completing the system, the equations of the bisectors are

$$\left(\frac{1}{2\beta^2}-\frac{1}{2\gamma^2}\right)x-\left(1+\frac{1}{2\gamma^2}\right)\;y+\left(1+\frac{1}{2\beta^2}\right)\;z+\left(\frac{1}{\beta}-\frac{1}{\gamma}\right)w=0,$$

$$\left(1+\frac{1}{2\gamma^2}\right)\;x+\left(\frac{1}{2\gamma^2}-\frac{1}{2\alpha^2}\right)y-\left(1+\frac{1}{2\alpha^2}\right)\;z+\left(\frac{1}{\gamma}-\frac{1}{\alpha}\right)w=0,$$

$$-\left(1+\frac{1}{2\beta^2}\right)\;x+\left(1+\frac{1}{2\alpha^2}\right)\;y+\left(\frac{1}{2\alpha^2}-\frac{1}{2\beta^2}\right)z+\left(\frac{1}{\alpha}-\frac{1}{\beta}\right)w=0.$$

In order to verify the geometrical construction, it only remains to show that each bisector touches two tactors. Consider the bisector and tactor

$$-\left(1+\frac{1}{2\beta^2}\right)x+\left(1+\frac{1}{2\alpha^2}\right)y+\left(\frac{1}{2\alpha^2}-\frac{1}{2\beta^2}\right)z+\left(\frac{1}{\alpha}-\frac{1}{\beta}\right)w=0,$$

$$-2\alpha\left(a\alpha-\sqrt{-2bc}\right)x+(c+a)\,y+(b+a)\,z+\left(2a\alpha-\sqrt{-2bc}\right)w=0\,;$$

and represent these for a moment by

$$\lambda x+\mu y+\nu z+\rho w=0,\quad \lambda' x+\mu' y+\nu' z+\rho' w=0\,;$$

if Δ be the same as before, and Δ' the like function of $\lambda', \mu', \nu', \rho'$, also if

$$\Phi = \lambda\lambda' + \mu\mu' + \nu\nu' - (\mu\nu' + \mu'\nu) - (\nu\lambda' + \nu'\lambda) - (\lambda\mu' + \lambda'\mu) - 2\rho\rho',$$

then

$$\Delta^2 = \left(2 + \frac{1}{\alpha\beta}\right)^2,$$

$$\Delta'^2 = (2a\alpha^2 - 2\alpha\sqrt{-2bc} + b + c)^2,$$

$$\Phi = a\alpha^2\left(2 + \frac{1}{\alpha\beta}\right)^2 - 2\alpha\sqrt{-2bc}\left(2 + \frac{1}{\alpha\beta}\right) + c\left(2 + \frac{1}{\beta^2}\right);$$

and the condition of contact $\Delta\Delta' = \Phi$ (taking the proper sign for the radicals) becomes

$$\left(2 + \frac{1}{\alpha\beta}\right)(2a\alpha^2 - 2\alpha\sqrt{-2bc} + b + c) = a\alpha^2\left(2 + \frac{1}{\alpha\beta}\right)^2 - 2\alpha\sqrt{-2bc}\left(2 + \frac{1}{\alpha\beta}\right) + c\left(2 + \frac{1}{\beta^2}\right);$$

or reducing,

$$a\alpha - b\beta + c\,\frac{\alpha - \beta}{2\alpha\beta + 1} = 0,$$

an equation which is evidently not altered by the interchange of a, α and b, β. The conditions, in order that each bisector may touch two tactors, reduce themselves to the three equations,

$$a\alpha - b\beta + c\,\frac{\alpha - \beta}{2\alpha\beta + 1} = 0,$$

$$a\,\frac{\beta - \gamma}{2\beta\gamma + 1} + b\beta - c\gamma = 0,$$

$$-a\alpha + b\,\frac{\gamma - \alpha}{2\alpha\gamma + 1} + c\gamma = 0,$$

which are satisfied by the values above found for the quantities a, b, c. The possibility and truth of the geometrical construction are thus demonstrated.

§ 4.

Let it be in the first instance proposed to find the equation of a section touching all or any of the sections $x = 0$, $y = 0$, $z = 0$ of the surface of the second order,

$$ax^2 + by^2 + cz^2 + 2fyz + 2gzx + 2hxy + pw^2 = 0.$$

Any section whatever of this surface may be written in the form

$$(a\lambda + h\mu + g\nu)\,x + (h\lambda + b\mu + f\nu)\,y + (g\lambda + f\mu + c\nu)\,z + \sqrt{-p}\,\nabla w = 0,$$

where

$$\nabla^2 = a\lambda^2 + b\mu^2 + c\nu^2 + 2f\mu\nu + 2g\nu\lambda + 2h\lambda\mu - K,$$

and λ, μ, ν are indeterminate. And considering any other section represented by a like equation,

$$(a\lambda' + h\mu' + g\nu')\, x + (h\lambda' + b\mu' + f\nu')\, y + (g\lambda' + f\mu' + c\nu')\, z + \sqrt{-p}\,\nabla' w = 0,$$

where

$$\nabla'^2 = a\lambda'^2 + b\mu'^2 + c\nu'^2 + 2f\mu'\nu' + 2g\nu'\lambda' + 2h\lambda'\mu' - K,$$

it may be shown by means of the lemma previously given, that the condition of contact is

$$a\lambda\lambda' + b\mu\mu' + c\nu\nu' + f(\mu\nu' + \mu'\nu) + g(\nu\lambda' + \nu'\lambda) + h(\lambda\mu' + \lambda'\mu) \pm K = \nabla\nabla'.$$

Suppose that λ', μ', ν' satisfy the equations

$$\nabla' = 0,$$

$$h\lambda' + b\mu' + f\nu' = 0,$$

$$g\lambda' + f\mu' + c\nu' = 0,$$

so that the last-mentioned section becomes $x = 0$; and observing that the first of the above equations may be transformed into

$$a\lambda' + h\mu' + g\nu' = \frac{K}{\lambda'},$$

it is easy to obtain $\lambda' = \sqrt{\mathfrak{A}}$, $\mu' = \dfrac{\mathfrak{H}}{\sqrt{\mathfrak{A}}}$, $\nu' = \dfrac{\mathfrak{G}}{\sqrt{\mathfrak{A}}}$. The condition of contact thus becomes

$$\frac{K}{\sqrt{\mathfrak{A}}}\lambda \pm K = 0;$$

and taking the under sign, $\lambda = \sqrt{\mathfrak{A}}$, so that if in the above written equation we establish all or any of the equations $\lambda = \sqrt{\mathfrak{A}}$, $\mu = \sqrt{\mathfrak{B}}$, $\nu = \sqrt{\mathfrak{C}}$, we have the equation of a section touching all or the corresponding sections of the sections

$$x = 0, \quad y = 0, \quad z = 0.$$

In particular we have for a solution of the problem of tactions, the following equation of the section touching $x = 0$, $y = 0$, $z = 0$, viz.

$$(a\sqrt{\mathfrak{A}} + h\sqrt{\mathfrak{B}} + g\sqrt{\mathfrak{C}})\, x + (h\sqrt{\mathfrak{A}} + b\sqrt{\mathfrak{B}} + f\sqrt{\mathfrak{C}})\, y + (g\sqrt{\mathfrak{A}} + f\sqrt{\mathfrak{B}} + c\sqrt{\mathfrak{C}})\, z$$

$$+ \frac{\sqrt{-p}}{\sqrt{K}}\sqrt{2(\sqrt{\mathfrak{B}\mathfrak{C}} - \mathfrak{F})(\sqrt{\mathfrak{A}\mathfrak{C}} - \mathfrak{G})(\sqrt{\mathfrak{A}\mathfrak{B}} - \mathfrak{H})} = 0.$$

Anticipating the use of a notation the value of which will subsequently appear, or putting

$$\mathrm{f} = \sqrt[4]{\mathfrak{A}}\sqrt{\sqrt{\mathfrak{B}\mathfrak{C}} - \mathfrak{F}}, \quad \mathrm{g} = \sqrt[4]{\mathfrak{B}}\sqrt{\sqrt{\mathfrak{A}\mathfrak{C}} - \mathfrak{G}}, \quad \mathrm{h} = \sqrt[4]{\mathfrak{C}}\sqrt{\sqrt{\mathfrak{A}\mathfrak{B}} - \mathfrak{H}}, \quad J = \sqrt{2}\,\sqrt[4]{\mathfrak{A}\mathfrak{B}\mathfrak{C}},$$

values which give

$$K^2 = -\mathrm{f}^4 - \mathrm{g}^4 - \mathrm{h}^4 + 2\mathrm{g}^2\mathrm{h}^2 + 2\mathrm{h}^2\mathrm{f}^2 + 2\mathrm{f}^2\mathrm{g}^2 - \frac{4\mathrm{f}^2\mathrm{g}^2\mathrm{h}^2}{J^2},$$

the equation of the section in question is

$$\frac{\mathrm{f}^2}{\sqrt{\mathfrak{A}}}(-\mathrm{f}^2+\mathrm{g}^2+\mathrm{h}^2)\,x + \frac{\mathrm{g}^2}{\sqrt{\mathfrak{B}}}(\mathrm{f}^2-\mathrm{g}^2+\mathrm{h}^2)\,y + \frac{\mathrm{h}^2}{\sqrt{\mathfrak{C}}}(\mathrm{f}^2+\mathrm{g}^2-\mathrm{h}^2)\,z + \frac{\mathrm{fgh}\sqrt{-p}\sqrt{K}}{J}\,w = 0.$$

I proceed to investigate a transformation of the equation for the section with an indeterminate parameter λ, which touches the two sections $y=0$, $z=0$. We have

$$a\nabla^2 = (a\lambda + h\mu + g\nu)^2 + (\mathfrak{C}\mu^2 + \mathfrak{B}\nu^2 - 2\mathfrak{F}\mu\nu) - \mathfrak{B}\mathfrak{C} + \mathfrak{F}^2;$$

or putting for μ and ν their values $\sqrt{\mathfrak{B}}$, $\sqrt{\mathfrak{C}}$ in the second term,

$$a\nabla^2 = (a\lambda + h\mu + g\nu)^2 + (\sqrt{\mathfrak{B}\mathfrak{C}} - \mathfrak{F})^2;$$

and introducing instead of λ an indeterminate quantity X, such that

$$a\lambda + h\mu + g\nu = (\sqrt{\mathfrak{B}\mathfrak{C}} - \mathfrak{F})\,X,$$

we have

$$a\nabla^2 = (\sqrt{\mathfrak{B}\mathfrak{C}} - \mathfrak{F})\sqrt{1+X^2};$$

also introducing throughout X instead of λ, and completing the substitution of $\sqrt{\mathfrak{B}}$, $\sqrt{\mathfrak{C}}$, for μ, ν, the equation of the section touching $y=0$, $z=0$, becomes

$$(ax + hy + gz)\,X + y\sqrt{\mathfrak{C}} + z\sqrt{\mathfrak{B}} + w\sqrt{-ap}\sqrt{1+X^2} = 0.$$

It may be remarked here in passing, that this is a very convenient form for the demonstration of the theorem; "If two sections of a surface of the second order touch each other, and are also tangent sections (of the same class) to two fixed sections, then considering the planes through the axis of the fixed sections and the poles of the tangent sections, and also the tangent planes through this axis, the anharmonic ratio of the four planes is independent of the position of the moveable tangent sections;" where by the axis of the fixed sections is to be understood the line joining their poles.

The sections touching $z=0$, $x=0$, and $x=0$, $y=0$, are of course

$$x\sqrt{\mathfrak{C}} + (hx + by + fz)\,Y + z\sqrt{\mathfrak{A}} + w\sqrt{-bp}\sqrt{1+Y^2} = 0,$$

$$x\sqrt{\mathfrak{B}} + y\sqrt{\mathfrak{A}} + (gx + fy + cz)\,Z + w\sqrt{-cp}\sqrt{1+Z^2} = 0,$$

where

$$h\lambda' + b\mu' + f\nu' = (\sqrt{\mathfrak{C}\mathfrak{A}} - \mathfrak{G})\,Y,\quad \lambda' = \sqrt{\mathfrak{A}},\quad \mu' = \mu',\quad \nu' = \sqrt{\mathfrak{C}},$$

$$g\lambda'' + f\mu'' + c\nu'' = (\sqrt{\mathfrak{A}\mathfrak{B}} - \mathfrak{H})\,Z,\quad \lambda'' = \sqrt{\mathfrak{A}},\quad \mu'' = \sqrt{\mathfrak{B}},\quad \nu'' = \nu''.$$

The conditions of contact of the sections represented by the above written equations would be perhaps most simply obtained directly from the lemma, but it is proper to deduce it from the formula for contact used in the present memoir. If for shortness

$$\Phi(\pm) = a\lambda'\lambda'' + b\mu'\mu'' + c\nu'\nu'' + f(\mu'\nu'' + \mu''\nu') + g(\nu'\lambda'' + \nu''\lambda') + h(\lambda'\mu'' + \lambda''\mu') \pm K,$$

where the symbol $\Phi(\pm)$ is used in order to mark the essentially different character of the results corresponding to the different values of the ambiguous sign, then

$$\begin{aligned} bc\Phi(-) = & f(h\lambda' + b\mu' + f\nu')(g\lambda'' + f\mu'' + c\nu''), \\ & + (\mathfrak{A}\nu' - \mathfrak{G}\lambda')(g\lambda'' + f\mu'' + c\nu''), \\ & + (\mathfrak{A}\mu'' - \mathfrak{H}\lambda'')(h\lambda' + b\mu' + f\nu'), \\ & + \nu'\mu''(-\mathfrak{A}f) + \nu'\lambda'' f\mathfrak{H} + \lambda'\mu'' f\mathfrak{G} + \lambda'\lambda''(K - f\mathfrak{F}) \\ & - \mathfrak{A}K - f^2K. \end{aligned}$$

$$\begin{aligned} = & f(h\lambda' + b\mu' + f\nu')(g\lambda'' + f\mu'' + c\nu'') \\ & + \sqrt{\mathfrak{A}}(\sqrt{\mathfrak{AC}} - \mathfrak{G})(g\lambda'' + f\mu'' + c\nu'') \\ & + \sqrt{\mathfrak{A}}(\sqrt{\mathfrak{AB}} - \mathfrak{H})(h\lambda' + b\mu' + f\nu') \\ & + f(-\mathfrak{A}\sqrt{\mathfrak{BC}} + \mathfrak{H}\sqrt{\mathfrak{CA}} + \mathfrak{G}\sqrt{\mathfrak{AB}} - \mathfrak{AF} - (\mathfrak{GH} - \mathfrak{AF})) \end{aligned}$$

$$\begin{aligned} = & f(h\lambda' + b\mu' + f\nu')(g\lambda'' + f\mu'' + c\nu'') \\ & + \sqrt{\mathfrak{A}}(\sqrt{\mathfrak{AC}} - \mathfrak{G})(g\lambda'' + f\mu'' + c\nu'') \\ & + \sqrt{\mathfrak{A}}(\sqrt{\mathfrak{AB}} - \mathfrak{H})(h\lambda' + b\mu' + f\nu') \\ & - f(\sqrt{\mathfrak{AC}} - \mathfrak{G})(\sqrt{\mathfrak{AB}} - \mathfrak{H}), \end{aligned}$$

that is, $\quad bc\Phi(-) = (\sqrt{\mathfrak{AC}} - \mathfrak{G})(\sqrt{\mathfrak{AB}} - \mathfrak{H})\{fYZ + \sqrt{\mathfrak{A}}(Y+Z) - f\}.$

What, however, is really required[1], is the value of $\Phi(+)$; to find this, we have

$$\begin{aligned} bc\Phi(+) &= bc\Phi(-) + 2bcK \\ &= (\sqrt{\mathfrak{AC}} - \mathfrak{G})(\sqrt{\mathfrak{AB}} - \mathfrak{H})\{fYZ + \sqrt{\mathfrak{A}}(Y+Z) + f\} \\ &\quad + 2bcK - 2f(\sqrt{\mathfrak{AC}} - \mathfrak{G})(\sqrt{\mathfrak{AB}} - \mathfrak{H}), \end{aligned}$$

the second line of which is

$$2(\sqrt{\mathfrak{AC}} - \mathfrak{G})(\sqrt{\mathfrak{AB}} - \mathfrak{H})\left\{\frac{bcK}{K^2bc}(\sqrt{\mathfrak{AC}} + \mathfrak{G})(\sqrt{\mathfrak{AB}} + \mathfrak{H}) - f\right\}$$

$$= \frac{2(\sqrt{\mathfrak{AC}} - \mathfrak{G})(\sqrt{\mathfrak{AB}} - \mathfrak{H})}{K}\{(\sqrt{\mathfrak{AC}} + \mathfrak{G})(\sqrt{\mathfrak{AB}} + \mathfrak{H}) - \mathfrak{GH} + \mathfrak{AF}\}$$

$$= 2(\sqrt{\mathfrak{AC}} - \mathfrak{G})(\sqrt{\mathfrak{AB}} - \mathfrak{H})\sqrt{\mathfrak{A}}\theta,$$

[1] It may be shown without difficulty that the $(-)$ sign would imply that the sections touching $z=0$, $x=0$, and $x=0$, $y=0$, were sections touching $x=0$ at the same point. By taking the $(-)$ sign in each equation we should have the solution of the problem "to determine three sections of a surface of the second order, the two sections of each pair touching one of three given sections at the same point," which is not without interest; the solution may be completed without any difficulty.

where

$$\theta = \frac{1}{K}(\sqrt{\mathfrak{ABC}} + \mathfrak{F}\sqrt{\mathfrak{A}} + \mathfrak{G}\sqrt{\mathfrak{B}} + \mathfrak{H}\sqrt{\mathfrak{C}});$$

and consequently

$$bc\Phi(+) = (\sqrt{\mathfrak{AC}} - \mathfrak{G})(\sqrt{\mathfrak{AB}} - \mathfrak{H})\{fYZ + \sqrt{\mathfrak{A}}(Y+Z) + f + 2\theta\sqrt{\mathfrak{A}}\},$$

a reduction, which on account of its peculiarity, I have thought right to work out in full.

The condition of contact is

$$\Phi(+) = \nabla'\nabla'' = \frac{1}{\sqrt{bc}}(\sqrt{\mathfrak{AC}} - \mathfrak{G})(\sqrt{\mathfrak{AB}} - \mathfrak{H})\sqrt{1+Y^2}\sqrt{1+Z^2}.$$

Hence finally, the condition in order that the sections

$$x\sqrt{\mathfrak{C}} + (hx + by + fz)Y + z\sqrt{\mathfrak{A}} + w\sqrt{-bp}\sqrt{1+Y^2} = 0,$$

$$x\sqrt{\mathfrak{B}} + y\sqrt{\mathfrak{A}} + (gx + fy + cz)Z + w\sqrt{-cp}\sqrt{1+Z^2} = 0,$$

(the former of which is a section touching $z=0$, $x=0$, and the latter a section touching $x=0$, $y=0$) may touch, is

$$fYZ + \sqrt{\mathfrak{A}}(Y+Z) + (f + 2\theta\sqrt{\mathfrak{A}}) - \sqrt{bc}\sqrt{1+Y^2}\sqrt{1+Z^2} = 0.$$

The preceding researches show that the solution of Steiner's extension of Malfatti's problem depends on a system of equations, such as the system mentioned at the commencement of the following section.

§ 5.

Consider the system of equations

$$\alpha + \beta(Y+Z) + \gamma YZ + \delta\sqrt{1+Y^2}\sqrt{1+Z^2} = 0,$$

$$\alpha' + \beta'(Z+X) + \gamma' ZX + \delta'\sqrt{1+Z^2}\sqrt{1+X^2} = 0,$$

$$\alpha'' + \beta''(X+Y) + \gamma'' XY + \delta''\sqrt{1+X^2}\sqrt{1+Y^2} = 0;$$

these equations may, it will be seen, be solved by quadratics only, when the coefficients satisfy the relations

$$\frac{\beta}{\gamma-\alpha} = \frac{\beta'}{\gamma'-\alpha'} = \frac{\beta''}{\gamma''-\alpha''},$$

$$\frac{\beta^2+\gamma^2-\delta^2}{\gamma^2-\alpha^2} = \frac{\beta'^2+\gamma'^2-\delta'^2}{\gamma'^2-\alpha'^2} = \frac{\beta''^2+\gamma''^2-\delta''^2}{\gamma''^2-\alpha''^2}.$$

It may be remarked that these equations are satisfied by

$$\beta = 0,\quad \beta' = 0,\quad \beta'' = 0,\quad \gamma = \delta,\quad \gamma' = \delta',\quad \gamma'' = \delta'',$$

or if we write

$$\frac{\alpha}{\gamma} = -l,\quad \frac{\alpha'}{\gamma'} = -m,\quad \frac{\alpha''}{\gamma''} = -n,$$

the equations become by a simple reduction,

$$\begin{aligned} Y^2 + Z^2 + 2l\,YZ &= l^2 - 1,\\ Z^2 + X^2 + 2mZX &= m^2 - 1,\\ X^2 + Y^2 + 2n\,XY &= n^2 - 1, \end{aligned}$$

which are equivalent to the equations discussed in my paper "On a system of Equations connected with Malfatti's Problem and on another Algebraical System," *Cambridge and Dublin Mathematical Journal*, t. IV. [1849] pp. 270—275, [79]; the solution might have been effected by the direct method, which I shall here adopt, of eliminating any one of the variables between the two equations into which it enters, and combining the result with the third equation.

Writing the second and third equations under the form

$$A' + B'X + C'\sqrt{1+X^2} = 0,$$

$$A'' + B''X + C''\sqrt{1+X^2} = 0,$$

the result of the elimination may be presented in the form

$$A'A'' + B'B'' - C'C'' = \sqrt{A'^2 + B'^2 - C'^2}\,\sqrt{A''^2 + B''^2 - C''^2},$$

which is most easily obtained by writing $X = \tan\phi$ and operating with the symbol $\cos^{-1}$; but if the rationalized equations be represented by

$$\lambda' + 2\mu'X + \nu'X^2 = 0 \text{ and } \lambda'' + 2\mu''X + \nu''X^2 = 0,$$

the form

$$4(\lambda'\nu' - \mu'^2)(\lambda''\nu'' - \mu''^2) = (\lambda'\nu'' + \lambda''\nu' - 2\mu'\mu'')^2$$

leads easily to the result in question. The values which enter are

$$\begin{aligned} A' &= \alpha' + \beta'Z, & A'' &= \alpha'' + \beta''Y,\\ B' &= \beta' + \gamma'Z, & B'' &= \beta'' + \gamma''Y,\\ C' &= \delta'\sqrt{1+Z^2}, & C'' &= \delta''\sqrt{1+Y^2}; \end{aligned}$$

whence, in the first place, by the equation connecting Y, Z,

$$C'C'' = -\frac{\delta'\delta''}{\delta}\{\alpha + \beta(Y+Z) + \delta YZ\}.$$

It is obviously convenient that $A'A''+B'B''$ should be symmetrical with respect to Y and Z, and this will be the case if

$$\alpha'\beta''+\beta'\gamma''=\alpha''\beta'+\beta''\gamma',$$

that is, if
$$\beta'(\gamma''-\alpha'')=\beta''(\gamma'-\alpha');$$

or assuming that the equations are symmetrically related to the system, we have the first set of relations between the coefficients, relations which are satisfied by

$$\alpha=\gamma+2\phi\beta,\quad \alpha'=\gamma'+2\phi\beta',\quad \alpha''=\gamma''+2\phi\beta'',$$

and the values of α, α', α'' will be considered henceforth as given by these conditions. We have

$$A'A''+B'B''-C'C''=\alpha'\alpha''+\beta'\beta''+(\gamma'\beta''+\gamma''\beta'+2\phi\beta'\beta'')(Y+Z)+(\beta'\beta''+\gamma'\gamma'')\,YZ$$
$$+\frac{\delta'\delta''}{\delta}\{\alpha+\beta(Y+Z)+\gamma YZ\}.$$

The quantities $A'^2+B'^2-C'^2$, $A''^2+B''^2-C''^2$ are quadratic functions of Z and Y respectively, and with proper relations between the coefficients, we may assume

$$(A'^2+B'^2-C'^2)(A''^2+B''^2-C''^2)=l^2s^2\{U^2+k[(\alpha+\beta(Y+Z)+\gamma YZ)^2-\delta^2(1+Y^2)(1+Z^2)]\},$$

in which U is a linear function of $Y+Z$ and YZ, and k and ls are constants. The first side must, in the first place, be symmetrical with respect to Y and Z, or

$$\alpha'^2+\beta'^2-\delta'^2,\quad (\alpha'+\gamma')\beta',\quad \beta'^2+\gamma'^2-\delta'^2$$

must be proportional to

$$\alpha''^2+\beta''^2-\delta''^2,\quad (\alpha''+\gamma'')\delta'',\quad \beta''^2+\gamma''^2-\delta''^2.$$

But since

$$(\alpha'+\gamma')\beta',\quad (\alpha''+\gamma'')\beta''$$

are proportional to

$$\gamma'^2-\alpha'^2,\quad \gamma''^2-\alpha''^2,$$

it is only necessary that

$$\beta'^2+\gamma'^2-\delta'^2,\quad \beta''^2+\gamma''^2-\delta''^2$$

should be proportional to

$$\gamma'^2-\alpha'^2,\qquad \gamma''^2-\alpha''^2;$$

or since the equations are supposed symmetrically related to the system, we must have the second set of relations between the coefficients. Suppose

$$\frac{\beta^2+\gamma^2-\delta^2}{\gamma^2-\alpha^2}=\frac{\beta'^2+\gamma'^2-\delta'^2}{\gamma'^2-\alpha'^2}=\frac{\beta''^2+\gamma''^2-\delta''^2}{\gamma''^2-\alpha''^2}=-\frac{s}{\phi},$$

then since

$$\gamma^2-\alpha^2=-4(\gamma+\phi\beta)\phi\beta,\ \&c.,$$

we have

$$\delta^2 = \beta^2 + \gamma^2 - 4s(\gamma + \phi\beta)\beta$$
$$\delta'^2 = \beta'^2 + \gamma'^2 - 4s(\gamma' + \phi\beta')\beta'$$
$$\delta''^2 = \beta''^2 + \gamma''^2 - 4s(\gamma'' + \phi\beta'')\beta'',$$

and δ, δ', δ'' will be supposed henceforth to satisfy these equations.

We have next

$$A'^2 + B'^2 - C'^2 = 4(\gamma' + \phi\beta')\beta'(s + \phi + Z + sZ^2)$$
$$A''^2 + B''^2 - C''^2 = 4(\gamma'' + \phi\beta'')\beta''(s + \phi + Y + sY^2),$$

which may be simplified by writing

$$s = \frac{\mu - \phi}{1 + \mu^2}, \qquad \nu = \frac{1 + \mu\phi}{\mu - \phi},$$

where μ, ν are to be considered as given functions of s and ϕ. These values give

$$A'^2 + B'^2 - C'^2 = 4(\gamma' + \phi\beta')\beta's(Z + \mu)(Z + \nu),$$
$$A''^2 + B''^2 - C''^2 = 4(\gamma'' + \phi\beta'')\beta''s(Y + \mu)(Y + \nu).$$

Hence, putting for simplicity

$$l^2 = 4(\gamma' + \phi\beta')(\gamma'' + \phi\beta'')\beta'\beta'',$$

we have

$$4(Z + \mu)(Z + \nu)(Y + \mu)(Y + \nu) = U^2 + k[(\alpha + \beta(Y + Z) + \gamma YZ)^2 - \delta^2(1 + Y^2)(1 + Z^2)];$$

and the two sides have next to be expressed in terms of $Y + Z$ and YZ.

If for symmetry we write

$$\xi = 1, \quad \eta = Y + Z, \quad \zeta = YZ,$$

then

$$4(\mu^2\xi + \mu\eta + \zeta)(\nu^2\xi + \nu\eta + \zeta) + k\delta^2[(\xi - \zeta)^2 + \eta^2] = U^2 + k(\alpha\xi + \beta\eta + \gamma\zeta)^2;$$

and U is now to be considered a linear function of ξ, η, ζ.

The condition that the first side of the equation may divide into factors, gives an equation for determining k; since the condition is satisfied for $k = 0$ and $k = \infty$, the equation will be linear, and it is easily seen that the value is $k = \frac{1}{\delta^2}(\mu - \nu)^2$. In fact

$$4(\mu^2\xi + \mu\eta + \zeta)(\nu^2\xi + \nu\eta + \zeta)^2 + (\mu - \nu)^2[(\xi - \zeta)^2 + \eta^2]$$
$$= (2\mu\nu\xi + (\mu + \nu)\eta + 2\zeta)^2 + (\mu - \nu)^2(\xi + \zeta)^2;$$

hence

$$\{2\mu\nu\xi + (\mu + \nu)\eta + 2\zeta\}^2 - U^2 = \frac{(\mu - \nu)^2}{\delta^2}\{(\alpha\xi + \beta\eta + \gamma\zeta)^2 - \delta^2(\xi + \zeta)^2\},$$

and we may assume

$$2\mu\nu\xi + (\mu+\nu)\eta + 2\zeta + U = \frac{\mu-\nu}{\delta}\Lambda\,\{(\alpha\xi+\beta\eta+\gamma\zeta) - \delta(\xi+\zeta)\},$$

$$2\mu\nu\xi + (\mu+\nu)\eta + 2\zeta - U = \frac{\mu-\nu}{\delta}\frac{1}{\Lambda}\,\{(\alpha\xi+\beta\eta+\gamma\zeta) + \delta(\xi+\zeta)\},$$

subject to its being shown that

$$4\mu\nu\xi + 2(\mu+\nu)\eta + 4\zeta = \frac{\mu-\nu}{\delta}\left\{\left(\Lambda+\frac{1}{\Lambda}\right)(\alpha\xi+\beta\eta+\gamma\zeta) - \delta\left(\Lambda-\frac{1}{\Lambda}\right)(\xi+\zeta)\right\}$$

gives a constant value for Λ. The comparison of coefficients gives

$$4\mu\nu = \frac{\mu-\nu}{\delta}\left\{\left(\Lambda+\frac{1}{\Lambda}\right)\alpha - \left(\Lambda-\frac{1}{\Lambda}\right)\delta\right\},$$

$$2\mu+2\nu = \frac{\mu-\nu}{\delta}\left(\Lambda+\frac{1}{\Lambda}\right)\beta,$$

$$4 = \frac{\mu-\nu}{\delta}\left\{\left(\Lambda+\frac{1}{\Lambda}\right)\gamma - \left(\Lambda-\frac{1}{\Lambda}\right)\delta\right\};$$

the first and third of these give

$$4(1-\mu\nu) = \frac{\mu-\nu}{\delta}\left(\Lambda+\frac{1}{\Lambda}\right)(\gamma-\alpha),$$

which will be identical with the second, if

$$\frac{2(1-\mu\nu)}{\mu+\nu} = \frac{\beta}{\gamma-a} = -2\phi,$$

which follows at once from the equation

$$\nu = \frac{1+\mu\phi}{\mu-\phi}.$$

Forming next the two equations

$$\Lambda+\frac{1}{\Lambda} = \frac{2}{(\mu-\nu)\beta}(\mu+\nu)\delta,$$

$$\Lambda-\frac{1}{\Lambda} = \frac{2}{(\mu-\nu)\beta}\{(\mu+\nu)\gamma - 2\beta\},$$

these will be equivalent to a single equation if

$$(\mu+\nu)^2\delta^2 = \{(\mu+\nu)\gamma - 2\beta\}^2 + (\mu-\nu)^2\beta^2,$$

that is, if

$$(\mu+\nu)^2\delta^2 = (\mu+\nu)^2(\beta^2+\gamma^2) - 4(\mu+\nu)\beta\gamma - 4(\mu\nu-1)\beta^2;$$

or finally, if

$$\delta^2 = \beta^2 + \gamma^2 - 4s\beta\left(\gamma + \frac{\mu\nu - 1}{\mu + \nu}\beta\right) = \beta^2 + \gamma^2 - 4s(\gamma + \phi\beta)\beta,$$

which is in fact the case.

Writing the equations for

$$\Lambda + \frac{1}{\Lambda}, \quad \Lambda - \frac{1}{\Lambda},$$

in the form

$$\Lambda + \frac{1}{\Lambda} = \frac{2\delta}{(\mu - \nu)\beta s}\ \Lambda - \frac{1}{\Lambda} = \frac{2}{(\mu - \nu)\beta s}(\gamma - 2\beta s),$$

and substituting in

$$U = \frac{\mu - \nu}{2\delta}\left\{\left(\Lambda - \frac{1}{\Lambda}\right)(\alpha\xi + \beta\eta + \gamma\zeta) - \left(\Lambda + \frac{1}{\Lambda}\right)\delta^2(\xi + \zeta)\right\},$$

we have

$$U = \frac{1}{s\beta\delta}\{(\gamma - 2\beta s)(\alpha\xi + \beta\eta + \gamma\zeta) - \delta^2(\xi + \zeta)\}$$

$$= \frac{1}{s\beta\delta}\{(-\beta + 2s\gamma + 2\phi\gamma)\xi + (\gamma - 2s\beta)\eta + (-\beta + 2s\gamma + 4s\phi\beta)\zeta\};$$

and consequently, multiplying by

$$ls = 2\sqrt{(\gamma' + \phi\beta')(\gamma'' + \phi\beta'')\beta'\beta'')}$$

we have

$$\sqrt{A'^2 + B'^2 - C'^2}\sqrt{A''^2 + B''^2 - C''^2}$$

$$= \frac{2}{\delta}\sqrt{(\gamma' + \phi\beta')(\gamma'' + \phi\beta'')\beta'\beta''}\{(-\beta + 2s\gamma + 2\phi\gamma)\xi + (\gamma - 2s\beta)\eta + (-\beta + 2s\gamma + 4s\phi\beta)\zeta\},$$

or collecting the different terms which enter into the equation

$$A'A'' + B'B'' - C'C'' = \sqrt{A'^2 + B'^2 - C'^2}\sqrt{A''^2 + B''^2 - C''^2}$$

the result is

$$(\alpha'\alpha'' + \beta'\beta'')\xi + (\gamma'\beta'' + \gamma''\beta' + 2\phi\beta'\beta'')\eta + (\beta'\beta'' + \gamma'\gamma'')\zeta + \frac{\delta'\delta''}{\delta}(\alpha\xi + \beta\eta + \gamma\zeta)$$

$$-\frac{2}{\delta}\sqrt{(\gamma' + 2\phi\beta')(\gamma'' + 2\phi\beta'')\beta'\beta''}\{(-\beta + 2s\gamma + 2\phi\gamma)\xi + (\gamma - 2s\beta)\eta + (-\beta + 2s\gamma + 4s\phi\beta)\zeta\} = 0,$$

which, combined with the first equation written under the form

$$(\alpha\xi + \beta\eta + \gamma\zeta)^2 - \delta^2[(\xi - \zeta)^2 + \eta^2] = 0,$$

determines the ratios of ξ, η, ζ, that is, the values of $Y + Z$ and YZ.

§ 6.

The system of equations

$$(f + 2\theta\sqrt{\mathfrak{A}}) + \sqrt{\mathfrak{A}}\,(Y + Z) + fYZ - \sqrt{bc}\,\sqrt{1 + Y^2}\,\sqrt{1 + Z^2} = 0,$$

$$(g + 2\theta\sqrt{\mathfrak{B}}) + \sqrt{\mathfrak{B}}\,(Z + X) + gZX - \sqrt{ca}\,\sqrt{1 + Z^2}\,\sqrt{1 + X^2} = 0,$$

$$(h + 2\theta\sqrt{\mathfrak{C}}) + \sqrt{\mathfrak{C}}\,(X + Y) + hXY - \sqrt{ab}\,\sqrt{1 + X^2}\,\sqrt{1 + Y^2} = 0,$$

where

$$\theta = \frac{1}{K}(\sqrt{\mathfrak{A}\mathfrak{B}\mathfrak{C}} + \mathfrak{F}\sqrt{\mathfrak{A}} + \mathfrak{G}\sqrt{\mathfrak{B}} + \mathfrak{H}\sqrt{\mathfrak{C}}),$$

on which depends the solution of Steiner's extension of Malfatti's problem, is at once seen to belong to the class of equations treated of in the preceding section, and we have $\phi = \theta$, $s = 0$. The equations at the conclusion of the preceding section become

$$\{\sqrt{\mathfrak{B}\mathfrak{C}} + gh + 2\theta(g\sqrt{\mathfrak{C}} + h\sqrt{\mathfrak{B}}) + 4\theta^2\sqrt{\mathfrak{B}\mathfrak{C}}\}\,\xi + \{g\sqrt{\mathfrak{C}} + h\sqrt{\mathfrak{B}} + 2\theta\sqrt{\mathfrak{B}\mathfrak{C}}\}\,\eta + \{\sqrt{\mathfrak{B}\mathfrak{C}} + gh\}\,\zeta$$

$$- a\,[(f + 2\theta\sqrt{\mathfrak{A}})\,\xi + \sqrt{\mathfrak{A}}\eta + f\zeta] - \frac{2}{\sqrt{bc}}\sqrt{(g + \theta\sqrt{\mathfrak{B}})\,(h + \theta\sqrt{\mathfrak{C}})\,\sqrt{\mathfrak{B}\mathfrak{C}}}\,\{(\sqrt{\mathfrak{A}} - 2\theta f)\,\xi - f\eta + \sqrt{\mathfrak{A}}\zeta\} = 0,$$

$$\{(f + 2\theta\sqrt{\mathfrak{A}})\,\xi + \sqrt{\mathfrak{A}}\eta + f\zeta\}^2 - bc\,\{(\xi - \zeta)^2 + \eta^2\} = 0,$$

which may also be written

$$(\sqrt{\mathfrak{B}\mathfrak{C}} + \mathfrak{F})\,(\xi + \zeta) + (-a\sqrt{\mathfrak{A}} + g\sqrt{\mathfrak{C}} + h\sqrt{\mathfrak{B}} + 2\theta\sqrt{\mathfrak{B}\mathfrak{C}})\,(\eta + 2\theta\xi)$$

$$- \frac{2}{\sqrt{bc}}\sqrt{(g + \theta\sqrt{\mathfrak{B}})\,(h + \theta\sqrt{\mathfrak{C}})\,\sqrt{\mathfrak{B}\mathfrak{C}}}\,((\sqrt{\mathfrak{A}} - 2\theta f)\,\xi - f\eta + \sqrt{\mathfrak{A}}\zeta) = 0,$$

$$\{f(\xi + \zeta) + \sqrt{\mathfrak{A}}\,(\eta + 2\theta\xi)\}^2 - bc\,\{(\xi - \zeta)^2 + \eta^2\} = 0.$$

Hence observing that

$$g + \theta\sqrt{\mathfrak{B}} = \frac{1}{K}(\sqrt{\mathfrak{B}\mathfrak{C}} + \mathfrak{F})\,(\sqrt{\mathfrak{A}\mathfrak{B}} + \mathfrak{H});\quad h + \theta\sqrt{\mathfrak{C}} = \frac{1}{K}(\sqrt{\mathfrak{B}\mathfrak{C}} + \mathfrak{F})\,(\sqrt{\mathfrak{A}\mathfrak{C}} + \mathfrak{G});$$

$$- a\sqrt{\mathfrak{A}} + h\sqrt{\mathfrak{B}} + g\sqrt{\mathfrak{C}} + 2\theta\sqrt{\mathfrak{B}\mathfrak{C}} = \theta(\sqrt{\mathfrak{B}\mathfrak{C}} + \mathfrak{F}),$$

and putting for a moment

$$\lambda = \frac{1}{K}\sqrt{(\sqrt{\mathfrak{A}\mathfrak{C}} + \mathfrak{G})\,(\sqrt{\mathfrak{A}\mathfrak{B}} + \mathfrak{H})\,\sqrt{\mathfrak{B}\mathfrak{C}}},$$

and therefore

$$\sqrt{(g + \theta\sqrt{\mathfrak{B}})\,(h + \theta\sqrt{\mathfrak{C}})\,\sqrt{\mathfrak{B}\mathfrak{C}}} = (\sqrt{\mathfrak{B}\mathfrak{C}} + \mathfrak{F})\,\lambda,$$

the first equation divides by $(\sqrt{\mathfrak{BC}}+\mathfrak{F})$, and the result is

$$(\xi+\zeta)+\theta(\eta+2\theta\xi)-\frac{2\lambda}{\sqrt{bc}}\{\sqrt{\mathfrak{A}}\,(\xi+\zeta)-f(\eta+2\theta\xi)\}=0.$$

Also, by an easy transformation, the second equation becomes

$$-\{\sqrt{\mathfrak{A}}\,(\xi+\zeta)-f(\eta+2\theta\xi)\}^2+4bc\zeta\,\big((\xi+\zeta)+\theta(\eta+2\theta\xi)-(1+\theta^2)\,\xi\big)=0,$$

or putting

$$\xi+\zeta\quad+\theta(\eta+2\theta\xi)=\Theta,$$

$$\frac{1}{\sqrt{bc}}\big(\sqrt{\mathfrak{A}}\,(\xi+\zeta)-f(\eta+2\theta\xi)\big)=\Phi,$$

$$\zeta\qquad\qquad=\Psi,$$

the equations become

$$\Theta-2\lambda\Phi=0,$$

$$-\Phi^2+4\Psi\{\Theta-(1+\theta^2)\,\Psi\}=0;$$

hence eliminating Φ,

$$\left(2\Psi-\frac{\Theta}{1+\theta^2}\right)^2=\frac{\Theta^2}{(1+\theta^2)^2}\left(1-\frac{1+\theta^2}{4\lambda^2}\right),$$

or observing that

$$1+\theta^2=\frac{1}{K^2}(\sqrt{\mathfrak{BC}}+\mathfrak{F})(\sqrt{\mathfrak{CA}}+\mathfrak{G})(\sqrt{\mathfrak{AB}}+\mathfrak{H}),$$

and reducing, we obtain

$$\Psi=\frac{K^2\Theta}{(\sqrt{\mathfrak{BC}}+\mathfrak{F})(\sqrt{\mathfrak{CA}}+\mathfrak{G})(\sqrt{\mathfrak{AB}}+\mathfrak{H})}\left(1+\frac{\sqrt{(\sqrt{\mathfrak{BC}}-\mathfrak{F})\sqrt{\mathfrak{BC}}}}{\sqrt{2\mathfrak{BC}}}\right);$$

also $\Theta=2\lambda\Phi$ gives

$$\Phi=\frac{K\Theta}{2\sqrt{(\sqrt{\mathfrak{AC}}+\mathfrak{G})(\sqrt{\mathfrak{AB}}+\mathfrak{H})\sqrt{\mathfrak{BC}}}}.$$

Suppose

$$\sqrt{\mathfrak{BC}}+\mathfrak{F}=\alpha,\quad \sqrt{\mathfrak{BC}}-\mathfrak{F}=\alpha_{,},\quad \therefore\ \alpha\alpha_{,}=Ka,$$

$$\sqrt{\mathfrak{CA}}+\mathfrak{G}=\beta,\quad \sqrt{\mathfrak{CA}}-\mathfrak{G}=\beta_{,},\quad \beta\beta_{,}=Kb,$$

$$\sqrt{\mathfrak{AB}}+\mathfrak{H}=\gamma,\quad \sqrt{\mathfrak{AB}}-\mathfrak{H}=\gamma_{,},\quad \gamma\gamma_{,}=Kc;$$

then substituting

$$\Theta-\frac{\sqrt{2}\,\sqrt{\alpha+\alpha_{,}}}{\sqrt{\beta_{,}\gamma_{,}}}\sqrt{bc}\,\Phi=0,$$

$$\Theta-\frac{4bc}{\beta_{,}\gamma_{,}}(\alpha+\alpha_{,})\left(1-\frac{\sqrt{\alpha_{,}}}{\sqrt{\alpha+\alpha_{,}}}\right)\Psi=0,$$

that is,

$$\xi + \zeta + \theta(\eta + 2\theta\xi) - \frac{\sqrt{2}\sqrt{\alpha+\alpha_{\prime}}}{\sqrt{\beta_{\prime}\gamma_{\prime}}}\{\sqrt{\mathfrak{A}}(\xi+\zeta) - f(\eta+2\theta\xi)\} = 0,$$

$$\xi + \zeta + \theta(\eta + 2\theta\xi) - \frac{4bc(\alpha+\alpha_{\prime})}{\beta_{\prime}\gamma_{\prime}}\left(1 - \frac{\sqrt{\alpha_{\prime}}}{\sqrt{\alpha+\alpha_{\prime}}}\right)\zeta = 0;$$

these may be written

$$L\xi + M\eta + N\zeta = 0,$$

$$L'\xi + M'\eta + N'\zeta = 0,$$

where

$$L = 1 + 2\theta^2 - \frac{\sqrt{2}\sqrt{\alpha+\alpha_{\prime}}}{\sqrt{\beta_{\prime}\gamma_{\prime}}}(\sqrt{\mathfrak{A}} - 2\theta f), \quad M = \theta + \frac{\sqrt{2}\sqrt{\alpha+\alpha_{\prime}}}{\sqrt{\beta_{\prime}\gamma_{\prime}}}f, \quad N = 1 - \frac{\sqrt{2}\sqrt{\alpha+\alpha_{\prime}}}{\sqrt{\beta_{\prime}\gamma_{\prime}}}\sqrt{\mathfrak{A}},$$

$$L' = 1 + 2\theta^2 - \frac{4bc(\alpha+\alpha_{\prime})}{\beta_{\prime}\gamma_{\prime}}\left(1 - \frac{\sqrt{\alpha_{\prime}}}{\sqrt{\alpha+\alpha_{\prime}}}\right), \quad M' = \theta \quad , \quad N' = 1;$$

or since ξ, η, ζ are equal to 1, $Y+Z$, YZ respectively,

$$1 : Y+Z : YZ = MN' - M'N : NL' - N'L : LM' - L'M$$

$$= \quad -\frac{\sqrt{2}(\alpha+\alpha_{\prime})}{\sqrt{\beta_{\prime}\gamma_{\prime}}}(f + \theta\sqrt{\mathfrak{A}})$$

$$: \quad \frac{4bc(\alpha+\alpha_{\prime})}{\beta_{\prime}\gamma_{\prime}}\left(1 - \frac{\sqrt{2}\sqrt{\alpha+\alpha_{\prime}}}{\sqrt{\beta_{\prime}\gamma_{\prime}}}\sqrt{\mathfrak{A}}\right)\left(1 - \frac{\sqrt{\alpha_{\prime}}}{\sqrt{\alpha+\alpha_{\prime}}}\right) + \frac{\sqrt{2}\sqrt{\alpha+\alpha_{\prime}}}{\sqrt{\beta_{\prime}\gamma_{\prime}}}(f + \theta\sqrt{\mathfrak{A}})\,2\theta$$

$$: -\frac{4bc(\alpha+\alpha_{\prime})}{\beta_{\prime}\gamma_{\prime}}\left(\theta + \frac{\sqrt{2}\sqrt{\alpha+\alpha_{\prime}}}{\sqrt{\beta_{\prime}\gamma_{\prime}}}f\right)\left(1 - \frac{\sqrt{\alpha_{\prime}}}{\sqrt{\alpha+\alpha_{\prime}}}\right) + \frac{\sqrt{2}\sqrt{\alpha+\alpha_{\prime}}}{\sqrt{\beta_{\prime}\gamma_{\prime}}}(f + \theta\sqrt{\mathfrak{A}}).$$

Also

$$f + \theta\sqrt{\mathfrak{A}} = \frac{\beta\gamma}{K} = \frac{Kbc}{\beta_{\prime}\gamma_{\prime}},$$

whence

$$Y + Z = -\frac{2\sqrt{2}\sqrt{\alpha+\alpha_{\prime}}\sqrt{\beta_{\prime}\gamma_{\prime}}}{K}\left(1 - \frac{\sqrt{2}\sqrt{\alpha+\alpha_{\prime}}}{\sqrt{\beta_{\prime}\gamma_{\prime}}}\sqrt{\mathfrak{A}}\right)\left(1 - \frac{\sqrt{\alpha_{\prime}}}{\sqrt{\alpha+\alpha_{\prime}}}\right) - 2\theta,$$

$$YZ = \frac{2\sqrt{2}\sqrt{\alpha+\alpha_{\prime}}\sqrt{\beta_{\prime}\gamma_{\prime}}}{K}\left(\theta + \frac{\sqrt{2}\sqrt{\alpha+\alpha_{\prime}}f}{\sqrt{\beta_{\prime}\gamma_{\prime}}}\right)\left(1 - \frac{\sqrt{\alpha_{\prime}}}{\sqrt{\alpha+\alpha_{\prime}}}\right) - 1;$$

and by forming the analogous expressions for $Z+X$ and ZX, $X+Y$ and XY, the values of X, Y, Z may be determined. But the equations in question simplify themselves in a remarkable manner by the notation before alluded to.

Suppose

$$\mathrm{f}=\sqrt[4]{\mathfrak{A}}\sqrt{\sqrt{\mathfrak{BC}}-\mathfrak{F}},\quad \mathrm{g}=\sqrt[4]{\mathfrak{B}}\sqrt{\sqrt{\mathfrak{CA}}-\mathfrak{G}},\quad \mathrm{h}=\sqrt[4]{\mathfrak{C}}\sqrt{\sqrt{\mathfrak{AB}}-\mathfrak{H}},\quad J=\sqrt{2}\sqrt[4]{\mathfrak{ABC}},$$

these values give

$$\frac{K\sqrt{\mathfrak{A}}}{\sqrt{\mathfrak{BC}}}\,a=2\mathrm{f}^2\left(1-\frac{\mathrm{f}^2}{J^2}\right),$$

$$\frac{K\sqrt{bc}}{\sqrt{\mathfrak{A}}}=2\mathrm{gh}\sqrt{1-\frac{\mathrm{g}^2}{J^2}}\sqrt{1-\frac{\mathrm{h}^2}{J^2}},$$

$$\frac{Kf}{\sqrt{\mathfrak{A}}}=\mathrm{f}^2-\mathrm{g}^2-\mathrm{h}^2+\frac{2\mathrm{g}^2\mathrm{h}^2}{J^2},$$

$$K\theta=-\mathrm{f}^2-\mathrm{g}^2-\mathrm{h}^2+2J^2,$$

$$K^2=-\mathrm{f}^4-\mathrm{g}^4-\mathrm{h}^4+2\mathrm{g}^2\mathrm{h}^2+2\mathrm{h}^2\mathrm{f}^2+2\mathrm{f}^2\mathrm{g}^2-\frac{4\mathrm{f}^2\mathrm{g}^2\mathrm{h}^2}{J^2}.$$

Applying these results to the preceding formulæ and forming for that purpose the equations

$$2\sqrt{2}\sqrt{\alpha+\alpha_{,}}\sqrt{\beta_{,}\gamma_{,}}=4\mathrm{gh},\qquad \frac{\sqrt{2}\sqrt{\alpha+\alpha_{,}}}{\sqrt{\beta_{,}\gamma_{,}}}=\frac{J^2}{\sqrt{\mathfrak{A}}\mathrm{gh}},\qquad \frac{\sqrt{\alpha_{,}}}{\sqrt{\alpha+\alpha_{,}}}=\frac{\mathrm{f}}{J},$$

$$\mathrm{gh}K\theta+\frac{J^2}{\sqrt{\mathfrak{A}}}K\mathrm{f}=(J^2-\mathrm{gh})\left(\mathrm{f}^2-(\mathrm{g}-\mathrm{h})^2\right)-2\mathrm{gh}\,(\mathrm{g}-\mathrm{h})^2,$$

we have

$$K(Y+Z)+2K\theta=4\,(J^2-\mathrm{gh})\left(1-\frac{\mathrm{f}}{J}\right),$$

$$K^2YZ+K^2=\{(J^2-\mathrm{gh})\left(\mathrm{f}^2-(\mathrm{g}-\mathrm{h})^2\right)-2\mathrm{gh}\,(\mathrm{g}-\mathrm{h})^2\}\left(1-\frac{\mathrm{f}}{J}\right);$$

the former of which, combined with the similar equations for $Z+X$ and $X+Y$, gives for X, Y, Z the values to be presently stated, and these values will of course verify the second equation and the corresponding equations for ZX and XY.

Recapitulating the preceding notation, if $x=0$, $y=0$, $z=0$ are the equations of the given sections, $w=0$ the equation of the polar plane of their point of intersection with respect to the surface,

$$ax^2+by^2+cz^2+2fyz+2gzx+2hxy+pw^2=0$$

the equation of the surface, $\mathfrak{A}$, $\mathfrak{B}$, $\mathfrak{C}$, $\mathfrak{F}$, $\mathfrak{G}$, $\mathfrak{H}$, K as usual, and

$$\theta=\frac{1}{K}(\sqrt{\mathfrak{ABC}}+\mathfrak{F}\sqrt{\mathfrak{A}}+\mathfrak{G}\sqrt{\mathfrak{B}}+\mathfrak{H}\sqrt{\mathfrak{C}}),$$

then the equations of the required sections are

$$(ax + hy + gz) X + y\sqrt{\mathfrak{C}} + z\sqrt{\mathfrak{B}} + w\sqrt{-ap}\sqrt{1+X^2} = 0,$$

$$x\sqrt{\mathfrak{C}} + (hx + by + fz) Y + z\sqrt{\mathfrak{A}} + w\sqrt{-bp}\sqrt{1+Y^2} = 0,$$

$$x\sqrt{\mathfrak{B}} + y\sqrt{\mathfrak{A}} + (gx + fy + cz) Z + w\sqrt{-cp}\sqrt{1+Z^2} = 0,$$

where X, Y, Z are to be determined by the following equations,

$$(f + 2\theta\sqrt{\mathfrak{A}}) + \sqrt{\mathfrak{A}}(Y+Z) + fYZ - \sqrt{bc}\sqrt{1+Y^2}\sqrt{1+Z^2} = 0,$$

$$(g + 2\theta\sqrt{\mathfrak{B}}) + \sqrt{\mathfrak{B}}(Z+X) + gZX - \sqrt{ca}\sqrt{1+Z^2}\sqrt{1+X^2} = 0,$$

$$(h + 2\theta\sqrt{\mathfrak{C}}) + \sqrt{\mathfrak{C}}(X+Y) + hXY - \sqrt{ab}\sqrt{1+X^2}\sqrt{1+Y^2} = 0;$$

and the solution of which, putting

$$\mathrm{f} = \sqrt[4]{\mathfrak{A}}\sqrt{\sqrt{\mathfrak{BC}} - \mathfrak{F}},\quad \mathrm{g} = \sqrt[4]{\mathfrak{B}}\sqrt{\sqrt{\mathfrak{CA}} - \mathfrak{G}},\quad \mathrm{h} = \sqrt[4]{\mathfrak{C}}\sqrt{\sqrt{\mathfrak{AB}} - \mathfrak{H}},\quad J = \sqrt{2}\sqrt[4]{\mathfrak{ABC}},$$

is given by the equations

$$KX = \frac{2\mathrm{fgh}}{J} + (-\mathrm{f}+\mathrm{g}+\mathrm{h})^2 - 2(-\mathrm{f}+\mathrm{g}+\mathrm{h})J,$$

$$KY = \frac{2\mathrm{fgh}}{J} + (\mathrm{f}-\mathrm{g}+\mathrm{h})^2 - 2(\mathrm{f}-\mathrm{g}+\mathrm{h})J,$$

$$KZ = \frac{2\mathrm{fgh}}{J} + (\mathrm{f}+\mathrm{g}-\mathrm{h})^2 - 2(\mathrm{f}+\mathrm{g}-\mathrm{h})J.\ (^1)$$

Instead of the direct but very tedious process by which these values of X, Y, Z have been obtained, we may substitute the following *à posteriori* verification.

We have

$$K^2(1+X^2) = 4(-\mathrm{f}+\mathrm{g}+\mathrm{h})^2 J^2\left(1+\frac{\mathrm{f}}{J}\right)\left(1-\frac{\mathrm{g}}{J}\right)\left(1-\frac{\mathrm{h}}{J}\right),$$

$$K^2\sqrt{1+Y^2}\sqrt{1+Z^2} = 4(\mathrm{f}^2 - (\mathrm{g}^2-\mathrm{h}))J^2\left(1-\frac{\mathrm{f}}{J}\right)\sqrt{1-\frac{\mathrm{g}^2}{J^2}}\sqrt{1-\frac{\mathrm{h}^2}{J^2}},$$

$$K^2(1+YZ) = 4\left(1-\frac{\mathrm{f}}{J}\right)\{(J^2-\mathrm{gh})(\mathrm{f}^2-(\mathrm{g}-\mathrm{h})^2) - 2\mathrm{gh}(\mathrm{g}-\mathrm{h})^2\},$$

$$K(Y+Z) - 2\mathrm{f}^2 - 2\mathrm{g}^2 - 2\mathrm{h}^2 + 4J^2 = 4\left(1-\frac{\mathrm{f}}{J}\right)(J^2-\mathrm{gh}).$$

Putting also

$$\mathrm{f}^2 - \mathrm{g}^2 - \mathrm{h}^2 + \frac{2\mathrm{g}^2\mathrm{h}^2}{J} = (\mathrm{f}^2-(\mathrm{g}-\mathrm{h})^2) - \frac{2\mathrm{gh}(J^2-\mathrm{gh})}{J^2},$$

$$K^2 = (\mathrm{f}^2-(\mathrm{g}-\mathrm{h})^2)\left((\mathrm{g}+\mathrm{h})^2 - \mathrm{f}^2 - \frac{4\mathrm{g}^2\mathrm{h}^2}{J^2}\right) - \frac{4\mathrm{g}^2\mathrm{h}^2(\mathrm{g}-\mathrm{h})^2}{J^2},$$

[1] It is perhaps worth noticing that the value of the quantity λ previously made use of,

$$\lambda = \frac{\mathrm{f}^2}{Ka\sqrt{\mathfrak{A}}}\left\{\frac{2\mathrm{fgh}}{J} - \mathrm{g}^2 - \mathrm{h}^2 + (J+\mathrm{f}-\mathrm{g}-\mathrm{h})^2\right\}.$$

we have

$$\left(\mathrm{f}^2-\mathrm{g}^2-\mathrm{h}^2+\frac{2\mathrm{g}^2\mathrm{h}^2}{J^2}\right)K^2(1+YZ)$$

$$=4\left(1-\frac{\mathrm{f}}{J}\right)\left\{(\mathrm{f}^2-(\mathrm{g}-\mathrm{h})^2)\left[(J^2-\mathrm{gh})(\mathrm{f}^2-(\mathrm{g}-\mathrm{h})^2)-2\mathrm{gh}(\mathrm{g}-\mathrm{h})^2-2\mathrm{gh}\frac{(J^2-\mathrm{gh})^2}{J^2}\right]\right.$$

$$\left.+\frac{4\mathrm{g}^2\mathrm{h}^2(\mathrm{g}-\mathrm{h})^2(J^2-\mathrm{gh})}{J^2}\right\},$$

$$K^2\{K(Y+Z)-2\mathrm{f}^2-2\mathrm{g}^2-2\mathrm{h}^2+4J^2\}$$

$$=4\left(1-\frac{\mathrm{f}}{J}\right)\left\{\mathrm{f}^2-(\mathrm{g}-\mathrm{h})^2)\left[(J^2-\mathrm{gh})\left((\mathrm{g}+\mathrm{h})^2-\mathrm{f}^2-\frac{4\mathrm{g}^2\mathrm{h}^2}{J^2}\right)\right]-\frac{4\mathrm{g}^2\mathrm{h}^2(\mathrm{g}-\mathrm{h})^2(J^2-\mathrm{gh})}{J^2}\right\}.$$

Also, since

$$(\mathrm{f}^2-(\mathrm{g}-\mathrm{h})^2)+\left((\mathrm{g}+\mathrm{h})^2-\mathrm{f}^2-\frac{4\mathrm{g}^2\mathrm{h}^2}{J^2}\right)=4\mathrm{gh}\frac{(J^2-\mathrm{gh})}{J^2},$$

we have

$$\left(\mathrm{f}^2-\mathrm{g}^2-\mathrm{h}^2+\frac{2\mathrm{g}^2\mathrm{h}^2}{J^2}\right)K^2(1+YZ)+K^2\{K(Y+Z)-2\mathrm{f}^2-2\mathrm{g}^2-2\mathrm{h}^2\}$$

$$=4\left(1-\frac{\mathrm{f}}{J}\right)(\mathrm{f}^2-(\mathrm{g}-\mathrm{h})^2)\,2\mathrm{gh}J^2\left(1-\frac{\mathrm{g}^2}{J^2}\right)\left(1-\frac{\mathrm{h}^2}{J^2}\right),$$

and the values obtained above give also

$$2\mathrm{gh}\sqrt{1-\frac{\mathrm{g}^2}{J^2}}\sqrt{1-\frac{\mathrm{h}^2}{J^2}}K^2\sqrt{1+Y^2}\sqrt{1+Z^2}$$

$$=4\left(1-\frac{\mathrm{f}}{J}\right)(\mathrm{f}^2-(\mathrm{g}-\mathrm{h})^2)\,2\mathrm{gh}J^2\left(1-\frac{\mathrm{g}^2}{J^2}\right)\left(1-\frac{\mathrm{h}^2}{J^2}\right),$$

which shows that the relation between Y and Z is verified by the assumed values of these quantities, and the other two equations are of course also verified. The solution of the problem will be rendered more complete if the equations of the required sections and of the auxiliary sections made use of in the geometrical construction are expressed in terms of f, g, h, J.

§ 7.

First, to substitute in the equations of the required sections or resultors. Writing the first equation in the form

$$\frac{K^2}{2\sqrt{\mathfrak{BC}}}\left\{aXx+(hX+\sqrt{\mathfrak{C}})\,y+(gX+\sqrt{\mathfrak{B}})\,z+\sqrt{-ap}\sqrt{1+X^2}\,w\right\}=0,$$

the coefficient of x will be

$$\frac{\mathrm{f}^2}{\sqrt{\mathfrak{A}}}\left(1-\frac{\mathrm{f}^2}{J^2}\right)\left\{\frac{2\mathrm{fgh}}{J}+(-\mathrm{f}+\mathrm{g}+\mathrm{h})^2-2J(-\mathrm{f}+\mathrm{g}+\mathrm{h})\right\},$$

or, as it is convenient to write it,

$$\left(1+\frac{\mathrm{f}}{J}\right)\mathrm{f}(-\mathrm{f}+\mathrm{g}+\mathrm{h})\frac{\mathrm{f}}{\sqrt{\mathfrak{A}}}\left(1-\frac{\mathrm{f}}{J}\right)\left\{\frac{2\mathrm{fgh}}{J(-\mathrm{f}+\mathrm{g}+\mathrm{h})}-\mathrm{f}+\mathrm{g}+\mathrm{h}-2J\right\}.$$

The coefficient of y is

$$\frac{1}{2\sqrt{\mathfrak{B}}}\left\{\left(-\mathrm{f}^2-\mathrm{g}^2+\mathrm{h}^2+\frac{2\mathrm{g}^2\mathrm{h}^2}{J^2}\right)\left(\frac{2\mathrm{fgh}}{J^2}+(-\mathrm{f}+\mathrm{g}+\mathrm{h})^2-2J(-\mathrm{f}+\mathrm{g}+\mathrm{h})\right)\right.$$

$$\left.-\mathrm{f}^4-\mathrm{g}^4-\mathrm{h}^4+2\mathrm{g}^2\mathrm{h}^2+2\mathrm{h}^2\mathrm{f}^2+2\mathrm{f}^2\mathrm{g}^2-\frac{4\mathrm{f}^2\mathrm{g}^2\mathrm{h}^2}{J^2}\right\},$$

or, after all reductions,

$$\left(1-\frac{\mathrm{f}}{J}\right)\mathrm{f}(-\mathrm{f}+\mathrm{g}+\mathrm{h})\frac{\mathrm{h}}{\sqrt{\mathfrak{B}}}\left(1-\frac{\mathrm{g}}{J}\right)\left\{\frac{-2\mathrm{fgh}}{J(-\mathrm{f}+\mathrm{g}+\mathrm{h})}+\mathrm{f}-\mathrm{g}+\mathrm{h}+\frac{2J(\mathrm{f}^2+\mathrm{g}^2-\mathrm{h}^2)}{2\mathrm{fg}}\right\};$$

and similarly the coefficient of z is

$$\frac{1}{2\sqrt{\mathfrak{C}}}\left\{\left(-\mathrm{f}^2+\mathrm{g}^2-\mathrm{h}^2+\frac{2\mathrm{h}^2\mathrm{f}^2}{J^2}\right)\left(\frac{2\mathrm{fgh}}{J}+(-\mathrm{f}+\mathrm{g}+\mathrm{h})^2-2J(-\mathrm{f}+\mathrm{g}+\mathrm{h})\right)\right.$$

$$\left.-\mathrm{f}^4-\mathrm{g}^4-\mathrm{h}^4+2\mathrm{g}^2\mathrm{h}^2+2\mathrm{h}^2\mathrm{f}^2+2\mathrm{f}^2\mathrm{g}^2-\frac{4\mathrm{f}^2\mathrm{g}^2\mathrm{h}^2}{J^2}\right\},$$

or, after all reductions,

$$\left(1+\frac{\mathrm{f}}{J}\right)\mathrm{f}(-\mathrm{f}+\mathrm{g}+\mathrm{h})\frac{\mathrm{h}}{\sqrt{\mathfrak{C}}}\left(1-\frac{\mathrm{h}}{J}\right)\left\{\frac{-2\mathrm{fgh}}{J(-\mathrm{f}+\mathrm{g}+\mathrm{h})}+\mathrm{f}+\mathrm{g}-\mathrm{h}+\frac{2J(\mathrm{f}^2-\mathrm{g}^2+\mathrm{h}^2)}{2\mathrm{fh}}\right\}$$

and the coefficient of w is

$$\left(1+\frac{\mathrm{f}}{J}\right)\mathrm{f}(-\mathrm{f}+\mathrm{g}+\mathrm{h})\,2\sqrt{K}\sqrt{1-\frac{\mathrm{f}}{J}}\sqrt{1-\frac{\mathrm{g}}{J}}\sqrt{1-\frac{\mathrm{h}}{J}}\sqrt{-p}.$$

Hence, forming the equation of the resultor in question, and by means of it those of the other resultors, the equations of the resultors are

$$\left(\frac{2\mathrm{fgh}}{J(-\mathrm{f}+\mathrm{g}+\mathrm{h})}-\mathrm{f}+\mathrm{g}+\mathrm{h}-2J\right)\frac{\mathrm{f}}{\sqrt{\mathfrak{A}}}\left(1-\frac{\mathrm{f}}{J}\right)x$$

$$+\left(\frac{-2\mathrm{fgh}}{J(-\mathrm{f}+\mathrm{g}+\mathrm{h})}+\mathrm{f}-\mathrm{g}+\mathrm{h}+2J\frac{\mathrm{f}^2+\mathrm{g}^2-\mathrm{h}^2}{2\mathrm{fg}}\right)\frac{\mathrm{g}}{\sqrt{\mathfrak{B}}}\left(1-\frac{\mathrm{g}}{J}\right)y$$

$$+\left(\frac{-2\mathrm{fgh}}{J(-\mathrm{f}+\mathrm{g}+\mathrm{h})}+\mathrm{f}+\mathrm{g}-\mathrm{h}+2J\frac{\mathrm{f}^2-\mathrm{g}^2+\mathrm{h}^2}{2\mathrm{fh}}\right)\frac{\mathrm{h}}{\sqrt{\mathfrak{C}}}\left(1-\frac{\mathrm{h}}{J}\right)z$$

$$+2\sqrt{K}\sqrt{1-\frac{\mathrm{f}}{J}}\sqrt{1-\frac{\mathrm{g}}{J}}\sqrt{1-\frac{\mathrm{h}}{J}}\sqrt{-p}\,w=0,$$

$$\begin{aligned}&\left(\frac{2\mathrm{fgh}}{J(\mathrm{f}-\mathrm{g}+\mathrm{h})}-\mathrm{f}+\mathrm{g}+\mathrm{h}+2J\frac{\mathrm{f}^2+\mathrm{g}^2-\mathrm{h}^2}{2\mathrm{fg}}\right)\frac{\mathrm{f}}{\sqrt{\mathfrak{A}}}\left(1-\frac{\mathrm{f}}{J}\right)x\\
&+\left(\frac{2\mathrm{fgh}}{J(\mathrm{f}-\mathrm{g}+\mathrm{h})}+\mathrm{f}-\mathrm{g}+\mathrm{h}-2J\right)\frac{\mathrm{g}}{\sqrt{\mathfrak{B}}}\left(1-\frac{\mathrm{g}}{J}\right)y\\
&+\left(\frac{-2\mathrm{fgh}}{J(\mathrm{f}-\mathrm{g}+\mathrm{h})}+\mathrm{f}+\mathrm{g}-\mathrm{h}+2J\frac{-\mathrm{f}^2+\mathrm{g}^2+\mathrm{h}^2}{2\mathrm{gh}}\right)\frac{\mathrm{h}}{\sqrt{\mathfrak{C}}}\left(1-\frac{\mathrm{h}}{J}\right)z\\
&\qquad+2\sqrt{K}\sqrt{1-\frac{\mathrm{f}}{J}}\sqrt{1-\frac{\mathrm{g}}{J}}\sqrt{1-\frac{\mathrm{h}}{J}}\sqrt{-p}\,w=0\end{aligned}$$

$$\begin{aligned}&\left(\frac{-2\mathrm{fgh}}{J(\mathrm{f}+\mathrm{g}-\mathrm{h})}-\mathrm{f}+\mathrm{g}+\mathrm{h}+2J\frac{\mathrm{f}^2-\mathrm{g}^2+\mathrm{h}^2}{2\mathrm{fh}}\right)\frac{\mathrm{f}}{\sqrt{\mathfrak{A}}}\left(1-\frac{\mathrm{f}}{J}\right)x\\
&+\left(\frac{-2\mathrm{fgh}}{J(\mathrm{f}+\mathrm{g}-\mathrm{h})}+\mathrm{f}-\mathrm{g}+\mathrm{h}+2J\frac{-\mathrm{f}^2+\mathrm{g}^2+\mathrm{h}^2}{2\mathrm{gh}}\right)\frac{\mathrm{g}}{\sqrt{\mathfrak{B}}}\left(1-\frac{\mathrm{g}}{J}\right)y\\
&+\left(\frac{2\mathrm{fgh}}{J(\mathrm{f}+\mathrm{g}-\mathrm{h})}+\mathrm{f}+\mathrm{g}-\mathrm{h}-2J\right)\frac{\mathrm{h}}{\sqrt{\mathfrak{C}}}\left(1-\frac{\mathrm{h}}{J}\right)z\\
&\qquad+2\sqrt{K}\sqrt{1-\frac{\mathrm{f}}{J}}\sqrt{1-\frac{\mathrm{g}}{J}}\sqrt{1-\frac{\mathrm{h}}{J}}\sqrt{-p}\,w=0,\end{aligned}$$

values which might be somewhat simplified by writing ξ, η, ζ, ω instead of

$$\frac{\mathrm{f}}{\sqrt{\mathfrak{A}}}\left(1-\frac{\mathrm{f}}{J}\right)x,\quad \frac{\mathrm{g}}{\sqrt{\mathfrak{B}}}\left(1-\frac{\mathrm{g}}{J}\right)y,\quad \frac{\mathrm{h}}{\sqrt{\mathfrak{C}}}\left(1-\frac{\mathrm{h}}{J}\right)z,\quad 2\sqrt{1-\frac{\mathrm{f}}{J}}\sqrt{1-\frac{\mathrm{g}}{J}}\sqrt{1-\frac{\mathrm{h}}{J}}\sqrt{-p}\,w;$$

and it may be also remarked, that the coefficients as well of these formulæ as of those which follow may be elegantly expressed in terms of the parts of a triangle having f, g, h for its sides.

The equations of the separators are found by taking the differences two and two of the equations of the resultors (this requires to be verified *à posteriori*); thus subtracting the third equation from the second the result contains a constant factor,

$$\frac{1}{J(\mathrm{f}^2-(\mathrm{g}-\mathrm{h})^2)\mathrm{gh}}\{4\mathrm{f}^2\mathrm{g}^2\mathrm{h}^2-J(\mathrm{f}^2-(\mathrm{g}-\mathrm{h})^2)((\mathrm{g}+\mathrm{h})^2-\mathrm{f}^2)\},$$

equivalent to

$$\frac{1}{J(\mathrm{f}^2-(\mathrm{g}-\mathrm{h})^2)\mathrm{gh}}\left(4\mathrm{f}^2\mathrm{g}^2\mathrm{h}^2-J^2\left(K^2+\frac{4\mathrm{f}^2\mathrm{g}^2\mathrm{h}^2}{J^2}\right)\right)\quad\text{or}\quad\frac{-JK^2}{(\mathrm{f}^2-(\mathrm{g}-\mathrm{h})^2)\mathrm{gh}}.$$

Rejecting the factor in question and forming the analogous two equations, the equations of the separators are

$$\begin{aligned}-\frac{\mathrm{g}-\mathrm{h}}{\mathrm{f}}\frac{\mathrm{f}}{\sqrt{\mathfrak{A}}}\left(1-\frac{\mathrm{f}}{J}\right)x+\frac{\mathrm{g}}{\sqrt{\mathfrak{B}}}\left(1-\frac{\mathrm{g}}{J}\right)y-\frac{\mathrm{h}}{\sqrt{\mathfrak{C}}}\left(1-\frac{\mathrm{h}}{J}\right)z&=0,\\
-\frac{\mathrm{f}}{\sqrt{\mathfrak{A}}}\left(1-\frac{\mathrm{f}}{J}\right)x-\frac{\mathrm{h}-\mathrm{f}}{\mathrm{g}}\frac{\mathrm{g}}{\sqrt{\mathfrak{B}}}\left(1-\frac{\mathrm{g}}{J}\right)y+\frac{\mathrm{h}}{\sqrt{\mathfrak{C}}}\left(1-\frac{\mathrm{h}}{J}\right)z&=0,\\
\frac{\mathrm{f}}{\sqrt{\mathfrak{A}}}\left(1-\frac{\mathrm{f}}{J}\right)x-\frac{\mathrm{g}}{\sqrt{\mathfrak{B}}}\left(1-\frac{\mathrm{g}}{J}\right)y-\frac{\mathrm{f}-\mathrm{g}}{\mathrm{h}}\frac{\mathrm{h}}{\sqrt{\mathfrak{C}}}\left(1-\frac{\mathrm{h}}{J}\right)z&=0;\end{aligned}$$

and from the mode of formation of these equations it is obvious that the separators have a line in common.

The equations of the determinators being $x=0$, $y=0$, $z=0$, the equations of the tactors are

$$\sqrt{\mathfrak{B}}\,z - \sqrt{\mathfrak{C}}\,y = 0, \quad \sqrt{\mathfrak{C}}\,x - \sqrt{\mathfrak{A}}\,z = 0, \quad \sqrt{\mathfrak{A}}\,y - \sqrt{\mathfrak{B}}\,x = 0;$$

and if $\alpha x + \beta y + \gamma z + \delta w = 0$ be the equation of the tactor touching

$$x = 0, \ \sqrt{\mathfrak{C}}\,x - \sqrt{\mathfrak{A}}\,z = 0 \text{ and } \sqrt{\mathfrak{A}}\,y - \sqrt{\mathfrak{B}}\,x = 0,$$

the conditions of contact are

$$\mathfrak{A}\left(\mathfrak{A}\alpha^2 + \ldots \frac{K}{p}\delta^2\right) = (\mathfrak{A}\alpha + \mathfrak{H}\beta + \mathfrak{G}\gamma)^2,$$

$$2\sqrt{\mathfrak{AB}}\,(\sqrt{\mathfrak{AB}} - \mathfrak{H})\left(\mathfrak{A}\alpha^2 + \ldots \frac{K}{p}\delta^2\right) = \left\{(\sqrt{\mathfrak{AB}} - \mathfrak{H})(\alpha\sqrt{\mathfrak{A}} - \beta\sqrt{\mathfrak{B}}) + \gamma\,(\mathfrak{G}\sqrt{\mathfrak{B}} - \mathfrak{H}\sqrt{\mathfrak{C}})\right\}^2,$$

$$2\sqrt{\mathfrak{AC}}\,(\sqrt{\mathfrak{AC}} - \mathfrak{G})\left(\mathfrak{A}\alpha^2 + \ldots \frac{K}{p}\delta^2\right) = \left\{(\sqrt{\mathfrak{AC}} - \mathfrak{G})(\alpha\sqrt{\mathfrak{A}} - \gamma\sqrt{\mathfrak{C}}) + \beta\,(\mathfrak{H}\sqrt{\mathfrak{C}} - \mathfrak{F}\sqrt{\mathfrak{A}})\right\}^2,$$

whence

$$\frac{1}{\sqrt{\mathfrak{A}}}\sqrt{2\sqrt{\mathfrak{AB}}\,(\sqrt{\mathfrak{AB}} - \mathfrak{H})}\,(\mathfrak{A}\alpha + \mathfrak{H}\beta + \mathfrak{G}\gamma) =$$
$$(\sqrt{\mathfrak{AB}} - \mathfrak{H})\sqrt{\mathfrak{A}}\,\alpha - (\sqrt{\mathfrak{AB}} - \mathfrak{H})\sqrt{\mathfrak{B}}\,\beta + (\mathfrak{G}\sqrt{\mathfrak{B}} - \mathfrak{F}\sqrt{\mathfrak{A}})\,\gamma,$$

$$\frac{1}{\sqrt{\mathfrak{A}}}\sqrt{2\sqrt{\mathfrak{AC}}\,(\sqrt{\mathfrak{AC}} - \mathfrak{G})}\,(\mathfrak{A}\alpha + \mathfrak{H}\beta + \mathfrak{G}\gamma) =$$
$$(\sqrt{\mathfrak{AC}} - \mathfrak{G})\sqrt{\mathfrak{A}}\,\alpha + (\mathfrak{H}\sqrt{\mathfrak{C}} - \mathfrak{F}\sqrt{\mathfrak{A}})\,\beta - (\sqrt{\mathfrak{AC}} - \mathfrak{G})\,\gamma,$$

$$c\beta^2 + b\gamma^2 - 2f\beta\gamma + \frac{\mathfrak{A}}{p}\delta^2 = 0,$$

and putting for a moment

$$\mu = \sqrt{\mathfrak{AC}} - \mathfrak{G} - \sqrt{2\sqrt{\mathfrak{AC}}\,(\sqrt{\mathfrak{AC}} - \mathfrak{G})},$$

$$\nu = \sqrt{\mathfrak{AB}} - \mathfrak{H} - \sqrt{2\sqrt{\mathfrak{AB}}\,(\sqrt{\mathfrak{AB}} - \mathfrak{H})};$$

after some reductions, and observing that the ratios only of the quantities α, β, γ, δ are material, we obtain

$$\alpha = \frac{K}{\sqrt{\mathfrak{A}}}(K + h\nu + g\mu),$$

$$\beta = \frac{K}{\sqrt{\mathfrak{A}}}(\qquad b\nu + f\mu),$$

$$\gamma = \frac{K}{\sqrt{\mathfrak{A}}}(\qquad f\nu + c\mu),$$

$$\delta = \frac{K}{\sqrt{\mathfrak{A}}}\sqrt{-p\,(b\nu^2 + c\mu^2 + 2f\mu\nu)}:$$

and it is easily seen also that the coordinates of the point of contact are

$$x = 0, \quad y = \nu, \quad z = \mu, \quad w = -\frac{\delta}{p}\frac{\sqrt{\mathfrak{A}}}{K};$$

also

$$\mu = -\frac{J\mathrm{g}}{\sqrt{\mathfrak{B}}}\left(1 - \frac{\mathrm{g}}{J}\right), \quad \nu = -\frac{J\mathrm{h}}{\sqrt{\mathfrak{C}}}\left(1 - \frac{\mathrm{h}}{J}\right).$$

Hence substituting and introducing throughout the quantities f, g, h, J, also forming the analogous equations, the equations of the tactors are

$$\left\{\mathrm{f}^2(-\mathrm{f}^2 + \mathrm{g}^2 + \mathrm{h}^2) + (\mathrm{g} + \mathrm{h})J\left(\mathrm{f}^2 - (\mathrm{g} - \mathrm{h})^2 - \frac{2\mathrm{f}^2\mathrm{gh}}{J^2}\right)\right\}\frac{1}{\sqrt{\mathfrak{A}}}x$$

$$-\left\{\mathrm{f}^2 - (\mathrm{g} - \mathrm{h})^2 + \frac{2\mathrm{gh}(\mathrm{g} - \mathrm{h})}{J}\right\}J\frac{\mathrm{g}}{\sqrt{\mathfrak{B}}}\left(1 - \frac{\mathrm{g}}{J}\right)y$$

$$-\left\{\mathrm{f}^2 - (\mathrm{g} - \mathrm{h})^2 - \frac{2\mathrm{gh}(\mathrm{g} - \mathrm{h})}{J}\right\}J\frac{\mathrm{h}}{\sqrt{\mathfrak{C}}}\left(1 - \frac{\mathrm{h}}{J}\right)z$$

$$+ 2\sqrt{K}\sqrt{\mathrm{gh}\left(1 - \frac{\mathrm{g}}{J}\right)\left(1 - \frac{\mathrm{h}}{J}\right)(\mathrm{f}^2 - (\mathrm{g} - \mathrm{h})^2)}\;\sqrt{-p}\,w = 0,$$

$$-\left\{\mathrm{g}^2 - (\mathrm{h} - \mathrm{f})^2 + \frac{2\mathrm{hf}(\mathrm{h} - \mathrm{f})}{J}\right\}J\frac{\mathrm{f}}{\sqrt{\mathfrak{A}}}\left(1 - \frac{\mathrm{f}}{J}\right)x$$

$$+\left\{\mathrm{g}^2(\mathrm{f}^2 - \mathrm{g}^2 + \mathrm{h}^2) + (\mathrm{h} + \mathrm{f})J\left(\mathrm{g}^2 - (\mathrm{h} - \mathrm{f})^2 - \frac{2\mathrm{fg}^2\mathrm{h}}{J^2}\right)\right\}\frac{1}{\sqrt{\mathfrak{B}}}y$$

$$-\left\{\mathrm{g}^2 - (\mathrm{h} - \mathrm{f})^2 - \frac{2\mathrm{hf}(\mathrm{h} - \mathrm{f})}{J}\right\}J\frac{\mathrm{h}}{\sqrt{\mathfrak{C}}}\left(1 - \frac{\mathrm{h}}{J}\right)z$$

$$+ 2\sqrt{K}\sqrt{\mathrm{hf}\left(1 - \frac{\mathrm{h}}{J}\right)\left(1 - \frac{\mathrm{f}}{J}\right)(\mathrm{g}^2 - (\mathrm{h} - \mathrm{f})^2)}\;\sqrt{-p}\,w = 0,$$

$$-\left\{\mathrm{h}^2 - (\mathrm{f} - \mathrm{g})^2 + \frac{2\mathrm{fg}(\mathrm{f} - \mathrm{g})}{J}\right\}J\frac{\mathrm{f}}{\sqrt{\mathfrak{A}}}\left(1 - \frac{\mathrm{f}}{J}\right)x$$

$$-\left\{\mathrm{h}^2 - (\mathrm{f} - \mathrm{g})^2 + \frac{2\mathrm{fg}(\mathrm{f} - \mathrm{g})}{J}\right\}J\frac{\mathrm{g}}{\sqrt{\mathfrak{B}}}\left(1 - \frac{\mathrm{g}}{J}\right)y$$

$$+\left\{\mathrm{h}^2(\mathrm{f}^2 + \mathrm{g}^2 - \mathrm{h}^2) + (\mathrm{f} + \mathrm{g})J\left(\mathrm{h}^2 - (\mathrm{f} - \mathrm{g})^2 - \frac{2\mathrm{fgh}^2}{J^2}\right)\right\}\frac{1}{\sqrt{\mathfrak{C}}}z$$

$$+ 2\sqrt{K}\sqrt{\mathrm{fg}\left(1 - \frac{\mathrm{f}}{J}\right)\left(1 - \frac{\mathrm{g}}{J}\right)(\mathrm{h}^2 - (\mathrm{f} - \mathrm{g})^2)}\;\sqrt{-p}\,w = 0.$$

It is obvious, from the equations, that each separator passes through the point of contact of a tactor and determinator, it consequently only remains to be shown that each separator touches two tactors. Consider the tactor which has been represented by $\alpha x + \beta y + \gamma z + \delta w = 0$, the unreduced values of the coefficients give

$$\mathfrak{A}\alpha + \mathfrak{H}\beta + \mathfrak{G}\gamma = K^2 \sqrt{\mathfrak{A}},$$

$$\mathfrak{H}\alpha + \mathfrak{B}\beta + \mathfrak{F}\gamma = \frac{K^2}{\sqrt{\mathfrak{A}}}(\mathfrak{H} + \nu),$$

$$\mathfrak{G}\alpha + \mathfrak{F}\beta + \mathfrak{C}\gamma = \frac{K^2}{\sqrt{\mathfrak{A}}}(\mathfrak{G} + \mu),$$

$$\sqrt{\mathfrak{A}\alpha^2 + \ldots \frac{K}{p}\delta^2} = \frac{1}{\sqrt{\mathfrak{A}}}(\mathfrak{A}\alpha + \mathfrak{H}\beta + \mathfrak{G}\gamma) = K^2.$$

Represent for a moment the separator

$$\frac{\mathrm{f}}{\sqrt{\mathfrak{A}}}\left(1 - \frac{\mathrm{f}}{J}\right) x - \frac{\mathrm{g}}{\sqrt{\mathfrak{B}}}\left(1 - \frac{\mathrm{g}}{J}\right) y - \frac{\mathrm{f} - \mathrm{g}}{\mathrm{h}} \frac{\mathrm{h}}{\sqrt{\mathfrak{C}}}\left(1 - \frac{\mathrm{h}}{J}\right) z = 0$$

by $lx + my + nz + sw = 0$. Then putting $\mathfrak{A}l^2 + \ldots \frac{K}{p} s^2 = \square^2$, since

$$\mathfrak{A}\alpha l + \ldots + \frac{K}{p}\delta s = K^2 \left\{ l\sqrt{\mathfrak{A}} + \frac{m}{\sqrt{\mathfrak{A}}}(\mathfrak{H} + \nu) + \frac{n}{\sqrt{\mathfrak{A}}}(\mathfrak{G} + \mu) \right\}$$

$$= K^2 \left\{ \mathrm{f}\left(1 - \frac{\mathrm{f}}{J}\right) - \mathrm{g}\left(1 - \frac{2\mathrm{h}}{J}\right)\left(1 - \frac{\mathrm{g}}{J}\right) - (\mathrm{f} - \mathrm{g})\left(1 - \frac{2\mathrm{g}}{J}\right)\left(1 - \frac{\mathrm{h}}{J}\right) \right\}$$

$$= \frac{K^2}{J}\left\{ -(\mathrm{f} - \mathrm{g})^2 + \mathrm{h}(\mathrm{f} + \mathrm{g}) - \frac{2\mathrm{fgh}}{J} \right\},$$

the condition of contact becomes

$$\square = \frac{1}{J}\left\{ -(\mathrm{f} - \mathrm{g})^2 + \mathrm{h}(\mathrm{f} + \mathrm{g}) - \frac{2\mathrm{fgh}}{J} \right\};$$

or, forming the value of $\square^2$ and substituting,

$$\mathrm{f}^2\left(1 - \frac{\mathrm{f}}{J}\right)^2 + \mathrm{g}^2\left(1 - \frac{\mathrm{g}}{J}\right)^2 + (\mathrm{f} - \mathrm{g})^2\left(1 - \frac{\mathrm{h}}{J}\right)^2 + 2\left(1 - \frac{2\mathrm{f}^2}{J^2}\right)(\mathrm{f} - \mathrm{g})\,\mathrm{g}\left(1 - \frac{\mathrm{g}}{J}\right)\left(1 - \frac{\mathrm{h}}{J}\right)$$

$$- 2\left(1 - \frac{2\mathrm{g}^2}{J^2}\right)(\mathrm{f} - \mathrm{g})\,\mathrm{f}\left(1 - \frac{\mathrm{h}}{J}\right)\left(1 - \frac{\mathrm{f}}{J}\right) - 2\left(1 - \frac{2\mathrm{h}^2}{J^2}\right)\mathrm{fg}\left(1 - \frac{\mathrm{f}}{J}\right)\left(1 - \frac{\mathrm{g}}{J}\right)$$

$$= \frac{1}{J^2}\left(-(\mathrm{f} - \mathrm{g})^2 + \mathrm{h}(\mathrm{f} + \mathrm{g}) - \frac{2\mathrm{fgh}}{J} \right)^2,$$

which may be verified without difficulty, and thus the construction for the resultors is shown to be true.

§ 8

Several of the formulæ of the preceding sections of this memoir apply to any number of variables. Consider the surface (i.e. hypersurface)

$$ax^2 + by^2 + cz^2 + 2fyz + 2gzx + 2hxy \ldots + pt^2 = 0,$$

and the section (i.e. hypersection)

$$(a\lambda + h\mu + g\nu \ldots)\, x + (h\lambda + b\mu + f\nu \ldots)\, y + (g\lambda + f\mu + c\nu \ldots)\, z \ldots + \sqrt{-p}\, \nabla t = 0,$$

where

$$\nabla^2 = a\lambda^2 + b\mu^2 + c\nu^2 + 2f\mu\nu + 2g\nu\lambda + 2h\lambda\mu \ldots - K,$$

the condition of contact with any other section represented by a similar equation is

$$a\lambda\lambda' + b\mu\mu' + c\nu\nu' + f(\mu\nu' + \mu'\nu) + g(\nu\lambda' + \nu'\lambda) + h(\lambda\mu' + \lambda'\mu) \ldots \pm K = \nabla\nabla',$$

where K is the determinant formed with the coefficients $a, b, c, f, g, h, \ldots$ And consequently, by establishing all or any of the equations $\lambda = \sqrt{\mathfrak{A}}$, $\mu = \sqrt{\mathfrak{B}}$, $\nu = \sqrt{\mathfrak{C}}$, ... we have the condition in order that the section in question may touch all or the corresponding sections of the sections $x = 0$, $y = 0$, $z = 0$, ...

Let $\mathfrak{n}$ be the number of the variables $x, y, z \ldots$, then $K^{\mathfrak{n}-1} = \begin{vmatrix} \mathfrak{A} & \mathfrak{H} & \mathfrak{G} & \ldots \\ \mathfrak{H} & \mathfrak{B} & \mathfrak{F} & \\ \mathfrak{G} & \mathfrak{F} & \mathfrak{C} & \\ \vdots & & & \end{vmatrix}$,

also $K^{\mathfrak{n}-2} \{(a\lambda + h\mu + g\nu \ldots)\, x + (h\lambda + b\mu + f\nu \ldots)\, y + (g\lambda + f\mu + c\nu \ldots)\, z \ldots\}$

$$= - \begin{vmatrix} & x & y & z & \ldots \\ \lambda & \mathfrak{A} & \mathfrak{H} & \mathfrak{G} & \\ \mu & \mathfrak{H} & \mathfrak{B} & \mathfrak{F} & \\ \nu & \mathfrak{G} & \mathfrak{F} & \mathfrak{C} & \\ \vdots & & & & \end{vmatrix},$$

whence also

$$K^{\mathfrak{n}-2} (\nabla^2 + K) = - \begin{vmatrix} & \lambda & \mu & \nu & \ldots \\ \lambda & \mathfrak{A} & \mathfrak{H} & \mathfrak{G} & \\ \mu & \mathfrak{H} & \mathfrak{B} & \mathfrak{F} & \\ \nu & \mathfrak{G} & \mathfrak{F} & \mathfrak{C} & \\ \vdots & & & & \end{vmatrix} \text{ or } K^{\mathfrak{n}-2} \nabla^2 = - \begin{vmatrix} 1 & \lambda & \mu & \nu & \ldots \\ \lambda & \mathfrak{A} & \mathfrak{H} & \mathfrak{G} & \\ \mu & \mathfrak{H} & \mathfrak{B} & \mathfrak{F} & \\ \nu & \mathfrak{G} & \mathfrak{F} & \mathfrak{C} & \\ \vdots & & & & \end{vmatrix},$$

and the equation of the section in question becomes

$$-\begin{vmatrix} & x & y & z & \cdots \\ \lambda & \mathfrak{A} & \mathfrak{H} & \mathfrak{G} & \\ \mu & \mathfrak{H} & \mathfrak{B} & \mathfrak{F} & \\ \nu & \mathfrak{G} & \mathfrak{F} & \mathfrak{C} & \\ \vdots & & & & \end{vmatrix} + K^{\frac{1}{2}n-1}\sqrt{-p}\sqrt{-\begin{vmatrix} 1 & \lambda & \mu & \nu & \cdots \\ \lambda & \mathfrak{A} & \mathfrak{H} & \mathfrak{G} & \\ \mu & \mathfrak{H} & \mathfrak{B} & \mathfrak{F} & \\ \nu & \mathfrak{G} & \mathfrak{F} & \mathfrak{C} & \\ \vdots & & & & \end{vmatrix}}\; t = 0,$$

also the condition of contact with the corresponding section is

$$-\begin{vmatrix} \mp 1 & \lambda & \mu & \nu & \cdots \\ \lambda' & \mathfrak{A} & \mathfrak{H} & \mathfrak{G} & \\ \mu' & \mathfrak{H} & \mathfrak{B} & \mathfrak{F} & \\ \nu' & \mathfrak{G} & \mathfrak{F} & \mathfrak{C} & \\ \vdots & & & & \end{vmatrix} = \sqrt{-\begin{vmatrix} 1 & \lambda & \mu & \nu & \cdots \\ \lambda & \mathfrak{A} & \mathfrak{H} & \mathfrak{G} & \\ \mu & \mathfrak{H} & \mathfrak{B} & \mathfrak{F} & \\ \nu & \mathfrak{G} & \mathfrak{F} & \mathfrak{C} & \\ \vdots & & & & \end{vmatrix}}\;\sqrt{-\begin{vmatrix} 1 & \lambda' & \mu' & \nu' & \cdots \\ \lambda' & \mathfrak{A} & \mathfrak{H} & \mathfrak{G} & \\ \mu' & \mathfrak{H} & \mathfrak{B} & \mathfrak{F} & \\ \nu' & \mathfrak{G} & \mathfrak{F} & \mathfrak{C} & \\ \vdots & & & & \end{vmatrix}}\,.$$

In particular the equation of the sections which touches all the sections $x=0$, $y=0$, $z=0, \ldots,$ is

$$-\begin{vmatrix} & x & y & z & \cdots \\ \sqrt{\mathfrak{A}} & \mathfrak{A} & \mathfrak{H} & \mathfrak{G} & \\ \sqrt{\mathfrak{B}} & \mathfrak{H} & \mathfrak{B} & \mathfrak{F} & \\ \sqrt{\mathfrak{C}} & \mathfrak{G} & \mathfrak{F} & \mathfrak{C} & \\ \vdots & & & & \end{vmatrix} + K^{\frac{1}{2}n-1}\sqrt{-p}\sqrt{-\begin{vmatrix} 1 & \sqrt{\mathfrak{A}} & \sqrt{\mathfrak{B}} & \sqrt{\mathfrak{C}} & \cdots \\ \sqrt{\mathfrak{A}} & \mathfrak{A} & \mathfrak{H} & \mathfrak{G} & \\ \sqrt{\mathfrak{B}} & \mathfrak{H} & \mathfrak{B} & \mathfrak{F} & \\ \sqrt{\mathfrak{C}} & \mathfrak{G} & \mathfrak{F} & \mathfrak{C} & \\ \vdots & & & & \end{vmatrix}}\; t = 0.$$

Again, the equations of the section touching $y=0$, $z=0, \ldots$ and the sections touching $x=0$, $z=0, \ldots$ are

$$-\begin{vmatrix} & x & y & z & \cdots \\ \lambda & \mathfrak{A} & \mathfrak{H} & \mathfrak{G} & \\ \sqrt{\mathfrak{B}} & \mathfrak{H} & \mathfrak{B} & \mathfrak{F} & \\ \sqrt{\mathfrak{C}} & \mathfrak{G} & \mathfrak{F} & \mathfrak{C} & \\ \vdots & & & & \end{vmatrix} + K^{\frac{1}{2}n-1}\sqrt{-p}\sqrt{-\begin{vmatrix} 1 & \lambda & \sqrt{\mathfrak{B}} & \sqrt{\mathfrak{C}} & \cdots \\ \lambda & \mathfrak{A} & \mathfrak{H} & \mathfrak{G} & \\ \sqrt{\mathfrak{B}} & \mathfrak{H} & \mathfrak{B} & \mathfrak{F} & \\ \sqrt{\mathfrak{C}} & \mathfrak{G} & \mathfrak{F} & \mathfrak{C} & \\ \vdots & & & & \end{vmatrix}}\; t = 0,$$

$$-\begin{vmatrix} & x & y & z & \cdots \\ \sqrt{\mathfrak{A}} & \mathfrak{A} & \mathfrak{H} & \mathfrak{G} & \\ \mu & \mathfrak{H} & \mathfrak{B} & \mathfrak{F} & \\ \sqrt{\mathfrak{C}} & \mathfrak{G} & \mathfrak{F} & \mathfrak{C} & \\ \vdots & & & & \end{vmatrix} + K^{\frac{1}{2}n-1}\sqrt{-p}\sqrt{-\begin{vmatrix} 1 & \sqrt{\mathfrak{A}} & \mu & \sqrt{\mathfrak{C}} & \cdots \\ \sqrt{\mathfrak{A}} & \mathfrak{A} & \mathfrak{H} & \mathfrak{G} & \\ \mu & \mathfrak{H} & \mathfrak{B} & \mathfrak{F} & \\ \sqrt{\mathfrak{C}} & \mathfrak{G} & \mathfrak{F} & \mathfrak{C} & \\ \vdots & & & & \end{vmatrix}}\; t = 0,$$

and the condition of contact of these two sections is

$$-\left|\begin{array}{cccc} \mp 1 & \lambda & \sqrt{\mathfrak{B}} & \sqrt{\mathfrak{C}}\ldots \\ \sqrt{\mathfrak{A}} & \mathfrak{A} & \mathfrak{H} & \mathfrak{G} \\ \mu & \mathfrak{H} & \mathfrak{B} & \mathfrak{F} \\ \sqrt{\mathfrak{C}} & \mathfrak{G} & \mathfrak{F} & \mathfrak{C} \\ \vdots & & & \end{array}\right| = \sqrt{-\left|\begin{array}{cccc} 1 & \lambda & \sqrt{\mathfrak{B}} & \sqrt{\mathfrak{C}}\ldots \\ \lambda & \mathfrak{A} & \mathfrak{H} & \mathfrak{G} \\ \sqrt{\mathfrak{B}} & \mathfrak{H} & \mathfrak{B} & \mathfrak{F} \\ \sqrt{\mathfrak{C}} & \mathfrak{G} & \mathfrak{F} & \mathfrak{C} \\ \vdots & & & \end{array}\right|}\ \sqrt{-\left|\begin{array}{cccc} 1 & \sqrt{\mathfrak{A}} & \mu & \sqrt{\mathfrak{C}}\ldots \\ \sqrt{\mathfrak{A}} & \mathfrak{A} & \mathfrak{H} & \mathfrak{G} \\ \mu & \mathfrak{H} & \mathfrak{B} & \mathfrak{F} \\ \sqrt{\mathfrak{C}} & \mathfrak{G} & \mathfrak{F} & \mathfrak{C} \\ \vdots & & & \end{array}\right|}.$$

It would seem from the appearance of these equations that there should be some simpler method of obtaining the solution than the method employed in the previous part of this memoir.

115.

NOTE ON THE PORISM OF THE IN-AND-CIRCUMSCRIBED POLYGON.

[From the *Philosophical Magazine*, vol. VI. (1853), pp. 99—102.]

THE equation of a conic passing through the points of intersection of the conics

$$U = 0, \quad V = 0$$

is of the form

$$wU + V = 0,$$

where w is an arbitrary parameter. Suppose that the conic touches a given line, we have for the determination of w a quadratic equation, the roots of which may be considered as parameters for determining the line in question. Let one of the values of w be considered as equal to a given constant k, the line is always a tangent to the conic

$$kU + V = 0;$$

and taking $w = p$ for the other value of w, p is a parameter determining the particular tangent, or, what is the same thing, the point of contact of this tangent.

Suppose the tangent meets the conic $U = 0$ (which is of course the conic corresponding to $w = \infty$) in the points P, P', and let θ, ∞ be the parameters of the point P, and θ', ∞ the parameters of the point P'. It follows from my "Note on the Geometrical representation of the Integral $\int dx \div \sqrt{(x+a)(x+b)(x+c)}$," [113] (1) and from the theory of invariants, that if $\square\xi$ represent the "Discriminant" of $\xi U + V$

1 I take the opportunity of correcting an obvious error in the note in question, viz. $a^2+b^2+c^2-2bc-2ca-2ab$ is throughout written instead of (what the expression should be) $b^2c^2+c^2a^2+a^2b^2-2a^2bc-2b^2ca-2c^2ab$. [This correction is made, *ante* p. 55.]

(I now use the term discriminant in the same sense in which determinant is sometimes used, viz. the discriminant of a quadratic function $ax^2 + by^2 + cz^2 + 2fyz + 2gzx + 2hxy$ or $(a, b, c, f, g, h)(x, y, z)^2$, is the determinant $k = abc - af^2 - bg^2 - ch^2 + 2fgh$), and if

$$\Pi\xi = \int_\infty \frac{d\xi}{\sqrt{\Box\xi}},$$

then the following theorem is true, viz.

"If (θ, ∞), (θ', ∞) are the parameters of the points P, P' in which the conic $U=0$ is intersected by the tangent, the parameter of which is p, of the conic $kU + V = 0$, then the equations

$$\Pi\theta = \Pi p - \Pi k,$$

$$\Pi\theta' = \Pi p + \Pi k,$$

determine the parameters θ, θ' of the points in question." And again,—

"If the variable parameters θ, θ' are connected by the equation

$$\Pi\theta' - \Pi\theta = 2\Pi k,$$

then the line PP' will be a tangent to the conic $kU + V = 0$." Whence, also,—

"If the sides of a triangle inscribed in the conic $U=0$ touch the conics

$$k\ U + V = 0,$$

$$k'\ U + V = 0,$$

$$k''U + V = 0,$$

then the equation

$$\Pi k + \Pi k' + \Pi k'' = 0$$

must hold good between the parameters k, k', k''."

And, conversely, when this equation holds good, there are an infinite number of triangles inscribed in the conic $U = 0$, and the sides of which touch the three conics; and similarly for a polygon of any number of sides.

The algebraical equivalent of the transcendental equation last written down is

$$\begin{vmatrix} 1, & k\ , & \sqrt{\Box k} \\ 1, & k', & \sqrt{\Box k'} \\ 1, & k'', & \sqrt{\Box k''} \end{vmatrix} = 0;$$

let it be required to find what this becomes when $k = k' = k'' = 0$, we have

$$\sqrt{\Box k} = A + Bk + Ck^2 + \dots,$$

and substituting these values, the determinant divides by

$$\begin{vmatrix} 1, & k\,, & k^2 \\ 1, & k', & k'^2 \\ 1, & k'', & k''^2 \end{vmatrix},$$

the quotient being composed of the constant term C, and terms multiplied by k, k', k''; writing, therefore, $k=k'=k''=0$, we have $C=0$ for the condition that there may be inscribed in the conic $U=0$ an infinity of triangles circumscribed about the conic $V=0$; C is of course the coefficient of ξ^2 in $\sqrt{\square\xi}$, i.e. in the square root of the discriminant of $\xi U+V$; and since precisely the same reasoning applies to a polygon of any number of sides,—

THEOREM. The condition that there may be inscribed in the conic $U=0$ an infinity of n-gons circumscribed about the conic $V=0$, is that the coefficient of ξ^{n-1} in the development in ascending powers of ξ of the square root of the discriminant of $\xi U+V$ vanishes. [This and the theorem p. 90 are erroneous, see *post*, 116].

It is perhaps worth noticing that $n=2$, i.e. the case where the polygon degenerates into two coincident chords, is a case of exception. This is easily explained.

In particular, the condition that there may be in the conic[1]

$$ax^2+by^2+cz^2=0$$

an infinity of n-gons circumscribed about the conic

$$x^2+\ y^2+\ z^2=0,$$

is that the coefficient of ξ^{n-1} in the development in ascending powers of ξ of

$$\sqrt{(1+a\xi)(1+b\xi)(1+c\xi)}$$

vanishes; or, developing each factor, the coefficient of ξ^{n-1} in

$$(1+\tfrac{1}{2}a\xi-\tfrac{1}{8}a^2\xi^2+\tfrac{1}{16}a^3\xi^3-\tfrac{5}{64}a^4\xi^4+\&\text{c.})(1+\tfrac{1}{2}b\xi-\&\text{c.})(1+\tfrac{1}{2}c\xi-\&\text{c.})$$

vanishes.

Thus, for a triangle this condition is

$$a^2+b^2+c^2-2bc-2ca-2ab=0\,;$$

for a quadrangle it is

$$a^3+b^3+c^3-bc^2-b^2c-ca^2-c^2a-ab^2-a^2b+2abc=0,$$

which may also be written

$$(b+c-a)(c+a-b)(a+b-c)=0\,;$$

and similarly for a pentagon, &c.

[1] I have, in order to present this result in the simplest form, purposely used a notation different from that of the note above referred to, the quantities $ax^2+by^2+cz^2$ and $x^2+y^2+z^2$ being, in fact, interchanged.

Suppose the conics reduce themselves to circles, or write

$$U = \qquad x^2 + y^2 - R^2 = 0,$$
$$V = (x - a)^2 + y^2 - r^2 = 0;$$

R is of course the radius of the circumscribed circle, r the radius of the inscribed circle, and a the distance between the centres. Then

$$\xi U + V = (\xi + 1,\ \xi + 1,\ -\xi R^2 - r^2 + a^2,\ 0,\ -a,\ 0)(x,\ y,\ 1)^2,$$

and the discriminant is therefore

$$-(\xi + 1)^2 (\xi R^2 + r^2 - a^2) - (\xi + 1)\, a^2 = -(1 + \xi)\{r^2 + \xi (r^2 + R^2 - a^2) + \xi^2 R^2\}.$$

Hence,

THEOREM. The condition that there may be inscribed in the circle $x^2 + y^2 - R^2 = 0$ an infinity of n-gons circumscribed about the circle $(x - a)^2 + y^2 - r^2 = 0$, is that the coefficient of ξ^{n-1} in the development in ascending powers of ξ of

$$\sqrt{(1 + \xi)\{r^2 + \xi (r^2 + R^2 - a^2) + \xi^2 R^2\}}$$

may vanish.

Now

$$(A + B\xi + C\xi^2)^{\frac{1}{2}} = \sqrt{A}\left\{1 + \tfrac{1}{2}B\,\frac{\xi}{A} + (\tfrac{1}{2}AC - \tfrac{1}{8}B^2)\,\frac{\xi^2}{A^2} + \dots\right\},$$

or the quantity to be considered is the coefficient of ξ^{n-1} in

$$(1 + \tfrac{1}{2}\xi - \tfrac{1}{8}\xi^2 \dots)\left\{1 + \tfrac{1}{2}B\,\frac{\xi}{A} + (\tfrac{1}{2}AC - \tfrac{1}{8}B^2)\,\frac{\xi^2}{A^2} + \dots\right\},$$

where, of course,

$$A = r^2, \quad B = r^2 + R^2 - a^2, \quad C = R^2.$$

In particular, in the case of a triangle we have, equating to zero the coefficient of ξ^2,

$$(A - B)^2 - 4AC = 0;$$

or substituting the values of A, B, C,

$$(a^2 - R^2)^2 - 4r^2R^2 = 0,$$

that is

$$(a^2 - R^2 + 2Rr)(a^2 - R^2 - 2Rr) = 0;$$

the factor which corresponds to the proper geometrical solution of the question is

$$a^2 - R^2 + 2Rr = 0,$$

Euler's well-known relation between the radii of the circles inscribed and circumscribed in and about a triangle, and the distance between the centres. I shall not now discuss the meaning of the other factor, or attempt to verify the formulæ which have been given by Fuss, Steiner and Richelot, for the case of a polygon of 4, 5, 6, 7, 8, 9, 12, and 16 sides. See Steiner, *Crelle*, t. II. [1827] p. 289, Jacobi, t. III. [1828] p. 376; Richelot, t. V. [1830] p. 250; and t. XXXVIII. [1849] p. 353.

2 *Stone Buildings, July* 9, 1853.

116.

CORRECTION OF TWO THEOREMS RELATING TO THE PORISM OF THE IN-AND-CIRCUMSCRIBED POLYGON.

[From the *Philosophical Magazine*, vol. VI. (1853), pp. 376—377.]

THE two theorems in my "Note on the Porism of the in-and-circumscribed Polygon" (see August Number), [115], are erroneous, the mistake arising from my having inadvertently assumed a wrong formulæ for the addition of elliptic integrals. The first of the two theorems (which, in fact, includes the other as a particular case) should be as follows:—

THEOREM. The condition that there may be inscribed in the conic $U=0$ an infinity of n-gons circumscribed about the conic $V=0$, depends upon the development in ascending powers of ξ of the square root of the discriminant of $\xi U+V$; viz. if this square root be

$$A+B\xi+C\xi^2+D\xi^3+E\xi^4+F\xi^5+G\xi^6+H\xi^7+\dots,$$

then for $n=3$, 5, 7, &c. respectively, the conditions are

$$|\,C\,|=0,\qquad \begin{vmatrix} C, & D \\ D, & E \end{vmatrix}=0,\qquad \begin{vmatrix} C, & D, & E \\ D, & E, & F \\ E, & F, & G \end{vmatrix}=0,\ \text{\&c.};$$

and for $n=4$, 6, 8, &c. respectively, the conditions are

$$|\,D\,|=0,\qquad \begin{vmatrix} D, & E \\ E, & F \end{vmatrix}=0,\qquad \begin{vmatrix} D, & E, & F \\ D, & F, & G \\ F, & G, & H \end{vmatrix}=0,\ \text{\&c.}$$

The examples require no correction; since for the triangle and the quadrilateral respectively, the conditions are (as in the erroneous theorem) $C=0$, $D=0$.

The second theorem gives the condition in the case where the conics are replaced by the circles $x^2 + y^2 - R^2 = 0$ and $(x - a)^2 + y^2 - r^2 = 0$, the discriminant being in this case

$$- (1 + \xi) \{r^2 + \xi (r^2 + R^2 - a^2) + \xi^2 R^2\}.$$

As a very simple example, suppose that the circles are concentric, or assume $a = 0$; the square root of the discriminant is here

$$(1 + \xi) \sqrt{r^2 + R^2\xi}\,;$$

and putting for shortness $\dfrac{R^2}{r^2} = \alpha$, we may write

$$A + B\xi + \ldots = (1 + \xi) \sqrt{1 + \alpha\xi},$$

that is, $A = 1$, $B = \frac{1}{2}\alpha + 1$, $C = -\frac{1}{8}\alpha^2 + \frac{1}{2}\alpha^2$, $D = \frac{1}{16}\alpha^3 - \frac{1}{8}\alpha^2$, $E = -\frac{5}{128}\alpha^4 + \frac{1}{16}\alpha^3$, &c.;

thus in the case of the pentagon,

$$\begin{aligned} CE - D^2 &= \tfrac{1}{1024}\alpha^4 \{(\alpha - 4)(5\alpha - 8) - 4(\alpha - 2)^2\} \\ &= \tfrac{1}{1024}\alpha^4 (\alpha^2 - 12\alpha + 16), \end{aligned}$$

and the required condition therefore is

$$\alpha^2 - 12\alpha + 16 = 0.$$

It is clear that, in the case in question,

$$\frac{r}{R} = \cos 36^\circ = \tfrac{1}{4}(\sqrt{5} + 1),$$

that is,
$$\frac{R}{r} = \sqrt{5} - 1, \text{ or } (R + r)^2 - 5r^2 = 0,$$

viz.
$$(\sqrt{\alpha} + 1)^2 - 5 = 0, \text{ or } \alpha + 2\sqrt{\alpha} - 4 = 0,$$

the rational form of which is

$$\alpha^2 - 12\alpha + 16 = 0,$$

and we have thus a verification of the theorem for this particular case.

2 *Stone Buildings, Oct.* 10, 1853.

117.

NOTE ON THE INTEGRAL $\int dx \div \sqrt{(m-x)(x+a)(x+b)(x+c)}$.

[From the *Philosophical Magazine*, vol. VI. (1853), pp. 103—105.]

IF in the formulæ of my "Note on the Porism of the in-and-circumscribed Polygon," [115], it is assumed that

$$U = x^2 + y^2 + z^2 + \frac{1}{m}(ax^2 + by^2 + cz^2)$$

$$V = \qquad\qquad ax^2 + by^2 + cz^2,$$

and if a new parameter ω connected with the parameter w by the equation

$$w = \frac{\omega m}{m - \omega}$$

be made use of instead of w, then

$$wU + V = \frac{m}{m-\omega}\{\omega(x^2 + y^2 + z^2) + ax^2 + by^2 + cz^2\};$$

and thus the equation $wU + V = 0$, viz. the equation

$$\omega(x^2 + y^2 + z^2) + ax^2 + by^2 + cz^2 = 0,$$

is precisely of the same form as that considered in my "Note on the Geometrical Representation of the Integral $\int dx \div \sqrt{(x+a)(x+b)(x+c)}$," [113.] Moreover, introducing instead of ξ a quantity η, such that

$$\xi = \frac{m\eta}{m-\eta},$$

then

$$\frac{d\xi}{\sqrt{\square\xi}} = \frac{\sqrt{m}\,d\eta}{\sqrt{(m-\eta)(a+\eta)(b+\eta)(c+\eta)}}.$$

Also $\xi=\infty$ gives $\eta=m$, the integral to be considered is therefore

$$\Pi,\eta=\int_m \frac{\sqrt{m}\,d\eta}{\sqrt{(m-\eta)(a+\eta)(b+\eta)(c+\eta)}};$$

i.e. if in the paper last referred to the parameter ∞ had been throughout replaced by the parameter m, the integral

$$\Pi\eta=\int_\infty \frac{d\eta}{\sqrt{(a+\eta)(b+\eta)(c+\eta)}}$$

would have had to be replaced by the integral Π,η. It is, I think, worth while to reproduce for this more general case a portion of the investigations of the paper in question, for the sake of exhibiting the rational and integral form of the algebraical equation corresponding to the transcendental equation $\pm\Pi,k\pm\Pi,p\pm\Pi,\theta=0$. Consider the point ξ, η, ζ on the conic $m(x^2+y^2+z^2)+ax^2+by^2+cz^2=0$, the equation of the tangent at this point is

$$(m+a)\,\xi x+(m+b)\,\eta y+(m+c)\,\zeta z=0;$$

and if θ be the other parameter of this line, then the line touches

$$\theta(x^2+y^2+z^2)+ax^2+by^2+cz^2=0;$$

or we have

$$\frac{(m+a)^2\xi^2}{\theta+a}+\frac{(m+b)^2\eta^2}{\theta+b}+\frac{(m+c)^2\zeta^2}{\theta+c}=0;$$

and combining this with

$$(m+a)\,\xi^2+(m+b)\,\eta^2+(m+c)\,\zeta^2=0,$$

we have

$$\begin{aligned}\xi:\eta:\zeta=&\sqrt{b-c}\;\sqrt{a+\theta}\,\sqrt{b+m}\,\sqrt{c+m}\\ &:\sqrt{(c-a)}\,\sqrt{b+\theta}\,\sqrt{c+m}\,\sqrt{a+m}\\ &:\sqrt{(a-b)}\,\sqrt{c+\theta}\,\sqrt{a+m}\,\sqrt{b+m}\end{aligned}$$

for the coordinates of the point P. Substituting these for x, y, z in the equation of the line PP' (the parameters of which are p, k), viz. in

$$x\sqrt{b-c}\,\sqrt{(a+k)(a+p)}+y\sqrt{c-a}\,\sqrt{(b+k)(b+p)}+z\sqrt{a-b}\,\sqrt{c+k}\,\sqrt{c+p}=0,$$

we have

$$(b-c)\frac{\sqrt{(a+p)(a+k)(a+\theta)}}{\sqrt{a+m}}+(c-a)\frac{\sqrt{(b+p)(b+k)(b+\theta)}}{\sqrt{b+m}}$$

$$+(a-b)\frac{\sqrt{(c+p)(c+k)(c+\theta)}}{\sqrt{c+m}}=0,$$

which is to be replaced by

$$\frac{(a+p)(a+k)(a+\theta)}{a+m} = (\lambda + \mu a)^2,$$

$$\frac{(b+p)(b+k)(b+\theta)}{b+m} = (\lambda + \mu b)^2,$$

$$\frac{(c+p)(c+k)(c+\theta)}{c+m} = (\lambda + \mu c)^2.$$

These equations give, omitting the common factor $(a+m)(b+m)(c+m)$,

$$\begin{aligned}\lambda^2 = {} & m^2(abc + pk\theta) \\ & + m\{-abc(p+k+\theta) + pk\theta(a+b+c)\} \\ & + \{abc(k\theta + \theta p + kp) + pk\theta(bc+ca+ab)\},\end{aligned}$$

$$\begin{aligned}2\lambda\mu = {} & m^2\{-(bc+ca+ab) + (k\theta + \theta p + pk)\} \\ & + m\{-abc - pk\theta + (bc+ca+ab)(p+k+\theta) + (k\theta+\theta p + pk)(a+b+c)\} \\ & + \{abc(p+k+\theta) - pk\theta(a+b+c)\},\end{aligned}$$

$$\begin{aligned}\mu^2 = {} & m^2\{a+b+c+p+k+\theta\} \\ & + m\{(bc+ca+ab) - (k\theta+\theta p + pk)\} \\ & + abc + pk\theta;\end{aligned}$$

and substituting in $4\lambda^2 . \mu^2 - (2\lambda\mu)^2 = 0$, we have the relation required. To verify that the equation so obtained is in fact the algebraical equivalent of the transcendental equation, it is only necessary to remark, that the values of λ^2, μ^2 are unaltered, and that of $\lambda\mu$ only changes its sign when a, b, c, m and p, k, θ, $-m$ are interchanged; and so this change will not affect the equation obtained by substituting in the equation $4\lambda^2 . \mu^2 - (2\lambda\mu)^2 = 0$. Hence precisely the same equation would be obtained by eliminating L, M from

$$(k+a)(k+b)(k+c) = (L+Mk)^2(m-k),$$

$$(p+a)(p+b)(p+c) = (L+Mp)^2(m-p),$$

$$(\theta+a)(\theta+b)(\theta+c) = (L+M\theta)^2(\theta-p);$$

or, putting $(L+Mk)(m-k) = \alpha + \beta k + \gamma k^2$, by eliminating α, β, γ from

$$(m-k)(k+a)(k+b)(k+c) = (\alpha + \beta k + \gamma k^2)^2,$$

$$(m-p)(p+a)(p+b)(p+c) = (\alpha + \beta p + \gamma p^2)^2,$$

$$(m-\theta)(\theta+a)(\theta+b)(\theta+c) = (\alpha + \beta\theta + \gamma\theta^2)^2,$$

$$0 = (\alpha + \beta m + \gamma m^2)^2,$$

which by Abel's theorem show that p, k, θ are connected by the transcendental equation above mentioned.

2 *Stone Buildings, July* 9, 1853.

118.

ON THE HARMONIC RELATION OF TWO LINES OR TWO POINTS.

[From the *Philosophical Magazine*, vol. VI. (1853), pp. 105—107.]

THE "harmonic relation of a point and line with respect to a triangle" is well known and understood[1]; but the analogous relation between two lines with respect to a quadrilateral, or between two points with respect to a quadrangle, is not, I think, sufficiently singled out from the mass of geometrical theorems so as to be recognized when implicitly occurring in the course of an investigation. The relation in question, or some particular case of it, is of frequent occurrence in the *Traité des Propriétés Projectives*, [Paris, 1822], and is, in fact, there substantially demonstrated (see No. 163); and an explicit statement of the theorem is given by M. Steiner, Lehrsätze 24 and 25, *Crelle*, t. XIII. [1835] p. 212 (a demonstration is given, t. XIX. [1839] p. 227). The theorem containing the relation in question may be thus stated.

THEOREM *of the harmonic relation of two lines with respect to a quadrilateral.* "If on each of the three diagonals of a *quadrilateral* there be taken two points harmonically related with respect to the angles upon this diagonal, then if three of the points lie in a *line*, the other three points will also lie in a *line*,"—the two lines are said to be harmonically related with respect to the quadrilateral.

It may be as well to exhibit this relation in a somewhat different form. The three diagonals of the quadrilateral form a triangle, the sides of which contain the six angles of the quadrilateral; and considering three only of these six angles (one angle on each side), these three angles are points which either lie in a line, or else

[1] The relation to which I refer is contained in the theorem, "If on each side of a *triangle* there be taken two points harmonically related with respect to the angles on this side, then if three of these points lie in *a line*, the lines joining the other three points with the opposite angles of the triangle meet in a *point*,"—the line and point are said to be harmonically related with respect to the triangle.

are such that the lines joining them with the opposite angles of the triangle meet in a point. Each of these points is, with respect to the involution formed by the two angles of the triangle, and the two points harmonically related thereto, a double point; and we have thus the following theorem of the harmonic relation of two lines to a triangle and line, or else to a triangle and point.

THEOREM. "If on the sides of a *triangle* there be taken three points, which either lie in a *line*, or else are such that the lines joining them with the opposite angles of a triangle meet in a *point;* and if on each side of the triangle there be taken two points, forming with the two angles on the same side an involution having the first-mentioned point on the same side for a double point; then if three of the six points lie in a *line*, the other three of the six points will also lie in a *line*,"—the two lines are said to be harmonically related to the triangle and line, or (as the case may be) to the triangle and point.

The theorems with respect to the harmonic relation of two points are of course the reciprocals of those with respect to the harmonic relation of two lines, and do not need to be separately stated.

The preceding theorems are useful in (among other geometrical investigations) the porism of the in-and-circumscribed polygon.

2 *Stone Buildings, July* 9, 1853.

119.

ON A THEOREM FOR THE DEVELOPMENT OF A FACTORIAL.

[From the *Philosophical Magazine*, vol. VI. (1853), pp. 182—185.]

THE theorem to which I refer is remarkable for the extreme simplicity of its demonstration. Let it be required to expand the factorial $\overline{x-a}\,\overline{x-b}\,\overline{x-c}\ldots$ in the form

$$\overline{x-\alpha}\,\overline{x-\beta}\,\overline{x-\gamma}\ldots+B\,\overline{x-\alpha}\,\overline{x-\beta}\ldots+C\,\overline{x-\alpha}\ldots+D\ldots\ \text{\&c.}$$

We have first

$$x-a=\overline{x-\alpha}+\overline{\alpha-a};$$

multiply the two sides of this by $\overline{x-b}$; but in multiplying by this factor the term $\overline{x-\alpha}$, write the factor in the form $\overline{x-\beta}+\overline{\beta-b}$; and in multiplying the term $\overline{\alpha-a}$, write the factor in the form $\overline{x-\alpha}+\overline{\alpha-b}$; the result is obviously

$$\overline{x-a}\,\overline{x-b}=\overline{x-\alpha}\,\overline{x-\beta}\ +(\overline{\alpha-a}+\overline{\beta-b})\,\overline{x-\alpha}\ +\overline{\alpha-a}\,\overline{\alpha-b};$$

multiply this by $x-c$, this factor being in multiplying the quantity on the right-hand side written successively under the forms $\overline{x-\gamma}+\overline{\gamma-c}$, $\overline{x-\beta}+\overline{\beta-c}$, $\overline{a-\alpha}+\overline{\alpha-c}$; the result is

$$\begin{aligned}\overline{x-a}\,\overline{x-b}\,\overline{x-c}=&\qquad\qquad\overline{x-\alpha}\,\overline{x-\beta}\,\overline{x-\gamma}\\ &+(\overline{\alpha-a}+\overline{\beta-b}+\overline{\gamma-c})\,\overline{x-\alpha}\,\overline{x-\beta}\\ &+(\overline{\alpha-a}\,\overline{\alpha-b}+\overline{\alpha-a}\,\overline{\beta-c}+\overline{\beta-b}\,\overline{\beta-c})\,\overline{x-\alpha}\\ &+\overline{\alpha-a}\,\overline{\alpha-b}\,\overline{\alpha-c},\end{aligned}$$

which may be thus written,

$$(x-a)(x-b)(x-c)=$$

$$(x-\alpha)(x-\beta)(x-\gamma)+\left[\begin{matrix}\alpha, & \beta, & \gamma \\ a, & b, & c\end{matrix}\right]_1 \overline{x-\alpha}\,\overline{x-\beta}+\left[\begin{matrix}\alpha, & \beta, & \\ a, & b, & c\end{matrix}\right]_2 \overline{x-\alpha}+\left[\begin{matrix}\alpha & & \\ a, & b, & c\end{matrix}\right]_3.$$

Consider, for instance,

$$\left[\begin{matrix}\alpha, & \beta, & \\ a, & b, & c\end{matrix}\right]_2=\overline{\alpha-a}\,\overline{\alpha-b}+\overline{\alpha-a}\,\overline{\beta-c}+\overline{\beta-b}\,\overline{\beta-c};$$

then, paying attention in the first instance to the Greek letters only, it is clear that the terms on the second side contain the combinations two and two, with repetitions, of the Greek letters α, β, and these letters appear in each term in the alphabetical order. Each such combination may therefore be considered as derived from the primitive combination α, α by a change of one or both of the α's into β; and if we take (instead of the mere combination α, α) the complete first term $\overline{\alpha-a}\,\overline{\alpha-b}$, and simultaneously with the change of the α of either of the factors into β make a similar change in the Latin letter of the factor, we derive from the first term the other terms of the expression on the right-hand side of the expression. It is proper also to remark, that, paying attention to the Latin letters only, the different terms contain all the combinations two and two, without repetitions, of the letters a, b, c. The same reasoning will show that

$$\begin{aligned}\overline{x-a}\,\overline{x-b}\,\overline{x-c}\,\overline{x-d}= \quad & \overline{x-\alpha}\,\overline{x-\beta}\,\overline{x-\gamma}\,\overline{x-\delta} \\ & +\left[\begin{matrix}\alpha, & \beta, & \gamma, & \delta \\ a, & b, & c, & d\end{matrix}\right]_1 \overline{x-\alpha}\,\overline{x-\beta}\,\overline{x-\gamma} \\ & +\left[\begin{matrix}\alpha, & \beta, & \gamma, & \\ a, & b, & c, & d\end{matrix}\right]_2 \overline{x-\alpha}\,\overline{x-\beta} \\ & +\left[\begin{matrix}\alpha, & \beta & & \\ a, & b, & c, & d\end{matrix}\right]_3 \overline{x-\alpha} \\ & +\left[\begin{matrix}\alpha & & & \\ a, & b, & c, & d\end{matrix}\right]_4;\end{aligned}$$

where, for instance,

$$\begin{aligned}\left[\begin{matrix}\alpha, & \beta & & \\ a, & b, & c, & d\end{matrix}\right]_3 = \quad & (\alpha-a)(\alpha-b)(\alpha-c) \\ & +(\alpha-a)(\alpha-b)(\beta-d) \\ & +(\alpha-a)(\beta-c)(\beta-d) \\ & +(\beta-b)(\beta-c)(\beta-d), \text{ \&c.}\end{aligned}$$

It is of course easy, by the use of subscript letters and signs of summation, to present the preceding theorem under a more condensed form; thus writing

$$\begin{bmatrix} \alpha_1 & \alpha_2 & \dots & \alpha_r & & \\ a_1 & a_2 & \dots & a_r & \dots & a_{r+s} \end{bmatrix}_{s+1} = \Sigma\left\{\overline{\alpha_{k_s} - a_{k_s+s}} \, . \, \overline{\alpha_{k_{s-1}} - a_{k_{s-1}+s-1}} \dots \overline{\alpha_{k_0} - a_{k_0}}\right\}$$

where k_s, k_{s-1}, ... k_0 form a decreasing series (equality of successive terms *not excluded*) of numbers out of the system r, $\overline{r-1}$, ... 3, 2, 1; the theorem may be written in the form

$$\overline{x-a_1}\,\overline{x-a_2}\dots\overline{x-a_p} = S_{q}{}^{p}_{0}\begin{bmatrix} \alpha_1 & \alpha_2 & \dots & \alpha_{p-q+1} \\ a_1 & a_2 & \dots\dots & a_p \end{bmatrix}_q \overline{x-\alpha_1}\,\overline{x-\alpha_2}\dots\overline{x-\alpha_{p-q}};$$

but I think that a more definite idea of the theorem is obtained through the notation first made use of. It is clear that the above theorem includes the binomial theorem for positive integers, the corresponding theorem for an ordinary factorial, and a variety of other theorems relating to combinations.

Thus, for instance, if $C_q(a_1, \dots a_p)$ denote the combinations of a_1, ... a_p, q and q together without repetitions, and $H_q(a_1, \dots a_p)$ denote the combinations of a_1, ... a_p, q and q together with repetitions, then making all the α's vanish,

$$\overline{x-a_1}\dots\overline{x-a_p} = S_{q}{}^{p}_{0}(-)^q\, C_q(a_1, \dots a_p)\, x^{p-q};$$

and therefore

$$(x-a)^p = S_{q}{}^{p}_{0}(-)^q\, C_q(a, a \dots \text{plures})\, x^{p-q} = S_{q}{}^{p}_{0}(-)^q \frac{[p]^q}{[q]^q}\, a^q x^{p-q},$$

the ordinary binomial theorem for a positive and integral index p.

So making all the a's vanish,

$$x^p = S_{q}{}^{p}_{0} H_q(\alpha_1 \dots \alpha_{p-q+1})\, \overline{x-\alpha_1}\,\overline{x-\alpha_2}\dots\overline{x-\alpha_{p-q}}.$$

If m be any integer less than p, the coefficient of x^m on the right-hand side must vanish, that is, we must have identically

$$0 = S_{q}{}^{p-m}_{0}(-)^q\, C_{p-q-m}(\alpha_1, \alpha_2, \dots \alpha_{p-q})\, H_q(\alpha_1, \alpha_2, \dots \alpha_{p-q+1}).$$

So also

$$C_{p-m}(a_1, a_2, \dots a_p) = S_{q}{}^{p-m}_{0}(-)^q\, C_{p-q-m}(\alpha_1, \alpha_2, \dots \alpha_{p-q})\begin{bmatrix} \alpha_1 & \dots & \alpha_{p-q+1} \\ a_1 & \dots\dots\dots & a_p \end{bmatrix}_q.$$

Suppose

$$a_1 = 0, \quad a_2 = 1 \dots a_p = p-1; \quad \alpha_1 = k, \quad \alpha_2 = k-1, \dots \alpha_p = k-p+1,$$

then

$$\begin{bmatrix} \alpha_1, \dots & \alpha_{p-q+1} \\ a_1 & \dots\dots & a_p \end{bmatrix}_q = \begin{bmatrix} k & \overline{k-1} & \dots & k-p+q \\ 0 & 1 & \dots\dots\dots\dots & p-1 \end{bmatrix}_q = \frac{[p]^q}{[q]^q}[k]^q\, \overline{x-\alpha_1}\dots\overline{x-\alpha_{p-q}} = [x]^{p-q};$$

and hence

$$[x+k]^p = S_{q}{}^{p}_{0}\frac{[p]^q}{[q]^q}[k]^q\,[x]^{p-q},$$

the binomial theorem for factorials.

A preceding formula gives at once the theorem

$$H_q(0,\ 1, \dots p-q) = \frac{1}{[p-q]^{p-q}} \Delta^{p-q} 0^p.$$

It may be as well to remark, with reference to a demonstration frequently given of the binomial theorem, that in whatever way the binomial theorem is demonstrated for integer positive indices, it follows from what has preceded that it is quite as easy to demonstrate the corresponding theorem for the factorial $[m]^p$. But the theorem being true for the factorial $[m]^p$, it is at once seen that the product of the series for $(1+x)^m$ and $(1+x)^n$ is identical with the series for $(1+x)^{m+n}$, and thus it becomes *unnecessary* to employ for the purpose of proving this identity the so-called principle of the permanence of equivalent forms; a principle which however, in the case in question, may legitimately be employed.

120.

NOTE ON A GENERALIZATION OF A BINOMIAL THEOREM.

[From the *Philosophical Magazine*, vol. VI. (1853), p. 185.]

THE formula (*Crelle*, t. I. [1826] p. 367) for the development of the binomial $(x+\alpha)^n$, but which is there presented in a form which does not put in evidence the law of the coefficients, is substantially equivalent to the theorem given by me as one of the Senate House Problems in the year 1851, and which is as follows:—

"If $\{\alpha+\beta+\gamma\ldots\}^p$ denote the expansion of $(\alpha+\beta+\gamma\ldots)^p$, retaining those terms $N\alpha^a\beta^b\gamma^c\delta^d\ldots$ only in which $b+c+d\ldots$ is not greater than $p-1$, $c+d+\ldots$ is not greater than $p-2$, &c., then

$$\begin{aligned} x^n = {} & 1 && (x+\alpha)^n \\ & -\frac{n}{1} \; \{\alpha\}^1 && (x+\alpha+\beta)^{n-1} \\ & +\frac{n(n-1)}{1.2} \; \{\alpha+\beta\}^2 && (x+\alpha+\beta+\gamma)^{n-2} \\ & -\frac{n(n-1)(n-2)}{1.2.3} \; \{\alpha+\beta+\gamma\}^3 && (x+\alpha+\beta+\gamma+\delta)^{n-3}. \\ & + \text{\&c.''} \end{aligned}$$

The theorem is, I think, one of some interest.

121.

NOTE ON A QUESTION IN THE THEORY OF PROBABILITIES.

[From the *Philosophical Magazine*, vol. VI. (1853), p. 259.]

THE following question was suggested to me, either by some of Prof. Boole's memoirs on the subject of probabilities, or in conversation with him, I forget which; it seems to me a good instance of the class of questions to which it belongs.

Given the probability α that a cause A will act, and the probability p that A acting the effect will happen; also the probability β that a cause B will act, and the probability q that B acting the effect will happen; required the total probability of the effect.

As an instance of the precise case contemplated, take the following: say a day is called *windy* if there is at least w of wind, and a day is called *rainy* if there is at least r of rain, and a day is called *stormy* if there is at least W of wind, *or* if there is at least R of rain. The day may therefore be stormy because of there being at least W of wind, or because of there being at least R of rain, or on both accounts; but if there is less than W of wind *and* less than R of rain, the day will not be stormy. Then α is the probability that a day chosen at random will be windy, p the probability that a windy day chosen at random will be stormy, β the probability that a day chosen at random will be rainy, q the probability that a rainy day chosen at random will be stormy. The quantities λ, μ introduced in the solution of the question mean in this particular instance, λ the probability that a windy day chosen at random will be stormy by reason of the quantity of wind, or in other words, that there will be at least W of wind; μ the probability that a rainy day chosen at random will be stormy by reason of the quantity of rain, or in other words, that there will be at least R of rain.

The sense of the terms being clearly understood, the problem presents of course no difficulty. Let λ be the probability that the cause A acting will act efficaciously; μ the probability that the cause B acting will act efficaciously; then

$$p = \lambda + (1-\lambda)\,\mu\beta,$$
$$q = \mu + (1-\mu)\,\alpha\lambda,$$

which determine λ, μ; and the total probability ρ of the effect is given by

$$\rho = \lambda\alpha + \mu\beta - \lambda\mu\alpha\beta;$$

suppose, for instance, $\alpha = 1$, then

$$p = \lambda + (1-\lambda)\,\mu\beta, \quad q = \mu + \lambda - \lambda\mu, \quad \rho = \lambda + \mu\beta - \lambda\mu\beta,$$

that is, $\rho = p$, for p is in this case the probability that (acting a cause which is certain to act) the effect will happen, or what is the same thing, p is the probability that the effect will happen.

Machynlleth, August 16, 1853.

122.

ON THE HOMOGRAPHIC TRANSFORMATION OF A SURFACE OF THE SECOND ORDER INTO ITSELF.

[From the *Philosophical Magazine*, vol. VI. (1853), pp. 326—333.]

THE following theorems in plane geometry, relating to polygons of any number (odd or even) of sides, are well known.

"If there be a polygon of $(m+1)$ sides inscribed in a conic, and m of the sides pass through given points, the $(m+1)$th side will envelope a conic having double contact with the given conic." And "If there be a polygon of $(m+1)$ sides inscribed in a conic, and m of the sides touch conics having double contact with the given conic, the $(m+1)$th side will envelope a conic having double contact with the given conic." The second theorem of course includes the first, but I state the two separately for the sake of comparison with what follows.

As regards the corresponding theory in geometry of three dimensions, Sir W. Hamilton has given a theorem relating to polygons of an odd number of sides, which may be thus stated: "If there be a polygon of $(2m+1)$ sides inscribed in a surface of the second order, and $2m$ of the sides pass through given points, the $(2m+1)$th side will constantly touch two surfaces of the second order, each of them intersecting the given surface of the second order in the same four lines[1]."

[1] See *Phil. Mag.* vol. XXXV. [1849] p. 200. The form in which the theorem is exhibited by Sir W. Hamilton is somewhat different; the surface containing the angles is considered as being an ellipsoid, and the two surfaces touched by the last or $(2m+1)$th side of the polygon are spoken of as being an ellipsoid, and a hyperboloid of two sheets, having respectively double contact with the given ellipsoid: the contact is, in fact, a quadruple contact at the *same* four points; *real* as regards two of them in the case of the ellipsoid, and as regards the other two in the case of the hyperboloid of two sheets; and a quadruple contact is the coincidence of four generating lines belonging two and two to the two series of generating lines, these generating lines being of course (in the case considered by Sir W. Hamilton) all of them imaginary.

The entire theory depends upon what may be termed the transformation of a surface of the second order into itself, or analytically, upon the transformation of a quadratic form of four indeterminates into itself. I use for shortness the term transformation simply; but this is to be understood as meaning a homographic transformation, or in analytic language, a transformation by means of linear substitutions. It will be convenient to remark at the outset, that if two points of a surface of the second order have the relation contemplated in the data of Sir W. Hamilton's theorem (viz. if the line joining the two points pass through a fixed point), the transformation is, using the language of the *Recherches Arithmétiques*, an *improper* one, but that the relation contemplated in the conclusion of the theorem (viz. that of two points of a surface of the second order, connected by a line touching two surfaces of the second order each of them intersecting the given surface of the second order in the same four lines) depends upon a *proper* transformation; and that the circumstance that an *even* number of improper transformations is required in order to make a proper transformation (that this circumstance, I say), is the reason why the theorem applies to polygons in which an even number of sides pass through fixed points, that is, to polygons of an *odd* number of sides.

Consider, in the first place, two points of a surface of the second order such that the line joining them passes through a given point. Let x, y, z, w be current coordinates[1], and let the equation of the surface be

$$(a, \dots)(x, y, z, w)^2 = 0,$$

and take for the coordinates of the two points on the surface x_1, y_1, z_1, w_1 and x_2, y_2, z_2, w_2, and for the coordinates of the fixed point $\alpha, \beta, \gamma, \delta$. Write for shortness

$$(a, \dots)(\alpha, \beta, \gamma, \delta)^2 = p,$$

$$(a, \dots)(\alpha, \beta, \gamma, \delta)(x_1, y_1, z_1, w_1) = q_1,$$

then the coordinates x_2, y_2, z_2, w_2 are determined by the very simple formulæ

$$x_2 = x_1 - \frac{2\alpha}{p} q_1,$$

$$y_2 = y_1 - \frac{2\beta}{p} q_1,$$

$$z_2 = z_1 - \frac{2\gamma}{p} q_1,$$

$$w_2 = w_1 - \frac{2\delta}{p} q_1.$$

[1] Strictly speaking, it is the ratios of these quantities, e.g. $x : w$, $y : w$, $z : w$, which are the coordinates, and consequently, even when the point is given, the values x, y, z, w are essentially indeterminate to a factor *près*. So that in assuming that a point is given, we should write $x : y : z : w = \alpha : \beta : \gamma : \delta$; and that when a point is obtained as the result of an analytical process, the conclusion is necessarily of the form just mentioned: but when this is once understood, the language of the text may be properly employed. It may be proper to explain here a notation made use of in the text: taking for greater simplicity the case of forms of two variables, $(l, m)(x, y)$ means $lx + my$; $(a, b, c)(x, y)^2$ means $ax^2 + 2bxy + cy^2$; $(a, b, c)(\xi, \eta)(x, y)$ means $a\xi x + b(\xi y + \eta x) + c\eta y$. The system of coefficients may frequently be indicated by a single coefficient only: thus in the text $(a, \dots)(x, y, z, w)^2$ stands for the most general quadratic function of four variables.

In fact, these values satisfy identically the equations

$$\left\|\begin{matrix} x_2, & y_2, & z_2, & w_2 \\ x_1, & y_1, & z_1, & w_1 \\ \alpha, & \beta, & \gamma, & \delta \end{matrix}\right\| = 0,$$

that is, the point (x_2, y_2, z_2, w_2) will be a point in the line joining (x_1, y_1, z_1, w_1) and $(\alpha, \beta, \gamma, \delta)$. Moreover,

$$\begin{aligned}(a, \ldots)(x_2, y_2, z_2, w_2)^2 = \;& (a, \ldots)(x_1, y_1, z_1, w_1)^2 \\ & - \frac{4q_1}{p}(a, \ldots)(\alpha, \beta, \gamma, \delta)(x_1, y_1, z_1, w_1) \\ & + \frac{4q_1^2}{p^2}(a, \ldots)(\alpha, \beta, \gamma, \delta)^2 \\ = \;& (a, \ldots)(x_1, y_1, z_1, w_1)^2 - \frac{4q_1}{p}q_1 + \frac{4q_1^2}{p^2}p,\end{aligned}$$

that is,

$$(a, \ldots)(x_2, y_2, z_2, w_2)^2 = (a, \ldots)(x_1, y_1, z_1, w_1)^2;$$

so that x_1, y_1, z_1, w_1 being a point on the surface, x_2, y_2, z_2, w_2 will be so too. The equation just found may be considered as expressing that the linear equations are a transformation of the quadratic form $(a, \ldots)(x, y, z, w)^2$ into itself. If in the system of linear equations the coefficients on the right-hand side were arranged square-wise, and the determinant formed by these quantities calculated, it would be found that the value of this determinant is -1. The transformation is on this account said to be *improper*. If in a system of linear equations for the transformation of the form into itself the determinant (which is necessarily $+1$ or else -1) be $+1$, the transformation is in this case said to be *proper*.

We have next to investigate the theory of the proper transformations of a quadratic form of four indeterminates into itself. This might be done for the absolutely general form by means of the theory recently established by M. Hermite, but it will be sufficient for the present purpose to consider the system of equations for the transformation of the form $x^2 + y^2 + z^2 + w^2$ into itself given by me some years since. (*Crelle*, vol. XXXII. [1846] p. 119, [52]) [1].

I proceed to establish (by M. Hermite's method) the formulæ for the particular case in question. The thing required is to find x_2, y_2, z_2, w_2 linear functions of x_1, y_1, z_1, w_1, such that

$$x_2^2 + y_2^2 + z_2^2 + w_2^2 = x_1^2 + y_1^2 + z_1^2 + w_1^2.$$

Write

$$x_1 + x_2 = 2\xi, \quad y_1 + y_2 = 2\eta, \quad z_1 + z_2 = 2\zeta, \quad w_1 + w_2 = 2\omega;$$

[1] It is a singular instance of the way in which different theories connect themselves together, that the formulæ in question were generalizations of Euler's formulæ for the rotation of a solid body, and also are formulæ which reappear in the theory of quaternions; the general formulæ cannot be established by any obvious generalization of the theory of quaternions.

then putting $x_2 = 2\xi - x_1$, &c., the proposed equation will be satisfied if only

$$\xi^2 + \eta^2 + \zeta^2 + \omega^2 = \xi x_1 + \eta y_1 + \zeta z_1 + \omega w_1,$$

which will obviously be the case if

$$\begin{aligned} x_1 &= \quad \xi + \nu\eta - \mu\zeta + a\omega, \\ y_1 &= -\nu\xi + \eta + \lambda\zeta + b\omega, \\ z_1 &= \quad \mu\xi - \lambda\eta + \zeta + c\omega, \\ w_1 &= -a\xi - b\eta - c\zeta + \omega, \end{aligned}$$

where λ, μ, ν, a, b, c are arbitrary.

Write for shortness

$$a\lambda + b\mu + c\nu = \phi,\ 1 + \lambda^2 + \mu^2 + \nu^2 + a^2 + b^2 + c^2 + \phi^2 = k,$$

then we have

$$\begin{aligned} k\xi &= (1 + \lambda^2 + b^2 + c^2)\, x_1 + (\lambda\mu - \nu - ab - c\phi)\, y_1 + (\nu\lambda + \mu - ca + b\phi)\, z_1 + (b\nu - c\mu - a - \lambda\phi) w_1, \\ k\eta &= (\lambda\mu + \nu - ab + c\phi)\, x_1 + (1 + \mu^2 + c^2 + a^2)\, y_1 + (\mu\nu - \lambda - bc + a\phi)\, z_1 + (c\lambda - a\nu - b - \mu\phi) w_1, \\ k\zeta &= (\nu\lambda - \mu - ca - b\phi)\, x_1 + (\mu\nu + \lambda - bc + a\phi)\, y_1 + (1 + \nu^2 + a^2 + b^2)\, z_1 + (a\mu - b\lambda - c - \nu\phi) w_1, \\ kw &= (b\nu - c\mu + a + \lambda\phi)\, x_1 + (c\lambda - a\nu + b + \mu\phi)\, y_1 + (a\mu - b\nu + c + \nu\phi)\, z_1 + (1 + \lambda^2 + \mu^2 + \nu^2)\, w_1; \end{aligned}$$

and from these we obtain

$$\begin{aligned} kx_2 &= (1 + \lambda^2 + b^2 + c^2 - \mu^2 - \nu^2 - a^2 - \phi^2)\, x_1 + 2\,(\lambda\mu - \nu - ab - c\phi)\, y_1 + 2\,(\nu\lambda + \mu - ca + b\phi)\, z_1 \\ &\qquad + 2\,(b\nu - c\mu - a - \lambda\phi)\, w_1, \\ ky_2 &= 2\,(\lambda\mu + \nu - ab + c\phi)\, x_1 + (1 + \mu^2 + c^2 + a^2 - \nu^2 - \lambda^2 - b^2 - \phi^2)\, y_1 + 2\,(\mu\nu - \lambda - bc - a\phi)\, z_1 \\ &\qquad + 2\,(c\lambda - a\nu - b - \mu\phi)\, w_1, \\ kz_2 &= 2\,(\nu\lambda - \mu - ca - b\phi)\, x_1 + 2\,(\mu\nu + \lambda - bc + a\phi)\, y_1 + (1 + \nu^2 + a^2 + b^2 - \lambda^2 - \mu^2 - c^2 - \phi^2)\, z_1 \\ &\qquad + 2\,(a\mu - b\lambda - c - \nu\phi)\, w_1, \\ kw_2 &= 2\,(b\nu - c\mu + a + \lambda\phi)\, x_1 + 2\,(c\lambda - a\nu + b + \mu\phi)\, y_1 + 2\,(a\mu - b\nu + c + \nu\phi)\, z_1 \\ &\qquad + (1 + \lambda^2 + \mu^2 + \nu^2 - a^2 - b^2 - c^2 - \phi^2)\, w_1, \end{aligned}$$

values which satisfy identically $x_2^2 + y_2^2 + z_2^2 + w_2^2 = x_1^2 + y_1^2 + z_1^2 + w_1^2$.

Dividing the linear equations by k, and forming with the coefficients on the right-hand side of the equation so obtained a determinant, the value of this determinant is $+1$; the transformation is consequently a *proper* one. And conversely, what is very important, *every* proper transformation may be exhibited under the preceding form[1].

[1] The nature of the reasoning by which this is to be established may be seen by considering the analogous relation for two variables. Suppose that x_1, y_1 are linear functions of x and y such that $x_1^2 + y_1^2 = x^2 + y^2$; then if $2\xi = x + x_1$, $2\eta = y + y_1$, ξ, η will be linear functions of x, y such that $\xi^2 + \eta^2 = \xi x + \eta y$, or $\xi(\xi - x) + \eta(\eta - y) = 0$; $\xi - x$ must be divisible either by η or else by $\eta - y$. On the former supposition, calling the quotient ν, we have $x = \xi - \nu\eta$, and thence $y = \nu\xi + \eta$, leading to a transformation such as is considered in the text, and which is a proper transformation; the latter supposition leads to an improper transformation. The given transformation, assumed to be proper, exists and cannot be obtained from the second supposition; it must therefore be obtainable from the first supposition, i.e. it is a transformation which may be exhibited under a form such as is considered in the text.

Next considering the equations connecting x, y, z, w with ξ, η, ζ, ω, we see that

$$\begin{aligned} x_1^2+y_1^2+z_1^2+w_1^2 = &(\quad \xi+\nu\eta-\mu\zeta+a\omega)^2 \\ &+(-\nu\xi+\eta+\lambda\zeta+b\omega)^2 \\ &+(\quad \mu\xi-\lambda\eta+\zeta+c\omega)^2 \\ &+(-a\xi-b\eta-c\zeta+\omega)^2. \end{aligned}$$

We are thus led to the discussion (in connexion with the question of the transformation into itself of the form $x^2+y^2+z^2+w^2$) of the new form

$$\begin{aligned} &(\quad x+\nu y-\mu z+aw)^2 \\ &+(-\nu x+y+\lambda z+bw)^2 \\ &+(\quad \mu x-\lambda y+z+cw)^2 \\ &+(-ax-by-cz+w)^2; \end{aligned}$$

or, as it may also be written,

$$(x^2+y^2+z^2+w^2)+(\nu y-\mu z+aw)^2+(\lambda z-\nu x+bw)^2+(\mu x-\lambda y+cw)^2+(ax+by+cz)^2.$$

Represent for a moment the forms in question by U, V, and consider the surfaces $U=0$, $V=0$. If we form from this the surface $V+qU=0$, and consider the discriminant of the function on the left-hand side, then putting for shortness

$$\kappa=\lambda^2+\mu^2+\nu^2+a^2+b^2+c^2,$$

this discriminant is

$$\{(q+1)^2+\kappa(q+1)+\phi^2\}^2,$$

which shows that the surfaces intersect in four lines. Suppose the discriminant vanishes; we have for the determination of q a quadratic equation, which may be written

$$q^2+(2+\kappa)q+K=0;$$

let the roots of this equation be $q_{,}$, $q_{,,}$; then each of the functions $q_{,}U+V$, $q_{,,}U+V$ will break up into linear factors, and we may write

$$q_{,}U+V=R_{,}S_{,},$$

$$q_{,,}U+V=R_{,,}S_{,,}.$$

(U and V are of course linear functions of $R_{,}S_{,}$ and $R_{,,}S_{,,}$) forms which put in evidence the fact of the two surfaces intersecting in four lines.

The equations

$$x_1+x_2=2\xi,\quad y_1+y_2=2\eta,\quad z_1+z_2=2\zeta,\quad w_1+w_2=2\omega,$$

show that the point $(\xi, \eta, \zeta, \omega)$ lies in the line joining the points (x_1, y_1, z_1, w_1) and (x_2, y_2, z_2, w_2); and to show that this line touches the surface $V=0$, it is only necessary to form the equation of the tangent plane at the point $(\xi, \eta, \zeta, \omega)$ of the surface in question; this is

$$(x+\nu y-\mu z+aw)(\xi+\nu\eta-\mu\zeta+a\omega)+\ldots=0;$$

or what is the same thing,

$$(x+\nu y-\mu z+aw)\,x_1+\ldots=0,$$

which is satisfied by writing (x_1, y_1, z_1, w_1) for (x, y, z, w), that is, the tangent plane of the surface contains the point (x_1, y_1, z_1, w_1). We see, therefore, that the line through (x_1, y_1, z_1, w_1) and (x_2, y_2, z_2, w_2) touches the surface $V=0$ at the point $(\xi, \eta, \zeta, \omega)$.

Write now

$$a'=\frac{-\lambda}{\phi},\quad b'=\frac{-\mu}{\phi},\quad c'=\frac{-\nu}{\phi},\quad \lambda'=\frac{-a}{\phi},\quad \mu'=\frac{-b}{\phi},\quad \nu'=\frac{-c}{\phi};$$

if we derive from the coordinates x_1, y_1, z_1, w_1, by means of these coefficients $a', b', c', \lambda', \mu', \nu'$, new coordinates in the same way as x_2, y_2, z_2, w_2 were derived by means of the coefficients $a, b, c, \lambda, \mu, \nu$, the coordinates so obtained are $-x_2, -y_2, -z_2, -w_2$, i.e. we obtain the *very same* point (x_2, y_2, z_2, w_2) by means of the coefficients $(a, b, c, \lambda, \mu, \nu)$, and by means of the coefficients $(a', b', c', \lambda', \mu', \nu')$. Call $\xi', \eta', \xi', \omega'$ what ξ, η, ζ, ω become when the second system of coefficients is substituted for the first; the point $\xi', \eta', \zeta', \omega'$ will be a point on the surface $V'=0$, where

$$V'=\phi^2(x^2+y^2+z^2+w^2)$$
$$+(-cy+bz-\lambda w)^2+(-az+cx-\mu w)^2+(-bx+ay-\nu w)^2+(-\lambda x-\mu y-\nu z)^2;$$

and since

$$V+V'=\kappa(x^2+y^2+z^2+w^2),$$

and $V=0$ intersects the surface $x^2+y^2+z^2+w^2=0$ in four lines, the surface $V'=0$ will also intersect this surface in the same four lines. And it is, moreover, clear that the line joining the points (x_1, y_1, z_1, w_1) and (x_2, y_2, z_2, w_2) touches the surface $V'=0$ in the point $(\xi', \eta', \zeta', \omega')$. We thus arrive at the theorem, that when two points of a surface of the second order are so connected that the coordinates of the one point are linear functions of the coordinates of the other point, and the transformation is a proper one, the line joining the two points touches two surfaces of the second order, each of them intersecting the given surface of the second order in the same four lines. Any two points so connected may be said to be corresponding points, or simply a pair. Suppose the four lines and also a single pair is given, it is not for the determination of the other pairs necessary to resort to the two auxiliary surfaces of the second order; it is only necessary to consider each point of the surface as determined by the two generating lines which pass through it; then considering first one point of the given pair, and the point the corresponding point to which has to be determined, take through each of these points a generating line, and take also two generating lines out of the given system of four lines, the four generating lines in question being all of them of the same set, these four generating lines intersecting either of the other two generating lines of the given system of four lines in four points. Imagine the same thing done with the other point of the given pair and the required point, we should have another system of four points (two of them of course identical with two of the points of the first-mentioned system of four points); these two systems must have their anharmonic ratios the same, a condition which enables the determination of the generating line in question through the required point: the other generating line through the required point is of course determined in the same manner, and thus the required point (i.e. the point corresponding to any point of the surface taken at pleasure) is determined by means of the two generating lines through such required point.

It is of course to be understood that the points of each pair belong to two distinct systems, and that the point belonging to the one system is not to be confounded or interchanged with the point belonging to the other system. Consider, now, a point of the surface, and the line joining such point with its corresponding point, but let the corresponding point itself be altogether dropped out of view. There are two directions in which we may pass along the surface to a consecutive point, in such manner that the line belonging to the point in question may be intersected by the line belonging to the consecutive point. We have thus upon the surface two series of curves, such that a curve of each series passes through a point chosen at pleasure on the surface. The lines belonging to the curves of the one series generate a series of developables, the edges of regression of which lie on one of the surfaces intersecting the surface of the second order in the four given lines; the lines belonging to the curves of the other series generate a series of developables, the edges of regression of which lie on the other of the surfaces intersecting the surface in the four given lines; the general nature of the system may be understood by considering the system of normals of a surface of the second order. Consider, now, the surface of the second order as given, and also the two surfaces of the second order intersecting it in the same four lines; from any point of the surface we may draw to the auxiliary surfaces four different tangents; but selecting any one of these, and considering the other point in which it intersects the surface as the point corresponding to the first-mentioned point, we may, as above, construct the entire system of corresponding points, and then the line joining any two corresponding points will be a tangent to the two auxiliary surfaces; the system of tangents so obtained may be called a system of *congruent* tangents. Now if we take upon the surface three points such that the first and second are corresponding points, and that the second and third are corresponding points, then it is obvious that the third and first are corresponding points;—observe that the two auxiliary surfaces for expressing the correspondence between the first and second point, those for the second and third point, and those for the third and first point, meet the surface, the two auxiliary surfaces of each pair in the same four lines, but that these systems of four lines are different for the different pairs of auxiliary surfaces. The same thing of course applies to any number of corresponding points. We have thus, finally, the theorem, if there be a polygon of $(m+1)$ sides inscribed in a surface of the second order, and the first side of the polygon constantly touches two surfaces of the second order, each of them intersecting the surface of the second order in the same four lines (and the side belong always to the same system of congruent tangents), and if the same property exists with respect to the second, third, &c.... and mth side of the polygon, then will the same property exist with respect to the $(m+1)$th side of the polygon.

We may add, that, instead of satisfying the conditions of the theorem, any two consecutive sides of the polygon, or the sides forming any number of pairs of consecutive sides, may pass each through a fixed point. This is of course only a particular case of the improper transformation of a surface of a second order into itself, a question which is not discussed in the present paper.

123.

ON THE GEOMETRICAL REPRESENTATION OF AN ABELIAN INTEGRAL.

[From the *Philosophical Magazine*, vol. VI. (1853), pp. 414—418.]

THE equation of a surface passing through the curve of intersection of the surfaces

$$x^2 + y^2 + z^2 + w^2 = 0,$$
$$ax^2 + by^2 + cz^2 + dw^2 = 0,$$

is of the form

$$\vartheta (x^2 + y^2 + z^2 + w^2) + ax^2 + by^2 + cz^2 + dw^2 = 0,$$

where ϑ is an arbitrary parameter. Suppose that the surface touches a given plane, we have for the determination of ϑ a cubic equation the roots of which may be considered as parameters defining the plane in question. Let one of the values of ϑ be considered equal to a given quantity k, the plane touches the surface

$$k (x^2 + y^2 + z^2 + w^2) + ax^2 + by^2 + cz^2 + dw^2 = 0,$$

and the other two values of ϑ may be considered as parameters defining the particular tangent plane, or what is the same thing, determining its point of contact with the surface.

Or more clearly, thus:—in order to determine the position of a point on the surface

$$k (x^2 + y^2 + z^2 + w^2) + ax^2 + by^2 + cz^2 + dw^2 = 0\,;$$

the tangent plane at the point in question is touched by two other surfaces

$$p (x^2 + y^2 + z^2 + w^2) + ax^2 + by^2 + cz^2 + dw^2 = 0,$$
$$q (x^2 + y^2 + z^2 + w^2) + ax^2 + by^2 + cz^2 + dw^2 = 0\,;$$

and, this being so, p and q are the parameters by which the point in question is determined. We may for shortness speak of the surface

$$k(x^2+y^2+z^2+w^2)+ax^2+by^2+cz^2+dw^2=0$$

as the surface (k). It is clear that we shall then have to speak of

$$x^2+y^2+z^2+w^2=0$$

as the surface (∞).

I consider now a chord of the surface (∞) touching the two surfaces (k) and (k'); and I take θ, ϕ as the parameters of the one extremity of this chord; (p, q) as the parameters of the point of contact with the surface (k); p', q' as the parameters of the point of contact with the surface (k'); and θ', ϕ' as the parameters of the other extremity of the chord; the points in question may therefore be distinguished as the points $(\infty\,;\,\theta, \phi)$, $(k\,;\,p, q)$, $(k'\,;\,p', q')$, and $(\infty\,;\,\theta', \phi')$. The coordinates of the point $(\infty\,;\,\theta, \phi)$ are given by

$$\begin{aligned} x : y : z : w = \; & \sqrt{(a+\theta)(a+\phi)} \div \sqrt{(a-b)(a-c)(a-d)} \\ : \; & \sqrt{(b+\theta)(b+\phi)} \div \sqrt{(b-c)(b-d)(b-a)} \\ : \; & \sqrt{(c+\theta)(c+\phi)} \div \sqrt{(c-d)(c-a)(c-b)} \\ : \; & \sqrt{(d+\theta)(d+\phi)} \div \sqrt{(d-a)(d-b)(d-c)}\,; \end{aligned}$$

those of the point $(k\,;\,p, q)$ by

$$\begin{aligned} x : y : z : w = \; & \sqrt{(a+p)(a+q)} \div \sqrt{(a-b)(a-c)(a-d)}\,\sqrt{a+k} \\ : \; & \sqrt{(b+p)(b+q)} \div \sqrt{(b-c)(b-d)(b-a)}\,\sqrt{b+k} \\ : \; & \sqrt{(c+p)(c+q)} \div \sqrt{(c-d)(c-a)(c-b)}\,\sqrt{c+k} \\ : \; & \sqrt{(d+p)(d+q)} \div \sqrt{(d-a)(d-b)(d-c)}\,\sqrt{d+k}\,; \end{aligned}$$

and similarly for the other two points.

Consider, in the first place, the chord in question as a tangent to the two surfaces (k) and (k'). It is clear that the tangent plane to the surface (k) at the point $(k\,;\,p, q)$ must contain the point $(k'\,;\,p', q')$, and *vice versâ*. Take for a moment ξ, η, ζ, ω as the coordinates of the point $(k\,;\,p, q)$, the equation of the tangent plane to (k) at this point is

$$\Sigma\,(a+k)\,\xi x = 0\,;$$

or substituting for $\xi,\ldots$ their values

$$\Sigma\left(x\sqrt{(a+p)(a+q)}\,\sqrt{a+k} \div \sqrt{(a-b)(a-c)(a-d)}\right)=0\,;$$

or taking for $x,\ldots$ the coordinates of the point (k', p', q'), we have for the conditions that this point may lie in the tangent plane in question,

$$\Sigma\left(\sqrt{(a+p)(a+q)}\sqrt{(a+p')(a+q')}\sqrt{(a+k)} \div \sqrt{(a+k')}(a-b)(a-c)(a-d)\right)=0;$$

or under a somewhat more convenient form we have

$$\Sigma\left((b-c)(c-d)(d-b)\sqrt{(a+p)(a+q)}\sqrt{(a+p')(a+q')}\frac{\sqrt{a+k}}{\sqrt{a+k'}}\right)=0,$$

for the condition in order that the point (k', p', q') may lie in the tangent plane at $(k;\ p,\ q)$ to the surface (k). Similarly, we have

$$\Sigma\left((b-c)(c-d)(d-b)\sqrt{(a+p)(a+q)}\sqrt{(a+p')(a+q')}\frac{\sqrt{a+k'}}{\sqrt{a+k}}\right)=0,$$

for the condition in order that the point (k, p, q) may lie in the tangent plane at $(k';\ p',\ q')$ to the surface (k'). The former of these two equations is equivalent to the system of equations

$$\sqrt{(a+p)(a+q)(a+p')(a+q')}\ \sqrt{\frac{a+k}{a+k'}}=\lambda+\mu a+\nu a^2,$$

$$\vdots$$

and the latter to the system of equations

$$\sqrt{(a+p)(a+q)(a+p')(a+q')}\sqrt{\frac{a+k'}{a+k}}=\lambda'+\mu' a+\nu' a^2;$$

$$\vdots$$

where in each system a is to be successively replaced by b, c, d, and where λ, μ, ν and λ', μ', ν' are indeterminate. Now dividing each equation of the one system by the corresponding equation in the other system, we see that the equation

$$\frac{x+k}{x+k'}=\frac{\lambda+\mu x+\nu x^2}{\lambda'+\mu' x+\nu' x^2}$$

is satisfied by the values a, b, c, d of x; and, therefore, since the equation in x is only of the third order, that the equation in question must be *identically* true. We may therefore write

$$\lambda+\mu x+\nu x^2=(\rho x+\sigma)(x+k),\quad \lambda'+\mu' x+\nu' x^2=(\rho x+\sigma)(x+k'),$$

and the two systems of equations become therefore equivalent to the single system,

$$\sqrt{(a+p)(a+q)(a+p')(a+q')}=(\rho a+\sigma)\sqrt{(a+k)(a+k')},$$

$$\sqrt{(b+p)(b+q)(b+p')(b+q')}=(\rho b+\sigma)\sqrt{(b+k)(b+k')},$$

$$\sqrt{(c+p)(c+q)(c+p')(c+q')}=(\rho c+\sigma)\sqrt{(c+k)(c+k')},$$

$$\sqrt{(d+p)(d+q)(d+p')(d+q')}=(\rho d+\sigma)\sqrt{(d+k)(d+k')},$$

a set of equations which may be represented by the single equation

$$\psi(x+p)(x+q)(x+p')(x+q')-(\rho x+\sigma)^2(x+k)(x+k')=\chi(x-a)(x-b)(x-c)(x-d),$$

where x is arbitrary; or what is the same thing, writing $-x$ instead of x,

$$\chi(x+a)(x+b)(x+c)(x+d)+(\rho x-\sigma)^2(x-k)(x-k')=\psi(x-p)(x-q)(x-p')(x-q').$$

Hence, putting

$$\Pi x=\int\frac{dx}{\sqrt{(x+a)(x+b)(x+c)(x+d)(x-k)(x-k')}},$$

$$\Pi_{,}x=\int\frac{xdx}{\sqrt{(x+a)(x+b)(x+c)(x+d)(x-k)(x-k')}},$$

we see that the algebraical equations between p, q; p', q' are equivalent to the transcendental equations

$$\Pi p\pm\Pi q\pm\Pi p'\pm\Pi q'=\text{const.}$$

$$\Pi_{,}p\pm\Pi_{,}q\pm\Pi_{,}p'\pm\Pi_{,}q'=\text{const.}$$

The algebraical equations which connect θ, ϕ with p, q; p', q', may be exhibited under several different forms; thus, for instance, considering the point $(\infty\,;\,\theta,\,\phi)$ as a point in the line joining $(k;\,p,\,q)$ and $(k';\,p',\,q')$, we must have

$$\left\|\begin{matrix}\sqrt{(a+p)(a+q)}\div\sqrt{a+k}, & \sqrt{(b+p)(b+q)}\div\sqrt{b+k},\ldots\\ \sqrt{(a+p')(a+q')}\div\sqrt{a+k'}, & \sqrt{(b+p')(b+q')}\div\sqrt{b+k'}\\ \sqrt{(a+\theta)(a+\phi)}, & \sqrt{(b+\theta)(b+\phi)}\end{matrix}\right\|=0,$$

i.e. the determinants formed by selecting any three of the four columns must vanish; the equations so obtained are equivalent (as they should be) to two independent equations.

Or, again, by considering $(\infty\,;\,\theta,\,\phi)$ first as a point in the tangent plane at $(k;\,p,\,q)$ to the surface (k), and then as a point in the tangent plane at $(k';\,p',\,q')$ to the surface (k'), we obtain

$$\Sigma\left((b-c)(c-d)(d-b)\sqrt{(a+p)(a+q)}\sqrt{(a+k)}\sqrt{(a+\theta)(a+\phi)}\right)=0,$$

$$\Sigma\left((b-c)(c-d)(d-b)\sqrt{(a+p')(a+q')}\sqrt{(a+k')}\sqrt{(a+\theta)(a+\phi)}\right)=0.$$

Or, again, we may consider the line joining $(\infty\,;\,\theta,\,\phi)$ and $(k;\,p,\,q)$ or $(k';\,p',\,q')$, as touching the surfaces (k) and (k'); the formulæ for this purpose are readily obtained by means of the lemma,—

"The condition in order that the line joining the points (ξ, η, ζ, ω) and ($\xi', \eta', \zeta', \omega'$) may touch the surface

$$\mathrm{a}x^2 + \mathrm{b}y^2 + \mathrm{c}z^2 + \mathrm{d}w^2 = 0$$

is

$$\Sigma \mathrm{ab}\,(\xi\eta' - \xi'\eta)^2 = 0,$$

the summation extending to the binary combinations of a, b, c, d."

But none of all these formulæ appear readily to conduct to the transcendental equations connecting θ, ϕ with p, q; p', q'. Reasoning from analogy, it would seem that there exist transcendental equations

$$\pm\,\Pi\theta \pm \Pi\phi \pm \Pi p \pm \Pi p' = \text{const.}$$

$$\pm\,\Pi_{\prime}\theta \pm \Pi_{\prime}\phi \pm \Pi_{\prime}p \pm \Pi_{\prime}p' = \text{const.},$$

or the similar equations containing q, q', instead of p, p', into which these are changed by means of the transcendental equations between p, q, p', q'. If in these equations we write θ', ϕ' instead of θ, ϕ, it would appear that the functions Πp, $\Pi p'$, $\Pi_{\prime}p$, $\Pi_{\prime}p'$ may be eliminated, and that we should obtain equations such as

$$\pm\,\Pi\theta \pm \Pi\phi \pm \Pi\theta' \pm \Pi\phi' = \text{const.}$$

$$\pm\,\Pi_{\prime}\theta \pm \Pi_{\prime}\phi \pm \Pi_{\prime}\theta' \pm \Pi_{\prime}\phi' = \text{const.}$$

to express the relations that must exist between the parameters θ, ϕ and θ', ϕ' of the extremities of a chord of the surface

$$x^2 + y^2 + z^2 + w^2 = 0,$$

in order that this chord may touch the two surfaces

$$k\,(x^2 + y^2 + z^2 + w^2) + ax^2 + by^2 + cz^2 + dw^2 = 0,$$

$$k'\,(x^2 + y^2 + z^2 + w^2) + ax^2 + by^2 + cz^2 + dw^2 = 0.$$

The quantities k, k', it will be noticed, enter into the radical of the integrals Πx, $\Pi_{\prime}x$. This is a very striking difference between the present theory and the analogous theory relating to conics, and leads, I think, to the inference that the theory of the polygon inscribed in a conic, *and the sides of which touch conics intersecting the conic in the same four points,* cannot be extended to surfaces in such manner as one might be led to suppose from the extension to surfaces of the much simpler theory of the polygon inscribed in a conic, *and the sides of which touch conics having double contact with the conic.* (See my paper "On the Homographic Transformation of a surface of the second order into itself," [122]).

The preceding investigations are obviously very incomplete; but the connexion which they point out between the geometrical question and the Abelian integral involving the root of a function of the sixth order, may I think be of service in the theory of these integrals.

124.

ON A PROPERTY OF THE CAUSTIC BY REFRACTION OF THE CIRCLE.

[From the *Philosophical Magazine*, vol. VI. (1853), pp. 427—431.]

M. ST LAURENT has shown (*Gergonne*, vol. XVIII. [1827] p. 1), that in certain cases the caustic by refraction of a circle is identical with the caustic of reflexion of a circle (the reflecting circle and radiant point being, of course, properly chosen), and a very elegant demonstration of M. St Laurent's theorems is given by M. Gergonne in the same volume, p. 48. A similar method may be employed to demonstrate the more general theorem, that the same caustic by refraction of a circle may be considered as arising from *six* different systems of a radiant point, circle, and index of refraction. The demonstration is obtained by means of the secondary caustic, which is (as is well known) an oval of Descartes. Such oval has three foci, any one of which may be taken for the radiant point: whichever be selected, there can always be found two corresponding circles and indices of refraction. The demonstration is as follows:—

Let c be the radius of the refracting circle, μ the index of refraction; and taking the centre of the circle as origin, let ξ, η be the coordinates of the radiant point, the secondary caustic is the envelope of the circle

$$\mu^2(\overline{x-\alpha}^2+\overline{y-\beta}^2)-(\overline{\xi-\alpha}^2+\overline{\eta-\beta}^2)=0,$$

where α, β are parameters which vary subject to the condition

$$\alpha^2+\beta^2-c^2=0;$$

the equation of the variable circle may be written

$$\{\mu^2(x^2+y^2+c^2)-(\xi^2+\eta^2+c^2)\}-2(\mu^2x-\xi)\alpha-2(\mu^2y-\eta)\beta=0,$$

which is of the form

$$C + A\alpha + B\beta = 0;$$

the envelope is therefore

$$C^2 = c^2 (A^2 + B^2).$$

Hence substituting, we have for the equation of the envelope, i.e. for the secondary caustic,

$$\{\mu^2 (x^2 + y^2 + c^2) - (\xi^2 + \eta^2 + c^2)\}^2 = 4c^2 \{(\mu^2 x - \xi)^2 + (\mu^2 y - \eta)^2\},$$

which may also be written

$$\{\mu^2 (x^2 + y^2 - c^2) - (\xi^2 + \eta^2 - c^2)\}^2 = 4c^2\mu^2 (\overline{x - \xi}^2 + \overline{y - \eta}^2);$$

and this may perhaps be considered as the standard form.

To show that this equation belongs to a Descartes' oval, suppose for greater convenience $\eta = 0$, and write

$$\mu^2 (x^2 + y^2 - c^2) - \xi^2 + c^2 = 2c\mu \sqrt{(x - \xi)^2 + y^2};$$

multiplying this equation by $1 - \dfrac{1}{\mu^2}$, and adding to each side $c^2 \left(\mu - \dfrac{1}{\mu}\right)^2 + (x - \xi)^2 + y^2$, we have

$$\left(1 - \frac{1}{\mu^2}\right) \{\mu^2 (x^2 + y^2 - c^2) - \xi^2 + c^2\} + (x - \xi)^2 + y^2 + c^2 \left(\mu - \frac{1}{\mu}\right)^2$$

$$= (x - \xi)^2 + y^2 + 2c \left(\mu - \frac{1}{\mu}\right) \sqrt{(x - \xi)^2 + y^2} + c^2 \left(\mu - \frac{1}{\mu}\right)^2;$$

or reducing

$$\mu^2 \left\{\left(x - \frac{\xi}{\mu^2}\right)^2 + y^2\right\} = \left\{\sqrt{(x - \xi)^2 + y^2} + c\left(\mu - \frac{1}{\mu}\right)\right\}^2;$$

again, multiplying the same equation by $\dfrac{1}{\mu^2}\left(1 - \dfrac{c^2}{\xi^2}\right)$, and adding to each side

$$\frac{\xi^2}{\mu^2}\left(1 - \frac{c^2}{\xi^2}\right)^2 + \frac{c^2}{\xi^2}(\overline{x - \xi}^2 + y^2),$$

we have

$$\frac{1}{\mu^2}\left(1 - \frac{c^2}{\xi^2}\right) \{\mu^2 (x^2 + y^2 - c^2) - \xi^2 + c^2\} + \frac{c^2}{\xi^2}(\overline{x - \xi}^2 + y^2) + \frac{\xi^2}{\mu^2}\left(1 - \frac{c^2}{\xi^2}\right)^2$$

$$= \frac{c^2}{\xi^2}((x - \xi)^2 + y^2) + \frac{2c}{\mu}\left(1 - \frac{c^2}{\xi^2}\right) \sqrt{(x - \xi)^2 + y^2} + \frac{\xi^2}{\mu^2}\left(1 - \frac{c^2}{\xi^2}\right)^2;$$

or reducing,

$$\left(x - \frac{c^2}{\xi}\right)^2 + y^2 = \left\{\frac{c}{\xi}\sqrt{(x - \xi)^2 + y^2} + \frac{\xi}{\mu}\left(1 - \frac{c^2}{\xi^2}\right)\right\}^2.$$

Hence, extracting the square roots of each side of the equations thus found, we have the equation of the secondary caustic in either of the forms

$$\sqrt{\left(x-\frac{\xi}{\mu^2}\right)^2+y^2}=\frac{1}{\mu}\sqrt{(x-\xi)^2+y^2}+\frac{c}{\mu}\left(\mu-\frac{1}{\mu}\right),$$

$$\sqrt{\left(x-\frac{c^2}{\xi}\right)^2+y^2}=\frac{c}{\xi}\sqrt{(x-\xi)^2+y^2}+\frac{1}{\mu}\left(\xi-\frac{c^2}{\xi}\right);$$

to which are to be joined

$$\sqrt{\left(x-\frac{c^2}{\xi}\right)^2+y^2}=\frac{c\mu}{\xi}\sqrt{\left(x-\frac{\xi}{\mu^2}\right)^2+y^2}+\frac{\xi}{\mu}-\frac{c^2\mu}{\xi},$$

$$c\left(\mu-\frac{1}{\mu}\right)\sqrt{\left(x-\frac{c^2}{\xi}\right)^2+y^2}+\left(-\xi+\frac{c^2}{\xi}\right)\sqrt{\left(x-\frac{\xi}{\mu^2}\right)^2+y^2}+\left(\frac{\xi}{\mu}-\frac{c^2\mu}{\xi}\right)\sqrt{(x-\xi)^2+y^2}=0.$$

Write successively,

$$\xi'=\xi\,,\quad c'=c\,,\quad \mu'=\mu\,,\qquad (1)$$

$$\xi'=\frac{c^2}{\xi},\quad c'=\frac{c}{\mu},\quad \mu'=\frac{c}{\xi}\,,\qquad (\alpha)$$

$$\xi'=\frac{\xi}{\mu^2},\quad c'=\frac{c}{\mu},\quad \mu'=\frac{1}{\mu}\,,\qquad (\beta)$$

$$\xi'=\xi\,,\quad c'=\frac{\xi}{\mu},\quad \mu'=\frac{\xi}{c}\,,\qquad (\gamma)$$

$$\xi'=\frac{c^2}{\xi},\quad c'=c\,,\quad \mu'=\frac{c\mu}{\xi},\qquad (\delta)$$

$$\xi'=\frac{\xi}{\mu^2},\quad c'=\frac{\xi}{\mu},\quad \mu'=\frac{\xi}{c\mu}\,;\qquad (\epsilon)$$

or, what is the same thing,

$$\xi=\xi'\,,\quad c=c'\,,\quad \mu=\mu'\,,\qquad (1)$$

$$\xi=\frac{\xi'}{\mu'^2},\quad c=\frac{\xi'}{\mu'},\quad \mu=\frac{\xi'}{c'\mu'}\,,\qquad (\alpha)$$

$$\xi=\frac{\xi'}{\mu'^2},\quad c=\frac{c'}{\mu'},\quad \mu=\frac{1}{\mu'}\,,\qquad (\beta)$$

$$\xi=\xi'\,,\quad c=\frac{\xi'}{\mu'},\quad \mu=\frac{\xi'}{c'}\,,\qquad (\gamma)$$

$$\xi=\frac{c'^2}{\xi'},\quad c=c'\,,\quad \mu=\frac{c'\mu'}{\xi'}\,,\qquad (\delta)$$

$$\xi=\frac{c'^2}{\xi'},\quad c=\frac{c'}{\mu'},\quad \mu=\frac{c'}{\xi'}\,;\qquad (\epsilon)$$

or, again,

$$\begin{array}{llll}
\xi' = \xi\ , & \dfrac{c'^2}{\xi'} = \dfrac{c^2}{\xi}, & \dfrac{\xi'}{\mu'^2} = \dfrac{\xi}{\mu^2}, & (1) \\[2ex]
\xi' = \dfrac{c^2}{\xi}\ , & \dfrac{c'^2}{\xi'} = \xi\ , & \dfrac{\xi'}{\mu'^2} = \xi\ , & (\alpha) \\[2ex]
\xi' = \dfrac{\xi}{\mu^2}, & \dfrac{c'^2}{\xi'} = \dfrac{c^2}{\xi}, & \dfrac{\xi'}{\mu'^2} = \xi\ , & (\beta) \\[2ex]
\xi' = \xi\ , & \dfrac{c'^2}{\xi'} = \dfrac{\xi}{\mu^2}, & \dfrac{\xi'}{\mu'^2} = \dfrac{c^2}{\xi}, & (\gamma) \\[2ex]
\xi' = \dfrac{c^2}{\xi}\ , & \dfrac{c'^2}{\xi'} = \xi\ , & \dfrac{\xi'}{\mu^2} = \dfrac{\xi}{\mu^2}, & (\delta) \\[2ex]
\xi' = \dfrac{\xi}{\mu^2}, & \dfrac{c'^2}{\xi'} = \xi\ , & \dfrac{\xi'}{\mu'^2} = \dfrac{c^2}{\xi}\ ; & (\epsilon)
\end{array}$$

then, whichever system of values of ξ', c', μ' be substituted for ξ, c, μ, we have in each case identically the same secondary caustic, the effect of the substitution being simply to interchange the different forms of the equation; and we have therefore identically the same caustic. By writing

$$\begin{aligned}(\xi',\ c',\ \mu') &= \quad (\xi,\ c,\ \mu)\\ &= \alpha\,(\xi,\ c,\ \mu),\\ &\quad \&c.,\end{aligned}$$

α, β, γ, δ, ϵ will be functional symbols, such as are treated of in my paper "On the Theory of Groups as depending on the symbolic equation $\theta^n = 1$," [125], and it is easy to verify the equations

$$\begin{aligned}
1 &= \alpha\beta = \beta\alpha = \gamma^2 = \delta^2 = \epsilon^2,\\
\alpha &= \beta^2 = \delta\gamma = \epsilon\delta = \gamma\epsilon,\\
\beta &= \alpha^2 = \epsilon\gamma = \gamma\delta = \delta\epsilon,\\
\gamma &= \delta\alpha = \epsilon\beta = \beta\delta = \alpha\epsilon,\\
\delta &= \epsilon\alpha = \gamma\beta = \alpha\gamma = \beta\epsilon,\\
\epsilon &= \gamma\alpha = \delta\beta = \beta\gamma = \alpha\delta.
\end{aligned}$$

Suppose, for example, $\xi = -c$, i.e. let the radiant point be in the circumference; then in the fourth system $\xi' = -c$, $c' = -\dfrac{c}{\mu}$, (or, since c' is the radius of a circle, this radius may be taken $\dfrac{c}{\mu}$), $\mu' = -1$, or the new system is a reflecting system. This is one of M. St Laurent's theorems, viz.

THEOREM. The caustic by refraction of a circle when the radiant point is on the circumference, is the caustic by reflexion for the same radiant point, and a concentric circle the radius of which is the radius of the first circle divided by the index of refraction.

Again, if $\xi = -c\mu$, the fifth system gives $\xi' = \frac{c^2}{\xi'}$, $c' = c$, $\mu' = -1$, or the new system is in this case also a reflecting system. This is the other of M. St Laurent's theorems, viz.:—

THEOREM. The caustic by refraction of a circle when the distance of the radiant point from the centre is equal to the radius of the circle multiplied by the index of refraction, is the caustic by reflexion of the same circle for a radiant point which is the image of the first radiant point.

Of course it is to be understood that the image of a point means a point whose distance from the centre = square of radius ÷ distance.

2 *Stone Buildings, Nov.* 2, 1853.

125.

ON THE THEORY OF GROUPS, AS DEPENDING ON THE SYMBOLIC EQUATION $\theta^n = 1$.

[From the *Philosophical Magazine*, vol. VII. (1854), pp. 40—47.]

LET θ be a symbol of operation, which may, if we please, have for its operand, not a single quantity x, but a system $(x, y, \ldots)$, so that

$$\theta(x, y, \ldots) = (x', y', \ldots),$$

where $x', y', \ldots$ are any functions whatever of $x, y, \ldots$, it is not even necessary that $x', y', \ldots$ should be the same in number with $x, y, \ldots$. In particular x', y', &c. may represent a permutation of x, y, &c., θ is in this case what is termed a substitution; and if, instead of a set $x, y, \ldots$, the operand is a single quantity x, so that $\theta x = x' = fx$, θ is an ordinary functional symbol. It is not necessary (even if this could be done) to attach any meaning to a symbol such as $\theta \pm \phi$, or to the symbol 0, nor consequently to an equation such as $\theta = 0$, or $\theta \pm \phi = 0$; but the symbol 1 will naturally denote an operation which (either generally or in regard to the particular operand) leaves the operand unaltered, and the equation $\theta = \phi$ will denote that the operation θ is (either generally or in regard to the particular operand) equivalent to ϕ, and of course $\theta = 1$ will in like manner denote the equivalence of the operation θ to the operation 1. A symbol $\theta\phi$ denotes the compound operation, the performance of which is equivalent to the performance, first of the operation ϕ, and then of the operation θ; $\theta\phi$ is of course in general different from $\phi\theta$. But the symbols $\theta, \phi, \ldots$ are in general such that $\theta . \phi\chi = \theta\phi . \chi$, &c., so that $\theta\phi\chi$, $\theta\phi\chi\omega$, &c. have a definite signification independent of the particular mode of compounding the symbols; this will be the case even if the functional operations involved in the symbols θ, ϕ, &c. contain parameters such as the quaternion imaginaries i, j, k; but not if these functional operations contain parameters such as the imaginaries which enter into the theory of octaves, &c., and for which, e.g. $\alpha . \beta\gamma$ is something different from $\alpha\beta . \gamma$, a supposition which is altogether excluded from the present paper. The order of the factors of a product $\theta\phi\chi\ldots$ must of course be attended to, since even in the case of a product of two factors the order is material; it is very convenient to speak of the symbols θ, $\phi\ldots$ as the first or furthest, second, third, &c., and last or nearest factor. What precedes may be almost entirely summed up in the remark, that the distributive law has no application to the symbols $\theta\phi\ldots$; and that these symbols are not in general convertible, but are associative. It is easy to see that $\theta^0 = 1$, and that the index law $\theta^m . \theta^n = \theta^{m+n}$, holds for all positive or negative integer values, not excluding 0. It should be noticed also, that if $\theta = \phi$, then, whatever the symbols α, β may be, $\alpha\theta\beta = \alpha\phi\beta$, and conversely.

A set of symbols,

$$1,\ \alpha,\ \beta, \ldots$$

all of them different, and such that the product of any two of them (no matter in what order), or the product of any one of them into itself, belongs to the set, is said to be a *group*[1]. It follows that if the entire group is multiplied by any one of the symbols, either as further or nearer factor, the effect is simply to reproduce the group; or what is the same thing, that if the symbols of the group are multiplied together so as to form a table, thus:

Further factors

Nearer factors	1	α	β	..
1	1	α	β	..
α	α	α^2	$\beta\alpha$	
β	β	$\alpha\beta$	β^2	
	:			

that as well each line as each column of the square will contain all the symbols $1, \alpha, \beta, \ldots$. It also follows that the product of any number of the symbols, with or without repetitions, and in any order whatever, is a symbol of the group. Suppose that the group

$$1,\ \alpha,\ \beta,\ \ldots$$

contains n symbols, it may be shown that each of these symbols satisfies the equation

$$\theta^n = 1\,;$$

so that a group may be considered as representing a system of roots of this symbolic binomial equation. It is, moreover, easy to show that if any symbol α of the group

[1] The idea of a group as applied to permutations or substitutions is due to Galois, and the introduction of it may be considered as marking an epoch in the progress of the theory of algebraical equations.

satisfies the equation $\theta^r = 1$, where r is less that n, then that r must be a submultiple of n; it follows that when n is a prime number, the group is of necessity of the form

$$1, \alpha, \alpha^2, \ldots \alpha^{n-1}, (\alpha^n = 1);$$

and the same may be (but is not necessarily) the case, when n is a composite number. But whether n be prime or composite, the group, *assumed to be of the form in question*, is in every respect analogous to the system of the roots of the ordinary binomial equation $x^n - 1 = 0$; thus, when n is prime, all the roots (except the root 1) are prime roots; but when n is composite, there are only as many prime roots as there are numbers less than n and prime to it, &c.

The distinction between the theory of the symbolic equation $\theta^n = 1$, and that of the ordinary equation $x^n - 1 = 0$, presents itself in the very simplest case, $n = 4$. For, consider the group

$$1, \alpha, \beta, \gamma,$$

which are a system of roots of the symbolic equation

$$\theta^4 = 1.$$

There is, it is clear, at least one root β, such that $\beta^2 = 1$; we may therefore represent the group thus,

$$1, \alpha, \beta, \alpha\beta, (\beta^2 = 1);$$

then multiplying each term by α as further factor, we have for the group $1, \alpha^2, \alpha\beta, \alpha^2\beta$, so that α^2 must be equal either to β or else to 1. In the former case the group is

$$1, \alpha, \alpha^2, \alpha^3, (\alpha^4 = 1),$$

which is analogous to the system of roots of the ordinary equation $x^4 - 1 = 0$. For the sake of comparison with what follows, I remark, that, representing the last-mentioned group by

$$1, \alpha, \beta, \gamma,$$

we have the table

	1,	α,	β,	γ
1	1	α	β	γ
α	α	β	γ	1
β	β	γ	1	α
γ	γ	1	α	β

If, on the other hand, $\alpha^2=1$, then it is easy by similar reasoning to show that we must have $\alpha\beta=\beta\alpha$, so that the group in the case is

$$1,\ \alpha,\ \beta,\ \alpha\beta,\ (\alpha^2=1,\ \beta^2=1,\ \alpha\beta=\beta\alpha);$$

or if we represent the group by

$$1,\ \alpha,\ \beta,\ \gamma,$$

we have the table

	1	α	β	γ
1	1	α	β	γ
α	α	1	γ	β
β	β	γ	1	α
γ	γ	β	α	1

,

or, if we please, the symbols are such that

$$\begin{aligned} \alpha^2 &= \beta^2 = \gamma^2 = 1, \\ \alpha &= \beta\gamma = \gamma\beta, \\ \beta &= \gamma\alpha = \alpha\beta, \\ \gamma &= \alpha\beta = \beta\alpha; \end{aligned}$$

[and we have thus a group essentially distinct from that of the system of roots of the ordinary equation $x^4-1=0$].

Systems of this form are of frequent occurrence in analysis, and it is only on account of their extreme simplicity that they have not been expressly remarked. For instance, in the theory of elliptic functions, if n be the parameter, and

$$\alpha(n)=\frac{c^2}{n}\quad \beta(n)=-\frac{c^2+n}{1+n}\quad \gamma(n)=-\frac{c^2(1+n)}{c^2+n},$$

then α, β, γ form a group of the species in question. So in the theory of quadratic forms, if

$$\begin{aligned} \alpha(a,\ b,\ c) &= (c,\ \ \ b,\ a) \\ \beta(a,\ b,\ c) &= (a,\ -b,\ c) \\ \gamma(a,\ b,\ c) &= (c,\ -b,\ a); \end{aligned}$$

although, indeed, in this case (treating forms which are properly equivalent as identical) we have $\alpha=\beta$, and therefore $\gamma=1$, in which point of view the group is simply a group of two symbols 1, α, $(\alpha^2=1)$.

Again, in the theory of matrices, if I denote the operation of inversion, and tr that of transposition, (I do not stop to explain the terms as the example may be passed over), we may write

$$\alpha = I, \quad \beta = \text{tr}, \quad \gamma = I\,.\,\text{tr} = \text{tr}\,.\,I.$$

I proceed to the case of a group of six symbols,

$$1, \ \alpha, \ \beta, \ \gamma, \ \delta, \ \epsilon,$$

which may be considered as representing a system of roots of the symbolic equation

$$\theta^6 = 1.$$

It is in the first place to be shown that there is at least one root which is a prime root of $\theta^3 = 1$, or (to use a simpler expression) a root having the index 3. It is clear that if there were a prime root, or root having the index 6, the square of this root would have the index 3, it is therefore only necessary to show that it is impossible that *all* the roots should have the index 2. This may be done by means of a theorem which I shall for the present assume, viz. that if among the roots of the symbolic equation $\theta^n = 1$, there are contained a system of roots of the symbolic equation $\theta^p = 1$ (or, in other words, if among the symbols forming a group of the order there are contained symbols forming a group of the order p), then p is a submultiple of n. In the particular case in question, a group of the order 4 cannot form part of the group of the order 6. Suppose, then, that γ, δ are two roots of $\theta^6 = 1$, having each of them the index 2; then if $\gamma\delta$ had also the index 2, we should have $\gamma\delta = \delta\gamma$; and 1, γ, δ, $\delta\gamma$, which is part of the group of the order 6, would be a group of the order 4. It is easy to see that $\gamma\delta$ must have the index 3, and that the group is, in fact, 1, $\gamma\delta$, $\delta\gamma$, γ, δ, $\gamma\delta\gamma$, which is, in fact, one of the groups to be presently obtained; I prefer commencing with the assumption of a root having the index 3. Suppose that α is such a root, the group must clearly be of the form

$$1, \ \alpha, \ \alpha^2, \ \gamma, \ \alpha\gamma, \ \alpha^2\gamma, \ (\alpha^3 = 1);$$

and multiplying the entire group by γ as nearer factor, it becomes γ, $\alpha\gamma$, $\alpha^2\gamma$, γ^2, $\alpha\gamma^2$, $\alpha^2\gamma^2$; we must therefore have $\gamma^2 = 1$, α, or α^2. But the supposition $\gamma^2 = \alpha^2$ gives $\gamma^4 = \alpha^4 = \alpha$, and the group is in this case 1, γ, γ^2, γ^3, γ^4, γ^5 $(\gamma^6 = 1)$; and the supposition $\gamma^2 = \alpha$ gives also this same group. It only remains, therefore, to assume $\gamma^2 = 1$; then we must have either $\gamma\alpha = \alpha\gamma$ or else $\gamma\alpha = \alpha^2\gamma$. The former assumption leads to the group

$$1, \ \alpha, \ \alpha^2, \ \gamma, \ \alpha\gamma, \ \alpha^2\gamma, \ (\alpha^3 = 1, \ \gamma^2 = 1, \ \gamma\alpha = \alpha\gamma),$$

which is, in fact, analogous to the system of roots of the ordinary equation $x^6 - 1 = 0$; and by putting $\alpha\gamma = \lambda$, might be exhibited in the form 1, λ, λ^2, λ^3, λ^4, λ^5, $(\lambda^6 = 1)$, under which this system has previously been considered. The latter assumption leads to the group

$$1, \ \alpha, \ \alpha^2, \ \gamma, \ \alpha\gamma, \ \alpha^2\gamma, \ (\alpha^3 = 1, \ \gamma^2 = 1, \ \gamma\alpha = \alpha^2\gamma),$$

and we have thus two, and only two, essentially distinct forms of a group of six.

If we represent the first of these two forms, viz. the group

$$1, \ \alpha, \ \alpha^2, \ \gamma, \ \alpha\gamma, \ \alpha^2\gamma, \ (\alpha^3 = 1, \ \gamma^2 = 1, \ \gamma\alpha = \alpha\gamma)$$

by the general symbols

$$1,\ \alpha,\ \beta,\ \gamma,\ \delta,\ \epsilon,$$

we have the table

	1,	α,	β,	γ,	δ,	ε
1	1	α	β	γ	δ	ε
α	α	β	γ	δ	ε	1
β	β	γ	δ	ε	1	α
γ	γ	δ	ε	1	α	β
δ	δ	ε	1	α	β	γ
ε	ε	1	α	β	γ	δ

;

while if we represent the second of these two forms, viz. the group

$$1,\ \alpha,\ \alpha^2,\ \gamma,\ \alpha\gamma,\ \alpha^2\gamma,\ (\alpha^3=1,\ \gamma^2=1,\ \gamma\alpha=\alpha^2\gamma),$$

by the same general symbols

$$1,\ \alpha,\ \beta,\ \gamma,\ \delta,\ \epsilon,$$

we have the table

	1	α	β	γ	δ	ε
1	1	α	β	γ	δ	ε
α	α	β	1	ε	γ	δ
β	β	1	α	δ	ε	γ
γ	γ	δ	ε	1	α	β
δ	δ	ε	γ	β	1	α
ε	ε	γ	δ	α	β	1

;

or, what is the same thing, the system of equations is

$$\begin{aligned}
1 &= \beta\alpha = \alpha\beta = \gamma^2 = \delta^2 = \epsilon^2, \\
\alpha &= \beta^2 = \delta\gamma = \epsilon\delta = \gamma\epsilon, \\
\beta &= \alpha^2 = \epsilon\gamma = \gamma\delta = \delta\epsilon, \\
\gamma &= \delta\alpha = \epsilon\beta = \beta\delta = \alpha\epsilon, \\
\delta &= \epsilon\alpha = \gamma\beta = \alpha\gamma = \beta\epsilon, \\
\epsilon &= \gamma\alpha = \delta\beta = \beta\gamma = \alpha\delta.
\end{aligned}$$

An instance of a group of this kind is given by the permutation of three letters; the group

$$1,\ \alpha,\ \beta,\ \gamma,\ \delta,\ \epsilon$$

may represent a group of substitutions as follows:—

$$\begin{matrix} abc, & cab, & bca, & acb, & cba, & bac \\ abc & abc & abc & abc & abc & abc. \end{matrix}$$

Another singular instance is given by the optical theorem proved in my paper "On a property of the Caustic by refraction of a Circle, [124]."

It is, I think, worth noticing, that if, instead of considering α, β, &c. as symbols of operation, we consider them as quantities (or, to use a more abstract term, 'cogitables') such as the quaternion imaginaries; the equations expressing the existence of the group are, in fact, the equations defining the meaning of the product of two complex quantities of the form

$$w + a\alpha + b\beta + \dots;$$

thus, in the system just considered,

$$(w + a\alpha + b\beta + c\gamma + d\delta + e\epsilon)(w' + a'\alpha + b'\beta + c'\gamma + d'\delta + e'\epsilon) = W + A\alpha + B\beta + C\gamma + D\delta + E\epsilon,$$

where

$$\begin{aligned}
W &= ww' + ab' + a'b + cc' + dd' + ee', \\
A &= wa' + w'a + bb' + dc' + ed' + ce', \\
B &= wb' + w'b + aa' + ec' + cd' + de', \\
C &= wc' + w'c + da' + eb' + bd' + ae', \\
D &= wd' + w'd + ea' + cb' + ac' + be', \\
E &= we' + w'e + ca' + db' + bc' + ad'.
\end{aligned}$$

It does not appear that there is in this system anything analogous to the modulus $w^2 + x^2 + y^2 + z^2$, so important in the theory of quaternions.

I hope shortly to resume the subject of the present paper, which is closely connected, not only with the theory of algebraical equations, but also with that of the composition of quadratic forms, and the 'irregularity' in certain cases of the determinants of these forms. But I conclude for the present with the following two examples of groups of higher orders. The first of these is a group of eighteen, viz.

$$1,\ \alpha,\ \beta,\ \gamma,\ \alpha\beta,\ \beta\alpha,\ \alpha\gamma,\ \gamma\alpha,\ \beta\gamma,\ \gamma\beta,\ \alpha\beta\gamma,\ \beta\gamma\alpha,\ \gamma\alpha\beta,\ \alpha\beta\alpha,\ \beta\gamma\beta,\ \gamma\alpha\gamma,\ \alpha\beta\gamma\beta,\ \beta\gamma\beta\alpha,$$

where

$$\alpha^2=1,\ \beta^2=1,\ \gamma^2=1,\ (\beta\gamma)^3=1,\ (\gamma\alpha)^3=1,\ (\alpha\beta)^3=1,\ (\alpha\beta\gamma)^2=1,\ (\beta\gamma\alpha)^2=1,\ (\gamma\alpha\beta)^2=1\,;$$

and the other a group of twenty-seven, viz.

$$1,\ \alpha,\ \alpha^2,\ \gamma,\ \gamma^2,\ \gamma\alpha,\ \alpha\gamma,\ \gamma\alpha^2,\ \alpha^2\gamma,\ \gamma^2\alpha,\ \alpha\gamma^2,\ \gamma^2\alpha^2,\ \alpha^2\gamma^2,$$

$$\alpha\gamma\alpha,\ \alpha\gamma^2\alpha,\ \alpha^2\gamma\alpha,\ \alpha^2\gamma^2\alpha,\ \alpha\gamma\alpha^2,\ \alpha\gamma^2\alpha^2,\ \alpha^2\gamma\alpha^2,\ \alpha^2\gamma^2\alpha^2,\ \gamma\alpha\gamma^2,\ \gamma\alpha^2\gamma^2,\ \gamma^2\alpha\gamma,\ \gamma^2\alpha^2\gamma,\ \gamma^2\alpha\gamma\alpha^2,\ \gamma\alpha\gamma^2\alpha^2,$$

where

$$\alpha^3=1,\ \gamma^3=1,\ (\gamma\alpha)^3=1,\ (\gamma^2\alpha)^3=1,\ (\gamma\alpha^2)^3=1,\ (\gamma^2\alpha^2)^3=1.$$

It is hardly necessary to remark, that each of these groups is in reality perfectly symmetric, the omitted terms being, in virtue of the equations defining the nature of the symbols, identical with some of the terms of the group: thus, in the group of 18, the equations $\alpha^2=1$, $\beta^2=1$, $\gamma^2=1$, $(\alpha\beta\gamma)^2=1$ give $\alpha\beta\gamma=\gamma\beta\alpha$, and similarly for all the other omitted terms. It is easy to see that in the group of 18 the index of each term is 2 or else 3, while in the group of 27 the index of each term is 3.

2 *Stone Buildings*, *Nov.* 2, 1853.

126.

ON THE THEORY OF GROUPS, AS DEPENDING ON THE SYMBOLIC EQUATION $\theta^n = 1$.—SECOND PART.

[From the *Philosophical Magazine*, vol. VII. (1854), pp. 408—409.]

IMAGINE the symbols

$$L, \; M, \; N, \ldots$$

such that (L being any symbol of the system),

$$L^{-1}L, \; L^{-1}M, \; L^{-1}N, \ldots$$

is the *group*

$$1, \quad \alpha, \quad \beta, \ldots;$$

then, in the first place, M being any other symbol of the system, $M^{-1}L$, $M^{-1}M$, $M^{-1}N, \ldots$ will be the same group $1, \alpha, \beta, \ldots$. In fact, the system $L, M, N, \ldots$ may be written $L, L\alpha, L\beta \ldots$; and if e.g. $M = L\alpha$, $N = L\beta$ then

$$M^{-1}N = (L\alpha)^{-1} L\beta = \alpha^{-1} L^{-1} L\beta = \alpha^{-1}\beta,$$

which belongs to the group $1, \alpha, \beta, \ldots$.

Next it may be shown that

$$LL^{-1}, \; ML^{-1}, \; NL^{-1}, \ldots$$

is a group, although not in general the same group as $1, \alpha, \beta, \ldots$. In fact, writing $M = L\alpha$, $N = L\beta$, &c., the symbols just written down are

$$LL^{-1}, \; L\alpha L^{-1}, \; L\beta L^{-1}, \ldots$$

and we have e.g. $L\alpha L^{-1} . L\beta L^{-1} = L\alpha\beta L^{-1} = L\gamma L^{-1}$, where γ belongs to the group $1, \alpha, \beta$.

The system $L, M, N, \ldots$ may be termed a group-holding system, or simply a holder; and with reference to the two groups to which it gives rise, may be said to hold on the nearer side the group $L^{-1}L, L^{-1}M, L^{-1}N, \ldots$, and to hold on the further side the group $LL^{-1}, LM^{-1}, LN^{-1}, \ldots$ Suppose that these groups are one and the same group $1, \alpha, \beta \ldots$, the system $L, M, N, \ldots$ is in this case termed a symmetrical holder, and in reference to the last-mentioned group is said to hold such group symmetrically. It is evident that the symmetrical holder $L, M, N, \ldots$ may be expressed indifferently and at pleasure in either of the two forms $L, L\alpha, L\beta, \ldots$ and $L, \alpha L, \beta L$; i.e. we may say that the group is convertible with any symbol L of the holder, and that the group operating upon, or operated upon by, a symbol L of the holder, produces the holder. We may also say that the holder operated upon by, or operating upon, a symbol α of the group reproduces the holder.

Suppose now that the group

$$1, \alpha, \beta, \gamma, \delta, \epsilon, \zeta, \ldots$$

can be divided into a series of symmetrical holders of the smaller group

$$1, \alpha, \beta, \ldots;$$

the former group is said to be a multiple of the latter group, and the latter group to be a submultiple of the former group. Thus considering the two different forms of a group of six, and first the form

$$1, \alpha, \alpha^2, \gamma, \gamma\alpha, \gamma\alpha^2, (\alpha^3 = 1, \gamma^2 = 1, \alpha\gamma = \gamma\alpha),$$

the group of six is a multiple of the group of three, $1, \alpha, \alpha^2$ (in fact, $1, \alpha, \alpha^2$ and $\gamma, \gamma\alpha, \gamma\alpha^2$ are each of them a symmetrical holder of the group $1, \alpha, \alpha^2$); and so in like manner the group of six is a multiple of the group of two, $1, \gamma$ (in fact, $1, \gamma$ and $\alpha, \alpha\gamma$, and $\alpha, \alpha^2\gamma$ are each a symmetrical holder of the group $1, \gamma$). There would not, in a case such as the one in question, be any harm in speaking of the group of six as the product of the two groups $1, \alpha, \alpha^2$ and $1, \gamma$, but upon the whole it is, I think, better to dispense with the expression.

Considering, secondly, the other form of a group of six, viz.

$$1, \alpha, \alpha^2, \gamma, \gamma\alpha, \gamma\alpha^2 (\alpha^3 = 1, \gamma^2 = 1, \alpha\gamma = \gamma\alpha^2);$$

here the group of six is a multiple of the group of three, $1, \alpha, \alpha^2$ (in fact, as before, $1, \alpha, \alpha^2$ and $\gamma, \gamma\alpha, \gamma\alpha^2$, are each a symmetrical holder of the group $1, \alpha, \alpha^2$, since, as regards $\gamma, \gamma\alpha, \gamma\alpha^2$, we have $(\gamma, \gamma\alpha, \gamma\alpha^2) = \gamma(1, \alpha, \alpha^2) = (1, \alpha^2, \alpha)\gamma$). But the group of six is not a multiple of any group of two whatever; in fact, besides the group $1, \gamma$ itself, there is not any symmetrical holder of this group $1, \gamma$; and so, in like manner, with respect to the other groups of two, $1, \gamma\alpha$, and $1, \gamma\alpha^2$. The group of three, $1, \alpha, \alpha^2$, is therefore, in the present case, the only submultiple of the group of six.

It may be remarked, that if there be any number of symmetrical holders of the same group, $1, \alpha, \beta, \ldots$ then any one of these holders bears to the aggregate of the holders a relation such as the submultiple of a group bears to such group; it is proper to notice that the aggregate of the holders is not of necessity itself a holder.

127.

ON THE HOMOGRAPHIC TRANSFORMATION OF A SURFACE OF THE SECOND ORDER INTO ITSELF.

[From the *Philosophical Magazine*, vol. VII. (1854), pp. 208—212: continuation of **122**.]

I PASS to the improper transformation. Sir W. R. Hamilton has given (in the note, p. 723 of his Lectures on Quaternions [Dublin, 1853)] the following theorem:—If there be a polygon of $2m$ sides inscribed in a surface of the second order, and $(2m-1)$ of the sides pass through given points, then will the $2m$-th side constantly touch two cones circumscribed about the surface of the second order. The relation between the extremities of the $2m$-th side is that of two points connected by the general improper transformation; in other words, if there be on a surface of the second order two points such that the line joining them touches two cones circumscribed about the surface of the second order, then the two points are as regards the transformation in question a pair of corresponding points, or simply a pair. But the relation between the two points of a pair may be expressed in a different and much more simple form. For greater clearness call the surface of the second order U, and the sections along which it is touched by the two cones, θ, ϕ; the cones themselves may, it is clear, be spoken of as the cones θ, ϕ. And let the two points be P, Q. The line PQ touches the two cones, it is therefore the line of intersection of the tangent plane through P to the cone θ, and the tangent plane through P to the cone ϕ. Let one of the generating lines through P meet the section θ in the point A, and the other of the generating lines through P meet the section ϕ in the point B. The tangent planes through P to the cones θ, ϕ respectively are nothing else than the tangent planes to the surface U at the points A, B respectively. We have therefore at these points two generating lines meeting in the point P; the other two generating lines at the points A, B meet in like manner in the point Q. Thus P, Q are opposite angles of a skew quadrangle formed by four generating lines (or, what is the same thing, lying upon the surface of the second order), and having its other two angles, one of them on the section θ and the other on the section ϕ; and if we consider the side PA as belonging determinately to one or the other of the two systems of generating lines, then when P is given, the corresponding point Q is, it is clear, completely determined. What precedes may be recapitulated in the statement, that in the improper transformation of a surface of the second order into itself, we have, as corresponding points, the opposite angles of a skew quadrangle lying upon the surface, and having the other two opposite angles upon given plane sections of the surface. I may add, that attending only to the sections through the points of intersection of θ, ϕ, if the point P be situate anywhere in one of these sections, the point Q will be always situate in the other of these sections, i.e. the sections correspond to each other in pairs; in particular, the sections θ, ϕ are corresponding sections, so also are the sections Θ, Φ (each of them two generating lines) made by tangent planes of the surface. Any three pairs of sections form an involution; the two sections which are the sibiconjugates of the involution are of course such, that, if the point P be situate in either of these sections, the corresponding point Q will be situate in the same section. It may be noticed that when the two sections θ, ϕ coincide, the line joining the corresponding points passes through a fixed point, viz. the pole of the plane of the coincident sections; in fact the lines PQ and AB are in every case reciprocal polars, and in the present case the line AB lies in a fixed plane, viz. the plane of the coincident sections, the line PQ passes therefore through the pole of this plane. This agrees with the remarks made in the first part of the present paper.

The analytical investigation in the case where the surface of the second order is represented under the form $xy - zw = 0$ is so simple, that it is, I think, worth while to reproduce it here, although for several reasons I prefer exhibiting the final result in relation to the form $x^2 + y^2 + z^2 + w^2 = 0$ of the equation of the surface of the second order. I consider then the surface $xy - zw = 0$, and I take $(\alpha, \beta, \gamma, \delta)$, $(\alpha', \beta', \gamma', \delta')$ for the coordinates of the poles of the two sections θ, ϕ, and also (x_1, y_1, z_1, w_1), (x_2, y_2, z_2, w_2) as the coordinates of the points P, Q. We have of course $x_1y_1 - z_1w_1 = 0$, $x_2y_2 - z_2w_2 = 0$. The generating lines through P are obtained by combining the equation $xy - zw = 0$ of the surface with the equation $xy_1 + yx_1 - zw_1 - wz_1 = 0$ of the tangent plane at P. Eliminating x from these equations, and replacing in the result x_1 by its value $\dfrac{z_1w_1}{y_1}$, we have the equation

$$(yz_1 - zy_1)(yw_1 - wy_1) = 0.$$

We may if we please take $yz_1 - zy_1 = 0$, $xy_2 + yx_1 - zw_1 - wz_1 = 0$ as the equations of the line PA; this leads to

$$\left.\begin{aligned} yz_1 - zy_1 &= 0, \\ xy_1 + yx_1 - zw_1 - wz_1 &= 0, \end{aligned}\right\} \quad \left.\begin{aligned} yw_2 - wy_2 &= 0, \\ xy_2 + yx_2 - zw_2 - wz_2 &= 0, \end{aligned}\right\}$$

for the equations of the lines PA, QA respectively; and we have therefore the coordinates of the point A, coordinates which must satisfy the equation

$$\beta x + \alpha y - \delta z - \gamma w = 0$$

of the plane θ. This gives rise to the equation

$$y_2(\alpha y_1 - \delta z_1) - w_2(\gamma y_1 - \beta z_1) = 0.$$

We have in like manner

$$\left.\begin{array}{l} yw_1 - y_1 w = 0, \\ xy_1 + yx_1 - zw_1 - wz_1 = 0, \end{array}\right\} \quad \left.\begin{array}{l} yz_2 - zy_2 = 0, \\ xy_2 + yx_2 - zw_2 - wz_2 = 0, \end{array}\right\}$$

for the equations of the lines PB, QB respectively; and we may thence find the coordinates of the point B, coordinates which must satisfy the equation

$$\beta' x + \alpha' y - \delta' z - \gamma' w = 0$$

of the plane ϕ. This gives rise to the equation

$$y_2(\alpha' y_1 - \gamma' w_1) - z_2(\delta' y_1 - \beta' w_1).$$

It is easy, by means of these two equations and the equation $x_2 y_2 - z_2 w_2 = 0$, to form the system

$$\begin{aligned} x_2 &= (\alpha y_1 - \delta z_1)(\alpha' y_1 - \gamma' w_1), \\ y_2 &= (\gamma y_1 - \beta z_1)(\delta' y_1 - \beta' w_1), \\ z_2 &= (\gamma y_1 - \beta z_1)(\alpha' y_1 - \gamma' w_1), \\ w_2 &= (\alpha y_1 - \delta z_1)(\delta' y_1 - \beta' w_1); \end{aligned}$$

or, effecting the multiplications and replacing $z_1 w_1$ by $x_1 y_1$, the values of x_2, y_2, z_2, w_2 contain the common factor y_1, which may be rejected. Also introducing on the left-hand sides the common factor MM', where $M^2 = \alpha\beta - \gamma\delta$, $M'^2 = \alpha'\beta' - \gamma'\delta'$, the equations become

$$\begin{aligned} MM'x_2 &= \gamma'\delta x_1 + \alpha\alpha' y_1 - \alpha'\delta z_1 - \alpha\gamma' w_1, \\ MM'y_2 &= \beta\beta' x_1 + \gamma\delta' y_1 - \beta\delta' z_1 - \beta'\gamma w_1, \\ MM'z_2 &= \beta\gamma' x_1 + \gamma\alpha' y_1 - \beta\alpha' z_1 - \gamma\gamma' w_1, \\ MM'w_2 &= \beta'\delta x_1 + \alpha\delta' y_1 - \delta\delta' z_1 - \alpha\beta' w_1, \end{aligned}$$

values which give identically $x_2 y_2 - z_2 w_2 = x_1 y_1 - z_1 w_1$. Moreover, by forming the value of the determinant, it is easy to verify that the transformation is in fact an improper one. We have thus obtained the equations for the improper transformation of the surface $xy - zw = 0$ into itself. By writing $x_1 + iy_1$, $x_1 - iy_1$ for x_1, y_1, &c., we have the following system of equations, in which (a, b, c, d), (a', b', c', d') represent, as before, the coordinates of the poles of the plane sections, and $M^2 = a^2 + b^2 + c^2 + d^2$, $M'^2 = a'^2 + b'^2 + c'^2 + d'^2$, viz. the system[1]

[1] The system is very similar in form to, but is *essentially* different from, that which could be obtained from the theory of quaternions by writing

$$MM'(w_2 + ix_2 + jy_2 + kz_2) = (d + ia + jb + kc)(w + ix + jy + kz)(d' + ia' + jb' + kc');$$

the last-mentioned transformation is, in fact, *proper*, and not *improper*.

$$MM'x_2 = (aa' - bb' - cc' - dd')\,x_1 + (\ ab' + a'b + cd' - c'd)\,y_1$$
$$+ (\ ac' + a'c + db' - d'b)\,z_1 + (\ ad' + a'd + bc' - b'c)\,w_1,$$
$$MM'y_2 = (ab' + a'b - cd' + c'd)\,x_1 + (-aa' + bb' - cc' - dd')\,y_1$$
$$+ (\ bc' + b'c - da' + d'a)\,z_1 + (\ bd' + b'd - ac' + a'c)\,w,$$
$$MM'z_2 = (ac' + a'c - db' + d'b)\,x_1 + (\ bc' + b'c - ad' + a'd)\,y_1$$
$$+ (-aa' - bb' + cc' - dd')\,z_1 + (\ cd' + c'd - ba' + b'a)\,w_1,$$
$$MM'w_2 = (ad' + a'd - bc' + b'c)\,x_1 + (\ bd' + b'd - ca' + c'a)\,y_1$$
$$+ (\ cd' + c'd - ab' + a'b)\,z_1 + (-aa' - bb' - cc' + dd')\,w_1,$$

values which of course satisfy identically $x_2^2 + y_2^2 + z_2^2 + w_2^2 = x_1^2 + y_1^2 + z_1^2 + w_1^2$, and which belong to an improper transformation. We have thus obtained the improper transformation of the surface of the second order $x^2 + y^2 + z^2 + w^2 = 0$ into itself.

Returning for a moment to the equations which belong to the surface $xy - zw = 0$, it is easy to see that we may without loss of generality write $\alpha = \beta = \alpha' = \beta' = 0$; the equations take then the very simple form

$$MM'x_2 = \gamma'\delta x_1,\quad MM'y_2 = \gamma\delta' y_1,\quad MM'z_2 = -\gamma\gamma' w_1,\quad MM'w_2 = -\delta\delta' z_1,$$

where $MM' = \sqrt{-\gamma\delta}\,\sqrt{-\gamma'\delta'}$; and it thus becomes very easy to verify the geometrical interpretation of the formulæ.

It is necessary to remark, that, whenever the coordinates of the points Q are connected with the coordinates of the points B by means of the equations which belong to an improper transformation, the points P, Q have to each other the geometrical relation above mentioned, viz. there exist two plane sections θ, ϕ such that P, Q are the opposite angles of a skew quadrangle upon the surface, and having the other two opposite angles in the sections θ, ϕ respectively. Hence combining the theory with that of the proper transformation, we see that if A and B, B and $C, \ldots,$ M and N are points corresponding to each other properly or improperly, then will N and A be points corresponding to each other, viz. properly or improperly, according as the number of the improper pairs in the series A and B, B and $C, \ldots,$ M and N is even or odd; i.e. if all the sides but one of a polygon satisfy the geometrical conditions in virtue of which their extremities are pairs of corresponding points, the remaining side will satisfy the geometrical condition in virtue of which its extremities will be a pair of corresponding points, the pair being proper or improper according to the rule just explained.

I conclude with the remark, that we may by means of two plane sections of a surface of the second order obtain a proper transformation. For, if the generating lines through P meet the sections θ, ϕ in the points A, B respectively, and the remaining generating lines through A, B respectively meet the sections ϕ, θ respectively in B', A', and the remaining generating lines through B', A' respectively meet in a point P'; then will P, P' be a pair of corresponding points in a proper transformation. In fact, the generating lines through P meeting the sections θ, ϕ in the points A, B respectively, and the remaining generating lines through A, B respectively meeting as before in the point Q, then P and Q will correspond to each other improperly, and in like manner P' and Q will correspond to each other improperly; i.e. P and P' will correspond to each other properly. The relation between P, P' may be expressed by saying that these points are opposite angles of the skew hexagon $PAB'P'A'B$ lying upon the surface, and having the opposite angles A, A' in the section θ, and the opposite angles B, B' in the section ϕ. It is, however, clear from what precedes, that the points P, P' lie in a section passing through the points of intersection of θ, ϕ, and thus the proper transformation so obtained is not the general proper transformation.

2 *Stone Buildings, January* 11, 1854.

128.

DEVELOPMENTS ON THE PORISM OF THE IN-AND-CIRCUMSCRIBED POLYGON.

[From the *Philosophical Magazine*, vol. VII. (1854), pp. 339—345.]

I PROPOSE to develope some particular cases of the theorems given in my paper, "Correction of two Theorems relating to the Porism of the in-and-circumscribed Polygon" (*Phil. Mag.* vol. VI. (1853), [116]). The two theorems are as follows:

THEOREM. The condition that there may be inscribed in the conic $U=0$ an infinity of n-gons circumscribed about the conic $V=0$, depends upon the development in ascending powers of ξ of the square root of the discriminant of $\xi U+V$; viz. if this square root be

$$A+B\xi+C\xi^2+D\xi^3+E\xi^4+F\xi^5+G\xi^6+H\xi^7+\dots,$$

then for $n=3, 5, 7$, &c. respectively, the conditions are

$$\left|\,C\,\right|=0,\quad \begin{vmatrix} C, & D \\ D, & E \end{vmatrix}=0,\quad \begin{vmatrix} C, & D, & E \\ D, & E, & F \\ E, & F, & G \end{vmatrix}=0,\ \text{\&c.};$$

and for $n=4, 6, 8$, &c. respectively, the conditions are

$$\left|\,D\,\right|=0,\quad \begin{vmatrix} D, & E \\ E, & F \end{vmatrix}=0,\quad \begin{vmatrix} D, & E, & F \\ E, & F, & G \\ F, & G, & H \end{vmatrix}=0,\ \text{\&c.}$$

THEOREM. In the case where the conics are replaced by the two circles

$$x^2 + y^2 - R^2 = 0, \quad (x-a)^2 + y^2 - r^2 = 0,$$

then the discriminant, the square root of which gives the series

$$A + B\xi + C\xi^2 + D\xi^3 + E\xi^4 + \&c.,$$

is

$$(1+\xi)\{r^2 + \xi(r^2 + R^2 - a^2) + \xi^2 R^2\}.$$

Write for a moment

$$A + B\xi + C\xi^2 + D\xi^3 + E\xi^4 + \&c. = \sqrt{(1+a\xi)(1+b\xi)(1+c\xi)},$$

then

$$\begin{aligned}
A &= 1,\\
2B &= a + b + c,\\
-8C &= a^2 + b^2 + c^2 - 2bc - 2ca - 2ab,\\
16D &= a^3 + b^3 + c^3 - a^2(b+c) - b^2(c+a) - c^2(a+b) + 2abc,\\
-128E &= 5a^4 + 5b^4 + 5c^4 - 4a^3(b+c) - 4b^3(c+a) - 4c^3(a+b)\\
&\quad + 4a^2bc + 4b^2ca + 4c^2ab - 2b^2c^2 - 2c^2a^2 - 2a^2b^2,\\
&\quad \&c.
\end{aligned}$$

To adapt these to the case of the two circles, we have to write

$$r^2(1+a\xi)(1+b\xi)(1+c\xi) = (1+\xi)\{r^2 + \xi(r^2 + R^2 - a^2) + \xi^2 R^2\},$$

and therefore

$$\begin{aligned}
c &= 1,\\
r^2(a+b) &= r^2 + R^2 - a^2,\\
r^2 ab &= R^2;
\end{aligned}$$

values which after some reductions give

$$\begin{aligned}
A &= 1,\\
r^2 \,.\, 2B &= 2r^2 + R^2 - a^2,\\
-r^4 \,.\, 8C &= (R^2 - a^2)^2 - 4R^2r^2,\\
r^6 \,.\, 16D &= (R^2 - a^2)\{(R^2 - a^2)^2 - 2r^2(R^2 + a^2)\},\\
-r^8 \,.\, 128E &= 5(R^2 - a^2)^4 - 8(R^2 - a^2)^2(R^2 + 2r^2)r^2 + 16a^4r^4.
\end{aligned}$$

Hence also

$$\begin{aligned}
r^{12} \,.\, 1024(CE - D^2) = \{5(R^2 - a^2)^4 - 8(R^2 - a^2)^2(R^2 + 2r^2)r^2 + 16a^4r^4\}\{(R^2 - a^2)^2 - 4R^2r^2)\}\\
- 4\{(R^2 - a^2)^3 - 2(R^2 - a^2)(R^2 + a^2)r^2\}^2,
\end{aligned}$$

which after all reductions is

$$\begin{aligned}&(R^2-a^2)^6\\&-12R^2(R^2-a^2)^4r^2\\&+16R^2(R^2+2a^2)(R^2-a^2)^2r^4\\&-64R^2a^4r^6.\end{aligned}$$

Hence the condition that there may be, inscribed in the circle $x^2+y^2-R^2=0$ and circumscribed about the circle $(x-a)^2+y^2-r^2=0$, an infinity of n-gons, is for $n=3$, 4, 5, i.e. in the case of a triangle, a quadrangle and a pentagon respectively, as follows.

I. For the triangle, the relation is

$$(R^2-a^2)^2-4R^2r^2=0,$$

which is the completely rationalized form (the simple power of a radius being of course analytically a radical) of the well-known equation

$$a^2=R^2-2Rr,$$

which expresses the relation between the radii R, r of the circumscribed and inscribed circles, and the distance a between their centres.

II. For the quadrangle, the relation is

$$(R^2-a^2)^2-2r^2(R^2+a^2)=0,$$

which may also be written

$$(R+r+a)(R+r-a)(R-r+a)(R-r-a)-r^4=0.$$

(Steiner, *Crelle*, t. II. [1827] p. 289.)

III. For the pentagon, the relation is

$$(R^2-a^2)^6-12R^2(R^2-a^2)^4r^2+16R^2(R^2+2A^2)(R^2-a^2)^2r^4-64R^2a^4r^6=0,$$

which may also be written

$$(R^2-a^2)^2\{(R^2-a^2)^2-4R^2r^2\}^2-4R^2r^2\{(R^2-a^2)^2-4a^2r^2\}^2=0.$$

The equation may therefore be considered as the completely rationalized form of

$$(R^2-a^2)^3+2R(R^2-a^2)^2r-4R^2(R^2-a^2)r^2-8Ra^2r^3=0.$$

This is, in fact, the form given by Fuss in his memoir "De polygonis symmetrice irregularibus circulo simul inscriptis et circumscriptis," *Nova Acta Petrop.* t. XIII. [1802] pp. 166—189 (I quote from Jacobi's memoir, to be presently referred to). Fuss puts $R+a=p$, $R-a=q$, and he finds the equation

$$\frac{p^2q^2-r^2(p^2+q^2)}{r^2q^2-p^2(r^2+q^2)}=\pm\sqrt{\frac{q-r}{q+p}},$$

which, he remarks, is satisfied by $r=-p$ and $r=\frac{pq}{p+q}$, and that consequently the rationalized equation will divide by $p+r$ and $pq-r(p+q)$; and he finds, after the division,

$$p^3q^3+p^2q^2(p+q)r-pq(p+q)^2r^2-(p+q)(p-q)^2r^3=0,$$

which, restoring for p, q their values $R+a$, $R-a$, is the very equation above found.

The form given by Steiner (*Crelle*, t. II. p. 289) is

$$r(R-a)=(R+a)\sqrt{(R-r+a)(R-r-a)}+(R+a)\sqrt{(R-r-a)\,2R},$$

which, putting p, q instead of $R+a$, $R-a$, is

$$qr=p\sqrt{(p-r)(q-r)}+p\sqrt{(q-r)(q+p)};$$

and Jacobi has shown in his memoir, "Anwendung der elliptischen Transcendenten u. s. w.," *Crelle*, t. III. [1828] p. 376, that the rationalized equation divides (like that of Fuss) by the factor $pq-(p+q)r$, and becomes by that means identical with the rational equation given by Fuss.

In the case of two concentric circles $a=0$, and putting for greater simplicity $\frac{R^2}{r^2}=M$, we have

$$A+B\xi+C\xi^2+D\xi^3+E\xi^4+\&c.=(1+\xi)\sqrt{1+M\xi}.$$

This is, in fact, the very formula which corresponds to the general case of two conics having double contact. For suppose that the polygon is inscribed in the conic $U=0$, and circumscribed about the conic $U+P^2=0$, we have then to find the discriminant of $\xi U+U+P^2$, i.e. of $(1+\xi)U+P^2$. Let K be the discriminant of U, and let F be what the polar reciprocal of U becomes when the variables are replaced by the coefficients of P, or, what is the same thing, let $-F$ be the determinant obtained by bordering K (considered as a matrix) with the coefficients of P. The discriminant of $(1+\xi)U+P^2$ is $(1+\xi)^3K+(1+\xi)^2F$, i.e. it is

$$(1+\xi)^2\{K(1+\xi)+F\},\ =(K+F)(1+\xi)^2(1+M\xi),$$

where $M=\frac{K}{K+F}$; or, what is the same thing, M is the discriminant of U divided by the discriminant of $U+P^2$. And M having this meaning, the condition of there being inscribed in the conic $U=0$ an infinity of n-gons circumscribed about the conic $U+P^2=0$, is found by means of the series

$$A+B\xi+C\xi^2+D\xi^3+E\xi^4+\&c.=(1+\xi)\sqrt{1+M\xi}.$$

We have, therefore,

$$\begin{aligned}
A &= 1, \\
2B &= M + 2, \\
-8C &= M^2 - 4M, \\
16D &= M^3 - 2M^2, \\
-128E &= 5M^4 - 8M^3, \\
&\&c. \\
1024(CE - D^2) &= M^4(M^2 - 12M + 16), \\
&\&c.
\end{aligned}$$

Hence for the triangle, quadrangle and pentagon, the conditions are—

I. For the triangle,

$$M + 2 = 0.$$

II. For the quadrangle,

$$M - 4 = 0.$$

III. For the pentagon,

$$M^2 - 12M + 16 = 0;$$

and so on.

It is worth noticing, that, in the case of two conics having a 4-point contact, we have $F = 0$, and consequently $M = 1$. The discriminant is therefore $(1 + \xi)^3$, and as this does not contain any variable parameter, the conics cannot be determined so that there may be for a given value of n (nor, indeed, for any value whatever of n) an infinity of n-gons inscribed in the one conic, and circumscribed about the other conic.

The geometrical properties of a triangle, &c. inscribed in a conic and circumscribed about another conic, these two conics having double contact with each other,

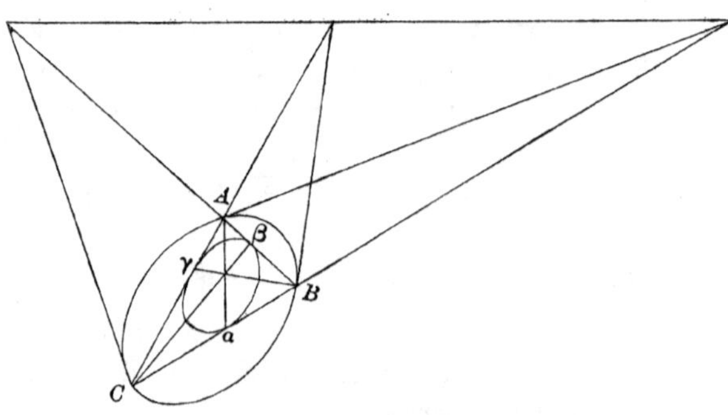

are at once obtained from those of the system in which the two conics are replaced by concentric circles. Thus, in the case of a triangle, if ABC be the triangle, and α, β, γ be the points of contact of the sides with the inscribed conic, then the tangents to the circumscribed conic at A, B, C meet the opposite sides BC, CA, AB in points lying in the chord of contact, the lines $A\alpha$, $B\beta$, $C\gamma$ meet in the pole of contact, and so on.

In the case of a quadrangle, if $ACEG$ be the quadrangle, and b, d, f, h the points of contact with the inscribed conic, then the tangents to the circumscribed

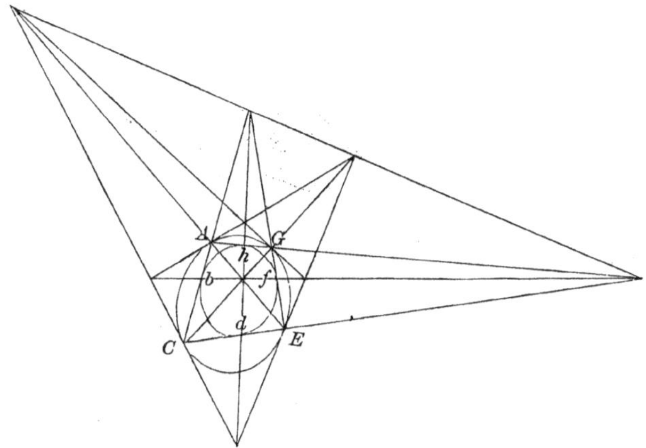

conic at the pair of opposite angles A, E and the corresponding diagonal CG, and in like manner the tangents at the pair of opposite angles C, G and the corresponding diagonal AE, meet in the chord of contact. Again, the pairs of opposite sides AC, EG, and the line dh joining the points of contact of the other two sides with the inscribed conic, and the pairs of opposite sides AG, CE, and the line bf joining the points of contact of the other two sides with the inscribed conic, meet in the chord of contact. The diagonals AE, CG, and the lines bf, dh through the points of contact of pairs of opposite sides with the inscribed conic, meet in the pole of contact, &c.

The beautiful systems of 'focal relations' for regular polygons (in particular for the pentagon and the hexagon), given in Sir W. R. Hamilton's Lectures on Quaternions, [Dublin, 1853] Nos. 379—393, belong, it is clear, to polygons which are inscribed in and circumscribed about two conics having double contact with each other. In fact, the focus of a conic is a point such that the lines joining such point with the circular points at infinity (i.e. the points in which a circle is intersected by the line infinity) are tangents to the conic. In the case of two concentric circles, these are to be considered as touching in the circular points at infinity; and consequently, when the concentric circles are replaced by two conics having double contact, the circular points at infinity are replaced by the points of contact of the two conics.

Thus, in the figure (which is simply Sir W. R. Hamilton's figure 81 put into

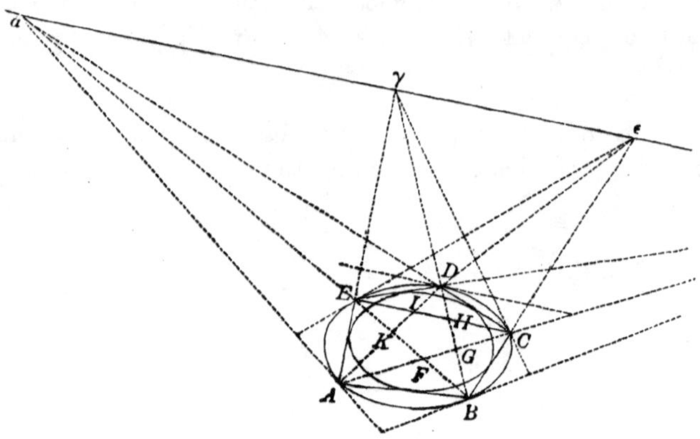

perspective), the system of relations

$$F,\ G\,(..)\,ABCI,$$
$$G,\ H(..)\,BCDK,$$
$$H,\ I\,(..)\,CDEF,$$
$$I,\ K\,(..)\,DEAG,$$
$$K,\ F(..)\,EABH,$$

will mean, F, $G\,(..)\,ABCI$, that there is a conic inscribed in the quadrilateral $ABCI$ such that the tangents to this conic through the points F and G pass two and two through the points of contact of the circumscribed and the inscribed conics, and similarly for the other relations of the system. As the figure is drawn, the tangents in question are of course (as the tangents through the foci in the case of the two concentric circles) imaginary.

2 *Stone Buildings, March* 7, 1854.

129.

ON THE PORISM OF THE IN-AND-CIRCUMSCRIBED TRIANGLE, AND ON AN IRRATIONAL TRANSFORMATION OF TWO TERNARY QUADRATIC FORMS EACH INTO ITSELF.

[From the *Philosophical Magazine*, vol. IX. (1855), pp. 513—517.]

THERE is an irrational transformation of two ternary quadratic forms each into itself, based upon the solution of the following geometrical problem,

Given that the line

$$lx + my + nz = 0$$

meets the conic

$$(a, b, c, f, g, h \between x, y, z)^2 = 0$$

in the point (x_1, y_1, z_1); to find the other point of intersection.

The solution is exceedingly simple. Take (x_2, y_2, z_2) for the coordinates of the other point of intersection, we must have identically with respect to x, y, z,

$$\begin{aligned}(a, \dots \between x, y, z)^2 . (\mathfrak{A}, \dots \between l, m, n)^2 - k(lx + my + nz)^2 \\ = (a, \dots \between x_1, y_1, z_1 \between x, y, z) . (a, \dots \between x_2, y_2, z_2 \between x, y, z)\end{aligned}$$

to a constant factor *près*.

Assume successively $x, y, z = \mathfrak{A}, \mathfrak{H}, \mathfrak{G};\ \mathfrak{H}, \mathfrak{B}, \mathfrak{F};\ \mathfrak{G}, \mathfrak{F}, \mathfrak{C}$; it follows that

$$\begin{aligned}x_2 : y_2 : z_2 = & \ y_1 z_1 \{\mathfrak{A}(\mathfrak{A}, \dots \between l, m, n)^2 - (\mathfrak{A}l + \mathfrak{H}m + \mathfrak{G}n)^2\} \\ : & \ z_1 x_1 \{\mathfrak{B}(\mathfrak{A}, \dots \between l, m, n)^2 - (\mathfrak{H}l + \mathfrak{B}m + \mathfrak{F}n)^2\} \\ : & \ x_1 y_1 \{\mathfrak{C}(\mathfrak{A}, \dots \between l, m, n)^2 - (\mathfrak{G}l + \mathfrak{F}m + \mathfrak{C}n)^2\};\end{aligned}$$

or, what is the same thing,

$$\begin{aligned} x_2 : y_2 : z_2 = \ & y_1 z_1 (bn^2 + cm^2 - 2fmn) \\ : \ & z_1 x_1 (cl^2 + an^2 - 2gnl) \\ : \ & x_1 y_1 (am^2 + bn^2 - 2hlm). \end{aligned}$$

It is not necessary for the present purpose, but it may be as well to give the corresponding solution of the problem:

Given that one of the tangents through the point (ξ, η, ζ) to the conic

$$(a, b, c, f, g, h \between x, y, z)^2 = 0$$

is the line $l_1 x + m_1 y + n_1 z = 0$; to find the equation to the other tangent.

Let $l_2 x + m_2 y + n_2 z = 0$ be the other tangent, then

$$\begin{aligned} (a, \ldots \between \xi, \eta, \zeta)^2 . (a, \ldots \between x, y, z)^2 - \{(a \ldots \between \xi, \eta, \zeta \between x, y, z)\}^2 \\ = (l_1 x + m_1 y + n_1 z)(l_2 x + m_2 y + n_2 z) \end{aligned}$$

to a constant factor *près*. Assume successively $y=0$, $z=0$; $z=0$, $x=0$; $x=0$, $y=0$; then we have

$$\begin{aligned} l_2 : m_2 : n_2 = \ & m_1 n_1 \{a (a, \ldots \between \xi, \eta, \zeta)^2 - (a\xi + h\eta + g\zeta)^2\} \\ : \ & n_1 \, l_1 \{b (a, \ldots \between \xi, \eta, \zeta)^2 - (h\xi + b\eta + f\zeta)^2\} \\ : \ & l_1 m_1 \{c (a, \ldots \between \xi, \eta, \zeta)^2 - (g\xi + f\eta + c\zeta)^2\}; \end{aligned}$$

or, as they may be more simply written,

$$\begin{aligned} l_2 : m_2 : n_2 = \ & m_1 n_1 (\mathfrak{B}\zeta^2 + \mathfrak{C}\eta^2 + 2\mathfrak{F}\eta\zeta) \\ : \ & n_1 \, l_1 (\mathfrak{C}\xi^2 + \mathfrak{A}\zeta^2 - 2\mathfrak{G}\zeta\xi) \\ : \ & l_1 m_1 (\mathfrak{A}\eta^2 + \mathfrak{B}\xi^2 - 2\mathfrak{H}\xi\eta). \end{aligned}$$

Returning now to the solution of the first problem, I shall for the sake of simplicity consider the formulæ obtained by taking for the equation of the conic,

$$\alpha x^2 + \beta y^2 + \gamma z^2 = 0.$$

We see, therefore, that if this conic be intersected by the line $lx + my + nz = 0$ in the points (x_1, y_1, z_1) and (x_2, y_2, z_2), then

$$\begin{aligned} x_2 : y_2 : z_2 = \ & y_1 z_1 (\gamma m^2 + \alpha n^2) \\ : \ & z_1 x_1 (\alpha n^2 + \beta l^2) \\ : \ & x_1 y_1 (\beta l^2 + \alpha m^2). \end{aligned}$$

We have, in fact, *identically*

$$ly_1z_1(\beta n^2+\gamma m^2)+mz_1x_1(\gamma l^2+\alpha n^2)+nx_1y_1(\alpha m^2+\beta l^2)$$
$$=(\alpha mnx_1+\beta nly_1+\gamma lmz_1)(lx_1+my_1+nz_1)-lmn(\alpha x_1^2+\beta y_1^2+\gamma z_1^2),$$

$$\alpha y_1^2z_1^2(\beta n^2+\gamma m^2)^2+\beta z_1^2x_1^2(\gamma l^2+\alpha n^2)^2+\gamma x_1^2y_1^2(\alpha m^2+\beta l^2)^2$$
$$=\alpha\beta\gamma\{-l^3x_1^3-m^3y_1^3-n^3z_1^3$$
$$+(my_1+nz_1)l^2x_1^2+(nz_1+lx_1)m^2y_1^2+(lx_1+my_1)n^2y_1^2-2lmnx_1y_1z_1\}(lx_1+my_1+nz_1)$$
$$-(l^4\beta\gamma x_1^2+m^4\gamma\alpha y_1^2+n^4\alpha\beta z_1^2)(\alpha x_1^2+\beta y_1^2+\gamma z_1^2);$$

which show that if $lx_1+my_1+nz_1=0$ and $\alpha x_1^2+\beta y_1^2+\gamma_1 z^2=0$, then also $lx_2+my_2+nz_2=0$ and $\alpha x_2^2+\beta y_2^2+\gamma z_2^2=0$: this is, of course, as it should be.

I shall now consider l, m, n as *given functions* of x_1, y_1, z_1 satisfying identically the equations

$$lx_1+my_1+nz_1=0,$$
$$l^2bc+m^2ca+n^2ab=0,$$

equations which express that $lx+my+nz=0$ is the tangent from the point (x_1, y_1, z_1) to the conic $ax^2+by^2+cz^2=0$. And I shall take for α, β, γ the following values, viz.

$$\alpha=ax_1^2+by_1^2+cz_1^2-a(x_1^2+y_1^2+z_1^2),$$
$$\beta=ax_1^2+by_1^2+cz_1^2-b(x_1^2+y_1^2+z_1^2),$$
$$\gamma=ax_1^2+by_1^2+cz_1^2-c(x_1^2+y_1^2+z_1^2);$$

so that x_1, y_1, z_1 continuing absolutely indeterminate, we have *identically* $\alpha x_1^2+\beta y_1^2+\gamma z_1^2=0$. Also taking Θ as a function of x_1, y_1, z_1, the value of which will be subsequently given, I write

$$x_2=\Theta y_1z_1(\beta n^2+\gamma m^2),$$
$$y_2=\Theta z_1x_1(\gamma l^2+\alpha n^2),$$
$$z_2=\Theta x_1y_1(\alpha m^2+\beta l^2);$$

so that x_1, y_1, z_1 are arbitrary, and x_2, y_2, z_2 are taken to be determinate functions of x_1, y_1, z_1. The point (x_2, y_2, z_2) is geometrically connected with the point (x_1, y_1, z_1) as follows, viz. (x_2, y_2, z_2) is the point in which the tangent through (x_1, y_1, z_1) to the conic $ax^2+by^2+cz^2=0$ meets the conic passing through the point (x_1, y_1, z_1) and the points of intersection of the conics $ax^2+by^2+cz^2=0$ and $x^2+y^2+z^2=0$. Consequently, in the particular case in which (x_1, y_1, z_1) is a point on the conic $x^2+y^2+z^2=0$, the point (x_2, y_2, z_2) is the point in which this conic is met by the tangent through (x_1, y_1, z_1) to the conic $ax^2+by^2+cz^2=0$.

It has already been seen that $lx_1+my_1+nz_1=0$ and $\alpha x_1^2+\beta y_1^2+\gamma z_1^2=0$ identically; consequently we have identically $lx_2+my_2+nz_2=0$ and $\alpha x_2^2+\beta y_2^2+\gamma z_2^2=0$. The latter equation, written under the form

$$(ax_1^2+by_1^2+cz_1^2)(x_2^2+y_2^2+z_2^2)-(x_1^2+y_1^2+z_1^2)(ax_2^2+by_2^2+cz_2^2)=0,$$

shows that if x_2, y_2, z_2 are such that $x_2^2+y_2^2+z_2^2=x_1^2+y_1^2+z_1^2$, then that also $ax_2^2+by_2^2+cz_2^2=ax_1^2+by_1^2+cz_1^2$. I proceed to determine Θ so that we may have $x_2^2+y_2^2+z_2^2=x_1^2+y_1^2+z_1^2$. We obtain immediately

$$\frac{1}{\Theta^2}(x_2^2+y_2^2+z_2^2)=(l^4x_1^2+m^4y_1^2+n^4z_1^2)(\alpha^2x_1^2+\beta^2y_1^2+\gamma^2z_1^2)$$
$$-(\alpha^2l^4x_1^4+\beta^2m^4y_1^4+\gamma^2n^4z_1^4-2\beta\gamma m^2n^2y_1^2z_1^2-2\gamma\alpha n^2l^2z_1^2x_1^2-2\alpha\beta l^2m^2x_1^2y_1^2);$$

write for a moment

$$ax_1^2+by_1^2+cz_1^2=p,\ x_1^2+y_1^2+z_1^2=q, \text{ so that } \alpha=p-aq,\ \beta=p-bq,\ \gamma=p-cq,$$

then

$$\alpha^2x_1^2+\beta^2y_1^2+\gamma^2z_1^2=qp^2-2p\,.\,pq+(a^2x_1^2+b^2y_1^2+c^2z_1^2)\,q^2,=q\,\{(a^2x_1^2+b^2y_1^2+c^2z_1^2)\,q-p^2\},$$
$$=q\,\{(b-c)^2\,y_1^2z_1^2+(c-a)^2\,z_1^2x_1^2+(a-b)^2\,x_1^2y_1^2\},$$

$$\alpha^2l^4x_1^4+\beta^2m^4y_1^4+\gamma^4n^4z_1^4-2\beta\gamma m^2n^2y_1^2z_1^2-2\gamma\alpha n^2l^2z_1^2x_1^2-2\alpha\beta l^2m^2x_1^2y_1^2$$
$$=\quad p^2\,\{l^4x_1^4+m^4y_1^4+n^4z_1^4-2m^2n^2y_1^2z_1^2-2n^2l^2z_1^2x_1^2-2l^2m^2x_1^2y_1^2\}$$
$$-\,2pq\,\{al^4x_1^4+bm^4y_1^4+cn^4z_1^4-(b+c)\,m^2n^2y_1^2z_1^2-(c+a)\,n^2l^2z_1^2x_1^2-(a+b)\,l^2m^2x_1^2y_1^2\}$$
$$+\quad q^2\,\{a^2l^4x_1^4+b^2m^4y_1^4+c^2n^4z_1^4-2bcm^2n^2y_1^2z_1^2-2can^2l^2z_1^2x_1^2-2abl^2m^2x_1^2y_1^2\},$$

the first line of which vanishes in virtue of the equation $lx_1+my_1+nz_1=0$; we have therefore

$$\frac{1}{\Theta^2}(x_2^2+y_2^2+z_2^2)\div(x_1^2+y_1^2+z_1^2)$$
$$=(l^4x_1^2+m^4y_1^2+n^4z_1^2)\,\{(b-c)^2\,y_1^2z_1^2+(c-a)^2\,z_1^2x_1^2+(a-b)^2\,x_1^2y_1^2\}$$
$$+\,2\,(ax_1^2+by_1^2+cz_1^2)\,\{al^4x_1^4+bm^4y_1^4+cn^4z_1^4-(b+c)\,m^2n^2y_1^2z_1^2-(c+a)\,n^2l^2z_1^2x_1^2-(a+b)\,l^2m^2x_1^2y_1^2\}$$
$$-\,(x_1^2+y_1^2+z_1^2)\,\{a^2l^4x_1^4+b^2m^4y_1^4+c^2n^4z_1^4-2bcm^2n^2y_1^2z_1^2-2can^2l^2z_1^2x_1^2-2abl^2m^2x_1^2y_1^2\}.$$

Hence reducing the function on the right-hand side, and putting

$$(x_2^2+y_2^2+z_2^2)\div(x_1^2+y_1^2+z_1^2)=1,$$

we have

$$\frac{1}{\Theta^2}=a^2l^4x_1^6+b^2m^4y_1^6+c^2n^4z_1^6$$
$$+\,(c^2m^4-2b^2m^2n^2)\,y_1^4z_1^2+(a^2n^4-2c^2n^2l^2)\,z_1^4x_1^2+(b^2l^4-2a^2l^2m^2)\,x_1^4y_1^2$$
$$+\,(b^2n^4-2c^2m^2n^2)\,y_1^2z_1^4+(c^2l^4-2a^2n^2l^2)\,z_1^2x_1^4+(a^2m^4-2b^2l^2m^2)\,x_1^2y_1^4$$
$$+\,\{l^4\,(b-c)^2+m^4\,(c-a)^2+n^4\,(a-b)^2$$
$$+\,2m^2n^2\,(bc-ca-ab)+2n^2l^2\,(-bc+ca-ab)+2l^2m^2\,(-bc-ca+ab)\}\,x_1^2y_1^2z_1^2.$$

The value of Θ might probably be expressed in a more simple form by means of the equations $lx_1+my_1+nz_1=0$ and $l^2bc+m^2ca+n^2ab=0$, even without solving these equations; but this I shall not at present inquire into.

Recapitulating, l, m, n are considered as functions of x_1, y_1, z_1 determined (to a common factor *près*) by the equations

$$lx_1 + my_1 + nz_1 = 0,$$
$$l^2bc + m^2ca + n^2ab = 0;$$

Θ is determined as above, and then writing

$$\alpha = ax_1^2 + by_1^2 + cz_1^2 - a(x_1^2 + y_1^2 + z_1^2),$$
$$\beta = ax_1^2 + by_1^2 + cz_1^2 - b(x_1^2 + y_1^2 + z_1^2),$$
$$\gamma = ax_1^2 + by_1^2 + cz_1^2 - c(x_1^2 + y_1^2 + z_1^2),$$

we have

$$x_2 = \Theta y_1 z_1 (\beta n^2 + \gamma m^2),$$
$$y_2 = \Theta z_1 x_1 (\gamma l^2 + \alpha n^2),$$
$$z_2 = \Theta x_1 y_1 (\alpha m^2 + \beta l^2);$$

and these values give

$$lx_2 + my_2 + nz_2 = 0,$$
$$x_2^2 + y_2^2 + z_2^2 = x_1^2 + y_1^2 + z_1^2,$$
$$ax_2^2 + by_2^2 + cz_2^2 = ax_1^2 + by_1^2 + cz_1^2.$$

In connexion with the subject I may add the following transformation, viz. if

$$3\sqrt{\alpha}\, x' = \sqrt{3\beta}\,(y - z) + \sqrt{(3\alpha - 2\beta)(x^2 + y^2 + z^2) + 2\beta(yz + zx + xy)},$$
$$\vdots$$

then reciprocally

$$3\sqrt{\beta}\, x = -\sqrt{3\alpha}\,(y' - z') + \sqrt{(3\beta - 2\alpha)(x'^2 + y'^2 + z'^2) + 2\alpha(y'z' + z'x' + x'y')}.$$
$$\vdots$$

$$x^2 + y^2 + z^2 = x'^2 + y'^2 + z'^2,$$
$$\beta(x^2 + y^2 + z^2 - yz - zx - xy) = \alpha(x'^2 + y'^2 + z'^2 - y'z' - z'x' - x'y').$$

Suppose $1 + \rho + \rho^2 = 0$, then

$$x^2 + y^2 + z^2 - yz - zx - xy = (x + \rho y + \rho^2 z)(x + \rho^2 + \rho z);$$

and in fact

$$3\sqrt{\alpha}\,(x' + \rho y' + \rho^2 z') = -\sqrt{3\beta}\,(1 + 2\rho)(x + \rho y + \rho^2 z),$$
$$3\sqrt{\alpha}\,(x' + \rho^2 y' + \rho z) = \sqrt{3\beta}\,(1 + 2\rho)(x + \rho^2 y + \rho z).$$

The preceding investigations have been in my possession for about eighteen months.

2 *Stone Buildings*, *April* 18, 1855.

130.

DEUXIÈME MÉMOIRE SUR LES FONCTIONS DOUBLEMENT PÉRIODIQUES.

[From the *Journal de Mathématiques Pures et Appliquées* (Liouville), tom. XIX. (1854), pp. 193—208: Sequel to Memoir t. X. (1845), **25.**]

Je vais essayer de développer ici les propriétés qui se rapportent aux transformations linéaires des périodes des fonctions yx, gx, Gx, Zx, dont je me suis occupé dans le Mémoire sur les fonctions doublement périodiques que j'ai donné dans ce Recueil en 1845. Avant d'entrer en matière, je remarque que partant des expressions

$$\Omega = \omega + \omega' i,$$
$$\Upsilon = \upsilon + \upsilon' i,$$

des deux périodes, où $i = \sqrt{-1}$, on obtient, en écrivant

$$\Omega^* = \omega - \omega' i,$$
$$\Upsilon^* = \upsilon - \upsilon' i,$$

les équations

$$\Omega\ \Upsilon^* = \omega\upsilon + \omega'\upsilon' - i(\omega\upsilon' - \omega'\upsilon),$$
$$\Omega^*\Upsilon = \omega\upsilon + \omega'\upsilon' + i(\omega\upsilon' - \omega'\upsilon),$$

au moyen desquelles et des valeurs

$$\beta = \frac{\pi i(\omega\upsilon' - \omega'\upsilon)}{\Omega\Upsilon \text{ mod. } (\omega\upsilon' - \omega'\upsilon)}, \qquad B = \frac{\pi(\omega\upsilon + \omega'\upsilon')}{\Omega\Upsilon \text{ mod. } (\omega\upsilon' - \omega'\upsilon)}$$

des quantités β, B, on déduit les formules

$$B + \beta = \frac{\pi\Omega^*}{\Omega \text{ mod. } (\omega\upsilon' - \omega'\upsilon)}, \qquad B - \beta = \frac{\pi\Upsilon^*}{\Upsilon \text{ mod. } (\omega\upsilon' - \omega'\upsilon)}.$$

Je ne fais attention qu'aux transformations qui correspondent à des entiers impairs et premiers, et je suppose, de plus, que la transformation soit toujours propre et régulière; c'est-à-dire qu'en écrivant

$$(2k+1)\,\Omega_, = \lambda\Omega + \mu\Upsilon = (\lambda,\ \mu),$$
$$(2k+1)\,\Upsilon_, = \nu\Omega + \rho\Upsilon = (\nu,\ \rho),$$

où $2k+1$ est un entier positif, impair et premier, et où λ, μ, ν, ρ sont des entiers tels, qu'au signe près, $\lambda\rho - \mu\nu$ soit égal à $2k+1$, je suppose

$$\lambda\rho - \mu\nu = 2k+1,$$

(condition pour que la transformation soit propre), et, en outre,

$$\lambda \equiv 1, \quad \mu \equiv 0, \ (\text{mod. } 2)$$
$$\nu \equiv 0, \quad \rho \equiv 1,$$

(condition pour que la transformation soit régulière).

On trouve tout de suite

$$\Omega = \quad \rho\Omega_, - \mu\Upsilon_, = (\rho,\ -\mu)_,\,,$$
$$\Upsilon = -\nu\Omega_, + \lambda\Upsilon_, = (-\nu,\ \lambda)_,\,;$$

j'écris aussi

$$\Omega_, = \omega_, + \omega_,' i, \quad \Omega_,^* = \omega_, - \omega_,' i,$$
$$\Upsilon_, = \upsilon_, + \upsilon_,' i, \quad \Upsilon_,^* = \upsilon_, - \upsilon_,' i,$$

et je suppose que $B_,$, $\beta_,$ soient des fonctions de $\omega_,$, $\upsilon_,$ telles que les fonctions B, β, de ω, υ.

Cela étant, je forme d'abord l'équation

$$(2k+1)\,(\omega_,\upsilon_,' - \omega_,'\upsilon_,) = \omega\upsilon' - \omega'\upsilon,$$

au moyen de laquelle l'équation

$$(B+\beta)\,\Omega = \frac{\pi}{\text{mod.}\,(\omega\upsilon' - \omega'\upsilon)}\,\Omega^*$$

se transforme en

$$(2k+1)(B+\beta)\,\Omega = \frac{\pi}{\text{mod.}\,(\omega_,\upsilon_,' - \omega_,'\upsilon_,)}\,\Omega^*.$$

De là

$$\{(2k+1)(B+\beta) - B_,\}\,\Omega = \frac{\pi}{\text{mod.}\,(\omega_,\upsilon_,' - \omega_,'\upsilon_,)}\left\{\Omega^* - \Omega\,\frac{\omega_,\upsilon_, + \omega_,'\upsilon_,'}{\Omega_,\Upsilon_,}\right\},$$

$$= \frac{\pi}{\text{mod.}\,(\omega_,\upsilon_,' - \omega_,'\upsilon_,)}\left\{\rho\Omega_,^* - \mu\Upsilon_,^* - \frac{\omega_,\upsilon_, + \omega_,'\upsilon_,'}{\Omega_,\Upsilon_,}\,(\rho\Omega_, - \mu\Upsilon_,)\right\},$$

$$= \frac{\pi}{\text{mod.}\,(\omega_,\upsilon_,' - \omega_,'\upsilon_,)}\left\{\rho\left(\Omega_,^* - \frac{\omega_,\upsilon_, + \omega_,'\upsilon_,'}{\Upsilon_,}\right) - \mu\left(\Upsilon_,^* - \frac{\omega_,\upsilon_, + \omega_,'\upsilon_,'}{\Omega_,}\right)\right\},$$

$$= \frac{\pi i\,(\omega_,\upsilon_,' - \omega_,'\upsilon_,)}{\Omega_,\Upsilon_,\,\text{mod.}\,(\omega_,\upsilon_,' - \omega_,'\upsilon_,)}\,(\rho\Omega_, + \mu\Upsilon_,),$$

ou enfin

$$\{\overline{2k+1}\,(B+\beta)-B_{,}\}\,\Omega = \quad \beta_{,}(\rho,\ \mu)_{,}\,;$$

et de même

$$\{\overline{2k+1}\,(B-\beta)-B_{,}\}\,\Omega = -\,\beta_{,}(\nu,\ \lambda)\,;$$

équations qui seront bientôt utiles.

Je suppose d'abord que $2k+1$ soit égal à l'unité, transformation que l'on peut nommer *triviale*. La fonction $\text{y}x$ est définie par l'équation

$$\text{y}x = \epsilon^{-\frac{1}{2}Bx^2}\,x\Pi\left\{1+\frac{x}{(m,\ n)}\right\},\quad \text{mod.}\,(m,\ n) < T,\ T=\infty\,;$$

dans $(m,\ n) = m\Omega + n\Upsilon$, les entiers $m,\ n$ doivent prendre toutes les valeurs positives ou négatives (le seul système $m=0,\ n=0$ excepté) qui satisfont à l'inégalité

$$\text{mod.}\,(m,\ n) < T,$$

dont le second membre T sera ensuite supposé infini. Soit $\text{y}_{,}x$ la fonction correspondante pour les périodes $\Omega_{,},\ \Upsilon_{,}\,$; on aura

$$\text{y}_{,}\,x = \epsilon^{-\frac{1}{2}B_{,}x^2}\,x\Pi\left\{1+\frac{x}{(m,\ n)_{,}}\right\},\quad \text{mod.}\,(m,\ n)_{,} < T,\qquad T=\infty\,.$$

Or

$$\begin{aligned}(m,\ n)_{,} &= m\Omega_{,} + n\Upsilon_{,}\,,\\ &= m\,(\lambda\Omega+\mu\Upsilon)+n\,(\nu\Omega+\rho\Upsilon),\\ &= (\lambda m+\nu n)\,\Omega+(\mu m+\rho n)\,\Upsilon,\\ &= m_{,}\Omega+n_{,}\Upsilon,\\ &= (m_{,},\ n_{,}).\end{aligned}$$

En écrivant, comme nous venons de le faire,

$$\begin{aligned}m_{,} &= \lambda m+\nu n,\\ n_{,} &= \nu m+\rho n,\end{aligned}$$

on voit tout de suite qu'à chaque système de valeurs entières de $m,\ n$, correspond un système, et un seul système, de valeurs entières de $m_{,},\ n_{,}$; et que de même à chaque système de valeurs entières de $m_{,},\ n_{,}$, correspond un système, et un seul système, de valeurs entières de $m,\ n$; de plus, les systèmes $m=0,\ n=0$ et $m_{,}=0,\ n_{,}=0$, correspondent l'un à l'autre. Il est donc permis d'écrire

$$x\Pi\left\{1+\frac{x}{(m,\ n)}\right\} = x\Pi\left\{1+\frac{x}{(m,\ n)_{,}}\right\},$$

les limites comme auparavant; car, à cause de

$$(m,\ n)_{,} = (m_{,},\ n_{,}),$$

la condition pour les limites, savoir :

$$\text{mod.}\,(m_,\ n)_, < T,\ \ T = \infty\,,$$

devient

$$\text{mod.}\,(m,\ n) < T,\ \ T = \infty\,.$$

Cela donne enfin l'équation

$$\mathrm{y}_{,}x = \epsilon^{-\frac{1}{2}(B_,-B)x^2}\,\mathrm{y}x\,;$$

et, au moyen de cette équation, on obtient une équation correspondante pour la transformation de l'une quelconque des fonctions $\mathrm{y}x$, $\mathrm{g}x$, Gx, Zx, définies par les équations

$$\mathrm{y}x = \epsilon^{-\frac{1}{2}Bx^2}\,.\,x\Pi\left\{1+\frac{x}{(m,\ n)}\right\},\quad \text{mod.}\,(m,\ n) < T,$$

$$\mathrm{g}x = \epsilon^{-\frac{1}{2}Bx^2}\,.\ \ \Pi\left\{1+\frac{x}{(\overline{m},\ n)}\right\},\quad \text{mod.}\,(\overline{m},\ n) < T,\quad T=\infty\,;$$

$$Gx = \epsilon^{-\frac{1}{2}Bx^2}\,.\ \ \Pi\left\{1+\frac{x}{(m,\ \overline{n})}\right\},\quad \text{mod.}\,(m,\ \overline{n}) < T,$$

$$Zx = \epsilon^{-\frac{1}{2}Bx^2}\,.\ \ \Pi\left\{1+\frac{x}{(\overline{m},\ \overline{n})}\right\},\quad \text{mod.}\,(\overline{m},\ \overline{n}) < T,$$

(équations dans lesquelles $\overline{m} = m + \frac{1}{2}$, $\overline{n} = n + \frac{1}{2}$). Je prends par exemple la fonction $\mathrm{g}x$, et j'écris dans l'équation entre $\mathrm{y}_{,}x$ et $\mathrm{y}x$, $x + \frac{1}{2}\Omega$ au lieu de x. Soit pour un moment $\rho = 2\rho' + 1$, $\mu = 2\mu'$; cela donne

$$x + \tfrac{1}{2}\Omega = x + \tfrac{1}{2}(\rho\Omega_, - \mu\Upsilon_,) = x + (\overline{\rho}', -\mu').$$

Donc

$$\mathrm{y}_{,}(x + \tfrac{1}{2}\Omega) = \epsilon^{\beta_,x(\overline{\rho}',\ \mu')_,}\,M_,\mathrm{g}_,x,$$

c'est-à-dire

$$\mathrm{y}_{,}(x + \tfrac{1}{2}\Omega) = \epsilon^{\frac{1}{2}\beta_,x(\rho,\ \mu)_,}\,M_,\mathrm{g}_,x\,;$$

de plus,

$$\mathrm{y}\ (x + \tfrac{1}{2}\Omega) = \epsilon^{\frac{1}{2}\beta\Omega x}\,M\mathrm{g}x.$$

Ces substitutions étant effectuées, les coefficients M, $M_,$ doivent être éliminés en écrivant $x = 0$; cela donne

$$\mathrm{g}_,x = \epsilon^{-\frac{1}{2}(B_,-B)\,x^2}\,\epsilon^{\frac{1}{2}x[(B+\beta-B_,)\,\Omega-\frac{1}{2}\beta_,(\rho,\ \mu)_,]}\,\mathrm{g}x,$$

ou enfin, au moyen d'une équation déjà trouvée,

$$\mathrm{g}_,x = \epsilon^{-\frac{1}{2}(B_,-B)x^2}\,\mathrm{g}x,$$

et de même pour les fonctions Gx, Zx.

Donc enfin, en représentant par Jx l'une quelconque des fonctions $\mathrm{y}x$, gx, Gx, Zx, on aura

$$J_{,}x = \epsilon^{-\frac{1}{2}(B_{,}-B)x^2} Jx,$$

où $J_{,}x$ est ce que devient Jx au moyen d'une transformation triviale (propre et régulière) des périodes.

Je passe à présent à la transformation pour un nombre impair et premier $(2k+1)$ quelconque; mais pour cela on a besoin de connaître la valeur de la fonction

$$u' = \Pi\left\{1 + \frac{x}{(m,\ n) + y}\right\}, \quad \text{mod. } \{(m,\ n) + y\} < T, \quad T = \infty,$$

où $y = a + bi$ est une quantité réelle ou imaginaire quelconque.

Soit u ce que devient u' en prenant pour la condition par rapport aux limites

$$\text{mod. } (m,\ n) < T, \quad T = \infty;$$

on trouve sans peine

$$u = \epsilon^{\frac{1}{2}Bx^2 + Bxy} \frac{\mathrm{y}\,(x+y)}{\mathrm{y}\,(y)}.$$

Pour trouver u', je forme l'équation

$$u \ :\ u' = \Pi\left\{1 + \frac{x}{(m,\ n)}\right\},$$

la limite inférieure du produit infini double étant

$$\text{mod. } \{(m,\ n) + y\} > T,$$

et la limite supérieure

$$\text{mod. } (m,\ n) < T, \quad T = \infty;$$

cela donne

$$\begin{aligned}\log u - \log u' &= x \sum \frac{1}{(m,\ n)+y} - \tfrac{1}{2} x^2 \sum \frac{1}{\{(m,\ n)+y\}^2} + \dots, \\ &= x \sum \frac{1}{(m,\ n)} - \tfrac{1}{2}(x^2 + 2yx) \sum \frac{1}{(m,\ n)^2} + \dots, \\ &= x \sum \frac{1}{(m,\ n)};\end{aligned}$$

car on peut démontrer que

$$\sum \frac{1}{(m,\ n)^2} = 0, \quad \sum \frac{1}{(m,\ n)^3} = 0, \ \text{\&c.}$$

Pour cela, observons que m et n étant infinis puisque T l'est, la première des sommes dont il s'agit peut se remplacer par l'intégrale double

$$I = \iint \frac{dm\,dn}{(m,\ n)^2},$$

laquelle (en écrivant $m = r\cos\theta$, $n = r\sin\theta$, ce qui donne, comme on sait, $dm\,dn = rdr\,d\theta$) devient

$$I = \iint \frac{dr\,d\theta}{r(\Omega\cos\theta + r\sin\theta)^2},$$

d'où

$$I = \iint \frac{(\log r)\,d\theta}{(\Omega\cos\theta + r\sin\theta)^2},$$

en prenant $(\log r)$ entre les limites convenables. Pour trouver ces limites, j'écris

$$(m,\ n) + y = r(\Omega\cos\theta + \Upsilon\sin\theta) + y\ ;$$

ce qui donne

$$\text{mod.}^2\,\{(m,\ n) + y\} = \{r(\Omega\cos\theta + \Upsilon\sin\theta) + y\}\,\{r(\Omega^*\cos\theta + \Upsilon^*\sin\theta) + y^*\},$$

savoir, à l'une des limites

$$r^2(\Omega\cos\theta + \Upsilon\sin\theta)(\Omega^*\cos\theta + \Upsilon^*\sin\theta)$$
$$+ r\{y^*(\Omega\cos\theta + \Upsilon\sin\theta) + y(\Omega^*\cos\theta + \Upsilon^*\sin\theta)\} + T^2 = 0\ ;$$

ou, en négligeant les puissances négatives de T,

$$r = \frac{T}{\sqrt{(\Omega\cos\theta + \Upsilon\sin\theta)(\Omega^*\cos\theta + \Upsilon^*\sin\theta)}}$$
$$- \tfrac{1}{2}\left\{\frac{y}{\Omega\cos\theta + \Upsilon\sin\theta} + \frac{y^*}{\Omega^*\cos\theta + \Upsilon^*\sin\theta}\right\},$$

et à l'autre limite,

$$r = \frac{T}{\sqrt{(\Omega\cos\theta + \Upsilon\sin\theta)(\Omega^*\cos\theta + \Upsilon^*\sin\theta)}}.$$

Or, en représentant ces deux équations par

$$r = R - \phi, \quad r = R,$$

on trouve, pour la valeur de $(\log r)$ entre les deux limites,

$$\log R - \log(R - \phi) = -\log\left(1 - \frac{\phi}{R}\right) = 0,$$

à cause de la valeur infinie de R. Ainsi la somme cherchée est nulle ; et il est tout clair que les sommes suivantes $\sum \frac{1}{(m,\ n)^3}$, &c., se réduisent de même à zéro.

Donc enfin,

$$\log u - \log u' = x\sum \frac{1}{(m,\ n)}.$$

Cela fait voir que

$$u' = \epsilon^{-kx}\, u,$$

le coefficient k étant donné au moyen de l'équation

$$k = \sum \frac{1}{(n,\ m)},$$

où la somme est prise, comme auparavant, entre les limites

$$\text{mod.}\ \{(m,\ n) + y\} > T, \quad \text{mod.}\ (m,\ n) < T, \quad T = \infty.$$

Mais il n'est pas permis d'écrire

$$k = \iint \frac{dm\, dn}{(m,\ n)}.$$

En effet, cette intégrale n'est que le premier terme d'une suite dont il faudrait, pour obtenir un résultat exact, prendre deux termes; le second terme de la suite serait une intégrale prise le long d'un contour, et il serait, ce me semble, très-difficile d'en trouver la valeur. Pour trouver la valeur de k, je remarque que k sera fonction linéaire des quantités T, y, y^*, $\frac{y^2}{T}$, &c., qui entrent dans les valeurs de r; donc, puisqu'en dernière analyse $T = \infty$, k ne peut être que de la forme $Ly + My^*$. Cela étant, en substituant pour u' sa valeur, je forme l'équation

$$\frac{\mathrm{y}(x+y)}{\mathrm{y}(y)} = \epsilon^{-\frac{1}{2}Bx^2}\, \epsilon^{(-By+Ly+My^*)x} \,.\, \Pi \left\{1 + \frac{x}{(m,\ n)+y}\right\},$$

$$\text{mod.}\ \{(m,\ n) + y\} < T, \quad T = \infty,$$

et j'écris successivement

$$y = \tfrac{1}{2}\Omega, \quad y = \tfrac{1}{2}\Upsilon,$$

ce qui donne pour les valeurs correspondantes du produit infini double $\epsilon^{-\frac{1}{2}Bx^2}.\, gx$ et $\epsilon^{-\frac{1}{2}Bx^2}.\, Gx$; en comparant les valeurs ainsi obtenues avec les équations qui donnent les valeurs de $\mathrm{y}(x + \frac{1}{2}\Omega)$, $\mathrm{y}(x + \frac{1}{2}\Upsilon)$, on trouve

$$L = 0, \quad M = \frac{\pi}{\text{mod.}\ (\omega\upsilon' - \omega'\upsilon)},$$

ou enfin,

$$\frac{\mathrm{y}(x+y)}{\mathrm{y}(y)} = \epsilon^{-\frac{1}{2}Bx^2}\, \epsilon^{\left(-By + \frac{\pi}{\text{mod.}\ (\omega\upsilon' - \omega'\upsilon)} y^*\right)x} \,.\, \Pi \left\{1 + \frac{x}{(m,\ n)+y}\right\},$$

$$\text{mod.}\ \{(m,\ n) + y\} < T, \quad T = \infty,$$

laquelle est l'équation qu'il s'agissait d'établir. Il est à peine nécessaire de faire la remarque que pour $y = 0$, on doit considérer à part le facteur $1 + \frac{x}{y}$, lequel multiplié par $\mathrm{y}(y)$ devient tout simplement x; l'équation subsiste donc dans ce cas.

En revenant au problème des transformations linéaires, partant des équations

$$(2k+1)\,\Omega_{,} = \lambda\Omega + \mu\Upsilon,$$
$$(2k+1)\,\Upsilon_{,} = \nu\Omega + \rho\Upsilon,$$

je suppose d'abord que les coefficients λ, ν ne satisfassent pas à la fois aux deux conditions

$$\lambda \equiv 0, \quad \nu \equiv 0, \quad \text{mod.}\,(2k+1),$$

et je prends p, q des entiers quelconques tels, que $\lambda p + \nu q$ ne soit pas $\equiv 0$, mod. $(2k+1)$.

Cela étant, soient

$$\lambda p + \nu q = p_{,},$$
$$\mu p + \rho q = q_{,},$$
$$(2k+1)\,\psi = p_{,}\Omega + q_{,}\Upsilon,$$

et, par conséquent,

$$\psi = p\Omega_{,} + q\Upsilon_{,}.$$

Je forme l'équation

$$(m_{,},\ n_{,}) + s\psi = (m,\ n)_{,},$$

savoir

$$m_{,}\Omega + n_{,}\Upsilon + s\psi = m\Omega_{,} + n\Upsilon_{,},$$

c'est-à-dire

$$\lambda m + \nu n - sp_{,} = (2k+1)\,m_{,},$$
$$\mu m + \nu n - sq_{,} = (2k+1)\,n_{,}\,,$$

ou, ce qui est la même chose,

$$m - sp = m_{,}\rho - n_{,}\nu\,,$$
$$n - sq = m_{,}\mu - n_{,}\lambda.$$

Or, $m_{,}$, $n_{,}$, s étant des entiers donnés, m, n seront aussi des entiers; de même, m, n étant des entiers donnés, on trouve de k à $-k$ un entier s qui donne $m_{,}$ un entier. Mais cela étant, $n_{,}$ sera aussi un entier; car autrement $n_{,}$ serait une fraction ayant pour dénominateur, lequel on voudrait, des nombres $2k+1$, λ, ν, ce qui est impossible à moins que

$$\lambda \equiv 0, \quad \nu \equiv 0, \quad \text{mod.}\,(2k+1).$$

Mais si ces équations avaient lieu, on trouverait d'abord s de manière à avoir $n_{,}$ entier, et alors, puisqu'on n'a pas aussi

$$\mu \equiv 0, \quad \rho \equiv 0, \quad \text{mod.}\,(2k+1)$$

(en effet, cela est impossible à cause de l'équation $\lambda\rho - \mu\nu = 2k+1$), on démontrerait, comme auparavant, pour $n_{,}$, que $m_{,}$ est entier. Donc, enfin, m, n étant des entiers donnés, on trouve pour $m_{,}$, $n_{,}$, s un système d'entiers tel que s soit compris de k à $-k$, et l'on voit sans peine qu'il n'y a qu'un seul système de cette espèce.

A présent, partant de l'équation

$$\frac{\mathrm{y}(x+y)}{\mathrm{y}(y)}=\epsilon^{-\frac{1}{2}Bx^2}\cdot\epsilon^{\left(-By+\frac{\pi}{\text{mod.}\,(\omega v'-\omega' v)}y^*\right)x}\cdot\Pi\left\{1+\frac{x}{(m_{,},\ n_{,})+y}\right\}$$

(et faisant attention à la particularité que présente le cas de $y=0$), j'écris successivement

$$y=0,\quad y=\pm\psi,\ldots,\quad y=\pm k\psi,$$

et je forme le produit des équations ainsi trouvées. Cela donne, à cause de $(m_{,},\ n_{,})+s\psi=(m,\ n)_{,}$,

$$\Pi\,\frac{\mathrm{y}(x+s\psi)}{\mathrm{y}(s\psi)}=\epsilon^{-\frac{1}{2}(2k+1)\,Bx^2}.\ x\Pi\left\{1+\frac{x}{(m,\ n)_{,}}\right\},$$

la condition, par rapport aux limites, étant

$$\text{mod.}\,(m,\ n)_{,}<T,\quad T=\infty.$$

Or

$$\mathrm{y}_{,}x=\epsilon^{-\frac{1}{2}B_{,}x^2}.\ x\Pi\left\{1+\frac{x}{(m,\ n)_{,}}\right\},$$

avec la même condition, par rapport aux limites; donc, enfin,

$$\mathrm{y}_{,}x=\epsilon^{-\frac{1}{2}[B_{,}-(2k+1)B]x^2}.\ \Pi\left\{\frac{\mathrm{y}(x+s\psi)}{\mathrm{y}(s\psi)}\right\},$$

où, dans le numérateur, s doit avoir toutes les valeurs entières depuis $s=-k$ jusqu'à $s=+k$, y compris $s=0$, et dans le dénominateur ces mêmes valeurs, hormis la valeur $s=0$.

Il est, à présent, facile de faire voir que cette propriété subsiste pour l'une quelconque des fonctions $\mathrm{y}x$, $\mathrm{g}x$, Gx, Zx; en effet, pour la démontrer pour $\mathrm{g}x$, j'écris $x+\frac{1}{2}\Omega$ au lieu de x; en prenant, pour un moment, $\rho=2\rho'+1$, $\mu=2\mu'$, cela donne

$$x+\tfrac{1}{2}\Omega=+(\bar{\rho}',\ -\mu'),$$

$$\mathrm{y}_{,}(x+\tfrac{1}{2}\Omega)=\epsilon^{\beta,x(\bar{\rho}',\ \mu')_{,}}M\mathrm{g}_{,}x=\epsilon^{\frac{1}{2}\beta,x(\rho,\ \mu)_{,}}M_{,}\mathrm{g}x,$$

c'est-à-dire

$$\frac{\mathrm{y}_{,}(x+\frac{1}{2}\Omega)}{\mathrm{y}_{,}(\frac{1}{2}\Omega)}=\epsilon^{\frac{1}{2}\beta,x(\rho,\ \mu)_{,}}\,\mathrm{g}_{,}x.$$

Or, on déduit de l'expression pour $\mathrm{y}_{,}x$,

$$\epsilon^{-\frac{1}{2}\beta,x(\rho,\ \mu)_{,}}.\ \frac{\mathrm{y}_{,}(x+\frac{1}{2}\Omega)}{\mathrm{y}_{,}(\frac{1}{2}\Omega)}$$

$$=\epsilon^{-\frac{1}{2}\beta,x(\rho,\ \mu)_{,}}\,\epsilon^{-\frac{1}{2}(B_{,}-\overline{2k+1}B)(x^2+\Omega x)}.\ \Pi\,\frac{\mathrm{y}(x+s\psi+\frac{1}{2}\Omega)}{\mathrm{y}(s\psi+\frac{1}{2}\Omega)},$$

$$=\epsilon^{-\frac{1}{2}\beta,x(\rho,\ \mu)_{,}}\,\epsilon^{-\frac{1}{2}(B_{,}-\overline{2k+1}B)x^2}\,\epsilon^{\frac{1}{2}(2k+1)\beta\Omega x}\,\Pi\,\frac{\mathrm{g}(x+s\psi)}{\mathrm{g}(s\psi)};$$

c'est-à-dire

$$\mathrm{g}_{,}x = \epsilon^{-\frac{1}{2}(B_{,}-\overline{2k+1}B)x^2}\, \epsilon^{\frac{1}{2}x\,\{-B_{,}\Omega+\overline{2k+1}\,(B+\beta)\,\Omega-\beta_{,}\,(\rho,\ \mu)_{,}\}}\, \Pi\, \frac{\mathrm{g}\,(x+s\psi)}{\mathrm{g}\,(s\psi)},$$

ou enfin, à cause de l'équation

$$-B_{,}\Omega + \overline{2k+1}\,(B+\beta)\,\Omega - \beta_{,}\,(\rho,\ \mu)_{,} = 0,$$

la valeur de $\mathrm{g}_{,}x$ est

$$\mathrm{g}_{,}x = \epsilon^{-\frac{1}{2}(B_{,}-\overline{2k+1}B)x^2}\,.\,\Pi\,\frac{\mathrm{g}\,(x+s\psi)}{\mathrm{g}\,(s\psi)}:$$

et en représentant, comme auparavant, l'une quelconque des fonctions yx, gx, Gx, Zx par Jx, on a l'équation

$$J_{,}x = \epsilon^{-\frac{1}{2}(B_{,}-\overline{2k+1}B)x^2}\,.\,\Pi\,\frac{J\,(x+s\psi)}{J\,(s\psi)},$$

équation dans laquelle s doit avoir, dans le numérateur, toutes les valeurs entières depuis $s=-k$ jusqu'à $s=k$, y compris $s=0$, et dans le dénominateur, ces mêmes valeurs, hormis la valeur $s=0$.

Je suppose que les valeurs de $p_{,}$, $q_{,}$ soient données (cela va sans dire que l'on ne doit pas avoir à la fois $p_{,}\equiv 0$, $q_{,}\equiv 0$, mod. $2k+1$), et je remarque que l'on a, pour déterminer λ, μ, ν, ρ, les conditions

$$\begin{aligned} \rho p_{,} - \nu q_{,} &\equiv 0, \qquad \text{mod. } (2k+1), \\ -\mu p_{,} + \lambda q_{,} &\equiv 0, \\[6pt] \lambda \equiv 1, \quad \mu &\equiv 0, \qquad \text{mod. } 2, \\ \nu \equiv 0, \quad \rho &\equiv 1, \\[6pt] \lambda\rho - \mu\nu &= 2k+1. \end{aligned}$$

Et cela étant, on aura ensuite, en rassemblant toutes les équations qui ont rapport à la transformation,

$$\begin{aligned} \rho p_{,} - \nu q_{,} &= (2k+1)\,p, \\ -\mu p_{,} + \lambda q_{,} &= (2k+1)\,q, \\ \Psi &= p_{,}\Omega + q_{,}\Upsilon, \\ (2k+1)\,\Omega_{,} &= \lambda\Omega + \mu\Upsilon\,, \\ (2k+1)\,\Upsilon_{,} &= \nu\,\Omega + \rho\Upsilon\,. \end{aligned}$$

Or, quoique les valeurs de λ, μ, ν, ρ ne soient pas complétement déterminées au moyen de ces conditions, cependant il est clair que la valeur de la fonction $J_{,}x$ ne dépend que des valeurs de $p_{,}$, $q_{,}$ (en effet, ces valeurs suffisent pour déterminer la quantité $\Psi = p_{,}\Omega + q_{,}\Upsilon$, de laquelle dépend la fonction $J_{,}x$). Les formes différentes de $J_{,}x$, pour les systèmes de valeurs de λ, μ, ν, ρ, qui correspondent à des valeurs données de $p_{,}$, $q_{,}$, doivent donc se dériver de l'une quelconque de ces formes, au moyen d'une transformation triviale des modules $\Omega_{,}$, $\Upsilon_{,}$. Il est, de plus, clair que les valeurs de $p_{,}$, $q_{,}$, qui sont égales à des multiples de $(2k+1)$ près, ne donnent qu'une seule valeur de $J_{,}x$. Je suppose d'abord que

$$p_{,} \equiv 0, \qquad \text{mod.}\,(2k+1),$$

on peut trouver un entier θ tel que

$$\theta p_{,} \equiv 1, \qquad \text{mod.}\,(2k+1)\,;$$

en prenant alors

$$\theta q_{,} \equiv q_{,}, \qquad \text{mod.}\,(2k+1),$$

cela donne

$$\theta\,(p_{,}\Omega + q_{,}\Upsilon_{,}) \equiv \Omega + q_{,}\Upsilon, \quad \text{mod.}\,(2k+1),$$

savoir

$$\theta\Upsilon \equiv \Omega + q_{,}\Upsilon, \quad \text{mod.}\,(2k+1).$$

Mais en donnant à s des valeurs entières quelconques, depuis $-k$ jusqu'à k, le système des valeurs de $s\psi$ est équivalent au système des valeurs de $s\theta\psi$, mod. $(2k+1)$; il est donc permis d'écrire, sans perte de généralité,

$$\Psi = \Omega + q_{,}\Upsilon.$$

De même pour

$$p_{,} \equiv 0, \quad \text{mod.}\,(2k+1),$$

on démontre que l'on peut donner à $q_{,}$ une valeur quelconque, sans changer pour cela la valeur de $J_{,}x$; il convient d'avoir $p_{,}$ impair et $q_{,}$ pair. J'écris donc, pour le premier cas, $2\mathrm{q}_{,}$ au lieu de $q_{,}$, et je suppose que, dans le deuxième cas, les valeurs de $p_{,}$, $q_{,}$ soient

$$p_{,} = 2k+1, \quad q_{,} = 2.$$

Cela donne:

Premier cas.

$$\Psi = \Omega + 2\mathrm{q}_{,}\Upsilon,$$

$\mathrm{q}_{,}$ un entier quelconque, y compris zéro, depuis $-k$ jusqu'à $+k$.

Deuxième cas.

$$\Psi = (2k+1)\,\Omega + 2\Upsilon\,;$$

le nombre des valeurs différentes de Ψ sera donc, en tout, $2k+2$.

On obtient tout de suite, pour le premier cas, le système d'équations

$$\begin{aligned} p_{,} &= 1, & q_{,} &= 2\mathrm{q}_{,}, \\ \lambda &= 1, & \mu &= 2\mathrm{q}_{,}, \\ \nu &= 0, & \rho &= (2k+1), \\ p &= 1, & q &= 0\,; \end{aligned}$$

$$\left\{ \begin{aligned} \psi &= \frac{1}{2k+1}(\Omega + 2\mathrm{q}_{,}\Upsilon), \\ \Omega_{,} &= \frac{1}{2k+1}(\Omega + 2\mathrm{q}_{,}\Upsilon) = \psi, \\ \Upsilon_{,} &= \Upsilon. \end{aligned} \right.$$

Le cas particulier le plus simple est celui de $q_{,} = 0$; cela donne

$$\psi = \Omega_{,} = \frac{1}{2k+1}\Omega, \qquad \Upsilon_{,} = \Upsilon,$$

et, de là,

$$\frac{\Omega_{,}}{\Upsilon_{,}} = \frac{1}{2k+1}\,\frac{\Omega}{\Upsilon};$$

et même le cas général se réduit à celui-ci, car, au moyen d'une transformation *triviale*, on obtiendrait

$$\Omega' = \Omega + 2\mathrm{q}_{,}\Upsilon, \qquad \Upsilon' = \Upsilon,$$

et puis

$$\psi = \Omega_{,} = \frac{1}{2k+1}\Omega', \qquad \Upsilon_{,} = \Upsilon',$$

et, de là,

$$\frac{\Omega_{,}}{\Upsilon_{,}} = \frac{1}{2k+1}\,\frac{\Omega'}{\Upsilon'}.$$

Les équations correspondantes pour le deuxième cas sont:

$$\begin{aligned} p_{,} &= 2k+1, & q_{,} &= 2, \\ \lambda &= 2k+1, & \mu &= 0, \\ \nu &= 0, & \rho &= 1, \\ p &= 1, & q &= 2, \end{aligned}$$

$$\psi = \frac{1}{2k+1}\{(2k+1)\,\Omega + 2\Upsilon\}$$

$$\Omega_{,} = \Omega,$$

$$\Upsilon_{,} = \frac{1}{2k+1}\Upsilon;$$

ce qui donne

$$\frac{\Omega_{,}}{\Upsilon_{,}} = (2k+1)\frac{\Omega}{\Upsilon}.$$

J'ajoute, sans m'arrêter pour les démontrer, quelques formules de transformation pour le nombre 2; je trouve d'abord

$$\Omega_{,} = \tfrac{1}{2}\Omega_{,}, \quad \Upsilon_{,} = \Upsilon,$$

$$\left\{\begin{aligned} \mathrm{y}_{,}x &= \epsilon^{-\frac{1}{4}(B_{,}-2B)x^2}\, \mathrm{y}x\, \mathrm{g}x, \\ \mathrm{g}_{,}x &= \epsilon^{-\frac{1}{2}(B_{,}-2B)x^2}\, \frac{\mathrm{g}(x-\frac{1}{4}\Omega)\, \mathrm{g}(x+\frac{1}{4}\Omega)}{\mathrm{g}^2(\frac{1}{4}\Omega)}, \\ G_{,}x &= \epsilon^{-\frac{1}{4}(B_{,}-2B)x^2}\, Gx\, Zx, \\ Z_{,}x &= \epsilon^{-\frac{1}{2}(B_{,}-2B)x^2}\, \frac{Z(x-\frac{1}{4}\Omega)\, Z(x+\frac{1}{4}\Omega)}{Z^2(\frac{1}{4}\Omega)}. \end{aligned}\right.$$

Ces équations donnent, en introduisant les fonctions elliptiques, ϕx, fx, Fx données au moyen de

$$\phi x = \frac{\mathrm{y}x}{Zx}, \qquad fx = \frac{gx}{Zx}, \qquad Fx = \frac{Gx}{Zx},$$

les équations

$$\frac{\phi_{,}x}{F_{,}x} = \phi x\, \frac{fx}{Fx},$$

$$f_{,}x = \frac{f(x-\frac{1}{4}\Omega)\, f(x+\frac{1}{4}\Omega)}{f^2(\frac{1}{4}\Omega)},$$

dont la seconde peut encore s'écrire sous la forme

$$f_{,}x = \frac{1 - c^2\, \dfrac{F^2(\frac{1}{4}\Omega)}{f^2(\frac{1}{4}\Omega)}\, \phi^2 x}{1 + e^2 c^2 \phi^2(\frac{1}{4}\Omega)\, \phi^2 x},$$

et les deux équations combinées ensemble conduisent sans peine à la valeur des modules $c_{,}$, $e_{,}$. On trouve en effet, en mettant comme à l'ordinaire $b^2 = c^2 + e^2$,

$$c_{,}^2 = 4bc,$$

$$e_{,}^2 = (b-c)^2,$$

et puis

$$\phi_{,}x = \frac{1 - c(c-b)\, \phi^2 x}{\phi x\, fx},$$

$$f_{,}x = \frac{1 - c(c+b)\, \phi^2 x}{1 - c(c-b)\, \phi^2 x},$$

$$F_{,}x = \frac{Fx}{1 - c(c-b)\, \phi^2 x},$$

formules qui correspondent à celles de la transformation de Lagrange. Les équations pour $\mathrm{y}_{,}x$, $Z_{,}x$ donnent encore une valeur de $\phi_{,}x$, laquelle, égalée à la valeur qui vient d'être trouvée, donne

$$\frac{\mathrm{y}x\, \mathrm{g}x\, Z^2(\frac{1}{4}\Omega)}{Z(x-\frac{1}{4}\Omega)\, Z(x+\frac{1}{4}\Omega)} = \frac{\phi x\, fx}{1 - c(c-b)\, \phi^2 x}.$$

On obtient tout de suite les formules pour la transformation analogue $\Omega_, = \Omega$, $\Upsilon_, = \frac{1}{2}\Upsilon$. Mais il faut de plus considérer le système

$$\Omega_, = \tfrac{1}{2}(\Omega - \Upsilon), \quad \Upsilon_, = \tfrac{1}{2}(\Omega + \Upsilon):$$

on aura alors

$$\left\{\begin{aligned}
\mathrm{y}_, x &= \epsilon^{-\frac{1}{2}(B_, - 2B)x^2}\, \mathrm{g}x\, Zx, \\
\mathrm{g}_, x &= \epsilon^{-\frac{1}{2}(B_, - 2B)x^2} \frac{\mathrm{y}\,(x + \frac{1}{4}\overline{\Omega - \Upsilon})\, \mathrm{y}\,(x - \frac{1}{4}\overline{\Omega - \Upsilon})}{-\mathrm{y}^2\,(\frac{1}{4}\overline{\Omega - \Upsilon})} \\
&= \epsilon^{-\frac{1}{2}(B_, - 2B)x^2} \frac{Z\,(x + \frac{1}{4}\overline{\Omega - \Upsilon})\, Z(x + \frac{1}{4}\overline{\Omega - \Upsilon})}{Z^2\,(\frac{1}{4}\overline{\Omega - \Upsilon})}, \\
G_, x &= \epsilon^{-\frac{1}{2}(B_, - 2B)x^2} \frac{\mathrm{y}\,(x + \frac{1}{4}\overline{\Omega + \Upsilon})\, \mathrm{y}\,(x - \frac{1}{4}\overline{\Omega + \Upsilon})}{-\mathrm{y}^2\,(\frac{1}{4}\overline{\Omega + \Upsilon})} \\
&= \epsilon^{-\frac{1}{2}(B - 2B)x^2} \frac{Z\,(x + \frac{1}{4}\overline{\Omega + \Upsilon})\, Z\,(x - \frac{1}{4}\overline{\Omega + \Upsilon})}{Z^2\,(\frac{1}{4}\overline{\Omega + \Upsilon})}, \\
Z_, x &= \epsilon^{-\frac{1}{2}(B_, - 2B)x^2}\, \mathrm{g}x\, Gx\,;
\end{aligned}\right.$$

et puis, en écrivant

$$c_,^2 = (e - ic)^2, \quad -e_,^2 = (e + ic)^2,$$

on obtient

$$\phi_, x = \frac{\phi x}{fx\, Fx},$$

$$f_, x = \frac{1 + ice\, \phi^2 x}{fx\, Fx},$$

$$F_, x = \frac{1 - ice\, \phi^2 x}{fx\, Fx};$$

$$\frac{1 + ice\, \phi^2 x}{fx\, Fx} = \frac{Z\,(x + \frac{1}{4}\overline{\Omega - \Upsilon})\, Z\,(x - \frac{1}{4}\overline{\Omega - \Upsilon})}{Z^2\,(\frac{1}{4}\overline{\Omega - \Upsilon})\, \mathrm{g}x\, Gx},$$

$$\frac{1 - ice\, \phi^2 x}{fx\, Fx} = \frac{Z\,(x + \frac{1}{4}\overline{\Omega + \Upsilon})\, Z\,(x - \frac{1}{4}\overline{\Omega + \Upsilon})}{Z^2\,(\frac{1}{4}\overline{\Omega + \Upsilon})\, \mathrm{g}x\, Gx},$$

$$\frac{1 + ice\, \phi^2 x}{1 - ice\, \phi^2 x} = \frac{Z\,(x + \frac{1}{4}\overline{\Omega - \Upsilon})\, Z\,(x - \frac{1}{4}\overline{\Omega - \Upsilon})\, Z^2\,(\frac{1}{4}\overline{\Omega + \Upsilon})}{Z\,(x + \frac{1}{4}\overline{\Omega + \Upsilon})\, Z\,(x - \frac{1}{4}\overline{\Omega - \Upsilon})\, Z^2\,(\frac{1}{4}\overline{\Omega - \Upsilon})}:$$

ou, au moins, ces formules seront exactes au signe de i près; car il serait peut-être difficile de déterminer quel est le signe qu'on doit donner à cette quantité.

131.

NOUVELLES RECHERCHES SUR LES COVARIANTS.

[From the *Journal für die reine und angewandte Mathematik* (Crelle), tom. XLVII. (1854), pp. 109—125.]

Je me sers de la notation

$$(a_0,\ a_1, \dots a_n)(x,\ y)^n$$

pour représenter la fonction

$$a_0 x^n + \frac{n}{1} a_1 x^{n-1} y + \dots + a_n y^n:$$

en supposant que les coefficients a_0', a_1' &c. soient donnés par l'équation

$$(a_0,\ a_1, \dots a_n)(\lambda x + \mu y,\ \lambda' x + \mu' y)^n = (a_0',\ a_1', \dots a_n')(x,\ y)^n,$$

supposée identique par rapport à x, y, soit $\phi(a_0,\ a_1, \dots a_n;\ x,\ y)$ une fonction des coefficients et des variables, telle que

$$\phi(a_0',\ a_1', \dots a_n';\quad x,\ y) = (\lambda\mu' - \lambda'\mu)^p\, \phi(a_0,\ a_1, \dots a_n;\quad \lambda x + \mu y,\ \lambda' x + \mu' y);$$

cette fonction ϕ sera généralement un *Covariant*, et dans le cas particulier où ϕ est fonction des seuls coefficients, un *Invariant* de la fonction donnée.

Je suppose d'abord que les nouveaux coefficients soient donnés par l'équation

$$(a_0,\ a_1, \dots a_n)(x + \lambda y,\ y)^n = (a_0',\ a_1', \dots a_n')(x,\ y)^n;$$

cela donne les relations

$$\begin{aligned} a_0' &= a_0, \\ a_1' &= a_1 + \lambda a_0, \\ a_2' &= a_2 + 2\lambda a_1 + \lambda^2 a_0, \\ &\text{\&c.} \end{aligned}$$

Il faut donc que le *covariant* ϕ satisfasse à l'équation

$$\phi(a_0', a_1', \ldots a_n'; \quad x, y) = \phi(a_0, a_1, \ldots a_n; \quad x + \lambda y, y),$$

laquelle peut aussi être écrite comme suit:

$$\phi(a_0', a_1', \ldots a_n'; \quad x - \lambda y, y) = \phi(a_0, a_1, \ldots a_n; \quad x, y). \quad \ldots\ldots\ldots\ldots(X)$$

De même, en faisant

$$(a_0, a_1, \ldots a_n)(x, \mu x + y)^n = (a_0^{\backprime}, a_1^{\backprime}, \ldots a_n^{\backprime})(x, y)^n,$$

ce qui donne

$$\begin{aligned} a_n^{\backprime} &= a_n, \\ a_{n-1}^{\backprime} &= a_{n-1} + \mu a_n, \\ a_{n-2}^{\backprime} &= a_{n-2} + 2\mu a_{n-1} + a_n, \\ &\&c. \end{aligned}$$

le *covariant* ϕ doit satisfaire aussi à l'équation

$$\phi(a_0^{\backprime}, a_1^{\backprime}, \ldots a_n^{\backprime}; \quad x, -\mu x + y) = \phi(a_0, a_1, \ldots a_n; \quad x, y); \quad \ldots\ldots(Y)$$

et réciproquement, toute fonction ϕ homogène par rapport aux coefficients et aussi par rapport aux variables, qui satisfait à ces équations (X, Y), sera un *covariant* de la fonction donnée.

Examinons d'abord l'équation (X) que je représente par $\phi' = \phi$. Soit pour le moment, $a_1' - a_1 = \lambda\alpha_1$, $a_2' - a_2 = \lambda\alpha_2$, &c., alors on aura, comme à l'ordinaire, l'équation symbolique

$$\phi' = e^{\lambda(\alpha_1\partial_{a_1} + \alpha_2\partial_{a_2} \ldots + \alpha_n\partial_{a_n} - y\partial_x)}\phi,$$

où les quantités a_1, a_2, &c., en tant qu'elles entrent dans α_1, α_2, &c., ne doivent pas être affectées par les symboles ∂_{a_1}, ∂_{a_2}, &c. de la différentiation. En substituant les valeurs de α_1, α_2, ..., et en ordonnant selon les puissances de λ, cette équation donne

$$\phi' = e^{\lambda\Box + \lambda^2\Box_1 \ldots + \lambda^n\Box_{n-1} - \lambda y\partial_x}\phi,$$

où les symboles $\Box$, $\Box_1$, &c. sont donnés par

$$\begin{aligned} \Box &= a_0\partial_{a_1} + 2a_1\partial_{a_2} \ldots + na_{n-1}\partial_{a_n}, \\ \Box_1 &= a_0\partial_{a_2} + 3a_1\partial_{a_3} \ldots + \frac{n\,.\,n-1}{1\,.\,2}a_{n-2}\partial_{a_n}, \\ &\vdots \\ \Box_{n-1} &= a_0\partial_{a_n}, \end{aligned}$$

et les quantités a_1, a_2, &c., en tant qu'elles entrent dans les symboles $\Box$, $\Box_1$, &c., ne doivent pas être affectées par les symboles ∂_{a_1}, ∂_{a_2}, &c. de la différentiation. Il est assez remarquable que l'équation symbolique peut aussi être écrite sous la forme plus simple

$$\phi' = e^{\lambda(\Box - y\partial_x)}\phi,$$

où les quantités a_1, a_2, ..., en tant qu'elles entrent dans le symbole $\underset{\bullet}{\square}$, sont censées affectées des symboles ∂_{a_1}, ∂_{a_2}, &c. de la différentiation; de manière que dans le développement, $\underset{\bullet}{\square}^2 . \phi$ par exemple, signifie $\underset{\bullet}{\square} . \underset{\bullet}{\square}\phi$, et ainsi de suite. Je ne m'arrête pas sur ce point, parce que pour ce que je vais démontrer de plus important, il suffit de faire attention à la *première* puissance de λ. D'ailleurs l'intelligibilité des équations dont il s'agit, sera facilitée en faisant les développements et en comparant les puissances correspondantes de λ. Cela donne par exemple:

$$\underset{\bullet}{\square}^2 = \underset{\bullet}{\square}^2 + 2\underset{\bullet}{\square}_1, \quad \underset{\bullet}{\square}^3 = \underset{\bullet}{\square}^3 + 3\underset{\bullet}{\square}\underset{\bullet}{\square}_1 + 6\underset{\bullet}{\square}_2, \text{ \&c.}$$

où les symboles $\underset{\bullet}{\square}^2$, $\underset{\bullet}{\square}^3$ &c. à gauche de ces équations dénotent la double, triple, &c. répétition de l'opération $\underset{\bullet}{\square}$, tandis qu'à côté droit des équations, les quantités a_1, a_2, ... &c., en tant qu'elles entrent dans les symboles $\underset{\bullet}{\square}$, $\underset{\bullet}{\square}_1$, &c. sont censées ne pas être affectées des symboles ∂_{a_1}, ∂_{a_2}, &c. de la différentiation. Dans la suite, si le contraire n'est pas dit, je me servirai des expressions $\underset{\bullet}{\square}^2$, $\underset{\bullet}{\square}^3$, &c. pour dénoter les répétitions de l'opération, et de même pour les combinaisons de deux ou de plusieurs symboles.

Cela étant, l'équation $\phi' = e^{\lambda(\underset{\bullet}{\square} - y\partial_x)}\,\phi = \phi$ donne

$$\phi = \left\{1 + \lambda\,(\underset{\bullet}{\square} - y\partial_x) + \frac{\lambda^2}{1 \,.\, 2}(\underset{\bullet}{\square} - y\partial_x)^2 + \ldots\right\}\phi,$$

où $(\underset{\bullet}{\square} - y\partial_x)^2 . \phi$ (je le répète) équivaut à $(\underset{\bullet}{\square} - y\partial_x) . (\underset{\bullet}{\square} - y\partial_x)\phi$; et ainsi de suite. Il faut d'abord que le coefficient de λ s'évanouisse, ce qui donne $(\underset{\bullet}{\square} - y\partial_x)\,\phi = 0$; et cette condition étant satisfaite, les coefficients des puissances supérieures s'évanouissent d'elles-mêmes; c'est-à-dire, l'équation (X) sera satisfaite en supposant que ϕ satisfait à l'équation aux différences partielles $(\underset{\bullet}{\square} - y\partial_x)\,\phi = 0$.

En posant

$$\begin{aligned}
\overset{\bullet}{\square} \quad &= a_n\partial_{a_{n-1}} + 2a_{n-1}\partial_{a_{n-2}} \ldots + na_1\partial_{a_0}, \\
\overset{\bullet}{\square}_1 \quad &= a_n\partial_{a_{n-2}} + 3a_{n-1}\partial_{a_{n-3}} \ldots + \frac{n\,(n-1)}{1 \,.\, 2}\,a_2\partial_{a_0}, \\
&\vdots \\
\overset{\bullet}{\square}_{n-1} &= a_n\partial_{a_0},
\end{aligned}$$

on fera un raisonnement analogue par rapport à l'équation (Y); et il sera ainsi démontré que ϕ doit satisfaire aussi à l'équation à différences partielles $(\overset{\bullet}{\square} - x\partial_y)\,\phi = 0$; donc enfin, on a le suivant

THÉORÈME. Tout *covariant* ϕ de la fonction

$$(a_0,\ a_1,\ \ldots\ a_n)(x,\ y)^n,$$

satisfait aux deux équations à différences partielles

$$(\underset{\bullet}{\square} - y\partial_x)\,\phi = 0, \quad (\overset{\bullet}{\square} - x\partial_y)\,\phi = 0, \quad \ldots\ldots\ldots\ldots\ldots\ldots\ldots (A)$$

où

$$\begin{aligned}
\underset{\bullet}{\square} &= a_0\partial_{a_1} + 2a_1\partial_{a_2} \ldots + na_{n-1}\partial_{a_n}, \\
\overset{\bullet}{\square} &= na_1\partial_{a_0} + (n-1)\,a_2\partial_{a_1} \ldots + a_n\partial_{a_{n-1}};
\end{aligned}$$

et réciproquement toute fonction, homogène par rapport aux coefficients et par rapport aux variables, qui satisfait à ces équations, est un *covariant* de la fonction donnée.

Par exemple, l'*invariant* $\phi = ac - b^2$ de la fonction $ax^2 + 2bxy + cy^2$ satisfait aux équations

$$(a\partial_b + 2b\partial_c)\phi = 0, \quad (2b\partial_a + c\partial_b)\phi = 0,$$

et le *covariant* $\phi = (ac - b^2)x^2 + (a\partial - bc)xy + (b\partial - c^2)y^2$ de la fonction $ax^3 + 3bx^2y + 3cxy^2 + dy^3$ satisfait aux équations

$$(a\partial_b + 2b\partial_c + 3c\partial_d - y\partial_x)\phi = 0, \quad (3d\partial_c + 2c\partial_b + b\partial_a - x\partial_y)\phi = 0.$$

Il est clair qu'en ne considérant que les fonctions qui restent les mêmes en prenant dans un ordre inverse les coefficients $a_0, a_1, \ldots a_n$ et les variables x, y, respectivement, les *covariants* seront définis par l'une ou l'autre des équations (A), et qu'il n'est plus nécessaire de considérer les deux équations. Cela posé, on trouve assez facilement les *covariants* par la méthode des coefficients indéterminés. Mais il y a à remarquer une circonstance de la plus grande importance dans cette théorie, savoir, que l'on obtient de cette manière un nombre d'équations plus grand qu'il n'en faut pour déterminer les coefficients dont il s'agit. Ces équations cependant, étant liées entre elles, se réduisent au nombre nécessaire d'équations indépendantes.

Cherchons par exemple pour la fonction $ax^3 + 3bx^2y + 3cxy^2 + dy^3$ un *invariant* ϕ de la forme

$$\phi = Aa^2d^2 + Babcd + Cac^3 + Cb^3d + Db^2c^2,$$

contenant les quatre coefficients indéterminés A, B, C, D. En substituant dans l'équation $(a\partial_b + 2b\partial_c + 3c\partial_d)\phi = 0$, on obtient

$$(3C + 2B)ab^2d + (3B + 6C + 2D)abc^2 + (6A + B)ac^2d + (3C + 4D)b^3c = 0\,;$$

or les *quatre* équations données par cette condition, se réduisent à *trois* équations indépendantes, de sorte qu'en faisant par exemple $A = -1$, les autres coefficients seront déterminés, et l'on obtient le résultat connu:

$$\phi = -a^2d^2 + 6abcd - 4ac^3 - 4b^3d + 3b^2c^2.$$

La circonstance mentionnée ci-dessus s'oppose à résoudre de la manière dont il s'agit, le problème de trouver le nombre des *invariants* d'un ordre donné: problème qui a toujours bravé mes efforts.

Avant d'entamer la solution des équations (A), je vais démontrer quelques propriétés générales des *covariants*, et des *invariants*. Pour abréger, je me servirai du mot *pesanteur*, en disant que les coefficients a_0, a_1, &c., ont respectivement les *pesanteurs* $0 - \frac{1}{2}n$, $1 - \frac{1}{2}n$, &c., que les variables x, y ont respectivement les pesanteurs $\frac{1}{2}$, $-\frac{1}{2}$, et que la *pesanteur* d'un produit est égale à la somme des *pesanteurs* des facteurs. Cela posé, je dis que tout *covariant* est composé de termes dont chacun à la pesanteur *zéro*. Pour démontrer cela, j'écris:

$$(\underset{\bullet}{\square} - y\partial_x)(\overset{\bullet}{\square} - x\partial_y) = \underset{\bullet}{\square}\overset{\bullet}{\square} - y\partial_x\overset{\bullet}{\square} - x\partial_y\underset{\bullet}{\square} + xy\,\partial_x\partial_y + y\partial_y,$$

$$(\overset{\bullet}{\square} - x\partial_y)(\underset{\bullet}{\square} - y\partial_x) = \overset{\bullet}{\square}\underset{\bullet}{\square} - y\partial_x\overset{\bullet}{\square} - x\partial_y\underset{\bullet}{\square} + xy\,\partial_x\partial_y + x\partial_x;$$

cela donne

$$(\underset{\bullet}{\square} - y\partial_x)(\overset{\bullet}{\square} - x\partial_y) - (\overset{\bullet}{\square} - x\partial_y)(\underset{\bullet}{\square} - y\partial_x) = \underset{\bullet}{\square}\overset{\bullet}{\square} - \overset{\bullet}{\square}\underset{\bullet}{\square} + y\partial_y - x\partial_x;$$

or, en faisant attention aux valeurs de $\underset{\bullet}{\square}$, $\overset{\bullet}{\square}$, savoir

$$\underset{\bullet}{\square}\overset{\bullet}{\square} = (\underset{\bullet}{\square}\overset{\bullet}{\square}) + na_0\partial_{a_0} + 2(n-1)a_1\partial_{a_1} \ldots + n\;1\;a_{n-1}\partial_{a_{n-1}},$$

$$\overset{\bullet}{\square}\underset{\bullet}{\square} = (\overset{\bullet}{\square}\underset{\bullet}{\square}) + \qquad n\,.\,1\,.\,a_1\partial_{a_1} \ldots + 2(n-1)a_{n-1}\partial_{a_{n-1}} + 1\,.\,na_n\partial_{a_n},$$

où, en formant les produits $(\underset{\bullet}{\square}\overset{\bullet}{\square})$, $(\overset{\bullet}{\square}\underset{\bullet}{\square})$, les quantités $a_0, a_1, \ldots a_n$ sont censées non affectées par les symboles $\partial_{a_0}, \partial_{a_1}, \ldots \partial_{a_n}$ de la différentiation, on en tire

$$\begin{aligned}\underset{\bullet}{\square}\overset{\bullet}{\square} - \overset{\bullet}{\square}\underset{\bullet}{\square} &= n\,a_0\partial_{a_0} + (n-2)a_1\partial_{a_1} \ldots - na_n\partial_{a_n}\\ &= -2\,\{(0-\tfrac{1}{2}n)a_0\partial_{a_0} + (1-\tfrac{1}{2}n)a_1\partial_{a_1} \ldots + (n-\tfrac{1}{2}n)a_n\partial_{a_n}\} = -2\Theta,\end{aligned}$$

en représentant par Θ l'expression symbolique entre les crochets. De là enfin on obtient:

$$(\underset{\bullet}{\square} - y\partial_x)(\overset{\bullet}{\square} - x\partial_y) - (\overset{\bullet}{\square} - x\partial_y)(\underset{\bullet}{\square} - y\partial_x) = -2\,(\Theta + \tfrac{1}{2}x\partial_x - \tfrac{1}{2}y\partial_y).$$

Or en supposant les deux parties de cette équation symbolique appliquées au *covariant* ϕ_1, la partie gauche de l'équation s'évanouit en vertu des équations (A) et l'équation se réduit à

$$(\Theta + \tfrac{1}{2}x\partial_x - \tfrac{1}{2}y\partial_y)\phi = 0; \quad \ldots\ldots\ldots\ldots\ldots\ldots\ldots\ldots\ldots\ldots \quad (B)$$

ce qui est une nouvelle équation à différences partielles, à laquelle satisfait le *covariant* ϕ. Il est aisé de voir que cette équation exprime le théorème énoncé ci-dessus, savoir que tout *covariant* est composé de termes de la pesanteur *zéro*.

Il suit de là, en considérant un *covariant*

$$\phi = (A_0,\ A_1,\ \ldots\ A_s)(x,\ y)^s$$

qu'un coefficient quelconque A_i aura la pesanteur $i - \tfrac{1}{2}s$, ou bien que les *pesanteurs* forment une progression arithmétique aux différences 1, et dont les termes extrêmes sont $-\tfrac{1}{2}s$, $+\tfrac{1}{2}s$.

Substituons maintenant cette valeur de ϕ dans les équations (A). La première équation donne d'abord:

$$\underset{\bullet}{\square}A_0 = 0,\quad \underset{\bullet}{\square}A_1 = A_0,\quad \underset{\bullet}{\square}A_2 = 2A_1,\ \ldots\ \underset{\bullet}{\square}A_s = sA_{s-1}. \quad \ldots\ldots\ldots\ldots(\alpha)$$

Cela est un système qui équivaut aux deux équations

$$\underset{\bullet}{\square}^s\,.\,A_s = 0,\quad \phi = y^s\,.\,e^{\underset{\bullet}{\square}\frac{x}{y}}\,.\,A_s. \quad \ldots\ldots\ldots\ldots\ldots\ldots\ldots\ldots(\alpha')$$

De même, la seconde équation donne

$$\dot{\square} A_s = 0,\quad \dot{\square} A_{s-1} = A_s,\quad \dot{\square} A_{s-2} = 2A_{s-1}, \ldots\quad \dot{\square} A_0 = sA_1: \quad \ldots\ldots\ldots(\beta)$$

système qui équivaut aux deux équations

$$\dot{\square}^{s+1} A_0 = 0,\quad \phi = x^s e^{\dot{\square}\frac{y}{x}} . A_0. \quad \ldots\ldots\ldots\ldots\ldots\ldots\ldots\ldots(\beta')$$

On voit que A_0 satisfait aux deux équations

$$\underset{\bullet}{\square} A_0 = 0,\quad \dot{\square}^{s+1} A_0 = 0, \quad \ldots\ldots\ldots\ldots\ldots\ldots\ldots\ldots\ldots(\gamma)$$

et en supposant que cette quantité soit connue, on trouve les autres coefficients $A_1, A_2, \ldots, A_s$ par la seule différentiation, au moyen des équations (β). Or cela étant, je dis que les équations (α) seront satisfaites d'elles-mêmes. En effet: des équations $\underset{\bullet}{\square} A_0 = 0$, $\dot{\square} A_0 = sA_1$ on tire $\dot{\square}\underset{\bullet}{\square} A_0 = 0$, $\underset{\bullet}{\square}\dot{\square} A_0 = s\underset{\bullet}{\square} A_1$, et de là $(\dot{\square}\underset{\bullet}{\square} - \underset{\bullet}{\square}\dot{\square}) A_0 = -s\underset{\bullet}{\square} A_1$.

Or nous avons déjà vu que $\dot{\square}\underset{\bullet}{\square} - \underset{\bullet}{\square}\dot{\square} = 2\Theta$, et l'équation (B) donne $\Theta . A_0 + \frac{1}{2}s . A_0 = 0$: donc l'équation $(\dot{\square}\underset{\bullet}{\square} - \underset{\bullet}{\square}\dot{\square}) A_0 = -s\underset{\bullet}{\square} A_1$ se réduit à $A_0 = \underset{\bullet}{\square} A_1$: équation du système (α). De la même manière on obtient les autres équations de ce système. On peut dire que l'on aurait pu déterminer également le coefficient A_s au moyen des équations

$$\dot{\square} A_s = 0,\quad \underset{\bullet}{\square}^s . A_s = 0, \quad \ldots\ldots\ldots\ldots\ldots\ldots\ldots\ldots\ldots\ldots(\delta)$$

et de là les coefficients $A_{s-1}, \ldots A_0$ par les équations (α).

Prenons par exemple un *covariant* $(A_0, A_1, A_2)(x, y)^2$ de la fonction cubique $ax^3 + 3bx^2y + 3cxy^2 + dy^3$. A_0 doit satisfaire aux deux équations

$$(a\partial_b + 2b\partial_c + 3c\partial_d) A_0 = 0,\quad (3b\partial_a + 2c\partial_b + d\partial_c)^3 A_0 = 0.$$

Ces équations sont en effet satisfaites en mettant $A_0 = ac - b^2$. On a donc les équations

$$2A_1 = (3b\partial_a + 2c\partial_b + d\partial_c) A_0,\quad A_2 = (3b\partial_a + 2c\partial_b + d\partial_c) A_1,$$

pour déterminer A_1, A_2; ce qui donne $2A_1 = ad - bc$, $A_2 = bd - c^2$, et on est conduit ainsi au *covariant* mentionné ci-dessus, savoir à

$$(ac - b^2)x^2 + (ad - bc)\, xy + (bd - c^2)\, y^2.$$

Soit maintenant

$$x^n \partial_{a_n} - x^{n-1} y\, \partial_{a_{n-1}} \ldots \pm y^n \partial_{a_0} = \Lambda,$$

on aura

$$\underset{\bullet}{\square}\Lambda = (\underset{\bullet}{\square}\Lambda),\quad \Lambda\underset{\bullet}{\square} = (\Lambda\underset{\bullet}{\square}) - y\frac{\partial\Lambda}{\partial x},$$

où dans ($\underset{\bullet}{\square}\Lambda$), ($\Lambda\underset{\bullet}{\square}$) les quantités a_0, a_1, ... sont censées non affectées par les symboles ∂_{a_1}, ∂_{a_2}, &c. de la différentiation. Cela donne

$$\underset{\bullet}{\square}\Lambda - \Lambda\underset{\bullet}{\square} = y\,\frac{d\Lambda}{\partial x}\,.$$

Or $\partial_x\Lambda - \Lambda\partial_x = \dfrac{d\Lambda}{\partial x}$, donc:

$$(\underset{\bullet}{\square} - y\partial_x)\,\Lambda = \Lambda\,(\underset{\bullet}{\square} - y\partial_x),$$

et de même:

$$(\dot{\square} - x\partial_y)\,\Lambda = \Lambda\,(\dot{\square} - x\partial_y).$$

Appliquons ces deux équations symboliques à un *covariant* ϕ. Les termes à droite s'évanouissent à cause des équations (A), et l'on obtient les deux équations

$$(\underset{\bullet}{\square} - y\partial_x)\,\Lambda\phi = 0, \quad (\dot{\square} - x\partial_y)\,\Lambda\phi = 0,$$

c'est-à-dire: $\Lambda\phi$ sera aussi un *covariant* de la fonction donnée. Par exemple de *l'invariant*

$$\phi = -a^2d^2 + 6abcd - 4ac^3 - 4b^3d + 3b^2c^2,$$

on tire le *covariant*

$$(x^3\partial_d - x^2y^2\partial_c + xy^2\partial_b - y^3\partial_a)\phi\,;$$

savoir:

$$\begin{aligned} &(-a^2d + 3abc - 2b^3\,)x^3 \\ -\,3\,(&\;abd - 2ac^2 + \;b^2c)x^2y \\ +\,3\,(&\;acd - 2b^2d + \;bc^2)xy^2 \\ -\;\;(&-ad^2 + 3acd - 2c^3\,)y^3\,; \end{aligned}$$

résultat déjà connu.

Essayons maintenant à intégrer les équations (A); savoir:

$$(\underset{\bullet}{\square} - y\partial_x)\phi = 0, \quad (\dot{\square} - x\partial_y)\phi = 0.$$

Pour intégrer la première, je reviens à une notation dont je me suis déjà servi dans ce mémoire et j'écris

$$\begin{aligned} a_0' &= a_0, \\ a_1' &= a_1 + \lambda a_0, \\ a_2' &= a_2 + 2\lambda a_1 + \lambda^2 a_0, \\ &\vdots \\ a_n' &= a_n + n\lambda a_{n-1} \ldots + \lambda^n a_0. \end{aligned}$$

En faisant $\lambda = -\frac{a_1}{a_0}$, ce qui donne $a_0' = 0$, on voit sans peine que l'on satisfera à l'équation, en mettant pour ϕ une fonction quelconque de quantités $a_0', a_1', \ldots a_n'$, $x + \lambda y$, y; le nombre de ces quantités étant $n+2$. Et cela est la solution générale de l'équation.

Ce résultat doit être substituée dans la seconde équation, savoir dans $(\dot{\square} - x\partial_y)\phi = 0$. Pour cela, imaginons que les quantités $a_0, a_1, \ldots a_n, x, y$ soient exprimées en fonction de $a_0', a_2', \ldots a_n', x, y$ et a_1; puisque ϕ est fonction des seules quantités $a_0', a_2' \ldots a_n'$, x, y, l'équation résultante doit être satisfaite, quelle que soit la valeur de a_1. Or on trouve que cette équation résultante a la forme $L + Ma_1 = 0$: donc il faut qu'on ait à la fois les deux équations $L = 0$, $M = 0$. (Je renvoie à une note les détails de la réduction.) En dernière analyse, et en remettant dans les équations $L = 0$, $M = 0$ les quantités $a_0, a_2, \ldots, a_n$ au lieu de $a_0', a_2', \ldots, a_n'$, je trouve les résultats suivants très simples, savoir, en écrivant

$$\bar{\Theta} = (0 - \tfrac{1}{2}n)a_0\partial_{a_0} + \quad (2 - \tfrac{1}{2}n)a_2\partial_{a_2} + (3 - \tfrac{1}{2}n)a_3\partial_{a_3} \ldots + (n - \tfrac{1}{2}n)a_n\partial_{a_n}:$$

$$\cdot\underset{\bullet}{\square} = \quad 3a_2\partial_{a_3} + \quad 4a_3\partial_{a_4} \ldots \quad + na_{n-1}\partial_{a_n},$$

$$\cdot\dot{\square} = (n-2)a_3\partial_{a_2} + (n-3)a_4\partial_{a_3} \ldots \quad + a_n\partial_{a_{n-1}}.$$

Les équations dont il s'agit sont

$$\{(n-1)a_2(\cdot\underset{\bullet}{\square} - y\partial_x) - a_0(\cdot\underset{\bullet}{\square} - x\partial_y)\}\bar{\phi} = 0, \quad \ldots\ldots\ldots\ldots\ldots\ldots(C)$$

$$(\bar{\Theta} + \tfrac{1}{2}x\partial_x - \tfrac{1}{2}y\partial_y)\bar{\phi} = 0, \quad \ldots\ldots\ldots\ldots\ldots\ldots(D)$$

et il y a à remarquer qu'on obtient l'équation (C) en éliminant entre les équations (A) le terme $\partial_{a_1}\phi$; et puis, en mettant $a_1 = 0$, on tire l'équation (D) de l'équation (B), en y mettant de même $a_1 = 0$. Il y a à remarquer aussi que la fonction $\bar{\phi}$ qui satisfait aux équations (C, D), est ce que devient un *covariant* quelconque ϕ, en y mettant $a_1 = 0$. On obtient d'abord la valeur générale en changeant $a_0, a_2, \ldots, a_n$ en $a_0', a_2', \ldots, a_n'$, et en mettant après pour ces quantités leurs valeurs en termes de $a_0, a_1, a_2, \ldots, a_n$. La solution du problème des *covariants* serait donc effectuée si l'on pourrait intégrer les équations (C, D).

Or la quantité a_0 entre dans l'équation (C) comme constante, et l'on voit sans peine que cette équation pourra être intégrée en mettant $a_0 = 1$; puis, en écrivant dans le résultat $\frac{a_2}{a_0}, \frac{a_3}{a_0}, \ldots \frac{a_n}{a_0}$ au lieu de $a_2, a_3, \ldots a_n$, et en multipliant par une puissance quelconque de a_0, le résultat ainsi obtenu, serait composé de termes de la même *pesanteur;* et en choisissant convenablement la puissance de a_0, on pourrait faire en sorte que ces termes fussent de la *pesanteur* zéro. Mais l'équation (D) ne fait qu'exprimer que la fonction $\bar{\phi}$ est composée de termes de la *pesanteur* zéro; le résultat obtenu de la manière dont il s'agit, satisfera donc par lui-même à l'équation (D), et il est permis de ne faire attention qu'à l'équation (C). Dans la pratique on intégrera cette équation en ayant soin de faire en sorte que les solutions soient de la *pesanteur* zéro, ce qui peut être effectué en multipliant par une puissance convenablement choisie de a_0. Et puisqu'en faisant abstraction de cette quantité a_0, l'équation (C) contient $n+1$ quantités variables, savoir $a_2, a_3, \ldots, a_n, x, y$, la fonction $\overline{\phi}$ sera une fonction arbitraire de n quantités; et en supposant que cette fonction ne contienne pas les variables x, y (cas auquel $\overline{\phi}$ serait ce que deviendrait un *invariant* quelconque en y mettant $a_1=0$), $\overline{\phi}$ sera une fonction arbitraire de $n-2$ quantités.

La même chose sera évidemment vrai, si l'on rétablit la valeur générale de a_1: donc tout *invariant* sera une fonction d'un nombre $n-2$ d'*invariants,* que l'on pourra prendre pour primitifs; et tout *covariant* sera une fonction de ces *invariants* primitifs de la fonction donnée (laquelle est évidemment un de ses propres *covariants*), et d'un autre *covariant* que l'on peut prendre pour primitif. Cela ne prouve nullement (ce qui est néanmoins vrai pour les *invariants,* à ce que je crois) que tout invariant est une fonction rationnelle et intégrale de $n-2$ *invariants* convenablement choisis, et que tout *covariant* est une fonction rationnelle et intégrale (ce qui en effet n'est pas vrai) de ces *invariants,* de la fonction donnée, et d'un *covariant* convenablement choisi.

Le cas $n=2$ fait dans cette théorie une exception. On sait qu'il existe dans ce cas un *invariant,* savoir $ac-b^2$ qui, selon la théorie générale, ne doit pas exister, et il n'existe pas de *covariant,* hormis la fonction donnée elle-même. Or cette particularité peut être aisément expliquée.

Le cas $n=3$ rentre, comme cela doit être, dans la théorie générale. En effet, il existe dans ce cas un *invariant,* savoir la fonction $-a^2d^2+6abcd+4ac^3-4b^3d+3b^2c^2$ ci-dessus trouvée, et tout *covariant* de la fonction peut être exprimé par cet *invariant* de la fonction donnée elle-même, et par le *covariant* $(ac-b^2)x^2+(ad-bc)xy+(bd-c^2)y^2$ ci-dessus trouvé. Il en est ainsi par exemple pour le *covariant* de troisième ordre par rapport aux variables et aux coefficients; car en représentant par Φ le *covariant* dont il s'agit, par H le *covariant* du second ordre, par u la fonction donnée $ax^3+3bx^2y+3cxy^2+dy^3$ et par ∇ l'*invariant,* on obtient l'équation identique $\Phi^2+\Box u^2=-4H^3$. Je fais mention de cette équation, parce que je crois qu'elle n'est pas généralement connue.

Je vais donner maintenant quelques exemples des équations (C et D). Soit d'abord $n=3$, et supposons que $\overline{\phi}$ ne contienne pas les variables x, y: $\overline{\phi}$ sera une fonction de a, c, d, et les équations reviendront à

$$(6c^2\partial_d-ad\partial_c)\overline{\phi}=0, \quad (-3a\partial_a+c\partial_c+3d\partial_d)\overline{\phi}=0.$$

Les quantités ac^3, a^2d^2, dont chacune est de la pesanteur *zéro,* satisfont par là à la seconde équation, et en mettant $\overline{\phi}=Aa^2d^2+Cac^3$, on obtient $4A-C=0$, en vertu de la première équation; ou en faisant $A=-1$, cela donne $C=-4$; de là on tire $\overline{\phi}=-a^2d^2-4ac^3$, et la solution générale est $\overline{\phi}=F(-a^2d^2-4ac^3)$, F étant une fonction quelconque. La formule plus générale $\overline{\phi}=F(a, -a^2d^2-4ac^3)$ satisferait sans doute à la première équation, mais pour que cette valeur satisfasse à la seconde équation, il faut que la quantité a, en tant qu'elle n'est pas contenue dans $-a^2d^2-4ac^3$, disparaisse. Ainsi la valeur donnée ci-dessus, savoir $\overline{\phi}=F(-a^2d^2-4ac^3)$, est la solution la plus générale des deux équations.

Écrivons a, $c-\frac{b^2}{a}$, $d-\frac{3bc}{a}+\frac{2b^3}{a^2}$ au lieu de a, c, d, et ϕ au lieu de $\overline{\phi}$, nous obtenons:

$$\phi=F(-a^2d^2+6abcd-4ac^3-4b^3d+3b^2c^2);$$

ce qui est l'expression la plus générale des *invariants* de la fonction $ax^3+3bx^2y+3cxy^2+y^3$, et l'on voit que tous ces *invariants* sont fonctions d'une seule quantité que nous avons prise ci-dessus pour l'*invariant* de la fonction de troisième ordre dont il s'agit.

Soit encore $n=4$, $\overline{\phi}$ sera une fonction de a, c, d, e qui satisfait aux équations

$$\{2ad\partial_c+(ae-9c^2)\,\partial_d-12cd\partial_e\}\,\overline{\phi}=0,$$

$$\{-2a\partial_a+d\partial_d+2e\partial_e\}\,\overline{\phi}=0,$$

dont la solution générale est $\overline{\phi}=F(ae+3c^2,\ ace-ad^2-c^3)$, F étant une fonction quelconque. On voit par là qu'il n'existe que les *invariants* indépendants $ae-4cd+3c^2$, $ace+2bcd-ad^2-b^2e-c^3$. Ce résultat est connu depuis longtemps.

Soit enfin $n=5$, $\overline{\phi}$ sera une fonction de a, c, d, e, f qui satisfait aux équations

$$\{3ad\partial_c+(2ae-12c^2)\,\partial_d+(af-16cd)\,\partial_e-20ce\partial_f\}\,\overline{\phi}=0,$$

$$\{-\tfrac{5}{2}a\partial_a-\tfrac{1}{2}c\partial_c+\tfrac{1}{2}d\partial_d+\tfrac{3}{2}e\partial_e+\tfrac{5}{2}f\partial_f\}\,\overline{\phi}=0.$$

On sait qu'il y en a une solution de quatrième ordre par rapport aux quantités a, c, d, e, f; et en prenant la fonction la plus générale dont les termes ont la *pesanteur* zéro, on aura:

$$\overline{\phi}=Aa^2f^2+Bacdf+Cace^2+Dad^2e+Ec^3e+Fc^2d^2:$$

fonction qui satisfait d'elle-même à la seconde équation. En substituant cette valeur dans la première équation, on trouvera que les coefficients A, B, &c. doivent satisfaire à ces *sept* équations:

$$2B+2C-40A=0,\quad 3B+D=0,\quad 3C+4D=0,\quad -12B+E=0,$$

$$9E-24D+4F-32C-20B=0,\quad 6F-16D=0,\quad -24F-16E=0,$$

qui se réduisent cependant (ce que l'on n'aurait pas facilement deviné par la forme des équations) à *cinq* équations indépendantes. En faisant donc $A=1$, on trouve aisément les autres coefficients B, C, &c. et on obtient ainsi:

$$\overline{\phi}=a^2f^2+4acdf+16ace^2-12ad^2e+48c^3e-32c^2d^2:$$

valeur qui peut être tirée d'une formule présentée dans mon mémoire sur les hyperdéterminants, [16], en y faisant $b=0$.

J'ai donné cet exemple pour faire voir qu'il serait impossible de déduire du nombre supposé connu des coefficients indéterminés qui correspondent à un ordre donné, le nombre des *invariants* de ce même ordre. Il est donc inutile de pousser plus loin cette discussion.

Note 1 *sur l'intégration des équations* (*A*).

En écrivant comme ci-dessus:

$$\underset{\bullet}{\Box} = a_0\partial_{a_1} + 2a_1\partial a_{a_2} \dots \qquad + na_{n-1}\partial_{a_n},$$

$$\overset{\bullet}{\Box} = na_1\partial_{a_0} + (n-1)a_2\partial_{a_1} \dots + a_n\partial_{a_{n-1}},$$

il s'agit de trouver une quantité ϕ, fonction de a_0, a_1, ... a_n, x et y qui satisfasse à la fois aux équations

$$(\underset{\bullet}{\Box} - y\partial_x)\,\phi = 0,$$

$$(\overset{\bullet}{\Box} - x\partial_y)\,\phi = 0.$$

Pour intégrer ces équations, j'écris, comme plus haut:

$$a_0' = a_0,$$

$$a_1' = a_1 + \lambda a_0,$$

$$a_2' = a_2 + 2\lambda a_1 + \lambda^2 a_0,$$

$$\vdots$$

$$a_n' = a_n + n\lambda a_{n-1} \dots \qquad + \lambda^n a_0,$$

et aussi $x' = x - \lambda y$, $y' = y$. Cela posé, je fais remarquer d'abord que $\dfrac{da_1'}{d\lambda} = a_0'$, $\dfrac{da_2'}{d\lambda} = 2a_1'$, et ainsi de suite. En considérant λ comme fonction quelconque de a_0, a_1, ... a_n, et en supposant que ϕ soit une fonction de a_0', a_1', ... a_n', x', y', on parvient assez facilement à l'équation identique $(\underset{\bullet}{\Box} - y\partial_x)\phi = (1 + \underset{\bullet}{\Box}\lambda)\,(\underset{\bullet}{\Box}' - y'\partial_{x'})\phi$, où $\underset{\bullet}{\Box}'$ est ce que devient $\underset{\bullet}{\Box}$, en y écrivant a_0', a_1', ... a_n' au lieu de a_0, a_1, ... a_n.

Nous pouvons donc satisfaire à la première équation, en déterminant λ au moyen de $1 + \underset{\bullet}{\Box}\lambda = 0$: équation qui serait satisfaite en écrivant $\lambda = -\dfrac{a_1}{a_0}$, ou, si l'on veut, en déterminant λ par $a_1' = 0$. Donc, en supposant toujours que λ ait cette valeur, ϕ sera une fonction quelconque de a_0', a_2', ... a_n', x', y', c'est-à-dire d'un nombre $n + 2$ de quantités. Ce sera donc là (comme on aurait pu facilement prévoir), la solution générale de la première équation. Or en considérant ϕ comme fonction de a_0', a_2', ... a_n', x', y', ou, si l'on veut, de a_0', a_1', a_2', ... a_n', x', y' (où $a_1' = a_1 + \lambda a_0 = 0$), et en substituant cette valeur dans l'équation $(\dot{\square} - x\partial_y)\phi = 0$, on voit d'abord que la variation de la quantité λ fournit au résultat le terme

$$\left(na_1 \frac{d\lambda}{da_0} + (n-1)a_2 \frac{d\lambda}{da_1}\right)(\underset{\bullet}{\square}' - y'\partial_{x'})\,\phi\,;$$

et puisque $na \dfrac{d\lambda}{da_0} + (n-1)a_2 \dfrac{d\lambda}{da_1}$ se réduit à $n\dfrac{a_1{}^2}{a_0{}^2} - (n-1)\dfrac{a_0 a_2}{a_0{}^2}$, ou enfin à $\lambda^2 - \dfrac{(n-1)a_2'}{a_0}$, ce terme devient

$$\left(\lambda^2 - \frac{(n-1)\,a_2'}{a_0}\right)(\underset{\bullet}{\square}' - y'\partial_{x'}).$$

Le terme $-x'\partial_y\,.\,\phi$ se réduit à $-(x'+\lambda y')(-\lambda\partial_{x'} + \partial_{y'})\phi$, savoir à

$$(-x'\partial_{y'} + \lambda^2 y'\partial_{x'} + \lambda x'\partial_{x'} - \lambda y'\partial_{y'})\,\phi,$$

et en mettant pour un moment

$$\begin{aligned} M = \qquad & na_1(\partial_{a_0'} + \lambda\partial_{a_1'} \;\ldots\ldots + \lambda^n\,\partial_{a_n'}) \\ +\,(n-1)\,a_2\,(& \qquad\quad \partial_{a_1'} \;\ldots\ldots + n\lambda^{n-1}\,\partial_{a_n'}) \\ & \;\vdots \\ +\,a_n\,(& \qquad\qquad \partial_{a_{n-1}'} + n\lambda\,\partial_{a_n'}), \end{aligned}$$

nous obtenons

$$(\dot{\square} - x\partial_y)\,\phi = M\phi + \left(\lambda^2 - \frac{(n-1)a_2'}{a_0}\right)(\underset{\bullet}{\square}' - y'\partial_{x'})\phi + (-x'\partial_{y'} + \lambda^2 y'\partial_{x'} + \lambda x'\partial_{x'} - \lambda y'\partial_{y'})\,\phi,$$

c'est-à-dire

$$(\dot{\square} - x\partial_y)\phi = (M - x'\partial_{y'})\phi + \lambda^2\underset{\bullet}{\square}'\phi - \frac{(n-1)a_2'}{a_0'}(\underset{\bullet}{\square}' - y'\partial_{x'})\,\phi + \lambda(x'\partial_{x'} - y'\partial_{y'})\phi.$$

Or en supposant que $\dot{\square}'$ est ce que devient $\dot{\square}$ en y écrivant a_0', a_1', ... a_n' au lieu de a_0, a_1, ... a_n, et en posant

$$\Theta' = (0 - \tfrac{1}{2}n)a_0'\partial_{a_0'} + (1 - \tfrac{1}{2}n)\,a_1'\partial_{a_1'} + \ldots\,(n - \tfrac{1}{2}n)a_n'\partial_{a_n'},$$

on obtient, après avoir fait une réduction un peu pénible:

$$M\phi + \lambda^2\underset{\bullet}{\square}'\phi = \dot{\square}'\phi + 2\lambda\Theta'\phi,$$

(en effet les coefficients de $\partial_{a_0'}\phi$, $\partial_{a_1'}\phi$ &c. aux deux côtés de cette équation deviennent les mêmes après des réductions convenables.) Donc enfin on a

$$(\dot{\square} - x\partial_y)\phi = (\dot{\square}' - x'\partial_{y'})\phi - \frac{(n-1)a_2'}{a_0'}(\underset{\bullet}{\square}' - y'\partial_{x'})\phi + 2\lambda\,(\Theta' + \tfrac{1}{2}x'\partial_{x'} - \tfrac{1}{2}y'\partial_{y'})\phi = 0,$$

ou bien, puisque cette équation doit être satisfaite indépendamment de la quantité λ (qui seule contient a_1), elle se décompose dans les deux équations

$$\{a_0'(\dot{\square}' - x'\partial_{y'}) - (n-1)\,a_2'\,(\underset{\bullet}{\square}' - y'\partial_{x'})\}\,\phi = 0,$$

$$\{\Theta' + \tfrac{1}{2}x'\partial_{x'} - \tfrac{1}{2}y'\partial_{y'}\}\,\phi = 0,$$

lesquelles, en y mettant d'abord $a_1' = 0$, puis en remettant $a_0, a_2, \ldots, a_n, x, y$ au lieu de $a_0', a_2', \ldots, a_n, x', y'$, et en écrivant $\overline{\phi}$, $\overline{\Theta}$, $\cdot\underset{\bullet}{\square}$, $\cdot\overset{\bullet}{\square}$ au lieu de ϕ, Θ, $\underset{\bullet}{\square}$, $\overset{\bullet}{\square}$, donnent en effet les équations (C, D) dont je me suis servi dans le texte.

Note 2.

Je vais résumer dans cette note quelques formules qui feront voir la liaison qui existe entre les *invariants* d'une fonction de n-ième ordre et de la fonction de $(n-1)$ième ordre que l'on obtient en réduisant à zéro le coefficient de y^n, et en supprimant le facteur x.

Il convient pour cela de considérer une fonction telle que

$$(a_0, a_1, \ldots a_n)(x, y)_n = a_0 x^n + a_1 x^{n-1} y \ldots + a_n y^n,$$

dans laquelle n'entrent plus les coefficients numériques du binôme $(1+x)^n$.

Écrivons

$$(a_0, a_1, \ldots a_n)(x, y)_n = a_0 (x - \alpha_1 y)(x - \alpha_2 y) \ldots (x - \alpha_n y);$$

je tâche d'abord à représenter les *invariants* au moyen des racines $\alpha_1, \alpha_2, \ldots, \alpha_n$, et j'étends pour le moment le terme *invariant* à toute fonction, symétrique ou non, des racines qui ait la propriété caractéristique des *invariants:* fonctions qui jusqu'ici ont été considérées tacitement comme rationnelles par rapport aux coefficients.

Mettons d'abord

$$\nabla = a_0^{2n-2}(\alpha_1 - \alpha_2)^2 (\alpha_1 - \alpha_3)^2 \ldots (\alpha_{n-1} - \alpha_n)^2;$$

cette quantité ∇ qui, égalée à zéro, exprime l'égalité de deux racines, et que je vais désormais nommer le *Discriminant* de la fonction, sera une fonction rationnelle des coefficients, et d'un *invariant* proprement dit. Mais de plus, toute fonction telle que $(\alpha_1 - \alpha_2)^n (\alpha_1 - \alpha_3)^n, \ldots$, dans laquelle la somme des indices des facteurs qui contiennent α_1, celle des indices des facteurs qui contiennent α_2, &c. sont égales, sera un *invariant;* et en réunissant ces fonctions, pour trouver une somme en fonction symétrique des racines, on obtiendra des *invariants* proprement dits. Cela soit dit en passant. Pour le moment il suffit de prendre les invariants les plus simples, savoir ceux de la forme

$$\frac{(\alpha_1 - \alpha_2)(\alpha_3 - \alpha_4)}{(\alpha_1 - \alpha_3)(\alpha_2 - \alpha_4)},$$

lesquels en effet sont des rapports *anharmoniques* de quatre racines, prises à volonté. Soient $Q_1, Q_2, \ldots, Q_{n-3}$ la fonction qui vient d'être écrite et les fonctions que l'on en tire en mettant $\alpha_5, \alpha_6, \ldots, \alpha_n$ au lieu de α_4. Les fonctions $\nabla, Q_1, Q_2, \ldots, Q_{n-3}$ seront des *invariants* indépendants, et le nombre de ces *invariants* est $n-2$. Donc, tout autre *invariant* sera une fonction des quantités ∇, Q_1, Q_2, ..., Q_{n-3}. Soit maintenant $a_n = 0$, et α_n la racine qui devient égale à zéro. Les quantités Q_1, Q_2, ..., Q_{n-4} seront toujours des rapports *anharmoniques* de quatre racines de l'équation du $(n-1)$ième ordre. Il n'y aura que la seule quantité Q_{n-3} qui change de forme, et elle ne sera pas un *invariant* de la fonction du $(n-1)$ième ordre. On voit aussi d'abord que le *discriminant* ∇ se réduit à $a^2_{n-1}\nabla_0$, en exprimant par ∇_0 le *discriminant* de la fonction du $(n-1)$ième ordre. (C'est je crois M. Joachimsthal qui a le premier remarqué cette circonstance.) Donc, en supposant $a_n = 0$, l'*invariant* de la fonction du n-ième ordre deviendra une fonction de $a^2_{n-1}\nabla_0$, Q_1, Q_2, ... Q_{n-4} et d'une quantité X qui n'est pas un *invariant* de la fonction du $(n-1)$ième ordre, mais qui sera toujours la même quel que soit l'invariant dont il s'agit. En considérant les *invariants* proprement dits de la fonction du $(n-1)$ième ordre, on peut former avec ces *invariants* des quotients I_1, I_2, ..., I_{n-4} du degré zéro par rapport aux coefficients. Nous pouvons remplacer par ces quotients les quantités Q_1, Q_2, ..., Q_{n-4}, et dire que l'*invariant* de la fonction du n-ième ordre, en mettant $a_n = 0$, deviendra une fonction des quantités $a^2_{n-1}\nabla_0$, I_1, I_2, ..., I_{n-4} et X.

Ces théorèmes auront, je crois, quelque utilité pour les recherches ultérieures: je les laisse à côté maintenant, et veux présenter une méthode assez simple pour calculer les *discriminants*.

Pour cela je remarque que les équations (A), en changeant, comme nous venons de le faire, les valeurs des coefficients, donnent pour les *invariants*:

$$(na_0\partial_{a_1} + (n-1)a_1\partial_{a_2} \ldots + a_{n-1}\partial_{a_n})\phi = 0,$$

$$(a_1\partial_{a_0} + 2a_2\partial_{a_1} \ldots + na_n\partial_{a_{n-1}})\phi = 0;$$

et ces équations seront satisfaites en mettant pour ϕ le *discriminant* ∇. Or, pour $a_n = 0$, la fonction ∇ devient $a^2_{n-1}\nabla_0$, ou, si l'on veut, $-a^2_{n-1}\nabla_0$; donc ∇ sera généralement de la forme

$$\nabla = -a^2_{n-1}\nabla_0 + Ba_n + Ca_n^2 + \ldots,$$

où a_n^{n-1} est la puissance la plus élevée de a_n. Donc, en supposant que ∇_0 soit connu, et en mettant la première des équations écrites ci-dessus sous la forme $(F + a_{n-1}\partial_{a_n})\nabla = 0$, où $F = na_0\partial_{a_1} + (n-1)\,a_1\partial_{a_2} \ldots + 2a_{n-2}\partial_{a_{n-1}}$, on obtiendra par la seule différentiation les coefficients B, C, &c. En effet, cette équation donne

$$a_{n-1}B = F(a^2_{n-1}\nabla_0), \quad 2a_{n-1}C = -F(B), \quad 3a_{n-1}D = -F(C);$$

et ainsi de suite.

En supposant par exemple $n = 3$, considérons la fonction du troisième ordre

$$\alpha x^3 + \beta x^2 y + \gamma xy^2 + \delta y^3:$$

le *discriminant* de $\alpha x^2 + \beta xy + \gamma y^2$ sera $4\alpha\gamma - \beta^2$. Nous avons alors

$$\nabla = -\gamma^2(4\alpha\gamma - \beta^2) + B\delta + C\delta^2,$$

et en mettant $F = 3\alpha\partial_\beta + 2\beta\partial_\gamma$, B, C seront donnés par

$$\gamma B = F(4\alpha\gamma^3 - \beta^2\gamma^2), \quad 2\gamma C = -F(B),$$

c'est-à-dire $B = 18\alpha\beta\gamma - 4\beta^3$, $C = -27\alpha^2$, et de là:

$$\nabla = -27\alpha^2\delta^2 + 18\alpha\beta\gamma\delta - 4\alpha\gamma^3 - 4\beta^3\delta + \beta^2\gamma^2:$$

valeur qui correspond en effet à la forme ordinaire

$$\nabla = -a^2d^2 + 6abcd - 4ac^3 - 4b^3d + 3b^2c^2,$$

en changeant d'une manière convenable les coefficients.

Londres, Stone Buildings, 23 *Févr.* 1852.

132.

RÉPONSE À UNE QUESTION PROPOSÉE PAR M. STEINER (Aufgabe 4, Crelle t. XXXI. (1846) p. 90).

[From the *Journal für die reine und angewandte Mathematik* (Crelle), tom. L. (1855), pp. 277—278.]

En partant des deux théorèmes :

I. Qu'il existe au moins une surface du second ordre qui touche neuf plans donnés quelconques ;

II. Que le lieu d'intersection de trois plans rectangles qui touchent une surface du second ordre est une sphère concentrique avec la surface, tandis que pour le paraboloïde cette sphère se réduit à un plan,

M. Steiner suppose le cas d'un parallélepipède rectangle, ou même d'un cube P et d'un point quelconque D, par lequel passent trois plans rectangles. Les six plans du parallélepipède P et les trois plans qui passent par le point D seront touchés d'une surface F du second ordre (I.), et les huit angles E du parallélepipède P et le point D doivent donc se trouver tous les neuf sur la surface d'une sphère, ou dans un plan (II.). Les huit angles E sont en effet situés sur la surface d'une sphère, déterminée par eux ; mais le point D étant arbitraire, ce point en général ne sera pas situé sur cette surface sphérique, de manière que les neuf points $8E$ et D ne seront situés, ni dans une surface sphérique, ni dans un plan ; ce qui ne s'accorde pas avec le théorème II. Cela étant, M. Steiner dit, qu'il y a à prouver que la contradiction n'est qu'apparente, et que tout cela n'affaiblit pas la validité générale des deux théorèmes.

Il s'agit de savoir ce que devient dans le cas supposé par M. Steiner la surface du second ordre qui touche les six plans du parallélepipède P et les trois plans qui passent par le point D. Cette surface sera en effet *la conique selon laquelle l'infini, considéré comme plan, est coupé par un cône déterminé, près la position du sommet.* En effet, menons par un point quelconque de l'espace trois plans parallèles aux plans du parallélepipède P, et par le point D trois autres plans parallèles à ces plans. Ces *six* plans seront touchés (en vertu d'un théorème connu) par un cône déterminé du second ordre, et on peut dire que ce cône, quelle que soit la position de son sommet, rencontre l'infini, considéré comme plan, dans une seule et même conique (cela n'est en effet autre chose que de dire que deux droites parallèles rencontrent l'infini, considéré comme plan, dans un seul et même point). Le cône dont il s'agit aura la propriété d'être touché par une infinité de systèmes de trois plans rectangles. En effet: le plan passant par le sommet, et perpendiculaire à la droite d'intersection de deux plans tangents quelconques sera un plan tangent du cône; les plans d'un tel système seront aussi des plans tangents de la conique mentionnée ci-dessus: donc le sommet du cône sera le point d'intersection de trois plans rectangles de la conique; et ce sommet étant un point entièrement indéterminé, le lieu de l'intersection des trois plans tangents rectangles de la conique, sera de même absolument indéterminé, ou si l'on veut, ce lieu sera l'espace entier près les points à une distance infinie. La contradiction apparente dont M. Steiner parle, a par conséquent son origine dans l'indétermination qui a lieu dans le cas dont il s'agit. Dans tout autre cas, le point d'intersection des trois plans rectangles de la surface du second ordre est parfaitement déterminé, et les théorèmes I. et II. sont tous deux légitimes.

133.

SUR UN THÉORÈME DE M. SCHLÄFLI.

[From the *Journal für die reine und angewandte Mathematik* (Crelle), tom. L. (1855), pp. 278—282.]

On lit dans (§ 13) d'un mémoire très intéressant de M. Schläfli intitulé "Über die Resultante eines Systems mehrerer algebraischer Gleichungen" (*Mém. de l'Acad. de Vienne,* t. IV. [1852]) un très beau théorème sur les *Résultants.*

Pour faire voir plus clairement en quoi consiste ce théorème, je prends un cas particulier. Soit

$$U = ax^3 + 3bx^2y + 3cxy^2 + dy^3 = (a,\ b,\ c,\ d)(x,\ y)^3,$$

$$V = \alpha x^2 + 2\beta xy + \gamma y^2 \qquad = (\alpha,\ \beta,\ \gamma)(x,\ y)^2.$$

Je fais $p = x^2$, $q = xy$, $r = y^2$, et je forme les opérateurs

$$\mathfrak{A} = \xi\partial_a + \tfrac{1}{3}\eta\partial_b + \tfrac{1}{3}\zeta\partial_c,$$

$$\mathfrak{B} = \tfrac{1}{3}\xi\partial_b + \tfrac{1}{3}\eta\partial_c + \zeta\partial_d,$$

lesquels, opérant sur U, donnent

$$x(p\xi + q\eta + r\zeta)\,; \quad y(p\xi + q\eta + r\zeta).$$

L'opérateur

$$\mathfrak{C} = \xi\partial_\alpha + \tfrac{1}{2}\eta\partial_\beta + \zeta\partial_\gamma,$$

opérant sur V, donne

$$p\xi + q\eta + r\zeta.$$

Cela étant, soit $\phi=0$ le *résultant* des équations $U=0$, $V=0$, c'est-à-dire l'équation que l'on obtient en éliminant x, y entre les équations $U=0$, $V=0$, ou autrement dit, soit ϕ le résultant des fonctions U, V. Pour fixer les idées j'écris la valeur de ce résultant comme suit:

$$\phi = \begin{vmatrix} & a, & 3b, & 3c, & d \\ a, & 3b, & 3c, & d & \\ & & \alpha, & 2\beta, & \gamma \\ & \alpha, & 2\beta, & \gamma & \\ \alpha, & 2\beta, & \gamma & & \end{vmatrix}.$$

Je suppose que les opérateurs $\mathfrak{A}$, $\mathfrak{B}$, $\mathfrak{C}$ opèrent sur le résultant ϕ, ce qui donne les fonctions

$$\mathfrak{A}\phi,\quad \mathfrak{B}\phi,\quad \mathfrak{C}\phi,$$

ou en écrivant pour $\mathfrak{A}$, $\mathfrak{B}$, $\mathfrak{C}$ leurs valeurs:

$$\begin{aligned} &(\ \xi\partial_a + \tfrac{1}{3}\eta\partial_b + \tfrac{1}{3}\zeta\partial_c)\,\phi, \\ &(\tfrac{1}{3}\xi\partial_b + \tfrac{1}{3}\eta\partial_c + \ \zeta\partial_d)\,\phi, \\ &(\ \xi\partial_\alpha + \tfrac{1}{2}\eta\partial_\beta + \ \zeta\partial_\gamma)\,\phi, \end{aligned}$$

et en considérant ces expressions comme des fonctions de ξ, η, ζ, j'en forme le résultant Φ, savoir

$$\Phi = \begin{vmatrix} \partial_a\phi, & \tfrac{1}{3}\partial_b\phi, & \tfrac{1}{3}\partial_c\phi \\ \tfrac{1}{3}\partial_b\phi, & \tfrac{1}{3}\partial_c\phi, & \partial_d\phi \\ \partial_\alpha\phi, & \tfrac{1}{2}\partial_\beta\phi, & \partial_\gamma\phi \end{vmatrix}.$$

Ce résultant Φ contiendra le carré de ϕ *comme facteur;* c'est ce qui donne, dans le cas particulier dont il s'agit, le théorème de M. Schläfli.

Généralement, en supposant que l'on ait autant de fonctions U, V, W, ... que d'indéterminées x, y, z, ..., on peut supposer que p, q, ... soient des monômes $x^l y^m z^n$, ... du même degré λ (il n'est pas nécessaire d'avoir la série entière de ces monômes), et on peut former des opérateurs $\mathfrak{A}$, $\mathfrak{B}$, &c. en même nombre que celui des monômes p, q, ... avec les indéterminées ξ, η, ..., tels que ces opérateurs $\mathfrak{A}$, $\mathfrak{B}$, ..., opérant sur les fonctions U, V, W, ... (chacun sur la fonction à laquelle il appartient), donnent $t(p\xi+q\eta\ldots)^{\mu}$, $t'(p\xi+q\eta\ldots)^{\mu'}$, &c.; t, t', &c. étant des monômes de la forme $x^f y^g z^h$ ....

Cela étant, soit ϕ le résultant des fonctions U, V, W, ...; en opérant sur ce résultant ϕ avec les opérateurs $\mathfrak{A}$, $\mathfrak{B}$, ... et en formant ainsi les fonctions $\mathfrak{A}\phi$, $\mathfrak{B}\phi$, ..., soit Φ le résultant de ces expressions considérées comme des fonctions de ξ, η, &c. Φ contiendra une puissance de ϕ *comme facteur*, et en supposant que μ ne soit plus petit qu'aucun autre des indices μ, μ', ...; $\pi=\mu\mu'\ldots$; et $\sigma=\frac{\pi}{\mu}+\frac{\pi}{\mu'}+\ldots$, l'indice de cette puissance sera au moins $\sigma-\frac{\pi}{\mu}$. Voilà le théorème général de M. Schläfli.

La démonstration donnée dans le mémoire cité est, on ne peut plus, simple et élégante. Elle repose d'abord sur un théorème connu (démontré au reste § 6) qui peut être énoncé ainsi; savoir, en supposant que les équations $U=0$, $V=0, \ldots$ soient satisfaites, on aura (près un facteur indépendant de $\xi, \eta, \ldots$):

$$\mathfrak{A}\phi = t\,(p\xi + q\eta \ldots)^{\mu}, \quad \mathfrak{B}\phi = t'\,(p\xi + q\eta \ldots)^{\mu'}, \text{ \&c.}$$

Puis, elle est fondée sur le théorème démontré (§ 12), savoir: le résultant des fonctions

$$\begin{array}{l} t\,(p\xi + q\eta \ldots)^{\mu} + f\,(\xi, \eta \ldots), \\ t'\,(p\xi + q\eta \ldots)^{\mu'} + f'\,(\xi, \eta \ldots), \\ \vdots \end{array}$$

(où $f, f', \ldots$ sont des polynômes de degrés $\mu, \mu', \ldots$ en ξ, η, &c., et $p, q, \ldots, t, t', \ldots$ des constantes quelconques) sera, en supposant que μ ne soit plus petit qu'aucun autre des indices $\mu, \mu', \ldots$, et en posant $\pi = \mu\mu' \ldots$, tout au plus du degré $\frac{\pi}{\mu}$ par rapport aux quantités t, t', &c. Voici cette démonstration, qui suppose aussi que le résultant ϕ soit *indécomposable*. Supposons que les coefficients de $U, V, W, \ldots$ soient assujettis à la seule condition d'être tels que le résultant ϕ soit un infiniment petit du premier ordre, il sera permis de supposer que tous ces coefficients des indéterminées $x, y, \ldots$ ne diffèrent des valeurs qui satisfont aux équations $U=0$, $V=0$, $W=0, \ldots$ que par des incréments infiniment petits du premier ordre; le résultant ϕ sera un infiniment petit du premier ordre, mais toute autre fonction des coefficients, à moins qu'elle ne contienne une puissance de ϕ comme facteur, aura une valeur finie, et toute fonction des coefficients infiniment petite de l'ordre k contiendra ϕ^k *comme facteur*. Dans cette supposition les équations $\mathfrak{A}\phi = 0$, $\mathfrak{B}\phi = 0$, &c. deviendront:

$$\begin{array}{l} t\,(p\xi + q\eta \ldots)^{\mu} + f\,(\xi, \eta \ldots) = 0, \\ t'\,(p\xi + q\eta \ldots)^{\mu'} + f'\,(\xi, \eta \ldots) = 0, \\ \vdots \end{array}$$

où $f, f', \ldots$ sont des polynômes de degrés $\mu, \mu', \ldots$ dont les coefficients sont des infiniment petits du premier ordre. En supposant toujours que μ ne soit plus petit qu'aucun autre des indices $\mu, \mu', \ldots$ et en posant $\pi = \mu\mu' \ldots$, $\sigma = \frac{\pi}{\mu} + \frac{\pi}{\mu'} \ldots$, le resultant Φ du système sera tout au plus du degré $\frac{\pi}{\mu}$ par rapport aux quantités finies $t, t', \ldots$. Le degré par rapport à tous les coefficients est σ; le degré par rapport aux coefficients de $f, f', \ldots$ sera donc au moins $\sigma - \frac{\pi}{\mu}$; c'est-à-dire, ce résultant sera un infiniment petit de l'ordre $\sigma - \frac{\pi}{\mu}$, ou enfin, Φ contiendra $\phi^{\sigma - \pi : \mu}$ *comme facteur*. Or les coefficients de $U, V, W, \ldots$ (assujettis à la seule condition ci-dessus mentionnée) étant d'ailleurs arbitraires, on voit sans peine qu'il est permis de faire abstraction de la *condition*, et que Φ contiendra en général cette même puissance $\phi^{\sigma - \pi : \mu}$ *comme facteur;* ce qu'il s'agissait de démontrer.

Rien n'empêche que Φ ne contienne une plus haute puissance que $\phi^{\sigma-\pi:\mu}$ comme facteur, ou que Φ ne s'évanouisse identiquement. On peut même assigner de plus près que l'a fait M. Schläfli, des cas où Φ s'évanouit identiquement. Soient $m, m', m'', \ldots$ les degrés de $U, V, W, \ldots$ par rapport à $x, y, z, \ldots$, $p = mm'm'' \ldots$, $s = \frac{p}{m} + \frac{p}{m'} + \frac{p}{m''} \ldots$, ϕ sera du degré $\frac{p}{m}$ par rapport aux coefficients de U. Soient aussi $\mu_{,}, \mu_{,,} \ldots$ les degrés de celles des fonctions $\mathfrak{A}\phi$, $\mathfrak{B}\phi, \ldots$, pour lesquelles les opérateurs $\mathfrak{A}$, $\mathfrak{B}, \ldots$ contiennent des différentielles par rapport aux coefficients de U, $\rho = \frac{\pi}{\mu_{,}} + \frac{\pi}{\mu_{,,}} \ldots$: pour ces fonctions les coefficients seront du degré $\frac{p}{m} - 1$ par rapport aux coefficients de U; pour les autres ils seront du degré $\frac{p}{m}$. Φ sera donc du degré $\left(\frac{p}{m} - 1\right)\rho + \frac{p}{m}(\sigma - \rho), = \frac{p}{m}\sigma - \rho$, par rapport aux coefficients dé U, et $\Phi \div \phi^{\sigma-\pi:\mu}$ sera du degré $\frac{p}{m}\sigma - \rho - \frac{p}{m}\left(\sigma - \frac{\pi}{\mu}\right)$, c'est-à-dire du degré $\frac{p}{m} \cdot \frac{\pi}{\mu} - \rho$ par rapport aux coefficients de U. De même, en supposant que les lettres $m', \rho', \ldots$ aient rapport à V, &c., $\Phi \div \phi^{\sigma-\pi:\mu}$ sera du degré $\frac{p}{m'} \cdot \frac{\pi}{\mu} - \rho'$, &c. par rapport aux coefficients de V, &c. Si l'un quelconque des nombres $\frac{p}{m} \cdot \frac{\pi}{\mu} - \rho$, $\frac{p}{m'} \cdot \frac{\pi}{\mu} - \rho'$, &c. est *négatif*, et à plus forte raison, si leur somme $s \cdot \frac{\pi}{\mu} - \sigma$ est *negative*, Φ doit s'évanouir identiquement. En particulier, en supposant que le nombre des fonctions $\mathfrak{A}\phi$, $\mathfrak{B}\phi, \ldots$ (c'est-à-dire le nombre des indéterminés $\xi, \eta, \ldots$) soit ν, on aura $\sigma > \nu \frac{\pi}{\mu}$, et par cette raison Φ s'évanouira identiquement si $\frac{\pi}{\mu}(\sigma - \nu)$ est négatif, c'est-à-dire si $\nu > \sigma$. Je ne parlerai pas ici des cas examinés par M. Schläfli, où Φ contient comme facteur une plus haute puissance que $\phi^{\sigma-\pi:\mu}$.

134.

REMARQUES SUR LA NOTATION DES FONCTIONS ALGÉBRIQUES.

[From the *Journal für die reine und angewandte Mathematik* (Crelle), tom. L. (1855), pp. 282—285.]

Je me sers de la notation

$$\left|\begin{array}{llll} \alpha\,, & \beta\,, & \gamma\,, & \ldots \\ \alpha', & \beta', & \gamma', & \ldots \\ \alpha'', & \beta'', & \gamma'', & \ldots \\ \vdots & & & \end{array}\right|$$

pour représenter ce que j'appelle une *matrice;* savoir un *système* de quantités rangées en forme de *carré,* mais d'ailleurs tout à fait *indépendantes* (je ne parle pas ici des *matrices rectangulaires*). Cette notation me parait très commode pour la théorie des équations *linéaires;* j'écris par exemple

$$(\xi,\ \eta,\ \zeta,\ \ldots) = (\left|\begin{array}{llll} \alpha\,, & \beta\,, & \gamma\,, & \ldots \\ \alpha', & \beta', & \gamma', & \ldots \\ \alpha'', & \beta'', & \gamma'', & \ldots \\ \vdots & & & \end{array}\right|)(x,\ y,\ z,\ \ldots)$$

pour représenter le système des équations

$$\begin{aligned} \xi &= \alpha\, x + \beta\, y + \gamma\, z\, \ldots, \\ \eta &= \alpha' x + \beta' y + \gamma' z\, \ldots, \\ \zeta &= \alpha'' x + \beta'' y + \gamma'' z\, \ldots, \\ &\vdots \end{aligned}$$

On obtient par là l'équation:

$$(x, y, z, \ldots) = \left(\begin{vmatrix} \alpha, & \beta, & \gamma, & \ldots \\ \alpha', & \beta', & \gamma', & \ldots \\ \alpha'', & \beta'', & \gamma'', & \ldots \\ \vdots & & & \end{vmatrix}^{-1} \right)(\xi, \eta, \zeta, \ldots),$$

qui représente le système d'équations qui donne $x, y, z, \ldots$ en termes de $\xi, \eta, \zeta, \ldots$, et on se trouve ainsi conduit à la notation

$$\begin{vmatrix} \alpha, & \beta, & \gamma, & \ldots \\ \alpha', & \beta', & \gamma', & \ldots \\ \alpha'', & \beta'', & \gamma'', & \ldots \\ \vdots & & & \end{vmatrix}^{-1}$$

de la matrice *inverse.* Les termes de cette matrice sont des fractions, ayant pour *dénominateur* commun le *déterminant* formé avec les termes de la matrice originale; les *numérateurs* sont les déterminants mineurs formés avec les termes de cette même matrice en supprimant l'une quelconque des lignes et l'une quelconque des colonnes.

Soit encore

$$(x, y, z, \ldots) = \left(\begin{vmatrix} a, & b, & c, & \ldots \\ a', & b', & c', & \ldots \\ a'', & b'', & c'', & \ldots \\ \vdots & & & \end{vmatrix} \right)(x, y, z, \ldots),$$

on peut écrire:

$$(\xi, \eta, \zeta, \ldots) = \left(\begin{vmatrix} \alpha, & \beta, & \gamma, & \ldots \\ \alpha', & \beta', & \gamma', & \ldots \\ \alpha'', & \beta'', & \gamma'', & \ldots \\ \vdots & & & \end{vmatrix} \begin{vmatrix} a, & b, & c, & \ldots \\ a', & b', & c', & \ldots \\ a'', & b'', & c'', & \ldots \\ \vdots & & & \end{vmatrix} \right)(x, y, z, \ldots),$$

et l'on parvient ainsi à l'idée d'une *matrice composée,* par ex.

$$\begin{vmatrix} \alpha, & \beta, & \gamma, & \ldots \\ \alpha', & \beta', & \gamma', & \ldots \\ \alpha'', & \beta'', & \gamma'', & \ldots \\ \vdots & & & \end{vmatrix} \begin{vmatrix} a, & b, & c, & \ldots \\ a', & b', & c', & \ldots \\ a'', & b'', & c'', & \ldots \\ \vdots & & & \end{vmatrix}.$$

On voit d'abord que la valeur de cette matrice composée est

$$\begin{vmatrix} (\alpha, \beta, \gamma, \ldots)(a, a', a'', \ldots), & (\alpha, \beta, \gamma, \ldots)(b, b', b'', \ldots), & \ldots \\ (\alpha', \beta', \gamma', \ldots)(a, a', a'', \ldots), & (\alpha', \beta', \gamma', \ldots)(b, b', b'', \ldots), & \ldots \\ \vdots & & \end{vmatrix}$$

où $(\alpha, \beta, \gamma, \ldots)(a, a', a'', \ldots) = \alpha a + \beta a' + \gamma a'' + \ldots$. Il faut faire attention, dans la composition des matrices, de combiner les *lignes* de la matrice à gauche avec les *colonnes* de la matrice à droite, pour former les *lignes* de la matrice composée. Il y aurait bien des choses à dire sur cette théorie de matrices, laquelle doit, il me semble, précéder la théorie de *Déterminants*.

Une notation semblable peut être employée dans la théorie des fonctions *quadratiques*. En effet, on peut dénoter par

$$\left(\begin{vmatrix} \alpha, & \beta, & \gamma, & \ldots \\ \alpha', & \beta', & \gamma', & \ldots \\ \alpha'', & \beta'', & \gamma'', & \ldots \\ \vdots & & & \end{vmatrix}\right)(\xi, \eta, \zeta)(x, y, z)$$

la fonction *lineo-linéaire*

$$\begin{aligned} &(\alpha\ \xi + \beta\ \eta + \gamma\ \zeta \ldots)\, x \\ +\, &(\alpha' \xi + \beta' \eta + \gamma' \zeta \ldots)\, y \\ +\, &(\alpha'' \xi + \beta'' \eta + \gamma'' \zeta \ldots)\, z \\ &\vdots \end{aligned}$$

et de là par

$$\left(\begin{vmatrix} a, & h, & g, & \ldots \\ h, & b, & f, & \ldots \\ g, & f, & c, & \ldots \\ \vdots & & & \end{vmatrix}\right)(x, y, z, \ldots)^2$$

la fonction *quadratique*

$$ax^2 + by^2 + cz^2 + 2fyz + 2gzx + 2hxy \ldots$$

que je représente aussi par

$$(a, b, c, \ldots f, g, h, \ldots)(x, y, z, \ldots)^2.$$

Je remarque qu'en général je représente une fonction rationnelle et intégrale, homogène et des *degrés* m, m', &c., par rapport aux indéterminées x, y, &c., x', y', &c., de la manière suivante:

$$(\ \Diamond\ )(x, y, \ldots)^m (x', y', \ldots)^{m'} \ldots.$$

Une fonction rationnelle et intégrale, homogène et du degré m par rapport aux deux indéterminées x, y sera donc représentée par

$$(\ \Diamond\ )(x, y)^m.$$

En introduisant dans cette notation les *coefficients*, j'écris par exemple

$$(a,\ b,\ c,\ d \between x,\ y)^3,$$

pour représenter la fonction

$$ax^3 + 3bx^2y + 3cxy^2 + dy^3,$$

tandis que je me sers de la notation

$$(a,\ b,\ c,\ d \,\overset{\frown}{\between}\, x,\ y)^3,$$

pour représenter la fonction

$$ax^3 + bx^2y + cxy^2 + dy^3,$$

et de même pour les fonctions d'un degré quelconque. J'ai trouvé cette distinction très commode.

[In the foregoing Paper as here printed, except in the expression in the second line of this page, $)($ is used instead of $\overset{\frown}{)(}$: it appears by a remark (*Crelle*, t. LI, errata) that the manuscript had the interlaced parentheses $\between$. Moreover in the manuscript $(\,|\;|\,)$ was used for a Matrix, which was thus distinguished from a Determinant, but in the absence of any real ambiguity, no alteration has been made in this respect. In the reprint of subsequent papers from Crelle, the arrowhead $\overset{\frown}{)(}$ or $\overset{\frown}{\between}$ is used instead of $(\smile)$.]

135.

NOTE SUR LES COVARIANTS D'UNE FONCTION QUADRATIQUE, CUBIQUE, OU BIQUADRATIQUE À DEUX INDÉTERMINÉES.

[From the *Journal für die reine und angewandte Mathematik* (Crelle), tom. L. (1855), pp. 285—287.]

La théorie d'une fonction à deux indéterminées d'un degré quelconque, par exemple

$$(\ \lozenge\ )(x,\ y)^m,$$

dépend du système des *covariants* de la fonction, lequel est censé contenir la fonction elle-même.

Pour une fonction *quadratique* le système de covariants est

$$(a,\ b,\ c)(x,\ y)^2,$$
$$ac - b^2.$$

Pour la fonction *cubique*, le système est

$$(a,\ b,\ c,\ d)(x,\ y)^3,$$
$$(ac - b^2,\ ad - bc,\ bd - c^2)(x,\ y)^2,$$
$$(-a^2d + 3abc - 2b^3,\ -abd + 2ac^2 - b^2c,\ acd - 2b^2d + bc^2,\ ad^2 - 3bcd + 2c^3)(x,\ y)^3,$$
$$-a^2d^2 + 6abcd - 4ac^3 - 4b^3d + 3b^2c^2,$$

fonctions lesquelles, en supposant qu'on les représente par U, H, Φ, $\square$, satisfont identiquement à l'équation

$$\Phi^2 + \square U^2 = -4H^3.$$

Pour la fonction *biquadratique*, le système est

$$(a,\ b,\ c,\ d,\ e)(x,\ y)^4,$$

$$ae - 4bd + 3c^2,$$

$$(ac - b^2,\ 2ad - 2bc,\ ae + 2bd - 3c^2,\ 2be - 2cd,\ ce - d^2)(x,\ y)^4,$$

$$ace + 2bcd - ad^2 - b^2e - c^3,$$

$$\left.\begin{cases} -a^2d + 3abc - 2b^3, \\ -a^2e - 2abd + 9ac^2 - 6b^2c, \\ -5abe + 15acd - 10b^2d, \\ +10a^2d - 10b^2e, \\ +5ade + 10bd^2 - 15bce, \\ +ae^2 + 2bde - 9c^2e + 6cd^2, \\ +be^2 - 3cde + 2d^3, \end{cases}\right\}(x,\ y)^6,$$

et ces fonctions, en supposant qu'on les représente par U, I, H, J, Φ, satisfont identiquement à l'équation

$$JU^3 - IU^2H + 4H^3 = -\Phi^2.$$

J'ajoute à ce système la fonction

$$\begin{aligned} I^3 - 27J^2 = {} & a^3e^3 - 12a^2bde^2 - 18a^2c^2e^2 + 54a^2cd^2e - 27a^2d \\ & + 54ab^2ce^2 - 6ab^2d^2e - 180abc^2de + 108abcd^3 + 81ac^4e \\ & - 54ac^3d^2 - 27b^4e^2 - 64b^3d^3 + 108b^3cde - 54b^2c^3e \\ & + 36b^2c^2d^2, \end{aligned}$$

qui est le *discriminant* de la fonction biquadratique.

Pour donner une application de ces formules, soit proposé de résoudre une équation quadratique, cubique ou biquadratique, ou autrement dit : de trouver un *facteur linéaire* de la fonction quadratique, cubique, ou biquadratique.

Il est assez singulier que pour la fonction quadratique la solution est en quelque sorte plus compliquée que pour les deux autres. En effet, il n'existe pas de solution symétrique, à moins qu'on n'introduise des quantités arbitraires et superflues; savoir, on trouve pour facteur linéaire de $(a,\ b,\ c)(x,\ y)^2$ l'expression

$$(a,\ b,\ c)(\alpha,\ \beta)(x,\ y) + \sqrt{-\square}\,.\,(\beta x - \alpha y),$$

où $(a,\ b,\ c)(\alpha,\ \beta)(x,\ y)$ dénote $a\alpha x + b\,(\alpha y + \beta x) + c\beta y$.

Pour la fonction *cubique*, l'équation $\Phi^2 + \square U^2 = -4H^3$ fait voir que les deux fonctions $\Phi + U\sqrt{-\square}$, $\Phi - U\sqrt{-\square}$ sont l'une et l'autre des cubes parfaits. L'expression

$$\sqrt[3]{\{\tfrac{1}{2}(\Phi + U\sqrt{-\square})\}} - \sqrt[3]{\{\tfrac{1}{2}(\Phi - U\sqrt{-\square})\}}$$

sera donc une fonction linéaire de x, y; et puisque cette fonction s'évanouit pour $U=0$, elle ne sera autre chose que l'un des facteurs linéaires de $(a, b, c, d)(x, y)^3$.

Pour la fonction *biquadratique*, en partant de l'équation

$$JU^3 - IU^2H + 4H^3 = -\Phi^2,$$

j'écris

$$M = \frac{I^3}{4J^2},$$

et je mets l'équation sous la forme

$$(1,\ 0, -M,\ M)(IH,\ JU)^3 = -\tfrac{1}{4} I^3 \Phi^2.$$

Donc, en supposant que ϖ_1, ϖ_2, ϖ_3 soient les racines de l'équation

$$(1,\ 0, -M,\ M)(\varpi,\ 1)^3 = 0,$$

ou plus simplement de l'équation

$$\varpi^3 - M(\varpi - 1) = 0,$$

ces expressions $IH - \varpi_1 JU$, $IH - \varpi_2 JU$, $IH - \varpi_3 JU$ seront toutes trois des carrés de fonctions quadratiques. L'expression

$$(\varpi_2 - \varpi_3)\sqrt{(IH - \varpi_1 JU)} + (\varpi_3 - \varpi_1)\sqrt{(IH - \varpi_2 JU)} + (\varpi_1 - \varpi_2)\sqrt{(IH - \varpi_3 JU)}$$

sera donc une fonction quadratique, et on voit sans peine qu'elle sera le carré d'une fonction *linéaire*. Or cette expression s'évanouit pour $U=0$; donc ce sera précisément le carré de l'un quelconque des facteurs linéaires de $(a, b, c, d, e)(x, y)^4$.

L'équation identique pour les covariants d'une fonction biquadratique donne lieu aussi (remarque que je dois à M. Hermite) à une transformation très élégante de *l'intégrale elliptique* $\int dx \div \sqrt{(a,\ b,\ c,\ d,\ e)(x,\ 1)^4}$.

136.

SUR LA TRANSFORMATION D'UNE FONCTION QUADRATIQUE EN ELLE-MÊME PAR DES SUBSTITUTIONS LINÉAIRES.

[From the *Journal für die reine und angewandte Mathematik* (Crelle), tom. L. (1855), pp. 288—299.]

Il s'agit de trouver les transformations linéaires d'une fonction quadratique $(\lozenge)(x, y, z, \ldots)^2$ *en elle-même*, c'est-à-dire de trouver pour $(x, y, z, \ldots)$ des fonctions linéaires de $x, y, z, \ldots$ telles que

$$(\lozenge)(\mathrm{x}, \mathrm{y}, \mathrm{z}, \ldots)^2 = (\lozenge)(x, y, z, \ldots)^2.$$

En représentant la fonction quadratique par

$$(\lozenge)(x, y, z, \ldots)^2 = \left|\begin{matrix} a, & h, & g, \ldots \\ h, & b, & f, \ldots \\ g, & f, & c, \ldots \\ \vdots \end{matrix}\right|(x, y, z, \ldots)^2,$$

la solution qu'a donnée M. Hermite de ce problème peut être résumée dans la seule équation

$$(\mathrm{x}, \mathrm{y}, \mathrm{z}, \ldots) =$$

$$\left(\left|\begin{matrix} a, & h, & g, \ldots \\ h, & b, & f, \ldots \\ g, & f, & c, \ldots \\ \vdots \end{matrix}\right|^{-1} \left|\begin{matrix} a, & h-\nu, & g+\mu, \ldots \\ h+\nu, & b, & f-\lambda, \ldots \\ g-\mu, & f+\lambda, & c\ , \ldots \\ \vdots \end{matrix}\right| \quad \left|\begin{matrix} a, & h+\nu, & g-\mu, \ldots \\ h-\nu, & b, & f+\lambda, \ldots \\ g+\mu, & f-\lambda, & c\ , \ldots \\ \vdots \end{matrix}\right|^{-1} \left|\begin{matrix} a, & h, & g, \ldots \\ h, & b, & f, \ldots \\ g, & f, & c, \ldots \\ \vdots \end{matrix}\right|\right)(x, y, z, \ldots),$$

où $\lambda, \mu, \nu, \ldots$ sont des quantités quelconques.

En effet, pour démontrer que cela est une solution, on n'a qu'à reproduire dans un ordre inverse le procédé de M. Hermite. En introduisant les quantités auxiliaires (ξ, η, ζ, ...), on peut remplacer l'équation par les deux équations

$$\left(\begin{vmatrix} a, & h, & g, \dots \\ h, & b, & f, \dots \\ g, & f, & c, \dots \\ \vdots \end{vmatrix}\right)(x, y, z, \dots) = \left(\begin{vmatrix} a, & h+\nu, & g-\mu, \dots \\ h-\nu, & b, & f+\lambda, \dots \\ g+\mu, & f-\lambda, & c, \dots \\ \vdots \end{vmatrix}\right)(\xi, \eta, \zeta, \dots)$$

$$\left(\begin{vmatrix} a, & h, & g, \dots \\ h, & b, & f, \dots \\ g, & f, & c, \dots \\ \vdots \end{vmatrix}\right)(\mathrm{x}, \mathrm{y}, \mathrm{z}, \dots) = \left(\begin{vmatrix} a, & h-\nu, & g+\mu, \dots \\ h+\nu, & b, & f-\lambda, \dots \\ g-\mu, & f+\lambda, & c, \dots \\ \vdots \end{vmatrix}\right)(\xi, \eta, \zeta, \dots)$$

qui donnent tout de suite d'abord

$$(\lozenge)(x, y, z, \dots)(\xi, \eta, \zeta, \dots) = (\lozenge)(\xi, \eta, \zeta, \dots)^2,$$

et puis

$$x + \mathrm{x} = 2\xi, \quad y + \mathrm{y} = 2\eta, \quad z + \mathrm{z} = 2\zeta, \quad \&c.$$

On obtient par là:

$$\begin{aligned}(\lozenge)(\mathrm{x}, \mathrm{y}, \mathrm{z}, \dots)^2 &= (\lozenge)(2\xi - \mathrm{x}, 2\eta - \mathrm{y}, 2\zeta - \mathrm{z}, \dots)^2, \\ &= 4(\lozenge)(\xi, \eta, \zeta, \dots)^2 - 4(\lozenge)(\xi, \eta, \zeta, \dots)(x, y, z, \dots) \\ &\quad + (\lozenge)(x, y, z, \dots)^2,\end{aligned}$$

c'est-à-dire l'équation

$$(\lozenge)(\mathrm{x}, \mathrm{y}, \mathrm{z}, \dots)^2 = (\lozenge)(x, y, z, \dots)^2,$$

qu'il s'agissait de vérifier.

Je remarque que la transformation est toujours *propre*. En effet, le déterminant de transformation est

$$\begin{vmatrix} a, h, g \dots \\ h, b, f \dots \\ g, f, c \dots \\ \vdots \end{vmatrix}^{-1} \begin{vmatrix} a, & h-\nu, & g+\mu \dots \\ h+\nu, & b, & f-\lambda \dots \\ g-\mu, & f+\lambda, & c \quad \dots \\ \vdots \end{vmatrix} \begin{vmatrix} a, & h+\nu, & g-\mu \dots \\ h-\nu, & b, & f+\lambda \dots \\ g+\mu, & f-\lambda, & c \quad \dots \\ \vdots \end{vmatrix}^{-1} \begin{vmatrix} a, h, g \dots \\ h, b, f \dots \\ g, f, c \dots \\ \vdots \end{vmatrix}.$$

Or les déterminants qui entrent dans les deux termes moyens, ne contiennent l'un ou l'autre que les puissances *paires* de λ, μ, ν, ... Donc ces deux déterminants sont égaux, et les quatre termes du produit sont *réciproques* deux à deux; le déterminant de transformation est donc $+1$, et la transformation est propre.

Pour obtenir une transformation *impropre*, il faut considérer une fonction quadratique qui contient outre les indéterminées x, y, z, ... une indéterminée θ, et puis réduire à zéro les coefficients de tous les termes dans lesquels entre cette indéterminée θ. Les valeurs de x, y, z, ... ne contiendront pas θ, et en représentant par ϑ l'indéterminée que l'on doit ajouter à la suite x, y, z, ..., la valeur de ϑ sera, comme on voit sans peine, $\vartheta = -\theta$; le déterminant de transformation pour la forme aux indéterminées $x, y, z, \ldots, \theta$ sera $+1$, et ce déterminant sera le produit du déterminant de transformation pour la forme aux indéterminées $x, y, z, \ldots$ multiplié par -1. Le déterminant de transformation pour la forme aux indéterminées $x, y, z, \ldots$ sera donc -1, c'est-à-dire, la transformation sera *impropre.*

Au lieu de la formule de transformation ci-dessus, on peut se servir des formules

$$(\xi, \eta, \zeta, \ldots) = \left(\begin{vmatrix} a\ , & h+\nu, & g-\mu, \ldots \\ h-\nu, & b\ , & f+\lambda, \ldots \\ g+\mu, & f-\lambda, & c\ , \ldots \\ \vdots \end{vmatrix}^{-1} \begin{vmatrix} a, & h, & g, \ldots \\ h, & b, & f, \ldots \\ g, & f, & c, \ldots \\ \vdots \end{vmatrix} \right)(x, y, z, \ldots),$$

$$\mathrm{x} = 2\xi - x, \quad \mathrm{y} = 2\eta - y, \quad \mathrm{z} = 2\zeta - z, \ldots.$$

Par exemple, en supposant que la forme à transformer soit

$$ax^2 + by^2 + cz^2 + \&c.,$$

on aura

$$(\xi, \eta, \zeta, \ldots) = \left(\begin{vmatrix} a, & \nu, & -\mu, \ldots \\ -\nu, & b, & \lambda, \ldots \\ \mu, & -\lambda, & c, \ldots \\ \vdots \end{vmatrix}^{-1} \right)(ax, by, cz, \ldots),$$

$$\mathrm{x} = 2\xi - x, \quad \mathrm{y} = 2\eta - y, \quad \mathrm{z} = 2\zeta - z, \ \&c.,$$

de manière qu'en posant

$$\begin{vmatrix} a, & \nu, & -\mu, \ldots \\ -\nu, & b, & \lambda, \ldots \\ \mu, & -\lambda, & c, \ldots \\ \vdots \end{vmatrix} = k,$$

$$\begin{vmatrix} a, & \nu, & -\mu, \ldots \\ -\nu, & b, & \lambda, \ldots \\ \mu, & -\lambda, & c, \ldots \\ \vdots \end{vmatrix}^{-1} = \frac{1}{k} \begin{vmatrix} A\ , & B\ , & C\ , \ldots \\ A', & B', & C', \ldots \\ A'', & B'', & C'', \ldots \\ \vdots \end{vmatrix},$$

on aura

$$(\mathrm{x}, \mathrm{y}, \mathrm{z} \ldots) = \frac{1}{k} \left(\begin{vmatrix} 2A - \frac{k}{a}, & 2B\ , & 2C & \ldots \\ 2A'\ , & 2B' - \frac{k}{b}, & 2C' & \ldots \\ 2A''\ , & 2B''\ , & 2C'' - \frac{k}{c} & \ldots \\ \vdots \end{vmatrix} \right)(ax, by, cz, \ldots),$$

ce qui est l'équation pour la transformation propre en elle-même, de la fonction $ax^2+by^2+cz^2+\&c.$ On en déduira, comme dans le cas général, la formule pour la transformation impropre. On trouvera des observations sur cette formule dans le mémoire "Recherches ultérieures sur les déterminants gauches" [137].

Je reviens à l'équation générale

$$(\lozenge)(\mathrm{x},\ \mathrm{y},\ \mathrm{z},\ \ldots)^2=(\lozenge)(x,\ y,\ z,\ \ldots)^2,$$

et je suppose seulement que x, y, z, ... soient des fonctions linéaires de x, y, z, ... qui satisfont à cette équation sans supposer rien davantage par rapport à la forme de la solution. Cela étant, je forme les fonctions linéaires $\mathrm{x}-sx$, $\mathrm{y}-sy$, $\mathrm{z}-sz$, &c., où s est une quantité quelconque, et je considère la fonction

$$(\lozenge)(\mathrm{x}-sx,\ \ \mathrm{y}-sy,\ \ \mathrm{z}-sz\ \ldots)^2,$$

laquelle, en la développant, devient

$$(1+s^2)(\lozenge)(x,\ y,\ z\ \ldots)^2-2s(\lozenge)(x,\ y,\ z\ \ldots)(\xi,\ \eta,\ \zeta\ \ldots);$$

et en développant de la même manière la fonction quadratique

$$(\lozenge)\left(\mathrm{x}-\frac{1}{s}x,\ \ \mathrm{y}-\frac{1}{s}y,\ \ \mathrm{z}-\frac{1}{s}z,\ \ldots\right)^2,$$

on obtient l'équation identique

$$(\lozenge)(\mathrm{x}-sx,\ \mathrm{y}-sy,\ \mathrm{z}-sz,\ \ldots)^2=s^2\ .\ (\lozenge)\left(\mathrm{x}-\frac{1}{s}x,\ \ \mathrm{y}-\frac{1}{s}y,\ \ \mathrm{z}-\frac{1}{s}z,\ \ldots\right)^2.$$

Soit $\square$ le déterminant formé avec les coefficients de fonctions linéaires $\mathrm{x}-sx$, $\mathrm{y}-sy$, $\mathrm{z}-sz$, &c. En supposant que le nombre des indéterminées x, y, z, &c., est n, $\square$ sera évidemment une fonction rationnelle et intégrale du degré n par rapport à s. Soit de même $\square'$ le déterminant formé avec les coefficients de

$$\mathrm{x}-\frac{1}{s}x,\ \ \mathrm{y}-\frac{1}{s}y,\ \ \mathrm{z}-\frac{1}{s}z,\ \&c.;$$

l'équation qui vient d'être trouvée, donne $\square^2=s^{2n}\square'^2$, c'est-à-dire $\square=\pm s^n\square'$. Cela fait voir que les coefficients du premier et du dernier terme, du second et de l'avant-dernier terme, &c., sont égaux, aux signes près. De plus, le coefficient de la plus haute puissance s^n est toujours ± 1, et on voit sans peine qu'en supposant d'abord que n soit *impair*, on a pour la transformation *propre:*

$$\square=(1,\ \ P,\ldots P,\ \ 1)(-s,\ 1)^n$$

et pour la transformation *impropre*

$$\square=(1,\ -P,\ \ldots\ P,\ -1)(-s,\ 1)^n:$$

équation qui peut être changée en celle-ci: $\square=-(1,\ P,\ \ldots\ P,\ 1)(s,\ 1)^n$. Puis, en supposant que n soit *pair*, on a pour la transformation *propre:*

$$\square=(1,\ \ P,\ldots P,\ \ 1)(-s,\ 1)^n$$

et pour la transformation *impropre:*

$$\square=(1,\ -P,\ \ldots\ P,\ -1)(-s,\ 1)^n,$$

le coefficient moyen étant dans ce cas égal à zéro. Ces théorèmes pour la forme du déterminant des fonctions linéaires $\mathrm{x}-sx$, $\mathrm{y}-sy$, $\mathrm{z}-sz$, ... sont dus à M. Hermite.

Il y a à remarquer que la forme $(\lozenge)(x, y, z\ldots)^2$ est tout à fait indéterminée; c'est-à-dire, on suppose seulement que x, y, z, ... soient des fonctions linéaires de $x, y, z, \ldots$, telles qu'il y ait une forme quadratique $(\lozenge)(x, y, z, \ldots)^2$ pour laquelle l'équation $(\lozenge)(\mathrm{x}, \mathrm{y}, \mathrm{z}\ldots)^2=(\lozenge)(x, y, z\ldots)^2$ est satisfaite.

Je regarde d'un autre point de vue ce problème de la transformation en elle-même, d'une fonction quadratique par des substitutions linéaires. Je suppose que x, y, z, &c. soient des fonctions linéaires données de $x, y, z, \ldots$, et je cherche une fonction linéaire de x, y, z, &c. qui, par la substitution de x, y, z, &c. au lieu de x, y, z, &c. se transforme en elle-même à un facteur près. Soit $(l, m, n, \ldots)(x, y, z, \ldots)$, cette fonction linéaire, il faut que $(l, m, n, \ldots)(\mathrm{x}, \mathrm{y}, \mathrm{z}, \ldots)$ soit identiquement $=s\,.\,(l, m, n, \ldots)(x, y, z, \ldots)$, ou, ce qui est la même chose, que $(l, m, n, \ldots)(\mathrm{x}-sx, \mathrm{y}-sy, \mathrm{z}-sz, \ldots)$ soit $=0$; c'est-à-dire, les quantités $l, m, n, \ldots$ seront déterminées par autant d'équations linéaires dont les coefficients sont précisément ceux de $\mathrm{x}-sx$, $\mathrm{y}-sy$, $\mathrm{z}-sz$, &c.; donc s sera déterminé si l'on rend égal à zéro le déterminant formé avec ces coefficients, et l, m, n, &c. se trouveront donnés rationnellement en termes de s. Cela étant, je suppose que les racines de l'équation en s soient $a, b, c, \ldots$, et ces différentes racines correspondront aux fonctions linéaires $\mathrm{x}_a, \mathrm{x}_b, \mathrm{x}_c, \ldots$ qui ont la propriété dont il s'agit. Soit $(\lozenge)(x, y, z, \ldots)^2$ une fonction quadratique qui se transforme en elle-même par la substitution de x, y, z, &c. au lieu de x, y, z, &c. Cette fonction peut être exprimée en fonction quadratique de $\mathrm{x}_a, \mathrm{x}_b, \mathrm{x}_c$, &c.; quantités qui, en substituant x, y, z, &c. au lieu de $x, y, z, \ldots$ deviennent $a\mathrm{x}_a, b\mathrm{x}_b, c\mathrm{x}_c, \ldots$.

Je prends les cas d'une fonction *binaire*, *ternaire*, &c., et d'abord le cas d'une fonction *binaire*.

En écrivant d'abord $(\lozenge)(x, y)^2=(A, B, C)(\mathrm{x}_a, \mathrm{x}_b)^2$, on doit obtenir identiquement $(A, B, C)(a\mathrm{x}_a, b\mathrm{x}_b)^2=(A, B, C)(\mathrm{x}_a, \mathrm{x}_b)^2$, c'est-à-dire $A(a^2-1)=0$, $B(ab-1)=0$, $C(b^2-1)=0$. Or la solution $A=B=C=0$ ne signifiant rien, on ne peut satisfaire à ces équations sans supposer des relations entre les quantités a, b; et pour obtenir une solution dans laquelle la fonction quadratique ne se réduit pas à un carré, il faut supposer, ou $ab-1=0$, ou $a^2-1=0$ et $b^2-1=0$. Le premier cas est celui de la transformation *propre*. Il donne

$$ab=1,\quad (\lozenge)(x, y)^2=l\,\mathrm{x}_a\mathrm{x}_b.$$

Le second cas est celui de la transformation *impropre*. Il donne

$$a=+1,\quad b=-1,\quad (\lozenge)(x, y)^2=l\,\mathrm{x}_a^2+m\,\mathrm{x}_b^2.$$

En passant au cas d'une fonction *ternaire*, soit

$$(\lozenge)(x, y, z)^2=(A, B, C, F, G, H)(\mathrm{x}_a, \mathrm{x}_b, \mathrm{x}_c)^2;$$

on doit avoir identiquement

$$(A, B, C, F, G, H)(a\mathrm{x}_a, b\mathrm{x}_b, c\mathrm{x}_c)^2=(A, B, C, F, G, H)(\mathrm{x}_a, \mathrm{x}_b, \mathrm{x}_c)^2,$$

c'est-à-dire $A(a^2-1)=0$, $B(b^2-1)=0$, $C(c^2-1)=0$, $F(bc-1)=0$, $G(ca-1)=0$, $H(ab-1)=0$, et on voit que pour obtenir une solution dans laquelle la fonction quadratique ne se réduit pas à un carré, ou à une fonction de deux indéterminées, il faut supposer par exemple $a^2-1=0$, $bc-1=0$. On a donc dans le cas d'une fonction ternaire:

$$a^2=1, \quad bc=1, \quad (\ \lozenge\ )(x,\ y,\ z)^2 = l\,\mathrm{x}_a{}^2 + m\,\mathrm{x}_b\mathrm{x}_c.$$

La transformation sera *propre*, ou *impropre*, selon que $a=+1$ ou $a=-1$.

Dans le cas d'une fonction *quaternaire*, on obtient pour la transformation *propre:*

$$ab=cd=1, \quad (\ \lozenge\ )(x,\ y,\ z,\ w)^2 = l\,\mathrm{x}_a\mathrm{x}_b + m\,\mathrm{x}_c\mathrm{x}_d,$$

et pour la transformation *impropre:*

$$a=+1, \quad b=-1, \quad cd=1, \quad (\ \lozenge\ )(x,\ y,\ z,\ w)^2 = l\,\mathrm{x}_a{}^2 + m\,\mathrm{x}_b{}^2 + n\,\mathrm{x}_c\mathrm{x}_d.$$

Dans le cas d'une fonction *quinaire* on obtient

$$a^2=1, \quad bc=de=1, \quad (\ \lozenge\ )(x,\ y,\ z,\ w,\ t)^2 = l\,\mathrm{x}_a{}^2 + m\,\mathrm{x}_b\mathrm{x}_c + n\,\mathrm{x}_d\mathrm{x}_e$$

et la transformation est propre ou impropre, selon que $a=+1$ ou $a=-1$; et ainsi de suite.

Cette méthode a des difficultés dans le cas où l'équation en s a des racines *égales*. Je n'entre pas ici dans ce sujet.

Dans les formules qu'on vient de trouver, on peut considérer les coefficients l, m, &c. comme des quantités *arbitraires*. Mais en supposant que la fonction quadratique soit *donnée*, ces coefficients deviennent *déterminés*. On les trouvera par la formule suivante que je ne m'arrête pas à démontrer.

Soient α, β, γ, &c. les coefficients de la fonction linéaire x_a, α', β', γ', &c. les coefficients de la fonction linéaire x_b, et ainsi de suite; alors, dans les différentes formules qui viennent d'être données, le coefficient d'un terme $\mathrm{x}_a{}^2$ à droite sera

$$\frac{-k}{(\dagger)(\alpha,\ \beta,\ \gamma,\ \ldots)},$$

et le coefficient d'un terme $\mathrm{x}_a\mathrm{x}_b$ à gauche sera

$$\frac{-k}{(\dagger)(\alpha,\ \beta,\ \gamma,\ \ldots)(\alpha',\ \beta',\ \gamma',\ \ldots)},$$

où k dénote le *discriminant* de la fonction quadratique à gauche, et où les coefficients des fonctions quadratiques des dénominateurs sont les coefficients inverses de cette même fonction quadratique à gauche[1].

[1] Je profite de cette occasion pour remarquer concernant ces recherches que les formules données dans la note sur les fonctions du second ordre (t. XXXVIII. [1848] p. 105) [71] pour les cas de trois et de quatre indéterminées, sont exactes, mais que je m'étais trompé dans la forme générale du théorème. [This correction is indicated vol. I. p. 589.]

L'application de la méthode à la forme binaire $(a, b, c)(x, y)^2$ donne lieu aux développements suivants.

J'écris $\mathrm{x} = \alpha x + \beta y$, $\mathrm{y} = \gamma x + \delta y$, et je représente par $(l, m)(x, y)$ une fonction linéaire qui par cette substitution est transformée en *elle-même*, au facteur s près. Nous aurons donc

$$(l, m)(\alpha x + \beta y, \gamma x + \delta y) = s\,(l, m)(x, y);$$

l'équation pour s sera

$$s^2 - s\,(\delta + \alpha) + \alpha\delta - \beta\gamma = 0;$$

laquelle peut aussi être écrite comme suit:

$$(1, -\delta - \alpha, \alpha\delta - \beta\gamma)(s, 1)^2 = 0.$$

Soient s', s'' les racines de cette équation. (Il est à peine nécessaire de remarquer que s', s'', et plus bas P, Q, sont ici ce que dans les formules générales j'ai représenté par a, b et x_a, x_b. De même les équations $\rho = s's''$, $\rho = s'^2$, $\rho = s''^2$, obtenues après, correspondent aux équations $ab = 1$, $a^2 = 1$, $b^2 = 1$.) On aura

$$s' + s'' = -\delta - \alpha, \qquad s's'' = \alpha\delta - \beta\gamma,$$

et les coefficients l, m seront déterminés rationnellement par s.

Mais on peut aussi déterminer ces coefficients par l'équation

$$l : m = l\alpha + m\gamma : l\beta + m\delta,$$

qui peut être écrite sous la forme

$$(\beta, \delta - \alpha, -\gamma)(l, m)^2 = 0,$$

et en éliminant entre cette équation et l'équation $lx + my = 0$ les quantités l, m, on voit que les fonctions linéaires $lx + my$ sont les facteurs de la fonction quadratique $(\beta, \delta - \alpha, -\gamma)(y, -x)^2$, ou, ce qui est la même chose, de la fonction quadratique

$$(\gamma, \delta - \alpha, -\beta)(x, y)^2;$$

je représente ces facteurs par P, Q et je remarque encore que l'équation en s aura des racines égales si

$$(\delta - \alpha)^2 + 4\beta\gamma = 0,$$

et que dans ce cas, et *exclusivement* dans ce cas, les fonctions P, Q ne forment qu'une seule et même fonction linéaire.

Je suppose maintenant que la fonction $(a, b, c)(x, y)^2$ se transforme en elle-même par la substitution $\alpha x + \beta y$, $\gamma x + \delta y$ au lieu de x, y, ou, ce qui est ici plus commode, je suppose que les deux fonctions sont égales à un facteur près, et j'écris

$$(a, b, c)(\alpha x + \beta y, \gamma x + \delta y)^2 = \rho\,(a, b, c)(x, y)^2.$$

En développant cette équation, on obtient

$$\left.\begin{array}{lllll} & x^2\,(a,\ b,\ c)(\alpha^2-\rho, & 2\alpha\gamma, & \gamma^2 &) \\ + & 2\,xy\,(a,\ b,\ c)(\alpha\beta, & \alpha\delta+\beta\gamma-\rho, & \gamma\delta &) \\ + & y^2\,(a,\ b,\ c)(\beta^2, & 2\beta\delta, & \delta^2-\rho &) \end{array}\right\}=0.$$

Voilà trois équations linéaires pour déterminer par les quantités α, β, γ, δ, considérées comme données, les coefficients (a, b, c) de la fonction quadratique. Les coefficients de ces équations linéaires sont

$$\begin{array}{lll} \alpha^2-\rho, & 2\alpha\gamma, & \gamma^2, \\ \alpha\beta, & \alpha\delta+\beta\gamma-\rho, & \gamma\delta, \\ \beta^2, & 2\beta\delta, & \delta^2-\rho. \end{array}$$

Le système inverse par lequel on trouve les valeurs de a, b, c, est

$$\begin{array}{lll} \delta^2(\alpha\delta-\beta\gamma)-(\alpha\delta+\beta\gamma+\delta^2)\,\rho+\rho^2, & -\beta\delta\,(\alpha\delta-\beta\gamma)+\alpha\beta\rho, & \beta^2(\alpha\delta-\beta\gamma)-\beta^2\rho, \\ -2\gamma\delta\,(\alpha\delta-\beta\gamma)+2\alpha\gamma\rho, & \alpha^2\delta^2-\beta^2\gamma^2-(\delta^2+\alpha^2)\,\rho+\rho^2, & -2\alpha\beta\,(\alpha\delta-\beta\gamma)+2\beta\delta\rho, \\ \gamma^2(\alpha\delta-\beta\gamma)+\gamma^2\rho, & -\alpha\gamma\,(\alpha\delta-\beta\gamma)+\gamma\delta\rho, & \alpha^2(\alpha\delta-\beta\gamma)-(\alpha\delta+\beta\gamma+\alpha^2)\,\rho+\rho^2, \end{array}$$

et le déterminant, égalé à zéro, donne

$$(\alpha\delta-\beta\gamma-\rho)\,\{(\alpha\delta-\beta\gamma+\rho)^2-\rho\,(\alpha+\delta)^2\}=0:$$

équation dont les racines sont

$$\rho=\alpha\delta-\beta\gamma,\quad \rho=\{\tfrac{1}{2}(\alpha+\delta)\pm\tfrac{1}{2}\sqrt{(\alpha-\delta)^2+4\beta\gamma}\}^2.$$

En comparant ces valeurs avec celles de s', s'', on voit que les racines de l'équation en ρ sont

$$\rho=s's'',\quad \rho=s'^2,\quad \rho=s''^2,$$

et nous allons voir que ces valeurs de ρ donnent en général les valeurs PQ, P^2, Q^2, pour la fonction quadratique.

Soit d'abord $\rho=\alpha\delta-\beta\gamma\,(=s's'')$, et posons pour abréger $\alpha\delta-\beta\gamma-\rho=\phi$, le système inverse devient:

$$\begin{array}{lll} (\delta^2-\rho)\,\phi-\beta\rho\,.\,2\gamma, & -\beta\delta\phi-\beta\rho\,(\delta-\alpha), & \beta^2\phi+\beta\rho\,.\,2\beta, \\ -2\gamma\delta\phi-\rho\,(\delta-\alpha)\,2\gamma, & (\alpha\delta+\beta\gamma-\rho)\,\phi-\ \rho\,(\delta-\alpha)^2, & -2\alpha\beta\phi+\rho\,(\delta-\alpha)\,2\beta, \\ -\gamma^2\phi+\gamma\rho\,.\,2\gamma, & -\alpha\gamma\phi+\gamma\rho\,(\delta-\alpha), & (\alpha^2-\rho)\,\phi-\gamma\rho\,.\,2\beta, \end{array}$$

et en mettant $\phi=0$, les termes de chaque ligne (en omettant un facteur) deviennent γ, $\frac{1}{2}(\delta-\alpha)$, β. On obtient ainsi dans ce cas, pour la fonction quadratique $(a, b, c)(x, y)^2$ la valeur

$$(\gamma,\ \delta-\alpha,\ -\beta)(x,\ y)^2,$$

qui est en effet le produit PQ des fonctions linéaires.

Il y a à remarquer qu'en supposant $(\delta-\alpha)^2+4\beta\gamma=0$, ce qui est le cas pour lequel ρ sera une racine *triple*, il n'y aura pas de changement à faire dans ce résultat. La fonction quadratique est, comme auparavant, le produit PQ des fonctions linéaires;

seulement ces deux fonctions linéaires dans le cas actuel sont identiques, de manière que la fonction quadratique se réduit à P^2.

Soit ensuite

$$\rho = \{\tfrac{1}{2}(\alpha+\delta) \pm \tfrac{1}{2}\sqrt{(\alpha-\delta)^2+4\beta\gamma}\}^2 \ (= s'^2 \text{ ou } s''^2);$$

en écrivant $\rho = s^2$ et en mettant pour abréger $\alpha\delta - \beta\gamma - s(\delta+\alpha) + s^2 = \chi$, le système inverse devient

$$\begin{array}{lll}
(\delta^2+s^2)\chi + s(\delta-s)^2(\delta+\alpha), & & -\beta\delta\chi - \beta s(\delta-s)(\delta+\alpha), \\
-2\gamma\delta\chi - 2\gamma s(\delta-s)(\delta+\alpha), & \{s^2+s(\delta+\alpha)+\alpha\delta+\beta\gamma\}\chi + 2\beta\gamma s(\delta+\alpha), & \\
\gamma^2\chi + \gamma^2 s(\delta+\alpha), & & -\alpha\gamma\chi - \gamma s(\alpha-s)(\delta+\alpha), \\
 & & \beta^2\chi + \beta^2 s(\delta+\alpha), \\
 & & -2\alpha\beta\chi - 2\beta s(\alpha-s)(\delta+\alpha), \\
 & & (\alpha^2+s^2)\chi + s(\alpha-s)^2(\delta+\alpha).
\end{array}$$

Donc, en écrivant $\chi = 0$ et en omettant le facteur $s(\delta+\alpha)$, le système inverse devient

$$\begin{array}{ccc}
(\delta-s)^2, & -\beta(\delta-s), & \beta^2, \\
\gamma(\delta-s), & \beta\gamma, & -\beta(\alpha-s), \\
\gamma^2, & -\gamma(\beta-s), & (\alpha-s)^2,
\end{array}$$

et les quantités dans chaque ligne sont dans le rapport $l^2 : lm : m^2$, de manière que la fonction quadratique est dans ce cas égale à P^2 ou Q^2. Cela suppose que $\delta+\alpha$ ne soit égal à zéro. En faisant pour le moment $\rho=1$, on en tire la conclusion qu'à moins de supposer $\delta+\alpha=0$, il n'existe pas de fonction quadratique binaire proprement dite (fonction non carrée) qui par la substitution *impropre* $\alpha x+\beta y$, $\gamma x+\delta y$ pour x, y, se transforme en elle-même. L'équation $\delta+\alpha=0$ donne $\rho=\alpha\delta-\beta\gamma$, qui est une racine double de l'équation cubique. On remarquera en passant par rapport à la signification de l'équation $\delta+\alpha=0$, que l'on a en général:

$$\begin{aligned}
&(\alpha,\ \beta)(\alpha x+\beta y,\ \gamma x+\delta y) : (\gamma,\ \delta)(\alpha x+\beta y,\ \gamma x+\delta y) \\
&= (\alpha^2+\beta\gamma)x + \beta(\delta+\alpha)y : \gamma(\delta+\alpha)x + (\delta^2+\beta\gamma)y,
\end{aligned}$$

et de là, qu'en supposant $\delta+\alpha=0$, on a

$$(\alpha,\ \beta)(\alpha x+\beta y,\ \gamma x+\delta y) : (\gamma,\ \delta)(\alpha x+\beta y,\ \gamma x+\delta y) = x,\ y.$$

Cela revient à dire qu'en faisant deux fois la substitution $\alpha x+\beta y$, $\gamma x+\delta y$ au lieu de x, y, on retrouve les quantités x, y, ou que la substitution est *périodique* du second ordre. Il y a aussi à remarquer que dans le cas dont il s'agit, savoir pour $\delta+\alpha=0$, on a $s''=-s'$, et que les deux fonctions linéaires P, Q restent parfaitement déterminées.

Nous venons de voir qu'il n'existe pas de transformation impropre d'une fonction quadratique binaire proprement dite, à moins que $\delta+\alpha$ ne soit pas $=0$. Mais en supposant $\delta+\alpha=0$, on voit que les coefficients des équations pour a, b, c deviennent

$$\begin{array}{ccc}
-\beta\gamma, & -\gamma(\delta-\alpha), & \gamma^2, \\
\alpha\beta, & \alpha(\delta-\alpha), & -\alpha\gamma, \\
\beta^2, & \beta(\delta-\alpha), & \beta\gamma,
\end{array}$$

c'est-à-dire: les coefficients de chaque équation sont dans le rapport de

$$\beta, \quad \delta - \alpha, \quad -\gamma,$$

de manière qu'en supposant que les coefficients a, b, c satisfont à la seule équation

$$(a, b, c)(\beta, \delta - \alpha, -\gamma) = 0,$$

où $\alpha, \beta, \gamma, \delta$ sont des quantités quelconques, telles que $\delta + \alpha = 0$, on aura

$$(a, b, c)(\alpha x + \beta y, \gamma x + \delta y)^2 = -(\alpha\delta - \beta\gamma)(a, b, c)(x, y)^2.$$

Ce n'est là qu'un cas particulier de l'équation identique

$$(a, b, c)(\alpha x + \beta y, \gamma x + \delta y)^2 + (\alpha\delta - \beta\gamma) \cdot (a, b, c)(x, y)^2 =$$
$$(\delta + \alpha) \cdot \big(a\alpha + b\gamma, b(\delta + \alpha), b\beta + c\delta\big)(x, y)^2 \quad + (\beta, \delta - \alpha, -\gamma)(a, b, c) \cdot (\beta, \delta - \alpha, -\gamma)(y, -x)^2.$$

Il faut remarquer qu'en supposant toujours l'équation

$$(a, b, c)(\beta, \delta - \alpha, -\gamma) = 0,$$

la fonction quadratique $(a, b, c)(x, y)^2$, *en supposant qu'elle se réduise à un carré*, est comme auparavant P^2 ou Q^2, c'est-à-dire le carré de l'une des fonctions linéaires. En effet: en écrivant $(a, b, c)(x, y)^2 = (lx + my)^2$, l'équation entre l, m serait évidemment $(\beta, \delta - \alpha, -\gamma)(l, m)^2 = 0$, de manière que l, m auraient les mêmes valeurs qu'auparavant. J'ajoute que tout ce qui précède par rapport à l'équation

$$(a, b, c)(\alpha x + \beta y, \gamma x + \delta y)^2 = \rho\,(a, b, c)(x, y)^2$$

fait voir qu'à moins que la fonction quadratique ne soit un carré, on aura toujours $\rho = \pm(\alpha\delta - \beta\gamma)$; ce qu'on savait déjà dès le commencement, et ce qui peut être démontré comme à l'ordinaire, en égalant les discriminants $(ac - b^2)(\alpha\delta - \beta\gamma)^2$ et $(ac - b^2)\rho^2$ des deux côtés. Je fais enfin $\rho = 1$, ce qui donne l'équation

$$(a, b, c)(\alpha x + \beta y, \gamma x + \delta y)^2 = (a, b, c)(x, y)^2,$$

et (en faisant abstraction du cas où la fonction quadratique est un carré) je tire de ce qui précède les résultats connus, savoir, que l'on a:

1. Pour la transformation propre:

$$\alpha\delta - \beta\gamma = 1,$$
$$a : 2b : c = \gamma : \delta - \alpha : -\beta.$$

2. Pour la transformation impropre:

$$\alpha\delta - \beta\gamma = -1, \quad \delta + \alpha = 0,$$
$$a\beta + b(\delta - \alpha) - c\gamma = 0.$$

Je crois que cette discussion a été utile pour compléter la théorie algébrique de la forme binaire $(a, b, c)(x, y)^2$.

137.

RECHERCHES ULTÉRIEURES SUR LES DÉTERMINANTS GAUCHES.

[From the *Journal für die reine und angewandte Mathematik* (Crelle), tom. L. (1855), pp. 299—313: Continuation of the Memoir t. XXXII. (1846) and t. XXXVIII. (1849); **52** and **69.**]

J'AI déjà donné une formule pour le développement d'un *déterminant gauche.* En prenant, pour fixer les idées, un cas particulier, soit

$$\overline{12345 \mid 12345} = \begin{vmatrix} 11, & 12, & 13, & 14, & 15 \\ 21, & 22, & 23, & 24, & 25 \\ 31, & 32, & 33, & 34, & 35 \\ 41, & 42, & 43, & 44, & 45 \\ 51, & 52, & 53, & 54, & 55 \end{vmatrix},$$

(où $12 = -21$, &c., tandis que les quantités 11, 22, &c. ne s'évanouissent pas). Cette formule peut être écrite comme suit:

$$\begin{aligned}
\overline{12345 \mid 12345} = \; & 11 \,.\, 22 \,.\, 33 \,.\, 44 \,.\, 55 \\
& + 11 \,.\, 22 \,.\, 33 \,.\, (45)^2 \\
& + 11 \,.\, 22 \,.\, 44 \,.\, (35)^2 \\
& + 11 \,.\, 22 \,.\, 55 \,.\, (34)^2 \\
& + 11 \,.\, 33 \,.\, 44 \,.\, (25)^2 \\
& + 11 \,.\, 33 \,.\, 55 \,.\, (24)^2 \\
& + 11 \,.\, 44 \,.\, 55 \,.\, (23)^2 \\
& + 22 \,.\, 33 \,.\, 44 \,.\, (15)^2 \\
& + 22 \,.\, 33 \,.\, 55 \,.\, (14)^2 \\
& + 22 \,.\, 44 \,.\, 55 \,.\, (13)^2 \\
& + 33 \,.\, 44 \,.\, 55 \,.\, (12)^2 \\
& + 11 \,.\, (2345)^2 \\
& + 22 \,.\, (1345)^2 \\
& + 33 \,.\, (1245)^2 \\
& + 44 \,.\, (1235)^2 \\
& + 55 \,.\, (1234)^2.
\end{aligned}$$

Les expressions 12, 1234, &c. à droite sont ici des *Pfaffians*. On a

$$12 = 12,$$

$$1234 = 12\,.\,34 + 13\,.\,42 + 14\,.\,23$$

et en écrivant encore un terme, pour mieux présenter la loi:

$$\begin{aligned} 123456 = {} & 12\,.\,34\,.\,56 + 13\,.\,45\,.\,62 + 14\,.\,56\,.\,23 + 15\,.\,62\,.\,34 + 16\,.\,23\,.\,45 \\ & + 12\,.\,35\,.\,64 + 13\,.\,46\,.\,25 + 14\,.\,52\,.\,36 + 15\,.\,63\,.\,42 + 16\,.\,24\,.\,53 \\ & + 12\,.\,36\,.\,45 + 13\,.\,42\,.\,56 + 14\,.\,53\,.\,62 + 15\,.\,64\,.\,23 + 16\,.\,25\,.\,34. \end{aligned}$$

J'ai trouvé récemment une formule analogue pour le développement d'un *déterminant gauche* bordé, tel que

$$\overline{\alpha 1234 \mid \beta 1234} = \begin{vmatrix} \alpha\beta, & \alpha 1, & \alpha 2, & \alpha 3, & \alpha 4 \\ 1\beta, & 11, & 12, & 13, & 14 \\ 2\beta, & 21, & 22, & 23, & 24 \\ 3\beta, & 31, & 32, & 33, & 34 \\ 4\beta, & 41, & 42, & 43, & 44 \end{vmatrix};$$

cette formule est:

$$\begin{aligned} \overline{\alpha 1234 \mid \beta 1234} = {} & \alpha\beta\,.\,11\,.\,22\,.\,33\,.\,44 \\ & + \alpha\beta\,.\,12\,.\,12\,.\,33\,.\,44 \\ & + \alpha\beta\,.\,13\,.\,13\,.\,22\,.\,44 \\ & + \alpha\beta\,.\,14\,.\,14\,.\,22\,.\,33 \\ & + \alpha\beta\,.\,23\,.\,23\,.\,11\,.\,44 \\ & + \alpha\beta\,.\,24\,.\,24\,.\,11\,.\,33 \\ & + \alpha\beta\,.\,34\,.\,34\,.\,11\,.\,22 \\ & + \alpha\beta\,.\,1234\,.\,1234 \\ & + \alpha 1\,.\,\beta 1\,.\,22\,.\,33\,.\,44 \\ & + \alpha 2\,.\,\beta 2\,.\,11\,.\,33\,.\,44 \\ & + \alpha 3\,.\,\beta 3\,.\,11\,.\,22\,.\,44 \\ & + \alpha 4\,.\,\beta 4\,.\,11\,.\,22\,.\,33 \\ & + \alpha 123\,.\,\beta 123\,.\,44 \\ & + \alpha 124\,.\,\beta 124\,.\,33 \\ & + \alpha 134\,.\,\beta 134\,.\,22 \\ & + \alpha 234\,.\,\beta 234\,.\,11. \end{aligned}$$

Il est à peine nécessaire de remarquer que dans les *Pfaffians* à droite, où entrent des symboles tels que 1α, $\beta 1$, &c., qui ne se trouvent pas dans le déterminant dont il s'agit, il faut écrire $1\alpha = -\alpha 1$, $\beta 1 = -1\beta$, &c. Le symbole $\beta\alpha$ ne se trouve ni dans le déterminant, ni au côté droit. Cependant il convient de poser $\beta\alpha = -\alpha\beta$; car cela étant, il serait permis de transformer les *Pfaffians*, en écrivant par exemple $\alpha\beta 12 = -\beta\alpha 12$.

Je remarque en passant que, si avant de poser l'équation $\beta\alpha = -\alpha\beta$, on suppose que les deux symboles α, β deviennent identiques (si par exemple on écrit $\alpha = \beta = 5$), on aurait par exemple

$$\alpha\beta\,.\,12 = \alpha\beta\,.\,12 + \alpha 1\,.\,2\beta + \alpha 2\,.\,\beta 1 = 55\,.\,12 + 51\,.\,25 + 52\,.\,51 = 55\,.\,12, \text{ \&c.},$$

et on retrouverait ainsi la formule pour le développement de $\overline{12345 \mid 12345}$.

La nouvelle formule peut être appliquée immédiatement au développement des *déterminants mineurs*. En effet, en se servant de la notation des *matrices*, on aura

$$\left|\begin{matrix} 11, & 12, & 13 \\ 21, & 22, & 23 \\ 31, & 32, & 33 \end{matrix}\right|^{-1} = \frac{1}{\overline{123 \mid 123}} \left|\begin{matrix} +\overline{23 \mid 23}, & -\overline{13 \mid 23}, & -\overline{12 \mid 32} \\ -\overline{23 \mid 13}, & +\overline{13 \mid 13}, & -\overline{21 \mid 31} \\ -\overline{32 \mid 12}, & -\overline{31 \mid 21}, & +\overline{12 \mid 12} \end{matrix}\right|,$$

$$\left|\begin{matrix} 11, & 12, & 13, & 14 \\ 21, & 22, & 23, & 24 \\ 31, & 32, & 33, & 34 \\ 41, & 42, & 43, & 44 \end{matrix}\right|^{-1} = \frac{1}{\overline{1234 \mid 1234}} \left|\begin{matrix} +\overline{234 \mid 234}, & -\overline{134 \mid 234}, & -\overline{124 \mid 324}, & -\overline{123 \mid 423} \\ -\overline{234 \mid 134}, & +\overline{134 \mid 134}, & -\overline{214 \mid 314}, & -\overline{213 \mid 413} \\ -\overline{324 \mid 124}, & -\overline{314 \mid 214}, & +\overline{214 \mid 214}, & -\overline{312 \mid 412} \\ -\overline{423 \mid 123}, & -\overline{413 \mid 213}, & -\overline{412 \mid 312}, & +\overline{123 \mid 123} \end{matrix}\right|$$

et ainsi de suite. On suppose toujours que chaque terme de la matrice à droite soit divisé par le dénominateur commun. On voit que les déterminants *mineurs* qui correspondent à des termes tels que $\alpha\alpha$, sont des déterminants *gauches ordinaires*, avec le signe $+$, tandis que les déterminants mineurs qui correspondent à des termes tels que $\alpha\beta$, sont des déterminants *gauches bordés* tels que $\overline{\beta\ldots \mid \alpha\ldots}$, avec le signe $-$.

Pour donner des exemples de la vérification de ces formules, je remarque que l'on doit avoir

$$\begin{aligned} \overline{123 \mid 123} = {} & \;\;\; 11\,.\,\overline{23 \mid 23} \\ & -12\,.\,\overline{23 \mid 13} \\ & -13\,.\,\overline{32 \mid 12}: \end{aligned}$$

équation qui peut aussi être écrite sous la forme

$$\begin{aligned} \overline{123 \mid 123} = {} & \;\;\; 11\,.\,\overline{23 \mid 23} \\ & +21\,.\,\overline{23 \mid 13} \\ & +31\,.\,\overline{32 \mid 12}. \end{aligned}$$

En effet, en développant les deux côtés, on obtient:

$$11.22.33 + 11.(23)^2 + 22.(13)^2 + 33.(12)^2 = 11.(22.33 + (23)^2) + 21.(21.33 + 23.13\dagger) + 31.(31.22 + 32.12\dagger).$$

On voit que les deux termes marqués par un † se détruisent et que l'équation est *identique*. On doit avoir de même,

$$\overline{1234 \mid 1234} = 11 \,.\, \overline{234 \mid 234} - 12 \,.\, \overline{234 \mid 134} - 13 \,.\, \overline{324 \mid 124} - 14 \,.\, \overline{423 \mid 123},$$

ou, ce qui est la même chose:

$$\overline{1234 \mid 1234} = 11 \,.\, \overline{234 \mid 234} + 21 \,.\, \overline{234 \mid 134} + 31 \,.\, \overline{324 \mid 124} + 41 \,.\, \overline{423 \mid 123};$$

c'est-à-dire, en développant des deux côtés:

$$11.22.33.44 + 11.22.(34)^2 + 11.33.(24)^2 + 11.44.(23)^2 + 22.33.(14)^2 + 22.44.(13)^2 + 33.44.(12)^2 + (1234)^2 =$$

$$11\,[22 \,.\, 33 \,.\, 44 + 22\,(34)^2 + 33\,(42)^2 + 44\,(23)^2] + 21\,[21 \,.\, 33 \,.\, 44 + 2134 \,.\, 34^* + 23 \,.\, 13 \,.\, 44\dagger + 24 \,.\, 14 \,.\, 33\dagger] + 31\,[31 \,.\, 22 \,.\, 44 + 3124 \,.\, 24^* + 32 \,.\, 12 \,.\, 44\dagger + 34 \,.\, 14 \,.\, 22\dagger] + 41\,[41 \,.\, 22 \,.\, 33 + 4123 \,.\, 23^* + 42 \,.\, 12 \,.\, 33\dagger + 43 \,.\, 13 \,.\, 22\dagger].$$

Cette expression est en effet identique, comme on le voit en observant que les six termes marqués par un † se détruisent deux à deux, et que les trois termes marqués par un (*) sont ensemble équivalents à $(1234)^2$.

Je remarque que le nombre des termes du développement du déterminant gauche est toujours une *puissance de* 2, et que de plus, ce nombre se réduit à la moitié, en réduisant à zéro un terme quelconque $\alpha\alpha$. Mais outre cela, le déterminant prend dans cette supposition la forme de déterminant d'un ordre inférieur de l'unité. Je considère par exemple le déterminant gauche $\overline{123 \mid 123}$. En y faisant $33 = 0$ et en accentuant, pour y mettre plus de clarté, tous les symboles, on trouve

$$\overline{123 \mid 123'} = 11' \,.\, (23')^2 + 22' \,.\, (13')^2.$$

De là, en écrivant

$$11 = 13'.11', \quad 12 = 11'.23',$$
$$22 = 13'.22',$$

on obtient

$$\overline{12 \mid 12} = 11 \,.\, 22 + (12)^2$$
$$= 11'.\{22'.(13')^2 + 11'.(23')^2\},$$

c'est-à-dire

$$\overline{12 \mid 12} = 11' \,.\, \overline{123 \mid 123'}.$$

On a de même

$$\overline{1234 \mid 1234'} = 11' \,.\, 22' \,.\, (34')^2 + 11' \,.\, 33' \,.\, (24')^2 + 22'33'\,(14')^2 + (1234')^2$$

et de là, en écrivant

$$11 = 14'.11', \quad 12 = 11'.24', \quad 23 = 1234',$$
$$22 = 14'.22', \quad 13 = 11'.34',$$
$$33 = 14'.33',$$

on obtient

$$\overline{123 \mid 123} = 11 \,.\, 22 \,.\, 33 + 11 \,.\, (23)^2 + 22 \,.\, (31)^2 + 33\,(12)^2$$
$$= 11' \,.\, 14'\,\{22' \,.\, 33' \,.\, (14')^2 + (1234')^2 + 11' \,.\, 22' \,.\, (34')^2 + 11' \,.\, 33' \,.\, (24')^2\},$$

c'est-à-dire,

$$\overline{123 \mid 123} = 11' \,.\, 14' \,.\, \overline{1234 \mid 1234'}.$$

De même

$$\begin{aligned}\overline{12345 \mid 12345'} = {} & 11' \,.\, 22' \,.\, 33' \,.\, (45')^2 \\ & + 11' \,.\, 22' \,.\, 44' \,.\, (35')^2 \\ & + 11' \,.\, 33' \,.\, 44' \,.\, (25')^2 \\ & + 22' \,.\, 33' \,.\, 44' \,.\, (15')^2 \\ & + 11' \,.\, (2345')^2 \\ & + 22' \,.\, (1345')^2 \\ & + 33' \,.\, (1245')^2 \\ & + 44' \,.\, (1235')^2.\end{aligned}$$

Or il est permis d'écrire

$$11 = 15'.11', \quad 12 = 11'.25', \quad 23 = 1235', \quad 1234 = 2345'.11'.15',$$
$$22 = 15'.22', \quad 13 = 11'.35', \quad 24 = 1245',$$
$$33 = 15'.33', \quad 14 = 11'.45', \quad 34 = 1345',$$
$$44 = 15'.44'.$$

En effet, les quantités à gauche ne sont liées entre elles que par la seule équation $1234 = 12 \,.\, 34 + 13 \,.\, 42 + 14 \,.\, 23$ qui est satisfaite identiquement par les valeurs à substituer pour les quantités qui y entrent. Cela étant, on trouve d'abord:

$$\overline{1234 \mid 1234} = 11'\,(15')^2\,\overline{12345 \mid 12345'}.$$

Je prends encore un exemple. On a

$$\begin{aligned}\overline{123456 \mid 123456'} = {} & 11' \,.\, 22' \,.\, 33' \,.\, 44' \,.\, (56')^2 \\ & + 11' \,.\, 22' \,.\, 33' \,.\, 55' \,.\, (46')^2 \\ & + 11' \,.\, 22' \,.\, 44' \,.\, 55' \,.\, (36')^2 \\ & + 11' \,.\, 33' \,.\, 44' \,.\, 55' \,.\, (26')^2 \\ & + 22' \,.\, 33' \,.\, 44' \,.\, 55' \,.\, (16')^2 \\ & + 11' \,.\, 22' \,.\, (3456')^2 \\ & + 11' \,.\, 33' \,.\, (2456')^2 \\ & + 11' \,.\, 44' \,.\, (2356')^2 \\ & + 11' \,.\, 55' \,.\, (2346')^2 \\ & + 22' \,.\, 33' \,.\, (1456')^2 \\ & + 22' \,.\, 44' \,.\, (1356')^2 \\ & + 22' \,.\, 55' \,.\, (1346')^2 \\ & + 33' \,.\, 44' \,.\, (1256')^2 \\ & + 33' \,.\, 55' \,.\, (1246')^2 \\ & + 44' \,.\, 55' \,.\, (1236')^2 \\ & + (123456')^2.\end{aligned}$$

Ici, il est permis d'écrire:

$$\begin{array}{lllll} 11 = 16'.11', & 12 = 11'.26', & 23 = 1236', & 34 = 1346', & 45 = 1456', \\ 22 = 16'.22', & 13 = 11'.36', & 24 = 1246', & 35 = 1356', & \\ 33 = 16'.33', & 14 = 11'.46', & 25 = 1256', & & \\ 44 = 16'.44', & 15 = 11'.56', & & & \\ 55 = 16'.55', & & & & \end{array}$$

$$\begin{array}{ll} 1234 = 2346'.11'.16', & 2345 = 123456'.16', \\ 1235 = 2356'.11'.16', & \\ 1245 = 2456'.11'.16', & \\ 1345 = 3456'.11'.16\,; & \end{array}$$

car les valeurs des quantités à gauche satisfont identiquement aux équations qui ont lieu entre ces mêmes quantités. Par exemple l'équation $1234 = 12.34 + 13.42 + 14.23$ devient $2346'.16' = 26'.1346' + 63'.1246' + 46'.1236'$.

Or l'expression à droite devient, en développant:

$$\begin{aligned} & 26'(13'.46' + 14'.63' + 16'.34') \\ & + 63'(12'.46' + 14'.62' + 16'.24') \\ & + 46'(12'.36' + 13'.62' + 16'.23'), \end{aligned}$$

et les termes qui contiennent le facteur 16′, donnent ensemble 16′. 2346′, les autres termes se détruisent deux à deux. On obtient enfin, en effectuant la substitution :

$$\overline{12345 \mid 12345} = 11' \,.\, (16')^3\, \overline{123456 \mid 123456'};$$

et ainsi de suite.

Je fais les mêmes substitutions dans les *matrices inverses*, en supprimant cependant la dernière ligne et la dernière colonne de chaque matrice. On trouve ainsi, en y ajoutant les équations ci-dessus trouvées par rapport aux déterminants :

$$11' \,.\, \overline{123 \mid 123'} = \overline{12 \mid 12},$$

$$\frac{1}{13' \,.\, \overline{123 \mid 123'}} \begin{vmatrix} +\overline{23 \mid 23'}, & -\overline{13 \mid 23'} \\ -\overline{23 \mid 13'}, & +\overline{13 \mid 13'} \end{vmatrix} = \frac{1}{\overline{12 \mid 12}} \begin{vmatrix} -\overline{2 \mid 2} + \dfrac{\overline{12 \mid 12}}{11}, & -\overline{1 \mid 2} \\ +\overline{2 \mid 1}, & +\overline{1 \mid 1} \end{vmatrix},$$

$$11' \,.\, 14' \,.\, \overline{1234 \mid 1234'} = \overline{123 \mid 123},$$

$$\frac{1}{14' \,.\, \overline{1234 \mid 1234'}} \begin{vmatrix} +\overline{234 \mid 234}, & -\overline{134 \mid 234}, & -\overline{124 \mid 324} \\ -\overline{234 \mid 124}, & +\overline{134 \mid 134}, & -\overline{214 \mid 314} \\ -\overline{324 \mid 124}, & -\overline{314 \mid 214}, & +\overline{124 \mid 124} \end{vmatrix} =$$

$$\frac{1}{\overline{123 \mid 123}} \begin{vmatrix} -\overline{23 \mid 23} + \dfrac{\overline{123 \mid 123}}{11}, & -\overline{13 \mid 23}, & -\overline{12 \mid 32} \\ +\overline{23 \mid 13}, & +\overline{13 \mid 13}, & -\overline{21 \mid 31} \\ +\overline{32 \mid 12}, & -\overline{31 \mid 21}, & +\overline{12 \mid 12} \end{vmatrix},$$

$$11' \,.\, (15')^2 \,.\, \overline{12345 \mid 12345} = \overline{1234 \mid 1234},$$

$$\frac{1}{15' \,.\, \overline{12345 \mid 12345'}} \begin{vmatrix} +\overline{2345 \mid 2345'}, & -\overline{1345 \mid 2345'}, & -\overline{1245 \mid 3245'}, & -\overline{1235 \mid 4235'} \\ -\overline{2345 \mid 1345'}, & +\overline{1345 \mid 1345'}, & -\overline{2145 \mid 3145'}, & -\overline{2135 \mid 4135'} \\ -\overline{3245 \mid 1245'}, & -\overline{3145 \mid 2145'}, & +\overline{1245 \mid 1245'}, & -\overline{3125 \mid 4125'} \\ -\overline{4235 \mid 1235'}, & -\overline{4135 \mid 2135'}, & -\overline{4125 \mid 3125'}, & +\overline{1235 \mid 1235'} \end{vmatrix} =$$

$$\frac{1}{\overline{1234 \mid 1234}} \begin{vmatrix} -\overline{234 \mid 234} + \dfrac{\overline{1234 \mid 1234}}{11}, & -\overline{134 \mid 234}, & -\overline{124 \mid 324}, & -\overline{123 \mid 423} \\ +\overline{234 \mid 134}, & +\overline{134 \mid 134}, & -\overline{214 \mid 314}, & -\overline{213 \mid 413} \\ +\overline{324 \mid 124}, & -\overline{314 \mid 214}, & +\overline{124 \mid 124}, & -\overline{312 \mid 423} \\ +\overline{423 \mid 123}, & -\overline{413 \mid 213}, & -\overline{412 \mid 312}, & +\overline{123 \mid 123} \end{vmatrix},$$

et ainsi de suite.

Il est bon de changer un peu la forme de ces équations. On en déduit sans peine:

$$\frac{1}{13' \, . \, \overline{123 \mid 123'}} \left| \begin{array}{ll} 2 \, . \, \overline{23 \mid 23'} - \dfrac{\overline{123 \mid 123'}}{11}, & -2 \, . \, \overline{13 \mid 23'} \\ -2 \, . \, \overline{23 \mid 13'}, & +2 \, . \, \overline{13 \mid 13'} - \dfrac{\overline{123 \mid 123'}}{22} \end{array} \right|$$

$$= \frac{1}{\overline{12 \mid 12}} \left| \begin{array}{ll} -2 \, . \, \overline{2 \mid 2} + \dfrac{\overline{12 \mid 12}}{11}, & -2 \, . \, \overline{1 \mid 2} \\ +2 \, . \, \overline{2 \mid 1}, & +2 \, . \, \overline{1 \mid 1} - \dfrac{\overline{12 \mid 12}}{22} \end{array} \right|,$$

$$\frac{1}{14' \, . \, \overline{1234 \mid 1234'}} \left| \begin{array}{lll} 2 \, . \, \overline{234 \mid 234'} - \dfrac{\overline{1234 \mid 1234'}}{11'}, & -2 \, . \, \overline{134 \mid 234'}, & -2 \, . \, \overline{124 \mid 324'} \\ -2 \, . \, \overline{234 \mid 134'}, & 2 \, . \, \overline{134 \mid 134'} - \dfrac{\overline{1234 \mid 1234'}}{22'}, & -2 \, . \, \overline{214 \mid 314'} \\ -2 \, . \, \overline{324 \mid 124'}, & -2 \, . \, \overline{314 \mid 214'}, & 2 \, . \, \overline{124 \mid 124'} - \dfrac{\overline{1234 \mid 12}}{33} \end{array} \right|$$

$$= \frac{1}{\overline{123 \mid 123}} \left| \begin{array}{lll} -2 \, . \, \overline{23 \mid 23} - \dfrac{\overline{123 \mid 123}}{11}, & -2 \, . \, \overline{13 \mid 23}, & -2 \, . \, \overline{12 \mid 32} \\ +2 \, . \, \overline{23 \mid 13}, & +2 \, . \, \overline{13 \mid 13} - \dfrac{\overline{123 \mid 123}}{22}, & -2 \, . \, \overline{21 \mid 31} \\ +2 \, . \, \overline{32 \mid 12}, & -2 \, . \, \overline{31 \mid 21}, & +2 \, . \, \overline{12 \mid 12} - \dfrac{\overline{123 \mid 123}}{33} \end{array} \right|,$$

et ainsi de suite.

Les formules que je viens de présenter, peuvent être appliquées aussitôt à la solution de la question suivante: "Trouver x_1, x_2, x_3, &c., fonctions linéaires de x_1, x_2, x_3, &c. telles que

$$11 \, \mathrm{x}_1^2 + 22 \, \mathrm{x}_2^2 + 33 \, \mathrm{x}_3^2 + \&\text{c.} = 11 \, x_1^2 + 22 \, x_2^2 + 33 \, x_3^2 + \&\text{c.},\text{"}$$

c'est-à-dire: transformer une fonction quadratique $11x_1^2 + 22x_2^2 + 33x_3^2$ + &c. *en elle-même* par des substitutions linéaires. Il suffira d'écrire la solution pour le cas de trois indéterminées: on satisfait identiquement à l'équation

$$11 \, \mathrm{x}_1^2 + 22 \, \mathrm{x}_2^2 + 33 \, \mathrm{x}_3^2 = 11 \, x_1^2 + 22 \, x_2^2 + 33 \, x_3^2$$

en écrivant

$$(\mathrm{x}_1, \quad \mathrm{x}_2, \quad \mathrm{x}_3) = \frac{1}{\overline{123 \mid 123}} \times$$

$$\left| \begin{array}{lll} +2 \, . \, \overline{23 \mid 23} - \dfrac{\overline{123 \mid 123}}{11}, & -2 \, . \, \overline{13 \mid 23}, & -2 \, . \, \overline{12 \mid 32} \\ -2 \, . \, \overline{23 \mid 13}, & +2 \, . \, \overline{13 \mid 13} - \dfrac{\overline{123 \mid 123}}{22}, & -2 \, . \, \overline{21 \mid 31} \\ -2 \, . \, \overline{32 \mid 12}, & -2 \, . \, \overline{31 \mid 21}, & +2 \, . \, \overline{12 \mid 12} - \dfrac{\overline{123 \mid 123}}{33} \end{array} \right| (11x_1, \; 22x_2, \; 33x_3).$$

Voilà la transformation *propre.* On en tire la transformation *impropre* de $11x_1^2+22x_2^2$ *en elle-même* en écrivant $33=0$; car, cela posé, les valeurs de x_1, x_2 ne contiennent pas x_3, et l'on n'a plus besoin de la valeur de x_3. On obtient ainsi la solution suivante; savoir, on satisfait identiquement à l'équation

$$11'\,\mathrm{x}_1^2+22'\,\mathrm{x}_2^2=11'\,x_1^2+22'\,x_2^2$$

en écrivant

$$(\mathrm{x}_1,\ \mathrm{x}_2)=\frac{1}{\overline{123\mid 123'}}\left|\begin{array}{ll} 2\,.\,\overline{23\mid 23'}-\dfrac{\overline{123\mid 123'}}{11'}, & -2\,.\,\overline{13\mid 23'} \\ -2\,.\,\overline{23\mid 13'}, & 2\,.\,\overline{13\mid 13'}-\dfrac{\overline{123\mid 123'}}{22} \end{array}\right|(11'x_1,\ 22'x_2),$$

ce qui est une transformation *impropre.* Mais en y faisant la substitution $11=13'.11'$, $22=13'.22'$, $12=11'.23'$, on réduit la solution à celle-ci, savoir on satisfait identiquement à l'équation $11\mathrm{x}_1^2+22\mathrm{x}_2^2=11x_1^2+22x_2^2$ en écrivant

$$(\mathrm{x}_1,\ \mathrm{x}_2)=\frac{1}{\overline{12\mid 12}}\left|\begin{array}{ll} -2\,.\,\overline{2\mid 2}+\dfrac{\overline{12\mid 12}}{11}, & -2\,.\,\overline{1\mid 2} \\ +2\,.\,\overline{2\mid 1}, & +2\,.\,\overline{1\mid 1}-\dfrac{\overline{12\mid 12}}{22} \end{array}\right|(11x_1,\ 22x_2),$$

ce qui est encore une transformation *impropre,* qui correspond de plus près à la formule pour la transformation *propre;* la seule différence est que les signes des termes de la première colonne de la matrice en sont changés.

En introduisant des lettres simples a, b, &c. à la place des symboles 11, 22, &c., je considère d'abord la transformation *propre*

$$a\mathrm{x}^2+b\mathrm{y}^2=ax^2+by^2.$$

Ici, en écrivant

$$\left|\begin{array}{ll} 11, & 12 \\ 21, & 22 \end{array}\right|=\left|\begin{array}{ll} a, & \nu \\ -\nu, & b \end{array}\right|,$$

la formule donne

$$(\mathrm{x},\ \mathrm{y})=\frac{1}{ab+\nu^2}\left|\begin{array}{ll} ab-\nu^2, & -2\nu b \\ 2\nu a\quad, & ab-\nu^2 \end{array}\right|(x,\ y).$$

La transformation *impropre*

$$a\mathrm{x}^2+b\mathrm{y}^2=ax^2+by^2$$

s'obtient au moyen de la formule donnée plus bas pour la transformation *propre* de la fonction $ax^2+by^2+cz^2$ *en elle-même.* En y écrivant $c=0$, on obtient

$$(\mathrm{x},\ \mathrm{y})=\frac{1}{a\lambda^2+b\mu^2}\left|\begin{array}{ll} a\lambda^2-b\mu^2, & 2\lambda\mu b \\ 2\lambda\mu a\quad, & -a\lambda^2+b\mu^2 \end{array}\right|(x,\ y).$$

J'ai déjà fait voir de quelle manière cette formule se rattache à la formule pour la transformation *propre;* la différence entre les formes de ces transformations dans ce cas simple est assez frappante.

Pour obtenir la transformation propre

$$a\mathrm{x}^2 + b\mathrm{y}^2 + c\mathrm{z}^2 = ax^2 + by^2 + cz^2,$$

j'écris

$$\left|\begin{matrix} 11, & 12, & 13 \\ 21, & 22, & 23 \\ 31, & 32, & 33 \end{matrix}\right| = \left|\begin{matrix} a, & \nu, & -\mu \\ -\nu, & b, & \lambda \\ \mu, & -\lambda, & c \end{matrix}\right|;$$

cette formule donne

$$(\mathrm{x},\ \mathrm{y},\ \mathrm{z}) = \frac{1}{abc + a\lambda^2 + b\mu^2 + c\nu^2} \times$$

$$\left|\begin{matrix} abc + a\lambda^2 - b\mu^2 - c\nu^2, & 2(\lambda\mu - c\nu)\, b, & 2(\nu\lambda + b\mu)\, c \\ 2(\lambda\mu + c\nu)\, a, & abc - a\lambda^2 + b\mu^2 - c\nu^2, & 2(\mu\nu - a\lambda)\, c \\ 2(\nu\lambda - b\mu)\, a, & 2(\mu\nu + a\lambda)\, b, & abc - a\lambda^2 - b\mu^2 - c\nu^2 \end{matrix}\right|(x,\ y,\ z).$$

La transformation *impropre en elle-même*

$$a\mathrm{x}^2 + b\mathrm{y}^2 + c\mathrm{z}^2 = ax^2 + by^2 + cz^2$$

peut être tirée de la transformation *propre en elle-même* de la fonction donnée ci-après $ax^2 + by^2 + cz^2 + dw^2$; en y écrivant $d = 0$, on obtient

$$(\mathrm{x},\ \mathrm{y},\ \mathrm{z}) = \frac{1}{bc\rho^2 + ca\sigma^2 + ab\tau^2 + \phi^2} \times$$

$$\left|\begin{matrix} -bc\rho^2 + ca\sigma^2 + ab\tau^2 - \phi^2, & -2b\tau\phi - 2bc\rho\sigma, & 2c\sigma\phi - 2bc\tau\rho \\ 2a\tau\phi - 2ac\rho\sigma, & bc\rho^2 - ca\sigma^2 + ab\tau^2 - \phi^2, & -2c\rho\phi - 2ca\sigma\tau \\ -2a\sigma\phi - 2ab\rho\tau, & 2b\rho\phi - 2ab\sigma\tau, & bc\rho^2 + ca\sigma^2 - ab\tau^2 - \phi^2 \end{matrix}\right|(x,\ y,\ z).$$

Pour vérifier que cette expression n'est en effet autre chose que la formule pour la transformation *propre,* en y changeant les signes de tous les termes, j'écris dans la formule pour la transformation propre, $a = b = c = \omega$. On a ainsi pour la transformation propre

$$\mathrm{x}^2 + \mathrm{y}^2 + \mathrm{z}^2 = x^2 + y^2 + z^2,$$

l'équation

$$(\mathrm{x},\ \mathrm{y},\ \mathrm{z}) = \frac{1}{\omega^2 + \lambda^2 + \mu^2 + \nu^2}\left|\begin{matrix} \omega^2 + \lambda^2 - \mu^2 - \nu^2, & 2\lambda\mu - 2\nu\omega, & 2\nu\lambda + 2\mu\omega \\ 2\lambda\mu + 2\nu\omega, & \omega^2 - \lambda^2 + \mu^2 - \nu^2, & 2\mu\nu - 2\lambda\omega \\ 2\nu\lambda - 2\mu\omega, & 2\mu\nu + 2\lambda\omega, & \omega^2 - \lambda^2 + \mu^2 - \nu^2 \end{matrix}\right|(x,\ y,\ z),$$

et en écrivant dans la formule pour la transformation impropre, $a=b=c=1$, $d=0$, et λ, μ, ν, $-\omega$ au lieu de ρ, σ, τ, ϕ, on obtient pour la transformation impropre

$$\mathrm{x}^2+\mathrm{y}^2+\mathrm{z}^2=x^2+y^2+z^2,$$

l'équation

$$(\mathrm{x},\ \mathrm{y},\ \mathrm{z})=\frac{1}{\omega^2+\lambda^2+\mu^2+\nu^2}\times$$

$$\left|\begin{array}{ccc} -\omega^2-\lambda^2+\mu^2+\nu^2, & -2\lambda\mu+2\nu\omega\ , & -2\nu\lambda-2\mu\omega \\ -2\lambda\mu-2\nu\omega\ , & -\omega^2+\lambda^2-\mu^2+\nu^2, & 2\mu\nu-2\lambda\omega \\ -2\nu\lambda+2\mu\omega\ , & -2\mu\nu-2\lambda\omega\ , & -\omega^2+\lambda^2-\mu^2-\nu^2 \end{array}\right|(x,\ y,\ z).$$

Pour obtenir la transformation propre

$$a\mathrm{x}^2-b\mathrm{y}^2+c\mathrm{z}^2+d\mathrm{w}^2=ax^2+by^2+cz^2+dw^2,$$

j'écris

$$\left|\begin{array}{cccc} 11, & 12, & 13, & 14 \\ 21, & 22, & 23, & 24 \\ 31, & 32, & 33, & 34 \\ 41, & 42, & 43, & 44 \end{array}\right| = \left|\begin{array}{cccc} a, & \nu\ , & -\mu, & \rho \\ -\nu, & b\ , & \lambda, & \sigma \\ \mu, & -\lambda, & c\ , & \tau \\ -\rho, & -\sigma, & -\tau, & d \end{array}\right|;$$

cela donne d'abord, en mettant pour abréger,

$$\phi=\lambda\rho+\mu\sigma+\nu\tau,$$

la valeur du déterminant

$$\overline{1234\ |\ 1234}=abcd+bc\rho^2+ca\sigma^2+ab\tau^2+ad\lambda^2+bd\mu^2+cd\nu^2+\phi^2$$

(ce que je représente par k).

J'ajoute aussi la valeur de la *matrice* inverse

$$\left|\begin{array}{cccc} 11, & 12, & 13, & 14 \\ 21, & 22, & 23, & 24 \\ 31, & 32, & 33, & 34 \\ 41, & 42, & 43, & 44 \end{array}\right|^{-1},$$

savoir:

$$\frac{1}{k}\left|\begin{array}{cc} bcd+b\tau^2+c\sigma^2+d\lambda^2, & -cd\nu-\tau\phi+d\lambda\mu-c\rho\sigma, \\ cd\nu+\tau\phi+d\lambda\mu-c\rho\sigma, & acd+b\tau^2+c\rho^2+d\mu^2, \\ -bd\mu-\sigma\phi+d\lambda\nu-b\rho\tau, & ad\lambda+\rho\phi+d\mu\nu-a\sigma\tau, \\ bc\rho+\lambda\phi+c\nu\sigma-b\mu\tau, & ac\sigma+\mu\phi-c\nu\rho+a\lambda\tau, \end{array}\right.$$

$$\left.\begin{array}{cc} bd\mu+\sigma\phi+d\lambda\nu-b\rho\tau\ , & -bc\rho-\lambda\phi+c\nu\sigma-b\mu\tau \\ -ad\lambda-\rho\phi+d\mu\nu-a\sigma\tau, & -ac\sigma-\mu\phi-c\nu\rho+a\lambda\tau \\ abd+a\sigma^2+b\rho^2+d\nu^2\ , & -ab\tau-\nu\phi+b\mu\rho-a\lambda\sigma \\ ab\tau+\nu\phi+b\mu\rho-a\lambda\sigma, & abc+a\lambda^2+b\mu^2+c\nu^2 \end{array}\right|.$$

On a pour la transformation, l'équation (x, y, z, w) =

$$\frac{1}{k}\left|\begin{array}{ll}
abcd - bc\rho^2 + ca\sigma^2 + ab\tau^2 + ad\lambda^2 - bd\mu^2 - cd\nu^2 - \phi^2\ , & 2b\,(-\,cd\nu - \tau\phi + d\lambda\mu - c\rho\sigma)\ , \\
2a\,(cd\nu + \tau\phi + d\lambda\mu - c\rho\sigma)\ , & abcd + bc\rho^2 - ca\sigma^2 + ab\tau^2 - ad\lambda^2 + bd\mu^2 - cd\nu^2 - \phi^2\ , \\
2a\,(-\,bd\mu - \sigma\phi + d\lambda\nu - b\rho\tau), & 2b\,(ad\lambda + \rho\phi + d\mu\nu - a\sigma\tau)\ , \\
2a\,(bc\rho + \lambda\phi + c\nu\sigma - b\mu\tau)\ , & 2b\,(ac\sigma + \mu\phi - c\nu\rho + a\lambda\tau)\ ,
\end{array}\right.$$

$$\left.\begin{array}{ll}
2c\,(bd\mu + \sigma\phi + d\lambda\nu - b\rho\tau)\ , & 2d\,(-\,bc\rho - \lambda\phi + c\nu\sigma - b\mu\tau) \\
2c\,(-\,ad\lambda - \rho\phi + d\mu\nu - a\sigma\tau)\ , & 2d\,(-ac\sigma - \mu\phi - c\nu\rho + a\lambda\tau) \\
abcd + bc\rho^2 + ca\sigma^2 - ab\tau^2 - ad\lambda^2 - bd\mu^2 + cd\nu^2 - \phi^2\ , & 2d\,(-\,ab\tau - \nu\phi + b\mu\rho - a\lambda\sigma) \\
2c\,(ab\tau + \nu\phi + b\mu\rho - a\lambda\sigma)\ , & abcd - bc\rho^2 - ca\sigma^2 - ab\tau^2 + ad\lambda^2 + b\delta\mu^2 + cd\iota^2 - \phi^2
\end{array}\right|(x,\ y,\ z,\ w).$$

Je suppose que l'on ait $a = b = c = d = \omega$, et j'écris $\psi = -\frac{\phi}{\omega}$, c'est-à-dire $\psi = -\frac{\lambda\rho + \mu\sigma + \nu\tau}{\omega}$ ou $\lambda\rho + \mu\sigma + \nu\tau + \psi\omega = 0$. En faisant cette substitution, on trouve d'abord $k = \omega^2 R$, où

$$R = \lambda^2 + \mu^2 + \nu^2 + \psi^2 + \rho^2 + \sigma^2 + \tau^2 + \omega^2,$$

et puis pour la transformation propre

$$\mathrm{x}^2 + \mathrm{y}^2 + \mathrm{z}^2 + \mathrm{w}^2 = x^2 + y^2 + z^2 + w^2,$$

l'équation (x, y, z, w) =

$$\frac{1}{R}\left|\begin{array}{ll}
-\rho^2 + \sigma^2 + \tau^2 + \omega^2 + \lambda^2 - \mu^2 - \nu^2 - \psi^2, & -\,2\omega\nu + 2\tau\psi + 2\lambda\mu - 2\rho\sigma, \\
2\omega\nu - 2\tau\psi + 2\lambda\mu - 2\rho\sigma, & \rho^2 - \sigma^2 + \tau^2 + \omega^2 - \lambda^2 + \mu^2 - \nu^2 - \psi^2, \\
-\,2\omega\mu + 2\sigma\psi + 2\lambda\nu - 2\rho\tau, & 2\omega\lambda - 2\rho\psi + 2\mu\nu - 2\sigma\tau, \\
2\omega\rho - 2\lambda\psi + 2\nu\sigma - 2\mu\tau, & 2\omega\sigma - 2\mu\psi - 2\nu\rho + 2\lambda\tau,
\end{array}\right.$$

$$\left.\begin{array}{ll}
2\omega\mu - 2\sigma\psi + 2\lambda\nu - 2\rho\tau\ , & -\,2\omega\rho + 2\lambda\psi + 2\nu\sigma - 2\mu\tau \\
-\,2\omega\lambda + 2\rho\psi + 2\mu\nu - 2\sigma\tau\ , & -\,2\omega\sigma + 2\mu\psi - 2\nu\rho + 2\lambda\tau \\
\rho^2 + \sigma^2 - \tau^2 + \omega^2 - \lambda^2 - \mu^2 + \nu^2 - \psi^2, & -\,2\omega\tau + 2\nu\psi + 2\mu\rho - 2\lambda\sigma \\
2\omega\tau - 2\nu\psi + 2\mu\rho - 2\lambda\sigma\ , & -\,\rho^2 - \sigma^2 - \tau^2 + \omega^2 + \lambda^2 + \mu^2 + \nu^2 - \psi^2
\end{array}\right|(x,\ y,\ z,\ w).$$

On peut changer la forme de cette expression, en y écrivant

$$\lambda = \tfrac{1}{2}(\alpha - \alpha'),\quad \mu = \tfrac{1}{2}(\beta - \beta'),\quad \nu = \tfrac{1}{2}(\gamma - \gamma'),\quad \psi = \tfrac{1}{2}(\delta - \delta'),$$

$$\rho = \tfrac{1}{2}(\alpha + \alpha'),\quad \sigma = \tfrac{1}{2}(\beta + \beta'),\quad \tau = \tfrac{1}{2}(\gamma + \gamma'),\quad \omega = \tfrac{1}{2}(\delta + \delta');$$

cela donne

$$\alpha^2+\beta^2+\gamma^2+\delta^2-\alpha'^2-\beta'^2-\gamma'^2-\delta'^2=0,$$
$$R=\tfrac{1}{2}(\alpha^2+\beta^2+\gamma^2+\delta^2+\alpha'^2+\beta'^2+\gamma'^2+\delta'^2),$$

de manière qu'en écrivant

$$M=\alpha^2+\beta^2+\gamma^2+\delta^2,$$
$$M'=\alpha'^2+\beta'^2+\gamma'^2+\delta'^2,$$

on obtient

$$R=\tfrac{1}{2}(M+M')=\sqrt{(MM')},$$

et la formule pour la transformation devient

$$(\mathrm{x},\ \mathrm{y},\ \mathrm{z},\ \mathrm{w})=\frac{1}{\sqrt{(MM')}}\left|\begin{array}{llll} -\alpha\alpha'+\beta\beta'+\gamma\gamma'+\delta\delta', & -\alpha\beta'-\beta\alpha'-\gamma\delta'+\delta\gamma', & -\alpha\gamma'+\beta\delta'-\gamma\alpha'-\delta\beta', & \alpha\delta'+\beta\gamma'-\gamma\beta'+\delta\alpha' \\ -\alpha\beta'-\beta\alpha'+\gamma\delta'+\delta\gamma', & \alpha\alpha'-\beta\beta'+\gamma\gamma'+\delta\delta', & -\alpha\delta'-\beta\gamma'-\gamma\beta'+\delta\alpha', & -\alpha\gamma'+\beta\delta'+\gamma\alpha'+\delta\beta' \\ -\alpha\gamma'-\beta\delta'-\gamma\alpha'+\delta\beta', & \alpha\delta'-\beta\gamma'-\gamma\beta'-\delta\alpha', & \alpha\alpha'+\beta\beta'-\gamma\gamma'+\delta\delta', & \alpha\beta'-\beta\alpha'+\gamma\delta'+\delta\gamma' \\ -\alpha\delta'+\beta\gamma'-\gamma\beta'-\delta\alpha', & -\alpha\gamma'-\beta\delta'+\gamma\alpha'-\delta\beta', & \alpha\beta'-\beta\alpha'-\gamma\delta'-\delta\gamma', & -\alpha\alpha'-\beta\beta'-\gamma\gamma'+\delta\delta' \end{array}\right|(x,\ y,\ z,\ w).$$

On voit donc que même sans supposer l'équation $M=M'$, cette formule donne la transformation propre

$$\mathrm{x}^2+\mathrm{y}^2+\mathrm{z}^2+\mathrm{w}^2=x^2+y^2+z^2+w^2.$$

Cette solution est à peu près de la même forme que la solution *impropre* donnée par Euler dans son mémoire "Problema algebraicum ob affectiones prorsus singulares memorabile" Nov. Comm. Petrop., t. XV. 1770, p. 75, et Comm. Arith. collectae, [4to. Petrop. 1849], t. I. p. 427. Je remarque aussi que cette même solution peut être déduite de la théorie des *Quaternions*. En effet, i, j, k étant des quantités imaginaires telles que $i^2=j^2=k^2=-1$, $jk=-kj=i$, $ki=-ik=j$, $ij=-ji=k$, on obtient, en effectuant la multiplication:

$$(\mathrm{x}i+\mathrm{y}j+\mathrm{z}k+\mathrm{w})=-\frac{1}{\sqrt{(MM')}}(\alpha i+\beta j+\gamma k+\delta)(xi+yj+zk+w)(\alpha' i+\beta' j+\gamma' k+\delta'),$$

x, y, z, w ayant les mêmes valeurs que dans la dernière formule de transformation.

En changeant les signes des termes de la quatrième colonne, on en tire pour la transformation impropre

$$\mathrm{x}^2+\mathrm{y}^2+\mathrm{z}^2+\mathrm{w}^2=x^2+y^2+z^2+w^2,$$

la formule suivante plus symétrique :

$$(\mathrm{x},\ \mathrm{y},\ \mathrm{z},\ \mathrm{w}) = \frac{1}{\sqrt{(MM')}} \left| \begin{matrix} -\alpha\alpha' + \beta\beta' + \gamma\gamma' + \delta\delta', & -\alpha\beta' - \beta\alpha' - \gamma\delta' + \delta\gamma', \\ -\alpha\beta' - \beta\alpha' + \gamma\delta' - \delta\gamma', & \alpha\alpha' - \beta\beta' + \gamma\gamma' + \delta\delta', \\ -\alpha\gamma' - \beta\delta' - \gamma\alpha' + \delta\beta', & \alpha\delta' - \beta\gamma' - \gamma\beta' - \delta\alpha', \\ -\alpha\delta' + \beta\gamma' - \gamma\beta' - \delta\alpha', & -\alpha\gamma' - \beta\delta' + \gamma\alpha' - \delta\beta', \end{matrix} \right.$$

$$\left. \begin{matrix} -\alpha\gamma' + \beta\delta' - \gamma\alpha' - \delta\beta', & -\alpha\delta' - \beta\gamma' + \gamma\beta' - \delta\alpha' \\ -\alpha\delta' - \beta\gamma' - \gamma\beta' + \delta\alpha', & \alpha\gamma' - \beta\delta' - \gamma\alpha' - \delta\beta' \\ \alpha\alpha' + \beta\beta' - \gamma\gamma' + \delta\delta', & -\alpha\beta' + \beta\alpha' - \gamma\delta' - \delta\gamma' \\ \alpha\beta' - \beta\alpha' - \gamma\delta' - \delta\gamma', & \alpha\alpha' + \beta\beta' + \gamma\gamma' - \delta\delta' \end{matrix} \right| (x,\ y,\ z,\ w).$$

Ces formules pour la transformation, tant propre qu'impropre, de la fonction $x^2 + y^2 + z^2 + w^2$ *en elle-même*, sont utiles dans la théorie des polygones inscrits dans une surface du second ordre.

138.

RECHERCHES SUR LES MATRICES DONT LES TERMES SONT DES FONCTIONS LINÉAIRES D'UNE SEULE INDÉTERMINÉE.

[From the *Journal für die reine und angewandte Mathematik* (Crelle), tom. L. (1855), pp. 313—317.]

Je pose la matrice

$$\begin{vmatrix} A\ , & B\ , & C\ , \dots \\ A', & B', & C', \dots \\ A'', & B'', & C'', \dots \\ \vdots & & \end{vmatrix}$$

dont les termes (n^2 en nombre) sont des fonctions linéaires d'une quantité s, et je considère le déterminant formé avec cette matrice, et les déterminants mineurs formés en supprimant un nombre quelconque des lignes et un nombre égal de colonnes de la matrice. En supprimant une *seule* ligne et une *seule* colonne, on obtient les *premiers mineurs;* en supprimant *deux* lignes et *deux* colonnes, on obtient les *seconds mineurs;* et ainsi de suite. Cela étant, je suppose que la quantité s a été trouvée en égalant à zéro le déterminant formé avec la matrice donnée; ce déterminant sera une fonction de s du n-ième degré qui généralement ne contiendra pas de facteurs multiples. On voit donc qu'un facteur simple du déterminant ne peut pas entrer comme facteur dans les premiers mineurs (c'est-à-dire dans *tous* les premiers mineurs); mais en supposant que le déterminant ait des facteurs multiples, un facteur multiple du déterminant peut entrer comme facteur (simple ou multiple) dans les premiers mineurs, ou dans les mineurs d'un ordre plus élevé. Il importe de trouver le degré selon lequel un facteur multiple du déterminant peut entrer comme facteur des premiers mineurs, ou des mineurs d'un ordre quelconque donné.

Cela se fait très facilement au moyen d'une propriété générale des déterminants; si les mineurs du $(r+1)$ième ordre contiennent le facteur $(s-a)^a$ (c'est-à-dire, si tous les mineurs de cet ordre contiennent le facteur $(s-a)^{\alpha}$, mais non pas tous les facteurs $(s-a)^{\alpha+1}$); et si de même les mineurs du r-ième ordre contiennent le facteur $(s-a)^{\beta}$; alors les mineurs du $(r-1)$ième ordre contiendront *au moins* le facteur $(s-a)^{2\beta-\alpha}$. Autrement dit: les mineurs du $(r-1)$ième ordre contiendront le facteur $(s-a)^{\gamma}$ où $\gamma > 2\beta-\alpha$, ou, ce qui est la même chose, $\alpha-2\beta+\gamma \nless 0$; c'est-à-dire: en formant la suite des indices des puissances selon lesquelles le facteur $(s-a)$ entre dans les mineurs premiers, seconds, &c. (il va sans dire que cette suite sera une suite *décroissante*), les différences secondes seront *positives* [c'est-à-dire non *négatives*]. Je représente par $\alpha, \beta, \gamma, \ldots$ la suite dont il s'agit; je suppose, pour fixer les idées, que δ soit le dernier terme qui ne s'évanouisse pas, et j'écris

$$\begin{array}{cccccc} \alpha, & \beta, & \gamma, & \delta, & 0, & 0, \ldots \\ \alpha-\beta, & \beta-\gamma, & \gamma-\delta, & \delta, & 0, \ldots & \\ \alpha-2\beta+\gamma, & \beta-2\gamma+\delta, & \gamma-2\delta, & \delta, & 0, \ldots; & \end{array}$$

ici, quel que soit le nombre des termes, tous les nombres de la troisième ligne seront positifs, et en représentant ces nombres par $\int, \int', \int''$, &c., on obtient:

$$\begin{aligned} \alpha &= \int + 2\int' + 3\int'' + 4\int''' + \ldots, \\ \beta &= \quad\;\; \int' + 2\int'' + 3\int''' + \ldots, \\ \gamma &= \qquad\quad\;\; \int'' + 2\int''' + \ldots, \\ \delta &= \qquad\qquad\qquad\; \int''' + \ldots, \\ &\vdots \end{aligned}$$

Il y a ici à considérer que le nombre α, indice de la puissance selon laquelle le facteur $(s-a)$ entre dans le déterminant, est donné; il sera donc permis de prendre pour $\int, \int', \int'', \ldots$ des valeurs entières et positives quelconques (zéro y compris) qui satisfont à la première équation; les autres équations donnent alors les valeurs de β, γ, δ, &c. On forme de cette manière une table des particularités que peut présenter un facteur multiple $(s-a)^{\alpha}$ du déterminant; cette table est composée des symboles $\alpha, \beta, \gamma, \ldots$, et les nombres $\alpha, \beta, \ldots$ de chaque symbole font voir le degré selon lequel le facteur $(s-a)$ entre dans les déterminants, dans les mineurs premiers, seconds, &c. Or il est très facile de former, au moyen des tables pour $\alpha=1$, $\alpha=2, \ldots$ $\alpha=k$, la table pour $\alpha=k+1$. On a par exemple pour $\alpha=1$, $\alpha=2$, $\alpha=3$, $a=4$ les tables suivantes:

Pour $\alpha=1$, 1.
Pour $\alpha=2$, 2, 21.
Pour $\alpha=3$, 3, 31, 321.
Pour $\alpha=4$, 4, 41, 42, 421, 4321.

De là on tire la table pour $\alpha=5$, savoir:

Pour $\alpha=5$, 5, 51, 52, 521, 531, 5321, 54321.

En effet, le premier terme est 5, et on obtient les autres termes en mettant le nombre 5 devant les symboles des tables pour $\alpha=1$, $\alpha=2$, $\alpha=3$, $\alpha=4$, en ayant seulement soin de supprimer les symboles 53, 54, 541, 542, 5421 pour lesquels le premier terme de la suite des différences secondes est négatif. On trouve de même pour $\alpha = 6$, la table suivante, savoir:

Pour $\alpha = 6$, 6, 61, 62, 621, 63, 631, 6321, 642, 6421, 64321, 654321;

et ainsi de suite. Les nombres des symboles pour $\alpha = 1, 2, 3, 4, 5, 6, 7, 8$, &c. sont 1, 2, 3, 5, 7, 11, 15, 22, [30, 42, 56], &c.; ce sont les coefficients des puissances x^1, x^2, x^3 &c. dans le développement de

$$(1-x)^{-1}\,(1-x^2)^{-1}\,(1-x^3)^{-1}\,(1-x^4)^{-1}\,(1-x^5)^{-1}\,\ldots\ \&c.$$

fonctions qui se présentent, comme on sait, dans la théorie de la *partition des nombres.*

Maintenant, au lieu de considérer un seul facteur du déterminant, je considère *tous* les facteurs: par exemple pour $n = 4$, le déterminant peut avoir un facteur double $(s-a)^2$, et un autre facteur double $(s-b)^2$; il peut de plus arriver que le facteur $(s-a)$ soit facteur simple des premiers mineurs, mais que le facteur $(s-b)$ n'entre pas dans les premiers mineurs. Le symbole qui correspond au facteur $(s-a)$ sera 21, et le symbole qui correspond au facteur $(s-b)$ sera 2. En combinant ces deux symboles, on aura le symbole composé $\begin{array}{|c|}\hline 21\\ 2\\ \hline\end{array}$, qui dénote que le déterminant a deux facteurs doubles de la classe dont il s'agit. Je forme de ces symboles composés des tables pour $n = 1$, $n = 2$, $n = 3$, $n = 4$, &c. On a:

Pour $n = 1$: $\begin{array}{|c|}\hline 1\\ \hline\end{array}$,

Pour $n = 2$: $\begin{array}{|c|}\hline 1\\ 1\\ \hline\end{array}$, $\begin{array}{|c|}\hline 2\\ \hline\end{array}$, $\begin{array}{|c|}\hline 21\\ \hline\end{array}$,

Pour $n = 3$: $\begin{array}{|c|}\hline 1\\ 1\\ 1\\ \hline\end{array}$, $\begin{array}{|c|}\hline 2\\ 1\\ \hline\end{array}$, $\begin{array}{|c|}\hline 21\\ 1\\ \hline\end{array}$, $\begin{array}{|c|}\hline 3\\ \hline\end{array}$, $\begin{array}{|c|}\hline 31\\ \hline\end{array}$, $\begin{array}{|c|}\hline 321\\ \hline\end{array}$,

Pour $n = 4$:

$\begin{array}{|c|}\hline 1\\ 1\\ 1\\ 1\\ \hline\end{array}$, $\begin{array}{|c|}\hline 2\\ 1\\ 1\\ \hline\end{array}$, $\begin{array}{|c|}\hline 21\\ 1\\ 1\\ \hline\end{array}$, $\begin{array}{|c|}\hline 3\\ 1\\ \hline\end{array}$, $\begin{array}{|c|}\hline 321\\ 1\\ \hline\end{array}$, $\begin{array}{|c|}\hline 2\\ 2\\ \hline\end{array}$, $\begin{array}{|c|}\hline 21\\ 2\\ \hline\end{array}$, $\begin{array}{|c|}\hline 4\\ \hline\end{array}$, $\begin{array}{|c|}\hline 41\\ \hline\end{array}$, $\begin{array}{|c|}\hline 42\\ \hline\end{array}$, $\begin{array}{|c|}\hline 421\\ \hline\end{array}$, $\begin{array}{|c|}\hline 4321\\ \hline\end{array}$,

et ainsi de suite.

Pour donner encore un exemple du sens de ces symboles, le symbole $\boxed{\begin{array}{l}321\\1\end{array}}$ dénote que le déterminant a un facteur $(s-a)$ qui entre comme facteur *triple* dans le déterminant, comme facteur *double* dans les premiers mineurs, et comme facteur *simple* dans les seconds mineurs; l'autre facteur du déterminant est un facteur simple $(s-b)$. Les nombres des symboles pour $n=1$, 2, 3, 4, 5, 6, 7, 8, 9, 10, 11, &c. sont 1, 3, 6, 14, 27, 58, 111, 223, 424, 817, 1527, &c.; ces nombres sont les coefficients de x^1, x^2, x^3, &c. dans le développement de

$$(1-x)^{-1}(1-x^2)^{-2}(1-x^3)^{-3}(1-x^4)^{-5}(1-x^5)^{-7}(1-x^6)^{-11}(1-x^7)^{-15}(1-x^8)^{-22}(1-x^9)^{-30}\ldots \text{ \&c.}$$

où les indices 1, 2, 3, 5, 7, 11, &c. forment la suite qui se présente dans la théorie de la partition des nombres, dont j'ai parlé plus haut. Il est très facile de démontrer qu'il en est ainsi.

Les résultats que je viens de présenter sont en partie dus à M. Sylvester (voyez son mémoire "An enumeration of the contacts of lines and surfaces of the second order," *Philosophical Magazine*, [vol. I. (1851), pp. 18—20]). En effet, M. Sylvester commence par étendre à des fonctions d'un nombre quelconque d'indéterminées l'idée géométrique des contacts des courbes et des surfaces. En considérant les deux équations quadratiques $U=0$, $V=0$, il forme le discriminant de la fonction quadratique $U+sV$, et il cherche dans quel degré chaque facteur de ce discriminant peut entrer comme facteur dans les mineurs premiers, seconds, &c. Le discriminant de M. Sylvester est un déterminant symétrique; mais cela ne change rien à la question, et je n'ai fait que reproduire l'analyse de M. Sylvester, en donnant cependant l'algorithme pour la formation des symboles, et de plus la loi pour le nombre des symboles. M. Sylvester donne pour $n=2$, 3, 4, 5, 6, des nombres qui, en ajoutant à chacun le nombre 2, pour embrasser deux cas extrêmes qui ne sont pas comptés, seraient 3, 6, 14, 26, 58. Il se trouve dans le nombre 26 une erreur de calcul; ce nombre devrait être 27, et en suppléant le premier terme 1, on a la suite trouvée plus haut, savoir 1, 3, 6, 14, 27, 58, &c.; il y a de même une erreur de calcul dans les nombres donnés par M. Sylvester pour $n=7$ et $n=8$.

Mais tout cela s'applique à une autre théorie géométrique, savoir à la théorie des figures *homographes*. Pour fixer les idées, je ne considère que les figures dans le plan. En supposant que x, y, z soient les coordonnées d'un point, et en prenant pour $(\mathrm{x}, \mathrm{y}, \mathrm{z})$ des fonctions linéaires de (x, y, z) on aura $(\mathrm{x}, \mathrm{y}, \mathrm{z})$ comme coordonnées d'un point *homographe* au point (x, y, z). En cherchant les points qui sont homographes chacun à soi-même, on est conduit aux équations $\mathrm{x}-sx=0$, $\mathrm{y}-sy=0$, $\mathrm{z}-sz=0$. Les quantités à gauche $\mathrm{x}-sx$, $\mathrm{y}-sy$, $\mathrm{z}-sz$ sont des fonctions linéaires de x, y, z, ayant pour coefficients des fonctions linéaires de s. On a ainsi une matrice dont les termes sont des fonctions linéaires de s; la théorie entière se rattache aux propriétés de cette matrice. Pour le cas général de *l'homographie* ordinaire, on a le symbole

$\boxed{\begin{array}{l}1\\1\\1\end{array}}$, pour l'*homologie,* le symbole $\boxed{\begin{array}{l}21\\1\end{array}}$; les autres symboles $\boxed{\begin{array}{l}2\\1\end{array}}$, $\boxed{3}$, $\boxed{31}$ se rapportent à des cas moins généraux, et le symbole $\boxed{321}$ au cas de l'identité complète des deux figures; y compris ce cas-limite de l'identité complète, il existe pour le *plan* 6 espèces d'homographie; pour l'*espace ordinaire* il existe 14 espèces d'homographie. Je reviendrai à cette théorie à une autre occasion.

Londres, le 24 *Mai* 1854.

139.

AN INTRODUCTORY MEMOIR UPON QUANTICS.

[From the *Philosophical Transactions of the Royal Society of London*, vol. CXLIV. *for the year* 1854, pp. 244—258. Received April 20,—Read May 4, 1854.]

1. THE term Quantics is used to denote the entire subject of rational and integral functions, and of the equations and loci to which these give rise; the word "quantic" is an adjective, meaning *of such a degree*, but may be used substantively, the noun understood being (unless the contrary appear by the context) function; so used the word admits of the plural "quantics."

The quantities or symbols to which the expression "degree" refers, or (what is the same thing) in regard to which a function is considered as a quantic, will be spoken of as "facients." A quantic may always be considered as being, in regard to its facients, homogeneous, since to render it so, it is only necessary to introduce as a facient unity, or some symbol which is to be ultimately replaced by unity; and in the cases in which the facients are considered as forming two or more distinct sets, the quantic may, in like manner, be considered as homogeneous in regard to each set separately.

2. The expression "an equation," used without explanation, is to be understood as meaning the equation obtained by putting any quantic equal to zero. I make no absolute distinction between the words "degree" and "order" as applied to an equation or system of equations, but I shall in general speak of the order rather than the degree. The equations of a system may be independent, or there may exist relations of connexion between the different equations of the system; the subject of a system of equations so connected together is one of extreme complexity and difficulty. It will be sufficient to notice here, that in any system whatever of equations, assuming only that the equations are not more than sufficient to determine the ratios of the facients, and joining to the system so many linear equations between the facients as will render the ratios of the facients determinate, the order of the system is the same thing as the order of the equation which determines any one of these ratios; it is clear that for a single equation the order so determined is nothing else than the order of the equation.

3. An equation or system of equations represents, or is represented by a locus. This assumes that the facients depend upon quantities $x, y, \ldots$ the coordinates of a point in space; the entire series of points, the coordinates of which satisfy the equation or system of equations, constitutes the locus. To avoid complexity, it is proper to take the facients themselves as coordinates, or at all events to consider these facients as linear functions of the coordinates; this being the case, the order of the locus will be the order of the equation, or system of equations.

4. I have spoken of the *coordinates* of a *point* in *space*. I consider that there is an ideal space of any number of dimensions, but of course, in the ordinary acceptation of the word, space is of three dimensions; however, the plane (the space of ordinary plane geometry) is a space of two dimensions, and we may consider the line as a space of one dimension. I do not, it should be observed, say that the only idea which can be formed of a space of two dimensions is the plane, or the only idea which can be formed of space of one dimension is the line; this is not the case. To avoid complexity, I will take the case of plane geometry rather than geometry of three dimensions; it will be unnecessary to speak of space, or of the number of its dimensions, or of the plane, since we are only concerned with space of two dimensions, viz. the plane; I say, therefore, simply that x, y, z are the coordinates of a point (strictly speaking, it is the ratios of these quantities which are the coordinates, and the quantities x, y, z themselves are indeterminates, i.e. they are only determinate to a common factor *près*, so that in assuming that the coordinates of a point are α, β, γ, we mean only that $x:y:z=\alpha:\beta:\gamma$, and we never as a result obtain $x, y, z=\alpha, \beta, \gamma$, but only $x : y : z=\alpha : \beta : \gamma$; but this being once understood, there is no objection to speaking of x, y, z as coordinates). Now the notions of coordinates and of a point are merely relative; we may, if we please, consider $x : y : z$ as the parameters of a curve containing two variable parameters; such curve becomes of course determinate when we assume $x : y : z=\alpha : \beta : \gamma$, and this very curve is nothing else than the point whose coordinates are α, β, γ, or as we may for shortness call it, the point (α, β, γ). And if the coordinates (x, y, z) are connected by an equation, then giving to these coordinates the entire system of values which satisfy the equation, the locus of the points corresponding to these values is the locus representing or represented by the equation; this of course fixes the notion of a curve of any order, and in particular the notion of a line as the curve of the first order.

The theory includes, as a very particular case, the ordinary theory of reciprocity in plane geometry; we have only to say that the word "point" shall mean "line," and the word "line" shall mean "point," and that expressions properly or primarily applicable to a point and a line respectively shall be construed to apply to a line and a point respectively, and any theorem (assumed of course to be a purely descriptive one) relating to points and lines will become a corresponding theorem relating to lines and points; and similarly with regard to curves of a higher order, when the ideas of reciprocity applicable to these curves are properly developed.

5. A quantic of the degrees $m, m' \ldots$ in the sets $(x, y \ldots)$, $(x', y' \ldots)$ &c. will for the most part be represented by a notation such as

$$(*\between x, y \ldots \overset{m}{\between} x', y' \ldots \overset{m'}{)} \ldots),$$

where the mark * may be considered as indicative of the absolute generality of the quantic; any such quantic may of course be considered as the sum of a series of terms $x^\alpha y^\beta \dots x'^{\alpha'} y'^{\beta'} \dots$, &c. of the proper degrees in the different sets respectively, each term multiplied by a coefficient; these coefficients may be mere numerical multiples of single letters or elements such as $a, b, c, \dots$, or else functions (in general rational and integral functions) of such elements; this explains the meaning of the expression "the elements of a quantic": in the case where the coefficients are mere numerical multiples of the elements, we may in general speak indifferently of the elements, or of the coefficients. I have said that the coefficients may be numerical multiples of single letters or elements such as $a, b, c, \dots$; by the appropriate numerical coefficient of a term $x^\alpha y^\beta \dots x'^{\alpha'} y'^{\beta'} \dots$, I mean the coefficient of this term in the expansion of

$$(x + y \dots)^m (x' + y' \dots)^{m'} \dots);$$

and I represent by the notation

$$(a, b, \dots \between x, y, \dots)^m (x', y', \dots)^{m'} \dots),$$

a quantic in which each term is multiplied as well by its appropriate numerical coefficient as by the literal coefficient or element which belongs to it in the set $(a, b, \dots)$ of literal coefficients or elements. On the other hand, I represent by the notation

$$(a, b, \dots \overset{\vee}{\between} x, y, \dots)^m (x', y', \dots)^{m'} \dots),$$

a quantic in which each term is multiplied only by the literal coefficient or element which belongs to it in the set $(a, b, \dots)$ of literal coefficients or elements. And a like distinction applies to the case where the coefficients are functions of the elements $(a, b, \dots)$.

6. I consider now the quantic

$$(* \between x, y, \dots)^m (x', y', \dots)^{m'} \dots),$$

and selecting any two facients of the same set, e.g. the facients x, y, I remark that there is always an operation upon the elements, tantamount as regards the quantic to the operation $x\partial_y$; viz. if we differentiate with respect to each element, multiply by proper functions of the elements and add, we obtain the same result as by differentiating with ∂_y and multiplying by x. The simplest example will show this as well as a formal proof; for instance, as regards $3ax^2 + bxy + 5cy^2$ (the numerical coefficients are taken haphazard), we have $\frac{1}{3}b\partial_a + 10c\partial_b$ tantamount to $x\partial_y$; as regards $a(x - \alpha y)(x - \beta y)$, we have $-a(\alpha + \beta)\partial_a + \alpha^2\partial_\alpha + \beta^2\partial_\beta$ tantamount to $x\partial_y$, and so in any other case. I represent by $\{x\partial_y\}$ the operation upon the elements tantamount to $x\partial_y$, and I write down the series of operations

$$\{x\partial_y\} - x\partial_y, \dots \quad \{x'\partial_{y'}\} - x'\partial_{y'}, \dots$$

where x, y are considered as being successively replaced by every permutation of two different facients of the set $(x, y, \dots)$; x', y' as successively replaced by every permutation of two different facients of the set $(x', y', \dots)$, and so on; this I call an entire system, and

I say that it is made up of partial systems corresponding to the different facient sets respectively; it is clear from the definition that the quantic is reduced to zero by each of the operations of the entire system. Now, besides the quantic itself, there are a variety of other functions which are reduced to zero by each of the operations of the entire system; any such function is said to be a covariant of the quantic, and in the particular case in which it contains only the elements, an invariant. (It would be allowable to define as a covariant *quoad any set or sets*, a function which is reduced to zero by each of the operations of the corresponding partial system or systems, but this is a point upon which it is not at present necessary to dwell.)

7. The definition of a covariant may however be generalized in two directions: we may instead of a single quantic consider two or more quantics; the operations $\{x\partial_y\}$, although represented by means of the same symbols x, y have, as regards the different quantics, different meanings, and we may form the sum $\Sigma\{x\partial_y\}$, where the summation refers to the different quantics: we have only to consider in place of the system before spoken of, the system

$$\Sigma\{x\partial_y\} - x\partial_y, \ldots ; \ \Sigma\{x'\partial_{y'}\} - x'\partial_{y'}, \ldots \text{ \&c. \&c.,}$$

and we obtain the definition of a covariant of two or more quantics.

Again, we may consider in connexion with each set of facients any number of new sets, the facients in any one of these new sets corresponding each to each with those of the original set; and we may admit these new sets into the covariant. This gives rise to a sum $S\{x\partial_y\}$, where the summation refers to the entire series of corresponding sets. We have in place of the system spoken of in the original definition, to consider the system

$$\{x\partial_y\} - S(x\partial_y), \ldots \ \{x'\partial_{y'}\} - S(x'\partial_{y'}), \ldots \text{ \&c. \&c.,}$$

or if we are dealing with two or more quantics, then the system

$$\Sigma\{x\partial_y\} - S(x\partial_y), \ldots ; \ \Sigma\{x'\partial_{y'}\} - S(x'\partial_{y'}), \ldots \text{ \&c. \&c.,}$$

and we obtain the generalized definition of a covariant.

8. A covariant has been defined simply as a function reduced to zero by each of the operations of the entire system. But in dealing with given quantics, we may without loss of generality consider the covariant as a function of the like form with the quantic, i.e. as being a rational and integral function homogeneous in regard to the different sets separately, and as being also a rational and integral function of the elements. In particular in the case where the coefficients are mere numerical multiples of the elements, the covariant is to be considered as a rational and integral function homogeneous in regard to the different sets separately, and also homogeneous in regard to the coefficients or elements. And the term "covariant" includes, as already remarked, "invariant."

It is proper to remark, that if the same quantic be represented by means of different sets of elements, then the symbols $\{x\partial_y\}$ which correspond to these different forms of the same quantic are mere transformations of each other, i.e. they become in virtue of the relations between the different sets of elements identical.

9. What precedes is a return to and generalization of the method employed in the first part of the memoir published in the *Camb. Math. Jour.*, t. IV. [1845], and *Camb. and Dubl. Math. Jour.*, t. I. [1846], under the title "On Linear Transformations," [13 and 14], and Crelle, t. XXX. [1846], under the title "Mémoire sur les Hyperdéterminants," [*16], and which I shall refer to as my original memoir. I there consider in fact the invariants of a quantic

$$(*\between x_1,\ x_2 \ldots x_m \between y_1,\ y_2 \ldots y_m) \ldots),$$

linear in regard to n sets each of them of m facients, and I represent the coefficients of a term $x_r y_s z_t \ldots$ by $rst \ldots$; there is no difficulty in seeing that α, β being any two different numbers out of the series $1, 2, \ldots m$, the operation $\{x_\beta \partial_{x_\alpha}\}$ is identical with the operation

$$\Sigma\Sigma \ldots \left(\alpha st \ldots \frac{d}{d\beta st \ldots}\right),$$

where the summations refer to $s, t, \ldots$ which pass respectively from 1 to m, both inclusive; and the condition that a function, assumed to be an invariant, i.e. to contain only the coefficients, may be reduced to zero by the operation $\{x_\beta \partial_{x_\alpha}\} - x_\beta \partial_{x_\alpha}$, is of course simply the condition that such function may be reduced to zero by the operation $\{x_\beta \partial_{x_\alpha}\}$; the condition in question is therefore the same thing as the equation

$$\Sigma\Sigma \ldots \left(\alpha st \ldots \frac{d}{d\beta st}\right) u = 0$$

of my original memoir.

10. But the definition in the present memoir includes also the method made use of in the second part of my original memoir. This method is substantially as follows: consider for simplicity a quantic $U=$

$$(*\between x,\ y, \ldots)^m$$

containing only the single set $(x, y \ldots)$, and let $U_1, U_2 \ldots$ be what the quantic becomes when the set $(x, y \ldots)$ is successively replaced by the sets $(x_1, y_1, \ldots)$, $(x_2, y_2, \ldots)$, ... the number of these new sets being equal to or greater than the number of facients in the set. Suppose that $A, B, C, \ldots$ are any of the determinants

$$\left\| \begin{matrix} \partial_{x_1}, & \partial_{x_2}, & \partial_{x_3}, & \ldots \\ \partial_{y_1}, & \partial_{y_2}, & \partial_{y_3}, & \\ \vdots & & & \end{matrix} \right\|,$$

then forming the derivative

$$A^p B^q C^r \ldots U_1 U_2 \ldots,$$

where $p, q, r \ldots$ are any positive integers, the function so obtained is a covariant involving the sets $(x_1, y_1, \ldots)$, $(x_2, y_2, \ldots)$ &c.; and if after the differentiations we replace these sets by the original set $(x, y, \ldots)$, we have a covariant involving only the original set $(x, y, \ldots)$ and of course the coefficients of the quantic. It is in fact easy to show that any such derivative is a covariant according to the definition given in this Memoir. But to do this some preliminary explanations are necessary.

11. I consider any two operations P, Q, involving each or either of them differentiations in respect of variables contained in the other of them. It is required to investigate the effect of the operation $P . Q$, where the operation Q is to be in the first place performed upon some operand Ω, and the operation P is then to be performed on the operand $Q\Omega$. Suppose that P involves the differentiations $\partial_a, \partial_b, \ldots$ in respect of variables $a, b, \ldots$ contained in Q and Ω, we must as usual in the operation P replace $\partial_a, \partial_b, \ldots$ by $\partial_a + \partial'_a, \partial_b + \partial'_b, \ldots$ where the unaccentuated symbols operate only upon Ω, and the accentuated symbols operate only upon Q. Suppose that P is expanded in ascending powers of the symbols $\partial'_a, \partial'_b, \ldots$, viz. in the form $P + P_1 + P_2 + \&c.$, we have first to find the values of P_1Q, P_2Q, &c., by actually performing upon Q as operand the differentiations $\partial'_a, \partial'_b \ldots$. The symbols PQ, P_1Q, P_2Q, &c. will then contain only the differentiations $\partial_a, \partial_b, \ldots$ which operate upon Ω, and the meaning of the expression being once understood, we may write

$$P . Q = PQ + P_1Q + P_2Q + \&c.$$

In particular if P be a linear function of $\partial_a, \partial_b, \ldots$, we have to replace P by $P + P_1$, where P_1 is the same function of $\partial'_a, \partial'_b, \ldots$ that P is of $\partial_a, \partial_b, \ldots$, and it is therefore clear that we have in this case

$$P . Q = PQ + P(Q),$$

where on the right-hand side in the term PQ the differentiations $\partial_a, \partial_b, \ldots$ are considered as not in anywise affecting the symbol Q, while in the term $P(Q)$ these differentiations, or what is the same thing, the operation P, is considered to be performed upon Q as operand.

Again, if Q be a linear function of $a, b, c, \ldots$, then $P_2Q = 0$, $P_3Q = 0$, &c., and therefore $P . Q = PQ + P_1Q$; and I shall in this case also (and consequently whenever $P_2Q = 0$, $P_3Q = 0$, &c.) write

$$P . Q = PQ + P(Q),$$

where on the right-hand side in the term PQ the differentiations $\partial_a, \partial_b, \ldots$ are considered as not in anywise affecting the symbol Q, while the term $P(Q)$ is in each case what has been in the first instance represented by P_1Q.

We have in like manner, if Q be a linear function of $\partial_a, \partial_b, \partial_c, \ldots$, or if P be a linear function of $a, b, c, \ldots$,

$$Q . P = QP + Q(P);$$

and from the two equations (since obviously $PQ = QP$) we derive

$$P . Q - Q . P = P(Q) - Q(P),$$

which is the form in which the equations are most frequently useful.

12. I return to the expression

$$A^pB^qC^r\dots U_1U_2\dots,$$

and I suppose that after the differentiations the sets $(x_1, y_1, \dots)$, $(x_2, y_2, \dots)$, &c. are replaced by the original set $(x, y, \dots)$. To show that the result is a covariant, we must prove that it is reduced to zero by an operation $\mathfrak{D} =$

$$\{x\partial_y\} - x\partial_y.$$

It is easy to see that the change of the sets $(x_1, y_1, \dots)$, $(x_2, y_2, \dots)$, &c. into the original set $(x, y, \dots)$ may be deferred until after the operation $\mathfrak{D}$, provided that $x\partial_y$ is replaced by $x_1\partial_{y_1} + x_2\partial_{y_2} + \dots$, or if we please by $Sx\partial_y$; we must therefore write $\mathfrak{D} = \{x\partial_y\} - Sx\partial_y$. Now in the equation

$$A \,.\, \mathfrak{D} - \mathfrak{D} \,.\, A = A\,(\mathfrak{D}) - \mathfrak{D}\,(A),$$

where, as before, $A\,(\mathfrak{D})$ denotes the result of the operation A performed upon $\mathfrak{D}$ as operand, and similarly $\mathfrak{D}\,(A)$ the result of the operation $\mathfrak{D}$ performed upon A as operand, we see first that $A\,(\mathfrak{D})$ is a determinant two of the lines of which are identical, it is therefore equal to zero; and next, since $\mathfrak{D}$ does not involve any differentiations affecting A, that $\mathfrak{D}\,(A)$ is also equal to zero. Hence $A \,.\, \mathfrak{D} - \mathfrak{D} \,.\, A = 0$ or A and $\mathfrak{D}$ are convertible. But in like manner $\mathfrak{D}$ is convertible with B, C, &c., and consequently $\mathfrak{D}$ is convertible with $A^pB^qC^r\dots$. Now $\mathfrak{D}\,U_1U_3\dots = 0$; hence

$$\mathfrak{D} \,.\, A^pB^qC^r\dots U_1U_2\dots = 0,$$

or $A^pB^qC^r\dots U_1U_2\dots$ is a covariant, the proposition which was to be proved.

13. I pass to a theorem which leads to another method of finding the covariants of a quantic. For this purpose I consider the quantic

$$(*\between x, y\dots\overset{m}{\between} x', y'\dots\overset{m'}{)}\dots),$$

the coefficients of which are mere numerical multiples of the elements $(a, b, c, \dots)$; and in connexion with this quantic I consider the linear functions $\xi x + \eta y \dots$, $\xi' x' + \eta' y' \dots$, which treating $(\xi, \eta, \dots)$, $(\xi', \eta', \dots)$, &c. as coefficients, may be represented in the form

$$(\xi, \eta, \dots\between x, y, \dots), \quad (\xi', \eta', \dots\between x', y', \dots), \dots$$

we may from the quantic (which for convenience I call U) form an operative quantic

$$(*\between \xi, \eta, \dots\overset{m}{\between} \xi', \eta', \dots\overset{m'}{)}\dots)$$

(I call this quantic Θ), the coefficients of which are mere numerical multiples of $\partial_a, \partial_b, \partial_c, \dots$, and which is such that

$$\Theta U = \overline{(\xi, \eta, \dots\between x, y, \dots)}|^m\, \overline{(\xi', \eta', \dots\between x', y', \dots)}|^{m'} \dots$$

i.e. a product of powers of the linear functions. And it is to be remarked that as regards the quantic Θ and its covariants or other derivatives, the symbols $\partial_a, \partial_b, \partial_c, \dots$ are to be considered as elements with respect to which we may differentiate, &c.

The quantic Θ gives rise to the symbols $\{\xi\partial_\eta\}$, &c. analogous to the symbols $\{x\partial_y\}$, &c. formed from the quantic U. Suppose now that Φ is any quantic containing as well the coefficients as all or any of the sets of Θ. Then $\{x\partial_y\}$ being a linear function of $a, b, c, \ldots$ the variables to which the differentiations in Φ relate, we have

$$\Phi \,.\, \{x\partial_y\} = \Phi \,\{x\partial_y\} + \Phi\,(\{x\partial_y\})\,;$$

again, $\{\eta\partial_\xi\}$ being a linear function of the differentiations with respect to the variables $\partial_a, \partial_b, \partial_c, \ldots$ in Φ, we have

$$\{\eta\partial_\xi\} \,.\, \Phi = \{\eta\partial_\xi\}\, \Phi + \{\eta\partial_\xi\}\,(\Phi)\,;$$

these equations serve to show the meaning of the notations $\Phi(\{x\partial_y\})$ and $\{\eta\partial_\xi\}(\Phi)$, and there exists between these symbols the singular equation

$$\Phi\,(\{x\partial_y\}) = \{\eta\partial_\xi\}\,(\Phi).$$

14. The general demonstration of this equation presents no real difficulty, but to avoid the necessity of fixing upon a notation to distinguish the coefficients of the different terms and for the sake of simplicity, I shall merely exhibit by an example the principle of such general demonstration. Consider the quantic

$$U = ax^3 + 3bx^2y + 3cy^2 + dy^3,$$

this gives

$$\Theta = \xi^3\partial_a + \xi^2\eta\partial_b + \xi\eta^2\partial_c + \eta^3\partial_d\,;$$

or if, for greater clearness, $\partial_a, \partial_b, \partial_c, \partial_d$ are represented by $\alpha, \beta, \gamma, \delta$, then

$$\Theta = \alpha\xi^3 + \beta\xi^2\eta + \gamma\xi\eta^2 + \delta\eta^3,$$

and we have

$$\{x\partial_y\} = 3b\partial_a + 2c\partial_b + d\partial_c,$$

and

$$\{\eta\partial_\xi\} = 3\alpha\partial_\beta + 2\beta\partial_\gamma + \gamma\partial_\delta.$$

Now considering Φ as a function of $\partial_a, \partial_b, \partial_c, \partial_d$, or, what is the same thing, of $\alpha, \beta, \gamma, \delta$, we may write

$$\Phi\,(\{x\partial_y\}) = \Phi\,(3b\alpha + 2c\beta + d\gamma)\,;$$

and if in the expression of Φ we write $\alpha + \partial_a$, $\beta + \partial_b$, $\gamma + \partial_c$, $\delta + \partial_d$ for $\alpha, \beta, \gamma, \delta$ (where only the symbols $\partial_a, \partial_b, \partial_c, \partial_d$ are to be considered as affecting a, b, c, d as contained in the operand $3b\alpha + 2c\beta + d\gamma$), and reject the first term (or term independent of $\partial_a, \partial_b, \partial_c, \partial_d$ in the expansion) we have the required value of $\Phi(\{x\delta_y\})$. This value is

$$(\partial_\alpha \Phi\, \partial_a + \partial_\beta \Phi\, \partial_b + \partial_\gamma \Phi\, \partial_c)\,(3b\alpha + 2c\beta + d\gamma)\,;$$

performing the differentiations $\partial_a, \partial_b, \partial_c, \partial_d$, the value is

$$(3\alpha\partial_\beta + 2\beta\partial_\gamma + \gamma\partial_\delta)\,\Phi,$$

i.e. we have

$$\Phi\,(\{x\partial_y\}) = \{\eta\partial_\xi\}\,(\Phi).$$

15. Suppose now that Φ is a covariant of Θ, then the operation Φ performed upon any covariant of U gives rise to a covariant of the system

$$(*\between x, y, \ldots \overset{m}{\between} x', y', \ldots \overset{m'}{)} \ldots),$$
$$(\xi, \eta, \ldots \between x, y, \ldots), \qquad (\xi', \eta', \ldots \between x', y', \ldots), \text{ \&c.}$$

To prove this it is to be in the first instance noticed, that as regards $(\xi, \eta, \ldots \between x, y, \ldots)$, &c. we have $\{x\partial_y\} = \eta\partial_\xi$, &c. Hence considering $\{x\partial_y\}$, &c. as referring to the quantic U, the operation $\Sigma \{x\partial_y\} - x\partial_y$ will be equivalent to $\{x\partial_y\} + \eta\partial_\xi - x\partial_y$, and therefore every covariant of the system must be reduced to zero by each of the operations

$$\mathfrak{D} = \{x\partial_y\} + \eta\partial_\xi - x\partial_y.$$

This being the case, we have

$$\mathfrak{D}\,.\,\Phi = \mathfrak{D}\Phi + \mathfrak{D}\,(\Phi),$$
$$\Phi\,.\,\mathfrak{D} = \Phi\mathfrak{D} + \Phi\,(\mathfrak{D}),$$

equations which it is obvious may be replaced by

$$\mathfrak{D}\,.\,\Phi = \mathfrak{D}\Phi + \eta\partial_\xi\,(\Phi),$$
$$\Phi\,.\,\mathfrak{D} = \Phi\mathfrak{D} + \Phi\,(\{x\partial_y\}),$$

and consequently (in virtue of the theorem) by

$$\mathfrak{D}\,.\,\Phi = \mathfrak{D}\Phi + \eta\partial_\xi\,(\Phi),$$
$$\Phi\,.\,\mathfrak{D} = \Phi\mathfrak{D} + \{\eta\partial_\xi\}\,(\Phi);$$

and we have therefore

$$\mathfrak{D}\,.\,\Phi - \Phi\,.\,\mathfrak{D} = -\,(\{\eta\partial_\xi\} - \eta\partial_\xi)\,(\Phi);$$

or, since Φ is a covariant of Θ, we have $\mathfrak{D}\,.\,\Phi = \Phi\,.\,\mathfrak{D}$. And since every covariant of the system is reduced to zero by the operation $\mathfrak{D}$, and therefore by the operation $\Phi\,.\,\mathfrak{D}$, such covariant will also be reduced to zero by the operation $\mathfrak{D}\,.\,\Phi$, or what is the same thing, the covariant operated on by Φ, is reduced to zero by the operation $\mathfrak{D}$ and is therefore a covariant, i.e. Φ operating upon a covariant gives a covariant.

16. In the case of a quantic such as $U =$

$$(*\between x, y \overset{m}{\between} x', y' \overset{m'}{)} \ldots),$$

we may instead of the new sets (ξ, η), $(\xi', \eta')\ldots$ employ the sets $(y, -x)$, $(y', -x')$, &c. The operative quantic Θ is in this case defined by the equation $\Theta U = 0$, and if Φ be, as before, any covariant of Θ, then Φ operating upon a covariant of U will give a covariant of U. The proof is nearly the same as in the preceding case; we have instead of the equation $\Phi\,(\{x\partial_y\}) = \{\eta\partial_\xi\}\,(\Phi)$ the analogous equation

$$\Phi\,(\{x\partial_y\}) = -\,\{x\partial_y\}\,(\Phi),$$

where on the left-hand side $\{x\partial_y\}$ refers to U, but on the right-hand side $\{x\partial_y\}$ refers to Θ, and instead of $\mathfrak{D} = \{x\partial_y\} + \eta\partial_\xi - x\partial_y$ we have simply $\mathfrak{D} = \{x\partial_y\} - x\partial_y$.

17. I pass next to the quantic

$$(\ast\between x, y)^m,$$

which I shall in general consider under the form

$$(a, b, \ldots b', a'\between x, y)^m,$$

but sometimes under the form

$$(a, b, \ldots b', a'\,\bar{\between}\, x, y)^m,$$

the former notation denoting, it will be remembered,

$$ax^m + \frac{m}{1} bx^{m-1}y \ldots + \frac{m}{1} b'xy^{m-1} + a'y^m,$$

and the latter notation

$$ax^m + bx^{m-1}y \ldots + b'xy^{m-1} + a'y^m.$$

But in particular cases the coefficients will be represented all of them by unaccentuated letters, thus $(a, b, c, d\between x, y)^3$ will be used to denote $ax^3 + 3bx^2y + 3cxy^2 + dy^3$, and $(a, b, c, d\,\bar{\between}\, x, y)^3$ will be used to denote $ax^3 + bx^2y + cxy^2 + dy^3$, and so in all similar cases.

Applying the general methods to the quantic

$$(a, b, \ldots b', a'\between x, y)^m,$$

we see that

$$\{y\partial_x\} = a\partial_b + 2b\partial_c \ldots + mb'\partial_{a'},$$

$$\{x\partial_y\} = mb\partial_a + (m-1c\partial_b \ldots + a'\partial_{b'};$$

in fact, with these meanings of the symbols the quantic is reduced to zero by each of the operations $\{y\partial_x\} - y\partial_x$, $\{x\partial_y\} - x\partial_y$; hence according to the definition any function which is reduced to zero by each of the last-mentioned operations is a covariant of the quantic. But in accordance with a preceding remark, the covariant may be considered as a rational and integral function, separately homogeneous in regard to the facients (x, y) and the coefficients $(a, b, \ldots b', a')$. If instead of the single set (x, y) the covariant contains the sets (x_1, y_1), (x_2, y_2), &c., then it must be reduced to zero by each of the operations $\{y\partial_x\} - Sy\partial_x$, $\{x\partial_y\} - Sx\partial_y$ (where $Sy\partial_x = y_1\partial_{x_1} + y_2\partial_{x_2} + \ldots$), but I shall principally attend to the case in which the covariant contains only the set (x, y).

Suppose, for shortness, that the quantic is represented by U, and let U_1, $U_2, \ldots$ be what U becomes when the set (x, y) is successively replaced by the sets (x_1, y_1), (x_2, y_2), &c. Suppose moreover that $\overline{12} = \partial_{x_1}\partial_{y_2} - \partial_{x_2}\partial_{y_1}$, &c., then the function

$$\overline{12}^p\,\overline{13}^q\,\overline{23}^r \ldots U_1U_2U_3 \ldots,$$

in which, after the differentiations, the new sets (x_1, y_1), $(x_2, y_2), \ldots$ may be replaced by the original set (x, y), will be a covariant of the quantic U. And if the number of differentiations be such as to make the facients disappear, i.e. if the sum of all the indices $p, q, \ldots$ of the terms $\overline{12}$, &c. which contain the symbolic number 1, the sum of all the indices $p, r, \ldots$ of the terms which contain the symbolic number 2, and so on, be severally equal to the degree of the quantic, we have an invariant. The operative quantic Θ becomes in the case under consideration

$$\Theta = (\partial_{a'},\ -\partial_{b'},\ \ldots \pm \partial_a \between x,\ y)^m,$$

the signs being alternately positive and negative; in fact it is easy to verify that this expression gives identically $\Theta U = 0$, and any covariant of Θ operating on a covariant of U gives rise to a covariant of U.

18. But the quantic

$$(a,\ b, \ldots b',\ a' \between x,\ y)^m,$$

considered as decomposable into linear factors, i.e. as expressible in the form

$$a\,(x - \alpha y)\,(x - \beta y)\ldots,$$

gives rise to a fresh series of results. We have in this case

$$\begin{aligned} \{y\partial_x\} &= \qquad\qquad\qquad \partial_\alpha + \ \ \partial_\beta \ldots, \\ \{x\partial_y\} &= -(\alpha + \beta \ldots)\,a\partial_a + \alpha^2\partial_\alpha + \beta^2\partial_\beta + \ldots; \end{aligned}$$

in fact with these meanings of the symbols the quantic is reduced to zero by each of the operations $\{x\partial_y\} - x\partial_y$, $\{y\partial_x\} - y\partial_x$, and we have consequently the definition of the covariant of a quantic considered as expressed in the form $a\,(x-\alpha y)\,(x-\beta y)\ldots$. And it will be remembered that these and the former values of the symbols $\{x\partial_y\}$ and $\{y\partial_x\}$ are, when the same quantic is considered as represented under the two forms $(a,\ b,\ldots b',\ a' \between x,\ y)^m$ and $a\,(x-\alpha y)\,(x-\beta y)\ldots$, identical.

19. Consider now the expression

$$a^\theta\,(x-\alpha y)^j\,(x-\beta y)^k \ldots (\alpha-\beta)^p \ldots,$$

where the sum of the indices $j,\ p,\ldots$ of all the simple factors which contain α, the sum of the indices $k,\ p,\ldots$ of all the simple factors which contain β, &c. are respectively equal to the index θ of the coefficient a. The index θ and the indices p, &c. may be considered as arbitrary, nevertheless within such limits as will give positive values (0 inclusive) for the indices $j,\ k,\ldots$.

The expression in question is reduced to zero by each of the operations $\{x\partial_y\} - x\partial_y$, $\{y\partial_x\} - y\partial_x$; and this is of course also the case with the expressions obtained by interchanging in any manner the roots $\alpha,\ \beta,\ \gamma,\ldots$, and therefore with the expression

$$a^\theta\,\Sigma\,(x-\alpha y)^j\,(x-\beta y)^k \ldots (\alpha-\beta)^p \ldots,$$

where Σ denotes a summation with respect to all the different permutations of the roots $\alpha,\ \beta, \ldots$.

The function so obtained (which is of course a rational function of $(a, b, \ldots b', a')$) will be a covariant, and if we suppose $\mu = m\theta - 2Sp$, where Sp denotes the sum of all the indices p of the different terms $(\alpha - \beta)^p$, &c., then the covariant will be of the order μ (i.e. of the degree μ in the facients x, y), and of the degree θ in the coefficients.

20. In connexion with this covariant

$$a^\theta \Sigma (x - \alpha y)^p (x - \beta y)^k \ldots (\alpha - \beta)^p \ldots,$$

of the order μ and of the degree θ in the coefficients, of the quantic $U =$

$$a (x - \alpha y)(x - \beta y) \ldots,$$

consider the covariant

$$\Sigma (\overline{12}^p \ldots) V_1 V_2 \ldots V_m$$

of a quantic $V =$

$$(* \between x, y)^\phi,$$

in which, after the differentiations, the sets (x_1, y_1), (x_2, y_2), ... are replaced by the original set (x, y). The last-mentioned covariant will be of the order $m(\phi - \theta) + \mu$, and will be of the degree m in the coefficients; and in particular if $\phi = \theta$, i.e. if V be a quantic of the order θ, then the covariant will be of the order μ and of the degree m in the coefficients. Hence to a covariant of the degree θ in the coefficients, of a quantic of the order m, there corresponds a covariant of the degree m in the coefficients, of a quantic of the order θ; the two covariants in question being each of them of the same order μ. And it is proper to notice, that if we had commenced with the covariant of the quantic V, a reverse process would have led to the covariant of the quantic U. We may, therefore, say that the covariants of a given order and of the degree θ in the coefficients, of a quantic of the order m, correspond each to each with the covariants of the same order and of the degree m in the coefficients, of a quantic of the order θ; and in particular the invariants of the degree θ of a quantic of the order m, correspond each to each with the invariants of the degree m of a quantic of the order θ. This is the law of reciprocity demonstrated by M. Hermite, by a method which (I am inclined to think) is substantially identical with that here made use of, although presented in a very different form: the discovery of the law, considered as a law relating to the *number* of invariants, is due to Mr Sylvester. The precise meaning of the law, in the last-mentioned point of view, requires some explanation. Suppose that we know all the really independent invariants of a quantic of the order m, the law gives the number of invariants of the degree m of a quantic of the order θ (it is convenient to assume $\theta > m$), viz. of the invariants of the degree in question, which are linearly independent, or asyzygetic, i.e. such that there do not exist any merely numerical multiples of these invariants having the sum zero; but the invariants in question may and in general will be connected *inter se* and with the other invariants of the quantic to which they belong by non-linear equations: and in particular the system of invariants of the degree m will comprise all the invariants of that degree (if any) which are rational and integral functions of the invariants of lower degrees. The like observations apply to the system of covariants of a given order and of the degree m in the coefficients, of a quantic of the order θ.

21. The number of the really independent covariants of a quantic $(*\between x, y)^m$ is precisely equal to the order m of the quantic, i.e. any covariant is a function (generally an irrational function only expressible as the root of an equation) of any m independent covariants, and in like manner the number of really independent invariants is $m-2$; we may, if we please, take $m-2$ really independent invariants as part of the system of the m independent covariants; the quantic itself may be taken as one of the other two covariants, and any other covariant as the other of the two covariants; we may therefore say that every covariant is a function (generally an irrational function only expressible as the root of an equation) of $m-2$ invariants, of the quantic itself and of a given covariant.

22. Consider any covariant of the quantic

$$(a, b, \ldots b', a'\between x, y)^m,$$

and let this be of the order μ, and of the degree θ in the coefficients. It is very easily shown that $m\theta-\mu$ is necessarily even. In particular in the case of an invariant (i.e. when $\mu=0$) $m\theta$ is necessarily even[1]: so that a quantic of an odd order admits only of invariants of an even degree. But there is an important distinction between the cases of $m\theta-\mu$ evenly even and oddly even. In the former case the covariant remains unaltered by the substitution of (y, x), $(a', b', \ldots b, a)$ for (x, y), $(a, b, \ldots b', a')$; in the latter case the effect of the substitution is to change the sign of the covariant. The covariant may in the former case be called a symmetric covariant, and in the latter case a skew covariant. It may be noticed in passing, that the simplest skew invariant is M. Hermite's invariant of the degree 18 of a quantic of the order 5.

23. There is another very simple condition which is satisfied by every covariant of the quantic

$$(a, b, \ldots b', a'\between x, y)^m,$$

viz. if we consider the facients (x, y) as being respectively of the weights $\tfrac{1}{2}, -\tfrac{1}{2}$, and the coefficients $(a, b, \ldots b', a')$ as being respectively of the weights $-\tfrac{1}{2}m, -\tfrac{1}{2}m+1, \ldots \tfrac{1}{2}m-1, \tfrac{1}{2}m$, then the weight of each term of the covariant will be zero. This is the most elegant statement of the law, but to avoid negative quantities, the statement may be modified as follows:—if the facients (x, y) are considered as being of the weights 1, 0 respectively, and the coefficients $(a, b \ldots b', a')$ as being of the weights $0, 1, \ldots, m-1, m$ respectively, then the weight of each term of the covariant will be $\tfrac{1}{2}(m\theta+\mu)$.

[1] I may remark that it was only M. Hermite's important discovery of an invariant of the degree 18 of a quantic of the order 5, which removed an erroneous impression which I had been under from the commencement of the subject, that $m\theta$ was of necessity *evenly even*.

24. The preceding laws as to the form of a covariant have been stated here by way of anticipation, principally for the sake of the remark, that they so far define the form of a covariant as to render it in very many cases practicable with a moderate amount of labour to complete the investigations by means of the operation $\{x\partial_y\} - x\partial_y$ and $\{y\partial_x\} - y\partial_x$. In fact, for finding the covariants of a given order, and of a given degree in the coefficients, we may form the most general function of the proper order and degree in the coefficients, satisfying the prescribed conditions as to symmetry and weight: such function, if reduced to zero by one of the operations in question, will, on account of the symmetry, be reduced to zero by the other of the operations in question; it is therefore only necessary to effect upon it, e.g. the operation $\{x\partial_y\} - x\partial_y$, and to determine if possible the indeterminate coefficients in such manner as to render the result identically zero: of course when this cannot be done there is not any covariant of the form in question. It is moreover proper to remark, as regards invariants, that if an invariant be expanded in a series of ascending powers of the first coefficient a, and the first term of the expansion is known, all the remaining terms can be at once deduced by mere differentiations. There is one very important case in which the value of such first term (i.e. the value of the invariant when a is put equal to 0) can be deduced from the corresponding invariant of a quantic of the next inferior order; the case in question is that of the discriminant (or function which equated to zero expresses the equality of a pair of roots); for by Joachimsthal's theorem, if in the discriminant of the quantic $(a, b, \dots b', a' \between x, y)^m$ we write $a = 0$, the result contains b^2 as a factor, and divested of this factor is precisely the discriminant of the quantic of the order $m-1$ obtained from the given quantic by writing $a = 0$ and throwing out the factor x: this is in practice a very convenient method for the calculation of the discriminants of quantics of successive orders. It is also to be noticed as regards covariants, that when the first or last coefficient of any covariant (i.e. the coefficient of the highest power of either of the facients) is known, all the other coefficients can be deduced by mere differentiations.

Postscript added October 7th, 1854.—I have, since the preceding memoir was written, found with respect to the covariants of a quantic $(* \between x, y)^m$, that a function of any order and degree in the coefficients satisfying the necessary condition as to weight, and such that it is reduced to zero by one of the operations $\{x\partial_y\} - x\partial_y$, $\{y\partial_x\} - y\partial_x$, will of necessity be reduced to zero by the other of the two operations, i.e. it will be a covariant; and I have been thereby led to the discovery of the law for the number of asyzygetic covariants of a given order and degree in the coefficients; from this law I deduce as a corollary, the law of reciprocity of MM. Sylvester and Hermite. I hope to return to the subject in a subsequent memoir.

140.

RESEARCHES ON THE PARTITION OF NUMBERS.

[From the *Philosophical Transactions of the Royal Society of London*, vol. CXLV *for the year* 1855, pp. 127—140. Received April 14,—Read May 24, 1855.]

I PROPOSE to discuss the following problem: "To find in how many ways a number q can be made up of the elements $a, b, c, \ldots$ each element being repeatable an indefinite number of times." The required number of partitions is represented by the notation

$$P(a, b, c, \ldots)q,$$

and we have, as is well known,

$$P(a, b, c, \ldots)q = \text{coefficient } x^q \text{ in } \frac{1}{(1-x^a)(1-x^b)(1-x^c)\ldots},$$

where the expansion is to be effected in ascending powers of x.

It may be as well to remark that each element is to be considered as a separate and distinct element, notwithstanding any equalities which may exist between the numbers $a, b, c, \ldots$; thus, although $a = b$, yet $a + a + a +$ &c. and $a + a + b +$ &c. are to be considered as two different partitions of the number q, and so in all similar cases.

The solution of the problem is thus seen to depend upon the theory, to which I now proceed, of the expansion of algebraical fractions.

Consider an algebraical fraction $\quad \dfrac{\phi x}{fx}$,

where the denominator is the product of any number of factors (the same or different) of the form $1 - x^m$. Suppose in general that $[1 - x^m]$ denotes the irreducible factor of $1 - x^m$, i.e. the factor which, equated to zero, gives the prime roots of the equation $1 - x^m = 0$. We have

$$1 - x^m = \Pi[1 - x^{m'}],$$

30—2

where m' denotes any divisor whatever of m (unity and the number m itself not excluded). Hence, if a represent a divisor of one or more of the indices m, and k be the number of the indices of which a is a divisor, we have

$$fx = \Pi\,[1 - x^a]^k.$$

Now considering apart from the others one of the multiple factors $[1 - x^a]^k$, we may write $fx = [1 - x^a]^k f_, x$.

Suppose that the fraction $\dfrac{\phi x}{fx}$ is decomposed into simpler fractions, in the form

$$\frac{\phi x}{fx} = I\,(x)$$

$$+ (x\partial_x)^{k-1} \frac{\theta x}{[1 - x^a]} + (x\partial_x)^{k-2} \frac{\theta_1 x}{[1 - x^a]} \cdots + \frac{\theta_{k-1} x}{[1 - x^a]}$$

$$+ \&c.,$$

where $I(x)$ denotes the integral part, and the &c. refers to the fractional terms depending upon the other multiple factors such as $[1 - x^a]^k$. The functions θx are to be considered as functions with indeterminate coefficients, the degree of each such function being inferior by unity to that of the corresponding denominator; and it is proper to remark that the number of the indeterminate coefficients in all the functions θx together is equal to the degree of the denominator fx.

The term $(x\partial_x)^{k-1} \dfrac{\theta x}{[1 - x^a]}$ may be reduced to the form

$$\frac{gx}{[1 - x^a]^k} + \frac{g_1 x}{[1 - x^a]^{k-1}} + \&c.,$$

the functions gx being of the same degree as θx, and the coefficients of these functions being linearly connected with those of the function θx. The first of the foregoing terms is the only term on the right-hand side which contains the denominator $[1 - x^a]^k$; hence, multiplying by this denominator and then writing $[1 - x^a] = 0$, we find

$$\frac{\phi x}{f_1 x} = gx,$$

which is true when x is any root whatever of the equation $[1 - x^a] = 0$. Now by means of the equation $[1 - x^a] = 0$, $\dfrac{\phi x}{f_1 x}$ may be expressed in the form of a rational and integral function Gx, the degree of which is less by unity than that of $[1 - x^a]$. We have therefore $Gx = gx$, an equation which is satisfied by each root of $[1 - x^a] = 0$, and which is therefore an identical equation; gx is thus determined, and the coefficients of θx being linear functions of those of gx, the function θx may be considered as determined. And this being so, the function

$$\frac{\phi x}{fx} - (x\partial_x)^{k-1} \frac{\theta x}{[1 - x^a]}$$

will be a fraction the denominator of which does not contain any power of $[1-x^a]$ higher than $[1-x^a]^{k-1}$; and therefore $\theta_1 x$ can be found in the same way as θx, and similarly $\theta_2 x$, and so on. And the fractional parts being determined, the integral part may be found by subtracting from $\frac{\phi x}{fx}$ the sum of the fractional parts, so that the fraction $\frac{\phi x}{fx}$ can by a direct process be decomposed in the above-mentioned form.

Particular terms in the decomposition of certain fractions may be obtained with great facility. Thus m being a prime number, assume

$$\frac{1}{(1-x^2)(1-x^3)\dots(1-x^m)} = \&\text{c.} + \frac{\theta x}{[1-x^m]};$$

then observing that $(1-x^m)=(1-x)[1-x^m]$, we have for $[1-x^m]=0$,

$$\theta x = \frac{1}{(1-x)(1-x^2)\dots(1-x^{m-1})}.$$

Now u being any quantity whatever and x being a root of $[1-x^m]=0$, we have identically

$$[1-u^m]=(u-x)(u-x^2)\dots(u-x^{m-1});$$

and therefore putting $u=1$, we have $m=(1-x)(1-x^2)\dots(1-x^{m-1})$, and therefore

$$\theta x = \frac{1}{m},$$

whence

$$\frac{1}{(1-x^2)(1-x^3)\dots(1-x^m)} = \&\text{c.} + \frac{1}{m}\frac{1}{[1-x^m]}.$$

Again, m being as before a prime number, assume

$$\frac{1}{(1-x)(1-x^2)\dots(1-x^m)} = \&\text{c.} + \frac{\theta x}{[1-x^m]},$$

we have in this case for $[1-x^m]=0$,

$$\theta x = \frac{1}{(1-x)^2(1-x^2)\dots(1-x^{m-1})},$$

which is immediately reduced to $\theta x = \frac{1}{m}\frac{1}{1-x}$. Now

$$\frac{[1-u^m]}{u-x} = \frac{[1-u^m]-[1-x^m]}{u-x} = (1+u+\dots+u^{m-2})+(1+u\dots+u^{m-3})x\dots+(1+u)x^{m-3}+x^{m-2};$$

or putting $u=1$,

$$\frac{m}{1-x} = (m-1)+(m-2)x\dots+x^{m-2};$$

and substituting this in the value of θx, we find

$$\frac{1}{(1-x)(1-x^2)\dots(1-x^m)} = \&\text{c.} + \frac{1}{m^2}\frac{(m-1)+(m-2)x\dots+x^{m-2}}{[1-x^m]}.$$

The preceding decomposition of the fraction $\frac{\phi x}{fx}$ gives very readily the expansion ot the fraction in ascending powers of x. For, consider a fraction such as

$$\frac{\theta x}{[1-x^a]},$$

where the degree of the numerator is in general less by unity than that of the denominator; we have

$$1-x^a = [1-x^a]\,\Pi\,[1-x^{a'}],$$

where a' denotes any divisor of a (including unity, but not including the number a itself). The fraction may therefore be written under the form

$$\frac{\theta x \Pi\,[1-x^{a'}]}{1-x^a},$$

where the degree of the numerator is in general less by unity than that of the denominator, i.e. is equal to $a-1$. Suppose that b is any divisor of a (including unity, but not including the number a itself), then $1-x^b$ is a divisor of $\Pi\,[1-x^{a'}]$, and therefore of the numerator of the fraction. Hence representing this numerator by

$$A_0 + A_1 x \dots + A_{a-1}x^{a-1},$$

and putting $a=bc$, we have (corresponding to the case $b=1$)

$$A_0 + A_1 + A_2 \dots + A_{a-1} = 0,$$

and generally for the divisor b,

$$\begin{aligned} &A_0 \quad + A_b \quad \dots + A_{(c-1)b} \quad = 0, \\ &A_1 \quad + A_{b+1} \dots + A_{(c-1)b+1} = 0, \\ &\vdots \\ &A_{b-1} + A_{2b-1} \dots + A_{cb-1} \quad = 0. \end{aligned}$$

Suppose now that a_q denotes a circulating element to the period a, i.e. write

$$a_q = 1, \quad q = 0 \text{ (mod. } a),$$

$$a_q = 0 \text{ in every other case;}$$

a function such as

$$A_0 a_q + A_1 a_{q-1} \dots + A_{a-1} a_{q-a+1}$$

will be a circulating function, or circulator to the period a, and may be represented by the notation

$$(A_0,\ A_1,\ \dots A_{a-1}) \text{ circlor } a_q.$$

In the case however where the coefficients A satisfy, for each divisor b of the number a, the above-mentioned equations, the circulating function is what I call a prime circulator, and I represent it by the notation

$$(A_0,\ A_1, \dots A_{a-1}) \text{ pcr } a_q.$$

By means of this notation we have at once

$$\text{coefficient } x^q \text{ in } \qquad \frac{\theta x}{[1-x^a]} = \quad (A_0,\ A_1 \dots A_{a-1}) \text{ pcr } a_q,$$

and thence also

$$\text{coefficient } x^q \text{ in } (x\partial_x)^r \frac{\theta x}{[1-x^a]} = q^r (A_0,\ A_1 \dots A_{a-1}) \text{ pcr } a_q.$$

Hence assuming that in the fraction $\dfrac{\phi x}{fx}$ the degree of the numerator is less than that of the denominator (so that there is not any integral part), we have

$$\text{coefficient } x^q \text{ in } \frac{\phi x}{fx} = \Sigma\, q^r (A_0,\ A_1,\ \dots A_{a-1}) \text{ pcr } a_q;$$

or, if we wish to put in evidence the non-circulating part arising from the divisor $a=1$,

$$\text{coefficient } x^q \text{ in } \frac{\phi x}{fx} = Aq^{k-1} + Bq^{k-2} \dots + Lq + M$$

$$+ \Sigma\, q^r (A_0,\ A_1 \dots A_{a-1}) \text{ pcr } a_q;$$

where k denotes the number of the factors $1-x^m$ in the denominator fx, a is any divisor (unity excluded) of one or more of the indices m; and for each value of a r extends from $r=0$ to $r=k-1$, where k denotes the number of indices m of which a is a divisor. The particular results previously obtained show, that m being a prime number,

$$\text{coefficient } x^q \text{ in } \frac{1}{(1-x^2)(1-x^3)\dots(1-x^m)} = \&\text{c.} + \frac{1}{m}(\qquad 1, -1, \quad 0,\ 0, \dots) \text{ pcr } m_q,$$

and

$$\text{coefficient } x^q \text{ in } \frac{1}{(1-x)(1-x^2)\dots(1-x^m)} = \&\text{c.} + \frac{1}{m^2}(m-1, -1, -1, \quad \dots) \text{ pcr } m_q.$$

Suppose, as before, that the degree of ϕx is less than that of fx, and let the analytical expression above obtained for the coefficient of x^q in the expansion in ascending powers of x of the fraction $\dfrac{\phi x}{fx}$ be represented by Fq, it is very remarkable that if we expand $\dfrac{\phi x}{fx}$ in descending powers of x, then the coefficient of x^q in this new expansion (q is here of course negative, since the expansion contains only negative powers of x) is precisely equal to $-Fq$; this is in fact at once seen to be the case with respect to each of the partial fractions into which $\frac{\phi x}{fx}$ has been decomposed, and it is consequently the case with respect to the fraction itself[1]. This gives rise to a result of some importance. Suppose that ϕx and fx are respectively of the degrees N and D; it is clear from the form of fx that we have $f\left(\frac{1}{x}\right)=(-)^D x^{-D} fx$; and I suppose that ϕx is also such that $\phi\left(\frac{1}{x}\right)=(\pm)^N x^{-N}\phi x$; then writing $D-N=h$, and supposing that $\frac{\phi x}{fx}$ is expanded in descending powers of x, so that the coefficient of x^q in the expansion is $-Fq$, it is in the first place clear that the expansion will commence with the term x^{-h}, and we must therefore have

$$Fq=0$$

for all values of q from $q=-1$ to $q=-(h-1)$.

Consider next the coefficient of a term x^{-h-q}, where q is 0 or positive; the coefficient in question, the value of which is $-F(-h-q)$, is obviously equal to the coefficient of x^{h+q} in the expansion in ascending powers of x of $\dfrac{\phi\left(\frac{1}{x}\right)}{f\left(\frac{1}{x}\right)}$, i.e. to

$$(\pm)^N(-)^D \text{ coefficient } x^{h+q} \text{ in } \frac{x^h\phi x}{fx},$$

or what is the same thing, to

$$(\pm)^N(-)^D \text{ coefficient } x^q \text{ in } \frac{\phi x}{fx};$$

and we have therefore, q being zero or positive,

$$F(-h-q)=-(\pm)^N(-)^D Fq.$$

In particular, when $\phi x=1$, $\quad Fq=0$

for all values of q from $q=-1$ to $q=-(D-1)$; and q being 0 or positive,

$$F(-D-q)=(-)^{D-1}Fq.$$

The preceding investigations show the general form of the function $P(a, b, c, \ldots)q$, viz. that

$$P(a, b, c, \ldots)q=Aq^{k-1}+Bq^{k-2}\ldots+Lq+M+\Sigma q^r(A_0, A_1, \ldots A_{l-1}) \text{ per } l_q,$$

a formula in which k denotes the number of the elements $a, b, c, \ldots$ &c., and l is any divisor (unity excluded) of one or more of these elements; the summation in the case of each such divisor extends from $r=0$ to $r=k-1$, where k is the number of the elements $a, b, c, \ldots$ &c. of which l is a divisor; and the investigations indicate

[1] The property is a fundamental one in the general theory of developments.

how the values of the coefficients A of the prime circulators are to be obtained. It has been moreover in effect shown, that if $D = a + b + c + \ldots$, then, writing for shortness $P(q)$ instead of $P(a, b, c, \ldots)q$, we have

$$P(q) = 0$$

for all values of q from $q = -1$ to $q = -(D-1)$, and that q being 0 or positive,

$$P(-D-q) = (-)^{D-1} P(q);$$

these last theorems are however uninterpretable in the theory of partitions, and they apply only to the analytical expression for $P(q)$.

I have calculated the following particular results:—

$$P(1, 2)q = \frac{1}{4}\left\{2q + 3 + (1, -1) \text{ per } 2_q\right\}$$

$$P(1, 2, 3)q = \frac{1}{72}\left\{6q^2 + 36q + 47 + 9(1, -1) \text{ per } 2_q + 8(2, -1, -1) \text{ per } 3_q\right\}$$

$$P(1, 2, 3, 4)q = \frac{1}{288}\left\{2q^3 + 30q^2 + 135q + 175 + (9q + 45)(1, -1) \text{ per } 2_q + 32\ (1, 0, -1) \text{ per } 3_q + 36\ (1, 0, -1, 0) \text{ per } 4_q\right\}$$

$$P(1, 2, 3, 4, 5)q = \frac{1}{86400}\left\{30q^4 + 900q^3 + 9300q^2 + 38250q + 50651 + (1350q + 10125)\ (1, -1) \text{ per } 2_q + 3200\ (2, -1, -1) \text{ per } 3_q + 5400\ (1, 1, -1, -1) \text{ per } 4_q + 3456\ (4, -1, -1, -1, -1) \text{ per } 5_q\right\}$$

$$P(2)q = \frac{1}{2}\left\{1 + (1, -1) \text{ per } 2_q\right\}$$

$$P(2, 3)q = \frac{1}{12}\left\{2q + 5 + 3\ (1, -1) \text{ per } 2_q + 4(1, -1, 0) \text{ per } 3_q\right\}$$

$$P(2,\ 3,\ 4)\,q = \frac{1}{288}\left\{\begin{array}{l} 6\,q^2 + 54\,q + 107 \\ + (18\,q + 81)\,(1,\ -1)\ \text{per}\ 2_q \\ + 32\qquad (2,\ -1,\ -1)\ \text{per}\ 3_q \\ + 36\ \ (1,\ -1,\ -1,\ 1)\ \text{per}\ 4_q \end{array}\right\}$$

$$P(2,\ 3,\ 4,\ 5)\,q = \frac{1}{1440}\left\{\begin{array}{l} 2\,q^3 + 42\,q^2 + 267\,q + 497 \\ + (45\,q + 315)\,(1,\ -1)\ \text{per}\ 2_q \\ + 160\qquad (1,\ -1,\ 0)\ \text{per}\ 3_q \\ + 180\ \ (1,\ 0,\ -1,\ 0)\ \text{per}\ 4_q \\ + 288\ (1,\ -1,\ 0,\ 0,\ 0)\ \text{per}\ 5_q \end{array}\right\}$$

$$P(2,\ 3,\ 4,\ 5,\ 6)\,q = \frac{1}{172800}\left\{\begin{array}{l} 10\,q^4 + 400\,q^3 + 5550\,q^2 + 31000\,q + 56877 \\ + (450\,q^2 + 9000\,q + 39075)\,(1,\ -1)\ \text{per}\ 2_q \\ + 3200\,q\qquad (1,\ -1,\ 0)\ \text{per}\ 3_q \\ + 1600\qquad (21,\ -19,\ -2)\ \text{per}\ 3_q \\ + 10800\qquad (1,\ 0,\ -1,\ 0)\ \text{per}\ 4_q \\ + 6912\qquad (4,\ -1,\ -1,\ -1,\ -1)\ \text{per}\ 5_q \\ + 4800\qquad (1,\ -1,\ -2,\ -1,\ 1,\ 2)\ \text{per}\ 6_q \end{array}\right\}$$

$$P(1,\ 2,\ 3,\ 5)\,q = \frac{1}{720}\left\{\begin{array}{l} 4\,q^3 + 66\,q^2 + 324\,q + 451 \\ + 45\qquad (1,\ -1)\ \text{per}\ 2_q \\ + 80\qquad (1,\ -1,\ 0)\ \text{per}\ 3_q \\ + 144\,(1,\ 0,\ 0,\ 0,\ -1)\ \text{per}\ 5_q \end{array}\right\}$$

$$P(1,\ 2,\ 2,\ 3,\ 4)\,q = \frac{1}{6912}\left\{\begin{array}{l} 6\,q^4 + 144\,q^3 + 1194\,q^2 + 3960\,q + 4267 \\ + (54\,q^2 + 648\,q + 1701)\,(1,\ -1)\ \text{per}\ 2_q \\ + 256\qquad (2,\ -1,\ -1)\ \text{per}\ 3_q \\ + 432\qquad (1,\ 0,\ -1,\ 0)\ \text{per}\ 4_q \end{array}\right\}$$

$$P(8)\,q = \frac{1}{8}\left\{\begin{array}{l} 1 \\ + 1\qquad (1,\ -1)\ \text{per}\ 2_q \\ + 2\qquad (1,\ 0,\ -1,\ 0)\ \text{per}\ 4_q \\ + 8\,(1,\ 0,\ 0,\ 0,\ -1,\ 0,\ 0,\ 0)\ \text{per}\ 8_q \end{array}\right\}$$

$$P(7,\ 8)\,q \quad = \frac{1}{112}\left\{\begin{array}{l} 2q+43 \\ +\ 7 \qquad\qquad\qquad\qquad (1,\ -1)\ \text{pcr}\ 2_q \\ +14 \qquad\qquad\qquad (1,\ -1,\ -1,\ 1)\ \text{pcr}\ 4_q \\ +16\ (3,\ 2,\ 1,\ 0,\ -1,\ -2,\ -3)\ \text{pcr}\ 7_q \\ +56\ (0,\ -1,\ -1,\ 0,\ 0,\ 1,\ 1,\ 0)\ \text{pcr}\ 8_q \end{array}\right\},$$

which are, I think, worth preserving.

Received April 14,—Read May 3 and 10, 1855.

I proceed to discuss the following problem: "To find in how many ways a number q can be made up as a sum of m terms with the elements 0, 1, 2,... k, each element being repeatable an indefinite number of times." The required number of partitions is represented by

$$P(0,\ 1,\ 2,\ldots k)^m\,q,$$

and the number of partitions of q less the number of partitions of $q-1$ is represented by

$$P'(0,\ 1,\ 2,\ \ldots\ k)^m\,q.$$

We have, as is well known,

$$P(0,\ 1,\ 2,\ldots k)^m\,q = \text{coefficient } x^q z^m \text{ in } \frac{1}{(1-z)(1-xz)\ldots(1-x^k z)},$$

where the expansion is to be effected in ascending powers of z. Now

$$\frac{1}{(1-z)(1-xz)\ldots(1-x^k z)} = 1 + \frac{1-x^{k+1}}{1-x}\,z + \frac{(1-x^{k+1})(1-x^{k+2})}{(1-x)\ (1-x^2)}\,z^2 + \&\text{c.},$$

the general term being

$$\frac{(1-x^{k+1})(1-x^{k+2})\ldots(1-x^{k+m})}{(1-x)(1-x^2)\ldots\ (1-x^m)}\,z^m,$$

or, what is the same thing,

$$\frac{(1-x^{m+1})(1-x^{m+2})\ldots(1-x^{m+k})}{(1-x)(1-x^2)\ldots\ (1-x^k)}\,z^m,$$

and consequently

$$P(0,\ 1,\ 2,\ \ldots\ k)^m\,q = \text{coefficient } x^q \text{ in } \frac{(1-x^{m+1})(1-x^{m+2})\ldots(1-x^{m+k})}{(1-x)(1-x^2)\ldots\ (1-x^k)};$$

to transform this expression I make use of the equation

$$(1+xz)(1+x^2 z)\ldots(1+z^k z) = 1 + \frac{x(1-x^k)}{1-x}\,z + \frac{x^3(1-x^k)(1-x^{k-1})}{(1-x)\ (1-x^2)}\,z^2 + \&\text{c.},$$

where the general term is

$$x^{\frac{1}{2}s(s+1)}\frac{(1-x^k)(1-x^{k-1})\dots(1-x^{k-s+1})}{(1-x)(1-x^2)\dots(1-x^s)}z^s,$$

and the series is a finite one, the last term being that corresponding to $s=k$, viz. $x^{\frac{1}{2}k(k+1)}z^k$. Writing $-x^m$ for z, and substituting the resulting value of

$$(1-x^{m+1})(1-x^{m+2})\dots(1-x^{m+k})$$

in the formula for $P(0, 1, 2, \dots k)^m q$, we have

$$P(0, 1, 2, \dots k)^m q = \Sigma_s\left\{(-)^s \text{ coefficient } x^q \text{ in } \frac{x^{sm+\frac{1}{2}s(s+1)}}{(1-x)(1-x^2)\dots(1-x^s)(1-x)(1-x^2)\dots(1-x^{k+s})}\right\},$$

where the summation extends from $s=0$ to $s=k$; but if for any value of s between these limits $sm+\frac{1}{2}s(s+1)$ becomes greater than q, then it is clear that the summation need only be extended from $s=0$ to the last preceding value of s, or what is the same thing, from $s=0$ to the greatest value of s for which $q-sm-\frac{1}{2}s(s+1)$ is positive or zero.

It is obvious, that if $q>km$, then

$$P(0, 1, 2\dots k)^m q = 0;$$

and moreover, that if $\theta \not> \frac{1}{2}km$, then

$$P(0, 1, 2, \dots k)^m \theta = P(0, 1, 2, \dots k)^m . km-\theta,$$

so that we may always suppose $q \not> \frac{1}{2}km$. I write therefore $q=\frac{1}{2}(km-\alpha)$ where α is zero or a positive integer not greater than km, and is even or odd according as km is even or odd. Substituting this value of q and making a slight change in the form of the result, we have

$$P(0, 1, 2\dots k)^m \tfrac{1}{2}(km-\alpha) =$$

$$\Sigma_s\left\{(-)^s \text{ coeff. } x^{(\frac{1}{2}k-s)m} \text{ in } \frac{x^{\frac{1}{2}\alpha+\frac{1}{2}s(s+1)}}{(1-x)(1-x^2)\dots(1-x^s)(1-x)(1-x^2)(1-x^{k-s})}\right\},$$

where the summation extends from $s=0$ to the greatest value of s for which $(\frac{1}{2}k-s)m-\frac{1}{2}\alpha-\frac{1}{2}s(s+1)$ is positive or zero. But we may, if we please, consider the summation as extending, when k is even, from $s=0$ to $s=\frac{1}{2}k-1$, and when k is odd, from $s=0$ to $s=\frac{1}{2}(k-1)$; the terms corresponding to values of s greater than the greatest value for which $(\frac{1}{2}k-s)m-\frac{1}{2}\alpha-\frac{1}{2}s(s+1)$ is positive or zero being of course equal to zero. It may be noticed, that the fraction will be a proper one if $\alpha<(k-s)(k-s+1)$; or substituting for s its greatest value, the fraction will be a proper one for all values of s, if, when k is even, $\alpha<\frac{1}{4}k(k+2)$, and when k is odd, $\alpha<\frac{1}{4}(k+1)(k+3)$.

We have in a similar manner,

$$P'(0, 1, 2\dots k)^m q = \text{coefficient } x^q z^m \text{ in } \frac{1-x}{(1-z)(1-xz)\dots(1-x^k z)},$$

which leads to

$P'(0,\ 1,\ 2\ldots k)^m \tfrac{1}{2}(km-\alpha) =$

$$\Sigma_s \left\{(-)^s \text{ coeff. } x^{(\frac{1}{2}k-s)m} \text{ in } \frac{x^{\frac{1}{2}\alpha+s(s+1)}}{(1-x^2)\ldots(1-x^s)(1-x)(1-x^2)\ldots(1-x^{k-s})}\right\},$$

where the summation extends, as in the former case, from $s=0$ to the greatest value of s, for which $(\frac{1}{2}k-s)m-\frac{1}{2}\alpha-\frac{1}{2}s(s+1)$ is positive or zero, or, if we please, when k is even, from $s=0$ to $s=\frac{1}{2}k-1$, and when s is odd, from $s=0$ to $s=\frac{1}{2}(k-1)$. The condition, in order that the fraction may be a proper one for all values of s, is, when k is even, $\alpha+1<\frac{1}{4}k(k+2)$, and when k is odd, $\alpha+1<\frac{1}{4}(k+1)(k+3)$.

To transform the preceding expressions, I write when k is odd x^2 instead of x, and I put for shortness θ instead of $\frac{1}{2}k-s$ or $2(\frac{1}{2}k-s)$, and γ instead of $\frac{1}{2}\alpha+\frac{1}{2}s(s+1)$ or $\alpha+s(s+1)$; we have to consider an expression of the form

$$\text{coefficient } x^{\theta m} \text{ in } \frac{x^\gamma}{Fx},$$

where Fx is the product of factors of the form $1-x^a$. Suppose that a' is the least common multiple of a and θ, then $(1-x^{a'})\div(1-x^a)$ is an integral function of x, equal χx suppose, and $1\div(1-x^a)=\chi x\div(1-x^{a'})$. Making this change in all the factors of Fx which require it (i.e. in all the factors except those in which a is a multiple of θ), the general term becomes

$$\text{coefficient } x^{\theta m} \text{ in } \frac{x^\gamma Hx}{Gx},$$

where Gx is a product of factors of the form $1-x^{a'}$, in which a' is a multiple of θ, i.e. Gx is a rational and integral function of x^θ. But in the numerator $x^\gamma Hx$ we may reject, as not contributing to the formation of the coefficient of $x^{\theta m}$, all the terms in which the indices are not multiples of θ; the numerator is thus reduced to a rational and integral function of x^θ, and the general term is therefore of the form

$$\text{coefficient } x^{\theta m} \text{ in } \frac{\lambda(x^\theta)}{\kappa(x^\theta)},$$

or what is the same thing, of the form

$$\text{coefficient } x^m \text{ in } \frac{\lambda x}{\kappa x},$$

where κx is the product of factors of the form $1-x^a$, and λx is a rational and integral function of x. The particular value of the fraction depends on the value of s; and uniting the different terms, we have an expression

$$\text{coefficient } x^m \text{ in } S_s\ (-)^s\ \frac{\lambda x}{\kappa x},$$

which is equivalent to

$$\text{coefficient } x^m \text{ in } \frac{\phi x}{fx},$$

where fx is a product of factors of the form $1-x^a$, and ϕx is a rational and integral function of x. And it is clear that the fraction will be a proper one when each of the fractions in the original expression is a proper fraction, i.e. in the case of $P(0, 1, 2 \ldots k)^m \tfrac{1}{2}(km-\alpha)$, when for k even, $\alpha < \tfrac{1}{4}k(k+2)$, and for k odd, $\alpha < \tfrac{1}{4}(k+1)(k+3)$; and in the case of $P'(0, 1, 2 \ldots k)^m \tfrac{1}{2}(km-\alpha)$, when for k even, $\alpha+1 < \tfrac{1}{4}k(k+2)$, and for k odd, $\alpha+1 < \tfrac{1}{4}(k+1)(k+3)$.

We see, therefore, that

$$P(0, 1, 2 \ldots k)^m \tfrac{1}{2}(km-\alpha),$$

and

$$P'(0, 1, 2 \ldots k)^m \tfrac{1}{2}(km-\alpha),$$

are each of them of the form

$$\text{coefficient } x^m \text{ in } \frac{\phi x}{fx},$$

where fx is the product of factors of the form $1-x^a$, and up to certain limiting values of α the fraction is a proper fraction. When the fraction $\dfrac{\phi x}{fx}$ is known, we may therefore obtain by the method employed in the former part of this Memoir, analytical expressions (involving prime circulators) for the functions P and P'.

As an example, take

$$P(0, 1, 2, 3)^m \tfrac{3}{2}m,$$

which is equal to

$$\text{coefficient } x^{3m} \text{ in } \frac{1}{(1-x^2)(1-x^4)(1-x^6)}$$

$$-\text{ coefficient } x^m \text{ in } \frac{1}{(1-x^2)(1-x^2)(1-x^4)}.$$

The multiplier for the first fraction is

$$\frac{(1-x^6)(1-x^{12})}{(1-x^2)(1-x^4)},$$

which is equal to

$$1+x^2+2x^4+x^6+2x^8+x^{10}+x^{12}.$$

Hence, rejecting in the numerator the terms the indices of which are not divisible by 3, the first term becomes

$$\text{coefficient } x^{3m} \text{ in } \frac{1+x^6+x^{12}}{(1-x^6)(1-x^{12})(1-x^6)},$$

or what is the same thing, the first term is

$$\text{coefficient } x^m \text{ in } \frac{1+x^2+x^4}{(1-x^2)^2(1-x^4)};$$

and, the second term being

$$-\text{coefficient } x^m \text{ in } \frac{x^2}{(1-x^2)^2(1-x^4)},$$

we have
$$P(0,\ 1,\ 2,\ 3)^m\,\tfrac{3}{2}m = \text{coefficient } x^m \text{ in } \frac{1+x^4}{(1-x^2)^2(1-x^4)};$$

and similarly it may be shown, that

$$P(0,\ 1,\ 2,\ 3)^m\,\tfrac{1}{2}(3m-1) = \text{coefficient } x^m \text{ in } \frac{x+x^3}{(1-x^2)^2(1-x^4)}.$$

As another example, take

$$P'(0,\ 1,\ 2,\ 3,\ 4,\ 5)\,\tfrac{5}{2}m,$$

which is equal to

$$\text{coefficient } x^{5m} \text{ in } \frac{1}{(1-x^4)(1-x^6)(1-x^8)(1-x^{10})}$$

$$-\text{coefficient } x^{3m} \text{ in } \frac{x^2}{(1-x^2)(1-x^4)(1-x^6)(1-x^8)}$$

$$+\text{coefficient } x^m \text{ in } \frac{x^6}{(1-x^2)(1-x^4)(1-x^4)(1-x^6)}.$$

The multiplier for the first fraction is

$$\frac{(1-x^{20})(1-x^{30})(1-x^{40})}{(1-x^4)(1-x^6)(1-x^8)},$$

which is a function of x^2 of the order 36, the coefficients of which are

1, 0, 1, 1, 2, 1, 3, 2, 4, 3, 4, 4, 6, 4, 6, 5, 7, 5, 7, 5, 7, 5, 6, 4, 6, 4, 4, 3, 4, 2, 3, 1, 2, 1, 1, 0, 1,

and the first part becomes therefore

$$\text{coefficient } x^m \text{ in } \frac{1+x^2+4x^4+5x^6+7x^8+4x^{10}+3x^{12}}{(1-x^2)(1-x^4)(1-x^6)(1-x^8)}.$$

The multiplier for the second fraction is

$$\frac{(1-x^6)(1-x^{12})(1-x^{24})}{(1-x^2)(1-x^4)(1-x^8)},$$

which is a function of x^2 of the order 14, the coefficients of which are

1, 1, 2, 1, 3, 2, 3, 1, 3, 2, 3, 1, 2, 1, 1;

and the second term becomes

$$-\text{coefficient } x^m \text{ in } \frac{2x^2+2x^4+3x^6+x^8+x^{10}}{(1-x^2)^2(1-x^4)(1-x^8)};$$

and the third term is

$$\text{coefficient } x^m \text{ in } \frac{x^6}{(1-x^2)(1-x^4)^2(1-x^6)}.$$

Now the fractions may be reduced to a common denominator

$$(1-x^2)(1-x^4)(1-x^6)(1-x^8)$$

by multiplying the terms of the second fraction by $\frac{1-x^6}{1-x^2}(=1+x^2+x^4)$, and the terms of the third fraction by $\frac{1-x^8}{1-x^4}(=1+x^4)$; performing the operations and adding, the numerator and denominator of the resulting fraction will each of them contain the factor $1-x^2$; and casting this out, we find

$$P(0,\ 1,\ 2,\ 3,\ 4,\ 5)^m\ \tfrac{5}{2}m = \text{coefficient } x^m \text{ in } \frac{1-x^6+x^{12}}{(1-x^4)(1-x^6)(1-x^8)}.$$

I have calculated by this method several other particular cases, which are given in my "Second Memoir upon Quantics", [141], the present researches were in fact made for the sake of their application to that theory.

Received April 20,—Read May 3 and 10, 1855.

Since the preceding portions of the present Memoir were written, Mr Sylvester has communicated to me a remarkable theorem which has led me to the following additional investigations[1].

Let $\frac{\phi x}{fx}$ be a rational fraction, and let $(x-x_1)^k$ be a factor of the denominator fx, then if

$$\left\{\frac{\phi x}{fx}\right\}_{x_1}$$

denote the portion which is made up of the simple fractions having powers of $x-x_1$ for their denominators, we have by a known theorem

$$\left\{\frac{\phi x}{fx}\right\}_{x_1} = \text{coefficient } \frac{1}{z} \text{ in } \frac{1}{x-x_1-z}\,\frac{\phi(x_1+z)}{f(x_1+z)}.$$

Now by a theorem of Jacobi's and Cauchy's,

$$\text{coefficient } \frac{1}{z} \text{ in } Fz = \text{coefficient } \frac{1}{t} \text{ in } F(\psi t)\,\psi' t;$$

whence, writing $x_1+z=x_1e^{-t}$, we have

$$\left\{\frac{\phi x}{fx}\right\}_{x_1} = \text{coefficient } \frac{1}{t} \text{ in } \frac{x_1}{x_1-xe^t}\,\frac{\phi(x_1e^{-t})}{f(x_1e^{-t})}.$$

[1] Mr Sylvester's researches are published in the *Quarterly Mathematical Journal*, July 1855, [vol. I. pp. 141—152], and he has there given the general formula as well for the circulating as the non-circulating part of the expression for the number of partitions.—Added 23rd February, 1856.—A. C.

Now putting for a moment $x = x_1 e^\theta$, we have

$$\frac{1}{x_1 - xe^t} = \frac{1}{x_1(1-e^{\theta+t})} = \frac{1}{x_1(1-e^\theta)} + \partial_\theta \frac{1}{x_1(1-e^\theta)} + \dots,$$

and $\partial_\theta = x\partial_x$, whence

$$\frac{1}{x_1 - xe^t} = \frac{1}{x_1 - x} + \frac{t}{1} x\partial_x \frac{1}{x_1 - x} + \frac{t^2}{1\,.\,2}(x\partial_x)^2 \frac{1}{x_1 - x} + \dots,$$

the general term of which is

$$\frac{t^{s-1}}{\Pi(s-1)}(x\partial_x)^{s-1}\frac{1}{x_1 - x}.$$

Hence representing the general term of

$$\frac{x_1\phi(x_1 e^{-t})}{f(x_1 e^{-t})}$$

by $\chi x_1 t^{-s}$, so that

$$\chi x_1 = \text{coefficient } \frac{1}{t} \text{ in } t^{s-1}\frac{x_1\phi(x_1 e^{-t})}{f(x_1 e^{-t})},$$

we find, writing down only the general term,

$$\left\{\frac{\phi x}{fx}\right\}_{x_1} = \dots + \frac{1}{\Pi(s-1)}(x\partial_x)^{s-1}\frac{\chi x_1}{x_1 - x} + \dots,$$

where the value of χx_1 depends upon that of s, and where s extends from $s = 1$ to $s = k$.

Suppose now that the denominator is made up of factors (the same or different) of the form $1 - x^m$. And let a be any divisor of one or more of the indices m, and let k be the number of the indices of which a is a divisor. The denominator contains the divisor $[1 - x^a]^k$, and consequently if ρ be any root of the equation $[1 - x^a] = 0$, the denominator contains the factor $(\rho - x)^k$. Hence writing ρ for x_1 and taking the sum with respect to all the roots of the equation $[1 - x^a] = 0$, we find

$$\left\{\frac{\phi x}{fx}\right\}_{[1-x^a]} = \dots + \frac{1}{\Pi(s-1)}(x\partial_x)^{s-1} S\frac{\chi\rho}{\rho - x} + \dots$$

$$= \dots + \frac{1}{\Pi(s-1)}(x\partial_x)^{s-1}\frac{\theta x}{[1 - x^a]} + \dots,$$

where

$$\chi\rho = \text{coefficient } \frac{1}{t} \text{ in } t^{s-1}\frac{\rho\phi(\rho e^{-t})}{f(\rho e^{-t})},$$

and as before s extends from $s = 1$ to $s = k$. We have thus the actual value of the function θx made use of in the memoir.

A preceding formula gives

$$\left\{\frac{\phi x}{fx}\right\}_1 = \text{coefficient } \frac{1}{t} \text{ in } \frac{1}{1 - xe^t}\frac{\phi(e^{-t})}{f(e^{-t})},$$

which is a very simple expression for the non-circulating part of the fraction $\frac{\phi x}{fx}$. This is, in fact, Mr Sylvester's theorem above referred to.

141.

A SECOND MEMOIR UPON QUANTICS.

[From the *Philosophical Transactions of the Royal Society of London*, vol. CXLVI. *for the year*, 1856, pp. 101—126. Received April 14,—Read May 24, 1855.]

THE present memoir is intended as a continuation of my Introductory Memoir upon Quantics, t. CXLIV. (1854), p. 245, and must be read in connexion with it; the paragraphs of the two Memoirs are numbered continuously. The special subject of the present memoir is the theorem referred to in the Postscript to the Introductory Memoir, and the various developments arising thereout in relation to the number and form of the covariants of a binary quantic.

25. I have already spoken of asyzygetic covariants and invariants, and I shall have occasion to speak of irreducible covariants and invariants. Considering in general a function u determined like a covariant or invariant by means of a system of partial differential equations, it will be convenient to explain what is meant by an asyzygetic integral and by an irreducible integral. Attending for greater simplicity only to a single set $(a, b, c, \ldots)$, which in the case of the covariants or invariants of a single function will be as before the coefficients or elements of the function, it is assumed that the system admits of integrals of the form $u = P$, $u = Q$, &c., or as we may express it, of integrals P, Q, &c., where P, Q, &c. are rational and integral homogeneous functions of the set $(a, b, c, \ldots)$, and moreover that the system is such that P, Q, &c. being integrals, $\phi(P, Q, \ldots)$ is also an integral. Then considering only the integrals which are rational and integral homogeneous functions of the set $(a, b, c, \ldots)$, integrals $P, Q, R, \ldots$ not connected by any linear equation or syzygy (such as $\lambda P + \mu Q + \nu R \ldots 0$), [1] are said to be asyzygetic; but in speaking of the asyzygetic integrals of a particular degree, it is implied that the integrals are a system such that every other integral of

[1] It is hardly necessary to remark, that the multipliers λ, μ, ν, ..., and generally any coefficients or quantities not expressly stated to contain the set $(a, b, c, \ldots)$, are considered as independent of the set, or to use a convenient word, are considered as "trivials."

the same degree can be expressed as a linear function (such as $\lambda P+\mu Q+\nu R\ldots$) of these integrals; and any integral P not expressible as a rational and integral homogeneous function of integrals of inferior degrees is said to be an irreducible integral.

26. Suppose now that A_1, A_2, A_3, &c. denote the number of asyzygetic integrals of the degrees 1, 2, 3, &c. respectively, and let α_1, α_2, α_3, &c. be determined by the equations

$$A_1=\alpha_1,$$
$$A_2=\tfrac{1}{2}\,\alpha_1(\alpha_1+1)+\alpha_2,$$
$$A_3=\tfrac{1}{6}\,\alpha_1(\alpha_1+1)(\alpha_1+2)+\alpha_1\alpha_2+\alpha_3,$$
$$A_4=\tfrac{1}{24}\,\alpha_1(\alpha_1+1)(\alpha_1+2)(\alpha_1+3)+\tfrac{1}{2}\,\alpha_1(\alpha_1+1)\,\alpha_2+\alpha_1\alpha_3+\tfrac{1}{2}\,\alpha_2(\alpha_2+1)+\alpha_4,\ \text{\&c.},$$

or what is the same thing, suppose that

$$1+A_1x+A_2x^2+\text{\&c.}=(1-x)^{-\alpha_1}(1-x^2)^{-\alpha_2}(1-x^3)^{-\alpha_3}\ldots;$$

a little consideration will show that α_r represents the number of irreducible integrals of the degree r less the number of linear relations or syzygies between the composite or non-irreducible integrals of the same degree. In fact the asyzygetic integrals of the degree 1 are necessarily irreducible, i.e. $A_1=\alpha_1$. Represent for a moment the irreducible integrals of the degree 1 by X, X', &c., then the composite integrals X^2, XX', &c., the number of which is $\tfrac{1}{2}\,\alpha_1(\alpha_1+1)$, must be included among the asyzygetic integrals of the degree 2; and if the composite integrals in question were asyzygetic, there would remain $A_2-\tfrac{1}{2}\,\alpha_1(\alpha_1+1)$ for the number of irreducible integrals of the degree 2; but if there exist syzygies between the composite integrals in question, the number to be subtracted from A_2 will be $\tfrac{1}{2}\,\alpha_1(\alpha_1+1)$ less the number of these syzygies, and we shall have $A_2-\tfrac{1}{2}\,\alpha_1(\alpha_1+1)$, i.e. α_2 equal to the number of the irreducible integrals of the degree 2 less the number of syzygies between the composite integrals of the same degree. Again, suppose that α_2 is negative $=-\beta_2$, we may for simplicity suppose that there are no irreducible integrals of the degree 2, but that the composite integrals of this degree, X^2, XX', &c., are connected by β_2 syzygies, such as $\lambda X^2+\mu XX'+\text{\&c.}=0$, $\lambda_1X^2+\mu_1XX'+\text{\&c.}=0$. The asyzygetic integrals of the degree 4 include X^4, X^3X', &c., the number of which is $\tfrac{1}{24}\,\alpha_1(\alpha_1+1)(\alpha_1+2)(\alpha_1+3)$; but these composite integrals are not asyzygetic, they are connected by syzygies which are augmentatives of the β_2 syzygies of the second degree, viz. by syzygies such as

$$(\lambda X^2+\mu XX'\ldots)\,X^2=0,\quad (\lambda X^2+\mu XX'\ldots)\,XX'=0,\ \text{\&c.}\ (\lambda_1X^2+\mu_1XX'\ldots)\,X^2=0,$$
$$(\lambda_1X^2+\mu_1XX'\ldots)\,XX'=0,\ \text{\&c.},$$

the number of which is $\tfrac{1}{2}\alpha_1(\alpha_1+1)\,\beta_2$. And these syzygies are themselves not asyzygetic, they are connected by secondary syzygies such as

$$\lambda_1(\lambda X^2+\mu XX'\ldots)\,X^2+\mu_1(\lambda X^2+\mu XX'\ldots)\,XX'+\text{\&c.}$$
$$-\lambda(\lambda_1X^2+\mu_1XX'\ldots)\,X^2-\mu(\lambda_1X^2+\mu_1XX'\ldots)\,XX'-\text{\&c.}=0,\ \text{\&c. \&c.},$$

the number of which is $\frac{1}{2}\beta_2(\beta_2-1)$. The real number of syzygies between the composite integrals X^4, X^3X', &c. (i.e. of the syzygies arising out of the β_2 syzygies between X^2, XX', &c.), is therefore $\frac{1}{2}\alpha_1(\alpha_1+1)\beta_2-\frac{1}{2}\beta_2(\beta_2-1)$, and the number of integrals of the degree 4, arising out of the integrals and syzygies of the degrees 1 and 2 respectively, is therefore

$$\tfrac{1}{24}\alpha_1(\alpha_1+1)(\alpha_1+2)(\alpha_1+3)-\tfrac{1}{2}\alpha_1(\alpha_1+1)\beta_2+\tfrac{1}{2}\beta_2(\beta_2-1);$$

or writing $-\alpha_2$ instead of β_2, the number in question is

$$\tfrac{1}{24}\alpha_1(\alpha_1+1)(\alpha_1+2)(\alpha_1+3)+\tfrac{1}{2}\alpha_1(\alpha_1+1)\alpha_2+\tfrac{1}{2}\alpha_2(\alpha_2+1).$$

The integrals of the degrees 1 and 3 give rise to $\alpha_1\alpha_3$ integrals of the degree 4; and if all the composite integrals obtained as above were asyzygetic, we should have

$$A_4-\tfrac{1}{24}\alpha_1(\alpha_1+1)(\alpha_1+2)(\alpha_1+3)-\tfrac{1}{2}\alpha_1(\alpha_1+1)\alpha_2-\tfrac{1}{2}\alpha_2(\alpha_2+1)-\alpha_1\alpha_3,$$

i.e. α_4 as the number of irreducible integrals of the degree 4; but if there exist any further syzygies between the composite integrals, then α_4 will be the number of the irreducible integrals of the degree 4 less the number of such further syzygies, and the like reasoning is in all cases applicable.

27. It may be remarked, that for any given partial differential equation, or system of such equations, there will be always a finite number ν such that given ν independent integrals every other integral is a function (in general an irrational function only expressible as the root of an equation) of the ν independent integrals; and if to these integrals we join a single other integral not a rational function of the ν integrals, it is easy to see that every other integral will be a rational function of the $\nu+1$ integrals; but every such other integral will not in general be a rational and integral function of the $\nu+1$ integrals; and [*incorrect*] there is not in general any finite number whatever of integrals, such that every other integral is a rational and integral function of these integrals, i.e. the number of irreducible integrals is in general infinite; and it would seem that this is in fact the case in the theory of covariants.

28. In the case of the covariants, or the invariants of a binary quantic, A_n is given (this will appear in the sequel) as the coefficient of x^n in the development, in ascending powers of x, of a rational fraction $\dfrac{\phi x}{fx}$, where fx is of the form

$$(1-x)^{\beta_1}(1-x^2)^{\beta_2}\ldots(1-x^k)^{\beta_k},$$

and the degree of ϕx is less than that of fx. We have therefore

$$1+A_1x+A_2x^2+\ldots=\frac{\phi x}{fx},$$

and consequently

$$\phi x=(1-x)^{\beta_1-\alpha_1}(1-x^2)^{\beta_2-\alpha_2}\ldots(1-x^k)^{\beta_k-\alpha_k}(1-x^{k+1})^{-\alpha_{k+1}}\ldots.$$

Now every rational factor of a binomial $1-x^m$ is the irreducible factor of $1-x^{m'}$, where m' is equal to or a submultiple of m. Hence in order that the series $\alpha_1, \alpha_2, \alpha_3, \ldots$ may terminate, ϕx must be made up of factors each of which is the irreducible factor of a binomial $1-x^m$, or if ϕx be itself irreducible, then ϕx must be the irreducible factor of a binomial $1-x^m$. Conversely, if ϕx be not of the form in question, the series $\alpha_1, \alpha_2, \alpha_3$, &c. will go on *ad infinitum*, and it is easy to see that there is no point in the series such that the terms beyond that point are all of them negative, i.e. there will be irreducible covariants or invariants of indefinitely high degrees; and the number of covariants or invariants will be infinite. The number of invariants is first infinite in the case of a quantic of the seventh order, or septimic; the number of covariants is first infinite in the case of a quantic of the fifth order, or quintic. [As is now well known, these conclusions are incorrect, the number of irreducible covariants or invariants is in every case finite.]

29. Resuming the theory of binary quantics, I consider the quantic

$$(a, b, \ldots b', a' \between x, y)^m.$$

Here writing

$$\{y\partial_x\} = a\partial_b + 2b\partial_c \ldots \qquad + mb'\partial_{a'}, \ = X,$$

$$\{x\partial_y\} = mb\partial_a + (m-1)\, c\partial_b \ldots + a'\partial_{b'}, \ = Y,$$

any function which is reduced to zero by each of the operations $X - y\partial_x$, $Y - x\partial_y$ is a covariant of the quantic. But a covariant will always be considered as a rational and integral function separately homogeneous in regard to the facients (x, y) and to the coefficients $(a, b, \ldots b', a')$. And the words order and degree will be taken to refer to the facients and to the coefficients respectively.

I commence by proving the theorem enunciated, No. 23. It follows at once from the definition, that the covariant is reduced to zero by the operation

$$\overline{X - y\partial_x}\,.\,\overline{Y - x\partial_y} - \overline{Y - x\partial_y}\,.\,\overline{X - y\partial_x},$$

which is equivalent to

$$X\,.\,Y - Y\,.\,X + y\partial_y - x\partial_x.$$

Now

$$X\,.\,Y = XY + X\,(Y)$$

$$Y\,.\,X = YX + Y(X),$$

where XY and YX are equivalent operations, and

$$X\,(Y) = \ 1ma\partial_a + 2\,(m-1)\,b\partial_b \ldots + m1b'\partial_{b'},$$

$$Y\,(X) = \qquad\qquad m1b\partial_b \ldots + 2\,(m-1)\,b'\partial_{b'} + 1ma'\partial_{a'},$$

whence

$$X\,(Y) - Y\,(X) = ma\partial_a + (m-2)\,b\partial_b \ldots - (m-2)\,b'\partial_b - ma'\partial_{a'}, \ = k \text{ suppose},$$

and the covariant is therefore reduced to zero by the operation

$$k + y\partial_y - x\partial_x.$$

Now as regards a term $a^\alpha b^\beta \ldots b'^{\beta'} a'^{\alpha'}\,.\,x^i y^j$, we have

$$k = m\alpha + (m-2)\,\beta \ldots, \ -(m-2)\,\beta' - m\alpha'$$

$$y\partial_y - x\partial_x = j - i\,;$$

and we see at once that for each term of the covariant we must have

$$m\alpha+(m-2)\beta\ldots-(m-2)\beta'-m\alpha'+j-i=0,$$

i.e. if (x, y) are considered as being of the weights $\frac{1}{2}, -\frac{1}{2}$ respectively, and $(a, b,\ldots b', a')$ as being of the weights $-\frac{1}{2}m, -\frac{1}{2}m+1, \ldots \frac{1}{2}m-1, \frac{1}{2}m$ respectively, then the weight of each term of the covariant is zero.

But if (x, y) are considered as being of the weights 1, 0 respectively, and $(a, b,\ldots b', a')$ as being of the weights $0, 1,\ldots m-1, m$ respectively, then writing the equation under the form

$$m(\alpha+\beta\ldots+\beta'+\alpha')+j+i-2(\beta+\ldots+\overline{m-1}\beta'+m\alpha'+i)=0,$$

and supposing that the covariant is of the order μ and of the degree θ, each term of the covariant will be of the weight $\frac{1}{2}(m\theta+\mu)$.

I shall in the sequel consider the weight as reckoned in the last-mentioned manner. It is convenient to remark, that as regards any function of the coefficients of the degree θ and of the weight q, we have

$$X.Y-Y.X=m\theta-2q.$$

30. Consider now a covariant

$$(A, B,\ldots B', A' \between x, y)^{\mu}$$

of the order μ and of the degree θ; the covariant is reduced to zero by each of the operations $X-y\partial_x$, $Y-x\partial_y$, and we are thus led to the systems of equations

$$\begin{aligned}
XA&=0,\\
XB&=\mu A,\\
XC&=(\mu-1)B,\\
&\vdots\\
XB'&=2C',\\
XA'&=B';
\end{aligned}$$

and

$$\begin{aligned}
YA&=B,\\
YB&=2C,\\
&\vdots\\
YC'&=(\mu-1)B',\\
YB'&=\mu A',\\
YA'&=0.
\end{aligned}$$

Conversely if these equations are satisfied the function will be a covariant.

I assume that A is a function of the degree θ and of the weight $\frac{1}{2}(m\theta-\mu)$, satisfying the condition

$$XA=0;$$

and I represent by YA, Y^2A, Y^3A, &c. the results obtained by successive operations with Y upon the function A. The function Y^sA will be of the degree θ and of the weight $\frac{1}{2}(m\theta-\mu)+s$. And it is clear that in the series of terms YA, Y^2A, Y^3A, &c., we must at last come to a term which is equal to zero. In fact, since m is the greatest weight of any coefficient, the weight of Y^s is at most equal to $m\theta$, and therefore if $\frac{1}{2}(m\theta-\mu)+s>m\theta$, or $s>\frac{1}{2}(m\theta+\mu)$, we must have $Y^sA=0$.

Now writing for greater simplicity XY instead of $X.Y$, and so in similar cases, we have, as regards Y^sA,

$$XY-YX=\mu-2s.$$

Hence

$$(XY-YX)A=\mu A,$$

and consequently

$$XYA=YXA+\mu A=\mu A.$$

Similarly

$$(XY-YX)\,YA=(\mu-2)\,YA,$$

and therefore

$$\begin{aligned}XY^2A&=YXYA+(\mu-2)\,YA\\&=\mu YA+(\mu-2)\,YA=2(\mu-1)\,YA.\end{aligned}$$

And again,

$$(XY-YX)\,Y^2A=(\mu-4)\,Y^2A,$$

and therefore

$$\begin{aligned}XY^3A&=YXY^2A+(\mu-4)\,Y^2A\\&=2(\mu-1)\,Y^2A+(\mu-4)\,Y^2A=3(\mu-2)\,Y^2A,\end{aligned}$$

or generally

$$XY^sA=s(\mu-s+1)\,Y^sA.$$

Hence putting $s=\mu+1$, $\mu+2$, &c., we have

$$\begin{aligned}&XY^{\mu+1}A=0,\\&XY^{\mu+2}A=-(\mu+2)\,1\,.\,Y^{\mu+1}A,\\&XY^{\mu+3}A=-(\mu+3)\,2\,.\,Y^{\mu+2}A,\\&\text{\&c.},\end{aligned}$$

equations which show that

$$Y^{\mu+1}A=0\,;$$

for unless this be so, i.e. if $Y^{\mu+1}A\neq 0$, then from the second equation $XY^{\mu+2}A\neq 0$, and therefore $Y^{\mu+2}A\neq 0$, from the third equation $XY^{\mu+3}\neq 0$, and therefore $X^{\mu+3}A\neq 0$, and so on *ad infinitum*, i.e. we must have $Y^{\mu+1}A=0$.

31. The suppositions which have been made as to the function A, give therefore the equations

$$\begin{aligned} XA &= 0, \\ XYA &= \mu A, \\ XY^2A &= 2(\mu-1)\,YA, \\ &\vdots \\ XY^{\mu}A &= \mu Y^{\mu+1}A, \\ Y^{\mu+1}A &= 0; \end{aligned}$$

and if we now assume

$$B = YA,\quad C = \tfrac{1}{2}YB, \ldots\ A^{\backprime} = \frac{1}{\mu}\,YB^{\backprime},$$

the system becomes

$$\begin{aligned} XA &= 0, \\ XB &= \mu A, \\ XC &= (\mu-1)\,B, \\ &\vdots \\ XA^{\backprime} &= B^{\backprime}, \\ YA' &= 0; \end{aligned}$$

so that the entire system of equations which express that $(A, B \ldots B^{\backprime}, A^{\backprime} \between x, y)^{\mu}$ is a covariant is satisfied; hence

THEOREM. Given a quantic $(a, b, \ldots b^{\backprime}, a^{\backprime} \between x, y)^m$; if A be a function of the coefficients of the degree θ and of the weight $\frac{1}{2}(m\theta - \mu)$ satisfying the condition $XA = 0$, and if $B, C, \ldots B^{\backprime}, A^{\backprime}$ are determined by the equations

$$B = YA,\quad C = \tfrac{1}{2}YB, \ldots\ A^{\backprime} = \frac{1}{\mu}\,YB^{\backprime},$$

then will

$$(A, B, \ldots B^{\backprime}, A^{\backprime} \between x, y)^{\mu}$$

be a covariant.

In particular, a function A of the degree θ and of the weight $\frac{1}{2}m\theta$, satisfying the condition $XA = 0$, will (also satisfy the equation $YA = 0$ and will) be an invariant.

32. I take now for A the most general function of the coefficients, of the degree θ and of the weight $\frac{1}{2}(m\theta - \mu)$; then XA is a function of the degree θ and of the weight $\frac{1}{2}(m\theta - \mu) - 1$, and the arbitrary coefficients in the function A are to be determined so that $XA = 0$. The number of arbitrary coefficients is equal to the number of terms in A, and the number of the equations to be satisfied is equal to the number of terms in XA; hence the number of the arbitrary coefficients which remains indeterminate is equal to the number of terms in A less the number of terms in XA; and since the covariant is completely determined when the leading coefficient is known,

the difference in question is equal to the number of the asyzygetic covariants, i.e. the number of the asyzygetic covariants of the order μ and the degree θ is equal to the number of terms of the degree θ and weight $\frac{1}{2}(m\theta-\mu)$, less the number of terms of the degree θ and weight $\frac{1}{2}(m\theta-\mu)-1$.

33. I shall now give some instances of the calculation of covariants by the method just explained. It is very convenient for this purpose to commence by forming the literal parts by Arbogast's Method of Derivations: we thus form tables such as the following:—

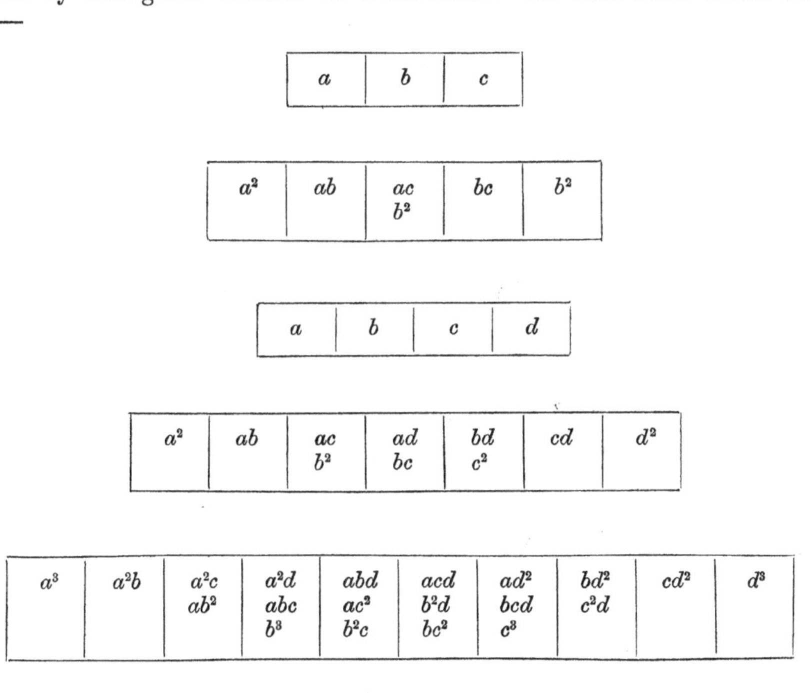

a	b	c

a^2	ab	ac b^2	bc	b^2

a	b	c	d

a^2	ab	ac b^2	ad bc	bd c^2	cd	d^2

a^3	a^2b	a^2c ab^2	a^2d abc b^3	abd ac^2 b^2c	acd b^2d bc^2	ad^2 bcd c^3	bd^2 c^2d	cd^2	d^3

a^4	a^3b	a^3c a^2b^2	a^3d a^2bc ab^3	a^2bd a^2c^2 ab^2c b^4	a^2cd ab^2d abc^2 b^3c	a^2d^2 $abcd$ ac^3 b^3d b^2c^2	abd^2 ac^2d b^2cd bc^3	acd^2 b^2d^2 bc^3d c^4	ad^3 bcd^2 c^3d	bd^3 c^2d^2	cd^3	d^4

a	b	c	d	e

a^2	ab	ac b^2	ad bc	ae bd c^2	be cd	bd c^2	cd	d^2

a^3	a^2b	a^2c	a^2d	a^2e	abe	ace	ade	ae^2	be^2	ce^2	de^2	e^3
		ab^2	abc	abd	acd	ad^2	bce	bde	cde	d^2e		
			b^3	ac^2	b^2d	b^2e	bd^2	c^2e	d^3			
				b^2c	bc^2	bcd	c^2d	cd^2				
						c^3						

34. Thus in the case of a cubic $(a, b, c, d \between x, y)^3$, the tables show that there will be a single invariant of the degree 4. Represent this by

$$\begin{aligned} &A a^2d^2 \\ &+ Babcd \\ &+ Cac^3 \\ &+ Db^3d \\ &+ Eb^2c^2, \end{aligned}$$

which is to be operated upon with $a\partial_b + 2b\partial_c + 3c\partial_d$. This gives

$+ B$		$+ 6A$	a^2cd
$+ 3D$	$+ 2B$		ab^2d
$+ 2E$	$+ 6C$	$+ 3B$	abc^2
	$+ 4E$	$+ 3D$	b^3c

i.e. $B + 6A = 0$, $3D + 2B = 0$, &c.; or putting $A = 1$, we find $B = -6$, $C = 4$, $D = 4$, $E = -3$, and the invariant is

$$\begin{aligned} &a^2d^2 \\ &- 6\,abcd \\ &+ 4\,ac^3 \\ &+ 4\,b^3d \\ &- 3\,b^2c^2. \end{aligned}$$

Again, there is a covariant of the order 3 and the degree 3. The coefficient of x^3 or leading coefficient is

$$\begin{aligned} &A\,a^2d \\ &+ B\,abc \\ &+ C\,b^3, \end{aligned}$$

which operated upon with $a\partial_b + 2b\partial_c + 3c\partial_d$, gives

$+ B$		$+ 3A$	a^2c
$+ 3C$	$+ 2B$		ab^2

i.e. $B + 3A = 0$, $3C + 2B = 0$; or putting $A = 1$, we have $B = -3$, $C = 2$, and the leading coefficient is

$$\begin{aligned} &a^2d \\ -\ &3\,abc \\ +\ &2\,b^3. \end{aligned}$$

The coefficient of x^2y is found by operating upon this with $(3b\partial_a + 2c\partial_b + d\partial_c)$, this gives

$+6$		-3	abd
	-6		ac^2
-9	$+12$		b^2c

i.e. the required coefficient of x^2y is

$$\begin{aligned} &3\,abd \\ -\ &6\,ac^2 \\ +\ &3\,b^2c\,; \end{aligned}$$

and by operating upon this with $\frac{1}{2}(3b\partial_a + 2c\partial_b + d\partial_c)$, we have for the coefficient of xy^2,

	$+3$	-6	acd
$+\frac{9}{2}$		$+\frac{3}{2}$	b^2d
-9	$+6$		bc^2

i.e. the coefficient of xy^2 is

$$\begin{aligned} -\ &3\,acd \\ +\ &6\,b^2d \\ -\ &3\,bc^2. \end{aligned}$$

Finally, operating upon this with $\frac{1}{3}(3b\partial_a + 2c\partial_b + d\partial_c)$, we have for the coefficient of y^3,

		-1	ad^2
-3	$+8$	-2	bcd
	-2		c^3

i.e. the coefficient of y^3 is

$$\begin{aligned} -\ &ad^2 \\ +\ &3\,bcd \\ -\ &2\,c^3, \end{aligned}$$

and the covariant is

(a^2d $+1$	abd $+3$	acd -3	ad^2 -1	
abc -3	ac^2 -6	b^2d $+6$	bcd $+3$	$\between x, y)^3$
b^3 $+2$	b^2c $+3$	bc^2 -3	c^3 -2	

[I now write the numerical coefficients after instead of before the literal terms.]

I have worked out the example in detail as a specimen of the most convenient method for the actual calculation of more complicated covariants[1].

35. The number of terms of the degree θ and of the weight q is obviously equal to the number of ways in which q can be made up as a sum of θ terms with the elements $(0, 1, 2, \ldots m)$, a number which is equal to the coefficient of $x^q z^\theta$ in the development of

$$\frac{1}{(1-z)(1-xz)(1-x^2z)\ldots(1-x^mz)};$$

and the number of the asyzygetic covariants of any particular degree for the quantic $(*\between x, y)^m$ can therefore be determined by means of this development. In the case of a cubic, for example, the function to be developed is

$$\frac{1}{(1-z)(1-xz)(1-x^2z)(1-x^3z)},$$

which is equal to

$$1 + z(1 + x + x^2 + x^3) + z^2(1 + x + 2x^2 + 2x^3 + 2x^4 + 2x^5 + x^6) + \&\text{c.},$$

where the coefficients are given by the following table; on account of the symmetry, the series of coefficients for each power of z is continued only to the middle term or middle of the series.

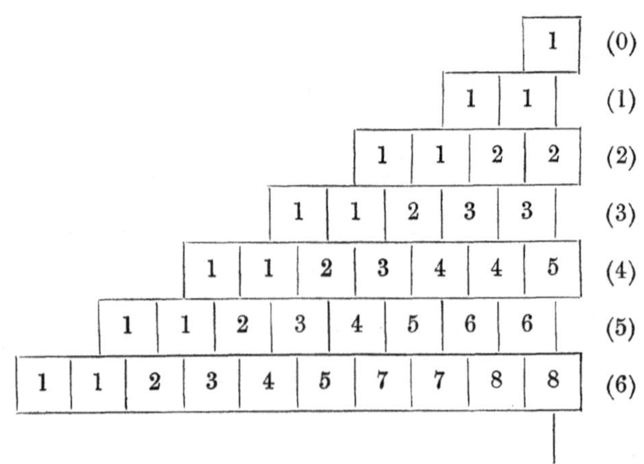

									1	(0)
								1	1	(1)
						1	1	2	2	(2)
					1	1	2	3	3	(3)
			1	1	2	3	4	4	5	(4)
		1	1	2	3	4	5	6	6	(5)
1	1	2	3	4	5	7	7	8	8	(6)

[1] Note added Feb. 7, 1856.—The following method for the calculation of an invariant or of the leading coefficient of a covariant, is easily seen to be identical in principle with that given in the text. Write down all the terms of the weight next inferior to that of the invariant or leading coefficient, and operate on each of these separately with the symbol

$$\text{ind.}\, b \,.\, \frac{b}{a} + 2\ \text{ind.}\, c \,.\, \frac{c}{b} + \ldots\ (m-1)\ \text{ind.}\, b' \,.\, \frac{b'}{a'},$$

where we are first to multiply by the fraction, rejecting negative powers, and then by the index of the proper letter in the term so obtained. Equating the results to zero, we obtain equations between the terms of the invariant or leading coefficient, and replacing in these equations each term by its numerical coefficient in the

and from this, by subtracting from each coefficient the coefficient which immediately precedes it, we form the table:

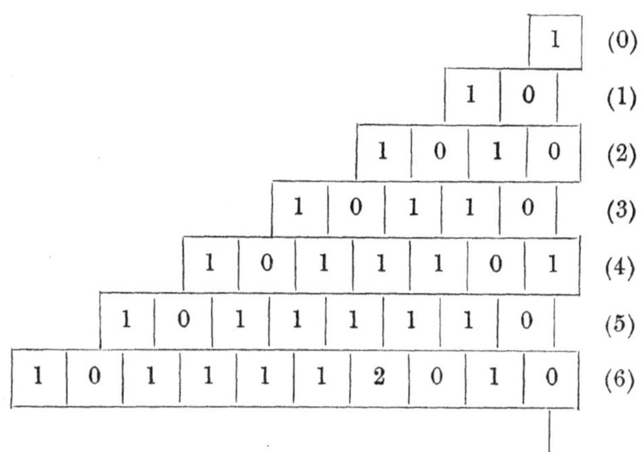

									1	(0)
								1	0	(1)
						1	0	1	0	(2)
				1	0	1	1	0		(3)
			1	0	1	1	1	0	1	(4)
		1	0	1	1	1	1	1	0	(5)
1	0	1	1	1	1	2	0	1	0	(6)

The successive lines fix the number and character of the covariants of the degrees 0, 1, 2, 3, &c. The line (0), if this were to be interpreted, would show that there is a single covariant of the degree 0; this covariant is of course merely the absolute constant unity, and may be excluded. The line (1) shows that there is a single covariant of the degree 1, viz. a covariant of the order 3; this is the cubic itself, which I represent by U. The line (2) shows that there are two asyzygetic covariants of the degree 2, viz. one of the order 6, this is merely U^2, and one of the order 2, this I represent by H. The line (3) shows that there are three asyzygetic covariants of the degree 3, viz. one of the order 9, this is U^3; one of the order 5, this is UH, and one of the order 3, this I represent by Φ. The line (4) shows that there are five asyzygetic covariants of the degree 4, viz. one of the order 12, this is U^4; one of the order 8, this is U^2H; one of the order 6, this is H^2; and one of the order 0, i.e. an invariant, this I represent by ∇. The line (5) shows that there are six asyzygetic covariants of the degree 5, viz. one of the order 15, this is U^5; one of the order 11, this is U^3H; one of the order 9, this is $U^2\Phi$; one of the order 7, this is UH^2; one of the order 5, this is $H\Phi$; and one of the order 3, this is ∇U. The line (6) shows that there are 8 asyzygetic covariants of the degree 6, viz. one of the order 18, this is U^6; one of the

invariant or leading coefficient, we have the equations of connexion of these numerical coefficients. Thus, for the discriminant of a cubic, the terms of the next inferior weight are a^2cd, ab^2d, abc^2, b^3c, and operating on each of these separately with the symbol

$$\text{ind. } b \,.\, \frac{b}{a} + 2 \text{ ind. } c \,.\, \frac{c}{b} + 3 \text{ ind. } d \,.\, \frac{d}{c},$$

we find

$abcd$		$+6\,a^2d^2$
$3\,b^3d$	$+2\,abcd$	
$2\,b^2c^2$	$+6\,ac^3$	$+3\,abcd$
	$+4\,b^2c^2$	$+3\,b^3d$

and equating the horizontal lines to zero, and assuming $a^2d^2=1$, we have $a^2d^2=1$, $abcd=-6$, $ac^3=4$, $b^3d=4$, $b^2c^2=-3$, or the value of the discriminant is that given in the text.

order 14, this is U^4H; one of the order 12, this is $U^3\Phi$; one of the order 10, this is U^2H^2; one of the order 8, this is $UH\Phi$; two of the order 6 (i.e. the three covariants H^3, Φ^2 and ∇U^2 are not asyzygetic, but are connected by a single linear equation or syzygy), and one of the order 2, this is ∇H. We are thus led to the irreducible covariants U, H, Φ, ∇ connected by a linear equation or syzygy between H^3, Φ^2 and ∇U^2, and this is in fact the complete system of irreducible covariants; ∇ is therefore the only invariant.

36. The asyzygetic covariants are of the form $U^pH^q\nabla^r$, or else of the form $U^pH^q\nabla^r\Phi$; and since U, H, ∇ are of the degrees 1, 2, 4 respectively, and Φ is of the degree 3, the number of asyzygetic covariants of the degree m of the first form is equal to the coefficient of x^m in $1 \div (1-x)(1-x^2)(1-x^4)$, and the number of the asyzygetic covariants of the degree m of the second form is equal to the coefficient of x^m in $x^3 \div (1-x)(1-x^2)(1-x^4)$. Hence the total number of asyzygetic covariants is equal to the coefficient of x^m in $(1+x^3) \div (1-x)(1-x^2)(1-x^4)$, or what is the same thing, in

$$\frac{1-x^6}{(1-x)(1-x^2)(1-x^3)(1-x^4)};$$

and conversely, if this expression for the number of the asyzygetic covariants of the degree m were established independently, it would follow that the irreducible invariants were four in number, and of the degrees 1, 2, 3, 4 respectively, but connected by an equation of the degree 6. As regards the invariants, every invariant is of the form ∇^p, i.e. the number of asyzygetic invariants of the degree m is equal to the coefficient of x^m in $\dfrac{1}{1-x^4}$, and conversely, from this expression it would follow that there was a single irreducible invariant of the degree 4.

37. In the case of a quartic, the function to be developed is:

$$\frac{1}{(1-z)(1-xz)(1-x^2z)(1-x^3z)(1-x^4z)},$$

and the coefficients are given by the table.

												1	(0)
										1	1	1	(1)
								1	1	2	2	3	(2)
						1	1	2	3	4	4	5	(3)
				1	1	2	3	5	5	7	7	8	(4)
		1	1	2	3	5	6	8	9	11	11	12	(5)
1	1	2	3	5	6	9	10	13	14	16	16	18	(6)

and subtracting from each coefficient the coefficient immediately preceding it, we have the table:

												1	(0)
										1	0	0	(1)
								1	0	1	0	1	(2)
						1	0	1	1	1	0	1	(3)
				1	0	1	1	2	0	2	0	1	(4)
		1	0	1	1	2	1	2	1	2	0	1	(5)
1	0	1	1	2	1	3	1	3	1	2	0	2	(6)

the examination of which will show that we have for the quartic the following irreducible covariants, viz. the quartic itself U; an invariant of the degree 2, which I represent by I; a covariant of the order 4 and of the degree 2, which I represent by H; an invariant of the degree 3, which I represent by J; and a covariant of the order 6 and the degree 3, which I represent by Φ; but that the irreducible covariants are connected by an equation of the degree 6, viz. there is a linear equation or syzygy between Φ^2, I^3H^3, I^2JH^2U, IJ^2HU^2 and J^3U^3; this is in fact the complete system of the irreducible covariants of the quartic: the only irreducible invariants are the invariants I, J.

38. The asyzygetic covariants are of the form $U^pI^qH^rJ^s$, or else of the form $U^pI^qH^rJ^s\Phi$, and the number of the asyzygetic covariants of the degree m is equal to the coefficient of x^m in $(1+x^3)\div(1-x)(1-x^3)^2(1-x^3)$, or what is the same thing, in

$$\frac{1-x^6}{(1-x)(1-x^2)^2(1-x^3)^2},$$

and the asyzygetic invariants are of the form I^pJ^q, and the number of the asyzygetic invariants of the degree m is equal to the coefficient of x^m in $1\div(1-x^2)(1-x^3)$. Conversely, if these formulæ were established, the preceding results as to the form of the system of the irreducible covariants or of the irreducible invariants, would follow.

39. In the case of a quintic, the function to be developed is

$$\frac{1}{(1-z)(1-xz)(1-x^2z)(1-x^3z)(1-x^4z)(1-x^5z)};$$

and the coefficients are given by the table:

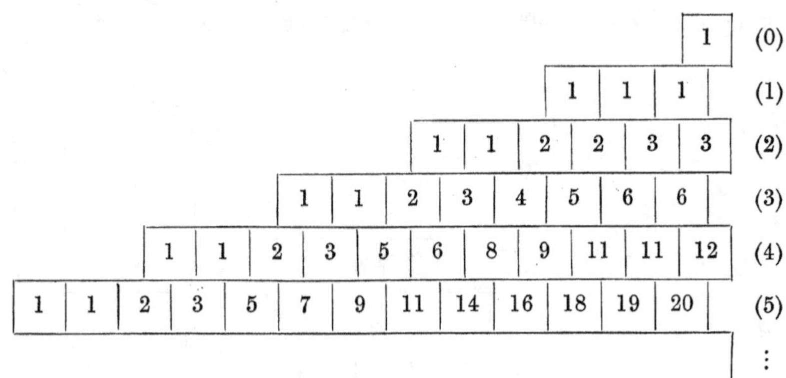

												1	(0)
										1	1	1	(1)
							1	1	2	2	3	3	(2)
					1	1	2	3	4	5	6	6	(3)
		1	1	2	3	5	6	8	9	11	11	12	(4)
1	1	2	3	5	7	9	11	14	16	18	19	20	(5)
													⋮

and subtracting from each coefficient the one which immediately precedes it, we have the table:

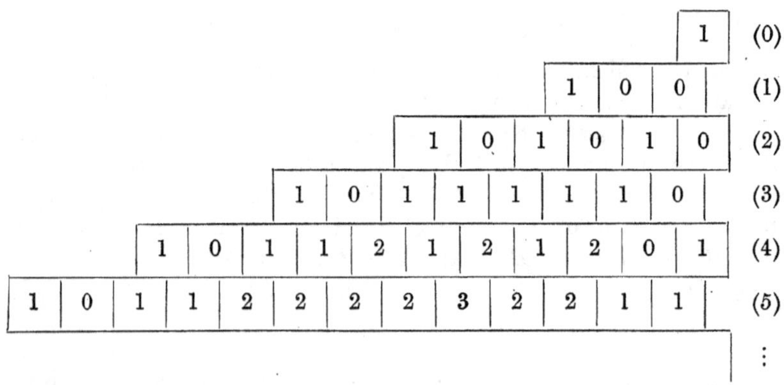

												1	(0)
										1	0	0	(1)
							1	0	1	0	1	0	(2)
					1	0	1	1	1	1	1	0	(3)
		1	0	1	1	2	1	2	1	2	0	1	(4)
1	0	1	1	2	2	2	2	3	2	2	1	1	(5)
													⋮

We thus obtain the following irreducible covariants, viz.:

Of the degree 1; a single covariant of the order 5, this is the quintic itself.

Of the degree 2; two covariants, viz. one of the order 6, and one of the order 2.

Of the degree 3; three covariants, viz. one of the order 9, one of the order 5, and one of the order 3.

Of the degree 4; three covariants, viz. one of the order 6, one of the order 4, and one of the order 0 (an invariant).

Of the degree 5; three covariants, viz. one of the order 7, one of the order 3, and one of the order 1 (a linear covariant).

These covariants are connected by a single syzygy of the degree 5 and of the order 11; in fact, the table shows that there are only two asyzygetic covariants of this degree and order; but we may, with the above-mentioned irreducible covariants of the degrees, 1, 2, 3 and 4, form three covariants of the degree 5 and the order 11; there is therefore a syzygy of this degree and order.

40. I represent the number of ways in which q can be made up as a sum of m terms with the elements 0, 1, 2, ... m, each element being repeatable an indefinite number of times by the notation

$$P(0,\ 1,\ 2, \ldots m)^\theta q,$$

and I write for shortness

$$P'(0,\ 1,\ 2, \ldots m)^\theta q = P(0,\ 1,\ 2 \ldots m)^\theta q - P(0,\ 1,\ 2 \ldots m)^\theta (q-1).$$

Then for a quantic of the order m, the number of asyzygetic covariants of the degree θ and of the order μ is

$$P'(0,\ 1,\ 2 \ldots m)^\theta \tfrac{1}{2}(m\theta - \mu).$$

In particular, the number of asyzygetic invariants of the degree θ is

$$P'(0,\ 1,\ 2 \ldots m)^\theta \tfrac{1}{2}m\theta.$$

To find the total number of the asyzygetic covariants of the degree θ, suppose first that $m\theta$ is even; then, giving to μ the successive values 0, 2, 4, ... $m\theta$, the required number is

$$\begin{aligned}
&P(\tfrac{1}{2}m\theta) \qquad\quad - P(\tfrac{1}{2}m\theta - 1)\\
&+ P(\tfrac{1}{2}m\theta - 1) - P(\tfrac{1}{2}m\theta - 2)\\
&\quad\vdots\\
&+ P(2) \qquad\qquad - P(1)\\
&+ P(1)\\
&= P(\tfrac{1}{2}m\theta),
\end{aligned}$$

i. e. when $m\theta$ is even, the number of the asyzygetic covariants of the degree θ is

$$P(0,\ 1,\ 2 \ldots m)^\theta \tfrac{1}{2}m\theta;$$

and similarly, when $m\theta$ is odd, the number of the asyzygetic covariants of the degree θ is

$$P(0,\ 1,\ 2, \ldots m)^\theta \tfrac{1}{2}(m\theta - 1).$$

But the two formulæ may be united into a single formula; for when $m\theta$ is odd $\tfrac{1}{2}m\theta$ is a fraction, and therefore $P(\tfrac{1}{2}m\theta)$ vanishes, and so when $m\theta$ is even $\tfrac{1}{2}(m\theta - 1)$ is a fraction, and $P\tfrac{1}{2}(m\theta - 1)$ vanishes; we have thus the theorem, that for a quantic of the order m:

The number of the asyzygetic covariants of the degree θ is

$$P(0,\ 1,\ 2 \ldots m)^\theta \tfrac{1}{2}m\theta + P(0,\ 1,\ 2, \ldots m)^\theta \tfrac{1}{2}(m\theta - 1).$$

41. The functions $P(\tfrac{1}{2}m\theta)$, &c. may, by the method explained in my "Researches on the Partition of Numbers," [140], be determined as the coefficients of x^θ in certain functions of x; I have calculated the following particular cases:—

Putting, for shortness,

$$P'(0,\ 1,\ 2, \ldots m)^\theta \tfrac{1}{2}m\theta = \text{coefficient } x^\theta \text{ in } \phi m,$$

then $$\phi 2 = \frac{1}{1-x^2},$$

$$\phi 3 = \frac{1}{1-x^4},$$

$$\phi 4 = \frac{1}{(1-x^2)(1-x^3)},$$

$$\phi 5 = \frac{1-x^6+x^{12}}{(1-x^4)(1-x^6)(1-x^8)},$$

$$\phi 6 = \frac{(1-x)(1+x-x^3-x^4-x^5+x^7+x^8)}{(1-x^2)^2(1-x^3)(1-x^4)(1-x^5)},$$

$$\phi 7 = \frac{1-x^6+2x^8-x^{10}+5x^{12}+2x^{14}+6x^{16}+2x^{18}+5x^{20}-x^{22}+2x^{24}-x^{26}+x^{32}}{(1-x^4)(1-x^6)(1-x^8)(1-x^{10})(1-x^{12})},$$

$$\phi 8 = \frac{(1-x)(1+x-x^3-x^4+x^6+x^7+x^8+x^9+x^{10}-x^{13}+x^{15}+x^{16})}{(1-x^2)^2(1-x^3)^2(1-x^4)(1-x^5)(1-x^7)};$$

$$P(0,\ 1,\ 2, \dots m)^\theta\, \tfrac{1}{2}m\theta = \text{coefficient of } x^\theta \text{ in } \psi m,$$

then $$\psi 2 = \frac{1}{(1-x)(1-x^2)},$$

$$\psi 3 = \frac{1+x^4}{(1-x^2)^2(1-x^4)},$$

$$\psi 4 = \frac{1-x+x^2}{(1-x)^2(1-x^2)(1-x^3)},$$

$$\psi 5 = \frac{1+x^2+6x^4+9x^6+12x^8+9x^{10}+6x^{12}+x^{14}+x^{16}}{(1-x^2)^2(1-x^4)(1-x^6)(1-x^8)};$$

$$P(0,\ 1,\ 2, \dots m)^\theta\, \tfrac{1}{2}(m\theta-1) = \text{coefficient of } x^\theta \text{ in } \psi_{,}m,$$

then $$\psi_{,}3 = \frac{x+x^3}{(1-x^2)^2(1-x^4)},$$

$$\psi_{,}5 = \frac{x+4x^3+8x^5+10x^7-10x^9+8x^{11}+4x^{13}+x^{15}}{(1-x^2)^2(1-x^4)(1-x^6)(1-x^8)}.$$

And from what has preceded, it appears that for a quantic of the order m, the number of asyzygetic covariants of the degree θ is for m even, coefficient x^θ in ψm, and for m odd, coefficient x^θ in $(\psi m + \psi_{,}m)$; and that the number of asyzygetic invariants of the degree θ is coefficient x^θ in ϕm. Attending first to the invariants:

42. For a quadric, the number of asyzygetic invariants of the degree θ is

$$\text{coefficient } x^\theta \text{ in } \frac{1}{1-x^2},$$

which leads to the conclusion that there is a single irreducible invariant of the degree 2.

43. For a cubic, the number of asyzygetic invariants of the degree θ is

$$\text{coefficient } x^{\theta} \text{ in } \frac{1}{1-x^4},$$

i.e. there is a single irreducible invariant of the degree 4.

44. For a quartic, the number of asyzygetic invariants of the degree θ is

$$\text{coefficient } x^{\theta} \text{ in } \frac{1}{(1-x^2)(1-x^3)},$$

i.e. there are two irreducible invariants of the degrees 2 and 3 respectively.

45. For a quintic, the number of asyzygetic invariants of the degree θ is

$$\text{coefficient } x^{\theta} \text{ in } \frac{1-x^6+x^{12}}{(1-x^4)(1-x^6)(1-x^8)}.$$

The numerator is the irreducible factor of $1-x^{36}$, i.e. it is equal to $(1-x^{36})(1-x^6) \div (1-x^{18})(1-x^{12})$; and substituting this value, the number becomes

$$\text{coefficient } x^{\theta} \text{ in } \frac{1-x^{36}}{(1-x^4)(1-x^8)(1-x^{12})(1-x^{18})},$$

i.e. there are in all four irreducible invariants, which are of the degrees 4, 8, 12 and 18 respectively; but these are connected by an equation of the degree 36, i.e. the square of the invariant of the degree 18 is a rational and integral function of the other three invariants; a result, the discovery of which is due to M. Hermite.

46. For a sextic, the number of asyzygetic invariants of the degree θ is

$$\text{coefficient } x^{\theta} \text{ in } \frac{(1-x)(1+x-x^3-x^4-x^5+x^7+x^8)}{(1-x^2)^2(1-x^3)(1-x^4)(1-x^5)}:$$

the second factor of the numerator is the irreducible factor $1-x^{30}$, i.e. it is equal to $(1-x^{30})(1-x^5)(1-x^3)(1-x^2) \div (1-x^{15})(1-x^{10})(1-x^6)(1-x)$; and substituting this value, the number becomes

$$\text{coefficient } x^{30} \text{ in } \frac{1-x^{30}}{(1-x^2)(1-x^4)(1-x^6)(1-x^{10})(1-x^{15})},$$

i.e. there are in all five irreducible invariants, which are of the degrees 2, 4, 6, 10 and 15 respectively; but these are connected by an equation of the degree 30, i.e. the square of the invariant of the degree 15 is a rational and integral function of the other four invariants.

47. For a septimic, the number of asyzygetic invariants of the degree θ is

$$\text{coefficient } x^{\theta} \text{ in } \frac{1-x^6+2x^8-x^{10}+5x^{12}+2x^{14}+6x^{16}+2x^{18}+5x^{20}-x^{22}+2x^{24}-x^{26}+x^{32}}{(1-x^4)(1-x^6)(1-x^8)(1-x^{10})(1-x^{12})},$$

the numerator is equal to

$$(1-x^6)(1-x^8)^{-2}(1-x^{10})(1-x^{12})^{-5}(1-x^{14})^{-4}\ldots,$$

where the series of factors does not terminate; hence [*incorrect*, see p. 253] the number of irreducible invariants is infinite; substituting the preceding value, the number of asyzygetic invariants of the degree θ is

$$\text{coefficient } x^\theta \text{ in } (1-x^4)^{-1}(1-x^8)^{-3}(1-x^{12})^{-6}(1-x^{14})^{-4}\ldots$$

The first four indices give the number of irreducible invariants of the corresponding degrees, i.e. there are 1, 3, 6 and 4 irreducible invariants of the degrees 4, 8, 12 and 14 respectively, but there is no reason to believe that the same thing holds with respect to the indices of the subsequent terms. To verify this it is to be remarked, that there are 1, 4, 10 and 4 asyzygetic invariants of the degrees in question respectively; there is therefore one irreducible invariant of the degree 4; calling this X_4, there is only one composite invariant of the degree 8, viz. X_4^2; there are therefore three irreducible invariants of this degree, say X_8, X_8', X_8''. The composite invariants of the degree 12 are four in number, viz. X_4^3, X_4X_8, X_4X_8', X_4X_8'', *and these cannot be connected by any syzygy*, for if they were so, X_4^2, X_8, X_8', X_8'' would be connected by a syzygy, or there would be less than 3 irreducible invariants of the degree 8. Hence there are precisely 6 irreducible invariants of the degree 12. And since the irreducible invariants of the degrees 4, 8 and 12 do not give rise to any composite invariant of the degree 14, there are precisely 4 irreducible invariants of the degree 14.

48. For an octavic, the number of the asyzygetic invariants of the degree θ is

$$\text{coefficient } x^\theta \text{ in } \frac{(1-x)(1+x-x^3-x^4+x^6+x^7+x^8+x^9+x^{10}-x^{12}-x^{13}+x^{15}+x^{16})}{(1-x^2)^2(1-x^3)^2(1-x^4)(1-x^5)(1-x^7)};$$

and the second factor of the numerator is

$$(1-x)^{-1}(1-x^2)(1-x^3)^{-1}(1-x^6)^{-1}(1-x^8)^{-1}(1-x^9)^{-1}(1-x^{10})^{-1}(1-x^{16})(1-x^{17})(1-x^{18})\ldots,$$

where the series of factors does not terminate, hence [*incorrect*] the number of irreducible invariants is infinite. Substituting the preceding value, the number of the asyzygetic invariants of the degree θ is

$$\text{coeff. } x^\theta \text{ in } (1-x^2)^{-1}(1-x^3)^{-1}(1-x^4)^{-1}(1-x^5)^{-1}(1-x^6)^{-1}(1-x^7)^{-1}(1-x^8)^{-1}(1-x^9)^{-1}(1-x^{10})^{-1}(1-x^{16})(1-x^{17})(1-x^{18})\ldots$$

There is certainly one, and only one irreducible invariant for each of the degrees 2, 3, 4, 5 and 6 respectively; but the formula does not show the number of the irreducible invariants of the degrees 7, &c.; in fact, representing the irreducible invariants of the degrees 2, 3, 4, 5 and 6 by X_2, X_3, X_4, X_5, X_6, these give rise to 3 composite invariants of the degree 7, viz. $X_2X_2X_3$, X_2X_5, X_3X_4, which may or may not be connected by a syzygy; if they are not connected by a syzygy, there will be a single irreducible invariant of the degree 7; but if they are connected by a syzygy, there will be two irreducible invariants of the degree 7; it is useless at present to pursue the discussion further.

Considering next the covariants,—

49. For a quadric, the number of asyzygetic covariants of the degree θ is

$$\text{coefficient } x^\theta \text{ in } \frac{1}{(1-x)(1-x^2)},$$

i.e. there are two irreducible covariants of the degrees 1 and 2 respectively; these are of course the quadric itself and the invariant.

50. For a cubic, the number of the asyzygetic covariants of the degree θ is

$$\text{coefficient } x^\theta \text{ in } \frac{(1+x)(1+x^2)}{(1-x^2)^2(1-x^4)}.$$

The first factor of the numerator is the irreducible factor of

$$1-x^2, = (1-x^2) \div (1-x),$$

and the second factor of the numerator is the irreducible factor of

$$1-x^4, = (1-x^4) \div (1-x^2);$$

substituting these values, the number is

$$\text{coefficient } x^\theta \text{ in } \frac{1-x^6}{(1-x)(1-x^2)(1-x^3)(1-x^4)},$$

i.e. there are 4 irreducible covariants of the degrees 1, 2, 3, 4 respectively; but these are connected by an equation of the degree 6; the covariant of the degree 1 is the cubic itself U, the other covariants are the covariants already spoken of and represented by the letters H, Φ and ∇ respectively (H is of the degree 2 and the order 3, Φ of the degree 3 and the order 3, and ∇ is of the degree 4 and the order 0, i.e. it is an invariant).

51. For a quartic, the number of the asyzygetic covariants of the degree θ is

$$\text{coefficient } x^\theta \text{ in } \frac{1-x+x^2}{(1-x)^2(1-x^2)(1-x^3)},$$

the numerator of which is the irreducible factor of $1-x^6$, i.e. it is equal to $(1-x^6)(1-x) \div (1-x^2)(1-x^3)$. Making this substitution, the number is

$$\text{coefficient } x^\theta \text{ in } \frac{1-x^6}{(1-x)(1-x^2)^2(1-x^3)^2},$$

i.e. there are five irreducible covariants, one of the degree 1, two of the degree 2, and two of the degree 3, but these are connected by an equation of the degree 6. The irreducible covariant of the degree 1 is of course the quartic itself U, the other irreducible covariants are those already spoken of and represented by I, H, J, Φ respectively (I is of the degree 2 and the order 0, and J is of the degree 3 and the order 0, i.e. I and J are invariants, H is of the degree 2 and the order 4, Φ is of the degree 3 and the order 6).

52. For a quintic, the number of irreducible covariants of the degree θ is

$$\text{coeff. } x^\theta \text{ in } \frac{1+x+x^2+4x^3+6x^4+8x^5+9x^6+10x^7+12x^8+10x^9+9x^{10}+8x^{11}+6x^{12}+4x^{13}+x^{14}+x^{15}+x^{16}}{(1-x^2)^2(1-x^4)(1-x^6)(1-x^8)},$$

the numerator of which is

$$(1+x)^2(1-x+2x^2+x^3+2x^4+3x^5+x^6+5x^7+x^8+3x^9+2x^{10}+x^{11}+2x^{12}-x^{13}+x^{14});$$

the first factor is $(1-x)^{-2}(1-x^2)^2$, the second factor is

$$(1-x)(1-x^2)^{-2}(1-x^3)^{-3}(1-x^4)^{-2}(1-x^5)^{-2}(1-x^6)^5(1-x^7)^5(1-x^8)^7(1-x^9)^1(1-x^{10})^{-9}(1-x^{11})^{-19}\ldots,$$

which does not terminate; hence [*incorrect*] the number of irreducible covariants is infinite. Substituting the preceding values, the expression for the number of the asyzygetic covariants of the degree θ is

$$\text{coeff. } x^\theta \text{ in } (1-x)^{-1}(1-x^2)^{-2}(1-x^3)^{-3}(1-x^4)^{-3}(1-x^5)^{-2}(1-x^6)^4(1-x^7)^5(1-x^8)^6(1-x^9)^1(1-x^{10})^{-9}(1-x^{11})^{-19}\ldots,$$

which agrees with a previous result: the numbers of irreducible covariants for the degrees 1, 2, 3, 4 are 1, 2, 3 and 3 respectively, and for the degree 5, the number of irreducible covariants is three, but there is one syzygy between the composite covariants of the degree in question; the difference $3-1=2$ is the index taken with its sign reversed of the factor $(1-x^5)^{-2}$.

53. I consider a system of the asyzygetic covariants of any particular degree and order of a given quantic, the system may of course be replaced by a system the terms of which are any linear functions of those of the original system, and it is necessary to inquire what covariants ought to be selected as most proper to represent the system of asyzygetic covariants; the following considerations seem to me to furnish a convenient rule of selection. Let the literal parts of the terms which enter into the coefficients of the highest power of x or leading coefficients be represented by M_α, M_β, M_γ,... these quantities being arranged in the natural or alphabetical order; the first in order of these quantities M, which enters into the leading coefficient of a particular covariant, may for shortness be called the leading term of such covariant, and a covariant the leading term of which is posterior in order to the leading term of another covariant, may be said to have a lower leading term.

It is clear, that by properly determining the multipliers of the linear functions we may form a covariant the leading term of which is lower than the leading term of any other covariant (the definition implies that there is but one such covariant); call this Θ. We may in like manner form a covariant such that its leading term is lower than the leading term of every other covariant except Θ_1; or rather we may form a system of such covariants, since if Φ_2 be a covariant having the property in question, $\Phi_2+k\Theta_1$ will have the same property, but k may be determined so that the covariant shall not contain the leading term of Θ_1, i.e. we may form a covariant Θ_2 such that its leading term is lower than the leading term of every other covariant excepting Θ_1, and that the leading term of Θ_1 does not enter into Θ_2; and there is but one such covariant, Θ_2. Again, we may form a covariant Θ_3 such that its leading term is lower than the leading term of every other covariant excepting Θ_1 and Θ_2, and that the leading terms of Θ_1 and Θ_2 do not either of them enter into Θ_3; and there is but one such covariant, Θ_3. And so on, until we arrive at a covariant the leading term of which is higher than the leading terms of the other covariants, and which does not contain the leading terms of the other covariants. We have thus a series of covariants Θ_1, Θ_2, Θ_3, &c. containing the proper number of terms, and which covariants may be taken to represent the asyzygetic covariants of the degree and order in question.

In order to render the covariants Θ definite as well numerically as in regard to sign, we may suppose that the covariant is in its least terms (i.e. we may reject numerical factors common to all the terms), and we may make the leading term positive. The leading term with the proper numerical coefficient (if different from unity) and with the proper power of x (or the order of the function) annexed, will, when the covariants of a quantic are tabulated, be sufficient to indicate, without any ambiguity whatever, the particular covariant referred to. I subjoin a table of the covariants of a quadric, a cubic and a quartic, and of the covariants of the degrees 1, 2, 3, 4 and 5 respectively of a quintic, and also two other invariants of a quintic.

[Except for the quantic itself, the algebraical sum of the numerical coefficients in any column is $=0$, viz. the sum of the coefficients with the sign $+$ is equal to that of the coefficients with the sign $-$, and I have as a numerical verification inserted at the foot of each column this sum with the sign $\pm$].

Covariant Tables (Nos. 1 to 26).

No. 1.

($a+1$	$b+2$	$c+1$	$\between x, y)^2$.

No. 2.

$ac+1$
b^2-1
± 1

The tables Nos. 1 and 2 are the covariants of a binary quadric. No. 1 is the quadric itself; No. 2 is the quadrinvariant, which is also the discriminant.

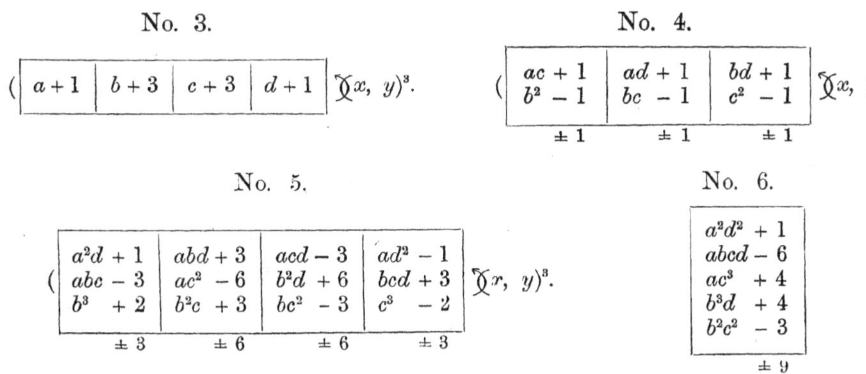

No. 3.

($a+1$	$b+3$	$c+3$	$d+1$	$\between x, y)^3$.

No. 4.

(			
$ac+1$	$ad+1$	$bd+1$	$\between x, y)^2$
b^2-1	$bc-1$	c^2-1	
± 1	± 1	± 1	

No. 5.

(				
a^2d+1	$abd+3$	$acd-3$	ad^2-1	$\between x, y)^3$.
$abc-3$	ac^2-6	b^2d+6	$bcd+3$	
b^3+2	b^2c+3	bc^2-3	c^3-2	
± 3	± 6	± 6	± 3	

No. 6.

a^2d^2+1
$abcd-6$
ac^3+4
b^3d+4
b^2c^2-3
± 9

The tables Nos. 3, 4, 5 and 6 are the covariants of a binary cubic. No. 3 is the cubic itself; No. 4 is the quadricovariant, or Hessian; No. 5 is the cubicovariant; No. 6 is the invariant, or discriminant. And if we write No. $3=U$, No. $4=H$, No. $5=\Phi$, No. $6=\nabla$, then identically, $$\Phi^2 - \nabla U^2 + 4H^3 = 0.$$

No. 7.

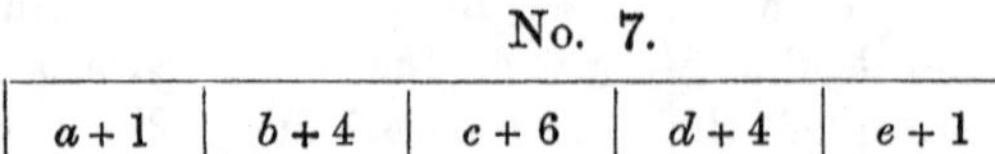

($a+1$	$b+4$	$c+6$	$d+4$	$e+1$	$)(x, y)^4$

No. 8.

$ae + 1$ $bd - 4$ $c^2 + 3$
± 4

No. 9.

($ac + 1$ $b^2 - 1$	$ad + 2$ $bc - 2$	$ae + 1$ $bd + 2$ $c^2 - 3$	$be + 2$ $cd - 2$	$bd + 1$ $c^2 - 1$	$)(x, y)^4$
± 1	± 2	± 3	± 2	± 1	

No. 10.

$ace + 1$ $ad^2 - 1$ $b^2e - 1$ $bcd + 2$ $c^3 - 1$
± 3

No. 11.

($a^2d + 1$ $abc - 3$ $b^3 + 2$	$a^2e + 1$ $abd + 2$ $ac^2 - 9$ $b^2c + 6$	$abe + 5$ $acd - 15$ $b^2d + 10$ bc^2 ...	ace ... $ad^2 - 10$ $b^2e + 10$ bcd ... c^3 ...	$ade - 5$ $bce + 15$ $bd^2 - 10$ c^2d ...	$ae^2 - 1$ $bde - 2$ $c^2e + 9$ $cd^2 - 6$	$be^2 - 1$ $cde + 3$ $d^3 - 2$	$)(x, y)^6$
± 3	± 9	± 15	± 10	± 15	± 9	± 3	

No. 12.

a^3e^3	$+\ 1$
a^2bde^2	$-\ 12$
$a^2c^2e^2$	$-\ 18$
a^2cd^2e	$+\ 54$
a^2d^4	$-\ 27$
ab^2ce^2	$+\ 54$
ab^2d^2e	$-\ 6$
abc^2de	$-\ 180$
$abcd^3$	$+\ 108$
ac^4e	$+\ 81$
ac^3d^2	$-\ 54$
b^4e^2	$-\ 27$
b^3cde	$+\ 108$
b^3d^3	$-\ 64$
b^2c^3e	$-\ 54$
$b^2c^2d^2$	$+\ 36$
bc^4d	...
c^6	...
	± 442

The tables Nos. 7, 8, 9, 10 and 11 are the irreducible covariants of a quartic. No. 7 is the quartic itself; No. 8 is the quadrinvariant; No. 9 is the quadricovariant, or Hessian; No. 10 is the cubinvariant; and No. 11 is the cubicovariant. The table No. 12 is the discriminant. And if we write No. $7 = U$, No. $8 = I$, No. $9 = H$, No. $10 = J$, No. $11 = \Phi$, No. $12 = \nabla$,

then the irreducible covariants are connected by the identical equation

$$JU^3 - IU^2H + 4H^3 + \Phi^2 = 0,$$

and we have

$$\nabla = I^3 - 27J^2.$$

[The Tables Nos. 13 to 24 which follow, and also Nos. 25 and 26 which are given in 143 relate to the binary quintic. I have inserted in the headings the capital letters A, B, ... L and also Q and Q′ by which I refer to these covariants of the quintic. A is the quintic itself, C is the Hessian, G is the quartinvariant, J a linear covariant: Q is the simplest octinvariant, and Q′ is the discriminant. As noticed in the original memoir we have $AI + BF - CE = 0$; and $Q' = G^2 - 128\,Q$, only the coefficient 128 was by mistake given as 1152.]

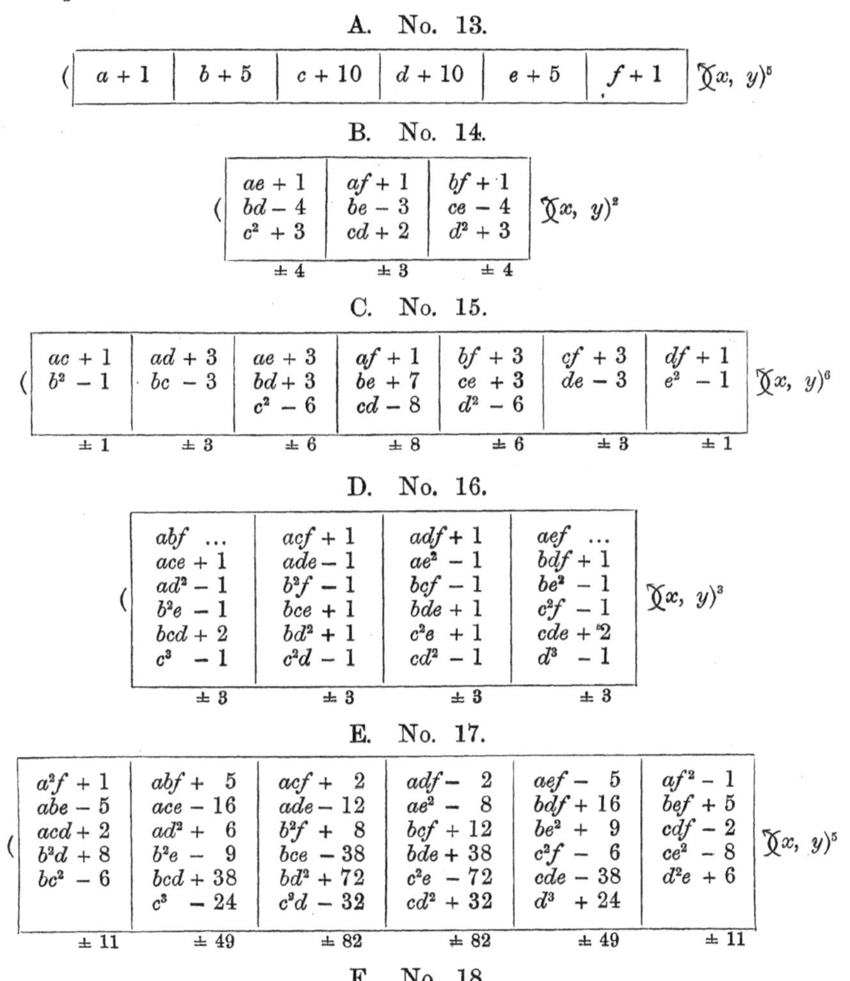

A. No. 13.

($a + 1$	$b + 5$	$c + 10$	$d + 10$	$e + 5$	$f + 1$	$)(x, y)^5$

B. No. 14.

($ae + 1$ $bd - 4$ $c^2 + 3$	$af + 1$ $be - 3$ $cd + 2$	$bf + 1$ $ce - 4$ $d^2 + 3$	$)(x, y)^2$
± 4	± 3	± 4	

C. No. 15.

($ac + 1$ $b^2 - 1$	$ad + 3$ $bc - 3$	$ae + 3$ $bd + 3$ $c^2 - 6$	$af + 1$ $be + 7$ $cd - 8$	$bf + 3$ $ce + 3$ $d^2 - 6$	$cf + 3$ $de - 3$	$df + 1$ $e^2 - 1$	$)(x, y)^6$
± 1	± 3	± 6	± 8	± 6	± 3	± 1	

D. No. 16.

(abf ... $ace + 1$ $ad^2 - 1$ $b^2e - 1$ $bcd + 2$ $c^3 - 1$	$acf + 1$ $ade - 1$ $b^2f - 1$ $bce + 1$ $bd^2 + 1$ $c^2d - 1$	$adf + 1$ $ae^2 - 1$ $bcf - 1$ $bde + 1$ $c^2e + 1$ $cd^2 - 1$	aef ... $bdf + 1$ $be^2 - 1$ $c^2f - 1$ $cde + 2$ $d^3 - 1$	$)(x, y)^3$
± 3	± 3	± 3	± 3	

E. No. 17.

($a^2f + 1$ $abe - 5$ $acd + 2$ $b^2d + 8$ $bc^2 - 6$	$abf + 5$ $ace - 16$ $ad^2 + 6$ $b^2e - 9$ $bcd + 38$ $c^3 - 24$	$acf + 2$ $ade - 12$ $b^2f + 8$ $bce - 38$ $bd^2 + 72$ $c^2d - 32$	$adf - 2$ $ae^2 - 8$ $bcf + 12$ $bde + 38$ $c^2e - 72$ $cd^2 + 32$	$aef - 5$ $bdf + 16$ $be^2 + 9$ $c^2f - 6$ $cde - 38$ $d^3 + 24$	$af^2 - 1$ $bef + 5$ $cdf - 2$ $ce^2 - 8$ $d^2e + 6$	$)(x, y)^5$
± 11	± 49	± 82	± 82	± 49	± 11	

F. No. 18.

($a^2d + 1$ $abc - 3$ $b^3 + 2$	$a^2e + 2$ $abd + 1$ $ac^2 - 12$ $b^2c + 9$	$a^2f + 1$ $abe + 11$ $acd - 34$ $b^2d + 16$ $bc^2 + 6$	$abf + 7$ $ace - 8$ $ad^2 - 34$ $b^2e + 29$ $bcd - 2$ $c^3 + 8$	$acf + 5$ $ade - 40$ $b^2f + 16$ $bce + 47$ $bd^2 - 44$ $c^2d + 16$	$adf - 5$ $ae^2 - 16$ $bcf + 40$ $bde - 47$ $c^2e + 44$ $cd^2 - 16$	$aef - 7$ $bdf + 8$ $be^2 - 29$ $c^2f + 34$ $cde + 2$ $d^3 - 8$	$af^2 - 1$ $bef - 11$ $cdf + 34$ $ce^2 - 16$ $d^2e - 6$	$bf^2 - 2$ $cef - 1$ $d^2f + 12$ $de^2 - 9$	$cf^2 - 1$ $def + 3$ $e^3 - 2$	$)(x, y)^9$
± 3	± 12	± 34	± 44	± 84	± 84	± 44	± 34	± 12	± 3	

G. No. 19.

$a^2f^2 + 1$
$abef - 10$
$acdf + 4$
$ace^2 + 16$
$ad^2e - 12$
$b^2df + 16$
$b^2e^2 + 9$
$bc^2f - 12$
$bcde - 76$
$bd^3 + 48$
$c^3e + 48$
$c^2d^2 - 32$
± 142

H. No. 20.

(| | | | |) $(x, y)^4$
|---|---|---|---|---|
| $a^2df + 1$ | $a^2ef + 2$ | $a^2f^2 + 1$ | $abf^2 + 2$ | $acf^2 + 1$ |
| a^2e^2 ... | $abdf - 4$ | $abef - 4$ | $acef - 4$ | $adef - 3$ |
| $abcf - 3$ | $abe^2 - 10$ | $acdf - 2$ | $ad^2f - 2$ | $ae^3 + 2$ |
| $abde - 5$ | $ac^2f - 2$ | $ace^2 + 4$ | $ade^2 + 4$ | b^2f^2 ... |
| $ac^2e + 10$ | $acde + 24$ | ad^2e ... | $b^2ef - 10$ | $bcef - 5$ |
| $acd^2 - 4$ | $ad^3 - 12$ | $b^2df + 4$ | $bcdf + 24$ | $bd^2f + 10$ |
| $b^3f + 2$ | $b^2cf + 4$ | $b^2e^2 - 9$ | $bce^2 + 16$ | $bde^2 - 5$ |
| $b^2ce - 5$ | $b^2de + 16$ | bc^2f ... | $bd^2e - 22$ | $c^2df - 4$ |
| $b^2d^2 + 14$ | $bc^2e - 22$ | $bcde + 50$ | $c^3f - 12$ | $c^2e^2 + 14$ |
| $bc^2d - 16$ | $bcd^2 - 4$ | $bd^3 - 36$ | $c^2de - 4$ | $cd^2e - 16$ |
| $c^4 + 6$ | $c^3d + 8$ | $c^3e - 36$ | $cd^3 + 8$ | $d^4 + 6$ |
| | | $c^2d^2 + 28$ | | |
| ± 33 | ± 54 | ± 87 | ± 54 | ± 33 |

I. No. 21.

(| | | | | | |) $(x, y)^6$.
|---|---|---|---|---|---|---|
| $a^2cf + 1$ | $a^2df + 2$ | a^2ef ... | a^2f^2 ... | abf^2 ... | $acf^2 - 2$ | $adf^2 - 1$ |
| $a^2de - 1$ | $a^2e^2 - 2$ | $abdf + 2$ | $abef$ ... | $acef - 2$ | $adef$ ... | $ae^2f + 1$ |
| $ab^2f - 1$ | $abcf - 10$ | $abe^2 - 2$ | $acdf$ ... | $ad^2f + 1$ | $ae^3 + 2$ | $bcf^2 + 1$ |
| $abce - 2$ | $abde + 10$ | $ac^2f - 1$ | $ace^2 - 20$ | $ade^2 + 1$ | $b^2f^2 + 2$ | $bdef + 2$ |
| $abd^2 + 4$ | ac^2e ... | $acde - 2$ | $ad^2e + 20$ | $b^2ef + 2$ | $bcef$ ... | $be^3 - 3$ |
| $ac^2d - 1$ | acd^2 ... | $ad^3 + 3$ | $b^2df + 20$ | $bcdf + 2$ | $bd^2f + 10$ | $c^2ef - 4$ |
| $b^3e + 3$ | $b^3f - 2$ | $b^2cf - 1$ | b^2e^2 ... | $bce^2 - 5$ | $bde^2 - 14$ | $cd^2f + 1$ |
| $b^2cd - 6$ | $b^2ce + 14$ | $b^2de + 5$ | $bc^2f - 20$ | $bd^2e - 1$ | $c^2df - 10$ | $cde^2 + 6$ |
| $bc^3 + 3$ | $b^2d^2 + 2$ | $bc^2e + 1$ | $bcde$ ... | $c^3f - 3$ | $c^2e^2 - 2$ | $d^3e - 3$ |
| | $bc^2d - 26$ | $bcd^2 - 9$ | $bd^3 - 20$ | $c^2de + 9$ | $cd^2e + 26$ | |
| | $c^4 + 12$ | $c^3d + 4$ | $c^3e + 20$ | $cd^3 - 4$ | $d^4 - 12$ | |
| | | | c^2d^2 ... | | | |
| ± 11 | ± 40 | ± 15 | ± 60 | ± 15 | ± 40 | ± 11 |

J. No. 22.

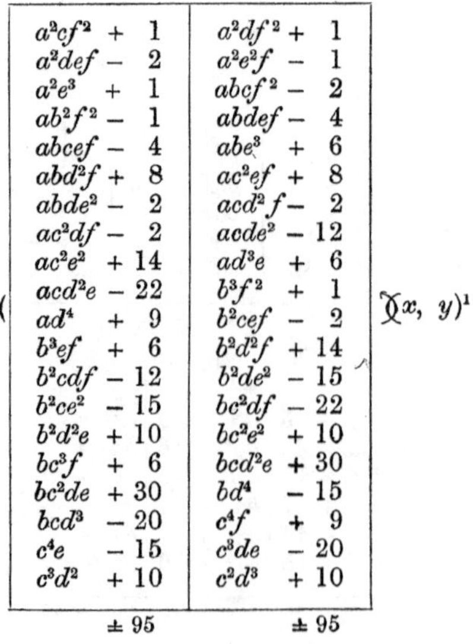

(	) $(x, y)^1$
$a^2cf^2 + 1$	$a^2df^2 + 1$
$a^2def - 2$	$a^2e^2f - 1$
$a^2e^3 + 1$	$abcf^2 - 2$
$ab^2f^2 - 1$	$abdef - 4$
$abcef - 4$	$abe^3 + 6$
$abd^2f + 8$	$ac^2ef + 8$
$abde^2 - 2$	$acd^2f - 2$
$ac^2df - 2$	$acde^2 - 12$
$ac^2e^2 + 14$	$ad^3e + 6$
$acd^2e - 22$	$b^3f^2 + 1$
$ad^4 + 9$	$b^2cef - 2$
$b^3ef + 6$	$b^2d^2f + 14$
$b^2cdf - 12$	$b^2de^2 - 15$
$b^2ce^2 - 15$	$bc^2df - 22$
$b^2d^2e + 10$	$bc^2e^2 + 10$
$bc^3f + 6$	$bcd^2e + 30$
$bc^2de + 30$	$bd^4 - 15$
$bcd^3 - 20$	$c^4f + 9$
$c^4e - 15$	$c^3de - 20$
$c^3d^2 + 10$	$c^2d^3 + 10$
± 95	± 95

K. No. 23.

a^2bf^2	...	a^2cf^2	$+\ 1$	a^2df^2	$-\ 1$	a^2ef^2	...
a^2cef	$+\ 1$	a^2def	$-\ 5$	a^2e^2f	$+\ 1$	$abdf^2$	$-\ 1$
a^2d^2f	$-\ 3$	a^2e^3	$+\ 4$	$abcf^2$	$+\ 5$	abe^2f	$+\ 1$
a^2de^2	$+\ 2$	ab^2f^2	$-\ 1$	$abdef$	$-\ 8$	ac^2f^2	$+\ 3$
ab^2ef	$-\ 1$	$abcef$	$+\ 8$	abe^3	$+\ 3$	$acdef$	$-\ 14$
$abcdf$	$+\ 14$	abd^2f	$+\ 11$	ac^2ef	$-\ 11$	ace^3	$+\ 8$
$abce^2$	$-\ 11$	$abde^2$	$-\ 17$	acd^2f	$+\ 11$	ad^3f	$+\ 9$
abd^2e	$-\ 1$	ac^2df	$-\ 11$	$acde^2$	$+\ 6$	ad^2e^2	$-\ 6$
ac^3f	$-\ 9$	ac^2e^2	$-\ 16$	ad^3e	$-\ 6$	b^2cf^2	$-\ 2$
ac^2de	$+\ 14$	acd^2e	$+\ 44$	b^3f^2	$-\ 4$	b^2def	$+\ 11$
acd^3	$-\ 6$	ad^4	$-\ 18$	b^2cef	$+\ 17$	b^2e^3	$-\ 9$
b^3df	$-\ 8$	b^3ef	$-\ 3$	b^2d^2f	$+\ 16$	bc^2ef	$+\ 1$
b^3e^2	$+\ 9$	b^2cdf	$-\ 6$	b^2de^2	$-\ 21$	bcd^2f	$-\ 14$
b^2c^2f	$+\ 6$	b^2ce^2	$+\ 21$	bc^2df	$-\ 44$	$bcde^2$	$+\ 16$
b^2cde	$-\ 16$	b^2d^2e	$-\ 5$	bc^2e^2	$+\ 5$	bd^3e	$-\ 3$
b^2d^3	$+\ 8$	bc^3f	$+\ 6$	bcd^2e	$+\ 39$	c^3df	$+\ 6$
bc^3e	$+\ 3$	bc^2de	$-\ 39$	bd^4	$-\ 12$	c^3e^2	$-\ 8$
bc^2d^2	$-\ 2$	bcd^3	$+\ 22$	c^4f	$+\ 18$	c^2d^2e	$+\ 2$
c^4d	...	c^4e	$+\ 12$	c^3de	$-\ 22$	cd^4	...
		c^3d^2	$-\ 8$	c^2d^3	$+\ 8$		
	$\pm\ 57$		$\pm\ 129$		$\pm\ 129$		$\pm\ 57$

$(\ \ \ )(x, y)^3$

L. No. 24.

a^3ef	...	a^3f^2	...	a^2bf^2	...	a^2cf^2	$-\ 1$	a^2df^2	$+\ 1$	a^2ef^2	...	a^2f^3	...	abf^3	...
a^2bdf	...	a^2bef	...	a^2cef	$-\ 3$	a^2def	$+\ 7$	a^2e^2f	$-\ 1$	$abdf^2$	$+\ 3$	$abef^2$	...	$acef^2$	...
a^2be^2	...	a^2cdf	$+\ 7$	a^2d^2f	$+\ 12$	a^2e^3	$-\ 6$	$abcf^2$	$-\ 7$	abe^2f	$-\ 3$	$acdf^2$	$-\ 7$	ad^2f^2	$-\ 2$
a^2c^2f	$+\ 2$	a^2ce^2	$-\ 10$	a^2de^2	$-\ 9$	ab^2f^2	$+\ 1$	$abdef$	$+\ 26$	ac^2f^2	$-\ 12$	ace^2f	$+\ 7$	ade^2f	$+\ 4$
a^2cde	$-\ 5$	a^2d^2e	$+\ 3$	ab^2ef	$+\ 3$	$abcef$	$-\ 26$	abe^3	$-\ 19$	$acdef$	$+\ 18$	ad^2ef	$+\ 7$	ae^4	$-\ 2$
a^2d^3	$+\ 3$	ab^2df	$-\ 7$	$abcdf$	$-\ 18$	abd^2f	$+\ 32$	ac^2ef	$-\ 32$	ace^3	$+\ 6$	ade^3	$-\ 7$	b^2ef^2	...
ab^2cf	$-\ 4$	ab^2e^2	$+\ 10$	$abce^2$	$-\ 18$	$abde^2$	$-\ 8$	acd^2f	$+\ 18$	ad^3f	$+\ 3$	b^2df^2	$+\ 10$	$bcdf^2$	$+\ 5$
ab^2de	$+\ 5$	abc^2f	$-\ 7$	abd^2e	$+\ 30$	ac^2df	$-\ 18$	$acde^2$	$+\ 53$	ad^2e^2	$-\ 15$	b^2e^2f	$-\ 10$	bce^2f	$-\ 5$
abc^2e	$+\ 5$	$abcde$	$-\ 8$	ac^3f	$-\ 3$	ac^2e^2	$+\ 6$	ad^3e	$-\ 39$	b^2cf^2	$+\ 9$	bc^2f^2	$-\ 3$	bd^2ef	$-\ 5$
$abcd^2$	$-\ 7$	abd^3	$+\ 9$	ac^2de	$+\ 45$	acd^2e	$+\ 52$	b^3f^2	$+\ 6$	b^2def	$+\ 18$	$bcdef$	$+\ 8$	bde^3	$+\ 5$
ac^3d	$+\ 1$	ac^3e	$+\ 22$	acd^3	$-\ 39$	ad^4	$-\ 39$	b^2cef	$+\ 8$	b^2e^3	$-\ 27$	bce^3	$-\ 2$	c^3f^2	$-\ 3$
b^4f	$+\ 2$	ac^2d^2	$-\ 19$	b^3df	$-\ 6$	b^3ef	$+\ 19$	b^2d^2f	$-\ 6$	bc^2ef	$-\ 30$	bd^3f	$-\ 22$	c^2def	$+\ 7$
b^3ce	$-\ 5$	b^3cf	$+\ 7$	b^3e^2	$+\ 27$	b^2cdf	$-\ 53$	b^2de^2	$-\ 20$	bcd^2f	$-\ 45$	bd^2e^2	$+\ 19$	c^2e^3	$+\ 2$
b^3d^2	$-\ 2$	b^3de	$+\ 2$	b^2c^2f	$+\ 15$	b^2ce^2	$+\ 20$	bc^2df	$+\ 45$	$bcde^2$	$+\ 87$	c^3ef	$-\ 9$	cd^3f	$-\ 1$
b^2c^2d	$+\ 8$	b^2c^2e	$-\ 19$	b^2cde	$-\ 87$	b^2d^2e	$-\ 25$	bc^2e^2	$+\ 25$	bd^3e	$-\ 12$	c^2d^2f	$+\ 19$	cd^2e^2	$-\ 8$
bc^4	$-\ 3$	b^2cd^2	$-\ 11$	b^2d^3	$+\ 6$	bc^3f	$+\ 39$	bcd^2e	$-\ 52$	c^3df	$+\ 39$	c^2de^2	$+\ 11$	d^4e	$+\ 3$
		bc^3d	$+\ 33$	bc^3e	$+\ 12$	bc^2de	$-\ 45$	bd^4	...	c^3e^2	$-\ 6$	cd^3e	$-\ 33$		
		c^5	$-\ 12$	bc^2d^2	$+\ 57$	bcd^3	$+\ 65$	c^4f	$+\ 39$	c^2d^2e	$-\ 57$	d^5	$+\ 12$		
				c^4d	$-\ 24$	c^4e	...	c^3de	$-\ 65$	cd^4	$+\ 24$				
						c^3d^2	$-\ 20$	c^2d^3	$+\ 20$						
	$\pm\ 26$		$\pm\ 93$		$\pm\ 207$		$\pm\ 241$		$\pm\ 241$		$\pm\ 207$		$\pm\ 93$		$\pm\ 26$

$(\ \ \ )(x, y)^7$

No. 25, $Q = +1\ a^3cdf^3 + \&c.$
No. 26, $Q' = +1\ a^4f^4 \ \ + \&c.$ } [see *post*, 143.]

142.

NUMERICAL TABLES SUPPLEMENTARY TO SECOND MEMOIR ON QUANTICS.

[Now first published (1889).]

In the present paper I arrange in a more compendious form and continue to a much greater extent the tables (first of each pair) given Nos. 35—39 of my Second Memoir on Quantics, 141, pp. 260—264, which relate to the cubic, the quartic and the quintic functions; and I give the like tables for the sextic, the septimic and the octavic functions respectively. The cubic table exhibits the coefficients of the several xz terms of the function $1 \div (1-z.1-xz.1-x^2z.1-x^3z)$, or, what is the same thing, it gives the number of partitions of a given number into a given number of parts, the parts being 0, 1, 2, 3, (repetitions admissible): or again, regarding the letters a, b, c, d, as having the weights 0, 1, 2, 3 respectively, it shows the number of literal terms of a given degree and given weight. And similarly for the quartic, quintic, sextic, septimic and octavic tables respectively, the parts of course being 0, 1,... up to 4, 5, 6, 7 or 8, and the letters being a, b, ... up to e, f, g, h or i. The extent of the tables is as follows:

cubic table extends to deg-weight				18—27
quartic	„	„	„	18—36
quintic	„	„	„	18—45
sextic	„	„	„	15—45
septimic	„	„	„	12—42
octavic	„	„	„	10—40

viz. for the quintic, the sextic and the octavic functions these are the deg-weights of the highest invariants respectively. I designate the Tables as the *ad*-, *ae*-, *af*-, *ag*-, *ah*- and *ai*-tables respectively.

It is to be noticed that in the several tables the lower part of each column is for shortness omitted; the column has to be completed by taking into it the series of bottom terms of each of the preceding columns: thus in the af- or quintic table the complete column for degree 3 would be

D	3
W	8-7
−0	6
−1	6
−2	5
−3	4
−4	3
−5	2
−6	1
−7	1

where the concluding terms 2, 1, 1 are the bottom terms of the three preceding columns respectively. And the meaning is that for degree 3, and weight 8, or 7, the number of terms is $=6$; for weight $7-1$, $=6$, the number of terms is $=6$; and similarly for weights 5, 4, 3, 2, 1, 0 the numbers are 5, 4, 3, 2, 1, 1; the numbers are those of the terms

W.	0	1	2	3	4	5	6	7	8
	a^3	a^2b	a^2c	a^2d	a^2e	a^2f	abf	acf	adf
			ab^2	abc	abd	abe	ace	ade	ae^2
				b^3	ac^2	acd	ad^2	b^2f	bcf
					b^2c	b^2d	b^2e	b^2ce	bde
						bc^2	bcd	bd^2	c^2e
							c^3	c^2d	cd^2
No.	1	1	2	3	4	5	6	6	6

The like remarks and explanations apply to the other tables.

ad-TABLE.

D	0	1	2	3	4	5	6	7	8	9	10	11	12	13	14	15	16	17	18	
W	0	2-1	3	5-4	6	8-7	9	11-10	12	14-13	15	17-16	18	20-19	21	23-22	24	26-25	27	
	1	1	2	3	5	6	8	10	13	15	18	21	25	28	32	36	41	45	50	−0
			2	3	4	6	8	10	12	15	18	21	24	28	32	36	40	45	50	−1
					4	5	7	9	12	14	17	20	24	27	31	35	40	44	49	−2
							7	8	11	13	17	19	23	26	31	34	39	43	49	−3
									10	12	15	18	22	25	29	33	38	42	47	−4
											14	16	20	23	28	31	36	40	46	−5
													19	21	26	29	35	38	44	−6
															24	27	32	36	42	−7
																	30	33	39	−8
																			37	−9

ae-TABLE.

D	0	1	2	3	4	5	6	7	8	9	10	11	12	13	14	15	16	17	18	
W	0	2	4	6	8	10	12	14	16	18	20	22	24	26	28	30	32	34	36	
	1	1	3	5	8	12	18	24	33	43	55	69	86	104	126	150	177	207	241	-0
		1	2	4	7	11	16	23	31	41	53	67	83	102	123	147	174	204	237	-1
			2	4	7	11	16	23	31	41	53	67	83	102	123	147	174	204	237	-2
				3	5	9	14	20	28	38	49	63	79	97	118	142	168	198	231	-3
					5	8	13	19	27	36	48	61	77	95	116	139	166	195	228	-4
						6	10	16	23	32	43	56	71	89	109	132	158	187	219	-5
							9	14	21	30	40	53	68	85	105	128	153	182	214	-6
								11	17	25	35	47	61	78	97	119	144	172	203	-7
									15	22	32	43	57	73	92	113	138	165	196	-8
										18	26	37	50	65	83	104	127	154	184	-9
											23	33	45	60	77	97	120	146	175	-10
												27	38	52	68	87	109	134	162	-11
													34	46	62	80	101	125	153	-12
														39	53	70	90	113	139	-13
															47	63	82	104	129	-14
																54	71	92	116	-15
																	64	83	106	-16
																		72	93	-17
																			84	-18

af-TABLE.

D	0	1	2	3	4	5	6	7	8	9	10	11	12	13	14	15	16	17	18	
W	0	3-2	5	8-7	10	13-12	15	18-17	20	23-22	25	28-27	30	33-32	35	38-37	40	43-42	45	
	1	1	3	6	12	20	32	49	73	102	141	190	252	325	414	521	649	795	967	-0
		1	3	6	11	19	32	48	71	101	141	188	249	322	414	518	645	791	966	-1
			2	5	11	18	30	46	70	98	137	184	247	317	408	511	641	783	957	-2
			2	4	9	16	29	43	66	93	134	178	240	309	402	501	630	770	948	-3
				3	8	14	25	39	63	88	127	170	233	299	390	488	619	754	930	-4
					6	11	23	35	57	81	121	161	222	286	379	472	601	734	912	-5
					5	9	19	30	52	74	111	150	212	272	362	453	583	711	886	-6
						7	16	26	45	66	103	139	197	256	346	433	559	684	860	-7
							12	21	40	58	92	126	184	239	325	409	536	655	827	-8
							10	17	33	50	83	114	168	220	306	385	507	623	795	-9
								13	28	43	72	101	154	202	283	359	480	590	756	-1
									22	35	63	89	137	183	262	333	448	554	719	-1
									18	29	53	77	123	165	238	306	419	519	677	-1
										23	45	66	107	146	217	280	386	482	637	-1
											36	55	94	129	194	253	356	446	593	-1
											30	46	80	112	174	228	324	409	553	-1
												37	68	97	152	203	295	374	509	-1
													56	82	134	180	264	339	469	-1
													47	69	115	157	237	306	427	-1
														57	99	137	209	273	389	-1
															83	117	185	243	350	-2
															70	100	160	214	315	-2
																84	139	188	279	-2
																	118	162	248	-2
																	101	140	217	-2
																		119	190	-2
																			163	-2
																			141	-2

ag-TABLE.

D	0	1	2	3	4	5	6	7	8	9	10	11	12	13	14	15	
W	0	3	6	9	12	15	18	21	24	27	30	33	36	39	42	45	
	1	1	4	8	18	32	58	94	151	227	338	480	676	920	1242	1636	−0
		1	3	8	16	32	55	94	147	227	332	480	668	920	1232	1635	−1
		1	3	7	16	30	55	90	146	221	330	471	664	907	1226	1617	−2
			2	7	14	29	51	88	139	217	319	464	648	896	1203	1601	−3
			2	5	13	25	48	81	134	205	310	446	634	870	1182	1565	−4
				4	10	23	42	76	123	196	293	431	608	847	1145	1533	−5
				3	9	19	39	68	116	182	280	408	587	813	1113	1483	−6
					6	16	32	61	103	169	258	387	553	780	1064	1435	−7
					5	12	28	52	94	152	241	359	525	737	1021	1373	−8
						10	22	46	81	139	218	335	488	699	965	1316	−9
						7	18	37	71	121	199	304	455	650	914	1244	−10
							13	31	59	107	175	278	415	607	852	1178	−11
							11	24	51	91	157	248	382	557	798	1102	−12
								19	40	78	134	222	341	512	733	1031	−13
								14	33	64	117	193	308	462	677	952	−14
									25	54	98	170	271	419	614	882	−15
									20	42	83	144	240	371	559	803	−16
										34	67	124	206	331	499	734	−17
										26	56	103	180	289	449	661	−18
											43	86	151	253	394	596	−19
											35	69	129	216	349	529	−20
												57	106	187	302	472	−21
												44	88	156	263	412	−22
													70	132	223	362	−23
													58	108	192	303	−24
														89	159	270	−25
														71	134	228	−26
															109	195	−27
															90	161	−28
																135	−29
																110	−30

ah-TABLE.

D	0	1	2	3	4	5	6	7	8	9	10	11	12	
W	0	4-3	7	11-10	14	18-17	21	25-24	28	32-31	35	39-38	42	
	1	1	4	10	24	49	94	169	289	468	734	1117	1656	-0
		1	4	10	23	48	94	166	285	464	734	1109	1646	-1
		1	3	9	23	46	90	162	282	454	722	1093	1634	-2
			3	8	20	43	88	155	272	441	709	1069	1605	-3
			2	7	19	39	81	146	263	424	686	1038	1572	-4
			2	5	16	35	76	136	247	403	663	1000	1524	-5
				4	14	30	68	125	233	379	629	957	1475	-6
				3	11	26	61	112	214	354	598	908	1410	-7
					9	21	52	100	197	325	558	856	1346	-8
					6	17	46	87	176	297	520	799	1271	-9
					5	13	37	75	158	268	477	742	1197	-10
						10	31	63	137	239	437	682	1114	-11
						7	24	53	120	210	392	623	1036	-12
							19	42	101	184	353	563	950	-13
							14	34	86	157	311	506	871	-14
							11	26	70	134	274	449	788	-15
								20	58	112	236	397	711	-16
								15	45	93	204	346	633	-17
									36	75	171	300	564	-18
									27	61	145	256	493	-19
									21	47	119	218	432	-20
										37	98	182	372	-21
										28	78	152	320	-22
											63	124	270	-23
											48	101	229	-24
											38	80	189	-25
												64	157	-26
												49	127	-27
													103	-28
													81	-29
													65	-30

ai-TABLE.

	0	1	2	3	4	5	6	7	8	9	10	
W	0	4	8	12	16	20	24	28	32	36	40	
	1	1	5	13	33	73	151	289	526	910	1514	-0
		1	4	12	31	71	147	285	519	902	1502	-1
		1	4	12	31	70	146	282	515	894	1492	-2
		1	3	11	28	66	139	272	499	873	1460	-3
			3	10	27	63	134	263	486	851	1430	-4
			2	8	23	57	123	247	461	816	1379	-5
			2	7	21	52	116	233	440	783	1331	-6
				5	17	45	103	214	409	738	1265	-7
				4	15	40	94	197	383	696	1214	-8
				3	11	33	81	176	348	645	1127	-9
					9	28	71	158	319	597	1057	-10
					6	22	59	137	284	543	974	-11
					5	18	51	120	255	495	900	-12
						13	40	101	221	441	816	-13
						10	33	86	194	394	742	-14
						7	25	70	164	345	662	-15
							20	58	141	302	593	-16
							14	45	116	258	519	-17
							11	36	97	222	457	-18
								27	77	185	393	-19
								21	63	156	340	-20
								15	48	127	286	-21
									38	104	243	-22
									28	82	200	-23
									22	66	167	-24
										50	134	-25
										39	109	-26
										29	85	-27
											68	-28
											51	-29
											40	-30

The numbers of each table are connected in several ways with those of the preceding tables. One of these connexions, which is of some importance, is best explained by an example: in the *af*-table, 8-20, the number of terms of degree 8 and weight 20 is 73; and we have $73 = 1 + 6 + 16 + 23 + 27$, viz. (see p. 288) these are the numbers of the terms in a^4, a^3, a^2, a^1, a^0 respectively: the complementary factors, (for example) of a^3 are bef^3, &c. terms in b, c, d, e, f of the degree 5 and weight 20, and (replacing therein each letter by that which immediately precedes it) these are in number equal to the terms in a, b, c, d, e of the degree 5 and weight $20 - 5, = 15$; thus the number 6 of the terms in question is that for the deg-weight 5-15 of the *ae*-table: and so 1, 6, 16, 23, 27 are the numbers in the *ae*-table for the deg-weights 4-16, 5-15, 6-14, 7-13 and 8-12 respectively, or (making a change rendered necessary by the abbreviated form of the tables) say for the deg-weights 4-0, 5-10, 6-14, 7-13 and 8-12.

143.

TABLES OF THE COVARIANTS M TO W OF THE BINARY QUINTIC: FROM THE SECOND, THIRD, FIFTH, EIGHTH, NINTH AND TENTH MEMOIRS ON QUANTICS.

[Arranged in the present form, 1889.]

THE binary quintic has in all (including the quintic itself and the invariants) 23 covariants, which I have represented by the capital letters, A, B, C, ... W (alternative forms of two of these are denoted by Q′ and S′). The covariants A, ... L, and also Q, Q′ were given in my Second Memoir on Quantics, and except Q and Q′ are reproduced in the present reprint thereof, 141; in all these I gave not only the literal terms actually presenting themselves, but also the terms with zero coefficients; in the other covariants however, or in most of them, the terms with zero coefficients were omitted. It is very desirable to have in every case the complete series of literal terms, and in the covariants as here printed they are accordingly inserted: the number of terms is in each case known beforehand by the foregoing *af*-table, 142, and any omission is thus precluded; by means of this *af*-table we have the numbers of terms as shown in the following list.

I have throughout (as was done in the Ninth and Tenth Memoirs) expressed the literal terms in a slightly different form from that employed in the Second Memoir: this is done in order to show at a glance in each column the set of terms which contain a given power of a, and in each such set the terms which contain a given power of b.

The numerical verifications are also given not only for the entire column but for each set of terms containing the same power of a; viz. in most cases, but not always, the positive and negative coefficients of a set have equal sums, which are shown by a number with the sign ± prefixed. The verification is in some cases given in regard to the subsets involving the same powers of a and b, here also the sums of the positive and negative coefficients are not in every case equal. The cases of inequality will be referred to at the end of this paper.

The whole series of covariants is as follows:

Mem.	No. of table.				deg-weight
2	13	A	=	$(1, 1, 1, 1, 1, 1 \between x, y)^5$	1 (0....5)
,,	14	B	=	$(3, 3, 3 \between x, y)^2$	2 (4.6)
,,	15	C	=	$(2, 2, 3, 3, 3, 2, 2 \between x, y)^6$	2 (2.....8)
,,	16	D	=	$(6, 6, 6, 6 \between x, y)^3$	3 (6..9)
,,	17	E	=	$(5, 6, 6, 6, 6, 5 \between x, y)^5$	3 (5....10)
,,	18	F	=	$(3, 4, 5, 6, 6, 6, 6, 5, 4, 3 \between x, y)^9$	3 (3........12)
,,	19	G	=	$(12 \between x, y)^0$, Invt.	4-10
,,	20	H	=	$(11, 11, 12, 11, 11 \between x, y)^4$	4 (8...12)
,,	21	I	=	$(9, 11, 11, 12, 11, 11, 9 \between x, y)^6$	4 (7.....13)
,,	22	J	=	$(20, 20 \between x, y)^1$	5 (12, 13)
,,	23	K	=	$(19, 20, 20, 19 \between x, y)^3$	5 (11..14)
,,	24	L	=	$(16, 18, 19, 20, 20, 19, 18, 16 \between x, y)^7$	5 (9......16)
8	83	M	=	$(32, 32, 32 \between x, y)^2$	6 (14.16)
,,	84	N	=	$(30, 32, 32, 32, 30 \between x, y)^4$	6 (13...17)
9	90	O	=	$(49, 49 \between x, y)^1$	7 (17, 18)
,,	91	P	=	$(46, 48, 49, 49, 48, 46 \between x, y)^5$	7 (15....20)
2	Q 25 Q′ 26	Q, Q′	=	$(73 \between x, y)^0$, Invt.	8-20
9	92	R	=	$(71, 73, 71 \between x, y)^2$	8 (19.21)
9 10	S 93 S 93 *bis*	S, S′	=	$(101, 102, 102, 101 \between x, y)^3$	9 (21..24)
9	94	T	=	$(190, 190 \between x, y)^1$	11 (27, 28)
3	29	U	=	$(252 \between x, y)^0$, Invt.	12-30
9	95	V	=	$(325, 325 \between x, y)^1$	13 (32, 33)
5	29A	W	=	$(967 \between x, y)^0$, Invt.	18-45

M. No. 83.

(

$a^3\ b^0ef^2$	...	$a^3\ b^0f^3$	...	$a^2\ b^1f^3$	...
$a^2\ b^1df^2$	...	$a^2\ b^1ef^2$	...	b^0cef^2	...
e^2f	...	b^0cdf^2	− 1	d^2f^2	− 1
$b^0c^2f^2$	− 1	ce^2f	+ 1	de^2f	+ 2
$cdef$	+ 5	d^2ef	+ 1	e^4	− 1
ce^3	− 3	de^3	− 1	$a^1\ b^2ef^2$	...
d^3f	− 3	$a^1\ b^2df^2$	+ 1	b^1cdf^2	+ 5
d^2e^2	+ 2	e^2f	− 1	ce^2f	− 5
$a^1\ b^2cf^2$	+ 2	$b^1c^2f^2$	+ 1	d^2ef	− 5
def	− 5	$cdef$	+ 6	de^3	+ 5
e^3	+ 3	ce^3	− 8	$b^0c^3f^2$	− 3
b^1c^2ef	− 5	d^3f	− 10	c^2def	+ 7
cd^2f	+ 7	d^2e^2	+ 11	c^2e^3	+ 2
cde^2	− 1	b^0c^3ef	− 10	cd^3f	− 1
d^3e	− 1	c^2d^2f	+ 11	cd^2e^2	− 8
b^0c^3df	− 1	c^2de^2	+ 18	d^4e	+ 3
c^3e^2	+ 6	cd^3e	− 28	$a^0\ b^3df^2$	− 3
c^2d^2e	− 8	d^5	+ 9	e^2f	+ 3
cd^4	+ 3	$a^0\ b^3cf^2$	− 1	$b^2c^2f^2$	+ 2
$a^0\ b^4f^2$	− 1	def	− 8	$cdef$	− 1
b^3cef	+ 5	e^3	+ 9	ce^3	− 3
d^2f	+ 2	b^2c^2ef	+ 11	d^3f	+ 6
de^2	− 3	cd^2f	+ 18	d^2e^2	− 4
b^2c^2df	− 8	cde^2	− 37	b^1c^3ef	− 1
c^2e^2	− 4	d^3e	+ 8	c^2d^2f	− 8
cd^2e	+ 7	b^1c^3df	− 28	c^2de^2	+ 7
d^4	− 1	c^3e^2	+ 8	cd^3e	+ 5
b^1c^4f	+ 3	c^2d^2e	+ 37	d^5	− 3
c^3de	+ 5	cd^4	− 17	b^0c^4df	+ 3
c^2d^3	− 4	b^0c^5f	+ 9	c^4e^2	− 1
b^0c^5e	− 3	c^4de	− 17	c^3d^2e	− 4
c^4d^2	− 2	c^3d^3	+ 8	c^2d^4	+ 2
	± 7		± 2		± 2
	21		57		22
	24		108		28
	± 52		± 167		± 52

$\between x, y)^2$

N. No. 84.

$a^3\ b^0df^2$	$-$ 1	$a^3\ b^0ef^2$	...	$a^3\ b^0f^3$	...	$a^2\ b^1f^3$	...	$a^2\ b^0cf^3$	$+$ 1
e^2f	$+$ 1	$a^2\ b^1df^2$	$-$ 4	$a^2\ b^1ef^2$	...	b^0cef^2	$+$ 4	def^2	$-$ 3
$a^2\ b^1cf^2$	$+$ 3	e^2f	$+$ 4	b^0cdf^2	...	d^2f^2	$-$ 4	e^3f	$+$ 2
def	$+$ 2	$b^0c^2f^2$	$+$ 4	ce^2f	$+$ 6	de^2f	$-$ 4	$a^1\ b^2f^3$	$-$ 1
e^3	$-$ 5	$cdef$	$-$ 8	d^2ef	$-$ 12	e^4	$+$ 4	b^1cef^2	$-$ 2
b^0c^2ef	$-$ 8	ce^3	$+$ 4	de^3	$+$ 6	$a^1\ b^2ef^2$	$-$ 4	d^2f^2	$+$ 8
cd^2f	$+$ 2	d^3f	...	$a^1\ b^2df^2$	$-$ 6	b^1cdf^2	$+$ 8	de^2f	$-$ 2
cde^2	$+$ 12	d^2e^2	...	e^2f	...	ce^2f	$-$ 16	e^4	$-$ 6
d^3e	$-$ 6	$a^1\ b^2cf^2$	$+$ 4	$b^1c^2f^2$	$+$ 12	d^2ef	$+$ 48	$b^0c^2df^2$	$-$ 2
$a^1\ b^3f^2$	$-$ 2	def	$+$ 16	$cdef$	...	de^3	$-$ 32	c^2e^2f	$+$ 6
b^2cef	$-$ 2	e^3	$-$ 24	ce^3	$-$ 36	$b^0c^3f^2$	...	cd^2ef	$-$ 20
d^2f	$-$ 6	b^1c^2ef	$-$ 48	d^3f	$+$ 48	c^2def	$-$ 40	cde^3	$+$ 12
de^2	$+$ 13	cd^2f	$+$ 40	d^2e^2	$-$ 12	c^2e^3	$+$ 56	d^4f	$+$ 9
b^1c^2df	$+$ 20	cde^2	$+$ 40	b^0c^3ef	$-$ 48	cd^3f	$+$ 8	d^3e^2	$-$ 6
c^2e^2	$+$ 4	d^3e	$-$ 24	c^2d^2f	...	cd^2e^2	$-$ 40	$a^0\ b^3ef^2$	$+$ 5
cd^2e	$-$ 52	b^0c^3df	$-$ 8	c^2de^2	$+$ 156	d^4e	$+$ 12	b^2cdf^2	$-$ 12
d^4	$+$ 24	c^3e^2	$+$ 56	cd^3e	$-$ 168	$a^0\ b^3df^2$	$-$ 4	ce^2f	$-$ 13
b^0c^4f	$-$ 9	c^2d^2e	$-$ 88	d^5	$+$ 54	e^2f	$+$ 24	d^2ef	$-$ 4
c^3de	$+$ 20	cd^4	$+$ 36	$a^0\ b^3cf^2$	$-$ 6	$b^2c^2f^2$	...	de^3	$+$ 15
c^2d^3	$-$ 10	$a^0\ b^4f^2$	$-$ 4	def	$+$ 36	$cdef$	$-$ 40	$b^1c^3f^2$	$+$ 6
$a^0\ b^4ef$	$+$ 6	b^3cef	$+$ 32	e^3	...	ce^3	$-$ 60	c^2def	$+$ 52
b^3cdf	$+$ 12	d^2f	$-$ 56	b^2c^2ef	$+$ 12	d^3f	$-$ 56	c^2e^3	$-$ 10
ce^2	$-$ 15	de^2	$+$ 60	cd^2f	$-$ 156	d^2e^2	$+$ 100	cd^3f	$-$ 20
d^2e	$+$ 10	b^2c^2df	$+$ 40	cde^2	...	b^1c^3ef	$+$ 24	cd^2e^2	$-$ 30
b^2c^3f	$+$ 6	c^2e^2	$-$ 100	d^3e	$+$ 60	c^2d^2f	$+$ 88	d^4e	$+$ 15
c^2de	$+$ 30	cd^2e	$-$ 80	b^1c^3df	$+$ 168	c^2de^2	$+$ 80	b^0c^4ef	$-$ 24
cd^3	$-$ 20	d^4	$+$ 60	c^3e^2	$-$ 60	cd^3e	$-$ 200	c^3d^2f	$+$ 10
b^1c^4e	$-$ 15	b^1c^4f	$-$ 12	c^2d^2e	...	d^5	$+$ 60	c^3de^2	$+$ 20
c^3d^2	$+$ 10	c^3de	$+$ 200	cd^4	$-$ 30	b^0c^4df	$-$ 36	c^2d^3e	$-$ 10
b^0c^5d	...	c^2d^3	$-$ 120	b^0c^5f	$-$ 54	c^4e^2	$-$ 60	cd^5	...
		b^0c^5e	$-$ 60	c^4de	$+$ 30	c^3d^2e	$+$ 120		
		c^4d^2	$+$ 40	c^3d^3	...	c^2d^4	$-$ 40		
	$\pm$ 1								
	19		$\pm$ 12		$\pm$ 12		$\pm$ 8		$\pm$ 3
	81		192		270		132		37
	62		432		306		496		123
	$\pm$ 163		$\pm$ 636		$\pm$ 588		$\pm$ 636		$\pm$ 163

$(\ldots \between x, y)^4$

O. No. 90.

Term	Coeff.	Term	Coeff.
$a^3\ b^0cf^3$	+ 1	$a^3\ b^0df^3$	− 1
def^2	− 4	e^2f^2	+ 1
e^3f	+ 3	$a^2\ b^1cf^3$	+ 4
$a^2\ b^2f^3$	− 1	def^2	+ 3
b^1cef^2	− 3	e^3f	− 7
d^2f^2	+ 16	$b^0c^2ef^2$	− 16
de^2f	+ 4	cd^2f^2	+ 6
e^4	− 15	cde^2f	+ 30
$b^0c^2df^2$	− 6	ce^4	− 8
c^2e^2f	+ 4	d^3ef	− 18
cd^2ef	− 22	d^2e^3	+ 6
cde^3	+ 26	$a^1\ b^3f^3$	− 3
d^4f	+ 9	b^2cef^2	− 4
d^3e^2	− 12	d^2f^2	− 4
$a^1\ b^3ef^2$	+ 7	de^2f	− 1
b^2cdf^2	− 30	e^4	+ 18
ce^2f	+ 1	$b^1c^2df^2$	+ 22
d^2ef	− 74	c^2e^2f	+ 74
de^3	+ 84	cd^2ef	− 160
$b^1c^3f^2$	+ 18	cde^3	− 32
c^2def	+ 160	d^4f	+ 81
c^2e^3	− 98	d^3e^2	+ 6
cd^3f	− 20	$b^0c^4f^2$	− 9
cd^2e^2	− 94	c^3def	+ 20
d^4e	+ 51	c^3e^3	− 112
b^0c^4ef	− 81	c^2d^3f	− 18
c^3d^2f	+ 18	$c^2d^2e^2$	+ 284
c^3de^2	+ 140	cd^4e	− 216
c^2d^3e	− 100	d^6	+ 54
cd^5	+ 18	$a^0\ b^4ef^2$	+ 15
$a^0\ b^4df^2$	+ 8	b^3cdf^2	− 26
e^2f	− 18	ce^2f	− 84
$b^3c^2f^2$	− 6	d^2ef	+ 98
$cdef$	+ 32	de^3	− 45
ce^3	+ 45	$b^2c^3f^2$	+ 12
d^3f	+ 112	c^2def	+ 94
d^2e^2	− 150	c^2e^3	+ 150
b^2c^3ef	− 6	cd^3f	− 140
c^2d^2f	− 284	cd^2e^2	− 50
c^2de^2	+ 50	d^4e	+ 15
cd^3e	+ 320	b^1c^4ef	− 51
d^5	− 120	c^3d^2f	+ 100
b^1c^4df	+ 216	c^3de^2	− 320
c^4e^2	− 15	c^2d^3e	+ 310
c^3d^2e	− 310	cd^5	− 90
c^2d^4	+ 130	b^0c^5df	− 18
b^0c^6f	− 54	c^5e^2	+ 120
c^5de	+ 90	c^4d^2e	− 130
c^4d^3	− 40	c^3d^4	+ 40
	± 4		± 1
	59		49
	497		559
	1003		954
	±1563		±1563

$)(x, y)^1$

P. No. 91.

$a^4\,b^0f^3$ …	$a^3\,b^2f^3$ …	$a^3\,b^0cf^3$ − 1	$a^3\,b^0df^3$ + 1	$a^3\,b^0ef^3$ …	$a^3\,b^0f^4$ …
$a^3\,b^1ef^2$ …	b^0cef^2 − 2	def^2 + 6	e^2f^2 − 1	$a^2\,b^1df^3$ + 2	$a^2\,b^1ef^3$ …
b^0cdf^2 + 1	d^2f^2 + 5	e^3f − 5	$a^2\,b^1cf^3$ − 6	e^2f^2 − 2	b^0cdf^3 − 1
ce^2f − 2	de^2f − 1	$a^2\,b^2f^3$ + 1	def^2 + 11	$b^0c^2f^3$ − 5	ce^2f^2 + 1
d^2ef + 2	e^4 − 2	b^1cef^2 − 11	e^3f − 5	$cdef^2$ + 17	d^2ef^2 + 3
de^3 − 1	$a^2\,b^2ef^2$ + 2	d^2f^2 − 4	$b^0c^2ef^2$ + 4	ce^3f − 7	de^3f − 5
$a^2\,b^2df^2$ − 1	b^1cdf^2 − 17	de^2f − 4	cd^2f^2 − 2	d^3f^2 − 4	e^5 + 2
e^2f + 2	ce^2f + 13	e^4 + 17	cde^2f + 4	d^2e^2f − 6	$a^1\,b^2df^3$ + 2
$b^1c^2f^2$ − 3	d^2ef − 32	$b^0c^2df^2$ + 2	ce^4 − 4	de^4 + 5	e^2f^2 − 2
$cdef$ − 6	de^3 + 32	c^2e^2f + 26	d^3ef − 10	$a^1\,b^2cf^3$ + 1	$b^1c^2f^3$ − 2
ce^3 + 13	$b^0c^3f^2$ + 4	cd^2ef − 2	d^2e^3 + 8	def^2 − 13	$cdef^2$ + 6
d^3f − 8	c^2def + 36	cde^3 − 40	$a^1\,b^3f^3$ + 5	e^3f + 12	ce^3f − 2
d^2e^2 + 2	c^2e^3 − 24	d^4f − 9	b^2cef^2 + 4	$b^1c^2ef^2$ + 32	d^3f^2 − 16
b^0c^3ef + 16	cd^3f − 10	d^3e^2 + 24	d^2f^2 − 26	cd^2f^2 − 36	d^2e^2f + 24
c^2d^2f − 2	cd^2e^2 − 16	$a^1\,b^3ef^2$ + 5	de^2f − 35	cde^2f − 42	de^4 − 10
c^2de^2 − 38	d^4e + 12	b^2cdf^2 − 4	e^4 + 42	ce^4 + 24	$b^0c^3ef^2$ + 8
cd^3e + 34	$a^1\,b^3df^2$ + 7	ce^2f + 35	$b^1c^2df^2$ + 2	d^3ef + 56	$c^2d^2f^2$ + 2
d^5 − 9	e^2f − 12	d^2ef − 26	c^2e^2f + 26	d^2e^3 − 34	c^2de^2f − 52
$a^1\,b^3cf^2$ + 5	$b^2c^2f^2$ + 6	de^3 − 22	cd^2ef + 72	$b^0c^3df^2$ + 10	c^2e^4 + 28
def + 2	$cdef$ + 42	$b^1c^3f^2$ + 10	cde^3 − 124	c^3e^2f − 54	cd^3ef + 52
e^3 − 12	ce^3 …	c^2def − 72	d^4f + 13	c^2d^2ef + 64	cd^2e^3 − 32
b^2c^2ef − 24	d^3f + 54	c^2e^3 − 106	d^3e^2 + 26	c^2de^3 + 46	d^5f − 18
cd^2f + 52	d^2e^2 − 91	cd^3f + 76	$b^0c^4f^2$ + 9	cd^4f − 37	d^4e^2 + 12
cde^2 + 7	b^1c^3ef − 68	cd^2e^2 + 210	c^3def − 76	cd^3e^2 − 50	$a^0\,b^3cf^3$ + 1
d^3e − 22	c^2d^2f − 64	d^4e − 99	c^3e^3 − 56	d^5e + 21	def^2 − 13
b^1c^3df − 52	c^2de^2 + 14	b^0c^4ef − 13	c^2d^3f + 10	$a^0\,b^4f^3$ + 2	e^3f + 12
c^3e^2 + 34	cd^3e + 204	c^3d^2f − 10	$c^2d^2e^2$ + 296	b^3cef^2 − 32	$b^2c^2ef^2$ − 2
c^2d^2e + 8	d^5 − 93	c^3de^2 + 128	cd^4e − 260	d^2f^2 + 24	cd^2f^2 + 38
cd^4 − 1	b^0c^4df + 37	c^2d^3e − 184	d^6 + 72	de^2f …	cde^2f − 7
b^0c^5f + 18	c^4e^2 + 86	cd^5 + 72	$a^0\,b^4ef^2$ − 17	e^4 …	ce^4 − 30
c^4de − 25	c^3d^2e − 208	$a^0\,b^4df^2$ + 4	b^3cdf^2 + 40	$b^2c^2df^2$ + 16	d^3ef − 34
c^3d^3 + 10	c^2d^4 + 86	e^2f − 42	ce^2f + 22	c^2e^2f + 91	d^2e^3 + 35
$a^0\,b^5f^2$ − 2	$a^0\,b^4cf^2$ − 5	$b^3c^2f^2$ − 8	d^2ef + 106	cd^2ef − 14	$b^1c^3df^2$ − 34
b^4cef + 10	def − 12	$cdef$ + 124	de^3 − 105	cde^3 − 105	c^3e^2f + 22
d^2f − 28	e^3 …	ce^3 + 105	$b^2c^3f^2$ − 24	d^4f − 86	c^2d^2ef − 8
de^2 + 30	b^3c^2ef + 34	d^3f + 56	c^2def − 210	d^3e^2 + 110	c^2de^3 + 50
b^3c^2df + 32	cd^2f − 46	d^2e^2 − 130	c^2e^3 + 130	$b^1c^4f^2$ − 12	cd^4f + 25
c^2e^2 − 35	cde^2 + 105	b^2c^3ef − 26	cd^3f − 128	c^3def − 204	cd^3e^2 − 70
cd^2e − 50	d^3e − 20	c^2d^2f − 296	cd^2e^2 + 170	c^3e^3 + 20	d^5e + 15
d^4 + 30	b^2c^3df + 50	c^2de^2 − 170	d^4e − 25	c^2d^3f + 208	$b^0c^5f^2$ + 9
b^2c^4f − 12	c^3e^2 − 110	cd^3e + 340	b^1c^4ef + 99	$c^2d^2e^2$ + 170	c^4def + 1
c^3de + 70	c^2d^2e − 170	d^5 − 60	c^3d^2f + 184	cd^4e − 250	c^4e^3 − 30
c^2d^3 − 40	cd^4 + 115	b^1c^4df + 260	c^3de^2 − 340	d^6 + 60	c^3d^3f − 10
b^1c^5e − 15	b^1c^5f − 21	c^4e^2 + 25	c^2d^3e + 150	b^0c^5ef + 93	$c^3d^2e^2$ + 40
c^4d^2 + 10	c^4de + 250	c^3d^2e − 150	cd^5 − 40	c^4d^2f − 86	c^2d^4e − 10
b^0c^6d …	c^3d^3 − 150	c^2d^4 …	b^0c^5df − 72	c^4de^2 − 115	cd^6 …
	b^0c^6e − 60	b^0c^6f − 72	c^5e^2 + 60	c^3d^3e + 150	
	c^5d^2 + 40	c^5de + 40	c^4d^2e …	c^2d^5 − 40	
		c^4d^3 …	c^3d^4 …		
± 3	± 5	± 6	± 1		
67	99	70	27	± 24	± 6
136	536	536	577	266	134
182	594	954	961	944	248
± 388	± 1234	± 1566	± 1566	± 1234	± 388

$\between (x, y)^5$

	Q. No. 25.	Q′. No. 26.
$a^4\ b^0f^4$	...	+ 1
$a^3\ b^1ef^3$	...	− 20
b^0cdf^3	+ 1	− 120
ce^2f^2	− 1	+ 160
d^2ef^2	− 3	+ 360
de^3f	+ 5	− 640
e^5	− 2	+ 256
$a^2\ b^2df^3$	− 1	+ 160
e^2f^2	+ 1	− 10
$b^1c^2f^3$	− 3	+ 360
$cdef^2$	+ 11	− 1640
ce^3f	− 5	+ 320
d^3f^2	+ 12	− 1440
d^2e^2f	− 30	+ 4080
de^4	+ 15	− 1920
$b^0c^3ef^2$	+ 12	− 1440
$c^2d^2f^2$	− 21	+ 2640
c^2de^2f	− 34	+ 4480
c^2e^4	+ 22	− 2560
cd^3ef	+ 78	− 10080
cd^2e^3	− 48	+ 5760
d^5f	− 27	+ 3456
d^4e^2	+ 18	− 2160
$a^1\ b^3cf^3$	+ 5	− 640
def^2	− 5	+ 320
e^3f	...	− 180
$b^2c^2ef^2$	− 30	+ 4080
cd^2f^2	− 34	+ 4480
cde^2f	+ 133	− 14920
ce^4	− 54	+ 7200
d^3ef	− 18	+ 960
d^2e^3	+ 3	− 600
$b^1c^3df^2$	+ 78	− 10080
c^3e^2f	− 18	+ 960
c^2d^2ef	− 220	+ 28480
c^2de^3	+ 106	− 16000
cd^4f	+ 93	− 11520
cd^3e^2	− 30	+ 7200
d^5e	− 9	...
$b^0c^5f^2$	− 27	+ 3456
c^4def	+ 93	− 11520
c^4e^3	− 38	+ 6400
c^3d^3f	− 42	+ 5120
$c^3d^2e^2$	+ 8	− 3200
c^2d^4e	+ 6	...
cd^6	...	...
$a^0\ b^5f^3$	− 2	+ 256
b^4cef^2	+ 15	+ 1920
d^2f^2	+ 22	− 2560
de^2f	− 54	+ 7200

	Q. No. 25.	Q′. No. 26.
$a^0\ b^4e^4$	+ 27	− 3375
$b^3c^2df^2$	− 48	+ 5760
c^2e^2f	+ 3	− 600
cd^2ef	+ 106	− 16000
cde^3	− 81	+ 9000
d^4f	− 38	+ 6400
d^3e^2	+ 38	− 4000
$b^2c^4f^2$	+ 18	− 2160
c^3def	− 30	+ 7200
c^3e^3	+ 38	− 4000
c^2d^3f	+ 8	− 3200
$c^2d^2e^2$	+ 25	+ 2000
cd^4e	− 57	...
d^6	+ 18	...
b^1c^5ef	− 9	...
c^4d^2f	+ 6	...
c^4de^2	− 57	...
c^3d^3e	+ 74	...
c^2d^5	− 24	...
$b^0\ c^6df$	...	...
c^6e^2	+ 18	...
c^5d^2e	− 24	...
c^4d^4	+ 8	...
	± 6 169 525 424 ±1124	± 128505

The sums for Q' are

1		= 1
776 −	780	= −4
21256 −	21250	= +6
68656 −	68660	= −4
37816 −	37815	= +1
128505 −	128505	= 0

R. No. 92.

$a^4\ b^0ef^3$ …	$a^0\ b^4ce^2f\ -\ 15$	$a^4\ b^0f^4$ …	$a^0\ b^4e^4$ …	$a^3\ b^1f^4$ …	$a^0\ b^3d^3ef\ +\ 2$
$a^3\ b^1df^3$ …	$d^2ef\ -\ 38$	$a^3\ b^1ef^3$ …	$b^3c^2df^2\ +\ 18$	b^0cef^3 …	$d^2e^3\ +\ 15$
e^2f^2 …	$de^3\ +\ 45$	b^0cdf^3 …	$c^2e^2f\ -\ 66$	$d^2f^3\ +\ 1$	$b^2c^3df^2\ -\ 32$
$b^0c^2f^3\ -\ 1$	$b^3c^3f^2\ +\ 3$	ce^2f^2 …	$cd^2ef\ +\ 20$	$de^2f^2\ -\ 2$	$c^3e^2f\ -\ 39$
$cdef^2\ +\ 6$	$c^2def\ +\ 102$	$d^2ef^2\ +\ 2$	cde^3 …	$e^4f\ +\ 1$	$c^2d^2ef\ -\ 24$
$ce^3f\ -\ 4$	$c^2e^3\ -\ 15$	$de^3f\ -\ 4$	$d^4f\ +\ 58$	$a^2\ b^2ef^3$ …	$c^2de^3\ +\ 175$
$d^3f^2\ -\ 3$	$cd^3f\ +\ 76$	$e^5\ +\ 2$	$d^3e^2\ -\ 50$	$b^1cdf^3\ -\ 6$	$cd^4f\ +\ 25$
$d^2e^2f\ +\ 1$	$cd^2e^2\ -\ 175$	$a^2\ b^2df^3$ …	$b^2c^4f^2\ -\ 6$	$ce^2f^2\ +\ 6$	$cd^3e^2\ -\ 120$
$de^4\ +\ 1$	$d^4e\ +\ 35$	e^2f^2 …	$c^3def\ +\ 72$	$d^2ef^2\ +\ 3$	$d^5e\ +\ 15$
$a^2\ b^2cf^3\ +\ 2$	$b^2c^4ef\ -\ 42$	$b^1c^2f^3\ -\ 2$	$c^3e^3\ +\ 50$	de^3f …	$b^1c^5f^2\ +\ 9$
$def^2\ -\ 6$	$c^3d^2f\ -\ 182$	$cdef^2$ …	$c^2d^3f\ -\ 156$	$e^5\ -\ 3$	$c^4def\ +\ 106$
$e^3f\ +\ 4$	$c^3de^2\ +\ 120$	$ce^3f\ +\ 4$	$c^2d^2e^2$ …	$b^0c^3f^3\ +\ 3$	$c^4e^3\ -\ 35$
$b^1c^2ef^2\ -\ 3$	$c^2d^3e\ +\ 150$	$d^3f^2\ -\ 14$	$cd^4e\ +\ 90$	$c^2def^2\ -\ 3$	$c^3d^3f\ -\ 60$
$cd^2f^2\ +\ 3$	$cd^5\ -\ 70$	$d^2e^2f\ +\ 30$	$d^6\ -\ 30$	$c^2e^3f\ -\ 6$	$c^3d^2e^2\ -\ 150$
$cde^2f\ -\ 18$	$b^1c^5df\ +\ 126$	$de^4\ -\ 18$	$b^1c^5ef\ -\ 24$	cd^3f^2 …	$c^2d^4e\ +\ 175$
$ce^4\ +\ 17$	$c^5e^2\ -\ 15$	$b^0c^3ef^2\ +\ 14$	$c^4d^2f\ +\ 94$	$cd^2e^2f\ +\ 3$	$cd^6\ -\ 45$
$d^3ef\ +\ 22$	$c^4d^2e\ -\ 175$	$c^2d^2f^2$ …	$c^4de^2\ -\ 90$	$cde^4\ +\ 6$	$b^0c^6ef\ -\ 36$
$d^2e^3\ -\ 21$	$c^3d^4\ +\ 75$	$c^2de^2f\ -\ 66$	c^3d^3e …	d^4ef …	$c^5d^2f\ +\ 21$
$b^0c^3df^2$ …	$b^0c^7f\ -\ 27$	$c^2e^4\ +\ 26$	$c^2d^5\ +\ 10$	$d^3e^3\ -\ 3$	$c^5de^2\ +\ 70$
$c^3e^2f\ +\ 13$	$c^6de\ +\ 45$	$cd^3ef\ +\ 56$	$b^0c^6df\ -\ 18$	$a^1\ b^3df^3\ +\ 4$	$c^4d^3e\ -\ 75$
$c^2d^2ef\ -\ 12$	$c^5d^3\ -\ 20$	$cd^2e^3\ -\ 18$	$c^6e^2\ +\ 30$	$e^2f^2\ -\ 4$	$c^3d^5\ +\ 20$
$c^2de^3\ -\ 21$		$d^5f\ -\ 18$	$c^5d^2e\ -\ 10$	$b^2c^2f^3\ -\ 1$	
$cd^4f\ -\ 3$		$d^4e^2\ +\ 6$	c^4d^4 …	$cdef^2\ +\ 18$	
$cd^3e^2\ +\ 32$		$a^1\ b^3cf^3\ +\ 4$		$ce^3f\ -\ 16$	
$d^5e\ -\ 9$		$def^2\ -\ 4$		$d^3f^2\ -\ 13$	
$a^1\ b^4f^3\ -\ 1$		e^3f …		$d^2e^2f\ -\ 3$	
b^3cef^2 …		$b^2c^2ef^2\ -\ 30$		$de^4\ +\ 15$	
$d^2f^2\ +\ 6$		$cd^2f^2\ +\ 66$		$b^1c^3ef^2\ -\ 22$	
$de^2f\ +\ 16$		cde^2f …		$c^2d^2f^2\ +\ 12$	
$e^4\ -\ 18$		$ce^4\ -\ 18$		$c^2de^2f\ +\ 18$	
$b^2c^2df^2\ -\ 3$		$d^3ef\ -\ 84$		$c^2e^4\ +\ 38$	
$c^2e^2f\ +\ 3$		$d^2e^3\ +\ 66$		$cd^3ef\ +\ 32$	
$cd^2ef\ -\ 18$		$b^1c^3df^2\ -\ 56$		$cd^2e^3\ -\ 102$	
$cde^3\ +\ 14$		$c^3e^2f\ +\ 84$		$d^5f\ -\ 18$	
$d^4f\ -\ 41$		c^2d^2ef …		$d^4e^2\ +\ 42$	
$d^3e^2\ +\ 39$		$c^2de^3\ -\ 20$		$b^0c^4df^2\ +\ 3$	
$b^1c^4f^2$ …		$cd^4f\ +\ 40$		$c^4e^2f\ +\ 41$	
$c^3def\ -\ 32$		$cd^3e^2\ -\ 72$		$c^3d^2ef\ -\ 84$	
$c^3e^3\ -\ 2$		$d^5e\ +\ 24$		$c^3de^3\ -\ 76$	
$c^2d^3f\ +\ 84$		$b^0c^5f^2\ +\ 18$		$c^2d^4f\ +\ 33$	
$c^2d^2e^2\ +\ 24$		$c^4def\ -\ 40$		$c^2d^3e^2\ +\ 182$	
$cd^4e\ -\ 106$		$c^4e^3\ -\ 58$		$cd^5e\ -\ 126$	
$d^6\ +\ 36$		c^3d^3f …		$d^7\ +\ 27$	
$b^0c^5ef\ +\ 18$		$c^3d^2e^2\ +\ 156$		$a^0\ b^4cf^3\ -\ 1$	
$c^4d^2f\ -\ 33$	$\pm\ 8$	$c^2d^4e\ -\ 94$	$\pm\ 4$	$def^2\ -\ 17$	$\pm\ 2$
$c^4de^2\ -\ 25$	93	$cd^6\ +\ 18$	136	$e^3f\ +\ 18$	21
$c^3d^3e\ +\ 60$	300	$a^0\ b^5f^3\ -\ 2$	476	$b^3c^2ef^2\ +\ 21$	465
$c^2d^5\ -\ 21$	780	$b^4cef^2\ +\ 18$	478	$cd^2f^2\ +\ 21$	693
$a^0\ b^5ef^2\ +\ 3$	± 1181	$d^2f^2\ -\ 26$	± 1094	$cde^2f\ -\ 14$	± 1181
$b^4cdf^2\ -\ 6$		$de^2f\ +\ 18$		$ce^4\ -\ 45$	

$)(x, y)^2.$

S. No. 93 *bis*; S′. No. 93. $(*\between x, y)^3$.

Coef. x^3	S	S′	Coef. x^3	S	S′	Coef. x^2y	S	S′	Coef. x^2y	S	S′
$a^4\ b^1f^4$	...	...	$a^1\ b^3d^3ef$	− 66	+ 528	$a^4\ b^0cf^4$	...	+ 9	$a^1\ b^2c^2e^4$	+ 66	+ 12960
b^0cef^3	...	+ 9	d^2e^3	+ 72	− 45	def^3	...	− 45	cd^3ef	+ 78	+ 18612
d^2f^3	...	+ 21	$b^2c^3df^2$	− 21	− 2592	e^3f^2	...	+ 36	cd^2e^3	− 186	− 18900
de^2f^2	...	− 78	c^3e^2f	− 96	− 9747	$a^3\ b^2f^4$	...	− 9	d^5f	+ 51	− 3888
e^4f	...	+ 48	c^2d^2ef	+ 36	− 8496	b^1cef^3	...	− 18	d^4e^2	− 9	+ 2970
$a^3\ b^2ef^3$	...	− 9	c^2de^3	+ 213	+ 26610	d^2f^3	...	+ 243	$b^1c^4df^2$	+ 111	+ 15228
b^1cdf^3	...	− 162	cd^4f	+ 120	+ 8544	de^2f^2	...	+ 9	c^4e^2f	− 78	− 4968
ce^2f^2	...	+ 99	cd^3e^2	− 303	− 16650	e^4f	...	− 216	c^3d^2ef	− 36	− 14544
d^2ef^2	...	+ 309	d^5e	+ 51	+ 720	$b^0c^2df^3$	− 3	− 351	c^3de^3	− 54	− 12960
de^3f	...	+ 12	$b^1c^5f^2$	+ 9	+ 972	$c^2e^2f^2$	+ 3	+ 144	c^2d^4f	− 96	+ 1296
e^5	...	− 240	c^4def	+ 174	+ 24624	cd^2ef^2	+ 24	+ 1836	$c^2d^3e^2$	+ 150	+ 22500
$b^0c^3f^3$	− 2	− 81	c^4e^3	− 36	− 5040	cde^3f	− 42	− 2592	cd^5e	+ 30	− 6480
c^2def^2	+ 15	+ 1026	c^3d^3f	− 204	− 15984	ce^5	+ 18	+ 1152	d^7	− 27	...
c^2e^3f	− 9	− 768	$c^3d^2e^2$	− 174	− 29340	d^4f^2	− 18	− 1458	$b^0c^6f^2$	− 27	− 3888
cd^3f^2	− 9	− 738	c^2d^4e	+ 330	+ 34320	d^3e^2f	+ 33	+ 2268	c^5def	+ 24	+ 5184
cd^2e^2f	− 6	− 564	cd^6	− 99	− 8640	d^2e^4	− 15	− 1008	c^5e^3	+ 54	+ 5760
cde^4	+ 9	+ 1056	b^0c^6ef	− 63	− 7776	$a^2\ b^3ef^3$	...	+ 63	c^4d^3f	+ 27	− 576
d^4ef	+ 9	+ 756	c^5d^2f	+ 66	+ 5184	b^2cdf^3	+ 6	− 234	$c^4d^2e^2$	− 93	− 9360
d^3e^3	− 7	− 696	c^5de^2	+ 99	+ 12960	ce^2f^2	− 6	− 18	c^3d^4e	+ 6	+ 2880
$a^2\ b^3df^3$	...	+ 120	c^4d^3e	− 147	− 14400	d^2ef^2	− 24	− 3231	c^2d^6	+ 9	...
e^2f^2	...	− 21	c^3d^5	+ 45	+ 3840	de^3f	+ 42	+ 4293	$a^0\ b^5cf^3$	+ 3	+ 288
$b^2c^2f^3$	+ 6	+ 486	$a^0\ b^6f^3$	+ 2	+ 192	e^5	− 18	− 972	def^2	− 30	− 3888
$cdef^2$	− 30	− 2160	b^5cef^2	− 15	− 1440	$b^1c^3f^3$	+ 3	+ 810	e^3f	+ 27	+ 3645
ce^3f	+ 18	+ 1023	d^2f^2	− 6	− 192	c^2def^2	− 78	− 3825	$b^4c^2ef^2$	...	+ 756[illegible]
d^3f^2	+ 9	+ 120	de^2f	− 18	− 1080	c^2e^3f	+ 69	+ 4032	cd^2f^2	+ 51	+ 7488
d^2e^2f	+ 6	− 1053	e^4	+ 27	+ 2025	cd^3f^2	+ 93	+ 7938	cde^2f	− 39	− 405[illegible]
de^4	− 9	+ 1314	$b^4c^2df^2$	+ 24	+ 1728	cd^2e^2f	− 51	− 9360	ce^4	− 27	− 6075
$b^1c^3ef^2$	− 15	− 1863	c^2e^2f	+ 51	+ 4410	cde^4	− 33	− 864	d^3ef	+ 60	− 432[illegible]
$c^2d^2f^2$	+ 21	+ 2538	cd^2ef	+ 102	+ 5280	d^4ef	− 57	− 1296	d^2e^3	− 45	+ 6075
c^2de^2f	− 6	+ 2340	cde^3	− 171	− 13500	d^3e^3	+ 54	+ 2700	$b^3c^3df^2$	− 39	− 712[illegible]
c^2e^4	+ 18	+ 672	d^4f	+ 6	− 4800	$b^0c^4ef^2$	+ 24	− 324	c^3e^2f	+ 45	+ 297[illegible]
cd^3ef	+ 30	+ 2820	d^3e^2	+ 18	+ 7800	$c^3d^2f^2$	− 36	− 2484	c^2d^2ef	− 108	+ 306[illegible]
cd^2e^3	− 51	− 7812	$b^3c^4f^2$	− 9	− 648	c^3de^2f	− 9	+ 6624	c^2de^3	+ 96	+ 1012[illegible]
d^5f	− 36	− 3024	c^3def	− 210	− 14040	c^3e^4	− 54	− 6912	cd^4f	− 111	+ 144[illegible]
d^4e^2	+ 39	+ 4572	c^3e^3	+ 43	+ 3075	c^2d^3ef	+ 24	− 4428	cd^3e^2	+ 147	− 1395[illegible]
$b^0c^4df^2$	− 3	− 324	c^2d^3f	− 120	+ 9120	$c^2d^2e^3$	+ 129	+ 12672	d^5e	− 30	+ 360[illegible]
c^4e^2f	+ 45	+ 3888	$c^2d^2e^2$	+ 345	+ 16350	cd^5f	+ 9	+ 1944	$b^2c^5f^2$	+ 9	+ 194[illegible]
c^3d^2ef	− 84	− 8748	cd^4e	− 87	− 19200	cd^4e^2	− 114	− 9072	c^4def	+ 6	− 162[illegible]
c^3de^3	− 63	− 4800	d^6	− 2	+ 4800	d^6e	+ 27	+ 1944	c^4e^3	− 48	− 450[illegible]
c^2d^4f	+ 45	+ 4248	b^2c^5ef	+ 72	+ 4860	$a^1\ b^4df^3$	− 3	+ 144	c^3d^3f	+ 234	− 36[illegible]
$c^2d^3e^2$	+ 150	+ 14520	c^4d^2f	+ 240	− 3240	e^2f^2	+ 3	− 243	$c^3d^2e^2$	− 150	+ 630[illegible]
cd^5e	− 117	− 11448	c^4de^2	− 192	− 8100	$b^3c^2f^3$	− 6	− 900	c^2d^4e	− 108	− 180[illegible]
d^7	+ 27	+ 2592	c^3d^3e	− 186	+ 9000	$cdef^2$	+ 108	+ 10620	cd^6	+ 57	...
$a^1\ b^4cf^3$	− 6	− 576	c^2d^5	+ 96	− 2400	ce^3f	− 96	− 8586	b^1c^6ef	+ 9	...
def^2	+ 15	+ 672	b^1c^6df	− 144	...	d^3f^2	− 21	− 864	c^5d^2f	− 141	...
e^3f	− 9	− 459	c^6e^2	+ 18	...	d^2e^2f	− 48	− 1215	c^5de^2	+ 87	...
$b^3c^2ef^2$	+ 30	+ 3456	c^5d^2e	+ 201	...	de^4	+ 63	+ 1215	c^4d^3e	+ 96	...
cd^2f^2	− 15	− 864	c^4d^4	− 87	...	$b^2c^3ef^2$	− 24	− 1836	c^3d^5	− 51	...
cde^2f	+ 24	+ 2094	b^0c^8f	+ 27	...	$c^2d^2f^2$	− 123	− 16812	b^0c^7df	+ 27	...
ce^4	− 45	− 3915	c^7de	− 45	...	c^2de^2f	+ 147	+ 6651	c^7e^2	− 18	...
			c^6d^3	+ 20	...				c^6d^2e	− 21	...
									c^5d^4	+ 12	...

For the Numerical Verifications for S see further pp. 304, 305.

	S	S′		S	S′
		± 78			± [illegible]
	± 33	3258		± 78	56[illegible]
	414	41253		480	430[illegible]
	1284	124524		927	1060[illegible]
	1292	68640		966	476[illegible]
	± 3023	± 237753		± 2451	± 2024[illegible]

S. No. 93 *bis*; S′. No. 93.

Coef. xy^2	S	S′
$a^4\, b^0df^4$	...	− 9
e^2f^3	...	+ 9
$a^3\, b^1cf^4$	..	+ 45
def^3	...	+ 18
e^3f^2	...	− 63
$b^0c^2ef^3$	...	− 243
cd^2f^3	+ 3	+ 351
cde^2f^2	− 6	+ 234
ce^4f	+ 3	− 144
d^3ef^2	− 3	− 810
d^2e^3f	+ 6	+ 900
de^5	− 3	− 288
$a^2\, b^3f^4$	...	− 36
b^2cef^3	...	− 9
d^2f^3	− 3	− 144
de^2f^2	+ 6	+ 18
e^4f	− 3	+ 243
$b^1c^2df^3$	− 24	− 1836
$c^2e^2f^2$	+ 24	+ 3231
cd^2ef^2	+ 78	+ 3825
cde^3f	− 108	− 10620
ce^5	+ 30	+ 3888
d^4f^2	− 24	+ 324
d^3e^2f	+ 24	+ 1836
d^2e^4	...	− 756
$b^0c^4f^3$	+ 18	+ 1458
c^3def^2	− 93	− 7938
c^3e^3f	+ 21	+ 864
$c^2d^3f^2$	+ 36	+ 2484
$c^2d^2e^2f$	+ 123	+ 16812
c^2de^4	− 51	− 7488
cd^4ef	− 111	− 15228
cd^3e^3	+ 39	+ 7128
d^6f	+ 27	+ 3888
d^5e^2	− 9	− 1944
b^4ef^3	...	+ 216
b^3cdf^3	+ 42	+ 2592
ce^2f^2	− 42	− 4293
d^2ef^2	− 69	− 4032
de^3f	+ 96	+ 8586
e^5	− 27	− 3645
$b^2c^3f^3$	− 33	− 2268
c^2def^2	+ 51	+ 9360
c^2e^3f	+ 48	+ 1215
cd^3f^2	+ 9	− 6624
cd^2e^2f	− 147	− 6651
cde^4	+ 39	+ 4050
d^4ef	+ 78	+ 4968
d^3e^3	− 45	− 2970
$b^1c^4ef^2$	+ 57	+ 1296
$c^3d^2f^2$	− 24	+ 4428
c^3de^2f	− 78	− 18612

Coef. xy^2	S	S′
$a^1\, b^1c^3e^4$	− 60	+ 4320
c^2d^3ef	+ 36	+ 14544
$c^2d^2e^3$	+ 108	− 3060
cd^5f	− 24	− 5184
cd^4e^2	− 6	+ 1620
d^6e	− 9	...
$b^0c^5df^2$	− 9	− 1944
c^5e^2f	− 51	+ 3888
c^4d^2ef	+ 96	− 1296
c^4de^3	+ 111	− 1440
c^3d^4f	− 27	+ 576
$c^3d^3e^2$	− 234	+ 360
c^2d^5e	+ 141	...
cd^7	− 27	...
$a^0\, b^5df^3$	− 18	− 1152
e^2f^2	+ 18	+ 972
$b^4c^2f^3$	+ 15	+ 1008
$cdef^2$	+ 33	+ 864
ce^3f	− 63	− 1215
d^3f^2	+ 54	+ 6912
d^2e^2f	− 66	− 12960
de^4	+ 27	+ 6075
$b^3c^3ef^2$	− 54	− 2700
$c^2d^2f^2$	− 129	− 12672
c^2de^2f	+ 186	+ 18900
c^2e^4	+ 45	− 6075
cd^3ef	+ 54	+ 12960
cd^2e^3	− 96	− 10125
d^5f	− 54	− 5760
d^4e^2	+ 48	+ 4500
$b^2c^4df^2$	+ 114	+ 9072
c^4e^2f	+ 9	− 2970
c^3d^2ef	− 150	− 22500
c^3de^3	− 147	+ 13950
c^2d^4f	+ 93	+ 9360
$c^2d^3e^2$	+ 150	− 6300
cd^5e	− 87	...
d^7	+ 18	...
$b^1c^6f^2$	− 27	− 1944
c^5def	− 30	+ 6480
c^5e^3	+ 30	− 3600
c^4d^3f	− 6	− 2880
$c^4d^2e^2$	+ 108	+ 1800
c^3d^4e	− 96	...
c^2d^6	+ 21	...
b^0c^7ef	+ 27	...
c^6d^2f	− 9	...
c^6de^2	− 57	...
c^5d^3e	+ 51	...
c^4d^5	− 12	...
		± 9
	± 12	1548
	426	45999
	912	62019
	1101	92853
	± 2451	± 202428

Coef. y^3	S	S′
$a^4\, b^0ef^4$	...	...
$a^3\, b^1df^4$	...	− 9
e^2f^3	...	+ 9
$b^0c^2f^4$	...	− 21
$cdef^3$	...	+ 162
ce^3f^2	...	− 120
d^3f^3	+ 2	+ 81
$d^2e^2f^2$	− 6	− 486
de^4f	+ 6	+ 576
e^6	− 2	− 192
$a^2\, b^2cf^4$	...	+ 78
def^3	...	− 99
e^3f^2	...	+ 21
$b^1c^2ef^3$	...	− 309
cd^2f^3	− 15	− 1026
cde^2f^2	+ 30	+ 2160
ce^4f	− 15	− 672
d^3ef^2	+ 15	+ 1863
d^2e^3f	− 30	− 3456
de^5	+ 15	+ 1440
$b^0c^3df^3$	+ 9	+ 738
$c^3e^2f^2$	− 9	− 120
$c^2d^2ef^2$	− 21	− 2538
c^2de^3f	+ 15	+ 864
c^2e^5	+ 6	+ 192
cd^4f^2	+ 3	+ 324
cd^3e^2f	+ 21	+ 2592
cd^2e^4	− 24	− 1728
d^5ef	− 9	− 972
d^4e^3	+ 9	+ 648
$a^1\, b^4f^4$	...	− 48
b^3cef^3	...	− 12
d^2f^3	+ 9	+ 768
de^2f^2	− 18	− 1023
e^4f	+ 9	+ 459
$b^2c^2df^3$	+ 6	+ 564
$c^2e^2f^2$	− 6	+ 1053
cd^2ef^2	+ 6	− 2340
cde^3f	− 24	− 2094
ce^5	+ 18	+ 1080
d^4f^2	− 45	− 3888
d^3e^2f	+ 96	+ 9747
d^2e^4	− 51	− 4410
$b^1c^4f^3$	− 9	− 756
c^3def^2	− 30	− 2820
c^3e^3f	+ 66	− 528
$c^2d^3f^2$	+ 84	+ 8748
$c^2d^2e^2f$	− 36	+ 8496
c^2de^4	− 102	− 5280
cd^4ef	− 174	− 24624
cd^3e^3	+ 210	+ 14040
d^6f	+ 63	+ 7776

Coef. y^3	S	S′
$a^1\, b^1d^5e^2$	− 72	− 4860
$b^0c^5ef^2$	+ 36	+ 3024
$c^4d^2f^2$	− 45	− 4248
c^4de^2f	− 120	− 8544
c^4e^4	− 6	+ 4800
c^3d^3ef	+ 204	+ 15984
$c^3d^2e^3$	+ 120	− 9120
c^2d^5f	− 66	− 5184
$c^2d^4e^2$	− 240	+ 3240
cd^6e	+ 144	...
d^8	− 27	...
$a^0\, b^5ef^3$	...	+ 240
b^4cdf^3	− 9	− 1056
ce^2f^2	+ 9	− 1314
d^2ef^2	− 18	− 672
de^3f	+ 45	+ 3915
e^5	− 27	− 2025
$b^3c^3f^3$	+ 7	+ 696
c^2def^2	+ 51	+ 7812
c^2e^3f	− 72	+ 45
cd^3f^2	+ 63	+ 4800
cd^2e^2f	− 213	− 26610
cde^4	+ 171	+ 13500
d^4ef	+ 36	+ 5040
d^3e^3	− 43	− 3075
$b^2c^4ef^2$	− 39	− 4572
$c^3d^2f^2$	− 150	− 14520
c^3de^2f	+ 303	+ 16650
c^3e^4	− 18	− 7800
c^2d^3ef	+ 174	+ 29340
$c^2d^2e^3$	− 345	− 16350
cd^5f	− 99	− 12960
cd^4e^2	+ 192	+ 8100
d^6e	− 18	...
$b^1c^5df^2$	+ 117	+ 11448
c^5e^2f	− 51	− 720
c^4d^2ef	− 330	− 34320
c^4de^3	+ 87	+ 19200
c^3d^4f	+ 147	+ 14400
$c^3d^3e^2$	+ 186	− 9000
c^2d^5e	− 201	...
cd^7	+ 45	...
$b^0c^7f^2$	− 27	− 2592
c^6def	+ 99	+ 8640
c^6e^3	+ 2	− 4800
c^5d^3f	− 45	− 3840
$c^5d^2e^2$	− 96	+ 2400
c^4d^5e	+ 87	...
c^3d^6	− 20	...
	± 8	± 828
	123	10920
	1071	79779
	1821	146226
	± 3023	± 237753

T. No. 94.

x coefficient.

$a^5\ b^0cf^5$		...
def^4		...
e^3f^3		...
$a^4\ b^2f^5$		...
b^1cef^4		...
d^2f^4		...
de^2f^3		...
e^4f^2		...
$b^0c^2df^4$	−	1
$c^2e^2f^3$	+	1
cd^2ef^3	+	7
cde^3f^2	−	12
ce^5f	+	5
d^4f^3	−	6
$d^3e^2f^2$	+	12
d^2e^4f	−	7
de^6	+	1
$a^3\ b^3ef^4$		...
b^2cdf^4	+	2
ce^2f^3	−	2
d^2ef^3	−	7
de^3f^2	+	12
e^5f	−	5
$b^1c^3f^4$	+	3
c^2def^3	−	30
$c^2e^3f^2$	+	21
cd^3f^3	+	44
$cd^2e^2f^2$	−	69
cde^4f	+	62
ce^6	−	28
d^4ef^2	−	6
d^3e^3f	−	8
d^2e^5	+	11
$b^0c^4ef^3$	−	6
$c^3d^2f^3$	−	11
$c^3de^2f^2$	+	96
c^3e^4f	−	64
$c^2d^3ef^2$	−	66
$c^2d^2e^3f$	−	29
c^2de^5	+	68
cd^5f^2	+	18
cd^4e^2f	+	75
cd^3e^4	−	78
d^6ef	−	27
d^5e^3	+	24
$a^2\ b^4df^4$	−	1
e^2f^3	+	1
$b^3c^2f^4$	−	8
$cdef^3$	+	46
ce^3f^2	−	30

$a^2\ b^3d^3f^3$	−	20
$d^2e^2f^2$	+	33
de^4f	−	48
e^6	+	27
$b^2c^3ef^3$	+	39
$c^2d^2f^3$	−	105
$c^2de^2f^2$	+	18
c^2e^4f	−	6
cd^3ef^2	+	114
cd^2e^3f	−	57
cde^5	+	12
d^5f^2	−	6
d^4e^2f	+	3
d^3e^4	−	12
$b^1c^4df^3$	+	90
$c^4e^2f^2$	−	198
$c^3d^2ef^2$	−	9
c^3de^3f	+	238
c^3e^5	+	116
$c^2d^4f^2$	−	6
$c^2d^3e^2f$	+	108
$c^2d^2e^4$	−	513
cd^5ef	−	294
cd^4e^3	+	513
d^7f	+	108
d^6e^2	−	153
$b^0c^6f^3$	−	27
c^5def^2	+	108
c^5e^3f	+	194
$c^4d^3f^2$	−	42
$c^4d^2e^2f$	−	663
c^4de^4	−	274
c^3d^4ef	+	570
$c^3d^3e^3$	+	914
c^2d^6f	−	153
$c^2d^5e^2$	−	1032
cd^7e	+	486
d^9	−	81
$a^1\ b^5cf^4$	+	7
def^3	−	16
e^3f^2	+	9
$b^4c^2ef^3$	−	53
cd^2f^3	+	104
cde^2f^2	−	150
ce^4f	+	117
d^3ef^2	−	48
d^2e^3f	+	138
de^5	−	108
$b^3c^3df^3$	−	82
$c^3e^2f^2$	+	315

x coefficient.

$a^1\ b^3c^2d^2ef^2$	+	153
c^2de^3f	−	390
c^2e^5	−	234
cd^4f^2	−	114
cd^3e^2f	−	308
cd^2e^4	+	735
d^5ef	+	208
d^4e^3	−	283
$b^2c^5f^3$	+	27
c^4def^2	−	396
c^4e^3f	−	337
$c^3d^3f^2$	+	222
$c^3d^2e^2f$	+	783
c^3de^4	+	880
c^2d^4ef	+	93
$c^2d^3e^3$	−	1986
cd^6f	−	240
cd^5e^2	+	1098
d^7e	−	144
$b^1c^6ef^2$	+	81
$c^5d^2f^2$	−	54
c^5de^2f	+	570
c^5e^4	−	148
c^4d^3ef	−	1116
$c^4d^2e^3$	−	527
c^3d^5f	+	474
$c^3d^4e^2$	+	1662
c^2d^6e	−	1185
cd^8	+	243
$b^0c^7df^2$		...
c^7e^2f	−	216
c^6d^2ef	+	369
c^6de^3	+	340
c^5d^4f	−	149
$c^5d^3e^2$	−	730
c^4d^5e	+	488
c^3d^7	−	102
$a^0\ b^7f^4$	−	2
b^6cef^3	+	20
d^2f^3	−	24
de^2f^2	+	72
e^4f	−	54
$b^5c^2df^3$	+	16
$c^2e^2f^2$	−	129
cd^2ef^2	−	108
cde^3f	+	72
ce^5	+	135
d^4f^2	+	84
d^3e^2f	−	112
d^2e^4		...

$a^0\ b^4c^4f^3$	−	6
c^3def^2	+	240
c^3e^3f	+	179
$c^2d^3f^2$	−	144
$c^2d^2e^2f$	+	306
c^2de^4	−	765
cd^4ef	+	28
cd^3e^3	+	280
d^6f	−	88
d^5e^2	+	40
$b^3c^5ef^2$	−	63
$c^4d^2f^2$	+	42
c^4de^2f	−	798
c^4e^4	+	175
c^3d^3ef	−	224
$c^3d^2e^3$	+	1365
c^2d^5f	+	368
$c^2d^4e^2$	−	1025
cd^6e	+	60
d^8	+	30
$b^2c^6df^2$		...
c^6e^2f	+	252
c^5d^2ef	+	798
c^5de^3	−	700
c^4d^4f	−	578
$c^4d^3e^2$	−	370
c^3d^5e	+	880
c^2d^7	−	240
$b^1c^8f^2$		...
c^7def	−	486
c^7e^3	+	60
c^6d^3f	+	312
$c^6d^2e^2$	+	645
c^5d^4e	−	735
c^4d^6	+	190
b^0c^9ef	+	81
c^8d^2f	−	54
c^8de^2	−	135
c^7d^3e	+	150
c^6d^5	−	40

± 26
436
3738
9116
6880

±20196

and see further p. 306.

T. No. 94.

y coefficient.

Term		
$a^5\ b^0df^5$		...
e^2f^4		...
$a^4\ b^1cf^5$		...
def^4		...
e^3f^3		...
$b^0c^2ef^4$		...
cd^2f^4	+	1
cde^2f^3	−	2
ce^4f^2	+	1
d^3ef^3	−	3
$d^2e^3f^2$	+	8
de^5f	−	7
e^7	+	2
$a^3\ b^3f^5$		...
b^2cef^4		...
d^2f^4	−	1
de^2f^3	+	2
e^4f^2	−	1
$b^1c^2df^4$	−	7
$c^2e^2f^3$	+	7
cd^2ef^3	+	30
cde^3f^2	−	46
ce^5f	+	16
d^4f^3	+	6
$d^3e^2f^2$	−	39
d^2e^4f	+	53
de^6	−	20
$b^0c^4f^4$	+	6
c^3def^3	−	44
$c^3e^3f^2$	+	20
$c^2d^3f^3$	+	11
$c^2d^2e^2f^2$	+	105
c^2de^4f	−	104
c^2e^6	+	24
cd^4ef^2	−	90
cd^3e^3f	+	82
cd^2e^5	−	16
d^6f^2	+	27
d^5e^2f	−	27
d^4e^2	+	6
$a^2\ b^4ef^4$		...
b^3cdf^4	+	12
ce^2f^3	−	12
d^2ef^3	−	21
de^3f^2	+	30
e^5f	−	9
$b^2c^3f^4$	−	12
c^2def^3	+	69
$c^2e^3f^2$	−	33
cd^3f^3	−	96

Term		
$a^2\ b^2cd^2e^2f^2$	−	18
cde^4f	+	150
ce^6	−	72
d^4ef^2	+	198
d^3e^3f	−	315
d^2e^5	+	129
$b^1c^4ef^3$	+	6
$c^3d^2f^3$	+	66
$c^3de^2f^2$	−	114
c^3e^4f	+	48
$c^2d^3ef^2$	+	9
$c^2d^2e^3f$	−	153
c^2de^5	+	108
cd^5f^2	−	108
cd^4e^2f	+	396
cd^3e^4	−	240
d^6ef	−	81
d^5e^3	+	63
$b^0c^5df^3$	−	18
$c^5e^2f^2$	+	6
$c^4d^2ef^2$	+	6
c^4de^3f	+	114
c^4e^5	−	84
$c^3d^4f^2$	+	42
$c^3d^3e^2f$	−	222
$c^3d^2e^4$	+	144
c^2d^5ef	+	54
$c^2d^4e^3$	−	42
cd^7f		...
cd^6e^2		...
d^8e		...
$a^1\ b^5df^4$	−	5
e^2f^3	+	5
$b^4c^2f^4$	+	7
$cdef^3$	−	62
ce^3f^2	+	48
d^3f^3	+	64
$d^2e^2f^2$	+	6
de^4f	−	117
e^6	+	54
$b^3c^3ef^3$	+	8
$c^2d^2f^3$	+	29
$c^2de^2f^2$	+	57
c^2e^4f	−	138
cd^3ef^2	−	238
cd^2e^3f	+	390
cde^5	−	72
d^5f^2	−	194
d^4e^2f	+	337
d^3e^4	−	179

y coefficient.

Term		
$a^1\ b^2c^4df^3$	−	75
$c^4e^2f^2$	−	3
$c^3d^2ef^2$	−	108
c^3de^3f	+	308
c^3e^5	+	112
$c^2d^4f^2$	+	663
$c^2d^3e^2f$	−	783
$c^2d^2e^4$	−	306
cd^5ef	−	570
cd^4e^3	+	798
d^7f	+	216
d^6e^2	−	252
$b^1c^6f^3$	+	27
c^5def^2	+	294
c^5e^3f	−	208
$c^4d^2e^2f$	−	93
$c^4d^3f^2$	−	570
c^4de^4	−	28
c^3d^4ef	+	1116
$c^3d^3e^3$	+	224
c^2d^6f	−	369
$c^2d^5e^2$	−	798
cd^7e	+	486
d^9	−	81
$b^0c^7ef^2$	−	108
$c^6d^2f^2$	+	153
c^6de^2f	+	240
c^6e^4	+	88
c^5d^3ef	−	474
$c^5d^2e^3$	−	368
c^4d^5f	+	149
$c^4d^4e^2$	+	578
c^3d^6e	−	312
c^2d^8	+	54
$a^0\ b^6cf^4$	−	1
def^3	+	28
e^3f^2	−	27
$b^5c^2ef^3$	−	11
cd^2f^3	−	68
cde^2f^2	−	12
ce^4f	+	108
d^3ef^2	−	116
d^2e^3f	+	234
de^5	−	135
$b^4c^3df^3$	+	78
$c^3e^2f^2$	+	12
$c^2d^2ef^2$	+	513
c^2de^3f	−	735
c^2e^5		...
cd^4f^2	+	274

Term		
$a^0\ b^4cd^3e^2f$	−	880
cd^2e^4	+	765
d^5ef	+	148
d^4e^3	−	175
$b^3c^5f^3$	−	24
c^4def^2	−	513
c^4e^3f	+	283
$c^3d^3f^2$	−	914
$c^3d^2e^2f$	+	1986
c^3de^4	−	280
c^2d^4ef	+	527
$c^2d^3e^3$	−	1365
cd^6f	−	340
cd^5e^2	+	700
d^7e	−	60
$b^2c^6ef^2$	+	153
$c^5d^2f^2$	+	1032
c^5de^2f	−	1098
c^5e^4	−	40
c^4d^3ef	−	1662
$c^4d^2e^3$	+	1025
c^3d^5f	+	730
$c^3d^4e^2$	+	370
c^2d^6e	−	645
cd^8	+	135
$b^1c^7df^2$	−	486
c^7e^2f	+	144
c^6d^2ef	+	1185
c^6de^3	−	60
c^5d^4f	−	488
$c^5d^3e^2$	−	880
c^4d^5e	+	735
c^3d^7	−	150
$b^0c^9f^2$	+	81
c^8def	−	243
c^8e^3	−	30
c^7d^3f	+	102
$c^7d^2e^2$	+	240
c^6d^4e	−	190
c^5d^6	+	40

± 12
395
1650
6511
11628

± 20196

and see further p. 306.

⅋$(x, y)^1$

U. No. 29.

Term	Sign	Coefficient
$a^6\ b^0f^6$		...
$a^5\ b^1ef^5$		...
b^0cdf^5		...
ce^2f^4		...
d^2ef^4		...
de^3f^3		...
e^5f^2		...
$a^4\ b^3f^5$		...
b^2df^5		...
e^2f^4		...
$b^1c^2f^5$		...
$cdef^4$		...
d^3f^4		...
$d^2e^2f^3$		...
de^4f^2		...
e^6f		...
$b^0c^3ef^4$		...
$c^2d^2f^4$	−	1
$c^2de^2f^3$	+	2
$c^2e^4f^2$	−	1
cd^3ef^3	+	6
$cd^2e^3f^2$	−	16
cde^5f	+	14
ce^7	−	4
d^5f^3	−	4
$d^4e^2f^2$	+	11
d^3e^4f	−	10
d^2e^6	+	3
$a^3\ b^3cf^5$		...
def^4		...
e^3f^3		...
$b^2c^2ef^4$		...
cd^2f^4	+	2
cde^2f^3	−	4
ce^4f^2	+	2
d^3ef^3	−	6
$d^2e^3f^2$	+	16
de^5f	−	14
e^7	+	4
$b^1c^3df^4$	+	6
$c^3e^2f^3$	−	6
$c^2d^2ef^3$	−	50
$c^2de^3f^2$	+	82
c^2e^5f	−	32
cd^4f^3	+	36
$cd^3e^2f^2$	−	30
cd^2e^4f	−	30
cde^6	+	24
d^5ef^2	−	28
d^4e^3f	+	50
$a^3\ b^1d^3e^5$	−	22
$b^0c^5f^4$	−	4
c^4def^3	+	36
$c^4e^3f^2$	−	16
$c^3d^3f^3$	−	22
$c^3d^2e^2f^2$	−	50
c^3de^4f	+	16
c^3e^6	+	16
$c^2d^4ef^2$	+	54
$c^2d^3e^3f$	+	46
$c^2d^2e^5$	−	60
cd^6f^2	−	6
cd^5e^2f	−	70
cd^4e^4	+	56
d^7ef	+	18
d^6e^3	−	14
$a^2\ b^5f^5$		...
b^4cef^4		...
d^2f^4	−	1
de^2f^3	+	2
e^4f^2	−	1
$b^3c^2df^4$	−	16
$c^2e^2f^3$	+	16
cd^2ef^3	+	82
cde^3f^2	−	132
ce^5f	+	50
d^4f^3	−	16
$d^3e^2f^2$	−	14
d^2e^4f	+	60
de^6	−	30
$b^2c^4f^4$	+	11
c^3def^3	−	30
$c^3e^3f^2$	−	14
$c^2d^3f^3$	−	50
$c^2d^2e^2f^2$	+	168
c^2de^4f	−	48
c^2e^6	−	4
cd^4ef^2	−	48
cd^3e^3f	−	2
cd^2e^5	+	6
d^6f^2	+	62
d^5e^2f	−	90
d^4e^4	+	39
$b^1c^5ef^3$	−	28
$c^4d^2f^3$	+	54
$c^4de^2f^2$	−	48
c^4e^4f	+	112
$c^3d^3ef^2$	+	82
$c^3d^2e^3f$	−	170
c^3de^5	−	104
$a^2\ b^1c^2d^5f^2$	−	108
$c^2d^4e^2f$	−	42
$c^2d^3e^4$	+	298
cd^6ef	+	242
cd^5e^3	−	294
d^8f	−	72
d^7e^2	+	78
$b^0c^6df^3$	−	6
$c^6e^2f^2$	+	62
$c^5d^2ef^2$	−	108
c^5de^3f	−	164
c^5e^5	−	24
$c^4d^4f^2$	+	63
$c^4d^3e^2f$	+	394
$c^4d^2e^4$	+	194
c^3d^5ef	−	324
$c^3d^4e^3$	−	440
c^2d^7f	+	78
$c^2d^6e^2$	+	428
cd^8e	−	180
d^{10}	+	27
$a^1\ b^6ef^4$		...
b^5cdf^4	+	14
ce^2f^3	−	14
d^2ef^3	−	32
de^3f^2	+	50
e^5f	−	18
$b^4c^3f^4$	−	10
c^2def^3	−	30
$c^2e^3f^2$	+	60
$cd^2e^2f^2$	−	48
cd^3f^3	+	16
cde^4f	+	38
ce^6	−	36
d^4ef^2	+	112
d^3e^3f	−	204
d^2e^5	+	102
$b^3c^4ef^3$	+	50
$c^3d^2f^3$	+	46
$c^3de^2f^2$	−	2
c^3e^4f	−	204
$c^2d^3ef^2$	−	170
c^2de^5	+	308
$c^2d^2e^3f$	+	42
cd^5f^2	−	164
cd^4e^2f	+	674
cd^3e^4	−	590
d^6ef	−	128
d^5e^3	+	138
$b^2c^6df^3$	−	70
$a^1\ b^2c^5e^2f^2$	−	90
$c^4d^2ef^2$	−	42
c^4de^3f	+	674
c^4e^5	−	4
$c^3d^4f^2$	+	394
$c^3d^2e^4$	−	652
$c^3d^3e^2f$	−	714
c^2d^5ef	−	498
$c^2d^4e^3$	+	1246
cd^7f	+	224
cd^6e^2	−	516
d^8e	+	48
$b^1c^7f^3$	+	18
c^6def^2	+	242
c^6e^3f	−	128
$c^5d^3f^2$	−	324
$c^5d^2e^2f$	−	498
c^5de^4	+	136
c^4d^4ef	+	1078
$c^4d^3e^3$	+	206
c^3d^6f	−	342
$c^3d^5e^2$	−	804
c^2d^7e	+	506
cd^9	−	90
$b^0c^8ef^2$	−	72
$c^7d^2f^2$	+	78
c^7de^2f	+	224
c^7e^4	+	16
c^6d^3ef	−	342
$c^6d^2e^3$	−	220
c^5d^5f	+	106
$c^5d^4e^2$	+	392
c^4d^6e	−	222
c^3d^8	+	40
$a^0\ b^7df^4$	−	4
e^2f^3	+	4
$b^6c^2f^4$	+	3
$cdef^3$	+	24
ce^3f^2	−	30
d^3f^3	+	16
$d^2e^2f^2$	−	4
de^4f	−	36
e^6	+	27
$b^5cd^3ef^2$	−	104
c^3ef^3	−	22
$c^2d^2f^3$	−	60
$c^2de^2f^2$	+	6
c^2e^4f	+	102
cd^2e^3f	+	308
cde^5	−	234
$a^0\ b^5d^5f^2$	−	24
d^4e^2f	−	4
d^3e^4	+	32
$b^4c^4df^3$	+	56
$c^4e^2f^2$	+	39
$c^3d^2ef^2$	+	298
c^3de^3f	−	590
c^3e^5	+	32
$c^2d^4f^2$	+	194
$c^2d^3e^2f$	−	652
$c^2d^2e^4$	+	713
cd^5ef	+	136
cd^4e^3	−	246
d^7f	+	16
d^6e^2	+	4
$b^3c^6f^3$	−	14
c^5def^2	−	294
c^5e^3f	+	138
$c^4d^3f^2$	−	440
$c^4d^2e^2f$	+	1246
c^4de^4	−	246
c^3d^4ef	+	206
$c^3d^3e^3$	−	868
c^2d^6f	−	222
$c^2d^5e^2$	+	550
cd^7e	−	56
d^9	−	4
$b^2c^7ef^2$	+	78
$c^6d^2f^2$	+	428
c^6de^2f	−	516
c^6e^4	+	4
c^5d^3ef	−	804
$c^5d^2e^3$	+	550
c^4d^5f	+	392
$c^4d^4e^2$	+	143
c^3d^6e	−	354
c^2d^8	+	83
$b^1c^8df^2$	−	180
c^8e^2f	+	48
c^7d^2ef	+	506
c^7de^3	−	56
c^6d^4f	−	222
$c^6d^3e^2$	−	354
c^5d^5e	+	330
c^4d^7	−	72
$b^0c^{10}f^2$	+	27
c^9def	−	90
c^9e^3	−	4
c^8d^3f	+	40
$c^8d^2e^2$	+	83
c^7d^4e	−	72
c^6d^6	+	16

±36, ±464, ±2608, ±7278, ±6878, together ±17264: and see further p. 307.

V. No. 95. $(*\between x, y)^1$.

x coefficient.

$a^6\ b^0cf^6$		...	$a^3\ b^4df^5$	−	2	$a^3\ b^0cd^6e^3$	+	876	$a^2\ b^1c^3d^3e^4$	+	2800
def^5		...	e^2f^4	+	2	d^9f	+	162	c^2d^6ef	+	6624
e^3f^4		...	$b^3c^2f^5$	−	16	d^8e^2	−	162	$c^2d^5e^3$	+	2052
$a^5\ b^2f^6$		...	$cdef^4$	+	32	$a^2\ b^5cf^5$	+	14	cd^8f	−	918
b^1cef^5		...	ce^3f^3		...	def^4	−	6	cd^7e^2	−	2304
d^2f^5		...	d^3f^4	−	8	e^3f^3	−	8	d^9e	+	486
de^2f^4		...	$d^2e^2f^3$	+	80	$b^4c^2ef^4$	−	50	c^7df^2		...
e^4f^3		...	de^4f^2	−	160	cd^2f^4	+	90	$b^0c^7e^2f^2$	+	504
$b^0c^2df^5$	−	2	e^6f	+	72	cde^2f^3	−	120	$c^6d^2ef^2$	−	576
$c^2e^2f^4$	+	2	$b^2c^3ef^4$	+	84	ce^4f^2	+	60	c^6de^3f	−	2288
cd^2ef^4	+	10	$c^2d^2f^4$	−	104	d^3ef^3	−	280	c^6e^5	+	1172
cde^3f^3	−	16	$c^2de^2f^3$	−	160	$d^2e^3f^2$	+	300	$c^5d^4f^2$	−	124
ce^5f^2	+	6	$c^2e^4f^2$	+	60	de^5f	+	216	$c^5d^3e^2f$	+	4336
d^4f^4	−	6	cd^3ef^3	+	320	e^7	−	216	$c^5d^2e^4$	−	2540
$d^3e^2f^3$	+	12	$cd^2e^3f^2$	+	80	$b^3c^3df^4$	−	160	c^4d^5ef	−	1912
$d^2e^4f^2$	−	10	cde^5f	−	496	$c^3e^2f^3$	−	80	$c^4d^4e^3$	+	2100
de^6f	+	6	ce^7	+	252	$c^2d^2ef^3$	+	1280	c^3d^7f	+	240
e^8	−	2	d^5f^3	−	72	$c^2de^3f^2$		...	$c^3d^6e^2$	−	1560
$a^4\ b^3ef^5$		...	$d^4e^2f^2$	−	420	c^2e^5f	−	312	c^2d^8e	+	810
b^2cdf^5	+	4	d^3e^4f	+	860	cd^4f^3	−	440	cd^{10}	−	162
ce^2f^4	−	4	d^2e^6	−	404	$cd^3e^2f^2$	−	2160	$a^1\ b^7f^5$	−	4
d^2ef^4	−	10	$b^1c^4df^4$	+	96	cd^2e^4f	+	1740	b^6cef^4	−	22
de^3f^3	+	16	$c^4e^2f^3$	−	120	cde^6	−	216	d^2f^4	−	26
e^5f^2	−	6	$c^3d^2ef^3$	−	560	d^5ef^2	+	2344	de^2f^3	+	76
$b^1c^3f^5$	+	6	$c^3de^3f^2$	+	160	d^4e^3f		3240	e^4f^2		...
c^2def^4	−	26	c^3e^5f	+	304	d^3e^5	+	1244	$b^5c^2df^4$	+	124
$c^2e^3f^3$	+	8	$c^2d^4f^3$	+	280	$b^2c^5f^4$	+	72	$c^2e^2f^3$	+	368
cd^3f^4	+	32	$c^2d^3e^2f^2$	+	1440	c^4def^3	−	240	cd^2ef^3	−	688
$cd^2e^2f^3$	−	116	$c^2d^2e^4f$	−	960	$c^4e^3f^2$	+	940	cde^3f^2	−	192
cde^4f^2	+	180	c^2de^6	−	376	$c^3d^3f^3$		...	ce^5f		...
ce^6f	−	78	cd^5ef^2	−	1296	$c^3d^2e^2f^2$	−	1320	d^4f^3	+	400
d^4ef^3	+	24	cd^4e^3f	+	80	c^3de^4f	−	2640	$d^3e^2f^2$	+	984
$d^3e^3f^2$	−	20	cd^3e^5	+	832	c^3e^6	+	908	d^2e^4f	−	2160
d^2e^5f	−	44	d^7f^2	+	432	$c^2d^4ef^2$	+	600	de^6	+	1080
de^7	+	34	d^6e^2f	−	72	$c^2d^3e^3f$	+	3360	$b^4c^4f^4$	−	60
$b^0c^4ef^4$	−	30	d^5e^4	−	240	$c^2d^2e^5$	−	168	c^3def^3	−	480
$c^3d^2f^4$	+	4	$b^0c^6f^4$	−	36	cd^6f^2	−	1656	$c^3e^3f^2$	−	1580
$c^3de^2f^3$	+	240	c^5def^3	+	288	cd^5e^2f	+	3408	$c^2d^3f^3$	+	40
$c^3e^4f^2$	−	130	$c^5e^3f^2$	−	56	cd^4e^4	−	3480	$c^2d^2e^2f^2$	+	2040
$c^2d^3ef^3$	−	160	$c^4d^3f^3$	−	140	d^7ef	−	1008	c^2de^4f	+	2910
$c^2d^2e^3f^2$	−	280	$c^4d^2e^2f^2$	−	480	d^6e^3	+	1224	c^2e^6	−	810
c^2de^5f	+	332	c^4de^4f	+	420	$b^1c^6ef^3$	−	144	cd^4ef^2	−	3420
c^2e^7	−	54	c^4e^6	−	276	$c^5d^2f^3$	+	108	cd^3e^3f	+	4800
cd^5f^3	+	24	$c^3d^4ef^2$	+	420	$c^5de^2f^2$	−	768	cd^2e^5	−	3510
$cd^4e^2f^2$	+	360	$c^3d^3e^3f$	−	1120	c^5e^4f	−	700	d^6f^2	−	1516
cd^3e^4f	−	320	$c^3d^2e^5$	+	1112	$c^4d^3ef^2$	+	900	d^5e^2f	+	2156
cd^2e^6	+	38	$c^2d^6f^2$	−	144	$c^4d^2e^3f$	+	8160	d^4e^4	−	430
d^6ef^2	−	108	$c^2d^5e^2f$	+	1620	c^4de^5	−	2148	$b^3c^5ef^3$	+	336
d^5e^3f	+	96	$c^2d^4e^4$	−	1620	$c^3d^5f^2$	+	912	$c^4d^2f^3$	−	40
d^4e^5	−	12	cd^7ef	−	864	$c^3d^4e^2f$	−	15060	$c^4de^2f^2$	+	2640

For the Numerical Verifications see p. 308.

V. No. 95 (continued).

x coefficient.

$a^1\ b^3c^4e^4f$	+	1840	$a^0\ b^7d^2ef^3$	+	184	$a^0\ b^2c^8ef^2$	−	594
$c^3d^3ef^2$	−	1280	de^3f^2	−	108	$c^7d^2f^2$	−	10296
$c^3d^2e^3f$	−	13360	e^5f		...	c^7de^2f	+	10080
c^3de^5	+	3200	$b^6c^3f^4$	+	18	c^7e^4	+	900
$c^2d^5f^2$	+	7312	c^2def^3	+	264	c^6d^3ef	+	19440
$c^2d^4e^2f$	−	2360	$c^2e^3f^2$	+	756	$c^6d^2e^3$	−	8800
$c^2d^3e^4$	+	3840	cd^3f^3	−	368	c^5d^5f	−	9160
cd^6ef	−	5344	$cd^2e^2f^2$	−	732	$c^5d^4e^2$	−	11900
cd^5e^3	+	2800	cde^4f	+	540	c^4d^6e	+	13900
d^8f	+	1956	ce^6		...	c^3d^8	−	3150
d^7e^2	−	1680	d^4ef^2	−	1172	$b^1c^9df^2$	+	3564
$b^2c^6df^3$	−	36	d^3e^3f	+	2520	c^9e^2f	−	1350
$c^6e^2f^2$	−	1296	d^2e^5	−	1350	c^8d^2ef	−	9540
$c^5d^2ef^2$	+	1668	$b^5c^4ef^3$	−	144	c^8de^3	−	750
c^5de^3f	−	1312	$c^3d^2f^3$	+	376	c^7d^4f	+	4260
c^5e^5	−	2060	$c^3de^2f^2$	−	1440	$c^7d^3e^2$	+	10800
$c^4d^4f^2$	−	8020	c^3e^4f	−	1530	c^6d^5e	−	9100
$c^4d^3e^2f$	+	15220	$c^2d^3ef^2$	+	6360	c^5d^7	+	2000
$c^4d^2e^4$	+	1180	$c^2d^2e^3f$	−	6000	$b^0c^{11}f^2$	−	486
c^3d^5ef	+	3712	c^2de^5	+	1350	$c^{10}def$	+	1620
$c^3d^4e^3$	−	8540	cd^5f^2	+	2344	$c^{10}e^3$	+	450
c^2d^7f	−	2952	cd^4e^2f	−	9260	c^9d^3f	−	720
$c^2d^6e^2$		...	cd^3e^4	+	7200	$c^9d^2e^2$	−	2250
cd^8e	+	3330	d^6ef	+	1720	c^8d^4e	+	1800
d^{10}	−	810	d^5e^3	−	1900	c^7d^6	−	400
$b^1c^8f^3$		...	$b^4c^5df^3$	−	168			
c^7def^2	−	576	$c^5e^2f^2$	+	648			
c^7e^3f	+	1824	$c^4d^2ef^2$	−	6420			
$c^6d^3f^2$	+	3792	c^4de^3f	+	9360			
$c^6d^2e^2f$	−	5808	c^4e^5	+	450			
c^6de^4	+	3240	$c^3d^4f^2$	−	10100			
c^5d^4ef	−	4768	$c^3d^3e^2f$	+	19920			
$c^5d^3e^3$	−	6240	$c^3d^2e^4$	−	10300			
c^4d^6f	+	2608	c^2d^5ef	+	4920			
$c^4d^5e^2$	+	12440	$c^2d^4e^3$	−	10100			
c^3d^7e	−	8160	cd^7f	−	3440			
c^2d^9	+	1620	cd^6e^2	+	7100			
$b^0c^9ef^2$	+	162	d^8e	−	750			
$c^8d^2f^2$	−	702	$b^3c^7f^3$	+	36			
c^8de^2f	−	90	c^6def^2	+	2988			
c^8e^4	−	1290	c^6e^3f	−	2880			
c^7d^3ef	+	1920	$c^5d^3f^2$	+	14688			
$c^7d^2e^3$	+	3640	$c^5d^2e^2f$	−	22740			
c^6d^5f	−	796	c^5de^4	+	600			
$c^6d^4e^2$	−	5340	c^4d^4ef	−	16520			
c^5d^6e	+	3100	$c^4d^3e^3$	+	23300			
c^4d^8	−	600	c^3d^6f	+	8760			
$a^0\ b^8ef^4$	+	18	$c^3d^5e^2$	−	5200			
b^7cdf^4	−	36	c^2d^7e	−	5400			
ce^2f^3	−	180	cd^9	+	1500			

V. No. 95 (continued).

y coefficient.

$a^6\ b^0df^6$	...	$a^3\ b^2c^2def^4$	− 116	$a^2\ b^4e^6f$	...	$a^2\ b^0c^5d^2e^3f$	+ 7312
e^2f^5	...	$c^2e^3f^3$	+ 80	$b^3c^3ef^4$	− 20	c^5de^5	+ 2344
$a^5\ b^1cf^6$	...	cd^3f^4	+ 240	$c^2d^2f^4$	− 280	$c^4d^5f^2$	− 124
def^5	...	$cd^2e^2f^3$	− 160	$c^2de^2f^3$	+ 80	$c^4d^4e^2f$	− 8020
e^3f^4	...	cde^4f^2	− 120	$c^2e^4f^2$	+ 300	$c^4d^3e^4$	− 10100
$b^0c^2ef^5$	...	ce^6f	+ 76	cd^3ef^3	+ 160	c^3d^6ef	+ 3792
cd^2f^5	− 2	d^4ef^3	− 120	$cd^2e^3f^2$	...	$c^3d^5e^3$	+ 14648
cde^2f^4	+ 4	$d^3e^3f^2$	− 80	cde^5f	− 192	c^2d^8f	− 702
ce^4f^3	− 2	d^2e^5f	+ 368	ce^7	− 108	$c^2d^7e^2$	− 10296
d^3ef^4	+ 6	de^7	− 180	d^5f^3	− 56	$c\ d^9e$	+ 3564
$d^2e^3f^3$	− 16	$b^1c^4ef^4$	+ 24	$d^4e^2f^2$	+ 940	d^{11}	− 486
de^5f^2	+ 14	$c^3d^2f^4$	− 160	d^3e^4f	− 1580	$a^1\ b^6cf^5$	+ 6
e^7f	− 4	$c^3de^2f^3$	+ 320	d^2e^6	+ 756	def^4	− 78
$a^4\ b^3f^6$	...	$c^3e^4f^2$	− 280	$b^2c^4df^4$	+ 360	e^3f^3	+ 72
b^2cef^5	...	$c^2d^3ef^3$	− 560	$c^4e^2f^3$	− 420	$b^5c^2ef^4$	− 44
d^2f^5	+ 2	$c^2d^2e^3f^2$	+ 1280	$c^3d^2ef^3$	+ 1440	cd^2f^4	+ 332
de^2f^4	− 4	c^2de^5f	− 688	$c^3de^3f^2$	− 2160	cde^2f^3	− 496
e^4f^3	+ 2	c^2e^7	+ 184	c^3e^5f	+ 984	ce^4f^2	+ 216
$b^1c^2df^5$	+ 10	cd^5f^3	+ 288	$c^2d^4f^3$	− 480	d^3ef^3	+ 304
$c^2e^2f^4$	− 10	$cd^4e^2f^2$	− 240	$c^2d^3e^2f^2$	− 1320	$d^2e^3f^2$	− 312
cd^2ef^4	− 26	cd^3e^4f	− 480	$c^2d^2e^4f$	+ 2040	de^5f	...
cde^3f^3	+ 32	cd^2e^6	+ 264	c^2de^6	− 732	e^7	...
ce^5f^2	− 6	d^6ef^2	− 144	cd^5ef^2	− 768	$b^4c^3df^4$	− 320
d^4f^4	− 30	d^5e^3f	+ 336	cd^4e^3f	+ 2640	$c^3e^2f^3$	+ 860
$d^3e^2f^3$	+ 84	d^4e^5	− 144	cd^3e^5	− 1440	$c^2d^2ef^3$	− 960
$d^2e^4f^2$	− 50	$b^0c^5df^4$	+ 24	d^7f^2	+ 504	$c^2de^3f^2$	+ 1740
de^6f	− 22	$c^5e^2f^3$	− 72	d^6e^2f	− 1296	c^2e^5f	− 2160
e^8	+ 18	$c^4d^2ef^3$	+ 280	d^5e^4	+ 648	cd^4f^3	+ 420
$b^0c^4f^5$	− 6	$c^4de^3f^2$	− 440	$b^1c^6f^4$	− 108	$cd^3e^2f^2$	− 2640
c^3def^4	+ 32	c^4e^5f	+ 400	c^5def^3	− 1296	cd^2e^4f	+ 2910
$c^3e^3f^3$	− 8	$c^3d^4f^3$	− 140	$c^5e^3f^2$	+ 2344	cde^6	+ 540
$c^2d^3f^4$	+ 4	$c^3d^3e^2f^2$	...	$c^4d^3f^3$	+ 420	d^5ef^2	− 700
$c^2d^2e^2f^3$	− 104	$c^3d^2e^4f$	+ 40	$c^4d^2e^2f^2$	+ 600	d^4e^3f	+ 1840
$c^2de^4f^2$	+ 90	c^3de^6	− 368	c^4de^4f	− 3420	d^3e^5	− 1530
c^2e^6f	− 26	$c^2d^5ef^2$	+ 108	c^4e^6	− 1172	$b^3c^5f^4$	+ 96
cd^4ef^3	+ 96	$c^2d^4e^3f$	− 40	$c^3d^4ef^2$	+ 900	c^4def^3	+ 80
$cd^3e^3f^2$	− 160	$c^2d^3e^5$	+ 376	$c^3d^3e^3f$	− 1280	$c^4e^3f^2$	− 3240
cd^2e^5f	+ 124	cd^7f^2	...	$c^3d^2e^5$	+ 6360	$c^3d^3f^3$	− 1120
cde^7	− 36	cd^6e^2f	− 36	$c^2d^6f^2$	− 576	$c^3d^2e^2f^2$	+ 3360
d^6f^3	− 36	cd^5e^4	− 168	$c^2d^5e^2f$	+ 1668	c^3de^4f	+ 4800
$d^5e^2f^2$	+ 72	d^8ef	...	$c^2d^4e^4$	− 6420	c^3e^6	+ 2520
d^4e^4f	− 60	d^7e^3	+ 36	cd^7ef	− 576	$c^2d^4ef^2$	+ 8160
d^3e^6	+ 18	$a^2\ b^5df^5$	+ 6	cd^6e^3	+ 2988	$c^2d^3e^3f$	− 13360
$a^3\ b^4ef^5$	...	e^2f^4	− 6	d^9f	+ 162	$c^2d^2e^5$	− 6000
b^3cdf^5	− 16	$b^4c^2f^5$	− 10	d^8e^2	− 594	cd^6f^2	− 2288
ce^2f^4	+ 16	$cdef^4$	+ 180	$b^0c^7ef^3$	+ 432	cd^5e^2f	− 1312
d^2ef^4	+ 8	ce^3f^3	− 160	$c^6d^2f^3$	− 144	cd^4e^4	+ 9360
de^3f^3	...	d^3f^4	− 130	$c^6de^2f^2$	− 1656	d^7ef	+ 1824
e^5f^2	− 8	$d^2e^2f^3$	+ 60	c^6e^4f	− 1516	d^6e^3	− 2880
$b^2c^3f^5$	+ 12	de^4f^2	+ 60	$c^5d^3ef^2$	+ 912	$b^2c^6ef^3$	− 72

For the Numerical Verifications see p. 308.

V. No. 95 (concluded).

y coefficient.

Term	Sign	Value
$a^1\ b^2c^5d^2f^3$	+	1620
$c^5de^2f^2$	+	3408
c^5e^4f	+	2156
$c^4d^3ef^2$	−	15060
$c^4d^2e^3f$	−	2360
c^4de^5	−	9260
$c^3d^5f^2$	+	4336
$c^3d^4e^2f$	+	15220
$c^3d^3e^4$	+	19920
c^2d^6ef	−	5808
$c^2d^5e^3$	−	22740
cd^8f	−	90
cd^7e^2	+	10080
d^9e	−	1350
$b^1c^7df^3$	−	864
$c^7e^2f^2$	−	1008
$c^6d^2ef^2$	+	6624
c^6de^3f	−	5344
c^6e^5	+	1720
$c^5d^4f^2$	−	1912
$c^5d^3e^2f$	+	3712
$c^5d^2e^4$	+	4920
c^4d^5ef	−	4768
$c^4d^4e^3$	−	16520
c^3d^7f	+	1920
$c^3d^6e^2$	+	19440
c^2d^8e	−	9540
cd^{10}	+	1620
$b^0c^9f^3$	+	162
c^8def^2	−	918
c^8e^3f	+	1956
$c^7d^3f^2$	+	240
$c^7d^2e^2f$	−	2952
c^7de^4	−	3440
c^6d^4ef	+	2608
$c^6d^3e^3$	+	8760
c^5d^6f	−	796
$c^5d^5e^2$	−	9160
c^4d^7e	+	4260
c^3d^9	−	720
$a^0\ b^8f^5$	−	2
$b^7c\,ef^4$	+	34
d^2f^4	−	54
de^2f^3	+	252
e^4f^2	−	216
$b^6c^2df^4$	+	38
$c^2e^2f^3$	−	404
cd^2ef^3	−	376
cde^3f^2	−	216
ce^5f	+	1080
$a^0\ b^6d^4f^3$	−	276
$d^3e^2f^2$	+	908
d^2e^4f	−	810
de^6		...
$b^5c^4f^4$	−	12
c^3def^3	+	832
$c^3e^3f^2$	+	1244
$c^2d^3f^3$	+	1112
$c^2d^2e^2f^2$	−	168
c^2de^4f	−	3510
c^2e^6	−	1350
cd^4ef^2	−	2148
cd^3e^3f	+	3200
cd^2e^5	+	1350
d^6f^2	+	1172
d^5e^2f	−	2060
d^4e^4	+	450
$b^4c^5ef^3$	−	240
$c^4d^2f^3$	−	1620
$c^4de^2f^2$	−	3480
c^4e^4f	−	430
$c^3d^3ef^2$	+	2800
$c^3d^2e^3f$	+	3840
c^3de^5	+	7200
$c^2d^5f^2$	−	2540
$c^2d^4e^2f$	+	1180
$c^2d^3e^4$	−	10300
cd^6ef	+	3240
cd^5e^3	+	600
d^8f	−	1290
d^7e^2	+	900
$b^3c^6df^3$	+	876
$c^6e^2f^2$	+	1224
$c^5d^2ef^2$	+	2052
c^5de^3f	+	2800
c^5e^5	−	1900
$c^4d^4f^2$	+	2100
$c^4d^3e^2f$	−	8540
$c^4d^2e^4$	−	10100
c^3d^5ef	−	6240
$c^3d^4e^3$	+	23300
c^2d^7f	+	3640
$c^2d^6e^2$	−	8800
cd^8e	−	750
d^{10}	+	450
$b^2c^8f^3$	−	162
c^7def^2	−	2304
c^7e^3f	−	1680
$c^6d^3f^2$	−	1560
$c^6d^2e^2f$		...
$a^0\ b^2c^6de^4$	+	7100
c^5d^4ef	+	12440
$c^5d^3e^3$	−	5200
c^4d^6f	−	5340
$c^4d^5e^2$	−	11900
c^3d^7e	+	10800
c^2d^9	−	2250
$b^1c^9ef^2$	+	486
$c^8d^2f^2$	+	810
c^8de^2f	+	3330
c^8e^4	−	750
c^7d^3ef	−	8160
$c^7d^2e^3$	−	5400
c^6d^5f	+	3100
$c^6d^4e^2$	+	13900
c^5d^6e	−	9100
c^4d^8	+	1800
$b^0c^{10}df^2$	−	162
$c^{10}e^2f$	−	810
c^9d^2ef	+	1620
c^9de^3	+	1500
c^8d^4f	−	600
$c^8d^3e^2$	−	3150
c^7d^5e	+	2000
c^6d^7	−	400

W, 29 A.

$a^9\ b^0f^9$		...	$a^6\ b^1ce^8f^2$	−	15	$a^5\ b^2c^2de^4f^4$	−	90	$a^5\ b^0c^3d^2e^7f$	+	1320
$a^8\ b^1ef^8$		...	d^5ef^5	+	10	$c^2e^6f^3$	+	30	c^3de^9	−	260
b^0cdf^8		...	$d^4e^3f^4$	−	35	cd^4ef^5	−	210	$c^2d^7f^4$	+	60
ce^2f^7		...	$d^3e^5f^3$	+	40	$cd^3e^3f^4$	+	120	$c^2d^6e^2f^3$	−	500
d^2ef^7		...	$d^2e^7f^2$	−	10	$cd^2e^5f^3$	+	360	$c^2d^5e^4f^2$	+	2235
de^3f^6		...	de^9f	−	10	cde^7f^2	−	420	$c^2d^4e^6f$	−	1995
e^5f^5		...	e^{11}	+	5	ce^9f	+	130	$c^2d^3e^8$	+	370
$a^7\ b^2df^8$		...	$b^0c^5f^7$	−	1	d^6f^5	−	5	cd^8ef^3	+	360
e^2f^7		...	c^4def^6	+	15	$d^5e^2f^4$	+	195	$cd^7e^3f^2$	−	1320
$b^1c^2f^8$		...	$c^4e^3f^5$	−	10	$d^4e^4f^3$	−	315	cd^6e^5f	+	1110
$cdef^7$		...	$c^3d^3f^6$		...	$d^3e^6f^2$	+	40	cd^5e^7	−	210
ce^3f^6		...	$c^3d^2e^2f^5$	−	90	d^2e^8f	+	165	$d^{10}f^3$	−	81
d^3f^7		...	$c^3de^4f^4$	+	120	de^{10}	−	75	$d^9e^2f^2$	+	270
$d^2e^2f^6$		...	$c^3e^6f^3$	−	40	$b^1c^5ef^6$	−	10	d^8e^4f	−	225
de^4f^5		...	$c^2d^4ef^5$	+	60	$c^4d^2f^6$	−	60	d^7e^6	+	45
e^6f^4		...	$c^2d^3e^3f^4$	+	30	$c^4de^2f^5$	+	210	$a^4\ b^6ef^7$		...
$b^0c^3ef^7$		...	$c^2d^2e^5f^3$	−	180	$c^4e^4f^4$	−	110	b^5cdf^7		...
$c^2d^2f^7$		...	$c^2de^7f^2$	+	120	$c^3d^3ef^5$		...	ce^2f^6		...
$c^2de^2f^6$		...	c^2e^9f	−	20	$c^3d^2e^3f^4$	+	60	d^2ef^6		...
$c^2e^4f^5$		...	cd^6f^5	−	15	$c^3de^5f^3$	−	360	de^3f^5		...
cd^3ef^6		...	$cd^5e^2f^4$	−	110	$c^3e^7f^2$	+	240	e^5f^4		...
$cd^2e^3f^5$		...	$cd^4e^4f^3$	+	265	$c^2d^5f^5$	+	30	$b^4c^3f^7$	−	10
cde^5f^4		...	$cd^3e^6f^2$	−	200	$c^2d^4e^2f^4$	−	210	c^2def^6	+	90
ce^7f^3		...	cd^2e^8f	+	65	$c^2d^3e^4f^3$	−	180	$c^2e^3f^5$	−	60
d^5f^6	+	1	cde^{10}	−	10	$c^2d^2e^6f^2$	+	1140	cd^3f^6	−	120
$d^4e^2f^5$	−	5	d^7ef^4	+	45	c^2de^8f	−	870	$cd^2e^2f^5$	+	90
$d^3e^4f^4$	+	10	$d^6e^3f^3$	−	100	c^2e^{10}	+	130	cde^4f^4		...
$d^2e^6f^3$	−	10	$d^5e^5f^2$	+	81	cd^6ef^4	+	310	ce^6f^3		...
de^8f^2	+	5	d^4e^7f	−	30	$cd^5e^3f^3$	−	240	d^4ef^5	+	110
$e^{10}f$	−	1	d^3e^9	+	5	$cd^4e^5f^2$	−	390	$d^3e^3f^4$	−	50
$a^6\ b^3cf^8$		...	$a^5\ b^5f^8$		...	cd^3e^7f	+	280	$d^2e^5f^3$	−	240
def^7		...	b^4cef^7		...	cd^2e^9	+	30	de^7f^2	+	280
e^3f^6		...	d^2f^7		...	d^8f^4	−	180	e^9f	−	90
$b^2c^2ef^7$		...	de^2f^6		...	$d^7e^2f^3$	+	300	$b^3c^4ef^6$	+	35
cd^2f^7		...	e^4f^5		...	$d^6e^4f^2$	−	120	$c^3d^2f^6$	−	30
cde^2f^6		...	$b^3c^2df^7$		...	d^5e^6f	+	30	$c^3de^2f^5$	−	120
ce^4f^5		...	$c^2e^2f^6$		...	d^4e^8	−	30	$c^3e^4f^4$	+	50
d^3ef^6		...	cd^2ef^6		...	$b^0c^6df^6$	+	15	$c^2d^3ef^5$	−	60
$d^2e^3f^5$		...	cde^3f^5		...	$c^6e^2f^5$	+	5	$c^2d^2e^3f^4$		...
de^5f^4		...	ce^5f^4		...	$c^5d^2ef^5$	−	30	$c^2de^5f^3$	+	360
e^7f^3		...	d^4f^6	+	10	$c^5de^3f^4$	−	270	$c^2e^7f^2$	−	210
$b^1c^3df^7$		...	$d^3e^2f^5$	−	40	$c^5e^5f^3$	+	196	cd^5f^5	+	270
$c^3e^2f^6$		...	$d^2e^4f^4$	+	60	$c^4d^4f^5$		...	$cd^4e^2f^4$	+	575
$c^2d^2ef^6$		...	de^6f^3	−	40	$c^4d^3e^2f^4$	+	225	$cd^3e^4f^3$	−	1700
$c^2de^3f^5$		...	e^8f^2	+	10	$c^4d^2e^4f^3$	+	615	$cd^2e^6f^2$	+	480
$c^2e^5f^4$		...	$b^2c^4f^7$	+	5	$c^4de^6f^2$	−	660	cde^8f	+	670
cd^4f^6	−	15	c^3def^6	−	60	c^4e^8f	+	45	ce^{10}	−	315
$cd^3e^2f^5$	+	60	$c^3e^3f^5$	+	40	$c^3d^5ef^4$	−	120	d^6ef^4	−	685
$cd^2e^4f^4$	−	90	$c^2d^3f^6$	+	90	$c^3d^4e^3f^3$	−	220	$d^5e^3f^3$	+	540
cde^6f^3	+	60	$c^2d^2e^2f^5$		...	$c^3d^3e^5f^2$	−	980	$d^4e^5f^2$	+	1515

For the Numerical Verifications see p. 309.

W, 29 A (continued).

$a^4\ b^3d^3e^7f$	−	2080	$a^4\ b^1cd^6e^6$	−	615	$a^3\ b^5d^5f^5$	−	196	$a^3\ b^3d^6e^6$	−	3880
d^2e^9	+	705	$d^{10}ef^2$	−	945	$d^4e^2f^4$	−	660	$b^2c^7ef^5$	−	300
$b^2c^5df^6$	+	110	d^9e^3f	+	900	$d^3e^4f^3$	+	1840	$c^6d^2f^5$	+	500
$c^5e^2f^5$	−	195	d^8e^5	+	45	$d^2e^6f^2$	−	1040	$c^6de^2f^4$	+	3810
$c^4d^2ef^5$	+	210	$b^0c^8ef^5$	+	180	de^8f	−	180	$c^6e^4f^3$	−	3710
$c^4de^3f^4$	−	575	$c^7d^2f^5$	−	60	e^{10}	+	216	$c^5d^3ef^4$	−	14040
$c^4e^5f^3$	+	660	$c^7de^2f^4$	−	1420	$b^4c^4df^6$	−	265	$c^5d^2e^3f^3$	+	16120
$c^3d^4f^5$	−	225	$c^7e^4f^3$	+	25	$c^4e^2f^5$	+	315	$c^5de^5f^2$	−	540
$c^3d^3e^2f^4$	+	1350	$c^6d^3ef^4$	+	780	$c^3d^2ef^5$	+	180	c^5e^7f	+	600
$c^3d^2e^4f^3$		...	$c^6d^2e^3f^3$	+	5760	$c^3de^3f^4$	+	1700	$c^4d^5f^4$	+	7020
$c^3de^6f^2$	−	1440	$c^6de^5f^2$	−	2945	$c^3e^5f^3$	−	1840	$c^4d^4e^2f^3$		...
c^3e^8f	−	75	c^6e^7f	+	1390	$c^2d^4f^5$	−	615	$c^4d^3e^4f^2$	−	1950
$c^2d^5ef^4$	−	1965	$c^5d^5f^4$		...	$c^2d^3e^2f^4$	−	1350	$c^4d^2e^6f$	−	17670
$c^2d^4e^3f^3$	+	6000	$c^5d^4e^2f^3$	−	7020	$c^2d^2e^4f^3$		...	c^4de^8	+	4170
$c^2d^3e^5f^2$	−	7050	$c^5d^3e^4f^2$	−	180	$c^2de^6f^2$	+	1560	$c^3d^6ef^3$	+	480
$c^2d^2e^7f$	+	3000	$c^5d^2e^6f$	−	1275	c^2e^8f	+	135	$c^3d^5e^3f^2$	−	31040
c^2de^9	+	265	c^5de^8	−	1110	cd^5ef^4	+	2210	$c^3d^4e^5f$	+	45180
cd^7f^4	+	1420	$c^4d^6ef^3$	+	3120	$cd^4e^3f^3$	−	4100	$c^3d^3e^7$	−	3160
$cd^6e^2f^3$	−	3810	$c^4d^5e^3f^2$	+	3900	$cd^3e^5f^2$	+	6000	$c^2d^8f^3$	−	140
$cd^5e^4f^2$	+	2310	$c^4d^4e^5f$	+	1240	cd^2e^7f	−	4880	$c^2d^7e^2f^2$	+	18000
cd^4e^6f	+	1795	$c^4d^3e^7$	+	3155	cde^9	+	990	$c^2d^6e^4f$	−	12180
cd^3e^8	−	1800	$c^3d^8f^3$	−	515	d^7f^4	−	25	$c^2d^5e^6$	−	13430
d^8ef^3	+	240	$c^3d^7e^2f^2$	−	2920	$d^6e^2f^3$	+	3710	cd^9ef^2	−	7200
$d^7e^3f^2$	+	30	$c^3d^6e^4f$	−	940	$d^5e^4f^2$	−	10755	cd^8e^3f	−	120
d^6e^5f	−	870	$c^3d^5e^6$	−	4300	d^4e^6f	+	9875	cd^7e^5	+	9960
d^5e^7	+	615	$c^2d^9ef^2$	+	675	d^3e^8	−	2845	$d^{11}f^2$	+	1890
$b^1c^7f^6$	−	45	$c^2d^8e^3f$	+	510	$b^3c^6f^6$	+	100	$d^{10}e^2f$	−	540
c^6def^5	−	310	$c^2d^7e^5$	+	2940	c^5def^5	+	240	d^9e^4	−	1710
$c^6e^3f^4$	+	685	$cd^{11}f^2$		...	$c^5e^3f^4$	−	540	$b^1c^8df^5$	−	360
$c^5d^3f^5$	+	120	$cd^{10}e^2f$	−	135	$c^4d^3f^5$	+	220	$c^8e^2f^4$	−	240
$c^5d^2e^2f^4$	+	1965	cd^9e^4	−	990	$c^4d^2e^2f^4$	−	6000	$c^7d^2ef^4$	+	5840
$c^5de^4f^3$	−	2210	$d^{12}ef$		...	$c^4de^4f^3$	+	4100	$c^7de^3f^3$	−	6560
$c^5e^6f^2$	−	960	$d^{11}e^3$	+	135	$c^4e^6f^2$	+	1340	$c^7e^5f^2$	+	8460
$c^4d^4ef^4$		...	$a^3\ b^7df^7$		...	$c^3d^4ef^4$	+	11700	$c^6d^4f^4$	−	3120
$c^4d^3e^3f^3$	−	11700	e^2f^6		...	$c^3d^3e^3f^3$		...	$c^6d^3e^2f^3$	−	480
$c^4d^2e^5f^2$	+	15435	$b^6c^2f^7$	+	10	$c^3d^2e^5f^2$	−	15240	$c^6d^2e^4f^2$	−	25880
c^4de^7f	−	2760	$cdef^6$	−	60	c^3de^7f	+	6960	c^6de^6f	−	1820
c^4e^9	+	555	ce^3f^5	+	40	c^3e^9	−	1620	c^6e^8	−	3620
$c^3d^6f^4$	−	780	d^3f^6	+	40	$c^2d^6f^4$	−	5760	$c^5d^5ef^3$	+	49680
$c^3d^5e^2f^3$	+	14040	$d^2e^2f^5$	−	30	$c^2d^5e^2f^3$	−	16120	$c^5d^4e^3f^2$		...
$c^3d^4e^4f^2$	−	10625	de^4f^4		...	$c^2d^4e^4f^2$	+	26700	$c^5d^3e^5f$	+	17520
$c^3d^3e^6f$	−	3220	e^6f^3		...	$c^2d^3e^6f$	−	5240	$c^5d^2e^7$	+	13500
$c^3d^2e^8$	−	570	$b^5c^3ef^6$	−	40	$c^2d^2e^8$	−	1640	$c^4d^7f^3$	−	120
$c^2d^7ef^3$	−	5840	$c^2d^2f^6$	+	180	cd^7ef^3	+	6560	$c^4d^6e^2f^2$	−	32280
$c^2d^6e^3f^2$	−	540	$c^2de^2f^5$	−	360	$cd^6e^3f^2$	+	7240	$c^4d^5e^4f$	−	46880
$c^2d^5e^5f$	+	5550	$c^2e^4f^4$	+	240	cd^5e^5f	−	24240	$c^4d^4e^6$	−	30040
$c^2d^4e^7$	+	1285	cd^3ef^5	+	360	cd^4e^7	+	11420	$c^3d^8ef^2$	+	12860
cd^9f^3	+	990	$cd^2e^3f^4$	−	360	d^9f^3	−	980	$c^3d^7e^3f$	+	32000
$cd^8e^2f^2$	+	3150	cde^5f^3		...	$d^8e^2f^2$	−	3420	$c^3d^6e^5$	+	46160
cd^7e^4f	−	3600	ce^7f^2		...	d^7e^4f	+	8100	$c^2d^{10}f^2$	−	2700

W, 29 A (continued).

$a^3\ b^1c^2d^9e^2f$	−	8820	$a^2\ b^6cd^3e^2f^4$	+	1440	$a^2\ b^4cd^5e^6$	−	18750	$a^2\ b^2c^3d^7e^4$	+	243000
$c^2d^8e^4$	−	34620	$cd^2e^4f^3$	−	1560	d^9ef^2	+	14115	$c^2d^{10}ef$	+	2340
$cd^{11}ef$	+	1080	cde^6f^2		...	d^8e^3f	−	23790	$c^2d^9e^3$	−	89550
$cd^{10}e^3$	+	12060	ce^8f		...	d^7e^5	+	8175	$cd^{12}f$	+	270
$d^{13}f$		...	d^5ef^4	+	960	$b^3c^7df^5$	+	1320	$cd^{11}e^2$	+	15120
$d^{12}e^2$	−	1620	$d^4e^3f^3$	−	1340	$c^7e^2f^4$	−	30	$d^{13}e$	−	810
$b^0c^{10}f^5$	+	81	$d^3e^5f^2$	−	2440	$c^6d^2ef^4$	+	540	$b^1c^{10}ef^4$	+	945
c^9def^4	−	990	d^2e^7f	+	4320	$c^6de^3f^3$	−	7240	$c^9d^2f^4$	−	675
$c^9e^3f^3$	+	980	de^9	−	1620	$c^6e^5f^2$	−	20390	$c^9de^2f^3$	+	7200
$c^8d^3f^4$	+	515	$b^5c^5f^6$	−	81	$c^5d^4f^4$	−	3900	$c^9e^4f^2$	−	14115
$c^8d^2e^2f^3$	+	140	c^4def^5	+	390	$c^5d^3e^2f^3$	+	31040	$c^8d^3ef^3$	−	12860
$c^8de^4f^2$	−	195	$c^4e^3f^4$	−	1515	$c^5d^2e^4f^2$	+	32370	$c^8d^2e^3f^2$	+	8220
c^8e^6f	−	5575	$c^3d^3f^5$	+	980	c^5de^6f	+	38820	c^8de^5f	+	150
$c^7d^4ef^3$	+	120	$c^3d^2e^2f^4$	+	7050	c^5e^8	+	9310	c^8e^7	+	6155
$c^7d^3e^3f^2$	−	800	$c^3de^4f^3$	−	6000	$c^4d^5ef^3$	−	49680	$c^7d^5f^3$	−	480
$c^7d^2e^5f$	+	22600	$c^3e^6f^2$	+	2440	$c^4d^4e^3f^2$		...	$c^7d^4e^2f^2$	+	26700
c^7de^7	+	7240	$c^2d^4ef^4$	−	15435	$c^4d^3e^5f$	−	91260	$c^7d^3e^4f$	+	63960
$c^6d^6f^3$		...	$c^2d^3e^3f^3$	+	15240	$c^4d^2e^7$	−	50550	$c^7d^2e^6$	−	6660
$c^6d^5e^2f^2$	−	1260	$c^2d^2e^5f^2$		...	$c^3d^7f^3$	+	800	$c^6d^6ef^2$		...
$c^6d^4e^4f$	−	42330	c^2de^7f	−	6480	$c^3d^6e^2f^2$	+	81840	$c^6d^5e^3f$	−	180600
$c^6d^3e^6$	−	34340	c^2e^9	+	1215	$c^3d^5e^4f$	+	360	$c^6d^4e^5$	−	71610
$c^5d^7ef^2$	+	480	cd^6f^4	+	2945	$c^3d^4e^6$	+	101450	$c^5d^8f^2$	−	4755
$c^5d^6e^3f$	+	48360	$cd^5e^2f^3$	+	540	$c^2d^8ef^2$	−	8220	$c^5d^7e^2f$	+	141240
$c^5d^5e^5$	+	73828	$cd^4e^4f^2$	−	795	$c^2d^7e^3f$	−	58080	$c^5d^6e^4$	+	219730
$c^4d^9f^2$	+	105	cd^3e^6f	−	4180	$c^2d^6e^5$	−	34300	c^4d^9ef	−	45130
$c^4d^8e^2f$	−	30265	cd^2e^8	+	4185	$cd^{10}f^2$	−	7590	$c^4d^8e^3$	−	240975
$c^4d^7e^4$	−	92290	d^7ef^3	−	8460	cd^9e^2f	+	41640	$c^3d^{11}f$	+	5580
$c^3d^{10}ef$	+	9540	$d^6e^3f^2$	+	20390	cd^8e^4	−	4650	$c^3d^{10}e^2$	+	128490
$c^3d^9e^3$	+	69220	d^5e^5f	−	16194	$d^{11}ef$	−	5580	$c^2d^{12}e$	−	34155
$c^2d^{12}f$	−	1215	d^4e^7	+	3765	$d^{10}e^3$	+	1980	cd^{14}	+	3645
$c^2d^{11}e^2$	−	30510	$b^4c^6ef^5$	+	120	$b^2c^9f^5$	−	270	$b^0c^{11}df^4$		...
$cd^{13}e$	+	7290	$c^5d^2f^5$	−	2235	c^8def^4	−	3150	$c^{11}e^2f^3$	−	1890
d^{15}	−	729	$c^5de^2f^4$	−	2310	$c^8e^3f^3$	+	3420	$c^{10}d^2ef^3$	+	2700
$a^2\ b^8cf^7$	−	5	$c^5e^4f^3$	+	10755	$c^7d^3f^4$	+	2920	$c^{10}de^3f^2$	+	7590
def^6	+	15	$c^4d^3ef^4$	+	10625	$c^7d^2e^2f^3$	−	18000	$c^{10}e^5f$	+	8256
e^3f^5	−	10	$c^4d^2e^3f^3$	−	26700	$c^7de^4f^2$	+	43800	$c^9d^4f^3$	−	105
$b^7c^2ef^6$	+	10	$c^4de^5f^2$	+	795	c^7e^6f	+	5030	$c^9d^3e^2f^2$	−	14360
cd^2f^6	−	120	c^4e^7f	−	10070	$c^6d^4ef^3$	+	32280	$c^9d^2e^4f$	−	43605
cde^2f^5	+	420	$c^3d^5f^4$	+	180	$c^6d^3e^3f^2$	−	81840	c^9de^6	−	12310
ce^4f^4	−	280	$c^3d^4e^2f^3$	+	1950	$c^6d^2e^5f$	−	85800	$c^8d^5ef^2$	+	4755
d^3ef^5	−	240	$c^3d^3e^4f^2$		...	c^6de^7	−	28710	$c^8d^4e^3f$	+	77790
$d^2e^3f^4$	+	210	$c^3d^2e^6f$	+	36510	$c^5d^6f^3$	+	1260	$c^8d^3e^5$	+	59835
de^5f^3		...	c^3de^8		...	$c^5d^5e^2f^2$		...	$c^7d^7f^2$		...
e^7f^2		...	$c^2d^6ef^3$	+	25880	$c^5d^4e^4f$	+	181980	$c^7d^6e^2f$	−	57060
$b^6c^3df^6$	+	200	$c^2d^5e^3f^2$	−	32370	$c^5d^3e^6$	+	153480	$c^7d^5e^4$	−	114960
$c^3e^2f^5$	−	40	$c^2d^4e^5f$	−	12180	$c^4d^7ef^2$	−	26700	c^6d^8ef	+	19020
$c^2d^2ef^5$	−	1140	$c^2d^3e^7$	−	9850	$c^4d^6e^3f$	−	41360	$c^6d^7e^3$	+	109660
$c^2de^4f^4$	−	480	cd^8f^3	+	195	$c^4d^5e^5$	−	306900	$c^5d^{10}f$	−	2481
$c^2e^6f^3$	+	1040	$cd^7e^2f^2$	−	43800	$c^3d^9f^2$	+	14360	$c^5d^9e^2$	−	56110
cd^4f^5	+	660	cd^6e^4f	+	72755	$c^3d^8e^2f$	−	16170	$c^4d^{11}e$	+	14895

W, 29 A (continued).

$a^2\ b^0c^3d^{13}$	−	1620	$a^1\ b^6d^8f^3$	+	5575	$a^1\ b^4cd^9e^3$	+	41250	$a^1\ b^1c^{11}e^3f^2$	+	5580
$a^1\ b^{10}f^7$	+	1	$d^7e^2f^2$	−	5030	$d^{12}f$	+	5445	$c^{10}d^3f^3$	−	9540
b^9cef^6	+	10	d^6e^4f	−	4255	$d^{11}e^2$	−	6525	$c^{10}d^2e^2f^2$	−	2340
d^2f^6	+	20	d^5e^6	+	2175	$b^3c^9ef^4$	−	900	$c^{10}de^4f$	+	20610
de^2f^5	−	130	$b^5c^6df^5$	−	1110	$c^8d^2f^4$	−	510	$c^{10}e^6$	−	4350
e^4f^4	+	90	$c^6e^2f^4$	+	870	$c^8de^2f^3$	+	120	$c^9d^4ef^2$	+	45130
$b^8c^2df^6$	−	65	$c^5d^2ef^4$	−	5550	$c^8e^4f^2$	+	23790	$c^9d^3e^3f$	−	92200
$c^2e^2f^5$	−	165	$c^5de^3f^3$	+	24240	$c^7d^3ef^3$	−	32000	$c^9d^2e^5$	−	25050
cd^2ef^5	+	870	$c^5e^5f^2$	+	16194	$c^7d^2e^3f^2$	+	58080	$c^8d^6f^2$	−	19020
cde^3f^4	−	670	$c^4d^4f^4$	−	1240	c^7de^5f	−	15440	$c^8d^5e^2f$	+	46050
ce^5f^3	+	180	$c^4d^3e^2f^3$	−	45180	c^7e^7	−	12500	$c^8d^4e^4$	+	138750
d^4f^5	−	45	$c^4d^2e^4f^2$	+	12180	$c^6d^5f^3$	−	48360	c^7d^7ef		...
$d^3e^2f^4$	+	75	c^4de^6f	−	66650	$c^6d^4e^2f^2$	+	41360	$c^7d^6e^3$	−	178200
$d^2e^4f^3$	−	135	c^4e^8	−	8550	$c^6d^3e^4f$	−	181600	c^6d^9f	−	1650
de^6f^2		...	$c^3d^5ef^3$	−	17520	$c^6d^2e^6$	−	18400	$c^6d^8e^2$	+	103950
e^8f		...	$c^3d^4e^3f^2$	+	91260	$c^5d^6ef^2$	+	180600	$c^5d^{10}e$	−	30250
$b^7c^4f^6$	+	30	$c^3d^3e^5f$		...	$c^5d^5e^3f$		...	c^4d^{12}	+	3600
c^3def^5	−	280	$c^3d^2e^7$	+	62100	$c^5d^4e^5$	+	289800	$b^0c^{13}ef^3$		...
$c^3e^3f^4$	+	2080	$c^2d^7f^3$	−	22600	$c^4d^8f^2$	−	77790	$c^{12}d^2f^3$	+	1215
$c^2d^3f^5$	−	1320	$c^2d^6e^2f^2$	+	85800	$c^4d^7e^2f$	−	87000	$c^{12}de^2f^2$	−	270
$c^2d^2e^2f^4$	−	3000	$c^2d^5e^4f$	−	148890	$c^4d^6e^4$	−	318500	$c^{12}e^4f$	−	5445
$c^2de^4f^3$	+	4880	$c^2d^4e^6$	+	1850	c^3d^9ef	+	92200	$c^{11}d^3ef^2$	−	5580
$c^2e^6f^2$	−	4320	cd^8ef^2	−	150	$c^3d^8e^3$	+	179500	$c^{11}d^2e^3f$	+	17520
cd^4ef^4	+	2760	cd^7e^3f	+	15440	$c^2d^{11}f$	−	17520	$c^{11}de^5$	+	8700
$cd^3e^3f^3$	−	6960	cd^6e^5	+	10350	$c^2d^{10}e^2$	−	69000	$c^{10}d^5f^2$	+	2481
$cd^2e^5f^2$	+	6480	$d^{10}f^2$	−	8256	$cd^{12}e$	+	15300	$c^{10}d^4e^2f$	−	10595
cde^7f		...	d^9e^2f	+	12210	d^{14}	−	1350	$c^{10}d^3e^4$	−	31150
ce^9		...	d^8e^4	−	7050	$b^2c^{10}df^4$	+	135	c^9d^6ef	+	1650
d^6f^4	−	1390	$b^4c^8f^5$	+	225	$c^{10}e^2f^3$	+	540	$c^9d^5e^3$	+	37950
$d^5e^2f^3$	−	600	c^7def^4	+	3600	$c^9d^2ef^3$	+	8820	c^8d^8f		...
$d^4e^4f^2$	+	10070	$c^7e^3f^3$	−	8100	$c^9de^3f^2$	−	41640	$c^8d^7e^2$	−	22275
d^3e^6f	−	12600	$c^6d^3f^4$	+	940	c^9e^5f	−	12210	c^7d^9e	+	6600
d^2e^8	+	4050	$c^6d^2e^2f^3$	+	12180	$c^8d^4f^3$	+	30265	c^6d^{11}	−	800
$b^6c^5ef^5$	−	30	$c^6de^4f^2$	−	72755	$c^8d^3e^2f^2$	+	16170	$a^0\ b^{11}ef^6$	−	5
$c^4d^2f^5$	+	1995	c^6e^6f	+	4255	$c^8d^2e^4f$	+	62025	$b^{10}cdf^6$	+	10
$c^4de^2f^4$	−	1795	$c^5d^4ef^3$	+	46880	c^8de^6	+	44225	ce^2f^5	+	75
$c^4e^4f^3$	−	9875	$c^5d^3e^3f^2$	−	360	$c^7d^5ef^2$	−	141240	d^2ef^5	−	130
$c^3d^3ef^4$	+	3220	$c^5d^2e^5f$	+	148890	$c^7d^4e^3f$	+	87000	de^3f^4	+	315
$c^3d^2e^3f^3$	+	5240	c^5de^7	+	38950	$c^7d^3e^5$	−	129000	e^5f^3	−	216
$c^3de^5f^2$	+	4180	$c^4d^6f^3$	+	42330	$c^6d^7f^2$	+	57060	$b^9c^3f^6$	−	5
c^3e^7f	+	12600	$c^4d^5e^2f^2$	−	181980	$c^6d^6e^2f$		...	c^2def^5	−	30
$c^2d^5f^4$	+	1275	$c^4d^4e^4f$		...	$c^6d^5e^4$	−	5250	$c^2e^3f^4$	−	705
$c^2d^4e^2f^3$	+	17670	$c^4d^3e^6$	−	220125	c^5d^8ef	−	46050	cd^3f^5	+	260
$c^2d^3e^4f^2$	−	36510	$c^3d^7ef^2$	−	63960	$c^5d^7e^3$	+	122800	$cd^2e^2f^4$	−	265
$c^2d^2e^6f$		...	$c^3d^6e^3f$	+	181600	$c^4d^{10}f$	+	10595	cde^4f^3	−	990
c^2de^8	−	6075	$c^3d^5e^5$	+	159000	$c^4d^9e^2$	−	88125	ce^6f^2	+	1620
cd^6ef^3	+	1820	$c^2d^9f^2$	+	43605	$c^3d^{11}e$	+	27300	d^4ef^4	−	555
$cd^5e^3f^2$	−	38820	$c^2d^8e^2f$	−	62025	c^2d^{13}	−	3375	$d^3e^3f^3$	+	1620
cd^4e^5f	+	66650	$c^2d^7e^4$	−	92500	$b^1c^{12}f^4$		...	$d^2e^5f^2$	−	1215
cd^3e^7	−	19800	$cd^{10}ef$	−	20610	$c^{11}def^3$	−	1080	de^7f		...

W, 29 A (concluded).

$a^0\ b^9e^9$		...
$b^8c^4ef^5$	+	30
$c^3d^2f^5$	−	370
$c^3de^2f^4$	+	1800
$c^3e^4f^3$	+	2845
$c^2d^3ef^4$	+	570
$c^2d^2e^3f^3$	+	1640
$c^2de^5f^2$	−	4185
c^2e^7f	−	4050
cd^5f^4	+	1110
$cd^4e^2f^3$	−	4170
$cd^3e^4f^2$		...
cd^2e^6f	+	6075
cde^8		...
d^6ef^3	+	3620
$d^5e^3f^2$	−	9310
d^4e^5f	+	8550
d^3e^7	−	3375
$b^7c^5df^5$	+	210
$c^5e^2f^4$	−	615
$c^4d^2ef^4$	−	1285
$c^4de^3f^3$	−	11420
$c^4e^5f^2$	−	3765
$c^3d^4f^4$	−	3155
$c^3d^3e^2f^3$	+	3160
$c^3d^2e^4f^2$	+	9850
c^3de^6f	+	19800
c^3e^8	+	3375
$c^2d^5ef^3$	−	13500
$c^2d^4e^3f^2$	+	50550
$c^2d^3e^5f$	−	62100
$c^2d^2e^7$		...
cd^7f^3	−	7240
$cd^6e^2f^2$	+	28710
cd^5e^4f	−	38950
cd^4e^6	+	25875
d^8ef^2	−	6155
d^7e^3f	+	12500
d^6e^5	−	7375
$b^6c^7f^5$	−	45
c^6def^4	+	615
$c^6e^3f^3$	+	3880
$c^5d^3f^4$	+	4300
$c^5d^2e^2f^3$	+	13430
$c^5de^4f^2$	+	18750
c^5e^6f	−	2175
$c^4d^4ef^3$	+	30040
$c^4d^3e^3f^2$	−	101450
$c^4d^2e^5f$	−	1850
c^4de^7	−	25875

$a^0\ b^6c^3d^6f^3$	+	34340
$c^3d^5e^2f^2$	−	153480
$c^3d^4e^4f$	+	220125
$c^3d^3e^6$		...
$c^2d^7ef^2$	+	6660
$c^2d^6e^3f$	+	18400
$c^2d^5e^5$	−	73375
cd^9f^2	+	12310
cd^8e^2f	−	44225
cd^7e^4	+	42500
$d^{10}ef$	+	4350
d^9e^3	−	5125
$b^5c^8ef^4$	−	45
$c^7d^2f^4$	−	2940
$c^7de^2f^3$	−	9960
$c^7e^4f^2$	−	8175
$c^6d^3ef^3$	−	46160
$c^6d^2e^3f^2$	+	34300
c^6de^5f	−	10350
c^6e^7	+	7375
$c^5d^5f^3$	−	73828
$c^5d^4e^2f^2$	+	306900
$c^5d^3e^4f$	−	159000
$c^5d^2e^6$	+	73375
$c^4d^6ef^2$	+	71610
$c^4d^5e^3f$	−	289800
$c^4d^4e^5$		...
$c^3d^8f^2$	−	59835
$c^3d^7e^2f$	+	129000
$c^3d^6e^4$	+	80500
c^2d^9ef	+	25050
$c^2d^8e^3$	−	80125
$cd^{11}f$	−	8700
$cd^{10}e^2$	+	19875
$d^{12}e$	−	1125
$b^4c^9df^4$	+	990
$c^9e^2f^3$	+	1710
$c^8d^2ef^3$	+	34620
$c^8de^3f^2$	+	4650
c^8e^5f	+	7050
$c^7d^4f^3$	+	92290
$c^7d^3e^2f^2$	−	243000
$c^7d^2e^4f$	+	92500
c^7de^6	−	42500
$c^6d^5ef^2$	−	219730
$c^6d^4e^3f$	+	318500
$c^6d^3e^5$	−	80500
$c^5d^7f^2$	+	114960
$c^5d^6e^2f$	+	5250
$c^5d^5e^4$		...

$a^0\ b^4c^4d^8ef$	−	138750
$c^4d^7e^3$	−	1250
$c^3d^{10}f$	+	31150
$c^3d^9e^2$	+	40000
$c^2d^{11}e$	−	18750
cd^{13}	+	2250
$b^3c^{11}f^4$	−	135
$c^{10}def^3$	−	12060
$c^{10}e^3f^2$	−	1980
$c^9d^3f^3$	−	69220
$c^9d^2e^2f^2$	+	89550
c^9de^4f	−	41250
c^9e^6	+	5125
$c^8d^4ef^2$	+	240975
$c^8d^3e^3f$	−	179500
$c^8d^2e^5$	+	80125
$c^7d^6f^2$	−	109660
$c^7d^5e^2f$	−	122800
$c^7d^4e^4$	+	1250
c^6d^7ef	+	178200
$c^6d^6e^3$		...
c^5d^9f	−	37950
$c^5d^8e^2$	−	37125
$c^4d^{10}e$	+	17875
c^3d^{12}	−	2125
$b^2c^{12}ef^3$	+	1620
$c^{11}d^2f^3$	+	30510
$c^{11}de^2f^2$	−	15120
$c^{11}e^4f$	+	6525
$c^{10}d^3ef^2$	−	128490
$c^{10}d^2e^3f$	+	69000
$c^{10}de^5$	−	19875
$c^9d^5f^2$	+	56110
$c^9d^4e^2f$	+	88125
$c^9d^3e^4$	−	40000
c^8d^6ef	−	103950
$c^8d^5e^3$	+	37125
c^7d^8f	+	22275
$c^7d^7e^2$		...
c^6d^9e	−	4125
c^5d^{11}	+	500
$b^1c^{13}df^3$	−	7290
$c^{13}e^2f^2$	+	810
$c^{12}d^2ef^2$	+	34155
$c^{12}de^3f$	−	15300
$c^{12}e^5$	+	1125
$c^{11}d^4f^2$	−	14895
$c^{11}d^3e^2f$	−	27300
$c^{11}d^2e^4$	+	18750
$c^{10}d^5ef$	+	30250

$a^0\ b^1c^{10}d^4e^3$	−	17875
c^9d^7f	−	6600
$c^9d^6e^2$	+	4125
c^8d^8e		...
c^7d^{10}		...
$b^0c^{15}f^3$	+	729
$c^{14}def^2$	−	3645
$c^{14}e^3f$	+	1350
$c^{13}d^3f^2$	+	1620
$c^{13}d^2e^2f$	+	3375
$c^{13}de^4$	−	2250
$c^{12}d^4ef$	−	3600
$c^{12}d^3e^3$	+	2125
$c^{11}d^6f$	+	800
$c^{11}d^5e^2$	−	500
$c^{10}d^7e$		...
c^9d^9		...

For the lower covariants the numerical verifications are given for the entire coefficient, but for the higher ones where the number of terms in a coefficient is considerable they are given separately for the different powers of a; and it is also interesting to consider them for the separate combinations of a and b. I recall that the positive and negative numerical coefficients are summed separately, so that ($\pm$ a number) means that the sum of the positive numerical coefficients is equal to the sum of the negative numerical coefficients and thus that the whole sum is $=0$.

It is to be observed that for the lower covariants the sums of the numerical coefficients do not vanish for the separate powers of a: thus in the invariant G, 141, the sums of the numerical coefficients for the terms in a^2, a^1, a^0 are $=1, -2, 1$ respectively.

As regards the invariants Q and Q'; for the first of these, Q, the sums of the numerical coefficients for the terms in a^4, a^3, a^2, a^1, a^0 are each of them $=0$, but this is not the case as regards Q'; in fact Q' is $=G^2+$ a multiple of Q; hence the sums for Q are the same as those for G^2, viz. they are $=1, -4, +6, -4, +1$ respectively. Like results present themselves in other cases, and they might probably be accounted for in a similar manner; we have a series of sums not each $=0$, but which are equal to a set of binomial coefficients taken with the signs + and − alternately and thus the sum of these sums is $=0$.

For R, S and S', I have given the sums for the different powers of a; and in regard to S I give here the following paragraphs from the Tenth Memoir on Quantics:—

I remark that I calculated the first two coefficients S_0, S_1, and deduced the other two, S_2 from S_1, and S_3 from S_0, by reversing the order of the letters (or which is the same thing, interchanging a and f, b and e, c and d) and reversing also the signs of the numerical coefficients. This process for S_2, S_3 is to a very great extent a verification of the values of S_0, S_1. For, as presently mentioned, the terms of S_0 form subdivisions such that in each subdivision the sum of the numerical coefficients is $=0$: in passing by the reversal process to the value of S_3, the terms are distributed into an entirely new set of subdivisions, and then in each of these subdivisions the sum of the numerical coefficients is found to be $=0$; and the like as regards S_1 and S_3.

If in the expressions for S_0, S_1, S_2, S_3 we first write $d=e=f=1$, thus in effect combining the numerical coefficients for the terms which contain the same powers in a, b, c, we find

$$\begin{aligned}S_0 = {} & a^3(-2c^3+6c^2-6c+2)\\ & +a^2\{b^2(6c^2-12c-6)+b(-15c^3+33c^2-21c+3)\\ & \qquad +b^0(42c^4-147c^3+195c^2-117c+27)\}\\ & +a\ \{b^4.0+b^3(30c^2-36c+6)+b^2(-117c^3+249c^2-183c+51)\\ & \qquad +b(9c^5+148c^4-378c^3+330c^2-99c)+b^0(-63c^6+165c^5-147c^4+45c^3)\}\end{aligned}$$

$$+ a^0 . \{b^6 . 2 + b^5(-15c+3) + b^4(75c^2 - 69c + 24) + b^3(-9c^4 - 167c^3 + 225c^2 - 87c - 2)$$
$$+ b^2(72c^5 + 48c^4 - 186c^3 + 96c^2) + b(-126c^6 + 201c^5 - 87c^4)$$
$$+ b^0(27c^8 - 45c^7 + 20c^6)\}$$

which for $c = 1$ becomes

$$= 2b^6 - 12b^5 + 30b^4 - 40b^3 + 30b^2 - 12b + 2, \text{ that is } 2(b-1)^6,$$

and for $b = 1$, becomes $= 0$.

$$S_2 = \quad a^3(0c^2 + 0c + 0)$$
$$+ a^2\{b^2(0c+0) + b(3c^3 - 9c^2 + 9c - 3) + b^0(24c^4 - 99c^3 + 153c^2 - 105c + 27)\}$$
$$+ a\ \{b^4 . 0 + b^3(-6c^2 + 12c - 6) + b^2(-24c^3 + 90c^2 - 108c + 42)$$
$$+ b(33c^4 - 90c^3 + 54c^2 + 30c - 27) + b^0(-27c^6 + 78c^5 - 66c^4 + 6c^3 + 9c^2)\}$$
$$+ a^0\{b^5(3c-3) + b^4(-15c+15) + b^3(6c^3 - 12c^2 + 36c - 30)$$
$$+ b^2(9c^5 - 42c^4 + 84c^3 - 108c^2 + 57c) + b(9c^6 - 54c^5 + 96c^4 - 51c^3)$$
$$+ b^0(9c^7 - 9c^6)\}$$

which for $c = 1$ becomes $= 0$.

$$S_3 = \quad a^3(0c + 0)$$
$$+ a^2\{b^2 . 0 + b(0c^2 + 0c + 0) + b^0(18c^4 - 72c^3 + 108c^2 - 72c + 18)\}$$
$$+ a\ \{b^3(0c+0) + b^2(-33c^3 + 99c^2 - 99c + 33) + b(57c^4 - 162c^3 + 144c^2 - 30c - 9)$$
$$+ b^0(-60c^5 + 207c^4 - 261c^3 + 141c^2 - 27c)\}$$
$$+ a^0\{b^5 . 0 + b^4(15c^2 - 30c + 15) + b^3(-54c^3 + 102c^2 - 42c - 6)$$
$$+ b^2(123c^4 - 297c^3 + 243c^2 - 87c + 18) + b(-27c^6 + 102c^4 - 96c^3 + 21c^2)$$
$$+ b^0(27c^7 - 60c^6 + 51c^5 - 12c^4)\}$$

which for $c = 1$ becomes $= 0$.

$$S_4 = \quad a^3 . 0$$
$$+ a^2\{b(0c+0) + b^0(0c^3 + 0c^2 + 0c + 0)\}$$
$$+ a\ \{b^3 . 0 + b^2(0c^2 + 0c + 0) + b(-9c^4 + 36c^3 - 54c^2 + 36c - 9)$$
$$+ b^0(36c^5 - 171c^4 + 324c^3 - 306c^2 + 144c - 27)\}$$
$$+ a^0\{b^4(0c+0) + b^3(7c^3 - 21c^2 + 21c - 7) + b^2(-39c^4 + 135c^3 - 171c^2 + 93c - 18)$$
$$+ b(66c^5 - 243c^4 + 333c^3 - 201c^2 + 45c)$$
$$+ b^0(-27c^7 + 101c^6 - 141c^5 + 87c^4 - 20c^3)\}$$

which for $c = 1$ becomes $= 0$.

It follows that for $c = d = e = f = 1$, the value of the covariant S is $= 2(b-1)^6 x^3$, which might be easily verified.

For T, U, V and W, I look at the sums for the different combinations of a and b.

Thus for T we have

x coefficient.

$a^4 b^0$		26	
			± 26
$a^3 b^2$		± 14	
b^1		141	
b^0		281	
			± 436
$a^2 b^4$		± 1	
b^3		106	
b^2		186	
b^1		1173	
b^0		2272	
			± 3738
$a^1 b^5$		± 16	
b^4		359	
b^3		1411	
b^2		3103	
b^1		3030	
b^0		1197	
			± 9116
$a^0 b^7$		− 2	
b^6	92	− 78	
b^5	307	− 349	
b^4	1073	− 1003	
b^3	2040	− 2110	
b^2	1930	− 1880	
b^1	1207	− 1221	
b^0	231	− 239	
			± 6880
			± 20196

y coefficient.

$a^4 b^0$	± 12	
		± 12
$a^3 b^2$	± 2	
b^1	112	
b^0	281	
		± 395
$a^2 b^3$	± 42	
b^2	546	
b^1	696	
b^0	366	
		± 1650
$a^1 b^5$	± 5	
b^4	179	
b^3	821	
b^2	2097	
b^1	2147	
b^0	1262	
		± 6511
$a^0 b^6$	± 28	
b^5	342	
b^4	1790	
b^3	3496	
b^2	3445	
b^1	2064	
b^0	463	
		± 11628
		± 20196

Observe here that in the x-coefficient for the terms in a^0 the successive sums are -2, $+14$, -42, $+70$, -70, $+42$, $-14+2$, which are the coefficients of $-2(\theta-1)^7$.

For U we have

$a^4 b^0$	$\pm$ 36	
		$\pm$ 36
$a^3 b^2$	$\pm$ 24	
b^1	198	
b^0	242	
		$\pm$ 464
$a^2 b^4$	$\pm$ 2	
b^3	208	
b^2	286	
b^1	866	
b^0	1246	
		$\pm$ 2608
$a^1 b^5$	$\pm$ 64	
b^4	328	
b^3	1258	
b^2	2586	
b^1	2186	
b^0	856	
		$\pm$ 7278
$a^0 b^7$	$\pm$ 4	
b^6	70	
b^5	448	
b^4	1488	
b^3	2140	
b^2	1678	
b^1	884	
b^0	166	
		$\pm$ 6878
		$\pm$ 17264

For V we have

x coefficient.

$a^5 b^0$		$\pm$ 36	
			$\pm$ 36
$a^4 b^2$		$\pm$ 20	
b^1		284	
b^0		1094	
			$\pm$ 1398
$a^3 b^4$		2	
b^3		184	
b^2		1656	
b^1		3624	
b^0		4898	
			$\pm$ 10364
$a^2 b^5$		$\pm$ 14	
b^4		666	
b^3		6608	
b^2		10512	
b^1		22042	
b^0		9162	
			$\pm$ 49004
$a^1 b^7$		$-$ 4	
b^6	76	$-$ 48	
b^5	2956	$-$ 3040	
b^4	11946	$-$ 11806	
b^3	23924	$-$ 24064	
b^2	25110	$-$ 25026	
b^1	25524	$-$ 25552	
b^0	8822	$-$ 8812	
			$\pm$ 98358
$a^0 b^8$	18		
b^7	184	$-$ 324	
b^6	4098	$-$ 3622	
b^5	19350	$-$ 20274	
b^4	42398	$-$ 41278	
b^3	51872	$-$ 52740	
b^2	44320	$-$ 43900	
b^1	20624	$-$ 20740	
b^0	3870	$-$ 3856	
			$\pm$ 186734
			$\pm$ 345894

y coefficient.

$a^5 b^0$		$\pm$ 24	
			$\pm$ 24
$a^4 b^2$		$\pm$ 4	
b^1		144	
b^0		436	
			$\pm$ 584
$a^3 b^3$		$\pm$ 24	
b^2		776	
b^1		2696	
b^0		1264	
			$\pm$ 4760
$a^2 b^5$		$\pm$ 6	
b^4		300	
b^3		2236	
b^2		8616	
b^1		15442	
b^0		33044	
			$\pm$ 59644
$a^1 b^6$		$\pm$ 78	
b^5		852	
b^4		8310	
b^3		30200	
b^2		56740	
b^1		39956	
b^0		17986	
			$\pm$ 154122
$a^0 b^8$		$-$ 2	
b^7	286	$-$ 270	
b^6	2026	$-$ 2082	
b^5	9360	$-$ 9248	
b^4	19760	$-$ 19900	
b^3	36442	$-$ 36330	
b^2	30340	$-$ 30396	
b^1	23426	$-$ 23410	
b^0	5120	$-$ 5122	
			$\pm$ 126760
			$\pm$ 345894

Here in the x-coefficient for a^1 the successive sums are -4, $+28$, -84, $+140$, -140, $+84$, -28, $+4$, which are the coefficients of $-4(\theta-1)^7$; and for a^0 the successive sums are 18, -140, $+476$, -924, $+1120$, -868, $+420$, -116, $+14$, which are the coefficients of $18(\theta-1)^8+4(\theta-1)^7$. In the y-coefficient the successive sums are -2, $+16$, -56, $+112$, -140, $+112$, -56, $+16$, -2, which are the coefficients of $-2(\theta-1)^8$.

Finally for W we have

$a^7 b^0$	± 16	
		± 16
$a^6 b$	± 175	
b^0	806	
		± 981
$a^5 b^3$	± 80	
b^2	1175	
b^1	2760	
b^0	6871	
		± 10886
$a^4 b^4$	± 570	
b^3	5200	
b^2	18005	
b^1	44720	
b^0	23810	
		± 92305
$a^3 b^6$	± 90	
b^5	2386	
b^4	26675	
b^3	84680	
b^2	107730	
b^1	199160	
b^0	240499	
		± 661220
$a^2 b^8$	± 15	
b^7	640	
b^6	8260	
b^5	59135	
b^4	182055	
b^3	341470	
b^2	699260	
b^1	612015	
b^0	304501	
		± 2207351

	+	−	
			± 2972759
$a^1 b^{10}$	+ 1		
b^9	120	− 130	
b^8	1125	− 1080	
b^7	30350	− 30470	
b^6	122400	− 122190	
b^5	332494	− 332746	
b^4	729150	− 728940	
b^3	880750	− 880870	
b^2	466935	− 466890	
b^1	363670	− 363680	
b^0	76116	− 76115	
			± 3003111
$a^0 b^{11}$	+	− 5	
b^{10}	400	− 346	
b^9	3500	− 3765	
b^8	26240	− 25460	
b^7	154030	− 155560	
b^6	409700	− 407600	
b^5	747985	− 750043	
b^4	745920	− 744480	
b^3	613100	− 613805	
b^2	311790	− 311560	
b^1	89215	− 89260	
b^0	9999	− 9995	
			± 3111879
			± 9087749

Here for the terms in a^1 the successive sums are

$$1,\ -10,\ +45,\ -120,\ +210,\ -252,\ +210,\ -120,\ +45,\ -10,\ +1,$$

which are the coefficients of $(\theta-1)^{10}$; and for the terms in a^0 the successive sums are

$$-5,\ +54,\ -265,\ +780,\ -1530,\ +2100,\ -2058,\ +1440,\ -705,\ +230,\ -45,\ +4,$$

which are the coefficients of $-5(\theta-1)^{11}-(\theta-1)^{10}$.

144.

A THIRD MEMOIR UPON QUANTICS.

[From the *Philosophical Transactions of the Royal Society of London*, vol. CXLVI. *for the year* 1856, pp. 627—647. Received March 13,—Read April 10, 1856.]

My object in the present memoir is chiefly to collect together and put upon record various results useful in the theories of the particular quantics to which they relate. The tables at the commencement relate to binary quantics, and are a direct sequel to the tables in my Second Memoir upon Quantics, vol. CXLVI. (1856), [141]. The definitions and explanations in the next part of the present memoir are given here for the sake of convenience, the further development of the subjects to which they relate being reserved for another occasion. The remainder of the memoir consists of tables and explanations relating to ternary quadrics and cubics.

Covariant and other Tables, Nos. 27 to 50 (Nos. 1 to 50 binary quantics)[1].

Nos. 27 to 29 are a continuation of the tables relating to the quintic

$$(a,\ b,\ c,\ d,\ e,\ f \between x,\ y)^5.$$

No. 27 gives the values of the different determinants of the matrix

$$\left(\begin{array}{cccccc} & a, & 4b, & 6c, & 4d, & e \\ a, & 4b, & 6c, & 4d, & e & \\ & b, & 4c, & 6d, & 4e, & f \\ b, & 4c, & 6d, & 4e, & f & \end{array}\right)$$

determinants which are represented by 1234, 1235, &c., where the numbers refer to

[1] The Tables 49 and 50 were inserted October 6, 1856.—A. C.

the different columns of the matrix. No. 28 gives the values of certain linear functions of these determinants, viz.

$$\begin{aligned}
L &= 1256 + 2345 - 2\,.\,1346,\\
L' &= 3\,.\,1256 - 1346,\\
8M &= -\,1345 + 2\,.\,1246,\\
8M' &= -\,2346 + 2\,.\,1356,\\
8N &= -\,1245 + 3\,.\,1236,\\
8N' &= -\,2356 + 3\,.\,1456,\\
80P &= L' - 3L = 5\,.\,1346 - 3\,.\,2345,\\
16P' &= -\,5L' - L = -\,16\,.\,1256 - 3\,.\,1346 - 2356.
\end{aligned}$$

At the end of the two tables there are given certain relations which exist between the terms of Tables 14, 16, 25, 26, 27 and 28.

No. 27.

1234.	1235.	1236.	1245.	1246.	1345.	1256.	2345.
a^2bf ...	a^2cf − 4	a^2df + 6	a^2df − 6	a^2ef + 4	a^2ef ...	a^2f^2 + 1	a^2f^2 ...
a^2ce − 16	a^2de + 24	a^2e^2 ...	a^2e^2 + 16	$abdf$ − 4	$abdf$ − 24	$abef$ − 2	$abef$ ...
a^2d^2 + 36	ab^2f + 4	$abcf$ − 22	$abcf$ + 6	abe^2 − 4	abe^2 + 64	$acdf$ − 16	$acdf$ + 20
ab^2e + 16	$abce$ − 84	$abde$ − 6	$abde$ − 26	ac^2f − 24	ac^2f + 24	ace^2 + 16	ace^2 − 80
$abcd$ − 152	abd^2 − 24	ac^2e + 16	ac^2e − 96	$acde$ + 24	$acde$ − 208	ad^2e + 16	ad^2e + 60
ac^3 + 96	ac^2d + 64	acd^2 ...	acd^2 + 96	ad^3 ...	ad^3 + 144	b^2df − 15	b^2df − 80
b^3d + 80	b^3e + 60	b^3f + 16	b^3f ...	b^2cf + 24	b^2cf ...	b^2e^2 ...	b^2e^2 + 240
b^2c^2 − 60	b^2cd − 40	b^2ce − 10	b^2ce + 90	b^2de − 20	b^2de − 40	bc^2f ...	bc^2f + 60
	bc^3 ...	b^2d^2 ...	b^2d^2 − 80	bc^2e ...	bc^2e + 60	$bcde$ ...	$bcde$ − 860
		bc^2d ...	bc^2d ...	bcd^2 ...	bcd^2 − 40	bd^3 ...	bd^3 + 960
		c^4 ...	c^4 ...	c^3d ...	c^3d ...	c^3e ..	c^3e + 960
						c^2d^2 ...	c^2d^2 − 320

1346.	2346.	1356.	2356.	1456.	2456.	3456.
a^2f^2 ...	abf^2 ...	abf^2 + 4	acf^2 − 6	acf^2 + 6	adf^2 − 4	aef^2 ...
$abef$ + 16	$acef$ − 24	$acef$ − 4	$adef$ + 6	$adef$ − 22	ae^2f + 4	bdf^2 − 16
$acdf$ − 36	ad^2f + 24	ad^2f − 24	ae^3 ...	ae^3 + 16	bcf^2 + 24	be^2f + 16
ace^2 − 16	ade^2 ...	ade^2 + 24	b^2f^2 + 16	b^2f^2 ...	$bdef$ − 84	c^2f^2 + 36
ad^2e + 36	b^2ef + 64	b^2ef − 4	$bcef$ − 26	$bcef$ − 6	be^3 + 60	$cdef$ − 152
b^2df − 16	$bcdf$ − 208	$bcdf$ + 24	b^2df − 96	bd^2f + 16	c^2ef − 24	ce^3 + 80
b^2e^2 ...	bce^2 − 40	bce^2 − 20	bde^2 + 90	bde^2 − 10	cd^2f + 64	d^3f + 96
bc^2f + 36	bd^2e + 60	bd^2e ...	c^2df + 96	c^2df ...	cde^2 − 40	d^2e^2 − 60
$bcde$ − 20	c^3f + 144	c^3f ...	c^2e^2 − 80	c^2e^2 ...	d^3e ...	
bd^3 ...	c^2de − 40	c^2de ...	cd^2e ...	cd^2e ...		
c^3e ...	cd^3 ...	cd^3 ...	d^4 ...	d^4 ...		
c^2d^2 ...						

No. 28.

N.	M.	L.	L′.	P.	P′.	M′.	N′.
a^2df + 3	a^2ef + 1	a^2f^2 + 1	a^2f^2 + 3	a^2f^2 ...	a^2f^2 − 1	abf^2 + 1	acf^2 + 3
a^2e^2 − 2	$abdf$ + 2	$abef$ − 34	$abef$ − 22	$abef$ + 1	$abef$ + 9	$acef$ + 2	$adef$ − 9
$abcf$ − 9	abe^2 − 9	$acdf$ + 76	$acdf$ − 12	$acdf$ − 3	$acdf$ − 1	ad^2f − 9	ae^3 + 6
$abde$ + 1	ac^2f − 9	ace^2 − 32	ace^2 + 64	ace^2 + 2	ace^2 − 18	ade^2 + 6	b^2f^2 − 2
ac^2e + 18	$acde$ + 32	ad^2e − 12	ad^2e − 36	ad^2e ...	ad^2e + 12	b^2ef − 9	$bcef$ + 1
acd^2 − 12	ad^3 − 18	b^2df − 32	b^2df + 64	b^2df + 2	b^2df − 18	$bcdf$ + 32	bd^2f + 18
b^3f + 6	b^2cf + 6	b^2e^2 + 225	b^2e^2 − 45	b^2e^2 − 9	b^2e^2 ...	bce^2 ...	bde^2 − 15
b^2ce − 15	b^2de ...	bc^2f − 12	bc^2f − 36	bc^2f ...	bc^2f + 12	bd^2e − 15	c^2df − 12
b^2d^2 + 10	bc^2e − 15	$bcde$ − 820	$bcde$ + 20	$bcde$ + 31	$bcde$ + 45	c^3f − 18	c^2e^2 + 10
bc^2d ...	bcd^2 + 10	bd^3 + 480	bd^3 ...	bd^3 − 18	bd^3 − 30	c^2de + 10	cd^2e ...
c^4 ...	c^3d ...	c^3e + 480	c^3e ...	c^3e − 18	c^3e − 30	cd^3 ...	d^4 ...
		c^2d^2 − 320	c^2d^2 ...	c^2d^2 + 12	c^2d^2 + 20		

If the coefficients of the table 14 are represented by $\tfrac{1}{2}A$, B, $\tfrac{1}{2}C$, viz. writing

$$A = 2\,(ae - 4bd + 3c^2),$$
$$B = \quad af - 3be + 2cd,$$
$$C = 2\,(bf - 4ce + 3d^2),$$

then we have the following relations between 1234, &c. and A, B, C, viz.

	$C\times$	$+B\times$	$+A\times$
1234 =	$+ 6\,a^2$	$- 12\,ab$	$+ 16\,ac - 10\,b^2$
1235 =	$+ 6\,ab$	$- 2\,ac - 10\,b^2$	$+ 6\,ad$
1236 =	$- 2\,ac + 8\,b^2$	$+ 6\,ad - 18\,bc$	$- 2\,df + 8\,e^2$
1245 =	$+ 18\,ac$	$- 6\,ad - 30\,bc$	$+ 8\,ae + 10\,bd$
1246 =	$+ 12\,bc$	$+ 4\,ae - 4\,bd - 24\,c^2$	$+ 4\,be + 8\,cd$
1345 =	$+ 24\,ad$	$- 8\,ae - 40\,bd$	$+ 4\,af + 20\,be$
1256 =	$- 1\,ae + 4\,bd + 3\,c^2$	$+ 1\,af + 5\,be - 18\,cd$	$- 1\,bf + 4\,ce + 3\,d^2$
2345 =	$+ 20\,ae + 40\,bd - 30\,c^2$	$- 80\,be + 20\,cd$	$+ 20\,bf + 40\,ce - 30\,d^2$
1346 =	$+ 4\,ae + 8\,bd + 6\,c^2$	$- 36\,cd$	$+ 4\,bf + 8\,ce + 6d^2$
2346 =	$+ 4\,af + 20\,be$	$- 8\,bf - 4\,ce$	$+ 24\,cf$
1356 =	$+ 4\,be + 8\,cd$	$+ 4\,bf - 4\,ce - 24\,d^2$	$+ 12\,de$
2356 =	$+ 8\,bf + 10\,ce$	$- 6\,cf - 30\,de$	$+ 18\,df$
1456 =	$+ 6\,ce$	$+ 6\,cf - 18\,de$	$- 2\,df + 8\,e^2$
2456 =	$+ 6\,cf$	$- 2\,df - 10\,e^2$	$+ 6\,ef$
3456 =	$+ 16\,df - 10\,e^2$	$- 12\,ef$	$+ 6\,f^2$

and the following relations between L, L', &c. and A, B, C, viz.

	$C\times$	$+B\times$	$+A\times$
$N =$	$- 3\,ac + 3\,b^2$	$+ 3\,ad - 3\,bc$	$- 1\,ae + 1\,bd$
$M =$	$- 3\,ad + 3\,bc$	$+ 3\,ae - 3\,c^2$	$- 1\,af + 1\,cd$
$L =$	$+ 11\,ae + 28\,bd - 39\,c^2$	$+ 1\,af - 75\,be + 74\,cd$	$+ 11\,bf + 28\,ce - 39\,d^2$
$L' =$	$- 7\,ae + 4\,bd + 3\,c^2$	$+ 3\,af + 15\,be - 18\,cd$	$- 7\,bf + 4\,ce + 3\,d^2$
$2P =$	$- 1\,ae - 2\,bd + 3\,c^2$	$+ 3\,be - 3\,cd$	$+ 1\,bf + 2\,ce - 3\,d^2$
$P' =$	$+ 3\,ae - 6\,bd + 3\,c^2$	$- 1\,af + 1\,cd$	$+ 3\,bf - 6\,ce + 3\,d^2$
$M' =$	$- 1\,af + 1\,cd$	$+ 3\,bf - 3\,d^2$	$- 3\,cf + 3\,de$
$N' =$	$- 1\,bf + 1\,ce$	$+ 3\,cf - 3\,de$	$- 3\,df + 3\,e^2$

We have also the following relations between L, L', &c. and a, b, c, d, e, f, viz.

$$\begin{array}{llllll} aP & -bM & +cN & & & =0, \\ aM'+bP' & -2cM+3dN & & & & =0, \\ aN'+2bM'-cL' & . & +3eN & & & =0, \\ 3bN' & . & - & dL'+2eM & +fN & =0, \\ & 3cN'-2dM'+ & & eP' & +fM & =0, \\ & & dN'- & eM' & +fP & =0. \end{array}$$

The quartinvariant No. 19 [G] is equal to

$$-AC+B^2,$$

i.e. it is in fact equal to -4 into the discriminant of the quintic No. 14, [A].

The octinvariant No. 25 [Q] is expressible in terms of the coefficients of Nos. 14 and 16, viz. A, B, C, as before, and $\tfrac{1}{3}\alpha$, β, γ, $\tfrac{1}{3}\delta$ the coefficients of No. 16, [D], i.e.

$$\begin{aligned} \alpha &= 3\,(ace-ad^2-b^2e+2bcd-c^3), \\ \beta &= acf-ade-b^2f+bd^2+bce-c^2d, \\ \gamma &= adf-ae^2-bcf+bde+c^2e-cd^2, \\ \delta &= 3\,(bdf-be^2+2cde-c^2f-d^3), \end{aligned}$$

then No. 25 is equal to

$$\begin{vmatrix} A, & B, & C \\ \alpha, & \beta, & \gamma \\ \beta, & \gamma, & \delta \end{vmatrix}.$$

The value of the discriminant No. 26, [Q′], is

$$(\text{No. }19)^2-128\text{ No. }25.\quad [\text{that is } Q'=G^2-128Q.]$$

We have also an expression for the discriminant in terms of L, L', &c., viz. three times the discriminant No. 26 is equal to

$$[\text{or say } 3Q'=]\ LL'+64MM'-64NN',$$

a remarkable formula, the discovery of which is due to Mr Salmon.

It may be noticed, that in the particular case in which the quintic has two square factors, if we write

$$(a,\ b,\ c,\ d,\ e,\ f\between x,\ y)^5=5\,\{(p,\ q,\ r\between x,\ y)^2\}^2\ .\ (\lambda,\ \mu\between x,\ y),$$

then

$$a = 5\lambda p^2, \qquad b = 4pq\lambda + p^2\mu, \qquad c = (2q^2 + pr)\lambda + 2pq\mu,$$
$$f = 5r^2\mu, \qquad e = r^2\lambda + 4qr\mu, \qquad d = 2qr\lambda + (2q^2 + pr)\mu;$$

and these values give

$$P = K(6q^2 - pr), \qquad P' = K(10q^2 - 15pr),$$
$$M = K \,.\, 10pq, \qquad M' = K \,.\, 10qr,$$
$$N = K \,.\, 5p^2, \qquad N' = K \,.\, 5r^2,$$

where the value of K is

$$8\,(p\mu^2 - 2q\mu\lambda + r\lambda^2)^2\,(pr - q^2)^2.$$

The table No. 29 is the invariant of the twelfth degree of the quintic, given in its simplest form, i.e. in a form not containing any power higher than the fourth of the leading coefficient a: this invariant was first calculated by M. Faa de Bruno.

No. 29. [See U. No. 29, p. 294.]

The tables Nos. 30 to 35 relate to a sextic. No. 30 is the sextic itself; No. 31 the quadrinvariant; Nos. 32 and 33 the quadricovariants (the latter of them the Hessian); No. 34 is the quartinvariant or catalecticant; and No. 35 is the sextinvariant in its best form, i.e. a form not containing any power higher than the second of the leading coefficient a.

No. 30.

(							
$a + 1$	$b + 6$	$c + 15$	$d + 20$	$e + 15$	$f + 6$	$g + 1$	$)(x, y)^6$

No. 31.

ag + 1 bf − 6 ce + 15 d^2 − 10
±16

No. 32.

(					
ae + 1 bd − 4 c^2 + 3	af + 2 be − 6 cd + 4	ag + 1 ce − 9 d^2 + 8	bg + 2 cf − 6 de + 4	cg + 1 df − 4 e^2 + 3	$)(x, y)^4$
±4	±6	±9	±6	±4	

No. 33.

(									
ac + 1 b^2 − 1	ad + 4 bc − 4	ae + 6 bd + 4 c^2 − 10	af + 4 be + 16 cd − 20	ag + 1 bf + 14 ce + 5 d^2 − 20	bg + 4 cf + 16 de − 20	cg + 6 df + 4 e^2 − 10	dg + 4 ef − 4	eg + 1 f^2 − 1	$)(x, y)^8$
±1	±4	±10	±20	±20	±20	±10	±4	±1	

No. 34.

$aceg$	+1
acf^2	−1
ad^2g	−1
$adef$	+2
ae^3	−1
b^2eg	−1
b^2f^2	+1
$bcdg$	+2
$bcef$	−2
bd^2f	−2
bde^2	+2
c^3g	−1
c^2df	+2
c^2e^2	+1
cd^2e	−3
d^4	+1

±12

No. 35.

$a^2d^2g^2$	+ 1	$acde^2f$	− 42	bc^2df^2	+ 60
a^2defg	− 6	ace^4	+ 12	bc^2e^2f	− 30
a^2df^3	+ 4	ad^4g	− 20	bcd^3g	+ 24
a^2e^3g	+ 4	ad^3ef	+ 24	bcd^2ef	− 84
$a^2e^2f^2$	− 3	ad^2e^3	− 8	$bcde^3$	+ 66
$abcdg^2$	− 6	b^3dg^2	+ 4	bd^4f	+ 24
$abcefg$	+ 18	b^3efg	− 12	bd^3e^2	− 24
$abcf^3$	− 12	b^3f^3	+ 8	c^4eg	+ 12
abd^2fg	+ 12	$b^2c^2g^2$	− 3	c^4f^2	− 27
$abde^2g$	− 18	b^2ce^2g	+ 30	c^3d^2g	− 8
abe^3f	+ 6	b^2cef^2	− 24	c^3def	+ 66
ac^3g^2	+ 4	b^2d^2eg	− 12	c^3e^3	− 8
ac^2e^2g	− 24	$b^2d^2f^2$	− 24	c^2d^3f	− 24
ac^2dfg	− 18	b^2de^2f	+ 60	$c^2d^2e^2$	− 39
ac^2ef^2	+ 30	b^2e^4	− 27	cd^4e	+ 36
acd^2eg	+ 54	bc^3fg	+ 6	d^6	− 8
acd^2f^2	− 12	bc^2deg	− 42		

±565

The sextinvariant may be thus represented by means of a determinant of the sixth order and of the quadrinvariant and quartinvariant.

$$5 \times \text{No. } 35 = \begin{vmatrix} & a, & 2b, & 3c, & 4d, & e \\ & b, & 2c, & 3d, & 4e, & f \\ & c, & 2d, & 3e, & 4f, & g \\ a, & 4b, & 3c, & 2d, & e & \\ b, & 4c, & 3d, & 2e, & f & \\ c, & 4d, & 3e, & 2f, & g & \end{vmatrix}$$

$$+ 4(ag - 6bf + 15ce - 10d^2) \begin{vmatrix} a, & b, & c, & d \\ b, & c, & d, & e \\ c, & d, & e, & f \\ d, & e, & f, & g \end{vmatrix}$$

The tables Nos. 36 and 37 relate to a septimic. No. 36 is the septimic itself; No. 37 the quartinvariant.

No. 36.

($a+1$	$b+7$	$c+21$	$d+35$	$e+35$	$f+21$	$g+7$	$h+1$	)($x, y)^7$

No. 37.

a^2h^2	− 1	bd^2h	− 40
$abgh$	+ 14	$bdeg$	− 50
$acfh$	− 18	bdf^2	− 360
acg^2	− 24	be^2f	+ 240
$adeh$	+ 10	c^2eg	− 360
$adfg$	+ 60	c^2f^2	− 81
ae^2g	− 40	cd^2g	+ 240
b^2fh	− 24	$cdef$	+ 990
b^2g^2	− 25	ce^3	− 600
$bcfg$	+ 234	d^3f	− 600
$bceh$	+ 60	d^2e^2	+ 375

±2223

The tables Nos. 38 to 45 relate to the octavic. No. 38 is the octavic itself; No. 39 the quadrinvariant; Nos. 40, 41 and 42 are the quadricovariants, the last of them being the Hessian; No. 43 is the cubinvariant; No. 44 the quartinvariant, and No. 45 the quintinvariant, which is also the catalecticant.

No. 38.

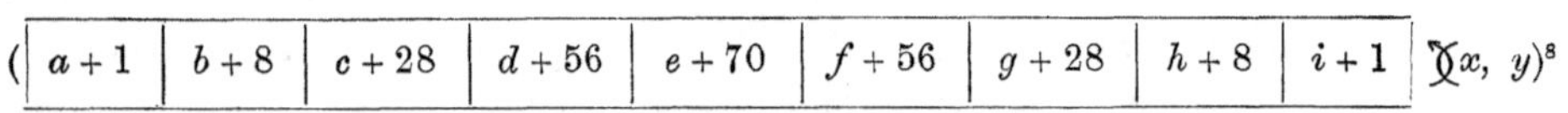

(a + 1	b + 8	c + 28	d + 56	e + 70	f + 56	g + 28	h + 8	i + 1) $(x, y)^8$

No. 39.

ai + 1
bh − 8
cg + 28
df − 56
e^2 + 35
±64

No. 40.

(ag + 1	ah + 2	ai + 1	bi + 2	ci + 1	
bf − 6	bg − 10	bh − 2	ch − 10	dh − 6	
ce + 15	cf + 18	cg − 8	dg + 18	eg + 15	) $(x, y)^4$.
d^2 − 10	de − 10	df + 34	ef − 10	f^2 − 10	
		e^2 − 25			
±16	±20	±35	±20	±16	

No. 41.

(ae + 1	af + 4	ag + 6	ah + 4	ai + 1	bi + 4	ci + 6	di + 4	ei + 1	
bd − 4	be − 12	bf − 8	bg + 8	bh + 12	ch + 8	dh − 8	eh − 12	fh − 4	
c^2 + 3	cd + 8	ce − 22	cf − 48	cg − 22	dg − 48	eg − 22	fg + 8	g^2 + 3	) $(x, y)^8$
		d^2 + 24	de + 36	df − 36	ef + 36	f^2 + 24			
				e^2 + 45					
±4	±12	±30	±48	±58	±48	±30	±12	±4	

No. 42.

+ 1	ad + 6	ae + 15	af + 20	ag + 15	ah + 6	ai + 1	bi + 6	ci + 15	di + 20	ei + 15	fi + 6	gi + 1
− 1	bc − 6	bd + 6	be + 50	bf + 90	bg + 78	bh + 34	ch + 78	dh + 90	eh + 50	fh + 6	gh − 6	h^2 − 1
		c^2 − 21	cd − 70	d^2 − 105	cf + 126	cg + 154	dg + 126	f^2 − 105	fg − 70	g^2 − 21		
					de − 210	df − 14	ef − 210					
						e^2 − 175						
±1	±6	±21	±70	±105	±210	±189	±210	±105	±70	±21	±6	±1

No. 43.

aei	+ 1
afh	− 4
ag^2	+ 3
bdi	− 4
beh	+ 12
bfg	− 8
c^2i	+ 3
cdh	− 8
ceg	− 22
cf^2	+ 24
d^2g	+ 24
def	− 36
e^3	+ 15

±82

No. 44.

$acgi$	− 1	$bcgh$	+ 3	cd^2i	− 2
ach^2	+ 1	$bdei$	+ 1	$cdeh$	− 23
$adfi$	+ 3	$bdfh$	− 10	$cdfg$	+ 27
$adgh$	− 3	bdg^2	+ 9	ce^3g	+ 19
ae^2i	− 2	be^2h	+ 11	cef^2	− 21
$aefh$	+ 1	$befg$	− 23	d^3h	+ 12
aeg^2	+ 3	bf^3	+ 12	d^2eg	− 21
af^2g	− 2	c^2ei	+ 3	d^2f^2	− 13
b^2gi	+ 1	c^2fh	+ 9	de^2f	+ 32
b^2h^2	− 1	c^2g^2	− 12	e^4	− 10
$bcfi$	− 3				

±147

No. 45.

$acegi$	+ 1	af^4	+ 1	$bdeg^2$	− 4	cd^2g^2	+ 1
$aceh^2$	− 1	b^2egi	− 1	bdf^2g	+ 2	$cdefg$	− 2
acf^2i	− 1	b^2eh^2	+ 1	be^3h	− 2	cdf^3	− 2
$acfgh$	+ 2	b^2fgh	− 2	be^2fg	+ 4	ce^3g	− 3
acg^3	− 1	b^2fi^2	+ 1	bef^3	− 2	ce^2dh	+ 4
ad^2gi	− 1	b^2g^3	+ 1	c^3gi	− 1	ce^2f^2	+ 3
ad^2h^2	+ 1	$bcdgi$	+ 2	c^3h^2	+ 1	d^4i	+ 1
$adefi$	+ 2	$bcdh^2$	− 2	c^2dfi	+ 2	d^3eh	− 2
$adegh$	− 2	$bcefi$	− 2	c^2dgh	− 2	d^3fg	− 2
adf^2h	− 2	$bcegh$	+ 2	c^2e^2i	+ 1	d^2e^2g	+ 3
$adfg^2$	+ 2	bcf^2h	+ 2	c^2efh	− 4	d^2ef^2	+ 3
ae^3i	− 1	$bcfg^2$	− 2	c^2eg^2	+ 2	de^3f	− 4
ae^2fh	+ 2	bd^2fi	− 2	c^2f^2g	+ 1	e^5	+ 1
ae^2g^2	+ 1	bd^2gh	+ 2	cd^2ei	− 3		
aef^2g	− 3	bde^2i	+ 2	cd^2fh	+ 2		

±56

If we write

$$\text{No. } 39 = I,$$
$$\text{No. } 43 = J,$$
$$\text{No. } 44 = K,$$
$$\text{No. } 45 = L,$$

then the determinant called the lambdaic, viz.

$$\begin{vmatrix} a, & b, & c, & d, & e-12\lambda \\ b, & c, & d, & e+3\lambda, & f \\ c, & d, & e-2\lambda, & f, & g \\ d, & e+3\lambda, & f, & g, & h \\ e-12\lambda, & f, & g, & h, & i \end{vmatrix}$$

is equal to

$$L + 2\lambda K + 3\lambda^2 J + 18\lambda^3 I - 2592\lambda^5.$$

Nos. 46 to 48 relate to the nonic. No. 46 is the nonic itself; Nos. 47 and 48 are the two quartinvariants, each of them in its best form, viz. No. 48 does not contain a^2, and No. 47 does not contain aci^2, the leading term of No. 48. The nonic is the lowest quantic with two quartinvariants.

No. 46.

(	$a+1$	$b+9$	$c+36$	$d+84$	$e+126$	$f+126$	$g+84$	$h+36$	$i+9$	$j+1$	$\between x, y)^9$.

No. 47.

a^2j^2	$-$ 1	bf^2h	...
$abij$	$+$ 18	bfg^2	$-$ 720
aci^2	...	c^2fj	$+$ 432
$achj$	$-$ 72	c^2gi	$-$ 1728
$adgj$	$+$ 168	c^2h^2	...
$adhi$	...	$cdej$	$-$ 720
$aefj$	$-$ 108	$cdfi$	$+$ 2160
$aegi$	$-$ 576	$cdgh$	$+$ 4608
aeh^2	$+$ 432	ce^2i	...
af^2i	$+$ 540	$cefh$	$-$ 2592
$afgh$	$-$ 720	ceg^2	$-$ 5760
ag^3	$+$ 320	cf^2g	$+$ 4320
b^2hj	...	d^3j	$+$ 320
b^2i^2	$-$ 81	d^2ei	$-$ 720
$bcgj$	...	d^2fh	$-$ 5760
$bchi$	$+$ 648	d^2g^2	$-$ 1536
$bdfj$	$-$ 576	$defg$	$+$ 14688
$bdgi$	$+$ 792	de^2h	$+$ 4320
bdh^2	$-$ 1728	df^3	$-$ 8640
$begh$	$+$ 2160	e^3g	$-$ 8640
be^2j	$+$ 540	e^2f^2	$+$ 5184
$befi$	$-$ 972		

± 41650

No. 48.

a^2j^2	...	bf^2h	$+$ 70
$abij$	...	bfg^2	$-$ 45
aci^2	$+$ 2	c^2fj	$+$ 27
$achj$	$-$ 2	c^2gi	$-$ 52
$adgj$	$+$ 7	c^2h^2	$+$ 25
$adhi$	$-$ 7	$cdej$	$-$ 45
$aefj$	$-$ 5	$cdfi$	$+$ 23
$aegi$	$-$ 22	$cdgh$	$+$ 22
aeh^2	$+$ 27	ce^3i	$+$ 70
af^2i	$+$ 25	$cefh$	$-$ 127
$afgh$	$-$ 45	ceg^2	$+$ 32
ag^3	$+$ 20	cf^2g	$+$ 25
b^2hj	$+$ 2	d^3j	$+$ 20
b^2i^2	$-$ 2	d^2ei	$-$ 45
$bcgj$	$-$ 7	d^2fh	$+$ 32
$bchi$	$+$ 7	d^2g^2	$+$ 47
$bdfj$	$-$ 22	$defg$	$+$ 85
$bdgi$	$+$ 74	de^2h	$+$ 25
bdh^2	$-$ 52	df^3	$-$ 50
$begh$	$+$ 23	e^3g	$-$ 50
be^2j	$+$ 25	e^2f^2	$+$ 30
$befi$	$-$ 73		

± 698

Nos 49, [49 A] and 50 relate to the dodecadic. No. 49 is the dodecadic itself; [No. 49 A, inserted in this place, but originally printed in the Fifth Memoir on Quantics, is the dodecadic quadricovariant], No. 50 is the cubinvariant. [The numerical coefficients in this last table as originally printed in the Third Memoir were altogether erroneous, and the table as here printed is in fact the table No. 50 *bis*, of the Fifth Memoir on Quantics.]

No. 49.

$a+1$	$b+12$	$c+66$	$d+220$	$e+495$	$f+792$	$g+924$	$h+792$	$i+495$	$j+220$	$k+66$	$l+12$	$m+1$	$\between x, y)$

No. 49 A.

(

$ag + 1$	$ah + 6$	$ai + 15$	$aj + 20$	$ak + 15$	$al + 6$	$am + 1$
$bf - 6$	$bg - 30$	$bh - 54$	$bi - 30$	$bj + 30$	$bk + 54$	$bl + 30$
$ce + 15$	$cf + 54$	$cg + 24$	$ch - 150$	$ci - 270$	$cj - 150$	$ck + 24$
$d^2 - 10$	$de - 30$	$df + 150$	$dg + 430$	$dh + 270$	$di - 270$	$dj - 430$
		$e^2 - 135$	$ef - 270$	$eg + 495$	$eh + 1080$	$ei + 495$
				$f^2 - 540$	$fg - 720$	$fh + 720$
						$g^2 - 840$
± 16	± 60	± 189	± 450	± 810	± 1140	± 1270

$bm + 6$	$cm + 15$	$dm + 20$	$em + 15$	$fm + 6$	$gm + 1$
$cl + 54$	$dl + 30$	$el - 30$	$fl - 54$	$gl - 30$	$hl - 6$
$dk - 150$	$ek - 270$	$fk - 150$	$gk + 24$	$hk + 54$	$ik + 15$
$ej - 270$	$fj + 270$	$gj + 430$	$hj + 150$	$ij - 30$	$j^2 - 10$
$fi + 1080$	$gi + 495$	$hi - 270$	$i^2 - 135$		
$gh - 720$	$h^2 - 540$				
± 1140	± 810	± 450	± 189	± 60	± 16

$)(x, y)^{12}$.

No. 50.

$agm + 1$	$cfl - 54$	$dhi + 270$
$ahl - 6$	$cgk + 24$	$e^2k - 135$
$aik + 15$	$chj + 150$	$efj + 270$
$aj^2 - 10$	$ci^2 - 135$	$egi + 495$
$bfm - 6$	$d^2m - 10$	$eh^2 - 540$
$bgl + 30$	$del + 30$	$f^2i - 540$
$bhk - 54$	$dfk + 150$	$fgh + 720$
$bij + 30$	$dgj - 430$	$g^3 - 280$
$cem + 15$		
		± 2200

Resuming now the general subject,—

54. The simplest covariant of a system of quantics of the form

$$(* \between x, y, \ldots)^m$$

(where the number of quantics is equal to the number of the facients of each quantic) is the functional determinant or *Jacobian*, viz. the determinant formed with the differential coefficients or derived functions of the quantics with respect to the several facients.

55. In the particular case in which the quantics are the differential coefficients or derived functions of a single quantic, we have a corresponding covariant of the single quantic, which covariant is termed the *Hessian*; in other words, the Hessian is the determinant formed with the second differential coefficients or derived functions of the quantic with respect to the several facients.

56. The expression, an *adjoint linear form*, is used to denote a linear function $\xi x + \eta y + \ldots$, or in the notation of quantics $(\xi, \eta, \ldots \between x, y, \ldots)$, having the same facients as

the quantic or quantics to which it belongs, and with indeterminate coefficients $(\xi, \eta, \ldots)$. The invariants of a quantic or quantics, and of an adjoint linear form, may be considered as quantics having $(\xi, \eta, \ldots)$ for facients, and of which the coefficients are of course functions of the coefficients of the given quantic or quantics. An invariant of the class in question is termed a contravariant of the quantic or quantics. The idea of a contravariant is due to Mr Sylvester.

In the theory of binary quantics, it is hardly necessary to consider the contravariants; for any contravariant is at once turned into an invariant by writing $(y, -x)$ for (ξ, η).

57. If we imagine, as before, a system of quantics of the form

$$(*\between x, y, \ldots)^m,$$

where the number of quantics is equal to the number of the facients in each quantic, the function of the coefficients, which, equated to zero, expresses the result of the elimination of the facients from the equations obtained by putting each of the quantics equal to zero, is said to be the *Resultant* of the system of quantics. The resultant is an invariant of the system of quantics.

And in the particular case in which the quantics are the differential coefficients, or derived functions of a single quantic with respect to the several facients, the resultant in question is termed the *Discriminant* of the single quantic; the discriminant is of course an invariant of the single quantic.

58. Imagine two quantics, and form the equations which express that the differential coefficients, or derived functions of the one quantic with respect to the several facients, are proportional to those of the other quantic. Join to these the equations obtained by equating each of the quantics to zero; we have a system of equations, one of which is contained in the others, and from which therefore the facients may be eliminated. The function which, equated to zero, expresses the result of the elimination is an invariant which (from its geometrical signification) might be termed the *Tactinvariant* of the two quantics, but I do not at present propose to consider this invariant except in the particular case where the system consists of a given quantic and of an adjoint linear form. In this case the tactinvariant is a contravariant of the given quantic, viz. the contravariant termed the *Reciprocant.*

59. Consider now a quantic

$$(*\between x, y, \ldots)^m,$$

and let the facients $x, y, \ldots$ be replaced by $\lambda x + \mu X$, $\lambda y + \mu Y, \ldots$ the resulting function may, it is clear, be considered as a quantic with the facients (λ, μ) and of the form

$$\left\{\begin{array}{l}(*\between x, y, \ldots)^m \\ (*\between x, y, \ldots)^{m-1}(X, Y, \ldots) \\ \vdots \\ (*\between X, Y, \ldots)^m\end{array}\right\}\between\lambda, \mu)^m.$$

The coefficients of this quantic are termed *Emanants*, viz., excluding the first coefficient, which is the quantic itself (but which might be termed the 0-th emanant), the other coefficients are the first, second, and last or ultimate emanants. The ultimate emanant is, it is clear, nothing else than the quantic itself, with $(X, Y, \ldots)$ instead of $(x, y, \ldots)$ for facients: the penultimate emanant is, in like manner, obtained from the first emanant by interchanging $(x, y, \ldots)$ with $(X, Y, \ldots)$, and similarly for the other emanants. The facients $(X, Y, \ldots)$ may be termed the *facients of emanation*, or simply the *new facients*. The theory of emanation might be presented in a more general form by employing two or more sets of emanating facients; we might, for example, write $\lambda x + \mu X + \nu X'$, $\lambda y + \mu Y + \nu Y', \ldots$ for $x, y, \ldots$, but it is not necessary to dwell upon this at present.

The invariants, in respect to the new facients, of any emanant or emanants of a quantic (i.e. the invariants of the emanant or emanants, considered as a function or functions of the new facients), are, it is easy to see, covariants of the original quantic, and it is in many cases convenient to define a covariant in this manner; thus the Hessian is the discriminant of the second or quadric emanant of the quantic.

60. If we consider a quantic

$$(a, b, \ldots \between x, y, \ldots)^m,$$

and an adjoint linear form, the operative quantic

$$(\partial_a, \partial_b, \ldots \between \xi, \eta, \ldots)^m$$

(which is, so to speak, a contravariant operator) is termed the *Evector*. The properties of the evector have been considered in the introductory memoir, and it has been in effect shown that the evector operating upon an invariant, or more generally upon a contravariant, gives rise to a contravariant. Any such contravariant, or rather such contravariant considered as so generated, is termed an *Evectant*. In the case of a binary quantic,

$$(a, b, \ldots \between x, y)^m,$$

the covariant operator

$$(\partial_a, \partial_b, \ldots \between y, -x)^m$$

may, if not with perfect accuracy, yet without risk of ambiguity, be termed the *Evector*, and a covariant obtained by operating with it upon an invariant or covariant, or rather such covariant considered as so generated, may in like manner be termed an *Evectant*.

61. Imagine two or more quantics of the same order,

$$(a, b, \ldots \between x, y)^m,$$
$$(\alpha, \beta, \ldots \between x, y)^m,$$
$$\vdots$$

we may have covariants such that for the coefficients of each pair of quantics the covariant is reduced to zero by the operators

$$a\partial_\alpha + b\partial_\beta + \ldots,$$
$$\alpha\partial_a + \beta\partial_b + \ldots.$$

Such covariants are called *Combinants,* and they possess the property of being invariantive, quoad the system, i.e. the covariant remains unaltered to a factor *près,* when each quantic is replaced by a linear function of all the quantics. This extremely important theory is due to Mr Sylvester.

Proceeding now to the theory of ternary quadrics and cubics,—

First for a ternary quadric, we have the following tables:—

Covariant and other Tables, Nos. 51 to 56 (a ternary quadric).

No. 51.

The quadric is represented by

$$(a,\ b,\ c,\ f,\ g,\ h \between x,\ y,\ z)^2,$$

which means

$$ax^2+by^2+cz^2+2fyz+2gzx+2hxy.$$

No. 52.

The first derived functions (omitting the factor 2) are—

$$\begin{aligned}&(a,\ h,\ g \between x,\ y,\ z),\\ &(h,\ b,\ f \between x,\ y,\ z),\\ &(g,\ f,\ c \between x,\ y,\ z).\end{aligned}$$

No. 53.

The operators which reduce a covariant to zero are

$$\begin{aligned}&(\ h,\ \ b,\ \ 2f \between \partial_g,\ \partial_f,\ \partial_c) - z\partial_y,\\ &(2g,\ \ f,\ \ c \between \partial_a,\ \partial_h,\ \partial_g) - x\partial_z,\\ &(\ a,\ \ 2h,\ \ g \between \partial_h,\ \partial_b,\ \partial_f) - y\partial_x,\\ &(\ g,\ \ 2f,\ \ c \between \partial_h,\ \partial_b,\ \partial_f) - y\partial_z,\\ &(\ a,\ \ h,\ \ 2g \between \partial_g,\ \partial_f,\ \partial_c) - z\partial_x,\\ &(2h,\ \ b,\ \ f \between \partial_a,\ \partial_h,\ \partial_g) - x\partial_y.\end{aligned}$$

No. 54.

The evector is

$$(\partial_a,\ \partial_b,\ \partial_c,\ \partial_f,\ \partial_g,\ \partial_h \between \xi,\ \eta,\ \zeta)^2.$$

No. 55.

The discriminant is

$$\begin{vmatrix} a, & h, & g \\ h, & b, & f \\ g, & f, & c \end{vmatrix}$$

which is equal to

$$abc - af^2 - bg^2 - ch^2 + 2fgh.$$

No. 56.

The reciprocant is

$$-\begin{vmatrix} & \xi, & \eta, & \zeta \\ \xi, & a, & h, & g \\ \eta, & h, & b, & f \\ \zeta, & g, & f, & c \end{vmatrix}$$

which is equal to

$$(bc - f^2,\ \ ca - g^2,\ \ ab - h^2,\ \ gh - af,\ \ hf - bg,\ \ fg - ch \between \xi,\ \eta,\ \zeta)^2.$$

The discriminant is, it will be noticed, the same function as the Hessian. The reciprocant is the evectant of the discriminant. The covariants are the quadric itself and the discriminant; the reciprocant is the only contravariant.

Next, for a ternary cubic, we have the following Tables:

Covariant and other Tables, Nos. 57 to 70 (a ternary cubic).

No. 57.

The cubic is $U =$

$$(a,\ b,\ c,\ f,\ g,\ h,\ i,\ j,\ k,\ l \between x,\ y,\ z)^3,$$

which means—

$$ax^3 + by^3 + cz^3 + 3fy^2z + 3gz^2x + 3hx^2y + 3iyz^2 + 3jzx^2 + 3kxy^2 + 6lxyz.$$

No. 58.

The first derived functions (omitting the factor 3) are

$$\begin{array}{l} (a,\ k,\ g,\ l,\ j,\ h \between x,\ y,\ z)^2, \\ (h,\ b,\ i,\ f,\ l,\ k \between x,\ y,\ z)^2, \\ (j,\ f,\ c,\ i,\ g,\ l \between x,\ y,\ z)^2. \end{array}$$

The second derived functions (omitting the factor 6) are

$$
\begin{array}{l}
(a,\ h,\ j \between x,\ y,\ z),\\
(k,\ b,\ f \between x,\ y,\ z),\\
(g,\ i,\ c \between x,\ y,\ z),\\
(l,\ f,\ i \between x,\ y,\ z),\\
(j,\ l,\ g \between x,\ y,\ z),\\
(h,\ k,\ l \between x,\ y,\ z).
\end{array}
$$

No. 59.

The operators which reduce a covariant to zero are

$$
\begin{array}{l}
(j,\ 3f,\ c,\ 2i,\ g,\ 2l \between \partial_h,\ \partial_b,\ \partial_i,\ \partial_f,\ \partial_l,\ \partial_k) - y\partial_z,\\
(a,\ k,\ 3g,\ 2l,\ 2j,\ h \between \partial_j,\ \partial_f,\ \partial_c,\ \partial_i,\ \partial_g,\ \partial_l) - z\partial_z,\\
(3h,\ b,\ i,\ f,\ 2l,\ 2k \between \partial_a,\ \partial_k,\ \partial_g,\ \partial_l,\ \partial_j,\ \partial_h) - x\partial_y,\\
(h,\ b,\ 3i,\ 2f,\ 2l,\ k \between \partial_j,\ \partial_f,\ \partial_c,\ \partial_i,\ \partial_g,\ \partial_l) - z\partial_y,\\
(3j,\ f,\ c,\ i,\ 2g,\ 2l \between \partial_a,\ \partial_k,\ \partial_g,\ \partial_l,\ \partial_j,\ \partial_h) - x\partial_z,\\
(a,\ 3k,\ g,\ 2l,\ j,\ 2h \between \partial_h,\ \partial_b,\ \partial_i,\ \partial_f,\ \partial_l,\ \partial_k) - y\partial_x.
\end{array}
$$

No. 60.

The evector is

$$(\partial_a,\ \partial_b,\ \partial_c,\ \partial_f,\ \partial_g,\ \partial_h,\ \partial_i,\ \partial_j,\ \partial_k,\ \partial_l \between \xi,\ \eta,\ \zeta)^3.$$

No. 61.

The Hessian is $HU =$

$$
- \left| \begin{array}{lll}
(a,\ h,\ j \between x,\ y,\ z), & (h,\ k,\ l \between x,\ y,\ z), & (j,\ l,\ g \between x,\ y,\ z)\\
(h,\ k,\ l \between x,\ y,\ z), & (k,\ b,\ f \between x,\ y,\ z), & (l,\ f,\ i \between x,\ y,\ z)\\
(j,\ l,\ g \between x,\ y,\ z), & (l,\ f,\ i \between x,\ y,\ z), & (g,\ i,\ c \between x,\ y,\ z)
\end{array} \right|
$$

which is equal to

$agk\ -1$	$bhi\ -1$	$cij\ -1$	$bch\ -1$	$acf\ -1$	$abg\ -1$	$bcj\ -1$	$ach\ -1$	$abi\ -1$	$abc\ -1$
$al^2\ +1$	$bl^2\ +1$	$cl^2\ +1$	$bgl\ +2$	$ai^2\ +1$	$afl\ +2$	$bg^2\ +1$	$afg\ -1$	$af^2\ +1$	$afi\ +1$
$gh^2\ +1$	$f^2h\ +1$	$fg^2\ +1$	$bij\ -1$	$chl\ +2$	$aik\ -1$	$cfh\ -1$	$ail\ +2$	$bgh\ -1$	$bgj\ +1$
$hjl\ -2$	$fkl\ -2$	$fkl\ -2$	$ck^2\ +1$	$cjk\ -1$	$bj^2\ +1$	$ckl\ +2$	$ch^2\ +1$	$bjl\ +2$	$chk\ +1$
$j^2k\ +1$	$ik^2\ +1$	$i^2j\ +1$	$f^2j\ +1$	$fgj\ +1$	$fhj\ -2$	$fij\ +1$	$fj^2\ +1$	$fjk\ -2$	$fgh\ -3$
			$fgk\ -2$	$g^2k\ +1$	$ghk\ +1$	$gik\ -2$	$gjk\ +1$	$gk^2\ +1$	$fjl\ +2$
			$fhi\ +1$	$ghi\ -2$	$h^2i\ +1$	$hi^2\ +1$	$hij\ -2$	$hki\ +1$	$gkl\ +2$
			$fl^2\ -1$	$gl^2\ -1$	$hl^2\ -1$	$il^2\ -1$	$jl^2\ -1$	$kl^2\ -1$	$hil\ +2$
									$ijk\ -3$
									$l^3\ -2$
± 3	± 3	± 3	± 5	± 5	± 5	± 5	± 5	± 5	± 9

$\between (x,\ y,\ z)^2.$

No. 62.

The quartinvariant is $S=$

$abcl$	-1	f^2j^2	$+1$
$abgi$	$+1$	$fghl$	$+3$
$acfk$	$+1$	$fgjk$	-1
af^2g	-1	$fhij$	-1
$afil$	$+1$	fjl^2	-2
ai^2k	-1	g^2k^2	$+1$
$bchj$	$+1$	$ghki$	-1
bg^2h	-1	gkl^2	-2
$bgjl$	$+1$	h^2i^2	$+1$
bij^2	-1	hil^2	-2
cf^2h	-1	$ijkl$	$+3$
$chkl$	$+1$	l^4	$+1$
cjk^2	-1		

± 16

No. 63.

The sextinvariant is $T=$

$a^2b^2c^2$	$+1$	acf^2hl	-24	$bcfhj^2$	-12	cfh^3i	-12	$fgjkl^2$	-12
a^2bcfi	-6	acf^2jk	-12	$bcgh^2l$	-24	cfh^2l^2	$+12$	fh^2i^2j	-12
a^2bi^3	$+4$	$acfgk^2$	-12	$bcghjk$	$+18$	$cfhjkl$	-60	$fhijl^2$	-12
a^2cf^3	$+4$	$acfhki$	$+18$	bch^2ij	-12	cfj^2k^2	$+24$	fij^2kl	$+36$
$a^2f^2i^2$	-3	$acfkl^2$	$+36$	$bchjl^2$	$+36$	$cghk^2l$	$+12$	fil^4	$+24$
ab^2cgj	-6	$acik^2l$	-24	bcj^2kl	-24	$cgjk^3$	-12	g^3k^3	$+8$
ab^2g^3	$+4$	af^3gj	-12	bfg^2hj	$+6$	ch^2ikl	$+12$	g^2hk^2i	-12
abc^2hk	-6	af^2g^2k	$+24$	$bfgj^2l$	$+12$	$chijh^2$	$+6$	$g^2k^2l^2$	-24
$abcfgh$	$+6$	af^2ghi	$+6$	$bfij^3$	-12	$chkl^3$	-12	gh^2ki^2	-12
$abcfjl$	$+12$	af^2gl^2	$+12$	bg^3hk	-12	cjk^2l^2	$+12$	$ghkil^2$	-12
$abcgkl$	$+12$	af^2ijl	$+12$	bg^2h^2i	$+24$	f^3j^3	$+8$	$gijk^2l$	$+36$
$abchil$	$+12$	$afgkil$	-60	bg^2hl^2	$+12$	$f^2g^2h^2$	-27	gkl^4	$+24$
$abcijk$	$+6$	$afhi^2l$	$+12$	bg^2jkl	$+12$	f^2gj^2k	-12	h^3i^3	$+8$
$abcl^3$	-20	afi^2jk	$+6$	$bghijl$	-60	f^2ghjl	$+36$	$h^2i^2l^2$	-24
$abfgij$	$+18$	$afil^3$	-12	$bgij^2k$	$+6$	f^2hij^2	-12	hil^4	$+24$
$abfg^2l$	-24	agk^2i^2	$+24$	$bgjl^3$	-12	$f^2j^2l^2$	-24	hi^2jkl	$+36$
abg^2ki	-12	ahi^3k	-12	bhi^2j^2	$+24$	fg^2hkl	$+36$	$i^2j^2k^2$	-27
$abghi^2$	-12	ai^2kl^2	$+12$	bij^2l^2	$+12$	fg^2jk^2	-12	$ijkl^3$	-36
$abgil^2$	$+36$	b^2cj^3	$+4$	$c^2h^2k^2$	-3	fgh^2il	$+36$	l^6	-8
abi^2jl	-24	$b^2g^2j^2$	-3	cf^2h^2j	$+24$	$fghijk$	-6		
ac^2k^3	$+4$	bc^2h^3	$+4$	$cfgh^2k$	$+6$	$fghl^3$	-36		

± 871

The discovery of the invariants S and T is due to Aronhold, the developed expressions were first obtained by Mr Salmon.

No. 64.

There is an octicovariant for which we may take

$$\Theta U = \begin{vmatrix} & \partial_x HU, & \partial_y HU, & \partial_z HU \\ \partial_x HU, & \tfrac{1}{6}\partial_x{}^2\ U, & \tfrac{1}{6}\partial_x\partial_y U, & \tfrac{1}{6}\partial_x\partial_z U \\ \partial_y HU, & \tfrac{1}{6}\partial_y\partial_x U, & \tfrac{1}{6}\partial_y{}^2\ U, & \tfrac{1}{6}\partial_y\partial_z U \\ \partial\ HU, & \tfrac{1}{6}\partial_z\partial_x U, & \tfrac{1}{6}\partial_y\partial_z U, & \tfrac{1}{6}\partial_z{}^2\ U \end{vmatrix} -,$$

or else

$$\Theta_{,} U = \begin{vmatrix} & \tfrac{1}{3}\partial_x U, & \tfrac{1}{3}\partial_y U, & \tfrac{1}{3}\partial_z U \\ \tfrac{1}{3}\partial_x U, & \partial_x^2 HU, & \partial_x \partial_y HU, & \partial_x\partial_z HU \\ \tfrac{1}{3}\partial_y U, & \partial_y \partial_x HU, & \partial_y^2 HU, & \partial_y\partial_z HU \\ \tfrac{1}{3}\partial_z U, & \partial_z \partial_x HU, & \partial_z \partial_y HU, & \partial_z^2 HU \end{vmatrix}$$

or else, what I believe is more simple, a function $\Theta_{,,}U$, which is a linear function of the last-mentioned two functions.

The relations between ΘU, $\Theta_{,}U$, $\Theta_{,,}U$ are

$$-\Theta_{,} U + 4\Theta U = T \,.\, U^2 - 24S \,.\, U \,.\, HU,$$

$$\Theta_{,,}U + 2\Theta U = T \,.\, U^2 - 10S \,.\, U \,.\, HU.$$

I have not worked out the developed expressions.

No. 65.

The cubicontravariant is $PU=$

bcl -1	acl -1	abl -1	ack $+1$	abi $+1$	bcj $+1$	abg $+1$	bch $+1$	acf $+1$	abc -1
bgi $+1$	agi $+1$	afk $+1$	afg -2	af^2 -1	bg^2 -1	afl $+1$	bgl $+1$	ai^2 -1	afi $+1$
cfk $+1$	chj $+1$	bhj $+1$	ail $+1$	bgh -2	cfh -2	aik -2	bij -2	chl $+1$	bgj $+1$
f^2g -1	g^2h -1	fh^2 -1	ch^2 -1	bjl $+1$	ckl $+1$	bj^2 -1	ck^2 -1	cjk -2	chk $+1$
fil $+1$	gjl $+1$	ckl $+1$	fj^2 $+2$	fhl $+3$	fgl $+3$	fhj -1	f^2j $+2$	fgj -1	fgh $+3$
i^2k -1	ij^2 -1	jk^2 -1	fhl $+3$	fjk -1	fij -1	ghk -1	fgk -1	g^2k $+2$	fjl -4
			gjk -1	gk^2 $+2$	gki -1	h^2i $+2$	fhi -1	ghi -1	gkl -4
			hij -1	hki -1	hi^2 $+2$	hl^2 -2	fl^2 -2	gl^2 -2	hil -4
			jl^2 -2	kl^2 -2	il^2 -2	jkl $+3$	ikl $+3$	ijl $+3$	ijk $+3$
									l^3 $+4$
± 3	± 3	± 3	± 7	± 7	± 7	± 7	± 7	± 7	± 13

$(\xi, \eta, \zeta)^3$.

No. 66.

The quintic contravariant is $QU=$

b^2c^2+1	a^2bc^2+1	a^2b^2c+1	a^2bci-3	ab^2cj-3	abc^2k-3	a^2bcf-3	ab^2cg-3	abc^2h-3	$abcfj+6$
$bcfi-6$	a^2cfi-3	a^2bfi-3	a^2cf^2+6	ab^2g^2+6	$abcfg+3$	a^2bi^2+6	$abcfl+6$	$abcgl+6$	$abcgk+6$
bi^3+4	a^2i^3+2	a^2f^3+2	a^2fi^2-3	$abcfh+3$	$abcil+6$	a^2f^2i-3	$abcik+3$	$abcij+3$	$abchi+6$
cf^3+4	$abcgj-6$	ab^2gj-3	$abcgh+3$	$abckl+6$	$abgi^2-6$	$abchl+6$	$abfgi+9$	abg^2i-6	$abcl^2-30$
f^2i^2-3	abg^3+4	$abchk-6$	$abcjl+6$	$abfij+9$	acf^2l-12	$abcjk+3$	abi^2l-12	ac^2k^2+6	$abfg^2-12$
$cgj-3$	ac^2hk-3	$abfgh+3$	$abgij+9$	$abfgl-24$	$acfki+9$	$abfgj+9$	acf^2k-6	acf^2j-6	$abgil+36$
g^3+2	$acfgh+3$	$abfjl+6$	abg^2l-12	$abgki-12$	af^2gi+3	abg^2k-6	af^3g-6	$acfgk-12$	abi^2j-12
$hk-3$	$acfjl+6$	$abgkl+6$	$acfhl-24$	$abhi^2-6$	afi^2l+6	$abghi-12$	af^2il+6	$acfhi+9$	acf^2h-12
$fgh+3$	$acgkl+6$	$abhil+6$	$acfjk-12$	$abil^2+18$	ai^3k-6	$abgl^2+18$	afi^2k+3	$acfl^2+18$	$acfkl+36$
$fjl+6$	$achil+6$	$abijk+3$	$acgk^2-6$	$acfk^2-6$	bc^2h^2+6	$abijl-24$	b^2cj^2+6	$acikl-24$	$acik^2-12$
$gkl+6$	$acijk+3$	abl^3-10	$achki+9$	af^3j-6	$bcfj^2-6$	$acfhk+9$	b^2g^2j-3	af^2g^2+12	af^2gl+12
$hil+6$	acl^3-10	ack^3+4	$ackl^2+18$	af^2gk+24	$bcghl-24$	ack^2l-12	$bcfhj-12$	$afgil-30$	af^2ij+6
$ijk+3$	$afgij+9$	af^2hl-12	af^2gj-18	af^2hi+3	$bcgjk+9$	af^2gh+3	$bcghk+9$	afi^2j+3	$afgki-30$
l^3-10	afg^2l-12	af^2jk-6	afg^2k+24	af^2l^2+6	$bchij-12$	af^2jl+6	bch^2i-6	$agki^2+24$	$afhi^2+6$
$gij+9$	ag^2ki-6	$afgk^2-6$	$afghi+6$	$afkil-30$	$bcjl^2+18$	$afgkl-30$	$bchl^2+18$	ahi^3-6	$afil^2-18$
g^2l-12	$aghi^2-6$	$afhki+9$	$afgl^2+12$	ak^2i^2+12	bfg^2j+3	$afhil+12$	$bcjkl-24$	ai^2l^2+6	ai^2kl+12
$ki-6$	$agil^2+18$	$ajkl^2+18$	$afijl+12$	b^2gj^2-3	bg^3k-6	$afijk+6$	bfg^2h+3	$bcghj+9$	$bcgh^2-12$
hi^2-6	ai^2jl-12	aik^2l-12	$agkil-30$	bch^2l-12	bg^2hi+24	afl^3-6	$bfgjl+12$	bcj^2l-12	$bchjl+36$
il^2+18	bcj^3+4	b^2j^3+2	ahi^2l+6	$bchjk+9$	bg^2l^2+6	agk^2i+24	$bfij^2-18$	bg^3h-6	bcj^2k-12
$jl-12$	bg^2j^2-3	bch^3+4	ai^2jk+3	$bfghj+6$	$bgijl-30$	ahi^2k-18	bg^2kl+6	bg^2jl+6	$bfgj^2+6$
k^3+2	c^2h^3+2	$bfhj^2-6$	ail^3-6	bfj^2l+6	bi^2j^2+12	$aikl^2+12$	$bghil-30$	$bgij^2+3$	bg^2hl+12
$hl-12$	$cfhj^2-6$	bgh^2l-12	$bchj^2-6$	bg^2hk-18	c^2hk^2-3	bch^2j-6	$bgijk+6$	c^2h^2k-3	bg^2jk+6
$jk-6$	cgh^2l-12	$bghjk+9$	bg^2hj+3	bgh^2i+24	cf^2hj+24	bfj^3-6	bgl^3-6	$cfgh^2+3$	$bghij-30$
gk^2-6	$cghjk+9$	bh^2ij-6	bgj^2l+6	$bghl^2+12$	$cfghk+6$	bg^2h^2+12	bhi^2j+24	$cfhjl-30$	$bgjl^2-18$
$hki+9$	ch^2ij-6	$bhjl^2+18$	bij^3-6	$bgjkl+12$	cfh^2i-18	$bghjl-30$	$bijl^2+12$	cfj^2k+24	bij^2l+12
kl^2+18	$chjl^2+18$	bj^2kl-12	cfh^2j+24	$bhijl-30$	$cfhl^2+12$	bgj^2k+3	cf^2h^2+12	$cghkl+12$	cfh^2l+12
k^2l-12	cj^2kl-12	ch^2k^2-3	cgh^2k+3	bij^2k+3	$cfjkl-30$	$bhij^2+24$	$cfhkl-30$	$cgjk^2-18$	$cfhjk-30$
$gj-6$	fg^2hj+3	f^2h^2j+12	ch^3i-6	bjl^3-6	cgk^2l+6	bj^2l^2+6	$cfjk^2+24$	ch^2il+6	$cghk^2+6$
g^2k+12	fgj^2l+6	fgh^2k+3	ch^2l^2+6	cfh^2k+3	$chikl+12$	cfh^3-6	cgk^3-6	$chijk+6$	ch^2ik+6
$ghi+3$	fij^3-6	fh^3i-6	$chjkl-30$	chk^2l+6	$cijk^2+3$	ch^2kl+6	$chik^2+3$	chl^3-6	$chkl^2-18$
gl^2+6	g^3hk-6	fh^2l^2+6	cj^2k^2+12	cjk^3-6	ckl^3-6	$chjk^2+3$	ck^2l^2+6	$cjkl^2+12$	cjk^2l+12
$ijl+6$	g^2h^2i+12	$fhjkl-30$	f^2j^3+12	f^2gh^2-27	f^2g^2h-27	f^2hj^2-6	f^3j^2+12	f^2gj^2-6	$f^2ghj+18$
$kil-30$	g^2hl^2+6	fj^2k^2+12	fg^2h^2-27	f^2j^2k-6	$f^2gjl+18$	fgh^2l+18	$f^2gjk-12$	fg^2hl+18	f^2j^2l-24
i^2l+6	g^2jkl+6	ghk^2l+6	fgj^2k-12	$f^2hjl+18$	f^2ij^2-6	$fghjk-3$	$f^2ghl+18$	fg^2jk-12	fg^2hk+18
$jk+3$	$ghijl-30$	gjk^3-6	$fghjl+36$	$fghkl+36$	fg^2kl+18	fh^2ij-12	$f^2hij-12$	$fghij-3$	fgh^2i+18
l^3-6	gij^2k+3	h^2ikl+6	$fhij^2-12$	$fgjk^2-12$	$fghil+36$	$fhjl^2-6$	f^2jl^2-24	$fgjl^2-6$	$fghl^2-54$
i^2+12	gjl^3-6	$hijk^2+3$	fj^2l^2-24	fh^2il+18	$fgijk-3$	fj^2kl+18	fg^2k^2-6	fij^2l+18	$fgjkl-12$
$k-6$	hi^2j^2+12	hkl^3-6	$g^2hkl+18$	$fhijk-3$	fgl^3-18	g^2hk^2-6	$fghik-3$	g^3k^2+12	$fhijl-12$
l^2+6	ij^2l^2+6	jk^2l^2+6	g^2jk^2-6	fhl^3-18	fhi^2j-12	gh^2ki-12	$fgkl^2-6$	$g^2hki-12$	fij^2k+18
			gh^2il+18	$fjkl^2-6$	$fijl^2-6$	$ghkl^2-6$	fh^2i^2-6	g^2kl^2-24	fjl^3+48
			$ghijk-3$	g^2k^3+12	g^2k^2i-6	gjk^2l+18	$fhil^2-6$	gh^2i^2-6	g^2k^2l-24
			ghl^3-18	ghk^2i-12	$ghki^2-12$	h^3i^2+12	$fijkl+36$	$ghil^2-6$	$ghkil-12$
			$gjki^2-6$	gk^2l^2-24	$gkil^2-6$	h^2il^2-24	fl^4+12	$gijkl+36$	$gijk^2+18$
			h^2i^2j-6	h^2ki^2-6	h^2i^3+12	hl^4+12	gik^2l+18	gl^4+12	gkl^3+48
			$hijl^2-6$	$hkil^2-6$	hi^2l^2-24	$hijkl+36$	hi^2kl+18	hi^2jl+18	h^2i^2l-24
			ij^2kl+18	ijk^2l+18	$i^2jkl+18$	ij^2k^2-27	i^2jk^2-27	i^2j^2k-27	hil^3+48
			jl^4+12	kl^4+12	il^4+12	jkl^3-18	ikl^3-18	jkl^3-18	hi^2jk+18
									$ijkl^2-54$
									l^5-24
± 145	± 145	± 145	± 282	± 282	± 282	± 282	± 282	± 282	± 486

$(\xi, \eta, \zeta)^3$

No. 67.

The reciprocant is $FU = (*\between \xi, \eta, \zeta)^6 =$

ξ^6.	η^6.	ζ^6.	$\eta^5\zeta$.	$\zeta^5\xi$.	$\xi^5\eta$.	$\eta\zeta^5$.	$\zeta\xi^5$.	$\xi\eta^5$.
b^2c^2 + 1	a^2c^2 + 1	a^2b^2 + 1	a^2ci − 6	ab^2j − 6	bc^2k − 6	a^2bf − 6	b^2cg − 6	ac^2h − 6
$bcfi$ − 6	$acgj$ − 6	$abhk$ − 6	$acgh$ + 6	$abfh$ + 6	$bcfg$ + 6	$abhl$ + 12	$bcfl$ + 12	$acgl$ + 12
bi^3 + 4	ag^3 + 4	ak^3 + 4	$acjl$ + 12	$abkl$ + 12	$bcil$ + 12	$abjk$ + 6	$bcik$ + 6	$acij$ + 6
cf^3 + 4	cj^3 + 4	bh^3 + 4	$agij$ + 18	afk^2 − 12	bgi^2 − 12	$afhk$ + 18	$bfgi$ + 18	ag^2i − 12
f^2i^2 − 3	g^2j^2 − 3	h^2k^2 − 3	ag^2l − 24	bh^2l − 24	cf^2l − 24	ak^2l − 24	bi^2l − 24	$cghj$ + 18
			chj^2 − 12	$bhjk$ + 18	$cfki$ + 18	bh^2j − 12	cf^2k − 12	cj^2l − 24
			g^2hj + 6	fh^2k + 6	f^2gi + 6	fh^3 − 12	f^3g − 12	g^3h − 12
			gj^2l + 12	hk^2l + 12	fi^2l + 12	h^2kl + 12	f^2il + 12	g^2jl + 12
			ij^3 − 12	jk^3 − 12	i^3k − 12	hjk^2 + 6	fi^2k + 6	gij^2 + 6
±9	±9	±9	±54	±54	±54	±54	±54	±54

$\eta^4\zeta^2$.	$\zeta^4\xi^2$.	$\xi^4\eta^2$.	$\eta^2\zeta^4$.	$\zeta^2\xi^4$.	$\xi^2\eta^4$.	$\eta^3\zeta^3$.	$\zeta^3\xi^3$.	$\xi^3\eta^3$.
a^2cf + 6	ab^2g + 6	bc^2h + 6	a^2bi + 6	b^2cj + 6	ac^2k + 6	a^2bc − 2	ab^2c − 2	abc^2 − 2
a^2i^2 + 9	$abfl$ − 12	$bcgl$ − 12	a^2f^2 + 9	b^2g^2 + 9	$acfg$ − 6	a^2fi − 18	$abfi$ + 6	$acfi$ + 6
$achl$ − 12	$abik$ − 6	$bcij$ − 6	$abgh$ − 6	$bcfh$ − 6	$acil$ − 12	$abgj$ + 6	af^3 − 4	ai^3 − 4
$acjk$ − 6	af^2k + 12	bg^2i + 12	$abjl$ − 12	$bckl$ − 12	agi^2 + 12	$achk$ + 6	b^2gj − 18	$bcgj$ + 6
$afgj$ − 18	b^2j^2 + 9	c^2k^2 + 9	$afhl$ − 36	$bfgl$ − 36	c^2h^2 + 9	$afgh$ + 18	$bchk$ + 6	bg^3 − 4
ag^2k + 12	$bfhj$ − 18	cf^2j + 12	$afjk$ − 18	$bfij$ − 18	cfj^2 + 12	$afjl$ + 36	$bfgh$ + 18	c^2hk − 18
$aghi$ − 18	$bghk$ − 18	$cfgk$ − 18	agk^2 + 12	$bgik$ − 18	$cghl$ − 36	$agkl$ − 48	$bfjl$ + 36	$cfgh$ + 18
agl^2 + 48	bh^2i + 12	$cfhi$ − 18	$aihk$ − 18	bhi^2 + 12	$cgjk$ − 18	$ahil$ + 36	$bgkl$ + 36	$cfjl$ − 48
$aijl$ − 36	bhl^2 + 48	cfl^2 + 48	akl^2 + 48	bil^2 + 48	$chij$ − 18	$aijk$ + 18	$bhil$ − 48	$cgkl$ + 36
ch^2j + 12	$bjkl$ − 36	$cikl$ − 36	bhj^2 + 12	cfk^2 + 12	cjl^2 + 48	al^3 − 32	$bijk$ + 18	$cijk$ + 18
fj^3 + 12	f^2h^2 − 3	f^2g^2 − 3	gh^2k − 6	f^3j + 12	fg^2j − 6	bj^3 − 4	bl^3 − 32	$chil$ + 36
g^2h^2 − 3	$fhkl$ − 24	$fgil$ − 24	fh^2j + 36	f^2gk + 36	g^3k + 12	ch^3 − 4	ck^3 − 4	cl^3 − 32
$ghjl$ − 24	fjk^2 + 36	fi^2j − 6	h^3i + 12	f^2hi − 6	g^2hi + 36	fhj^2 − 36	f^2hl + 12	fg^2l + 12
gj^2k − 6	gk^3 + 12	gi^2k + 36	h^2l^2 − 12	f^2l^2 − 12	g^2l^2 − 12	gh^2l + 12	f^2jk − 36	$fgij$ + 12
hij^2 + 36	hik^2 − 6	hi^3 + 12	$hjkl$ − 24	$fikl$ − 24	$gijl$ − 24	$ghjk$ + 12	fgk^2 − 36	g^2ki − 36
j^2l^2 − 12	k^2l^2 − 12	i^2l^2 − 12	j^2k^2 − 3	i^2k^2 − 3	i^2j^2 − 3	h^2ij − 36	$fhki$ + 12	ghi^2 − 36
						hjl^2 + 24	fkl^2 + 24	gil^2 + 24
						j^2kl + 12	ik^2l + 12	i^2jl + 12
±135	±135	±135	±135	±135	±135	±180	±180	±180

$\xi^4\eta\zeta$.	$\eta^4\zeta\xi$.	$\zeta^4\xi\eta$.	$\xi^3\eta^2\zeta$.	$\eta^3\zeta^2\xi$.	$\zeta^3\xi\eta^2$.	$\xi^3\eta\zeta^2$.	$\eta^3\zeta\xi^2$.	$\zeta^3\xi^2\eta$.	$\xi^2\eta^2\zeta^2$.
$bcfj\ -12$	$acfj\ -12$	$abgk\ -12$	$abci\ +6$	$abcj\ +6$	$abck\ +6$	$abcf\ +6$	$abcg\ +6$	$abch\ +6$	$abcl\ -24$
$bcgk\ +30$	$acgk\ -12$	$abhi\ -12$	$acf^2\ -12$	$abg^2\ -12$	$abfg\ -24$	$abi^2\ -12$	$acfl\ +36$	$abgl\ +36$	$abgi\ +6$
$bchi\ -12$	$achi\ +30$	$abij\ +30$	$afi^2\ +6$	$acfh\ -24$	$abil\ +36$	$af^2i\ +6$	$acik\ -24$	$abij\ -24$	$acfk\ +6$
$bcl^2\ -24$	$acl^2\ -24$	$abl^2\ -24$	$bcgh\ -24$	$ackl\ +36$	$af^2l\ +12$	$bchl\ +36$	$afgi\ -30$	$ack^2\ -12$	$af^2g\ +30$
$bfg^2\ -18$	$afg^2\ +24$	$af^2h\ -18$	$bcjl\ +36$	$afgl\ -60$	$afik\ -30$	$bcjk\ -24$	$ai^2l\ +12$	$af^2j\ -36$	$afil\ -48$
$bgil\ +12$	$agil\ +12$	$afkl\ +12$	$bg^2l\ +12$	$afij\ +54$	$bch^2\ -12$	$bfgj\ +54$	$bcj^2\ -12$	$afgk\ +12$	$ai^2k\ +30$
$bi^2j\ +24$	$ai^2j\ -18$	$aik^2\ +24$	$bgij\ -30$	$ahi^2\ -36$	$bfj^2\ -36$	$bg^2k\ -36$	$bg^2j\ +6$	$afhi\ +54$	$bchj\ +6$
$cf^2h\ +24$	$cgh^2\ -18$	$bgh^2\ +24$	$cfhl\ -60$	$agik\ +12$	$bghl\ -60$	$bghi\ +12$	$cfhj\ +12$	$afl^2\ +24$	$bg^2h\ +30$
$cfkl\ +12$	$chjl\ +12$	$bhjl\ +12$	$cfjk\ +12$	$ail^2\ +24$	$bgjk\ +54$	$bgl^2\ +24$	$cghk\ +54$	$aikl\ -60$	$bgjl\ -48$
$cik^2\ -18$	$cj^2k\ +24$	$bj^2k\ -18$	$cgk^2\ -36$	$bgj^2\ +6$	$bhij\ +12$	$bijl\ -60$	$ch^2i\ -36$	$bghj\ -30$	$bij^2\ +30$
$f^2gl\ +60$	$fgj^2\ -12$	$fh^2l\ +60$	$chik\ +54$	$ch^2l\ +12$	$bjl^2\ +24$	$cfhk\ -30$	$chl^2\ +24$	$bj^2l\ +12$	$cfh^2\ +30$
$f^2ij\ -12$	$g^2hl\ +60$	$fhjk\ -66$	$ckl^2\ +24$	$chjk\ -30$	$chk^2\ +6$	$ck^2l\ +12$	$cjkl\ -60$	$ch^2k\ +6$	$chkl\ -48$
$fgik\ -66$	$g^2jk\ -12$	$ghk^2\ -12$	$f^2gj\ -24$	$fghj\ +78$	$f^2hj\ +60$	$f^2gh\ -66$	$fg^2h\ -66$	$fgh^2\ -66$	$cjk^2\ +30$
$fhi^2\ -12$	$ghij\ -66$	$h^2ki\ -12$	$fg^2k\ +60$	$fj^2l\ -48$	$fghk\ +78$	$f^2jl\ -48$	$fgjl\ +72$	$fhjl\ -12$	$f^2j^2\ +24$
$fil^2\ -48$	$gjl^2\ -48$	$hkl^2\ -48$	$fghi\ +78$	$g^2hk\ -24$	$fh^2i\ -24$	$fgkl\ -12$	$fij^2\ -24$	$fj^2k\ +60$	$fghl\ +108$
$i^2kl\ +60$	$j^2il\ +60$	$jk^2l\ +60$	$fgl^2\ -96$	$gh^2i\ +60$	$fhl^2\ -96$	$fhil\ +72$	$g^2kl\ -48$	$ghkl\ +72$	$fgjk\ -114$
			$fijl\ +72$	$ghl^2\ -96$	$fjkl\ -12$	$fijk\ +78$	$ghil\ -12$	$gjk^2\ -24$	$fhij\ -114$
			$gikl\ -12$	$gjkl\ +72$	$gk^2l\ -48$	$fl^3\ +48$	$gijk\ +78$	$h^2il\ -48$	$fjl^2\ +24$
			$hi^2l\ -48$	$hijl\ -12$	$hikl\ +72$	$gik^2\ +60$	$gl^3\ +48$	$hijk\ +78$	$g^2k^2\ +24$
			$i^2jk\ -66$	$ij^2k\ -66$	$ijk^2\ -66$	$hi^2k\ -24$	$hi^2j\ +60$	$hl^3\ +48$	$ghki\ -114$
			$il^3\ +48$	$jl^3\ +48$	$kl^3\ +48$	$ikl^2\ -96$	$ijl^2\ -96$	$jkl^2\ -96$	$gkl^2\ +24$
									$h^2i^2\ +24$
									$hil^2\ +24$
									$ijkl\ +108$
									$l^4\ -48$
± 222	± 222	± 222	± 408	± 408	± 408	± 408	± 408	± 408	± 558

The preceding Tables contain the complete system [not so] of the covariants and contravariants of the ternary cubic, i.e. the covariants are the cubic itself U, the quartinvariant S, the sextinvariant T, the Hessian HU, and an octicovariant, say ΘU; the contravariants are the cubicontravariant PU, the quinticontravariant QU, and the reciprocant FU.

The contravariants are all of them evectants, viz. PU is the evectant of S, QU is the evectant of T, and the reciprocant FU is the evectant of QU, or what is the same thing, the second evectant of T.

The discriminant is a rational and integral function of the two invariants; representing it by R, we have $R = 64\ S^3 - T^2$.

If we combine U and HU by arbitrary multipliers, say α and 6β, so as to form the sum $\alpha U + 6\beta HU$, this is a cubic, and the question arises, to find the covariants and contravariants of this cubic: the results are given in the following Table:

No. 68.

$$\alpha U + 6\beta HU \qquad = \alpha U + 6\beta HU.$$

$$H(\alpha U + 6\beta HU) = \quad (0,\ 2S,\quad T,\quad 8S^2 \between \alpha,\ \beta)^3\, U$$

$$+ (1,\quad 0,\ -12S,\ -2T \between \alpha,\ \beta)^3\, HU.$$

$$P(\alpha U + 6\beta HU) = (1,\ 0,\ 12S,\ 4T \between \alpha,\ \beta)^3 PU$$
$$+ (0,\ 1,\ 0,\ -4S \between \alpha,\ \beta)^3 QU.$$
$$Q(\alpha U + 6\beta HU) = (0,\ 60S,\ 30T,\ 0,\ -120TS,\ -24T^2 + 576S^3 \between \alpha,\ \beta)^5 PU$$
$$+ (1,\ 0,\ 0,\ 10T,\ 240S^2,\ 24TS \between \alpha,\ \beta)^5 QU.$$
$$S(\alpha U + 6\beta HU) = (S,\ T,\ 24S^2,\ 4TS,\ T^2 - 48S^3 \between \alpha,\ \beta)^4.$$
$$T(\alpha U + 6\beta HU) = (T,\ 96S^2,\ 60TS,\ 20T^2,\ 240TS^2,\ -48T^2S + 4608S^4,\ -8T^3 + 576TS^3 \between \alpha,\ \beta)^6.$$
$$R(\alpha U + 6\beta HU) = [(1,\ 0,\ -24S,\ -8T,\ -48S^2 \between \alpha,\ \beta)^4]^3 R.$$
$$F(\alpha U + 6\beta HU) = (1,\ 0,\ -24S,\ -8T,\ -48S^2 \between \alpha,\ \beta)^4 FU$$
$$+ (0,\ 24,\ 0,\ 0,\ -48T \between \alpha,\ \beta)^4 (PU)^2$$
$$+ (0,\ 0,\ 24,\ 0,\ 96S \between \alpha,\ \beta)^4 PU . QU$$
$$+ (0,\ 0,\ 0,\ 8,\ 0 \between \alpha,\ \beta)^4 . (QU)^2.$$

We have, in like manner, for the covariants and contravariants of the cubic $6\alpha PU + \beta QU$, the following Table:

No. 69.

$$6\alpha PU + \beta QU = 6\alpha PU + \beta QU.$$
$$H(6\alpha PU + \beta QU) = (-2T,\ 48S^2,\ 18TS,\ T^2 + 16S^3 \between \alpha,\ \beta)^3 PU$$
$$+ (8S,\ T,\ -8S^2,\ -TS \between \alpha,\ \beta)^3 QU.$$
$$P(6\alpha PU + \beta QU) = (32S^2,\ 12TS,\ T^2 + 32S^3,\ 4TS^2 \between \alpha,\ \beta)^3 U$$
$$+ (4T,\ 96S^2,\ 12TS,\ T^2 - 32S^3 \between \alpha,\ \beta)^3 HU.$$

$$Q(6\alpha PU + \beta QU) = \left(\begin{array}{l} +384T S^2, \\ +120T^2S + 7680 S^4, \\ +\ 10T^3 + 3200TS^3, \\ +480T^2S^2, \\ +\ 30T^3S, \\ +\ 1T^4 - 24T^2S^3 + 512S^6 \end{array}\right\between \alpha,\ \beta)^5 U$$

$$+ \left(\begin{array}{l} -\ 24T^2 + 4608 S^3, \\ +1920T S^2, \\ +\ 480T^2S, \\ +\ 30T^3 + 1920TS^3, \\ +\ 120T^2S^2 + 7680 S^5, \\ -\ 6T^3S + 768TS^4 \end{array}\right\between \alpha,\ \beta)^5 HU.$$

$$S(6\alpha PU+\beta QU)=\left(\begin{array}{l} +\ \ 1T^2\ +192\ S^3, \\ +128T S^2, \\ +\ 18T^2S+384\ S^4, \\ +\ \ 1T^3\ +\ 64TS^3, \\ +\ \ 5T^2S^2-\ 64\ S^5 \end{array}\right)(\alpha,\ \beta)^4.$$

$$T(6\alpha PU+\beta QU)=\left(\begin{array}{l} -\ \ 8T^3\ +\ 4608TS^3, \\ +1920T^2S^2+73728\ S^5, \\ +\ 360T^3S\ +38400TS^4, \\ +\ \ 20T^4\ +\ 8960T^2S^3, \\ +\ 840T^3S^2+\ 7680TS^5, \\ +\ \ 36T^4S+\ \ 384T^2S^4+24576\ S^7, \\ +\ \ \ 1T^5\ -\ \ 40T^3S^2+\ 2560TS^6 \end{array}\right)(\alpha,\ \beta)^6.$$

$$6\alpha PU+\beta QU)=[(48S,\ 8T,\ -96S^2,\ -24TS,\ -T^2-16S^3)(\alpha,\ \beta)^4]^3\ R^2.$$

$$\begin{aligned} 6\alpha PU+\beta QU)=&\ (192\,S,\ 32T,\ -384\ S^2,\ -96T\,S,\ -4T^2-64S^3)(\alpha,\ \beta)^4\ \Theta U \\ &+(0,\ 512\ S^3,\ 192T\,S^2,\ 24T^2S,\ T^3)(\alpha,\ \beta)^4.\ U^2 \\ &+(1344S^2,\ 352TS,\ 24T^2-1152S^3,\ -288T\,S^2,\ -20T^2S+64\,S^4)(\alpha,\ \beta)^4\ U.HU \\ &+(48\,T,\ 0,\ 288T\,S,\ 24T^2+1536S^3,\ 144T\,S^2)(\alpha,\ \beta)^4\,(HU)^2. \end{aligned}$$

The tables for the ternary cubic become much more simple if we suppose that the cubic is expressed in Hesse's canonical form; we have then the following table:

No. 70.

$$\begin{aligned} U\ &= x^3+y^3+z^3+6lxyz. \\ S\ &= -l+l^4. \\ T\ &= 1-20l^3-8l^6. \\ R\ &= -(1+8l^3)^3. \\ HU &= l^2(x^3+y^3+z^3)-(1+2l^3)\,xyz. \\ \Theta U &= (1+8l^3)^2(y^3z^3+z^3x^3+x^3y^3) \\ &\quad+(-9l^6)(x^3+y^3+z^3)^2 \\ &\quad+(-2l-5l^4-20l^7)(x^3+y^3+z^3)\,xyz\,. \\ &\quad+(-15l^2-78l^5+12l^8)\,x^2y^2z^2. \\ \Theta_{\prime}U &= 4\,(1+8l^3)^2(y^3z^3+z^3x^3+x^3y^3) \\ &\quad+(-1-4l^3-4l^6)(x^3+y^3+z^3)^2 \\ &\quad+(4l+100l^4+112l^7)(x^3+y^3+z^3)\,xyz \\ &\quad+(48l^2+552l^5+48l^8)\,x^2y^2z^2. \end{aligned}$$

$$\begin{aligned}
\Theta_{\prime\prime} U = & -2(1+8l^3)^2(y^3z^3+z^3x^3+x^3y^3) \\
& +(1-10l^3)(x^3+y^3+z^3)^2 \\
& +(6l-180l^4+96l^7)(x^3+y^3+z^3)\,xyz \\
& +(6l^2-624l^5-192l^8)\,x^2y^2z^2. \\
PU = & -l(\xi^3+\eta^3+\zeta^3)+(-1+4l^3)\,\xi\eta\zeta. \\
QU = & (1-10l^3)(\xi^3+\eta^3+\zeta^3)-6l^2(5+4l^3)\,\xi\eta\zeta. \\
FU = & -4(1+8l^3)(\eta^3\zeta^3+\zeta^3\xi^3+\xi^3\eta^3) \\
& +(\xi^3+\eta^3+\zeta^3)^2 \\
& -24l^2(\xi^3+\eta^3+\zeta^3)\,\xi\eta\zeta \\
& -24l(1+2l^3)\,\xi^2\eta^2\zeta^2,
\end{aligned}$$

to which it is proper to join the following transformed expressions for ΘU, $\Theta_{\prime} U$, $\Theta_{\prime\prime} U$, viz.

$$\begin{aligned}
\Theta U = & (1+8l^3)^2(y^3z^3+z^3x^3+x^3y^3) \\
& +(-2l^3-l^6)\,U^2 \\
& +(2l-5l^4)\,U\,.\,HU \\
& +(-3l^2\quad)(HU). \\
\Theta_{\prime} U = & 4(1+8l^3)^2(y^3z^3+z^3x^3+x^3y^3) \\
& +\;(-1+12l^3+4l^6)\,U^2 \\
& +\;(-16l+4l^4\quad)\,U.HU \\
& +\;(-12l^2\quad)(HU)^2. \\
\Theta_{\prime\prime} U = & -2(1+8l^3)^2(y^3z^3+z^3x^3+x^3y^3) \\
& +\;(1-16l^3-6l^6)\,U^2 \\
& +\;(6l\quad)\,U.HU \\
& +\;(6l^2\quad)(HU)^2.
\end{aligned}$$

The last preceding table affords a complete solution of the problem to reduce a ternary cubic to its canonical form.

[I add to the present Memoir, in the notation hereof $(a, b, c, f, g, h, i, j, k, l \between x, y, z)^3$ for the ternary cubic, some formulæ originally contained in the paper "On Homogeneous Functions of the third order with three variables," (1846), but which on account of the difference of notation were omitted from the reprint, 35, of that paper.

Representing the determinant

$$-\begin{vmatrix} ax+hy+jz, & hx+ky+bz, & jx+ly+gz, & \xi \\ hx+ky+bz, & kx+by+fz, & lx+fy+iz, & \eta \\ jx+ly+gz, & lx+fy+iz, & gx+iy+cz, & \zeta \\ \xi\quad, & \eta\quad, & \zeta & \end{vmatrix}$$

by

$$(A,\ B,\ C,\ F,\ G,\ H \between x,\ y,\ z)^2$$

the values of A, B, C, F, G, H (equations (10) of 35) are

	A	B	C	F	G	H
ξ^2	gk $-\ l^2$	bi $-\ f^2$	cf $-\ i^2$	bc $-\ fi$	fg $+\ ck$ $-\ 2il$	ki $+\ bg$ $-\ 2fl$
η^2	ag $-\ j^2$	hi $-\ l^2$	cj $-\ g^2$	ij $+\ ch$ $-\ 2gl$	ca $-\ gj$	gh $+\ ai$ $-\ 2jl$
ζ^2	ak $-\ h^2$	bh $-\ k^2$	fj $-\ l^2$	hf $+\ bj$ $-\ 2kl$	jk $+\ af$ $-\ 2hl$	ab $-\ hk$
$\eta\zeta$	$2hj$ $-\ 2al$	$2kl$ $-\ 2hf$	$2gl$ $-\ 2ij$	l^2 $+\ gk$ $-\ hi$ $-\ fj$	gh $-\ ai$	jk $-\ af$
$\zeta\xi$	$2hl$ $-\ 2jk$	$2fk$ $-\ 2bl$	$2il$ $-\ 2fg$	ki $-\ bg$	l^2 $+\ hi$ $-\ fj$ $-\ gk$	hf $-\ bj$
$\xi\eta$	$2jl$ $-\ 2gh$	$2fl$ $-\ 2ki$	$2gi$ $-\ 2cl$	fg $-\ ck$	ij $-\ ch$	l^2 $+\ fj$ $-\ gk$ $-\ hi$

Moreover writing

$$FU = \begin{vmatrix} a, & k, & g, & l, & j, & h, & \xi \\ h, & b, & i, & f, & l, & k, & \eta \\ j, & f, & c, & i, & g, & l, & \zeta \\ 2\xi & . & . & . & \zeta & \eta & . \\ . & 2\eta & . & \zeta & . & \xi & . \\ . & . & 2\zeta & \eta & \xi & . & . \\ A, & B, & C, & F, & G, & H & . \end{vmatrix},$$

so that

$$FU = A\mathrm{a} + B\mathrm{b} + C\mathrm{c} + 2F\mathrm{f} + 2G\mathrm{g} + 2H\mathrm{h},$$

then the values of a, b, c, f, g, h (equations (13) of 35) are

	a	b	c	f	g	h
ξ^4	0	$2cf$ $-2i^2$	$2bi$ $-2f^2$	fi $-bc$	0	0
η^4	$2cj$ $-2g^2$		$2ag$ $-2j^2$	0	gj $-ca$	0
ζ^4	$2bh$ $-2k^2$	$2ak$ $-2h^2$		0	0	hk $-ab$
$\eta^3\zeta$	$8gl$ $-6ij$ $-2ch$	0	$4hj$ $-4al$	$2j^2$ $-2ag$	$3ai$ $-2jl$ $-gh$	ca $-gj$
$\zeta^3\xi$	$4fk$ $-4bl$	$8hl$ $-6jk$ $-2af$	0	ab $-hk$	$2k^2$ $-2bh$	$3bj$ $-2kl$ $-hf$
$\xi\eta^3$	0	$4gi$ $-4cl$	$8fl$ $-6ki$ $-2bg$	$3ck$ $-2il$ $-fg$	bc $-fi$	$2i^2$ $-2cf$
$\eta\zeta^3$	$8kl$ $-6hf$ $-2bj$	$4hj$ $-4al$	0	$2h^2$ $-2ak$	ab $-hk$	$3af$ $-2hl$ $-jk$
$\zeta\xi^3$	0	$8il$ $-6fg$ $-2ck$	$4fk$ $-4bl$	$3bg$ $-2fl$ $-ki$	$2f^2$ $-2bi$	bc $-fi$
$\xi\eta^3$	$4gi$ $-4cl$	0	$8jl$ $-6gh$ $-2ai$	ca $-gj$	$3ch$ $-2gl$ $-ij$	$2g^2$ $-2cj$
$\eta^2\zeta^2$	$6hi$ $+6fj$ $-4gk$ $-8l^2$	$2ag$ $-2j^2$	$2ak$ $-2h^2$	$4al$ $-4hj$	jk $+2hl$ $-3af$	gh $+2jl$ $-3ai$
$\zeta^2\xi^2$	$2bi$ $-2f^2$	$6fj$ $+6gk$ $-4hi$ $-8l^2$	$2bh$ $-2k^2$	hf $+2kl$ $-3bj$	$4bl$ $-4fk$	ki $+2fl$ $-3bg$
$\xi^2\eta^2$	$2cf$ $-2i^2$	$2cj$ $-2g^2$	$6gk$ $+6hi$ $-4fj$ $-8l^2$	ij $+2gl$ $-3ch$	fg $+2il$ $-3ck$	$4cl$ $-4gi$
$\xi^2\eta\zeta$	$2fi$ $-2bc$	$4ch$ $+4gl$ $-8ij$	$4bj$ $+4kl$ $-8hf$	$4l^2$ $+2hi$ $+2fj$ $-8gk$	$7ki$ $-6fl$ $-bg$	$7fg$ $-6il$ $-ck$
$\xi\eta^2\zeta$	$4ck$ $+4il$ $-8fg$	$2gj$ $-2ca$	$4af$ $+4hl$ $-8jk$	$7gh$ $-6jl$ $-ai$	$4l^2$ $+2fj$ $+2gk$ $-8hi$	$7ij$ $-6gl$ $-ch$
$\xi\eta\zeta^2$	$4bg$ $+4fl$ $-8ki$	$4ai$ $+4jl$ $+8gh$	$2hk$ $-2ab$	$7jk$ $-6hl$ $-af$	$7hf$ $-6kl$ $-bj$	$4l^2$ $+2gk$ $+2hi$ $-8fj$

Also if the discriminant be written

$$K(U) = \begin{vmatrix} a & k & g & l & j & h \\ h & b & i & f & l & k \\ j & i & c & i & g & l \\ \mathfrak{A} & \mathfrak{K} & \mathfrak{G} & \mathfrak{L} & \mathfrak{J} & \mathfrak{H} \\ \mathfrak{H} & \mathfrak{B} & \mathfrak{I} & \mathfrak{F} & \mathfrak{L} & \mathfrak{K} \\ \mathfrak{J} & \mathfrak{I} & \mathfrak{C} & \mathfrak{I} & \mathfrak{G} & \mathfrak{L} \end{vmatrix},$$

then the values of $\mathfrak{A}$, $\mathfrak{B}$, $\mathfrak{C}$, $\mathfrak{F}$, $\mathfrak{G}$, $\mathfrak{H}$, $\mathfrak{I}$, $\mathfrak{J}$, $\mathfrak{K}$, $\mathfrak{L}$ (equations (20) of 35) are

$$\begin{aligned}
\mathfrak{A} &= akg + 2hjl - al^2 - gh^2 - j^2k, \\
\mathfrak{B} &= bih + 2fkl - bl^2 - hf^2 - k^2i, \\
\mathfrak{C} &= cjf + 2gil - cl^2 - fg^2 - i^2j, \\
3\mathfrak{F} &= bch + bij - ck^2 + 2gfk - 2bgl + fl^2 - f^2j - fih, \\
3\mathfrak{G} &= caf + cjk - ai^2 + 2hgi - 2chl + gl^2 - g^2k - gjf, \\
3\mathfrak{H} &= abg + aki - bj^2 + 2fhj - 2afl + hl^2 - h^2i - hkg, \\
3\mathfrak{I} &= bcj + cfh - bg^2 + 2kig - 2ckl + jl^2 - i^2h - fij, \\
3\mathfrak{J} &= cak + agf - ch^2 + 2ijh - 2ail + kl^2 - j^2f - gjk, \\
3\mathfrak{K} &= abi + bhg - af^2 + 2jkf - 2bjl + il^2 - k^2g - hki, \\
6\mathfrak{L} &= abc + 3fgh + 3ijk + 2l^3 - afi - bgj - chk - 2lgk - 2lhi - 2lfj.
\end{aligned}$$

The equation $K(U) = R = 64S^3 - T^2$ would however afford a perhaps easier formula for the calculation of the discriminant.]

145.

A MEMOIR UPON CAUSTICS.

[From the *Philosophical Transactions of the Royal Society of London,* vol. CXLVII. *for the year* 1857, pp. 273—312. Received May 1,—Read May 8, 1856.]

THE following memoir contains little or nothing that can be considered new in principle; the object of it is to collect together the principal results relating to caustics *in plano,* the reflecting or refracting curve being a right line or a circle, and to discuss, with more care than appears to have been hitherto bestowed upon the subject, some of the more remarkable cases. The memoir contains in particular researches relating to the caustic by refraction of a circle for parallel rays, the caustic by reflexion of a circle for rays proceeding from a point, and the caustic by refraction of a circle for rays proceeding from a point; the result in the last case is not worked out, but it is shown how the equation in rectangular coordinates is to be obtained by equating to zero the discriminant of a rational and integral function of the sixth degree. The memoir treats also of the secondary caustic, or orthogonal trajectory of the reflected or refracted rays, in the general case of a reflecting or refracting circle and rays proceeding from a point; the curve in question, or rather *a* secondary caustic, is, as is well known, the Oval of Descartes or 'Cartesian': the equation is discussed by a method which gives rise to some forms of the curve which appear to have escaped the notice of geometers. By considering the caustic as the evolute of the secondary caustic, it is shown that the caustic, in the general case of a reflecting or refracting circle and rays proceeding from a point, is a curve of the sixth class only. The concluding part of the memoir treats of the curve which, when the incident rays are parallel, must be taken for the secondary caustic in the place of the Cartesian, which, for the particular case in question, passes off to infinity. In the course of the memoir, I reproduce a theorem first given, I believe, by me in the Philosophical Magazine, viz. that there are six different systems of a radiant point and refracting circle which give rise to identically the same caustic, [see *post*, XXVIII]. The memoir is divided into sections, each of which is to a considerable extent intelligible by itself, and the subject of each section is for the most part explained by the introductory paragraph or paragraphs.

I.

Consider a ray of light reflected or refracted at a curve, and suppose that ξ, η are the coordinates of a point Q on the incident ray, α, β the coordinates of the point G of incidence upon the reflecting or refracting curve, a, b the coordinates of a point N upon the normal at the point of incidence, x, y the coordinates of a point q on the reflected or refracted ray.

Write for shortness,

$$(b-\beta)(\xi-\alpha)-(a-\alpha)(\eta-\beta)=\nabla QGN,$$
$$(a-\alpha)(\xi-\alpha)+(b-\beta)(\eta-\beta)=\square QGN,$$

then ∇QGN is equal to twice the area of the triangle QGN, and if ξ, η instead of being the coordinates of a point Q on the incident ray were current coordinates, the equation $\nabla QGN=0$ would be the equation of the line through the points G and N, i.e. of the normal at the point of incidence; and in like manner the equation $\square QGN=0$ would be the equation of the line through G perpendicular to the line through the points G and N, i.e. of the tangent at the point of incidence.

We have

$$\overline{NG}^2=(a-\alpha)^2+(b-\beta)^2,$$
$$\overline{QG}^2=(\xi-\alpha)^2+(\eta-\beta)^2,$$

and therefore identically,

$$\overline{NG}^2\cdot\overline{QG}^2=\overline{\nabla QGN}^2+\overline{\square QGN}^2.$$

Suppose for a moment that ϕ is the angle of incidence and ϕ' the angle of reflexion or refraction; and let μ be the index of refraction (in the case of reflexion $\mu=-1$), then writing

$$(b-\beta)(x-\alpha)-(a-\alpha)(y-\beta)=\nabla qGN,$$
$$(a-\alpha)(x-\alpha)+(b-\beta)(y-\beta)=\square qGN,$$

and

$$\overline{qG}^2=(x-\alpha)^2+(y-\beta)^2,$$

we have

$$\sin\phi=\frac{\nabla QGN}{NG\,.\,GQ},\quad \sin\phi'=\frac{\nabla qGN}{NG\,.\,Gq};$$

and substituting these values in the equation

$$\sin^2\phi-\mu^2\sin^2\phi'=0,$$

we obtain

$$\overline{qG}^2\,\overline{\nabla QGN}^2 - \mu^2\,\overline{QG}^2\,\overline{\nabla qGN}^2 = 0,$$

an equation which is rational of the second order in x, y, the coordinates of a point q on the refracted ray; this equation must therefore contain, as a factor, the equation of the refracted ray; the other factor gives the equation of a line equally inclined to, but on the opposite side of the normal; this line (which of course has no physical existence) may be termed *the false refracted ray.* The caustic is *geometrically* the envelope of *the pair of rays,* and for finding the equation of the caustic it is obviously convenient to take the equation of the two rays conjointly in the form under which such equation has just been found, without attempting to break the equation up into its linear factors.

It is however interesting to see how the resolution of the equation may be effected; for this purpose multiply the equation by $\overline{NG}^2$, then reducing by means of a previous formula, the equation becomes

$$(\overline{\nabla qGN}^2 + \overline{\Box qGN}^2)\overline{\nabla QGN}^2 - \mu^2(\overline{\nabla QGN}^2 + \overline{\Box QGN}^2)\overline{\nabla qGN}^2 = 0,$$

which is equivalent to

$$\overline{\nabla qGN}^2(\mu^2\,\overline{\Box QGN}^2 + (\mu^2-1)\overline{\nabla QGN}^2) - \overline{\Box qGN}^2\,\overline{\nabla QGN}^2 = 0,$$

and the factors are

$$\nabla qGN\sqrt{\mu^2\,\overline{\Box QGN}^2 + (\mu^2-1)\overline{\nabla QGN}^2} \mp \Box qGN\,.\,\nabla QGN = 0\,;$$

it is in fact easy to see that these equations represent lines passing through the point G and inclined to GN at angles $\pm\phi'$, where ϕ' is given by the equations

$$\sin\phi = \mu\sin\phi',$$

$$\tan\phi = \frac{\nabla QGN}{\Box QGN},$$

and there is no difficulty in distinguishing in any particular case between the refracted ray and the false refracted ray.

In the case of reflexion $\mu = -1$, and the equations become

$$\nabla qGN\,.\,\Box QGN \mp \Box qGN\,.\,\nabla QGN = 0\,;$$

the equation

$$\nabla qGN\,.\,\Box QGN - \Box qGN\,.\,\nabla QGN = 0$$

is obviously that of the incident ray, which is what the false refracted ray becomes in the case of reflexion; and the equation

$$\nabla qGN\,.\,\Box QGN + \Box qGN\,.\,\nabla QGN = 0$$

is that of the reflected ray.

II.

But instead of investigating the nature of the caustic itself, we may begin by finding the secondary caustic or orthogonal trajectory of the refracted rays, i.e. a curve having the caustic for its evolute; suppose that the incident rays are all of them normal to a certain curve, and let Q be a point upon this curve, and considering the ray through the point Q, let G be the point of incidence upon the refracting curve; then if the point G be made the centre of a circle the radius of which is $\mu^{-1} . GQ$, the envelope of the circles will be the secondary caustic. It should be noticed that, if the incident rays proceed from a point, the most simple course is to take such point for the point Q. The remark, however, does not apply to the case where the incident rays are parallel; the point Q must here be considered as the point in which the incident ray is intersected by some line at right angles to the rays, and there is not in general any one line which can be selected in preference to another. But if the refracting curve be a circle, then the line perpendicular to the incident rays may be taken to be a diameter of the circle. To translate the construction into analysis, let ξ, η be the coordinates of the point Q, and α, β the coordinates of the point G, then ξ, η, α, β are in effect functions of a single arbitrary parameter; and if we write

$$\overline{GQ}^2 = (\xi - \alpha)^2 + (\eta - \beta)^2,$$

$$\overline{Gq}^2 = (x - \alpha)^2 + (y - \beta)^2,$$

then the equation

$$\mu^2\overline{Gq}^2 - \overline{GQ}^2 = 0,$$

where x, y are to be considered as current coordinates, and which involves of course the arbitrary parameter, is the equation of the circle, and the envelope is obtained in the usual manner. This is the well-known theory of Gergonne and Quetelet.

III.

There is however a simpler construction of the secondary caustic in the case of the reflexion of rays proceeding from a point. Suppose, as before, that Q is the radiant point, and let G be the point of incidence. On the tangent at G to the reflecting curve, let fall a perpendicular from Q, and produce it to an equal distance on the other side of the tangent; then if q be the extremity of the line so produced, it is clear that q is a point on the reflected ray Gq, and it is easy to see that the locus of q is the secondary caustic. Produce now QG to a point Q' such that $GQ' = QG$, it is clear that the locus of Q' will be a curve similar to and similarly situated with and twice the magnitude of the reflecting curve, and that the two curves have the point Q for a centre of similitude. And the tangent at Q' passes through the point q, i.e. q is the foot of the perpendicular let fall from Q upon the tangent at Q'; we have therefore the theorem due to Dandelin, viz.

If rays proceeding from a point Q are reflected at a curve, then the secondary caustic is the locus of the feet of the perpendiculars let fall from the point Q upon the tangents of a curve similar to and similarly situated with and twice the magnitude of the reflecting curve, and such that the two curves have the point Q for a centre of similitude.

IV.

If rays proceeding from a point Q are reflected at a line, the reflected rays will proceed from a point q situate on the perpendicular let fall from Q, and at an equal distance on the other side of the reflecting line. The point q may be spoken of as the image of Q; it is clear that if Q be considered as a variable point, then the locus of the image q will be a curve equal and similar but oppositely situated to the curve, the locus of Q, and which may be spoken of as the image of such curve. Hence it at once follows, that if the incidental rays are tangent, or normal, or indeed in any other manner related to a curve, then the reflected rays will be tangent, or normal, or related in a corresponding manner to a curve the image of the first-mentioned curve. The theory of the combined reflexions and refractions of a pencil of rays transmitted through a plate or prism, is, by the property in question, rendered very simple. Suppose, for instance, that a pencil of rays is refracted at the first surface of a plate or prism, and after undergoing any number of internal reflexions, finally emerges after a second refraction at the first or second surface; in order to find the caustic enveloped by the rays after the second refraction, it is only necessary to form the successive images of the first caustic corresponding to the different reflexions, and finally to determine the caustic for refraction in the case where the incident rays are the tangents of the caustic which is the last of the series of images; the problem is not in effect different from that of finding the caustic for refraction in the case where the incident rays are the tangents to the caustic after the first refraction, but the line at which the second refraction takes place is arbitrarily situate with respect to the caustic. Thus e.g. suppose the incident rays proceed from a point, the caustic after the first refraction is, it will be shown in the sequel, the evolute of a conic; for the complete theory of the combined reflexions and refractions of the pencil by a plate or prism, it is only necessary to find the caustic by refraction, where the incident rays are the normals of a conic, and the refracting line is arbitrarily situate with respect to the conic.

V.

Suppose that rays proceeding from a point Q are refracted at a line; and take the refracting line for the axis of y, the axis of x passing through the radiant point Q, and take the distance QA for unity. Suppose that the index of refraction μ is put equal to $\frac{1}{k}$. Then if ϕ be the angle of incidence and ϕ' the angle of refraction,

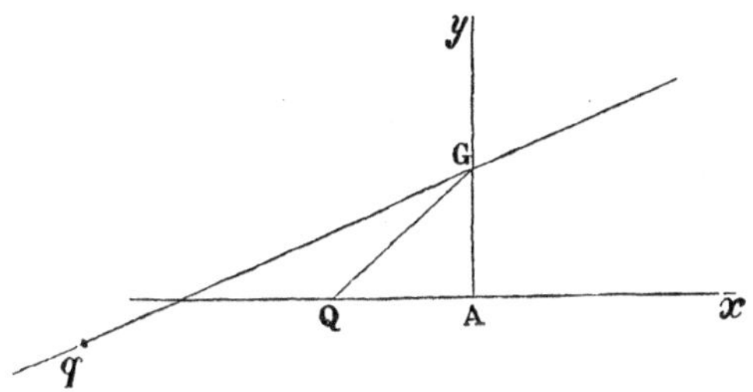

we have $\sin\phi' = k\sin\phi$, and the equation $y - x\tan\phi' = \tan\phi$ of the refracted ray becomes, putting for ϕ' its value,

$$y - \frac{k\sin\phi}{\sqrt{1-k^2\sin^2\phi}}\,x - \tan\phi = 0.$$

Differentiating with respect to the variable parameter and combining the two equations, we obtain, after a simple reduction,

$$kx = -\frac{(1-k^2\sin^2\phi)^{\frac{3}{2}}}{\cos^3\phi},$$

$$k'y = -\frac{k'^3\sin^3\phi}{\cos^3\phi},$$

where $k' = \sqrt{1-k^2}$; hence eliminating,

$$(kx)^{\frac{2}{3}} - (k'y)^{\frac{2}{3}} = 1,$$

which is the equation of the caustic. When the refraction takes place into a denser medium k is less than 1, and k'^2 is positive, the caustic is therefore the evolute of a hyperbola (see fig. 1); but when the refraction takes place in a rarer medium k is greater than 1, and k'^2 is negative, the caustic is therefore the evolute of an ellipse (see fig. 2). These results appear to have been first obtained by Gergonne. The conic (hyperbola or ellipse) is the secondary caustic, and as such may be obtained as follows.

VI.

The equation of the variable circle is

$$x^2 + (y-\tan\phi)^2 - k^2\sec^2\phi = 0;$$

or reducing, the equation is

$$x^2 + y^2 - 2y\tan\phi + k'^2\tan^2\phi - k^2 = 0:$$

whence, considering $\tan\phi$ as the variable parameter, the equation of the envelope is

$$k'^2(x^2+y^2-k^2) - y^2 = 0,$$

that is,

$$k'^2x^2 - k^2y^2 - k^2k'^2 = 0,$$

or

$$\frac{x^2}{k^2} - \frac{y^2}{k'^2} = 1$$

is the equation of the secondary caustic, or conic having the caustic for its evolute. The radiant point, it is clear, is a focus of the conic.

VII.

Let the equation of the refracted ray be represented by

$$Xx + Yy + Z = 0,$$

we have

$$X : Y : Z = \frac{-k \sin\phi}{\sqrt{1 - k^2 \sin^2\phi}} : 1 : -\tan\phi,$$

from which we obtain

$$\frac{k^2}{X^2} - \frac{k'^2}{Y^2} = \frac{1}{Z^2}$$

for the tangential equation of the caustic; or if we represent the equation of the refracted ray by

$$Xx + Yy - k = 0,$$

then we have

$$\frac{k^2}{X^2} - \frac{k'^2}{Y^2} = \frac{1}{k^2}$$

for the tangential equation of the caustic.

Fig. 1. Fig. 2.

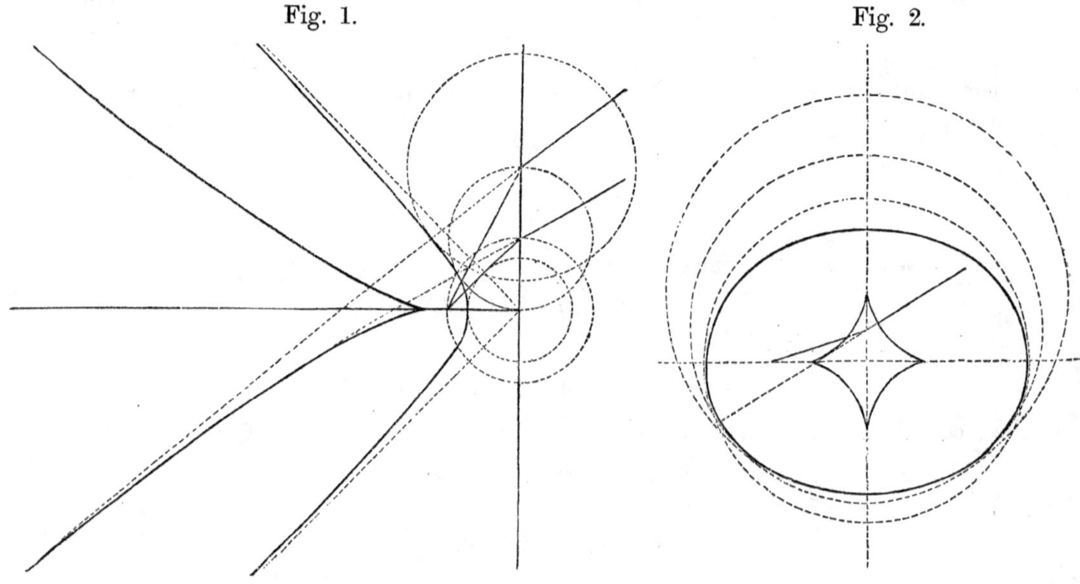

VIII.

If a ray be reflected at a circle; we may take a, b as the coordinates of the centre of the circle, and supposing as before that ξ, η are the coordinates of a point Q in the incident ray, α, β the coordinates of the point G of incidence, and x, y the coordinates of a point q in the reflected ray, the equation of the reflected ray, treating x, y as current coordinates, is

$$\{(b-\beta)(x-\alpha)-(a-\alpha)(y-\beta)\}\{(a-\alpha)(\xi-\alpha)+(b-\beta)(\eta-\beta)\}$$
$$+\{(a-\alpha)(x-\alpha)+(b-\beta)(y-\beta)\}\{(b-\beta)(\xi-\alpha)-(a-\alpha)(\eta-\beta)\}=0.$$

Write for shortness,

$$N_{q,G}=(b-\beta)(x-\alpha)-(a-\alpha)(y-\beta),$$
$$T_{q,G}=(a-\alpha)(x-\alpha)+(b-\beta)(y-\beta),$$

and similarly for $N_{Q,G}$, &c.; the equation of the reflected ray is

$$N_{q,G}T_{Q,G}+T_{q,G}N_{Q,G}=0.$$

Suppose that the reflected ray meets the circle again in G' and undergoes a second reflexion, and let x', y' be the coordinates of a point q' in the ray thus twice reflected. We see first (G' being a point in the first reflected ray) that

$$N_{G',G}T_{Q,G}+T_{G',G}N_{Q,G}=0.$$

Again, considering G as a point in the ray by the reflexion of which the second reflected ray arises, the equation of the second reflected ray is

$$N_{q',G'}T_{G,G'}+T_{q',G'}N_{G,G'}=0;$$

and from the form of the expressions $N_{q,G}$, $T_{q,G}$ it is clear that

$$N_{G,G'}=-N_{G',G},\ T_{G,G'}=+T_{G',G};$$

the equation for the second reflected ray may therefore be written under the form

$$N_{q',G'}T_{G',G}-T_{q',G'}N_{G',G}=0;$$

or reducing by a previous equation, we obtain finally for the equation of the second reflected ray,

$$N_{q',G'}T_{Q,G}+T_{q',G'}N_{Q,G}=0;$$

and in like manner the equation for the third reflected ray is

$$N_{q'',G''}T_{Q,G}+T_{q'',G''}N_{Q,G}=0,$$

and so on, the equation for the last reflected ray containing, it will be observed, the coordinates of the radiant point and of the first and last points of incidence (the coordinates of the last point of incidence can of course only be calculated from those of the radiant point and the first point of incidence, through the coordinates of the intermediate points of incidence), but not containing explicitly the coordinates of any of the intermediate points of incidence. The form is somewhat remarkable, but the result is really the same with that obtained by simple geometrical considerations, as follows.

IX.

Consider a ray reflected any number of times at a circle; and let $G_0G_,$ be the ray incident at $G_,$ and GG' the last reflected ray, the point at which the reflexion takes place or last point of incidence being G. Take the centre O of the circle for

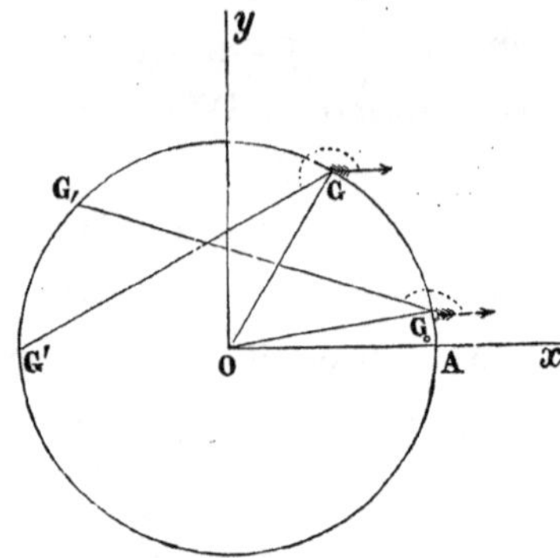

the origin, and any two lines Ox, Oy through the centre and at right angles to each other for axes, and let Ox meet the circle in the point A. Write

$$\angle AOG_0 = \theta_0, \quad \angle xG_0G_, = \psi_0,$$
$$\angle AOG = \theta, \quad \angle xGG' = \psi,$$
$$\angle G_0G_,O = \phi;$$

then the radius of the circle being taken as the centre of the circle, the equation of the reflected ray is

$$y - \sin\theta = \tan\psi\,(x - \cos\theta);$$

and if there have been n reflexions, then

$$\theta = \theta_0 + n(\pi - 2\phi) = \theta_0 + n\pi - 2n\phi,$$
$$\psi = \psi_0 - 2n\phi,$$

and therefore the equation of the reflected ray is

$$y\cos(\psi_0 - 2n\phi) - x\sin(\psi_0 - 2n\phi) + (-)^n \sin(\psi_0 - \theta_0) = 0.$$

X.

If a pencil of parallel rays is reflected any number of times at a circle, then taking AO for the direction of the incident rays, we may write $\theta_0 = \phi$, $\psi_0 = \pi$, and the equation of a reflected ray is

$$x\sin 2n\phi + y\cos 2n\phi = (-)^n \sin\phi;$$

differentiating with respect to the variable parameter, we find

$$x \cos 2n\phi - y \sin 2n\phi = (-)^n \frac{1}{2n} \cos\phi\,;$$

and these equations give

$$x = \frac{(-)^n}{4n}\left\{\quad (2n+1)\cos(2n-1)\phi - (2n-1)\cos(2n+1)\phi\right\},$$

$$y = \frac{(-)^n}{4n}\left\{-(2n+1)\sin(2n-1)\phi + (2n-1)\sin(2n+1)\phi\right\},$$

which may be taken for the equation of the caustic; the caustic is therefore an epicycloid: this is a well-known result.

XI.

If rays proceeding from a point upon the circumference are reflected any number of times at a circle, then taking the point A for the radiant point, we have $\alpha_0 = 0$, $\psi_0 = \pi - \phi$, and the equation of a reflected ray is

$$x \sin(2n+1)\phi + y \cos(2n+1)\phi = (-)^n \sin\phi\,;$$

differentiating with respect to the variable parameter, we find

$$x \cos(2n+1)\phi - y \sin(2n+1)\phi = (-)^n \frac{1}{2n+1} \sin\phi\,;$$

and these equations give

$$x = \frac{(-)^n}{2n+1}\left\{\quad (n+1)\cos 2n\phi - n\cos(2n+2)\phi\right\},$$

$$y = \frac{(-)^n}{2n+1}\left\{-(n+1)\sin 2n\phi + n\cos(2n+2)\phi\right\},$$

which may be taken as the equation of the caustic; the caustic is therefore in this case also an epicycloid: this is a well-known result.

XII.

Consider a pencil of parallel rays refracted at a circle; take the radius of the circle as unity, and let the incident rays be parallel to the axis of x, then if ϕ, ϕ' be the angles of incidence and refraction, and μ or $\frac{1}{k}$ be the index of refraction, so that $\sin\phi' = k\sin\phi$, the coordinates of the point of incidence are $\cos\phi$, $\sin\phi$, and the equation of the refracted ray is

$$y - \sin\phi = \tan(\phi - \phi')(x - \cos\phi),$$

i.e.

$$\cos(\phi - \phi')(y - \sin\phi) = \sin(\phi - \phi')(x - \cos\phi),$$

or

$$y\cos(\phi-\phi')-x\sin(\phi-\phi')=\sin\phi',$$

which may also be written

$$(y\cos\phi-x\sin\phi)\cos\phi'+(y\sin\phi+x\cos\phi-1)\sin\phi'=0.$$

Writing $k\sin\phi$, $\sqrt{1-k^2\sin^2\phi}$ instead of $\sin\phi'$, $\cos\phi'$, and putting for shortness

$$y\cos\phi-x\sin\phi=Y,$$
$$y\sin\phi+x\cos\phi=X,$$
$$\frac{k\sin\phi}{\sqrt{1-k^2\sin^2\phi}}=\Phi,$$

the equation of the refracted ray becomes

$$Y+\Phi(X-1)=0;$$

and differentiating with respect to the variable parameter ϕ, observing that

$$\frac{dY}{d\phi}=-X,\quad \frac{dX}{d\phi}=Y,$$
$$\frac{d\Phi}{d\phi}=\frac{k\cos\phi}{(1-k^2\sin^2\phi)^{\frac{3}{2}}}=\frac{\cot\phi}{1-k^2\sin^2\phi}\Phi,$$

we have

$$-X+\Phi\left(Y+\frac{\cot\phi\,(X-1)}{1-k^2\sin^2\phi}\right)=0,$$

and the combination of the two equations gives

$$Y=-\frac{\Phi(1-k^2\sin^2\Phi)}{\Phi\cot\phi-1},$$
$$X=\frac{\Phi\cot\phi-k^2\sin^2\phi}{\Phi\cot\phi-1},$$

and we have therefore

$$y=Y\cos\phi+X\sin\phi=\frac{k^2\sin^3\phi\,(\Phi\cot\phi-1)}{\Phi\cot\phi-1}=k^2\sin^3\phi,$$

$$x=X\cos\phi-Y\sin\phi=\frac{\Phi\left(\frac{1}{\sin\phi}-k^2\sin^3\phi\right)-k^2\sin^2\phi\cos\phi}{\Phi\cot\phi-1},$$

i.e.

$$x=\frac{\Phi(1-k^2\sin^4\phi)-k^2\sin^3\phi\cos\phi}{\Phi\cos\phi-\sin\phi};$$

or multiplying the numerator and denominator by $(1 - k^2 \sin^2 \phi)(\Phi \cos \phi + \sin \phi)$, the numerator becomes

$$\begin{aligned}
&(1 - k^2 \sin^2 \phi) \{\Phi^2 \cos \phi (1 - k^2 \sin^4 \phi) - k^2 \sin^4 \phi \cos \phi \\
&\qquad + \Phi \left(\sin \phi (1 - k^2 \sin^2 \phi) - k^2 \sin^3 \phi \cos \phi\right)\} \\
&= k^2 \sin^2 \phi \cos \phi \{(1 - k^2 \sin^4 \phi) - \sin^2 \phi (1 - k^2 \sin^2 \phi)\} \\
&\qquad + k \sin^2 \phi \sqrt{1 - k^2 \sin^2 \phi} (1 - k^2 \sin^2 \phi) \\
&= k^2 \sin^2 \phi \cos^3 \phi + k \sin^2 \phi (1 - k^2 \sin^2 \phi)^{\frac{3}{2}},
\end{aligned}$$

and the denominator becomes

$$\begin{aligned}
&k^2 \sin^2 \phi \cos^2 \phi - (1 - k^2 \sin^2 \phi) \sin^2 \phi \\
&= - k'^2 \sin^2 \phi,
\end{aligned}$$

if $k'^2 = 1 - k^2$.

Hence we have for the coordinates of the point of the caustic,

$$\begin{cases} k'^2 x = - k^2 \cos^3 \phi - k (1 - k^2 \sin^2 \phi)^{\frac{3}{2}}, \\ \quad y = \quad k^2 \sin^3 \phi\,; \end{cases}$$

and eliminating ϕ, we obtain for the equation of the caustic,

$$k'^2 x = - k^2 \{1 - k^{-\frac{4}{3}} y^{\frac{2}{3}}\}^{\frac{3}{2}} - k \{1 - k^{\frac{2}{3}} y^{\frac{2}{3}}\}^{\frac{3}{2}}\,;$$

or writing $\frac{1}{\mu}$ instead of k, we find

$$(1 - \mu^2) x = (1 - \mu^{\frac{4}{3}} y^{\frac{2}{3}})^{\frac{3}{2}} + \mu (1 - \mu^{-\frac{2}{3}} y^{\frac{2}{3}})^{\frac{3}{2}}$$

for the equation of the caustic by refraction of the circle, for parallel rays. The equation was first obtained by St Laurent.

XIII.

The discussion of the preceding equation presents considerable interest. In the first place to obtain the rational form write

$$\alpha = (1 - \mu^2) x, \ \beta = (1 - \mu^{\frac{4}{3}} y^{\frac{2}{3}})^{\frac{3}{2}}, \ \gamma = \mu (1 - \mu^{-\frac{2}{3}} y^{\frac{2}{3}})^{\frac{3}{2}};$$

this gives

$$\alpha^4 - 2\alpha^2 (\beta^2 + \gamma^2) + (\beta^2 - \gamma^2)^2 = 0,$$

and we have

$$\beta^2 = 1 - 3\mu^{\frac{4}{3}} y^{\frac{2}{3}} + 3\mu^{\frac{8}{3}} y^{\frac{4}{3}} - \mu^4 y^2,$$

$$\gamma^2 = \mu^2 - 3\mu^{\frac{4}{3}} y^{\frac{2}{3}} + 3\mu^{\frac{2}{3}} y^{\frac{4}{3}} - y^2,$$

and consequently

$$\beta^2 - \gamma^2 = (1 - \mu^2) \{1 - 3\mu^{\frac{2}{3}} y^{\frac{4}{3}} + (1 + \mu^2) y^2\}.$$

Hence dividing out by the factor $(1-\mu^2)^2$, the equation becomes

$$(1-\mu^2)^2 x^4 - 2\left(1+\mu^2 - 6\mu^{\frac{4}{3}}y^{\frac{2}{3}} + 3\mu^{\frac{2}{3}}(1+\mu^2)\,y^{\frac{4}{3}} - (1+\mu^4)\,y^2\right)2x^2 + \left(1 - 3\mu^{\frac{2}{3}}y^{\frac{4}{3}} + (1+\mu^2)\,y^2\right)^2 = 0\,;$$

or reducing and arranging,

$$\begin{aligned}(1-\mu^2)^2 x^4 - 2(1+\mu^2)\,x^2 + 2(1+\mu^4)\,x^2y^2 + 1 + 2(1+\mu^2)\,y^2 + (1+\mu^2)^2 y^4 \\ + (12\mu^{\frac{4}{3}}x^2 + 9\mu^{\frac{4}{3}}y^2)\,y^{\frac{2}{3}} - \left(6\mu^{\frac{2}{3}}(1+\mu^2)\,x^2 + 6\mu^{\frac{2}{3}} + 6\mu^{\frac{2}{3}}(1+\mu^2)\,y^2\right)y^{\frac{4}{3}} = 0,\end{aligned}$$

which is of the form

$$A + 3\mu^{\frac{4}{3}}By^{\frac{2}{3}} - 6\mu^{\frac{2}{3}}Cy^{\frac{4}{3}} = 0\,;$$

and the rationalized equation is

$$A^3 + 27\mu^4B^3y^2 - 216\mu^2C^3y^4 + 54\mu^2ABCy^2 = 0,$$

where the values of A, B, C may be written

$$\begin{aligned}A &= (x^2+y^2)\,\{(1-\mu^2)^2 x^2 + (1+\mu^2)^2 y^2\} - 2(1+\mu^2)(x^2-y^2) + 1,\\ B &= 4x^2 + 3y^2,\\ C &= (1+\mu^2)(x^2+y^2) + 1\,;\end{aligned}$$

the caustic is therefore a curve of the 12th order.

To find where the axis of x meets the curve, we have

$$y = 0,\ A_0^{\ 3} = 0,$$

where

$$\begin{aligned}A_0 &= (1-\mu^2)^2 x^4 - 2(1+\mu^2)\,x^2 + 1\\ &= \{(1-\mu)^2 x^2 - 1\}\,\{(1+\mu)^2 x^2 - 1\},\end{aligned}$$

i.e.

$$\begin{cases} y = 0, \\ x = \pm\dfrac{1}{1-\mu},\ x = \pm\dfrac{1}{1+\mu}, \end{cases}$$

or there are in all four points, each of them a point of triple intersection.

To find where the line ∞ meets the curve, we have

$$\infty,\ A'^3 = 0,$$

where

$$A' = (x^2+y^2)\,\{(1-\mu^2)^2 x^2 + (1+\mu^2)^2 y^2\},$$

i.e.

$$\begin{cases} \infty, \\ x = \pm\, iy,\ x = \pm\dfrac{1+\mu^2}{1-\mu^2}\,iy, \end{cases}$$

or the curve meets the line ∞ in four points, each of them a point of triple intersection: two of these points are the circular points at ∞.

To find where the circle $x^2+y^2=1$ meets the curve, this gives $x^2=1-y^2$, and thence

$$\begin{aligned}A &= \mu^2(\mu^2-4)+4(1+2\mu^2)y^2,\\ B &= 4-y^2,\\ C &= \mu^2+2,\end{aligned}$$

and the equation becomes

$$\{\mu^2(\mu^2-4)+4(1+2\mu^2)y^2\}^3+27\mu^4(4-y^2)^3y^2-216(\mu^2+2)^3y^4$$
$$+54\mu^2(\mu^2+2)y^2(4-y^2)\{\mu^2(\mu^2-4)+4(1+2\mu^2)y^2\}=0,$$

which is only of the eighth order; it follows that each of the circular points at ∞ (which have been already shown to be points upon the curve) are quadruple points of intersection of the curve and circle. The equation of the eighth order reduces itself to

$$(y^2-\mu^2)^3\{27\mu^4y^2+(\mu^2-4)^3\}=0;$$

the values of x corresponding to the roots $y=\pm\mu$ are obtained without difficulty, and those corresponding to the other roots are at once found by means of the identical equation

$$(\mu^2-4)^3+27\mu^4+(1-\mu^2)(\mu^2+8)^2=0;$$

we thus obtain for the coordinates of the points of intersection of the curve with the circle $x^2+y^2=1$, the values

$$\begin{cases}\infty,\\ x=\pm iy,\end{cases}\quad \begin{cases}x=\pm\sqrt{1-\mu^2}\\ y=\pm\mu,\end{cases}\quad \begin{cases}x=\pm\dfrac{(\mu^2+8)\sqrt{1-\mu^2}}{3\sqrt{3}\mu^2}\,i,\\[2ex] y=\pm\dfrac{(\mu^2-4)^{\frac{3}{2}}}{3\sqrt{3}\mu^2}\,i,\end{cases}$$

each of the points of the first system being a quadruple point of intersection, each of the points of the second system a triple point of intersection, and each of the points of the third system a single point of intersection.

Next, to find where the circle $x^2+y^2=\dfrac{1}{\mu^2}$ meets the curve; writing $x^2=\dfrac{1}{\mu^2}-y^2$, we obtain for y an equation of the eighth order, which after all reductions is

$$\left(y^2-\frac{1}{\mu^4}\right)^3\{27\mu^4y^2+(1-4\mu^2)^3\}=0,$$

and we have for the coordinates of the points of intersection,

$$\begin{cases}\infty, \\ x = \pm iy,\end{cases} \qquad \begin{cases} x = \pm \dfrac{1}{\mu}\sqrt{1-\dfrac{1}{\mu^2}}, \\ y = \pm \dfrac{1}{\mu^2}, \end{cases} \qquad \begin{cases} x = \pm \dfrac{(1+8\mu^2)\sqrt{1-\dfrac{1}{\mu^2}}}{3\sqrt{3}\mu}\, i, \\ y = \pm \dfrac{(1-4\mu^2)^{\frac{3}{2}}}{3\sqrt{3}\mu}\, i, \end{cases}$$

each of the points of the first system being a quadruple point of intersection, each of the points of the second system a triple point of intersection, and each of the points of the third system a single point of intersection.

The points of intersection with the axes of x, and the points of triple intersection with the circles $x^2+y^2=1$ and $x^2+y^2=\dfrac{1}{\mu^2}$, are all of them cuspidal points; the two circular points at ∞ are, I think, triple points, and the other two points of intersection with the line ∞, cuspidal points, but I have not verified this: assuming that it is so, there will be a reduction 54 accounted for in the class of the curve, but the curve is, in fact, as will be shown in the sequel, of the class 6; there is consequently a reduction 72 to be accounted for by other singularities of the curve.

XIV.

It is obvious from the preceding formulæ that the caustic stands to the circle, radius $\dfrac{1}{\mu}$, in a relation similar to that in which it stands to the circle, radius 1, i.e. to the refracting circle. In fact, the very same caustic would have been obtained if the circle radius $\dfrac{1}{\mu}$ had been taken for the refracting circle, the index of refraction being $\dfrac{1}{\mu}$ instead of μ. This may be shown very simply by means of the irrational form of the equation as follows.

The equation of the caustic by refraction of the circle, radius 1, index of refraction μ, is as we have seen

$$(1-\mu^2)\,x = (1-\mu^{\frac{4}{3}}y^{\frac{2}{3}})^{\frac{3}{2}} + \mu\,(1-\mu^{-\frac{2}{3}}y^{\frac{2}{3}})^{\frac{3}{2}};$$

hence the equation of the caustic by refraction of the circle radius c', index of refraction μ', is

$$(1-\mu'^2)\,\frac{x}{c'} = \left\{1-\mu'^{\frac{4}{3}}\left(\frac{y}{c'}\right)^{\frac{2}{3}}\right\}^{\frac{3}{2}} + \mu'\left\{1-\mu'^{-\frac{2}{3}}\left(\frac{y}{c'}\right)^{\frac{2}{3}}\right\}^{\frac{3}{2}},$$

or, what is the same thing,

$$(1-\mu'^2)\frac{x}{c'\mu'}=\left\{1-\mu'^{-\frac{2}{3}}c'^{-\frac{2}{3}}y^{\frac{2}{3}}\right\}^{\frac{3}{2}}+\frac{1}{\mu'}\left\{1-\mu'^{\frac{4}{3}}c'^{-\frac{2}{3}}y^{\frac{2}{3}}\right\}^{\frac{3}{2}},$$

which becomes identical with the equation of the first-mentioned caustic if $\mu'=c'=\dfrac{1}{\mu}$. Hence taking c instead of 1 as the radius of the first circle, we find,

THEOREM. The caustic by refraction for parallel rays of a circle, radius c, index of refraction μ, is the same curve as the caustic by refraction for parallel rays of a concentric circle, radius $\dfrac{c}{\mu}$, index of refraction $\dfrac{1}{\mu}$.

XV.

We may consequently in tracing the caustic confine our attention to the case in which the index of refraction is greater than unity. The circle, radius $\dfrac{c}{\mu}$, will in this case be within the refracting circle, and it is easy to see that if from the extremity of the diameter of the refracting circle perpendicular to the direction of the incident rays, tangents are drawn to the circle, radius $\dfrac{c}{\mu}$, the points of contact are the points of triple intersection of the caustic with the last-mentioned circle, and these points of intersection being, as already observed, cusps, the tangents in question are the tangents to the caustic at these cusps. The points of intersection with the axis of x are also cusps of the caustic, the tangents at these cusps coinciding with the axis of x: two of the last-mentioned cusps, viz. those whose distances from the centre are $\pm\dfrac{1}{\mu+1}$, lie within the circle, radius $\dfrac{c}{\mu}$, the other two of the same four cusps, viz. those whose distances from the centre are $\pm\dfrac{1}{\mu-1}$, lie without the circle, radius $\dfrac{c}{\mu}$; the last-mentioned two cusps lie without the refracting circle, when $\mu<2$, upon this circle, when $\mu=2$, and within it and therefore between the two circles, when $\mu>2$. The caustic is therefore of the forms in the annexed figures 3, 4, 5, in each

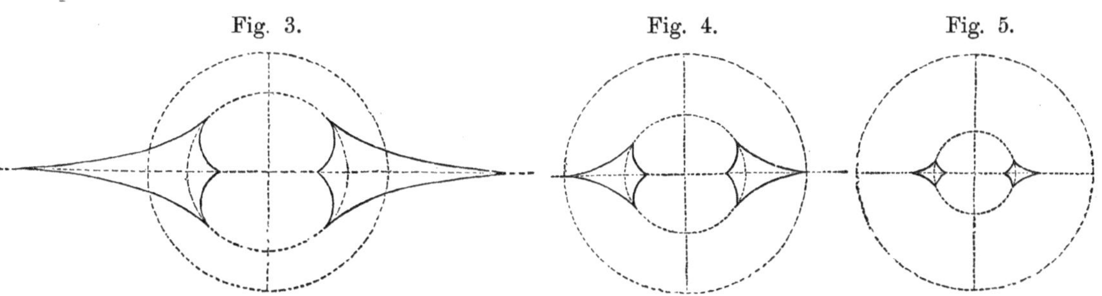
Fig. 3. Fig. 4. Fig. 5.

of which the outer circle is the refracting circle, and μ is > 1, but the three figures correspond respectively to the cases $\mu < 2$, $\mu = 2$ and $\mu > 2$. The same three figures will represent the different forms of the caustic when the inner circle is the refracting circle and μ is < 1, the three figures then respectively corresponding to the cases $\mu > \frac{1}{2}$, $\mu = \frac{1}{2}$, and $\mu < \frac{1}{2}$.

XVI.

To find the tangential equation, I retain k instead of its value $\frac{1}{\mu}$; the equation of the refracted ray then is

$$x(k\cos\phi - \sqrt{1-k^2\sin^2\phi}) + y(k\sin\phi + \cot\phi\sqrt{1-k^2\sin^2\phi}) - k = 0,$$

and representing this by

$$Xx + Yy - k = 0,$$

we have

$$X = k\cos\phi - \sqrt{1-k^2\sin^2\phi},$$

$$Y = k\sin\phi + \cot\phi\sqrt{1-k^2\sin^2\phi},$$

equations which give

$$X\cos\phi + Y\sin\phi = k,$$

$$X^2 + Y^2 = \frac{1}{\sin^2\phi},$$

and consequently

$$\sin\phi = \frac{1}{\sqrt{X^2+Y^2}}$$

$$\cos\phi = \frac{\sqrt{X^2+Y^2-1}}{\sqrt{X^2+Y^2}},$$

and we have

$$X\sqrt{X^2+Y^2-1} + Y - k\sqrt{X^2+Y^2} = 0,$$

which gives

$$(X^2+Y^2)(X^2-1-k^2) = -2kY\sqrt{X^2+Y^2};$$

or, dividing out by the factor $\sqrt{X^2+Y^2}$, the equation becomes

$$\sqrt{X^2+Y^2}(X^2-1-k^2) = -2kY,$$

from which

$$(X^2+Y^2)(X^2-1-k^2)^2 - 4k^2Y^2 = 0;$$

or reducing and arranging, we obtain

$$X^2(X^2-1-k^2)^2 + Y^2(X+1+k)(X+1-k)(X-1+k)(X-1-k) = 0$$

for the tangential equation of the caustic by refraction of a circle for parallel rays. The caustic is therefore of the class 6.

XVII.

Suppose next that rays proceeding from a point are reflected at a circle.

A very elegant solution of the problem is given by Lagrange in the *Mém. de Turin*; the investigation, as given by Mr P. Smith in a note in the *Cambridge and Dublin Mathematical Journal*, t. ii. [1847] p. 237, is as follows:

Let B be the radiant point, RBP an incident ray, and PS a reflected ray; CA a fixed radius; $ACP=\alpha$, $ACB=\epsilon$, reciprocal of $CB=c$, reciprocal of $CP=a$. The equations of the incident and reflected ray, where $u=\frac{1}{r}$, may be written

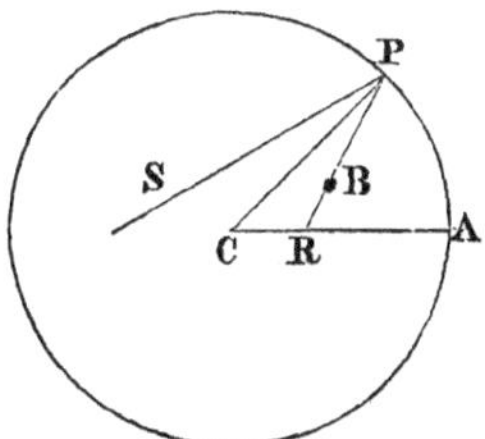

$$u=A\sin\theta+B\cos\theta;\ \text{incident ray},$$

$$u=A\sin(2\alpha-\theta)+B\cos(2\alpha-\theta);\ \text{reflected ray},$$

the conditions for determining A and B being

$$a=A\sin\alpha+B\cos\alpha,$$

$$c=A\sin\epsilon+B\cos\epsilon,$$

whence

$$A=\frac{a\cos\epsilon-c\cos\alpha}{\sin(\alpha-\epsilon)},\quad B=\frac{c\sin\alpha-a\sin\epsilon}{\sin(\alpha-\epsilon)}.$$

Substituting these values, the equation of the reflected ray becomes

$$a\sin(2\alpha-\theta-\epsilon)=u\sin(\alpha-\epsilon)+c\sin(\alpha-\theta),$$

from which and its differential with respect to the arbitrary parameter α, the equation of the caustic, or envelope of the reflected rays, will be found by eliminating α.

In this, α being the only quantity treated as variable in the differentiation, let

$$2\alpha-\theta-\epsilon=2\phi,$$

therefore

$$\alpha=\phi+\tfrac{1}{2}(\theta+\epsilon),$$

and the equation becomes

$$a \sin 2\phi = u \sin \{\phi + \tfrac{1}{2}(\theta - \epsilon)\} + c \sin \{\phi - \tfrac{1}{2}(\theta - \epsilon)\}.$$

Make

$$P = \frac{(u + c) \cos \frac{1}{2}(\theta - \epsilon)}{2a},$$

$$Q = \frac{(u - c) \sin \frac{1}{2}(\theta - \epsilon)}{2a},$$

also

$$x = \frac{1}{\cos \phi}, \quad y = \frac{1}{\sin \phi},$$

then the equation becomes

$$Px + Qy = 1,$$

with the condition

$$x^{-2} + y^{-2} = 1.$$

Hence

$$P = \lambda x^{-3}, \quad Q = \lambda y^{-3};$$

multiplying by x and y, and adding, we find $\lambda = 1$; therefore

$$x^{-2} = P^{\frac{2}{3}}, \quad y^{-2} = Q^{\frac{2}{3}}.$$

Hence

$$P^{\frac{2}{3}} + Q^{\frac{2}{3}} = 1;$$

or restoring the values of P and Q,

$$\{(u + c) \cos \tfrac{1}{2}(\theta - \epsilon)\}^{\frac{2}{3}} + \{(u - c) \sin \tfrac{1}{2}(\theta - \epsilon)\}^{\frac{2}{3}} = 1,$$

the equation of the caustic.

XVIII.

But the equation of the caustic for rays proceeding from a point and reflected at a circle may be obtained by a different method, as follows:

Take the centre of the circle for origin; let c be the radius of the circle, a, b the coordinates of the radiant point, α, β the coordinates of the point of incidence, x, y the coordinates of a point in the reflected ray. Then we have from the equation of the circle $\alpha^2 + \beta^2 = c^2$, and the equation of the reflected ray is by the general formula,

$$(b\alpha - a\beta)(\alpha x + \beta y - c^2) + (y\alpha - x\beta)(a\alpha + b\beta - c^2) = 0;$$

or arranging the terms in a different order,

$$(bx+ay)(\alpha^2-\beta^2)+2(by-ax)\alpha\beta-c^2(b+y)\alpha+c^2(a+x)\beta=0;$$

and writing herein $\alpha=c\cos\theta$, $\beta=c\sin\theta$, the equation becomes

$$(bx+ay)\cos 2\theta+(by-ax)\sin 2\theta-(b+y)c\cos\theta+(a+x)c\sin\theta=0,$$

where θ is a variable parameter.

Now in general to find the envelope of

$$A\cos 2\theta+B\sin 2\theta+C\cos\theta+D\sin\theta+E=0,$$

we may put $e^{i\theta}=z$, which gives the equation

$$(A-Bi)z^4+(C-Di)z^3+2Ez^2+(C+Di)z+(A+Bi)=0,$$

and equate the discriminant to zero: this gives

$$(4I)^3-27(-8J)^2=0,$$

where

$$4I=4(A^2+B^2)-(C^2+D^2)+\tfrac{4}{3}E^2,$$
$$-8J=A(C^2-D^2)+2BCD-\{8(A^2+B^2)+(C^2+D^2)\}\tfrac{1}{3}E+\tfrac{8}{27}E^3,$$

and consequently

$$\{4(A^2+B^2)-(C^2+D^2)+\tfrac{4}{3}E^2\}^3$$
$$-27\{A(C^2-D^2)+2BCD-\left(8(A^2+B^2)+(C^2+D^2)\right)\tfrac{1}{3}E+\tfrac{8}{27}E^3\}^2=0;$$

and substituting for A, B, C, D, E their values, we find

$$\{4(a^2+b^2)(x^2+y^2)-c^2\left((x+a)^2+(y+b)^2\right)\}^3-27(bx-ay)^2(x^2+y^2-a^2-b^2)^2=0,$$

for the equation of the caustic in the case of rays proceeding from a point and reflected at a circle: the equation was first obtained by St Laurent.

It will be convenient to consider the axis of x as passing through the radiant point; this gives $b=0$; and if we assume also $c=1$, the equation of the caustic becomes

$$\{(4a^2-1)(x^2+y^2)-2ax-a^2\}^3-27a^2y^2(x^2+y^2-a^2)^2=0.$$

XIX.

Reverting to the equation of the reflected ray, and putting, as before, $c=1$, $b=0$, this becomes

$$(-2a\cos\theta+1)x+\frac{a\cos 2\theta-\cos\theta}{\sin\theta}y+a=0;$$

differentiating with respect to θ, we have

$$(-2a\sin\theta)x+\frac{-a\cos\theta(1+2\sin^2\theta)+1}{\sin^2\theta}y=0;$$

and from these equations

$$x = \frac{a^2 \cos\theta\,(1 + 2\sin^2\theta) - a}{1 - 3a\cos 2\theta + 2a^2},$$

$$y = \frac{2a^2 \sin^3\theta}{1 - 3a\cos 2\theta + 2a^2},$$

which give the coordinates of a point of the caustic in terms of the angle θ which determines the position of the point of incidence. The values in question satisfy, as they should do, the equation

$$\{(4a^2 - 1)(x^2 + y^2) - 2ax - a^2\}^3 - 27a^2y^2(x^2 + y^2 - a^2)^2 = 0;$$

we have, in fact,

$$x^2 + y^2 \qquad - a^2 = \frac{4a^3(\cos\theta - a)^3}{(1 - 3a\cos 2\theta + 2a^2)^2},$$

$$(4a^2 - 1)(x^2 + y^2) - 2ax - a^2 = \frac{12a^4(\cos\theta - a)^2}{(1 - 3a\cos 2\theta + 2a^2)^2},$$

from which it is easy to derive the equation in question.

XX.

If we represent the equation of the reflected ray by

$$Xx + Yy + a = 0,$$

then we have

$$X = -2a\cos\theta + 1,$$

$$Y = \frac{a\cos 2\theta - \cos\theta}{\sin\theta},$$

and thence

$$(X - 1)^2 - 4a^2 = -4a^2\sin^2\theta,$$

$$X^2 + Y^2 \qquad = \frac{1}{\sin^2\theta}(1 - 2a\cos\theta + a^2),$$

$$X + a^2 \qquad = 1 - 2a\cos\theta + a^2,$$

and consequently

$$(X^2 + Y^2)\{(X - 1)^2 - 4a^2\} + 4a^2X + 4a^4 = 0,$$

or, what is the same thing,

$$\{X(X - 1) - 2a^2\}^2 + Y^2\{(X - 1)^2 - 4a^2\} = 0,$$

which may be considered as the tangential equation of the caustic by reflexion of a circle; or if we consider X, Y as the coordinates of a point, then the equation may be considered as that of the polar of the caustic. The polar is therefore a curve of the fourth order, having two double points defined by the equations $X(X - 1) - 2a^2 = 0$, $Y = 0$, and a third double point at infinity on the axis of Y, i.e. three double points in all; the number of cusps is therefore 0, and there are consequently 4 double tangents and 6 inflections, and the curve is of the class 6. And as Y is given as an explicit function of X, there is of course no difficulty in tracing the curve. We thus see that the caustic by reflexion of a circle is a curve of the order 6, and has 4 double points and 6 cusps (the circular points at infinity are each of them a cusp, so that the number of cusps at a finite distance is 4): this coincides with the conclusions which will be presently obtained by considering the equation of the caustic.

XXI.

The equation of the caustic by reflexion of a circle is

$$\{(4a^2-1)(x^2+y^2)-2ax-a^2\}^3-27a^2y^2(x^2+y^2-a^2)^2=0.$$

Suppose first that $y=0$, we have

$$\{(4a^2-1)x^2-2ax-a^2\}^3=0,$$

i.e.

$$x=\frac{-a}{2a+1},\quad x=\frac{a}{2a-1},$$

or the curve meets the axis of x in two points, each of which is a triple point of intersection.

Write next $x^2+y^2=a^2$, this gives

$$\{(4a^2-1)a^2-2ax-a^2\}^3=0,$$

and consequently

$$x=-a(1-2a^2),$$
$$y=\pm\,2a^2\sqrt{1-a^2},$$

or the curve meets the circle $x^2+y^2-a^2=0$ in two points, each of which is a triple point of intersection.

To find the nature of the infinite branches, we may write, retaining only the terms of the degrees six and five,

$$(4a^2-1)^3(x^2+y^2)^3-6(4a^2-1)^2a(x^2+y^2)^2x-27\,a^2y^2(x^2+y^2)^2=0\,;$$

and rejecting the factor $(x^2+y^2)^2$, this gives

$$(4a^2-1)^3x^2+\{(4a^2-1)^3-27a^2\}y^2-6(4a^2-1)^2ax=0\,;$$

or reducing,

$$(4a^2-1)^3x^2-(1-a^2)(8a^2+1)^2y^2-6(4a^2-1)^2ax=0\,;$$

and it follows that there are two asymptotes, the equations of which are

$$y=\frac{(4a^2-1)^{\frac{3}{2}}}{\sqrt{1-a^2}\,(8a^2+1)}\left\{x-\frac{3a}{4a^2-1}\right\}.$$

Represent for a moment the equation of one of the asymptotes by $y = A(x - \alpha)$, then the perpendicular from the origin or centre of the reflecting circle is $A\alpha \div \sqrt{1 + A^2}$, and

$$A\alpha = \frac{3a\sqrt{4a^2 - 1}}{\sqrt{1 - a^2}\,(1 + 8a^2)},$$

$$1 + A^2 = \frac{(1 - a^2)(1 + 8a^2)^2 + (4a^2 - 1)^3}{(1 - a^2)(1 + 8a^2)^2} = \frac{27a^2}{(1 - a^2)(1 + 8a^2)^2},$$

$$\sqrt{1 + A^2} = \frac{3\sqrt{3}\,a}{\sqrt{1 - a^2}\,(1 + 8a^2)},$$

and the perpendicular is $\frac{1}{\sqrt{3}}\sqrt{4a^2 - 1}$, which is less than a if only $a^2 < 1$, i.e. in every case in which the asymptote is real.

The tangents parallel and perpendicular to the axis of x are most readily obtained from the equation of the reflected ray, viz.

$$(-2a\cos\theta + 1)\,x + \frac{a\cos 2\theta - \cos\theta}{\sin\theta}\,y + a = 0\,;$$

the coefficient of x (if the equation is first multiplied by $\sin\theta$) vanishes if $\sin\theta = 0$, which gives the axis of x, or if $\cos\theta = \frac{1}{2a}$, which gives $y = \pm\frac{\sqrt{4a^2 - 1}}{2a}$, for the tangents parallel to the axis of x.

The coefficient of y vanishes if $a\cos 2\theta - \cos\theta = 0$; this gives

$$\cos\theta = \frac{1 \pm \sqrt{8a^2 + 1}}{4a}, \qquad \sin\theta = \frac{1}{8a^2}\left(4a^2 - 1 \mp \sqrt{8a^2 + 1}\right),$$

and the tangents perpendicular to the axis of x are thus given by

$$x = \frac{-2a}{1 \mp \sqrt{8a^2 + 1}}\,;$$

these tangents are in fact double tangents of the caustic. In order that the point of contact may be real, it is necessary that $\sin\theta$, $\cos\theta$ should be real; this will be the case for both values of the ambiguous sign if $a >$ or $= 1$, but only for the upper value if $a < 1$.

It has just been shown that for the tangents parallel to the axis of x, we have

$$y = \pm\frac{\sqrt{4a^2 - 1}}{2a},$$

the values of y being real for $a > \frac{1}{2}$: it may be noticed that the value $y = \frac{\sqrt{4a^2 - 1}}{2a}$

is greater, equal, or less than or to $y=2a^2\sqrt{1-a^2}$, according as $a >=$ or $< \frac{1}{\sqrt{2}}$; this depends on the identity $(4a^2-1)-16a^6(1-a^2)=(2a^2-1)^3(2a^2+1)$.

To find the points of intersection with the reflecting circle, $x^2+y^2-1=0$, we have

$$(3a^2-1-2ax)^3-27a^2(1-x^2)(1-a^2)^2=0;$$

or, reducing,

$$8a^3x^3+(-27a^4+18a^2-15)a^2x^2+(54a^4-36a^2+6)ax+(-27a^4+18a^2+1)=0,$$

i.e.

$$(ax-1)^2(8ax-27a^4+18a^2+1)=0.$$

The factor $(ax-1)^2$ equated to zero shows that the caustic touches the circle in the points $x=\frac{1}{a}$, $y=\pm\sqrt{1-\frac{1}{a^2}}$, i.e. in the points in which the circle is met by the polar of the radiant point, and which are real or imaginary according as $a>$ or <1. The other factor gives

$$x=\frac{27a^4-18a^2-1}{8a}.$$

Putting this value equal to ± 1, the resulting equation is $(a\mp 1)(27a^2+9a+1)=0$, and it follows that x will be in absolute magnitude greater or less than 1, i.e. the points in question will be imaginary or real, according as $a>1$ or $a<1$.

It is easy to see that the curve passes through the circular points at infinity, and that these points are cusps on the curve; the two points of intersection with the axis of x are cusps (the axis of x being the tangent), and the two points of intersection with the circle $x^2+y^2-a^2=0$ are also cusps, the tangent at each of the cusps coinciding with the tangent of the circle; there are consequently in all 6 cusps.

XXII.

To investigate the position of the double points we may proceed as follows: write for shortness $P=(4a^2-1)(x^2+y^2)-2ax-a^2$, $Q=ayS$, $S=x+y^2-a^2$; the equation of the caustic is

$$P^3-27Q^2=0;$$

hence, at a double point,

$$P^2\frac{dP}{dx}-18Q\frac{dQ}{dx}=0,$$

$$P^2\frac{dP}{dy}-18Q\frac{dQ}{dy}=0;$$

one of which equations may be replaced by

$$\frac{dP}{dx}\frac{dQ}{dy}-\frac{dP}{dy}\frac{dQ}{dx}=0.$$

Now

$$\frac{dP}{dx} = 2\{(4a^2-1)x - a\}, \quad \frac{dP}{dy} = 2(4a^2-1)y,$$

$$\frac{dQ}{dx} = 2axy, \quad \frac{dQ}{dy} = a(x^2+3y^2-a^2) = a(S+2y^2);$$

substituting these values in the last preceding equation, we find

$$\frac{(4a^2-1)x - a}{(4a^2-1)y} = \frac{2xy}{S+2y^2},$$

or, reducing,

$$(4a^2-1)x - a = \frac{2ay^2}{S};$$

and using this to simplify the equation

$$P^2\frac{dP}{dx} - 18Q\frac{dQ}{dx} = 0,$$

we have

$$P^2\frac{4ay^2}{S} - 18ayS \,.\, 2axy = 0,$$

i.e.

$$\frac{P^2}{S} - 9axS = 0,$$

and therefore

$$9x = \frac{P^2}{aS^2}.$$

Multiplying by P and writing for P^3 its value $27a^2y^2S^2$, we have

$$Px = 3ay^2,$$

and thence

$$P = \frac{3ay^2}{x}, \quad P^3 = \frac{27a^3y^6}{x^3} = 27a^2y^2S^2,$$

whence

$$\frac{ay^4}{x^3} = S^2, \quad \frac{a^{\frac{1}{2}}y^2}{x^{\frac{3}{2}}} = S, \quad \text{or} \quad \frac{y^2}{S} = \frac{x^{\frac{3}{2}}}{a^{\frac{1}{2}}};$$

and substituting in the equation

$$x = \frac{a}{4a^2-1}\left(1 + \frac{2y^2}{S}\right),$$

we find

$$x = \frac{a}{4a^2-1}\left(1 + \frac{2x}{a^{\frac{1}{2}}}\right),$$

or, rationalising,

$$4ax^3 - \{(4a^2-1)x - a\}^2 = 0,$$

or, what is the same thing,

$$(4ax-1)\left(x-a\left(a+i\sqrt{1-a^2}\right)^2\right)\left(x-a\left(a-i\sqrt{1-a^2}\right)^2\right)=0.$$

The factor $4ax-1$ equated to zero gives $x=\frac{1}{4a}$ from which y may be found, but the resulting point is not a double point; the other factors give each of them double points, and if we write

$$x=a\left(a+i\sqrt{1-a^2}\right)^2,$$

we find

$$y=\frac{2a^2i\left(a+i\sqrt{1-a^2}\right)^{\frac{3}{2}}}{\left(3a-i\sqrt{1-a^2}\right)^{\frac{1}{2}}},$$

values which, in fact, belong to one of the four double points. It is easy to seė that the points in question are always imaginary.

It may be noticed, by way of verification, that the preceding values of x, y give

$$(4a^2-1)(x^2+y^2)-2ax-a^2=\frac{12a^4}{1+8a^2}\left(1-4a^2-4ai\sqrt{1-a^2}\right),$$

$$x^2+y^2-a^2 \qquad =\frac{-4a^3}{1+8a^2}\left(3a+i\sqrt{1-a^2}\right),$$

$$y^2 \qquad =\frac{4a^4}{1+8a^2}\left(-1+14a^2-16a^4+2a(3-8a^2)\,i\sqrt{1-a^2}\right);$$

and if the quantities within ( ) on the right-hand side are represented by A, B, C, then

$$\frac{A}{B}=-\left(a+i\sqrt{1-a^2}\right),$$

$$\frac{C}{B}=-\left(a+i\sqrt{1-a^2}\right)^3,$$

whence we have identically,

$$\left(\frac{A}{B}\right)^3=\frac{C}{B},\quad\text{or}\quad A^3=B^2C,$$

by means of which it appears that the values of x, y satisfy, as they should do, the equation of the caustic; and by forming the expressions for $(4a^2-1)\,x-a$ and $x^2+3y^2-a^2$, it might be shown, *à posteriori*, that the point in question was a double point.

XXIII.

The equation

$$\{(4a^2-1)(x^2+y^2)-2ax-a^2\}^3-27a^2y^2(x^2+y^2-a^2)^2=0$$

becomes when $a=1$ (i.e. when the radiant point is in the circumference),

$$\{3y^2+(x-1)(3x+1)\}^3-27y^2(y^2+x^2-1)^2=0;$$

it is easy to see that this divides by $(x-1)^2$; and throwing out this factor, we have for the caustic the equation of the fourth order,

$$27y^4+18y^2(3x^2-1)+(x-1)(3x+1)^3=0.$$

XXIV.

The equation

$$\{(4a^2-1)(x^2+y^2)-2ax-a^2\}^3-27a^2y^2(x^2+y^2-a^2)^2=0$$

becomes when $a=\infty$ (i.e. in the case of parallel rays),

$$(4x^2+4y^2-1)^3-27y^2=0,$$

which may also be written

$$64x^6+48x^4(4y^2-1)+12x^2(4y^2-1)^2+(8y^2+1)^2(y^2-1)=0.$$

XXV.

It is now easy to trace the curve. Beginning with the case $a=\infty$, the curve lies wholly within the reflecting circle, which it touches at two points; the line joining the points of contact, being in fact the axis of y, divides the curve into two equal portions; the curve has in the present, as in every other case (except one limiting case), two cusps on the axis of x (see fig. 6). Next, if a be positive and >1, the general form of the curve is the same as before, only the line joining the points of contact with the reflecting circle divides the curve into unequal portions, that in the

Fig. 6. $a=\infty$. Fig. 7. $a>1$.

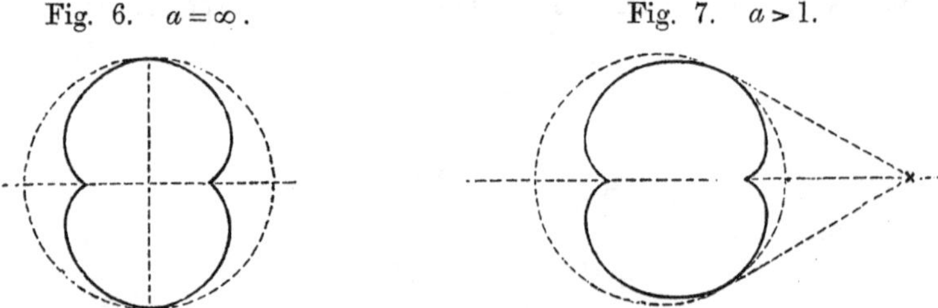

neighbourhood of the radiant point being the smaller of the two portions (see fig. 7). When $a=1$, the two points of contact with the reflecting circle unite together at the radiant point; the curve throws off, as it were, the two coincident lines $x=1$, and the order is reduced from 6 to 4. The curve has the form fig. 8, with only a single cusp on the axis of x. If a be further diminished, $a<1>\frac{1}{\sqrt{2}}$, the curve takes the form shown by fig. 9, with two infinite branches, one of them having simply a cusp on the axis of x, the other having a cusp on the axis of x, and a pair of cusps at its intersection with the circle through the radiant point; there are two asymptotes equally inclined to the axis of x. In the case $a=\frac{1}{\sqrt{2}}$, the form of the curve is nearly the same as before, only the cusps upon the circle through the radiant point lie on the axis of y (see fig. 10). The case $a<\frac{1}{\sqrt{2}}>\frac{1}{2}$ is shown, fig. 11. For $a=\frac{1}{2}$, the two asymptotes coincide with the axis of x; one of the branches of the curve has wholly disappeared, and the form of the other is modified by the coincidence of the asymptotes

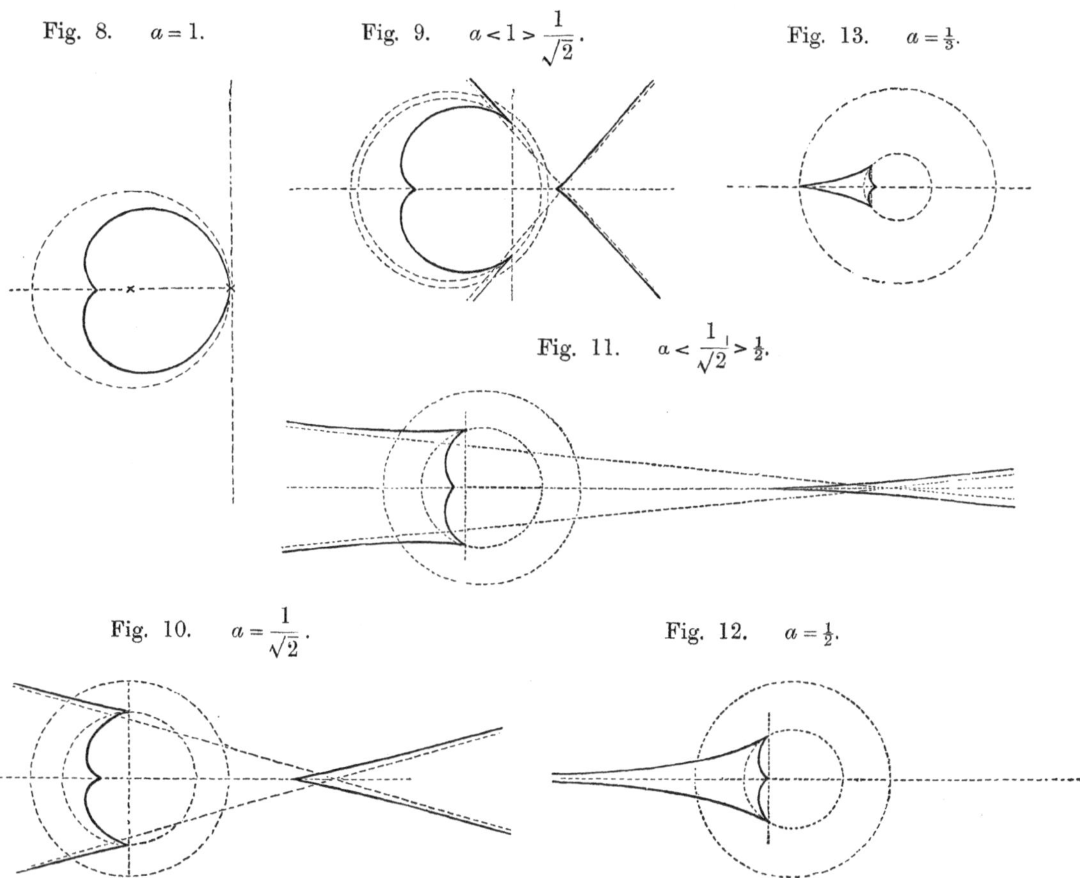

with the axis of x; it has in fact acquired a cusp at infinity on the axis of x (see fig. 12). When $a < \frac{1}{2}$, the curve consists of a single finite branch, with two cusps on the axis of x, and two cusps at the points of intersection with the circle through the radiant point; one of the last-mentioned cusps will be outside the reflecting circle as long as $a > \frac{1}{3}$; fig. 13 represents the case $a = \frac{1}{3}$, for which this cusp is upon the reflecting circle. For $a < \frac{1}{3}$, the curve lies wholly within the reflecting circle, one of the cusps upon the axis of x being always within, and the other always without the circle through the radiant point, and as a approaches O the curve becomes smaller and smaller, and ultimately disappears in a point. The case a negative is obviously included in the preceding one.

Several of the preceding results relating to the caustic by reflexion of a circle were obtained, and the curve is traced in a memoir by the Rev. Hamnet Holditch, *Quarterly Mathematical Journal*, t. I. [1857, pp. 93—111].

XXVI.

Suppose next that rays proceeding from a point are *refracted* at a circle. Take the centre of the circle as origin, let the radius be c, and take ξ, η as the coordinates of the radiant point, α, β the coordinates of the point of incidence, x, y the coordinates of a point in the refracted ray: then the general equation

$$-\overline{qG}^2\ \overline{\nabla QGN}^2 + \mu^2\ \overline{QG}^2\ \overline{\nabla qGN}^2 = 0$$

becomes, taking the centre of the circle as the point N on the normal, or writing $a = b,\ b = 0$,

$$-\{(x-\alpha)^2 + (y-\beta)^2\}\,(\beta\xi - \alpha\eta)^2 + \mu^2\{(\xi-\alpha)^2 + (\eta-\beta)^2\}\,(\beta x - \alpha y)^2 = 0\,;$$

or putting $\alpha^2 + \beta^2 = c^2$, and expanding,

$$\begin{aligned}
&\alpha^3\quad \{2\,(\eta^2 x - \mu^2 y^4 \xi)\}\\
&+\alpha^2\beta\ \{-4\,(\xi\eta x - \mu^2 xy\xi) + 2\,(\eta^2 y - \mu^2 y^2\eta)\}\\
&+\alpha\beta^2\ \{-4\,(\xi\eta y - \mu^2 xy\eta) + 2\,(\xi^2 x - \mu^2 x^2\xi)\}\\
&+\beta^3\quad \{2\,(\xi^2 y - \mu^2 x^2\eta)\}\\
&-\alpha^2\quad \{(x^2+y^2+c^2)\,\eta^2 - \mu^2\,(\xi^2+\eta^2+c^2)\,y^2\}\\
&+2\alpha\beta\,\{(x^2+y^2+c^2)\,\xi\eta - \mu^2\,(\xi^2+\eta^2+c^2)\,xy\}\\
&-\beta^2\quad \{(x^2+y^2+c^2)\,\xi^2 - \mu^2\,(\xi^2+\eta^2+c^2)\,x^2\}\\
&=0,
\end{aligned}$$

which may be represented by

$$A\alpha^3 + B\alpha^2\beta + C\alpha\beta^2 + D\beta^3 + F\alpha^2 + G\alpha\beta + H\beta^2 = 0.$$

Now $\alpha^2 + \beta^2 = c^2$, and we may write

$$\alpha = \tfrac{1}{2}c\left(z + \frac{1}{z}\right),\qquad \beta = -\tfrac{1}{2}ci\left(z - \frac{1}{z}\right).$$

The equation thus becomes

$$\begin{aligned}
&A\left(z+\frac{1}{z}\right)^3 - Bi\left(z+\frac{1}{z}\right)^2\left(z-\frac{1}{z}\right) - C\left(z+\frac{1}{z}\right)\left(z-\frac{1}{z}\right)^2 - Di\left(z-\frac{1}{z}\right)^3\\
&\qquad + \frac{2}{c}F\left(z+\frac{1}{z}\right)^2 - \frac{2}{c}Gi\left(z+\frac{1}{z}\right)\left(z-\frac{1}{z}\right) - \frac{2}{c}H\left(z-\frac{1}{z}\right)^2 = 0
\end{aligned}$$

or expanding,

$$\left.\begin{aligned}
&(A - Bi - C - Di) && z^3\\
&+\frac{2}{c}(F - Gi - H) && z^2\\
&+\ (3A - Bi + C + 3Di) && z\\
&+\frac{4}{c}(F + H) && \\
&+\ (3A + Bi + C - 3Di) && \frac{1}{z}\\
&+\frac{2}{c}(F + Gi - H) && \frac{1}{z^2}\\
&+\ (A + Bi - C + Di) && \frac{1}{z^3}
\end{aligned}\right\} = 0,$$

in which z may be considered as the variable parameter; hence the equation of the caustic may be obtained by equating to zero the discriminant of the above function of z; but the discriminant of a sextic function has not yet been calculated. The equation would be of the order 20, and it appears from the result previously obtained for parallel rays, that the equation must be of the order 12 at the least; it is, I think, probable that there is not any reduction of order in the general case. It is however practicable, as will presently be seen, to obtain the tangential equation of the caustic by refraction, and the curve is thus shown to be only of the class 6.

XXVII.

Suppose that rays proceeding from a point are refracted at a circle, and let it be required to find the equation of the secondary caustic: take the centre of the circle as origin, let c be the radius, ξ, η the coordinates of the radiant point, α, β the coordinates of a point upon the circle, μ the index of refraction; the secondary caustic will be the envelope of the circle,

$$\mu^2 \{(x-\alpha)^2 + (y-\beta)^2\} - \{(\xi-\alpha)^2 + (\eta-\beta)^2\} = 0,$$

where α, β are variable parameters connected by the equation $\alpha^2 + \beta^2 - c^2 = 0$; the equation of the circle may be written in the form

$$\mu^2 (x^2 + y^2 + c^2) - (\xi^2 + \eta^2 + c^2) - 2(\mu^2 x - \xi)\,\alpha - 2(\mu^2 y - \eta)\,\beta = 0.$$

But in general the envelope of $A\alpha + B\beta + C = 0$, where α, β are connected by the equation $\alpha^2 + \beta^2 - c^2 = 0$, is $c^2(A^2 + B^2) - C^2 = 0$, and hence in the present case the equation of the envelope is

$$\{\mu^2 (x^2 + y^2 + c^2) - (\xi^2 + \eta^2 + c^2)\}^2 = 4c^2 \{(\mu^2 x - \xi)^2 + (\mu^2 y - \eta)^2\},$$

which may also be written

$$\{\mu^2 (x^2 + y^2 - c^2) - (\xi^2 + \eta^2 - c^2)\}^2 = 4c^2\mu^2 \{(x-\xi)^2 + (y-\eta)^2\}.$$

If the axis of x be taken through the radiant point, then $\eta = 0$, and writing also $\xi = a$, the equation becomes

$$\{\mu^2 (x^2 + y^2 - c^2) - a^2 + c^2\}^2 = 4c^2\mu^2 \{(x-a)^2 + y^2\};$$

or taking the square root of each side,

$$\{\mu^2 (x^2 + y^2 - c^2) - a^2 + c^2\} = 2c\mu \sqrt{(x-a)^2 + y^2};$$

whence multiplying by $1 - \dfrac{1}{\mu^2}$ and adding on each side $c^2\left(\mu - \dfrac{1}{\mu}\right)^2 + (x-a)^2 + y^2$, we have

$$\mu^2 \left\{\left(x - \frac{a}{\mu^2}\right)^2 + y^2\right\} = \left\{\sqrt{(x-a)^2 + y^2} + c\left(\mu - \frac{1}{\mu}\right)\right\}^2,$$

or

$$\mu \sqrt{\left(x - \frac{a}{\mu^2}\right)^2 + y^2} = \sqrt{(x-a)^2 + y^2} + c\left(\mu - \frac{1}{\mu}\right),$$

which shows that the secondary caustic is the Oval of Descartes, or as it will be convenient to call it, the Cartesian.

It is proper to remark, that the Cartesian consists in general of two ovals, one of which is the orthogonal trajectory of the refracted rays, the other the orthogonal trajectory of the false refracted rays. In the case of reflexion, the secondary caustic is a Cartesian having a double point; this may be either a conjugate point, or a real double point arising from the union and intersection of the two ovals; the same secondary caustic may arise also from refraction, as will be presently shown.

XXVIII.

Reverting to the original form of the equation of the secondary caustic, multiplying by $\frac{1}{\mu^2}\left(1-\frac{c^2}{a^2}\right)$ and adding on each side $\frac{a^2}{\mu^2}\left(1-\frac{c^2}{a^2}\right)^2+\frac{c^2}{a^2}\{(x-a)^2+y^2\}$, the equation becomes

$$\left\{\left(x-\frac{c^2}{a}\right)^2+y^2\right\}=\left\{\frac{c}{a}\sqrt{(x-a)^2+y^2}+\frac{a}{\mu}\left(1-\frac{c^2}{a^2}\right)\right\}^2,$$

or extracting the square root,

$$\sqrt{\left(x-\frac{c^2}{a}\right)^2+y^2}=\frac{c}{a}\sqrt{(x-a)^2+y^2}+\frac{a}{\mu}\left(1-\frac{c^2}{a^2}\right).$$

Combining this with the former result, we see that the equation may be expressed indifferently in any one of the four forms,

$$\sqrt{\left(x-\frac{a}{\mu^2}\right)^2+y^2}=\frac{1}{\mu}\sqrt{(x-a)^2+y^2}+\frac{c}{\mu}\left(\mu-\frac{1}{\mu}\right),$$

$$\sqrt{\left(x-\frac{c^2}{a}\right)^2+y^2}=\frac{c}{a}\sqrt{(x-a)^2+y^2}+\frac{1}{\mu}\left(a-\frac{c^2}{a}\right),$$

$$\sqrt{\left(x-\frac{c^2}{a}\right)^2+y^2}=\frac{c\mu}{a}\sqrt{\left(x-\frac{a}{\mu^2}\right)^2+y^2}+\frac{a}{\mu}-\frac{c^2\mu}{a},$$

$$c\left(\mu-\frac{1}{\mu}\right)\sqrt{\left(x-\frac{c^2}{a}\right)^2+y^2}+\left(-a+\frac{c^2}{a}\right)\sqrt{\left(x-\frac{a}{\mu^2}\right)^2+y^2}+\left(\frac{a}{\mu}-\frac{c^2\mu}{a}\right)\sqrt{(x-a)^2+y^2}=0.$$

It follows, that if we write successively

$$a'=a,\quad c'=c,\quad \mu'=\mu \qquad (1)$$

$$a'=\frac{c^2}{a},\quad c'=\frac{c}{\mu},\quad \mu'=\frac{c}{a} \qquad (\alpha)$$

$$a'=\frac{a}{\mu^2},\quad c'=\frac{c}{\mu},\quad \mu'=\frac{1}{\mu} \qquad (\beta)$$

$$a'=a,\quad c'=\frac{a}{\mu},\quad \mu'=\frac{a}{c} \qquad (\gamma)$$

$$a'=\frac{c^2}{a},\quad c'=c,\quad \mu'=\frac{c\mu}{a} \qquad (\delta)$$

$$a'=\frac{a}{\mu^2},\quad c'=\frac{a}{\mu},\quad \mu'=\frac{a}{c\mu} \qquad (\epsilon),$$

or what is the same thing,

$$a = a', \quad c = c', \quad \mu = \mu' \qquad (1)$$

$$a = \frac{a'}{\mu'^2}, \quad c = \frac{a'}{\mu'}, \quad \mu = \frac{a'}{c'\mu'} \qquad (\alpha)$$

$$a = \frac{a'}{\mu'^2}, \quad c = \frac{c'}{\mu'}, \quad \mu = \frac{1}{\mu'} \qquad (\beta)$$

$$a = a', \quad c = \frac{a'}{\mu'}, \quad \mu = \frac{a'}{c'} \qquad (\gamma)$$

$$a = \frac{c'^2}{a'}, \quad c = c', \quad \mu = \frac{c'\mu'}{a'} \qquad (\delta)$$

$$a = \frac{c'^2}{a'}, \quad c = \frac{c'}{\mu'}, \quad \mu = \frac{c'}{a'} \qquad (\epsilon),$$

or what is again the same thing,

$$a' = a, \quad \frac{c'^2}{a'} = \frac{c^2}{a}, \quad \frac{a'}{\mu'^2} = \frac{a}{\mu^2} \qquad (1)$$

$$a' = \frac{c^2}{a}, \quad \frac{c'^2}{a'} = \frac{a}{\mu^2}, \quad \frac{\dot{a}'}{\mu'^2} = a \qquad (\alpha)$$

$$a' = \frac{a}{\mu^2}, \quad \frac{c'^2}{a'} = \frac{c^2}{a}, \quad \frac{a'}{\mu'^2} = a \qquad (\beta)$$

$$a' = a, \quad \frac{c'^2}{a'} = \frac{a}{\mu^2}, \quad \frac{a'}{\mu'^2} = \frac{c^2}{a} \qquad (\gamma)$$

$$a' = \frac{c^2}{a}, \quad \frac{c'^2}{a'} = a, \quad \frac{a'}{\mu'^2} = \frac{a}{\mu^2} \qquad (\delta)$$

$$a' = \frac{a}{\mu^2}, \quad \frac{c'^2}{a'} = a, \quad \frac{a'}{\mu'^2} = \frac{c^2}{a} \qquad (\epsilon),$$

we have in each case identically the same secondary caustic, and therefore also identically the same caustic; in other words, the same caustic is produced by six different systems of a radiant point and refracting circle. It is proper to remark that if we represent the six systems of equations by $(a', c', \mu') = (a, c, \mu)$, $(a', c', \mu') = \alpha(a, c, \mu)$, &c., then, α, β, γ, δ, ϵ will be functional symbols satisfying the conditions

$$1 = \alpha\beta = \beta\alpha = \gamma^2 = \delta^2 = \epsilon^2,$$

$$\alpha = \beta^2 = \delta\gamma = \epsilon\delta = \gamma\epsilon,$$

$$\beta = \alpha^2 = \gamma\delta = \delta\epsilon = \epsilon\gamma,$$

$$\gamma = \delta\alpha = \alpha\epsilon = \epsilon\beta = \beta\delta,$$

$$\delta = \epsilon\alpha = \alpha\gamma = \gamma\beta = \beta\epsilon,$$

$$\epsilon = \gamma\alpha = \alpha\delta = \delta\beta = \beta\gamma.$$

XXIX.

The preceding formulæ, which were first given by me in the *Philosophical Magazine*, December 1853, [124] include as particular cases a preceding theorem with respect to the caustic by refraction of parallel rays, and also two theorems of St Laurent, *Gergonne*, t. XVIII., [1827, pp. 1—19] viz. if we suppose first that $a=c$, i.e. that the radiant point is in the circumference of the refracting circle, then the system (α) shows that the same caustic would be obtained by writing c, $\frac{c}{\mu}$, 1 (or what is the same thing -1) in the place of c, c, μ, and we have

THEOREM. The caustic by refraction for a circle when the radiant point is in the circumference is also the caustic by reflexion for the same radiant point, and for a reflecting circle concentric with the refracting circle, but having its radius equal to the quotient of the radius of the refracting circle by the index of refraction.

Next, if we write $a=c\mu$, then the refracted rays all of them pass through a point which is a double point of the secondary caustic, the entire curve being in this case the orthogonal trajectory, not of the refracted rays, but of the false refracted rays; the formula (δ) shows that the same caustic is obtained by writing $\frac{c^2}{a}$, c, 1 (or what is the same thing -1) in the place of a, c, $\mu\left(=\frac{a}{c}\right)$, and we have

THEOREM. The caustic by refraction for a circle when the distance of the radiant point from the centre is to the radius of the circle in the ratio of the index of refraction to unity, is also the caustic by reflexion for the same circle considered as a reflecting circle, and for a radiant point the image of the former radiant point.

XXX.

The curve is most easily traced by means of the preceding construction; thus if we take the radiant point outside the refracting circle, and consider μ as varying from a small to a large value (positive or negative values of μ give the same curve), we see that when μ is small the curve consists of two ovals, one of them within and the other without the refracting circle (see fig. 14). As μ increases the exterior oval continually increases, but undergoes modifications in its form; the interior oval in the first instance diminishes until we arrive at a curve, in which the interior oval is reduced to a conjugate point (see fig. 15); then as μ continues to increase the interior oval reappears (see fig. 16), and at last connects itself with the exterior oval, so as to form a curve with a double point (see fig. 17); and as μ increases still further the curve again breaks up into an exterior and an interior oval (see fig. 18); and thenceforward as μ goes on increasing consists always of two ovals; the shape of the exterior oval is best perceived from the figures. An examination of the figures will also show how the same curves may originate from a different refracting circle and radiant point.

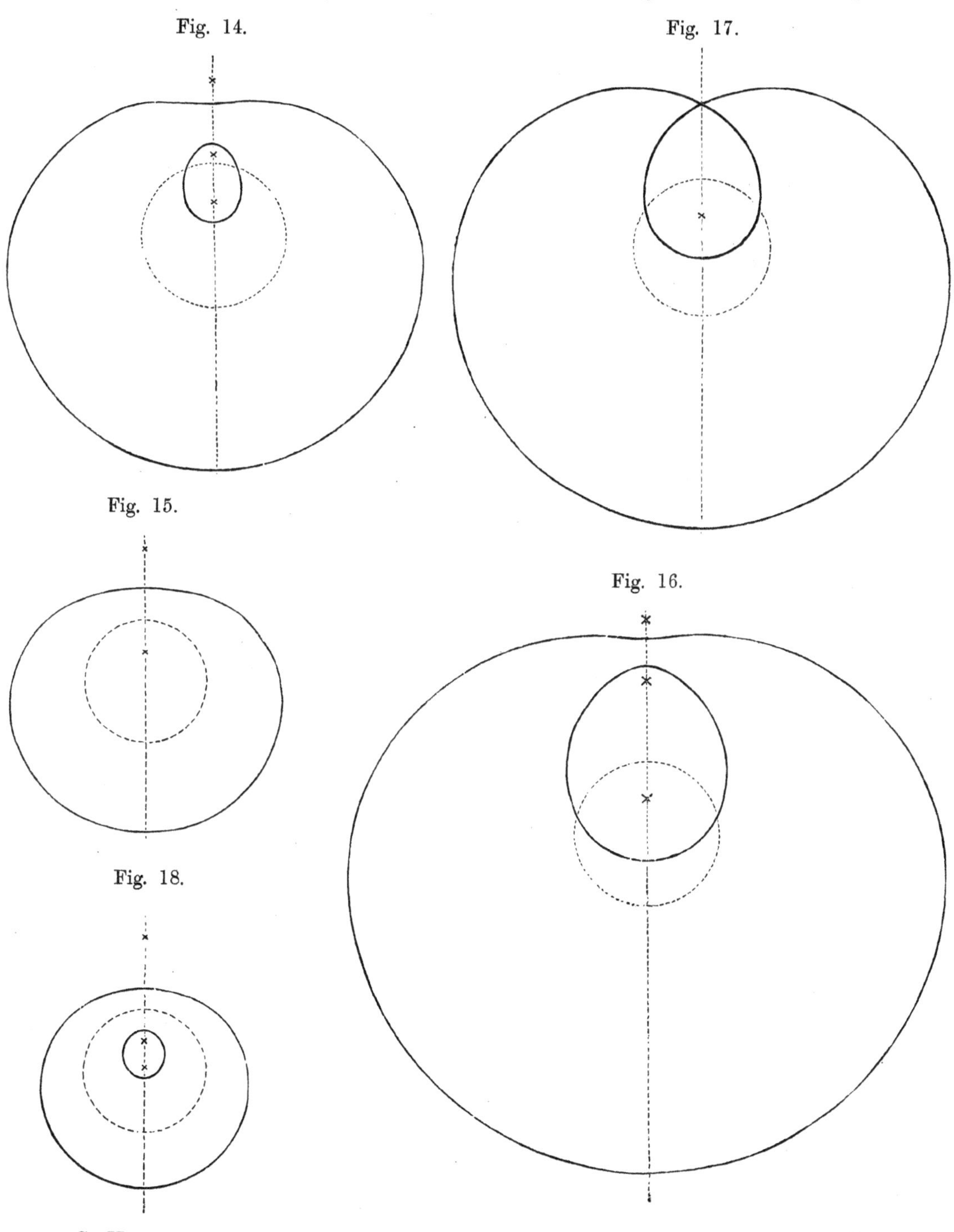
Fig. 14.
Fig. 17.
Fig. 15.
Fig. 16.
Fig. 18.

XXXI.

The theorem, "If a variable circle have its centre upon a circle S, and its radius proportional to the tangential distance of the centre from a circle C, the envelope is a Cartesian,"
is at once deducible from the theorem—

"If a variable circle have its centre upon a circle S and its radius proportional to the distance of the centre from a point C', the locus is a Cartesian,"

which last theorem was in effect given in discussing the theory of the secondary caustic. In fact, the locus of a point P such that its tangential distances from the circles C, C' are in a constant ratio, is a circle S. Conversely, if there be a circle C, and the locus of P be a circle S, then the circle C' may be found such that the tangential distances of P from the two circles are in a constant ratio, and the circle C' may be taken to be a point, i.e. if there be a circle C and the locus of P be a circle S, then a point C' may be found such that the tangential distance of P from the circle C is in a constant ratio to the distance from the point C'.

Hence treating P as the centre of the variable circle, it is clear that the variable circle is determined in the two cases by equivalent constructions, and the envelope is therefore the same in both cases.

XXXII.

The equation of the secondary caustic developed and reduced is

$$\mu^4(x^2+y^2)^2 - 2\mu^2\left(a^2+(\mu^2+1)c^2\right)(x^2+y^2) + 8c^2\mu^2ax + a^4 - 2a^2c^2(\mu^2+1) + (\mu^2-1)^2c^4 = 0,$$

or, what is the same thing,

$$\{\mu^2(x^2+y^2) - \left(a^2+(\mu^2+1)c^2\right)\}^2 + 8c^2\mu^2ax - 4c^2\left(c^2\mu^2+(\mu^2+1)a^2\right) = 0,$$

which may also be written

$$\left(x^2+y^2-\left(\frac{a^2}{\mu^2}+(1+\frac{1}{\mu^2})\,c^2\right)\right)^2 + \frac{8}{\mu^2}c^2ax - \frac{4c^2}{\mu^2}\left(c^2+(1+\frac{1}{\mu^2})\,a^2\right) = 0,$$

which is of the form

$$(x^2+y^2-\alpha)^2 + 16A\,(x-m) = 0\,;$$

and the values of the coefficients are

$$\alpha = \frac{1}{\mu^2}a^2 + (1+\frac{1}{\mu^2})\,c^2,$$

$$A = \frac{c^2a}{2\mu^2},$$

$$m = \frac{1}{2a}\left(c^2+(1+\frac{1}{\mu^2})\,a^2\right).$$

The equation just obtained should, I think, be taken as the standard form of the equation of the Cartesian, and the form of the equation shows that the Cartesian may be defined as the locus of a point, such that the fourth power of its tangential distance from a given circle is in a constant ratio to its distance from a given line.

XXXIII.

The Cartesian is a curve of the fourth order, symmetrical about a certain line which it intersects in four arbitrary points, and these points determine the curve. Taking the line in question (which may be called the axis) as the axis of x, and a line at right angles to it as the axis of y, let a, b, c, d be the values of x corresponding to the points of intersection with the axis, then the equation of the curve is

$$y^4 + y^2 [2x^2 - (a+b+c+d)\, x - \tfrac{1}{4}(a^2+b^2+c^2+d^2 - 2ab - 2ac - 2ad - 2bc - 2bd - 2cd)] \\ + (x-a)(x-b)(x-c)(x-d) = 0.$$

It is easy to see that the form of the equation is not altered by writing $x+\theta$ for x, and $a+\theta$, $b+\theta$, $c+\theta$, $d+\theta$ for a, b, c, d, we may therefore without loss of generality put $a+b+c+d=0$, and the equation of the curve then becomes

$$y^4 + y^2 (2x^2 + ab + ac + ad + bc + bd + cd) + (x-a)(x-b)(x-c)(x-d) = 0,$$

where

$$a+b+c+d=0;$$

the curve is in this case said to be referred to the centre as origin.

The last-mentioned equation may be written

$$(x^2+y^2)^2 + (ab+ac+ad+bc+bd+cd)(x^2+y^2) - (abc+abd+acd+bcd)\, x + abcd = 0,$$

or

$$\begin{aligned} &\{x^2+y^2+\tfrac{1}{2}(ab+ac+ad+bc+bd+cd)\}^2 \\ &\quad - (abc+abd+acd+bcd)\, x \\ &\quad - \tfrac{1}{4}\left\{\begin{array}{l} a^2b^2+a^2c^2+a^2d^2+b^2c^2+b^2d^2+c^2d^2 \\ +\, 2a^2bc + 2a^2bd + 2a^2cd + 2b^2ac + 2b^2ad + 2b^2cd \\ +\, 2c^2ab + 2c^2ad + 2c^2bd + 2d^2ab + 2d^2ac + 2d^2bc \\ +\, 2abcd \end{array}\right\} = 0, \end{aligned}$$

or observing that

$$\begin{aligned} &a^2bc + a^2bd + a^2cd + b^2ac + b^2ad + b^2cd \\ &+ c^2ab + c^2ad + c^2bd + d^2ab + d^2ac + d^2bc \\ &\quad = abc\,(a+b+c) + abd\,(a+b+d) + acd\,(a+c+d) + bcd\,(b+c+d) \\ &\quad = -4abcd, \end{aligned}$$

the equation becomes

$$\begin{aligned}&\{x^2 + y^2 + \tfrac{1}{2}(ab + ac + ad + bc + bd + cd)\}^2\\ &\quad - (abc + abd + acd + bcd)\,x\\ &\quad - \tfrac{1}{4}(a^2b^2 + a^2c^2 + a^2d^2 + b^2c^2 + b^2d^2 + c^2d^2 - 6abcd) = 0,\end{aligned}$$

which is of the form

$$(x^2 + y^2 - \alpha)^2 + 16A\,(x - m) = 0,$$

and, as already remarked, signifies that the fourth power of the tangential distance of a point in the curve from a given circle, is proportional to the distance of the same point from a given line. The circle in question (which may be called the dirigent circle) has for its equation

$$x^2 + y^2 + \tfrac{1}{2}(ab + ac + ad + bc + bd + cd) = 0\,;$$

the line in question, which may be called the directrix, has for its equation

$$x + \frac{a^2b^2 + a^2c^2 + a^2d^2 + b^2c^2 + b^2d^2 + c^2d^2 - 6abcd}{4\,(abc + abd + acd + bcd)} = 0\,;$$

the multiplier of the distance from the directrix is

$$abc + abd + acd + bcd.$$

It may be remarked that a, b, c, d being real, the dirigent circle is real; the equation may, in fact, be written

$$x^2 + y^2 = \tfrac{1}{8}[(a + b)^2 + (a + c)^2 + (a + d)^2 + (b + c)^2 + (b + d)^2 + (c + d)^2].$$

XXXIV.

Considering the equation of the Cartesian under the form

$$(x^2 + y^2 - \alpha)^2 + 16A\,(x - m) = 0,$$

the centre of the dirigent circle $x^2 + y^2 - \alpha = 0$ must be considered as a real point, but α may be positive or negative, i.e. the radius may be either a real or a pure imaginary distance: the coefficients A, m must be real, the directrix is therefore a real line. The equation shows that for all points of the curve $x - m$ is always negative or always positive, according as A is positive or negative, i.e. that the curve lies wholly on one side of the directrix, viz. on the same side with the centre of the dirigent circle if A is positive, but on the contrary side if A is negative. In the former case the curve may be said to be an 'inside' curve, in the latter an 'outside' curve. If $m = 0$, or the directrix passes through the centre of the dirigent circle, then the distinction between an inside curve and an outside curve no longer exists. It is clear that the curve touches the directrix in the points of intersection of this line and the dirigent circle, and that the points in question are the only points of intersection of the curve with the directrix or the dirigent circle; hence if the directrix and dirigent circle do not intersect, the curve does not meet either the directrix or the dirigent circle.

XXXV.

To discuss the equation

$$(x^2+y^2-\alpha)^2+16A\,(x-m)=0,$$

I write first $y=0$, which gives

$$x^4-2\alpha x^2+16Ax+\alpha^2-16Am=0$$

for the points of intersection with the axis of x. If this equation has equal roots, there will be a double point on the axis of x, and it is important to find the condition that this may be the case. The equation may be written in the form

$$(3,\ 0,\ -\alpha,\ 12A,\ 3\alpha^2-48Am \between x,\ 1)^4=0,$$

the condition for a part of equal roots is then at once seen to be

$$-(\alpha^2-12Am)^3+(\alpha^3-18Am\alpha+54A^2)^2=0\,;$$

or reducing and throwing out the factor A^2, this is

$$27A^2+2m\,(8m^2-9\alpha)\,A-\alpha^2\,(m^2-\alpha)=0.$$

This equation will give two equal values for A if

$$m^2\,(8m^2-9\alpha)^2+27\alpha^2\,(m^2-\alpha)=0,$$

an equation which reduces itself to

$$(4m^2-3\alpha)^3=0.$$

Hence, if $4m^2-3\alpha$ be negative, i.e. if $\alpha>\frac{4m^2}{3}$, the values of A will be imaginary, but if $4m^2-3\alpha$ be positive, or $\alpha<\frac{4m^2}{3}$, the values of A will be real. If $\alpha=\frac{4m^2}{3}$, then there will be two equal values of A, which in fact corresponds to a cusp upon the axis of x. Whenever the curve is real there will be at least two real points on the axis of x; and when $\alpha<\frac{4m^2}{3}$, but not otherwise, then for properly selected values of A there will be four real points on the axis of x.

Differentiating the equation of the curve, we have

$$\big((x^2+y^2-\alpha)\,x+4A\big)\,dx+(x^2+y^2-\alpha)\,ydy=0\,;$$

and if in this equation we put $dx=0$, we find $y=0$, or $x^2+y^2-\alpha=0$, i.e. that the points on the axis of x, and the points of intersection with the circle $x^2+y^2-\alpha=0$, are the only points at which the curve is perpendicular to the axis of x. To find the points at which the curve is parallel to the axis of x, we must write $dx=0$, this gives

$$(x^2+y^2-\alpha)\,x+4A=0,$$

and thence

$$x^2 + y^2 - \alpha = -\frac{4A}{x},$$

and

$$A + x^2(x - m) = 0:$$

this equation will have three real roots if $A < \frac{4m^3}{27}$, and only a single real root if $A > \frac{4m^3}{27}$; for $A = \frac{4m^3}{27}$, the equation in question will have a pair of equal roots. It is easy to see that there is always a single real root of the equation which gives rise to a real value of y, i.e. to a real point upon the curve; but, when the equation has three real roots, two of the roots may or may not give rise to real points upon the curve.

XXXVI.

It is now easy to trace the curve. First, when $m = 0$, or the directrix passes through the centre of the dirigent circle, the curve is here an oval bent in so as to have double contact with the directrix, and lying on the one or the other side of the directrix according to the sign of A. See fig. *a*.

Fig. *a*.

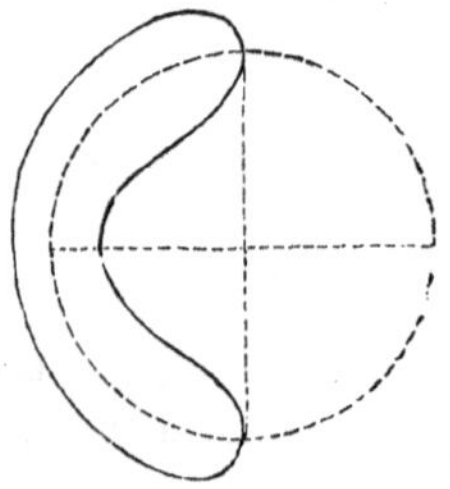

Fig. *b*.

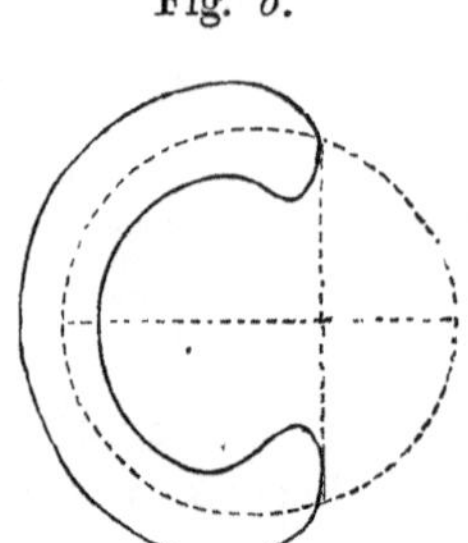

Fig. *c*.

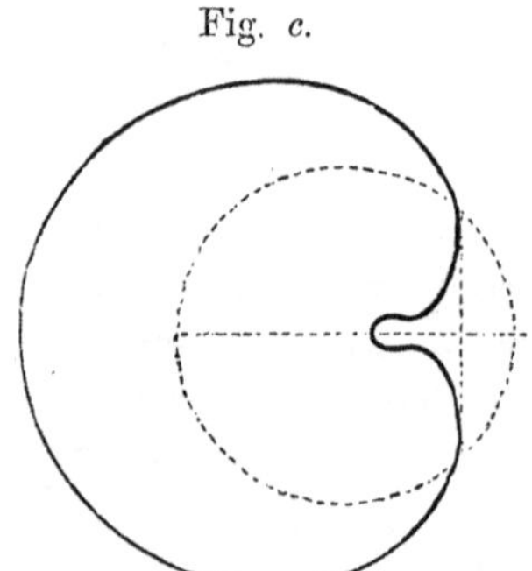

Fig. *d*.

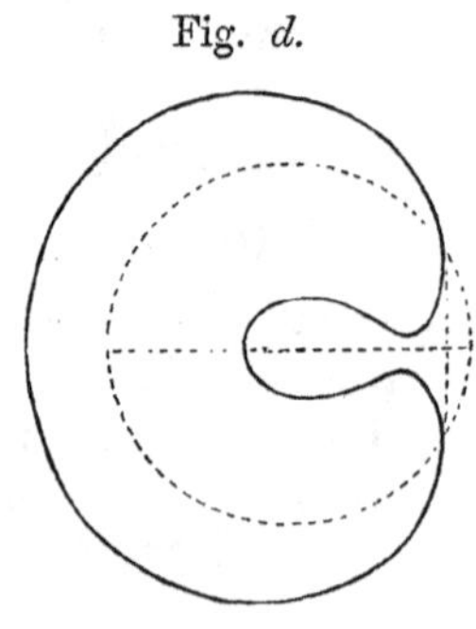

Next, when the directrix does not pass through the centre of the dirigent circle, it will be convenient to suppose always that m is positive, and to consider A as passing first from 0 to ∞ and then from 0 to $-\infty$, i.e. to consider first the different inside curves, and then the different outside curves. Suppose $\alpha > \frac{4m^2}{3}$, the inside curve is at first an oval, as in fig. *b*, where (attending to one side only of the axis) it will be noticed that there are three tangents parallel to the axis, viz. one for the convexity of the oval, and two for the concavity. For $A = \frac{4m^3}{27}$ the two tangents for the concavity come together, and give rise to a stationary tangent (i.e. a tangent at an inflection) parallel to the axis, and for $A > \frac{4m^3}{27}$ the two tangents for the concavity disappear. The outside curve is an oval (of course on the opposite side of, and) bent in so as to have double contact with the directrix.

Next, if $\alpha = \frac{4m^2}{3}$, the inside curve is at first an oval, as in fig. *c*, and there are, as before, three tangents parallel to the axis: for $A = \frac{4m^3}{27}$, the tangents for the concavity of the oval come to coincide with the axis, and are tangents at a cusp, and for $A > \frac{4m^3}{27}$ the cusp disappears, and there are not for the concavity of the oval any tangents parallel to the axis. The outside curve is an oval as before, but smaller and more compressed.

Next, $\alpha < \frac{4m^2}{3} > m^2$, then the inside curve is at first an oval, as in fig. *d*, and there are, as before, three tangents parallel to the axis; when A attains a certain value which is less than $\frac{4m^3}{27}$, the curve acquires a double point; and as A further increases, the curve breaks up into two separate ovals, and there are then only two tangents parallel to the axis, viz. one for the exterior oval and one for the interior oval. As A continues to increase, the interior oval decreases; and when A attains a certain value which is less than $\frac{4m^3}{27}$, the interior oval reduces itself to a conjugate point, and it afterwards disappears altogether. The outside curve is an oval as before, but smaller and more compressed.

Next, if the directrix touch the dirigent circle, i.e. if $\alpha = m^2$. Then the inside curve is at first composed of an exterior oval which touches the dirigent circle, and of an interior oval which lies wholly within the dirigent circle. As A increases the interior oval decreases, reduces itself to a conjugate point, and then disappears. The outside curve is an oval which always touches the dirigent circle, at first very small (it may be considered as commencing from a conjugate point corresponding to $A = 0$), but increasing as A increases negatively.

Next, when the directrix does not meet the dirigent circle, i.e. if $\alpha < m^2$. The inside curve consists at first of two ovals, an exterior oval lying without the dirigent circle, and an interior oval lying within the dirigent circle. As A increases the interior oval decreases, reduces itself to a conjugate point and disappears. The outside curve is at first imaginary, but when A attains a sufficiently large negative value, it makes its appearance as a conjugate point, and afterwards becomes an oval which gradually increases.

Next, when the dirigent circle reduces itself to a point, i.e. if $\alpha = 0$. The inside curve makes its appearance as a conjugate point (corresponding to $A = 0$), and as A increases it becomes an oval and continually increases. The outside curve comports itself as in the last preceding case.

Finally, when the dirigent circle becomes imaginary, or has for its radius a pure imaginary distance, i.e. if α is negative. The inside curve is at first imaginary, but when A attains a certain value it makes its appearance as a conjugate point, and as A increases becomes an oval and continually increases. The outside curve, as in the preceding two cases, comports itself in a similar manner.

The discussion, in the present section, of the different forms of the curve is not a very full one, and a large number of figures would be necessary in order to show completely the transition from one form to another. The forms delineated in the four figures were selected as forms corresponding to imaginary values of the parameters by means of which the equation of the curve is usually represented, e.g. the equations in Section XXVIII.

XXXVII.

It has been shown that for rays proceeding from a point and refracted at a circle, the secondary caustic is the Cartesian; the caustic itself is therefore the evolute of the Cartesian; this affords a means of finding the tangential equation of the caustic. In fact, the equation of the Cartesian is

$$(x^2 + y^2 - \alpha)^2 + 16A\,(x - m) = 0\,;$$

and if we take for the equation of the normal

$$X\xi + Y\eta + Z = 0,$$

(where ξ, η are current coordinates), then

$$\begin{aligned} X : Y : Z = &- y\,(x^2 + y^2 - \alpha) \\ &: \ x\,(x^2 + y^2 - \alpha) + 4A \\ &: \ 4Ay, \end{aligned}$$

equations which give

$$\begin{aligned} Z^2Yx &= Y(mZ^2 - AX^2), \\ -Z^2Yy &= Z^3 + X(mZ^2 - AX^2), \\ Z^4Y^2(x^2 + y^2 - \alpha) &= 4AZ^3XY^2, \end{aligned}$$

whence eliminating, we have

$$\{Z^3 + X(mZ^2 - AX^2)\}^2 + Y^2(mZ^2 - AX^2)^2 - Z^3Y^2(\alpha Z + 4AX) = 0,$$

where if, as before, c denotes the radius of the refracting circle, a the distance of the radiant point from the centre, and μ the index of refraction, we have

$$\begin{aligned} \alpha &= \frac{1}{\mu^2}a^2 + (1 + \frac{1}{\mu^2})\,c^2, \\ A &= \frac{c^2a}{2\mu^2}, \\ m &= \frac{1}{2a}\left(c^2 + (1 + \frac{1}{\mu^2})\,a^2\right). \end{aligned}$$

The above equation is the condition in order that the line $Xx + Yy + Z = 0$ may be a normal to the secondary caustic $(x^2 + y^2 - \alpha)^2 + 16A(x - m) = 0$, or it is the tangential equation of the caustic, which is therefore a curve of the class 6 only. The equation may be written in the more convenient form

$$Z^6 + 2Z^3X(mZ^2 - AX^2) + (X^2 + Y^2)(mZ^2 - AX^2)^2 - Z^3Y^2(\alpha Z + 4AX) = 0.$$

XXXVIII.

To compare the last result with that previously obtained for the caustic by reflexion, I write $\mu = -1$, and putting also $c = 1$ and $Z = a$ (for the equation of the reflected ray was assumed to be $Xx + Yy + a = 0$), we have

$$\alpha = a^2 + 2,\ A = \tfrac{1}{2}a,\ m = \frac{1}{2a}(1 + 2a^2),$$

and the equation becomes, after a slight reduction,

$$4a^4 + 4a^2X(2a^2 + 1 - X^2) + (X^2 + Y^2)(2a^2 + 1 - X^2)^2 - 4a^2Y^2(a^2 + 2 + 2X) = 0,$$

which may be written

$$\left(2a^2 + X(2a^2 + 1 - X^2)\right)^2 + Y^2\left(-4a^2 + 1 - 8a^2X - 2(2a^2 + 1)X^2 + X^4\right) = 0\,;$$

this divides out by the factor $(X + 1)^3$, and the equation then becomes,

$$(X^2 - X - 2a^2)^2 + Y^2\left((X - 1)^2 - 4a^2\right) = 0,$$

which agrees with the result before obtained.

XXXIX.

Again, to compare the general equation with that previously obtained for parallel rays refracted at a circle, we must write $\mu = \frac{1}{k}$, $c = 1$, $a = \infty$, $Z = k$ (for the equation of the refracted ray was taken to be $Xx + Yy + k = 0$); we have then

$$\alpha = 1 + k^2 + k^2a^2,\ A = \tfrac{1}{2}k^2a^2,\ m = \frac{1}{2a}\left(1 + (1 + k^2)\,a^2\right),$$

and, after the substitution, $a = \infty$. The equation becomes in the first instance

$$k^6 + 2k^3X\left\{\frac{1}{2a}\left(1 + (1 + k^2)\,a^2\right)k^2 - \tfrac{1}{2}k^2aX^2\right\} + (X^2 + Y^2)\left\{\frac{1}{2a}\left(1 + (1 + k^2)\,a^2\right)k^2 - \tfrac{1}{2}k^2aX^2\right\}^2$$
$$- k^3Y^2\left(1 + k^2 + k^2a^2 + 2k^2aX\right) = 0\,;$$

and then putting $a = \infty$, or, what is the same thing, attending only to the terms which involve a^2, and throwing out the constant factor k^4, we obtain

$$(X^2 + Y^2)(X^2 - 1 - k^2)^2 - 4k^2Y^2 = 0,$$

or

$$X^2(X^2 - 1 - k^2)^2 + Y^2(X + 1 + k)(X - 1 - k)(X + 1 - k)(X - 1 - k) = 0,$$

which agrees with the former result.

XL.

It was remarked that the ordinary construction for the secondary caustic could not be applied to the case of parallel rays (the entire curve would in fact pass off to an infinite distance), and that the simplest course was to measure the distance GQ from a line through the centre of the refracting circle perpendicular to the direction of the rays. To find the equation of the resulting curve, take the centre of the circle as the origin and the direction of the incident rays for the axis of x; let the radius of the circle be taken equal to unity, and let μ denote, as before, the index of refraction. Then if α, β are the coordinates of the point of incidence of a ray, we have $\alpha^2 + \beta^2 = 1$, and considering α, β as variable parameters connected by this equation, the required curve is the envelope of the circle,

$$\mu^2\left\{(x - \alpha)^2 + (y - \beta)^2\right\} - \alpha^2 = 0.$$

Write now $\alpha = \cos\theta$, $\beta = \sin\theta$, then multiplying the equation by -2, and writing $1 + \cos 2\theta$ instead of $2\cos^2\theta$, the equation becomes

$$1 + \cos 2\theta - 2\mu^2\left(x^2 + y^2 - 2x\cos\theta - 2y\sin\theta + 1\right) = 0,$$

which is of the form

$$A\cos 2\theta + B\sin 2\theta + C\cos\theta + D\sin\theta + E = 0,$$

and the values of the coefficients are

$$A = 1,$$
$$B = 0,$$
$$C = 4\mu^2 x,$$
$$D = 4\mu^2 y,$$
$$E = -2\mu^2 (x^2 + y^2) - 2\mu^2 + 1.$$

Substituting these values in the equation

$$\{12 (A^2 + B^2) - 3 (C^2 + D^2) + 4E^2\}^3$$
$$- \{27A (C^2 - D^2) + 54BCD - \left(72 (A^2 + B^2) + 9 (C^2 + D^2)\right) E + 8E^3\}^2 = 0,$$

the equation of the envelope is found to be

$$16 \{(1 - \mu^2 + \mu^4) - (\mu^2 + \mu^4)(x^2 + y^2) + \mu^4 (x^2 + y^2)^2\}^3$$
$$- \left\{ \begin{array}{l} 4 - 6\mu^2 - 6\mu^4 + 4\mu^6 \\ - (6\mu^2 + 3\mu^4 + 6\mu^6)(x^2 + y^2) - 27\mu^4 (x^2 - y^2) \\ - (6\mu^4 + 6\mu^6)(x^2 + y^2)^2 \\ + 4\mu^6 (x^2 + y^2)^3 \end{array} \right\}^2 = 0,$$

which is readily seen to be only of the 8th order. But to simplify the result, write first $(x^2 + y^2 - 1) + 1$, and $2x^2 - 1 - (x^2 + y^2 - 1)$ in the place of $x^2 + y^2$ and $x^2 - y^2$ respectively, the equation becomes

$$4\{(1 - \mu^2)^2 - \mu^2 (1 - \mu^2)(x^2 + y^2 - 1) + \mu^4 (x^2 + y^2 - 1)^2\}^3$$
$$- \left\{ \begin{array}{l} 2 (1 - \mu^2)^3 \\ - 3\mu^2 (1 - \mu^2)^2 (x^2 + y^2 - 1) - 27\mu^4 x^2 \\ - 3\mu^4 (1 - \mu^2)(x^2 + y^2 - 1)^2 \\ + 2\mu^6 (x^2 + y^2 - 1)^3 \end{array} \right\}^2 = 0.$$

Write for a moment $1 - \mu^2 = q$, $\mu^2 (x^2 + y^2 - 1) = \rho$, the equation becomes

$$4 (q^2 - q\rho + \rho^2)^3 - (2q^3 - 3q^2\rho - 3\rho^2 + 2\rho^3 - 27\mu^4 x^2)^2 = 0;$$

or developing,

$$4 (q^2 - q\rho + \rho^2)^3 - (2q^3 - 3q^2\rho - 3q\rho^2 + 2\rho^3)^2$$
$$+ 54 (2q^3 - 3q^2\rho - 3q\rho^2 + 2\rho^3) \mu^4 x^2 - 729\mu^8 x^4 = 0,$$

and reducing and dividing out by 27, this gives

$$q^2\rho^2 (\rho - q)^2 + 2 (\rho + q)(2\rho - q)(\rho - 2q) \mu^4 x^2 - 27\mu^8 x^4 = 0,$$

whence replacing q, ρ by their values, the required equation is

$$(1-\mu^2)^2(x^2+y^2-1)^2\left(\mu^2(x^2+y^2)-1\right)^2$$
$$+2\left(\mu^2(x^2+y^2)-2\mu^2+1\right)\left(2\mu^2(x^2+y^2)-\mu^2-1\right)\left(\mu^2(x^2+y^2)-2+\mu^2\right)x^2-27\mu^4x^4=0,$$

which is the equation of an orthogonal trajectory of the refracted rays.

In the case of reflexion, $\mu=-1$, and the equation becomes

$$4(x^2+y^2-1)^3-27x^2=0.$$

Comparing this with the equation of the caustic, it is easy to see,

THEOREM. In the case of parallel rays and a reflecting circle, there is a secondary caustic which is a curve similar to and double the magnitude of the caustic, the position of the two curves differing by a right angle.

XLI.

The entire system of the orthogonal trajectories of the refracted rays might in like manner be determined by finding the envelope of the circle (where, as before, α, β are variable parameters connected by the equation $\alpha^2+\beta^2=1$),

$$\mu^2\{(x-\alpha)^2+(y-\beta)^2\}-(\alpha+m)^2=0.$$

{The result, as far as I have worked it out, is as follows, viz.—

$$\left(3-12\left[m^2+2m\mu^2x+\mu^4(x^2+y^2)\right]+\left[1-2\mu^2+2m^2-2\mu^2(x^2+y^2)\right]^2\right)^3$$
$$-\left(\left[1-2\mu^2+2m^2-2\mu^2(x^2+y^2)\right]\left[9+18m^2+36m\mu^2x+18\mu^4(x^2+y^2)\right]\right.$$
$$\left.-54\left[m^2+2m\mu^2x+\mu^4(x^2-y^2)\right]-\left[1-2\mu^2+2m^2-2\mu^2(x^2+y^2)\right]^3\right)^2=0,$$

which, it is easy to see, is an equation of the order 8 only. Added Sept. 12.—A. C.}

146.

A MEMOIR ON CURVES OF THE THIRD ORDER.

[From the *Philosophical Transactions of the Royal Society of London*, vol. CXLVII. *for the year* 1857, pp. 415—446. Received October 30,—Read December 11, 1856.]

A CURVE of the third order, or cubic curve, is the locus represented by an equation such as $U = (*\between x, y, z)^3 = 0$; and it appears by my "Third Memoir on Quantics," [144], that it is proper to consider, in connexion with the curve of the third order $U = 0$, and its Hessian $HU = 0$ (which is also a curve of the third order), two curves of the third class, viz. the curves represented by the equations $PU = 0$ and $QU = 0$. These equations, I say, represent curves of the third class; in fact, PU and QU are contravariants of U, and therefore, when the variables x, y, z of U are considered as point coordinates, the variables ξ, η, ζ of PU and QU must be considered as line coordinates, and the curves will be curves of the third class. I propose (in analogy with the form of the word Hessian) to call the two curves in question the Pippian and Quippian respectively. [The curve $PU = 0$ is now usually called the Cayleyan.] A geometrical definition of the Pippian was readily found; the curve is in fact Steiner's curve R_0 mentioned in the memoir "Allgemeine Eigenschaften der algebraischen Curven," *Crelle*, t. XLVII. [1854] pp. 1—6, in the particular case of a basis-curve of the third order; and I also found that the Pippian might be considered as occurring implicitly in my "Mémoire sur les courbes du troisième ordre," *Liouville*, t. IX. [1844] pp. 285—293 [26] and "Nouvelles remarques sur les courbes du troisième ordre," *Liouville*, t. X. [1845] pp. 102—109 [27]. As regards the Quippian, I have not succeeded in obtaining a satisfactory geometrical definition; but the search after it led to a variety of theorems, relating chiefly to the first-mentioned curve, and the results of the investigation are contained in the present memoir. Some of these results are due to Mr Salmon, with whom I was in correspondence on the subject. The character of the results makes it difficult to develope them in a systematic order; but the results are given in such connexion one with another as I have been able to present them in. Considering the object of the memoir to be the establishment of a distinct geometrical theory of the Pippian, the leading results will be found summed up in the nine different definitions or modes of generation of the Pippian, given in the concluding number. In the course of the memoir I give some further developments relating to the theory in the memoirs in *Liouville* above referred to, showing its relation to the Pippian, and the analogy with theorems of Hesse in relation to the Hessian.

Article No. 1.—*Definitions, &c.*

1. It may be convenient to premise as follows:—Considering, in connexion with a curve of the third order or cubic, *a point,* we have:

(*a*) The *first or conic polar* of the point.

(*b*) The *second or line polar* of the point.

The meaning of these terms is well known, and they require no explanation.

Next, considering, in connexion with the cubic, *a line*—

(*c*) The first or conic polars of each point of the line meet in four points, which are the *four poles* of the line.

(*d*) The second or line polars of each point of the line envelope a conic, which is the *lineo-polar envelope* of the line.

And reciprocally considering, in connexion with a curve of the third class, *a line,* we have:

(*e*) The *first or conic pole* of the line.

(*f*) The *second or point-pole* of the line.

And considering, in connexion with the curve of the third class, *a point*—

(*g*) The first or conic poles of each line through the point touch four lines, which are the *four polars* of the point.

(*h*) The second or point poles of each line through the point generate a conic which is the *point-pole locus* of the point.

But I shall not have occasion in the present memoir to speak of these reciprocal figures, except indeed the first or conic pole of the line.

The term *conjugate poles* of a cubic is used to denote two points, such that the first or conic polar of either of them, with respect to the cubic, is a pair of lines passing through the other of them. Reciprocally, the term *conjugate polars* of a curve of the third class denotes two lines, such that the first or conic pole of either of them, with respect to the curve of the third class, is a pair of points lying in the other of them.

The expression, a *syzygetic cubic*, used in reference to two cubics, denotes a curve of the third order passing through the points of intersection of the two cubics; but in the present memoir the expression is in general used in reference to a single cubic, to denote a curve of the third order passing through the points of intersection of the cubic and its Hessian. As regards curves of the third class, I use in the memoir the full expression, a curve of the third class syzygetically connected with two given curves of the third class.

It is a well-known theorem, that if at the points of intersection of a given line with a given cubic tangents are drawn to the cubic, these tangents again meet the cubic in three points which lie in a line; such line is in the present memoir termed the *satellite line* of the given line, and the point of intersection of the two lines is termed the *satellite point* of the given line; the given line in reference to its satellite line or point is termed the *primary line.*

In particular, if the primary line be a tangent of the cubic, the satellite line coincides with the primary line, and the satellite point is the point of simple intersection of the primary line and the cubic.

Article No. 2.—*Group of Theorems relating to the Conjugate Poles of a Cubic.*

2. The theorems which I have first to mention relate to or originate out of the theory of the conjugate poles of a cubic, and may be conveniently connected together and explained by means of the accompanying figure.

The point E is a point of the Hessian; this being so, its first or conic polar, with respect to the cubic, will be a pair of lines passing through a point F of the

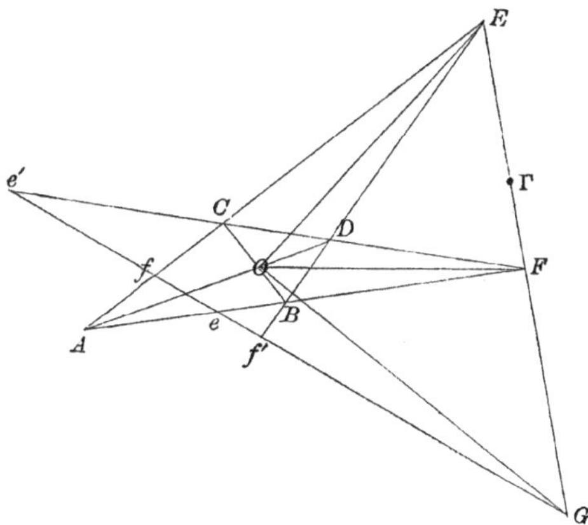

Hessian; and not only so, but the first or conic polar of the point F, with respect to the cubic will be a pair of lines passing through E. The pair of lines through

F are represented in the figure by FBA, FDC, and the pair of lines through E are represented by ECA, EDC, and the lines of the one pair meet the lines of the other pair in the points A, B, C, D. The point O, which is the intersection of the lines AD, BC, is a point of the Hessian, and joining EO, FO, these lines are tangents to the Hessian at the points E, F, that is, the points E, F are *corresponding points* of the Hessian, in the sense that the tangents to the Hessian at these points meet in a point of the Hessian. The two points E, F are, according to a preceding definition, conjugate poles of the cubic.

The line EF meets the Hessian in a third point G, and the points G, O are conjugate poles of the cubic. The first or conic polar of G, with respect to the cubic, is the pair of lines AOD, BOC meeting in O. The first or conic polar of O, with respect to the cubic, is the pair of lines GEF and $Gf'ef e'$ meeting in G. The four poles of the line EO, with respect to the cubic, are the points of intersection of the first or conic polars of the two points E and O, that is, the four poles in question are the points F, F, e, e'. Similarly, the four poles of the line FO, with respect to the cubic, are the points E, E, f, f'.

The line EF, that is, any line joining two conjugate poles of the cubic, is a tangent to the Pippian, and the point of contact Γ is the harmonic with respect to the points E, F (which are points on the Hessian) of G, the third point of intersection with the Hessian. Conversely, any tangent of the Pippian meets the Hessian in three points, two of which are conjugate poles of the cubic, and the point of contact is the harmonic, with respect to these two points, of the third point of intersection with the Hessian.

The line GO in the figure is of course also a tangent of the Pippian, and moreover the lines FBA, FDC (that is, the pair of lines which are the first or conic polar of E) and the lines ECA, EDB (that is, the pair of lines which are the first or conic polar of F) are also tangents to the Pippian. The point E represents *any* point of the Hessian, and the three tangents through E to the Pippian are the line EFG and the lines ECA, EDB; the line EFG is the line joining E with the conjugate pole F, and the lines ECA, EDB are the first or conic polar of this conjugate pole F with respect to the cubic. The figure shows that the line EO (the tangent to the Hessian at the point E) and the before-mentioned three lines (the tangents through E to the Pippian), are harmonically related, viz. the line EO the tangent of the Hessian, and the line EF one of the tangents to the Pippian, are harmonics with respect to the other two tangents to the Pippian. It is obvious that the tangents to the Pippian through the point F are in like manner the line GFE, and the pair of lines FBA, FBC, and that these lines are harmonically related to FO the tangent at F of the Hessian. And similarly, the tangents to the Pippian through the point O are the line GO and the lines AOD, BOC, and the tangents to the Pippian through the point G are the line GO and the lines GFE and $Gf'efe'$. Thus all the lines of the figure are tangents to the Pippian except the lines EO, FO, which are tangents to the Hessian. It may be added, that the lineo-polar envelope of the line EF with respect to the cubic is the pair of lines OE, OF.

It will be presently seen that the analytical theory leads to the consideration of a line IJ (not represented in the figure): the line in question is the polar of E (or F) with respect to the conic which is the first or conic polar of F (or E) with respect to *any* syzygetic cubic. The line IJ is a tangent of the Pippian, and moreover the lines EF and IJ are conjugate polars of a curve of the third class syzygetically connected with the Pippian and Quippian, and which is moreover such that its Hessian is the Pippian.

Article Nos. 3 to 19.—*Analytical investigations, comprising the proof of the theorems, Article No.* 2.

3. The analytical theory possesses considerable interest. Take as the equation of the cubic,

$$U = x^3 + y^3 + z^3 + 6lxyz = 0\,;$$

then the equation of the Hessian is

$$HU = l^2 (x^3 + y^3 + z^3) - (1 + 2l^3)\, xyz = 0\,;$$

and the equation of the Pippian in line coordinates (that is, the equation which expresses that $\xi x + \eta y + \zeta z = 0$ is a tangent of the curve) is

$$PU = -\, l (\xi^3 + \eta^3 + \zeta^3) + (-1 + 4l^3)\, \xi\eta\zeta = 0.$$

The equation of the Quippian in line coordinates is

$$QU = (1 - 10l^3)(\xi^3 + \eta^3 + \zeta^3) - 6l^2 (5 + 4l^3)\, \xi\eta\zeta = 0\,;$$

and the values of the two invariants of the cubic form are

$$S = -\, l + l^4,$$

$$T = 1 - 20l^3 - 8l^6,$$

values which give identically,

$$T^2 - 64S^3 = (1 + 8l^3)^3\,;$$

the last-mentioned function being in fact the discriminant.

4. Suppose now that $(X,\ Y,\ Z)$ are the coordinates of the point E, and $(X',\ Y',\ Z')$ the coordinates of the point F; then the equations which express that these points are conjugate poles of the cubic, are

$$XX' + l(YZ' + Y'Z) = 0,$$

$$YY' + l(ZX' + Z'X) = 0,$$

$$ZZ' + l(XY' + X'Y) = 0\,;$$

and by eliminating from these equations, first $(X',\ Y',\ Z')$, and then $(X,\ Y,\ Z)$, we find

$$l^2 (X^3 + Y^3 + Z^3) - (1 + 2l^3)\, XYZ = 0,$$

$$l^2 (X'^3 + Y'^3 + Z'^3) - (1 + 2l^3)\, X'Y'Z' = 0,$$

which shows that the points E, F are each of them points of the Hessian.

5. I may notice, in passing, that the preceding equations give rise to a somewhat singular *unsymmetrical* quadratic transformation of a cubic form. In fact, the second and third equations give $X' : Y' : Z' = YZ - l^2X^2 : l^2XY - lZ^2 : l^2ZX - lY^2$. And substituting these values for X', Y', Z' in the form

$$l^2(X'^3 + Y'^3 + Z'^3) - (1 + 2l^3)X'Y'Z',$$

the result must contain as a factor

$$l^2(X^3 + Y^3 + Z^3) - (1 + 2l^3)XYZ;$$

the other factor is easily found to be

$$-l^3(l^3(X^3 + Y^3 + Z^3) + 3lXYZ).$$

Several of the formulæ given in the sequel conduct in like manner to unsymmetrical transformations of a cubic form.

6. I remark also, that the last-mentioned system of equations gives, *symmetrically*,

$$\begin{aligned} &X'^2 : Y'^2 : Z'^2 : Y'Z' : Z'X' : X'Y' \\ &\quad = YZ - l^2X^2 : ZX - l^2Y^2 : XY - l^2Z^2 : l^2YZ - lX^2 : l^2ZX - lY^2 : l^2XY - lZ^2; \end{aligned}$$

and it is, I think, worth showing how, by means of these relations, we pass from the equation between X', Y', Z' to that between X, Y, Z. In fact, representing, for shortness, the foregoing relations by

$$X'^2 : Y'^2 : Z'^2 : Y'Z' : Z'X' : X'Y' = A : B : C : F : G : H,$$

we may write

$$X' = AF = GH, \quad Y' = BG = HF, \quad Z' = CH = FG, \quad ABC = FGH;$$

and thence

$$X'^3 = AF \,.\, G^2H^2, \quad Y'^3 = BG \,.\, H^2F^2, \quad Z'^3 = CH \,.\, F^2G^2, \quad X'Y'Z' = F^2G^2H^2;$$

hence

$$l^2(X'^3 + Y'^3 + Z'^3) - (1 + 2l^3)X'Y'Z' = FGH\{l^2(AGH + BHF + CFG) - (1 + 2l^3)FGH\}.$$

But we have

$$\begin{aligned} l^2(AGH + BHF + CFG) &= -(2l^5 + l^8)(X^3 + Y^3 + Z^3)XYZ + (l^4 + 2l^7)(Y^3Z^3 + Z^3X^3 + X^3Y^3), \\ -(1 + 2l^3)FGH \quad &= \quad (l^5 + 2l^8)(X^3 + Y^3 + Z^3)XYZ + (l^4 + 2l^7)(Y^3Z^3 + Z^3X^3 + X^3Y^3) \\ &\qquad + l^3(1 - l^3)(1 + 2l^3)X^2Y^2Z^2; \end{aligned}$$

and thence

$$\begin{aligned} &l^2(AGH + BHF + CFG) - (1 + 2l^3)FGH \\ &\qquad = -l^3(1 - l^3)\{l^2(X^3 + Y^3 + Z^3)XYZ - (1 + 2l^3)X^2Y^2Z^2\}; \end{aligned}$$

and finally,

$$\begin{aligned} l^2(X'^3 + Y'^3 + Z'^3) - (1 + 2l^3)X'Y'Z' &= l^5(-l + l^4)(lYZ - X^2)(lZX - Y^2)(lXY - Z^2)XYZ \\ &\quad \times \{l^2(X^3 + Y^3 + Z^3) - (1 + 2l^3)XYZ\}. \end{aligned}$$

We have also, identically,

$$ABC - FGH = \frac{1}{l}(-l + l^4) XYZ \{l^2 (X^3 + Y^3 + Z^3) - (1 + 2l^3) XYZ\},$$

which agrees with the relation $ABC - FGH = 0$.

7. Before going further, it will be convenient to investigate certain relations which exist between the quantities (X, Y, Z), (X', Y', Z'), connected as before by the equations

$$\begin{aligned} XX' + l(YZ' + Y'Z) &= 0, \\ YY' + l(ZX' + Z'X) &= 0, \\ ZZ' + l(XY' + X'Y) &= 0, \end{aligned}$$

and the quantities

$$\begin{aligned} \xi &= YZ' - Y'Z, & \alpha &= XX' = -\frac{1}{l}(YZ' + Y'Z), \\ \eta &= ZX' - Z'X, & \beta &= YY' = -\frac{1}{l}(ZX' + Z'X), \\ \zeta &= XY' - X'Y, & \gamma &= ZZ' = -\frac{1}{l}(XY' + X'Y). \end{aligned}$$

We have identically,

$$2XX'(YZ' - Y'Z) + (XY' + X'Y)(ZX' - Z'X) + (ZX' + Z'X)(XY' - X'Y) = 0;$$

or expressing in terms of ξ, η, ζ, α, β, γ the quantities which enter into this equation, and forming the analogous equations, we have

$$\begin{aligned} 2l\alpha\xi - \gamma\eta - \beta\zeta &= 0, \qquad \text{(A)} \\ -\gamma\xi + 2l\beta\eta - \alpha\zeta &= 0, \\ -\beta\xi - \alpha\eta + 2l\gamma\zeta &= 0. \end{aligned}$$

We have also

$$X^2Y'Z' - X'^2YZ = \tfrac{1}{2}\{-(XY' + X'Y)(ZX' - Z'X) + (ZX' + Z'X)(XY' - X'Y)\},$$

and thence in like manner,

$$\begin{aligned} X^2Y'Z' - X'^2YZ &= \frac{1}{2l}(\gamma\eta - \beta\zeta), \qquad \text{(B)} \\ Y^2Z'X' - Y'^2ZX &= \frac{1}{2l}(\alpha\zeta - \gamma\xi), \\ Z^2X'Y' - X'^2YZ &= \frac{1}{2l}(\beta\xi - \alpha\eta). \end{aligned}$$

Again, we have

$$\begin{aligned} (YZ' - Y'Z)^2 &= (YZ' + Y'Z)^2 - 4YY'ZZ', \\ (ZX' - Z'X)(XY' - X'Y) &= -(ZX' + Z'X)(XY' + X'Y) + 2XX'(YZ' + Y'Z); \end{aligned}$$

and thence

$$\begin{aligned}
\xi^2 &= \frac{1}{l^2}\alpha^2 - 4\beta\gamma, \\
\eta^2 &= \frac{1}{l^2}\beta^2 - 4\gamma\alpha, \\
\zeta^2 &= \frac{1}{l^2}\gamma^2 - 4\alpha\beta, \\
\eta\zeta &= -\frac{2}{l}\alpha^2 - \frac{1}{l^2}\beta\gamma, \\
\zeta\xi &= -\frac{2}{l}\beta^2 - \frac{1}{l^2}\gamma\alpha, \\
\xi\eta &= -\frac{2}{l}\gamma^2 - \frac{1}{l^2}\alpha\beta;
\end{aligned} \tag{C}$$

and conversely

$$\begin{aligned}
\frac{1}{l^2}(1+8l^3)\,\alpha^2 &= \xi^2 - 4l^2\eta\zeta, \\
\frac{1}{l^2}(1+8l^3)\,\beta^2 &= \eta^2 - 4l^2\zeta\xi, \\
\frac{1}{l^2}(1+8l^3)\,\gamma^2 &= \zeta^2 - 4l^2\xi\eta, \\
-\frac{1}{l^2}(1+8l^3)\,\beta\gamma &= 2l\xi^2 + \eta\zeta, \\
-\frac{1}{l^2}(1+8l^3)\,\gamma\alpha &= 2l\eta^2 + \zeta\xi, \\
-\frac{1}{l^2}(1+8l^3)\,\alpha\beta &= 2l\zeta^2 + \xi\eta.
\end{aligned} \tag{D}$$

8. It is obvious that

$$\xi x + \eta y + \zeta z = 0$$

is the equation of the line EF joining the two conjugate poles, and it may be shown that

$$\alpha x + \beta y + \gamma z = 0$$

is the equation of the line IJ, which is the polar of E with respect to a conic which is the first or conic polar of F with respect to *any* syzygetic cubic. In fact, the equation of a syzygetic cubic will be $x^3+y^3+z^3+6\lambda xyz=0$, where λ is arbitrary, and the equation of the line in question is

$$(X\partial_x + Y\partial_y + Z\partial_z)\,(X'\partial_x + Y'\partial_y + Z'\partial_z)\,(x^3+y^3+z^3+6\lambda xyz) = 0\,;$$

or developing,

$$\begin{aligned}
&XX'x + YY'y + ZZ'z \\
&+\lambda\,\{YZ' + Y'Z)\,x + (ZX' + Z'X)\,y + (XY' + X'Y)\,z\} = 0\,;
\end{aligned}$$

and the function on the left-hand side is

$$\left(1-\frac{\lambda}{l}\right)(\alpha x+\beta y+\gamma z),$$

which proves the theorem.

9. The equations (A) by the elimination of (ξ, η, ζ), give

$$-l(\alpha^3+\beta^3+\gamma^3)+(-1+4l^3)\alpha\beta\gamma=0,$$

which shows that the line IJ is a tangent of the Pippian: the proof of the theorem is given in this place because the relation just obtained between α, β, γ is required for the proof of some of the other theorems.

10. To find the coordinates of the point G in which the line EF joining two conjugate poles again meets the Hessian.

We may take for the coordinates of G,

$$uX+vX', \quad uY+vY', \quad uZ+vZ';$$

and, substituting in the equation of the Hessian, the terms containing u^3, v^3 disappear, and the ratio $u : v$ is determined by a simple equation. It thus appears that we may write

$$u=-3l^2(XX'^2+YY'^2+ZZ'^2)+(1+2l^3)(Y'Z'X+Z'X'Y+X'Y'Z),$$
$$v=\ \ 3l^2(X^2X'+Y^2Y'+Z^2Z')-(1+2l^3)(YZX'+ZXY'+XYZ');$$

hence introducing, as before, the quantities ξ, η, ζ, α, β, γ, we find

$$uX+vX'=3l^2(\gamma\eta-\beta\zeta)+(1+2l^3)(X^2Y'Z'-X'^2YZ);$$

but from the first of the equations (B),

$$X^2Y'Z'-X'^2YZ=\frac{1}{2l}(\gamma\eta-\beta\zeta),$$

and therefore the preceding value of $uX+vX'$ becomes

$$\left(3l^2-\frac{1+2l^3}{2l}\right)(\gamma\eta-\beta\zeta),$$

which is equal to

$$\frac{-1+4l^3}{2l}(\gamma\eta-\beta\zeta).$$

Hence throwing out the constant factor, we find, for the coordinates of the point G, the values

$$\gamma\eta-\beta\zeta, \quad \alpha\zeta-\gamma\xi, \quad \beta\xi-\alpha\eta.$$

11. To find the coordinates of the point O.

Consider O as the point of intersection of the tangents to the Hessian at the points E, F, then the coordinates of O are proportional to the terms of

$$\left\|\begin{matrix} 3l^2X^2-\overline{1+2l^3}YZ, & 3l^2Y^2-\overline{1+2l^3}ZX, & 3l^2Z^2-\overline{1+2l^3}XY \\ 3l^2X'^2-\overline{1+2l^3}Y'Z', & 3l^2Y'^2-\overline{1+2l^3}Z'X', & 3l^2Z'^2-\overline{1+2l^3}X'Y' \end{matrix}\right\|$$

Hence the x-coordinate is proportional to

$$(3l^2Y^2 - \overline{1+2l^3}ZX)(3l^2Z'^2 - \overline{1+2l^3}X'Y') - (3l^2Z^2 - \overline{1+2l^3}XY)\,3l^2Y'^2 - \overline{1+2l^3}Z'X'),$$

which is equal to

$$9l^4(Y^2Z'^2 - Y'^2Z^2) + 3l^2(1+2l^3)\,YY'(XY' - X'Y) + 3l^2(1+2l^3)\,ZZ'(ZX' - Z'X)$$
$$- (1+2l^3)^2\,XX'(YZ' - Y'Z);$$

or introducing, as before, the quantities ξ, η, ζ, α, β, γ, to

$$-9l^3\alpha\xi + 3l^2(1+2l^3)(\beta\zeta + \gamma\eta) - (1+2l^3)^2\alpha\xi,$$
$$= (-1 - 13l^3 - 4l^6)\,\alpha\xi + 2l^2(1+2l^3)(\beta\zeta + \gamma\eta).$$

But by the first of the equations (A) $\beta\zeta + \gamma\eta = 2l\alpha\xi$, and the preceding value thus becomes $(-1 - 7l^3 + 8l^6)\,\alpha\xi$. Hence throwing out the constant factor the coordinates of the point O are found to be

$$\alpha\xi,\quad \beta\eta,\quad \gamma\zeta.$$

12. The points G, O are conjugate poles of the cubic.

Take a, b, c for the coordinates of G, and a', b', c' for the coordinates of O, we have

$$a,\ b,\ c = \gamma\eta - \beta\zeta,\quad \alpha\zeta - \gamma\xi,\quad \beta\xi - \alpha\eta,$$
$$a',\ b',\ c' = \alpha\xi\quad,\quad \beta\eta\quad,\quad \gamma\zeta.$$

These values give $aa' + l(bc' + b'c)$

$$= \alpha\xi(\gamma\eta - \beta\zeta) + l\,\{\beta\eta(\beta\xi - \alpha\eta) + \gamma\zeta(\alpha\zeta - \gamma\xi)\}$$
$$= \xi\eta(\alpha\gamma + l\beta^2) + \eta^2(-l\alpha\beta) + \zeta^2(l\alpha\gamma) + \xi\zeta(-\alpha\beta - l\gamma^2);$$

or substituting for $\xi\eta$, η^2, ζ^2, $\xi\zeta$ their values in terms of α, β, γ, this is

$$\left(-\frac{2}{l}\gamma^2 - \frac{1}{l^2}\alpha\beta\right)(\alpha\gamma + l\beta^2)$$
$$+\left(\frac{\beta^2}{l^2} - 4\gamma\alpha\right)(-l\alpha\beta)$$
$$+\left(\frac{\gamma^2}{l^2} - 4\alpha\beta\right)(l\alpha\gamma)$$
$$+\left(-\frac{2}{l}\beta^2 - \frac{1}{l}\alpha\gamma\right)(-\alpha\beta - l\gamma^2),$$

which is identically equal to zero. Hence, completing the system, we find

$$aa' + l(bc' + b'c) = 0,$$
$$bb' + l(ca' + c'a) = 0,$$
$$cc' + l(ab' + a'b) = 0,$$

equations which show that O (as well as G) is a point of the Hessian, and that the points G, O are corresponding poles of the cubic.

13. The line EF joining a pair of conjugate poles of the cubic is a tangent of the Pippian[1].

In fact, the equations (A), by the elimination of α, β, γ, give

$$-l(\xi^3+\eta^3+\zeta^3)+(-1+4l^3)\,\xi\eta\zeta=0,$$

which proves the theorem.

14. To find the equation of the pair of lines through F, and to show that these lines are tangents of the Pippian.

The equation of the pair of lines considered as the first or conic polar of the conjugate pole E, is

$$X(x^2+2lyz)+Y(y^2+2lzx)+Z(z^2+2lxy)=0.$$

Let one of the lines be

$$\lambda x+\mu y+\nu z=0,$$

then the other is

$$\frac{X}{\lambda}x+\frac{Y}{\mu}y+\frac{Z}{\nu}z=0;$$

and we find

$$\begin{aligned} 2lX\mu\nu - \quad Y\nu^2 - \quad Z\mu^2 &= 0,\\ -\quad X\nu^2 + 2lY\nu\lambda - \quad Z\lambda^2 &= 0,\\ -\quad X\mu^2 - \quad Y\lambda^2 + 2lZ\mu\nu &= 0, \end{aligned}$$

any two of which determine the ratios λ, μ, ν.

The elimination of X, Y, Z gives

$$\begin{vmatrix} 2l\mu\nu, & -\nu^2, & -\mu^2 \\ -\nu^2, & 2l\nu\lambda, & -\lambda^2 \\ -\mu^2, & -\lambda^2, & 2l\lambda\mu \end{vmatrix}=0,$$

which is equivalent to

$$\lambda\mu\nu\,\{-l(\lambda^3+\mu^3+\nu^3)+(-1+4l^3)\,\lambda\mu\nu\}=0;$$

or, omitting a factor, to

$$-l(\lambda^3+\mu^3+\nu^3)+(-1+4l^3)\,\lambda\mu\nu=0,$$

which shows that the line in question is a tangent of the Pippian.

15. To find the equation of the pair of lines through O.

The equation of the pair of lines through E is in like manner

$$X'(x^2+2lyz)+Y'(y^2+2lzx)+Z'(z^2+2lxy)=0;$$

[1] Steiner's curve R_0, in the particular case of a cubic basis-curve, is according to definition the envelope of the line EF, that is, the curve R_0 in the particular case in question is the Pippian.

and combining this with the foregoing equation,

$$X(x^2+2lyz)+Y(y^2+2lzx)+Z(z^2+2lxy)=0$$

of the pair of lines through F, viz. multiplying the two equations by

$$X^2X'+Y^2Y'+Z^2Z',\quad -(XX'^2+YY'^2+ZZ'^2),$$

and adding, then if as before

$$a : b : c = \gamma\eta-\beta\zeta : \alpha\zeta-\gamma\xi : \beta\xi-\alpha\eta,$$

we find as the equation of a conic passing through the points A, B, C, D, the equation

$$a(x^2+2lyz)+b(y^2+2lzx)+c(z^2+2lxy)=0.$$

But putting, as before,

$$a' : b' : c' = \alpha\xi : \beta\eta : \gamma\zeta,$$

then a', b', c' are the coordinates of the point O, and the equations

$$\begin{aligned} aa'+l(bc'+b'c)&=0,\\ bb'+l(ca'+c'a)&=0,\\ cc'+l(ab'+a'b)&=0,\end{aligned}$$

show that the conic in question is in fact the pair of lines through the point O.

16. To find the coordinates of the point Γ, which is the harmonic of G with respect to the points E, F.

The coordinates of the point in question are

$$uX-vX',\quad uY-vY',\quad uZ-vZ',$$

where u, v have the values given in No. 10, viz.

$$\begin{aligned} u&=-3l^2(XX'^2+YY'^2+ZZ'^2)+(1+2l^3)(Y'Z'X+Z'X'Y+X'Y'Z),\\ v&=\;\;\,3l^2(X^2X'+Y^2Y'+Z^2Z')-(1+2l^3)(YZX'+ZXY'+XYZ');\end{aligned}$$

these values give

$$\begin{aligned} uX-vX'=&-3l^2\{2X^2X'^2+(XY'+X'Y)YY'+(XZ'+X'Z)ZZ'\}\\ &+(1+2l^3)\{(XY'+X'Y)(XZ'+X'Z)+XX'(YZ'+Y'Z\};\end{aligned}$$

and therefore

$$\begin{aligned} uX-vX'&=-3l^2\left\{2\alpha^2-\frac{2}{l}\beta\gamma\right\}+(1+2l^3)\left\{\frac{1}{l^2}\beta\gamma-\frac{1}{l}\alpha^2\right\}\\ &=\frac{1}{l^2}(1+8l^3)(-l\alpha^2+\beta\gamma);\end{aligned}$$

and consequently, omitting the constant factor, the coordinates of Γ may be taken to be

$$-l\alpha^2+\beta\gamma,\quad -l\beta^2+\gamma\alpha,\quad -l\gamma^2+\alpha\beta.$$

17. The line through two consecutive positions of the point Γ is the line EF.

The coordinates of the point Γ are

$$-l\alpha^2+\beta\gamma, \quad -l\beta^2+\gamma\alpha, \quad -l\gamma^2+\alpha\beta\,;$$

and it has been shown that the quantities α, β, γ satisfy the equation

$$-l(\alpha^3+\beta^3+\gamma^3)+(-1+4l^3)\,\alpha\beta\gamma=0.$$

Hence, considering α, β, γ as variable parameters connected by this equation, the equation of the line through two consecutive positions of the point Γ is

$$\left|\begin{array}{cccc} & -3l\alpha^2+(-1+4l^3)\,\beta\gamma, & -3l\beta^2+(-1+4l^3)\,\gamma\alpha, & -3l\gamma^2+(-1+4l^3)\,\alpha\beta \\ x, & -2l\alpha \;, & \gamma \;, & \beta \\ y, & \gamma \;, & -2l\beta \;, & \alpha \\ z, & \beta \;, & \alpha \;, & -2l\gamma \end{array}\right| = 0\,;$$

and representing this equation by

$$Lx+My+Nz=0,$$

we find

$$\begin{aligned} L = \; & (4l^2\beta\gamma-\alpha^2)\left(-3l\alpha^2+(-1+4l^3)\beta\gamma\right) \\ & +(\alpha\beta+2l\gamma^2)\left(-3l\beta^2+(-1+4l^3)\,\gamma\alpha\right) \\ & +(\alpha\gamma+2l\beta^2)\left(-3l\gamma^2+(-1+4l^3)\,\alpha\beta\right); \end{aligned}$$

or, multiplying out and collecting,

$$L=3l\alpha^4+(-1-8l^3)\,\alpha^2\beta\gamma+(-5l+8l^4)\,(\alpha\beta^3+\alpha\gamma^3)+(-16l^2+16l^5)\,\beta^2\gamma^2\,;$$

but the equation

$$-l(\alpha^3+\beta^3+\gamma^3)+(-1+4l^3)\,\alpha\beta\gamma=0$$

gives

$$3l\alpha^4=-3l(\alpha\beta^3+\alpha\gamma^3)+(-3+12l^3)\,\alpha^2\beta\gamma,$$

and we have

$$\begin{aligned} L &= (-4+4l^3)\,\alpha^2\beta\gamma+(-8l+8l^4)\,(\alpha\beta^3+\alpha\gamma^3)+(-16l^2+16l^5)\,\beta^2\gamma^2 \\ &= (-4+4l^3)\left(\alpha^2\beta\gamma+2l(\alpha\beta^3+\alpha\gamma^3)+4l^2\beta^2\gamma^2\right) \\ &= (-4+4l^3)\,(\alpha\gamma+2l\beta^2)\,(\alpha\beta+2l\gamma^2)\,; \end{aligned}$$

or, in virtue of the equations (D),

$$L=(-4+4l^3)\,l^2\zeta\xi\,.\,l^2\xi\eta=(-4+4l^3)\,l^4\xi^2\eta\zeta=(-4+4l^3)\,l^4\xi\eta\zeta\,.\,\xi.$$

Hence, omitting the common factor, we find $L:M:N=\xi:\eta:\zeta$, and the equation $Lx+My+Nz=0$ becomes

$$\xi x+\eta y+\zeta z=0,$$

which is the equation of the line EF, that is, the line through two consecutive positions of Γ is the line EF; or what is the same thing, the line EF touches the Pippian in the point Γ which is the harmonic of G with respect to the points E, F.

18. The lineo-polar envelope of the line EF, with respect to the cubic, is the pair of lines OE, OF.

The equation of the pair of lines OE, OF, considered as the tangents to the Hessian at the points E, F, is

$$\left.\begin{array}{l}\{(3l^2X^2 - \overline{1+2l^3}\,Y\,Z\,)\,x + (3l^2Y^2 - \overline{1+2l^3}\,Z\,X\,)\,y + (3l^2Z^2 - \overline{1+2l^3}\,X\,Y\,)\,z\} \\ \times\ \{(3l^2X'^2 - \overline{1+2l^3}\,Y'Z')\,x + (3l^2\,Y'^2 - \overline{1+2l^3}\,Z'X')\,y + (3l^2Z'^2 - \overline{1+2l^3}\,X'Y')\,z\}\end{array}\right\} = 0.$$

Here on the left-hand side the coefficient of x^2 is

$$9l^4X^2X'^2 - 3l^2\,(1+2l^3)\,(X^2Y'Z' + X'^2YZ) + (1+2l^3)^2\,YY'ZZ',$$

which is equal to

$$9l^4\alpha^2 - 3l^2\,(1+2l^3)\,(l^2\beta\gamma + \frac{1}{l}\,\alpha^2) + (1+2l^3)^2\,\beta\gamma,$$

that is

$$\frac{1}{l}\,(-l+l^4)\,\{3l\alpha^2 + 2\,(1+2l^3)\,\beta\gamma\}\,;$$

and the coefficient of yz is

$$9l^4\,(Y^2Z'^2 + Y'^2Z^2) - 3l^2\,(1+2l^3)\,\big(YY'\,(XY' + X'Y) + ZZ'\,(XZ' + X'Z)\big)$$
$$+ (1+2l^3)^2\,XX'\,(YZ' + Y'Z),$$

which is equal to

$$9l^4\left(\frac{1}{l^2}\,\alpha^2 - 2\beta\gamma\right) - 3l^2\,(1+2l^3)\left(-\frac{2}{l}\,\beta\gamma\right) + (1+2l^3)^2\,\alpha\left(-\frac{1}{l}\,\alpha\right),$$

that is

$$\frac{1}{l}\,(-l+l^4)\,\{(1-4l^3)\,\alpha^2 - 6l^2\beta\gamma\}.$$

Hence completing the system and throwing out the constant factor, the equation of the pair of lines is

$$\begin{array}{l}(3l\alpha^2 + 2\,(1+2l^3)\,\beta\gamma,\quad 3l\beta^2 + 2\,(1+2l^3)\,\gamma\alpha,\quad 3l\gamma^2 + 2\,(1+2l^3)\,\alpha\beta, \\ \qquad (1-4l^3)\,\alpha^2 - 6l^2\beta\gamma,\quad (1-4l^3)\,\beta^2 - 6l^2\gamma\alpha,\quad (1-4l^3)\,\gamma^2 - 6l^2\alpha\beta\between x,\ y,\ z)^2 = 0.\end{array}$$

But the equation of the line EF is $\xi x + \eta y + \zeta z = 0$, and the equation of its lineo-polar envelope is

$$\left|\begin{array}{cccc} & \xi, & \eta, & \zeta \\ \xi, & x, & lz, & ly \\ \eta, & lz, & y, & lx \\ \zeta, & ly, & x, & z \end{array}\right| = 0\,;$$

or expanding,

$$(yz - l^2x^2,\ \ zx - l^2y^2,\ \ xy - l^2z^2,\ \ l^2yz - lx^2,\ \ l^2zx - ly^2,\ \ l^2xy - lz^2 \between \xi,\ \eta,\ \zeta)^2 = 0\,;$$

or arranging in powers of x, y, z,

$$(-l^2\xi^2 - 2l\eta\zeta,\ \ -l^2\eta^2 - 2l\zeta\xi,\ \ -l^2\zeta^2 - 2l\xi\eta,\ \ \tfrac{1}{2}\xi^2 + l^2\eta\zeta,\ \ \tfrac{1}{2}\eta^2 + l^2\zeta\xi,\ \ \tfrac{1}{2}\zeta^2 + l^2\xi\eta \between x,\ y,\ z)^2 = 0:$$

and if in this equation we replace ξ^2, &c. by their values in terms of α, β, γ, as given by the equations (D), we obtain the equation given as that of the pair of lines OE, OF.

19. It remains to prove the theorem with respect to the connexion of the lines EF, IJ.

The equations (A) show that the two lines

$$\xi x + \eta y + \zeta z = 0,$$
$$\alpha x + \beta y + \zeta z = 0,$$

(where ξ, η, ζ and α, β, γ have the values before attributed to them) are conjugate polars with respect to the curve of the third class,

$$l(\xi^3 + \eta^3 + \zeta^3) - 3\xi\eta\zeta = 0,$$

in which equation ξ, η, ζ denote current line coordinates. The curve in question is of the form $APU + BQU = 0$. We have, in fact, identically,

$$3T\,.\,PU - 4S\,.\,QU = (1 + 8l^3)^2\,\{l(\xi^3 + \eta^3 + \zeta^3) - 3\xi\eta\zeta\}.$$

It is clear that the curve in question must have the curve $PU = 0$ for its Hessian; and in fact, in the formula of my Third Memoir, [144]

$$\begin{aligned} H(6\alpha PU + \beta QU) = &(-2T,\ 48S^2,\ 18TS,\ T^2 + 16S^3 \between \alpha,\ \beta)^3\,PU \\ &+ (\ \ 8S,\ \ T,\ -8S^2,\ \ -TS \between \alpha,\ \beta)^3\,QU, \end{aligned}$$

the coefficient of QU is

$$(8S\alpha + T\beta)\,(\alpha^2 - S\beta^2)\,;$$

and therefore, putting $\alpha = \tfrac{1}{2}T$, $\beta = -4S$, we find

$$H(3T\,.\,PU - 4S\,.\,QU) = -\tfrac{1}{4}\,(T^2 - 64S^3)^2\,PU.$$

Article No. 20.—*Theorem relating to the curve of the third class, mentioned in the preceding Article.*

20. The consideration of the curve $3T\,.\,PU - 4S\,.\,QU = 0$, gives rise to another geometrical theorem. Suppose that the line $(\xi,\ \eta,\ \zeta)$, that is, the line whose equation is $\xi x + \eta y + \zeta z = 0$, is with respect to this curve of the third class one of the four polars of a point $(X,\ Y,\ Z)$ of the Hessian, and that it is required to find the envelope of the line $\xi x + \eta y + \zeta z = 0$.

We have

$$X : Y : Z = l\xi^2 - \eta\zeta : l\eta^2 - \zeta\xi : l\zeta^2 - \xi\eta,$$

and X, Y, Z are to be eliminated from these equations, and the equation

$$l^2(X^3 + Y^3 + Z^3) - (1 + 2l^3) XYZ = 0$$

of the Hessian. We have

$$\begin{aligned} X^3 + Y^3 + Z^3 = {} & l^3(\xi^3 + \eta^3 + \zeta^3)^2 \\ & - 3l^2(\xi^3 + \eta^3 + \zeta^3)\,\xi\eta\zeta \\ & + 9l\ \xi^2\eta^2\zeta^2 \\ & - (1 + 2l^3)(\eta^3\zeta^3 + \zeta^3\xi^3 + \xi^3\eta^3), \\ XYZ = {} & l(\xi^3 + \eta^3 + \zeta^3)\,\xi\eta\zeta \\ & + \ (-1 + l^3)\,\xi^2\eta^2\zeta^2 \\ & - l^2(\eta^3\zeta^3 + \zeta^3\xi^3 + \xi^3\eta^3), \end{aligned}$$

and thence

$$\begin{aligned} HU = {} & l^5\ (\xi^3 + \eta^3 + \zeta^3)^2 \\ & - (l + 5l^4)(\xi^3 + \eta^3 + \zeta^3)\,\xi\eta\zeta \\ & + (1 + 10l^3 - 2l^6)\,\xi^2\eta^2\zeta^2\,; \end{aligned}$$

and equating the right-hand side to zero, we have the equation in line coordinates of the curve in question, which is therefore a curve of the sixth class in quadratic syzygy with the Pippian and Quippian.

Article No. 21.—*Geometrical definition of the Quippian.*

21. I have not succeeded in obtaining any good geometrical definition of the Quippian, and the following is only given for want of something better.

The curve

$$T.PU\{P\,6H(\alpha U + 6\beta HU)\} - P(6HU)\{T(\alpha U + 6\beta HU)\,.\,P(\alpha U + 6\beta HU)\} = 0,$$

which is derived in what may be taken to be a known manner from the cubic, is in general a curve of the sixth class. But if the syzygetic cubic $\alpha U + 6\beta HU = 0$ be properly selected, viz. if this curve be such that its Hessian breaks up into three lines, then both the Pippian of the cubic $\alpha U + 6\beta HU = 0$, and the Pippian of its Hessian will break up into the same three points, which will be a portion of the curve of the sixth class, and discarding these three points the curve will sink down to one of the third class, and will in fact be the Quippian of the cubic.

To show this we may take

$$\alpha U + 6\beta HU = x^3 + y^3 + z^3, = 0$$

as the equation of the syzygetic cubic satisfying the prescribed condition, for this value in fact gives

$$H(\alpha U + 6\beta HU) = -xyz, = 0,$$

a system of three lines. We find, moreover,

$$P(\alpha U + 6\beta HU) = P(x^3 + y^3 + z^3), = -\xi\eta\zeta$$

and

$$P\{6H(\alpha U + 6\beta HU)\} = P(-6xyz), = -4\xi\eta\zeta,$$

the latter equation being obtained by first neglecting all but the highest power of l in the expression of PU, and then writing $l = -1$: we have also $T(\alpha U + 6\beta HU) = 1$. Substituting the above values, the curve of the sixth class is

$$\xi\eta\zeta\{-4T \,.\, PU + P(6HU)\} = 0\,;$$

or throwing out the factor $\xi\eta\zeta$, we have the curve of the third class,

$$-4T \,.\, PU + P(6HU) = 0.$$

Now the general expression in my Third Memoir, viz.

$$P(\alpha U + 6\beta HU) = (\alpha^3 + 12S\alpha\beta^2 + 4T\beta^3)\, PU + (\alpha^2\beta - 4S\beta^3)\, QU,$$

putting $\alpha = 0$, $\beta = 1$, gives

$$P(6HU) = 4T \,.\, PU - 4S \,.\, QU,$$

or what is the same thing,

$$-4T \,.\, PU + P(6HU) = -4S \,.\, QU\,;$$

and the curve of the third class is therefore the Quippian $QU = 0$. It may be remarked, that for a cubic $U = 0$ the Hessian of which breaks up into three lines, the above investigation shows that we have $PU = -\xi\eta\zeta$, $P(6HU) = -4\xi\eta\zeta$, and $T = 1$, and consequently that $-4T.PU + P(6HU)$ ought to vanish identically; this in fact happens in virtue of the factor S on the right-hand side, the invariant S of a cubic of the form in question being equal to zero; the appearance of the factor S on the right-hand side is thus accounted for *à priori*.

Article No. 22.—*Theorem relating to a line which meets three given conics in six points in involution.*

22. The envelope of a line which meets three given conics, the first or conic polars of any three points with respect to the cubic, in six points in involution, is the Pippian.

It is readily seen that if the theorem is true with respect to the three conics,

$$\frac{dU}{dx} = 0, \quad \frac{dU}{dy} = 0, \quad \frac{dU}{dz} = 0,$$

it is true with respect to any three conics whatever of the form

$$\lambda \frac{dU}{dx} + \mu \frac{dU}{dy} + \nu \frac{dU}{dz} = 0,$$

that is, with respect to any three conics, each of them the first or conic polar of some point (λ, μ, ν) with respect to the cubic. Considering then these three conics, take $\xi x + \eta y + \zeta z = 0$ as the equation of the line, and let (X, Y, Z) be the coordinates of a point of intersection with the first conic, we have

$$\begin{aligned} \xi X + \eta Y + \zeta Z &= 0, \\ X^2 + 2lYZ \quad &= 0; \end{aligned}$$

and combining with these a linear equation

$$\alpha X + \beta Y + \gamma Z = 0,$$

in which (α, β, γ) are arbitrary quantities, we have

$$X : Y : Z = \gamma\eta - \beta\zeta : \alpha\zeta - \gamma\xi : \beta\xi - \alpha\eta;$$

and hence

$$(\gamma\eta - \beta\zeta)^2 + 2l(\alpha\zeta - \gamma\xi)(\beta\xi - \alpha\eta) = 0,$$

an equation in (α, β, γ) which is in fact the equation in line coordinates of the two points of intersection with the first conic. Developing and forming the analogous equations, we find

$$\begin{aligned} &(-2l\eta\zeta, \quad \zeta^2, \quad \eta^2, \quad -\eta\zeta - l\xi^2, \quad l\xi\eta, \quad l\xi\zeta \quad \between \alpha, \beta, \gamma)^2 = 0, \\ &(\zeta^2, \quad -2l\zeta\xi, \quad \xi^2, \quad l\xi\eta, \quad -\zeta\xi - l\eta^2, \quad l\eta\zeta \quad \between \alpha, \beta, \gamma)^2 = 0, \\ &(\eta^2, \quad \xi^2, \quad -2l\xi\eta, \quad l\zeta\xi, \quad l\eta\zeta, \quad -\xi\eta - l\zeta^2 \between \alpha, \beta, \gamma)^2 = 0, \end{aligned}$$

which are respectively the equations in line coordinates of the three pairs of intersections.

Now combining these equations with the equation $\gamma = 0$, we have the equations of the pairs of lines joining the points of intersection with the point $(x = 0, y = 0)$, and if the six points are in involution, the six lines must also be in involution, or the condition for the involution of the six points is

$$\begin{vmatrix} -2l\eta\zeta, & \zeta^2, & l\xi\zeta \\ \zeta^2, & -2l\zeta\eta, & l\eta\zeta, \\ \eta^2, & \xi^2, & -\xi\eta - l\zeta^2 \end{vmatrix} = 0,$$

that is,

$$4l^2\zeta^2\xi\eta(-\xi\eta - l\zeta^2) + l\eta^3\zeta^3 + l\xi^3\zeta^3 + 2l^2\xi^2\eta^2\zeta^2 + 2l^2\xi^2\eta^2\zeta^2 + \zeta^4(-\xi\eta - l\zeta^2) = 0;$$

or, reducing and throwing out the factor ζ^3, we find

$$-l(\xi^3 + \eta^3 + \zeta^3) + (-1 + 4l^3)\xi\eta\zeta = 0,$$

which shows that the line in question is a tangent of the Pippian.

It is to be remarked that any three conics whatever may be considered as the first or conic polars of three properly selected points with respect to a properly selected cubic curve. The theorem applies therefore to any three conics whatever, but in this case the cubic curve is not given, and the Pippian therefore stands merely for a curve of the third class, and the theorem is as follows, viz. the envelope of a line which meets any three conics in six points in involution, is a curve of the third class.

Article No. 23.—*Completion of the theory in Liouville, and comparison with analogous theorems of* HESSE.

In order to convert the foregoing theorem into its reciprocal, we must replace the cubic $U=0$ by a curve of the third class, that is we must consider the coordinates which enter into the equation as line coordinates; and it of course follows that the coordinates which enter into the equation $PU=0$ must be considered as point coordinates, that is we must consider the Pippian as a curve of the third order: we have thus the theorem; The locus of a point such that the tangents drawn from it to three given conics (the first or conic poles of any three lines with respect to a curve of the third class) form a pencil in involution, is the Pippian considered as a curve of the third order. This in fact completes the fundamental theorem in my memoirs in *Liouville* above referred to, and establishes the analogy with Hesse's results in relation to the Hessian; to show this I set out the two series of theorems as follows:

Hesse, in his memoirs On Curves of the Third Order and Curves of the Third Class, *Crelle*, tt. XXVIII. XXXVI. and XXXVIII. [1844, 1848, 1849], has shown as follows:

(α) The locus of a point such that its polars with respect to the three conics $X=0$, $Y=0$, $Z=0$ (or more generally its polars with respect to all the conics of the series $\lambda X+\mu Y+\nu Z=0$) meet in a point, is a curve of the third order $V=0$.

(β) Conversely, given a curve of the third order $V=0$, there exists a series of conics such that the polars with respect to all the conics of any point whatever of the curve $V=0$, meet in a point.

(γ) The equation of any one of the conics in question is

$$\lambda\frac{dU}{dx}+\mu\frac{dU}{dy}+\nu\frac{dU}{dz}=0,$$

that is, the conic is the first or conic polar of a point (λ, μ, ν) with respect to a certain curve of the third order $U=0$; and this curve is determined by the condition that its Hessian is the given curve $V=0$, that is, we have $V=HU$.

(δ) The equation $V=HU$ is solved by assuming $U=aV+bHV$, for we have then $H(aV+bHV)=AV+BHV$, where A, B are given cubic functions of a, b, and thence $V=HU=AV+BHV$, or $A=1$, $B=0$; the latter equation gives what is alone important, the ratio $a:b$; and it thus appears that there are three distinct series of conics, each of them having the above-mentioned relation to the given curve of the third order $V=0$.

In the memoirs in *Liouville* above referred to, I have in effect shown that—

(α') The locus of a point such that the tangents from it to three conics, represented in line coordinates by the equations $X=0$, $Y=0$, $Z=0$ (or more generally with respect to any three conics of the series $\lambda X+\mu Y+\nu Z=0$) form a pencil in involution, is a curve of the third order $V=0$.

(β') Conversely, given a curve of the third order $V=0$, there exists a series of conics such that the tangents from any point whatever of the curve to any three of the conics, form a pencil in involution.

Now, considering the coordinates which enter into the equation of the Pippian as point coordinates, and consequently the Pippian as a curve of the third order, I am able to add as follows:

(γ') The equation in line coordinates of any one of the conics in question is

$$\lambda\frac{dU}{d\xi}+\mu\frac{dU}{d\eta}+\nu\frac{dU}{d\zeta}=0,$$

that is, the conic is the first or conic polar of a line (λ, μ, ν) with respect to a certain curve of the third class $U=0$; and this curve is determined by the condition that its Pippian is the given curve of the third order $V=0$, that is, we have $V=PU$.

(δ') The equation $V=PU$ is solved by assuming $U=aPV+bQV$, for we have then $P(aPV+bQV)=AV+BHV$, where A and B are given cubic functions of a, b; and thence $V=PU=AV+BHV$, or $A=1$, $B=0$; the latter equation gives what is alone important, the ratio $a:b$; and it thus appears that there are three distinct curves of the third class $U=0$, and therefore (what indeed is shown in the Memoirs in *Liouville*) three distinct series of conics having the above-mentioned relation to the given curve of the third order $V=0$.

It is hardly necessary to remark that the preceding theorems, although precisely analogous to those of Hesse, are entirely distinct theorems, that is the two series are not connected together by any relation of reciprocity.

Article Nos. 24 to 28.—*Various investigations and theorems.*

24. Reverting to the theorem (No. 18), that the lineo-polar envelope of the line EF is the pair of lines OE, OF; the line EF is any tangent of the Pippian, hence the theorem includes the following one:

The lineo-polar envelope with respect to the cubic, of any tangent of the Pippian, is a pair of lines.

And conversely,

The Pippian is the envelope of a line such that the lineo-polar envelope of the line with respect to the cubic is a pair of lines.

It is I think worth while to give an independent proof. It has been shown that the equation of the lineo-polar envelope with respect to the cubic, of the line $\xi x + \eta y + \zeta z = 0$ (where ξ, η, ζ are arbitrary quantities), is

$$(-l^2\xi^2 - 2l\eta\zeta,\ \ -l^2\eta^2 - 2l\zeta\xi,\ \ -l^2\zeta^2 - 2l\xi\eta,\ \ \tfrac{1}{2}\xi^2 + l^2\eta\zeta,\ \ \tfrac{1}{2}\eta^2 + l^2\zeta\xi,\ \ \tfrac{1}{2}\zeta^2 + l^2\xi\eta \between x,\ y,\ x)^2 = 0\,;$$

and representing this equation by

$$\tfrac{1}{2}(a,\ b,\ c,\ f,\ g,\ h \between x,\ y,\ z)^2 = 0,$$

we find

$$\begin{aligned}
bc - f^2 &= \xi\,(-\xi^3 + 8l^3\eta^3 + 8l^3\zeta^3 + 12l^2\xi\eta\zeta),\\
ca - g^2 &= \eta\,(8l^3\xi^3 - \eta^3 + 8l^3\zeta^3 + 12l^2\xi\eta\zeta),\\
ab - h^2 &= \zeta\,(8l^3\xi^3 + 8l^3\eta^3 - \zeta^3 + 12l^2\xi\eta\zeta),
\end{aligned}$$

$$\begin{aligned}
gh - af &= \xi\,\big(2l^2(\xi^3 + \eta^3 + \zeta^3) + 4l(1 + 2l^3)\,\xi\eta\zeta\big) + (1 + 8l^3)\,\eta^2\zeta^2,\\
hf - bg &= \eta\,\big(2l^2(\xi^3 + \eta^3 + \zeta^3) + 4l(1 + 2l^3)\,\xi\eta\zeta\big) + (1 + 8l^3)\,\zeta^2\xi^2,\\
fg - ch &= \zeta\,\big(2l^2(\xi^3 + \eta^3 + \zeta^3) + 4l(1 + 2l^3)\,\xi\eta\zeta\big) + (1 + 8l^3)\,\xi^2\eta^2\,;
\end{aligned}$$

and after all reductions,

$$\begin{aligned}
&abc - af^2 - bg^2 - ch^2 + 2fgh\\
&\quad = [-l(\xi^3 + \eta^3 + \zeta^3) + (-1 + 4l^3)\,\xi\eta\zeta]^2 = (PU)^2,
\end{aligned}$$

or the condition in order that the conic may break up into a pair of lines is $PU = 0$.

25. The following formulæ are given in connexion with the foregoing investigation, but I have not particularly considered their geometrical signification. The lineo-polar envelope of an arbitrary line $\xi x + \eta y + \zeta z = 0$, with respect to the cubic

$$x^3 + y^3 + z^3 + 6lxyz = 0,$$

has been represented by

$$(a,\ b,\ c,\ f,\ g,\ h \between x,\ y,\ z)^2 = 0\,;$$

and if in like manner we represent the lineo-polar envelope of the same line, with respect to a syzygetic cubic

$$x^3 + y^3 + z^3 + 6l'xyz = 0,$$

by

$$(a',\ b',\ c',\ f',\ g',\ h' \between x,\ y,\ z)^2 = 0,$$

then we have

$$a'(bc-f^2)+b'(ca-g^2)+c'(ab-h^2)+2f'(gh-af)+2g'(hf-bg)+2h'(fg-ch)$$
$$\begin{aligned}= \quad & (l'^2+2l^2)(\xi^3+\eta^3+\zeta^3)^2 \\ & +(2l'+4l-32l^3l'+8l^4)(\xi^3+\eta^3+\zeta^3)\xi\eta\zeta \\ & +(24ll'^2+48l^4l'^2-72l^2l'+24l^3+3)\xi^2\eta^2\zeta^2,\end{aligned}$$

which may be verified by writing $l'=l$, in which case the right-hand side becomes as it should do, $3(PU)^2$. If $l'=-\dfrac{1+2l^3}{6l^2}$, that is, if the syzygetic cubic be the Hessian, then the formula becomes

$$a'(bc-f^2)+\&\text{c.}=\frac{1}{36l^4}\left\{\begin{aligned} & (1+4l^3+76l^6)(\xi^3+\eta^3+\zeta^3)^2 \\ & +12l^2(-1+26l^3+56l^6)(\xi^3+\eta^3+\zeta^3)\xi\eta\zeta \\ & +12l\ (2+57l^3+168l^6+16l^{10})\xi^2\eta^2\zeta^2\end{aligned}\right\}$$

which is equal to

$$\frac{1}{36l^4}\left\{\overline{QU}^2-24S.\overline{PU}^2\right\}.$$

26. The equation

$$(bc'+b'c-2ff', \ldots gh'+g'h-af'-a'f, \ldots \between \xi, \eta, \zeta)^2=0$$

is the equation in line coordinates of a conic, the envelope of the line which cuts harmonically the conics

$$(a, b, c, f, g, h\between x, y, z)^2=0,$$
$$(a', b', c', f', g', h'\between x, y, z)^2=0;$$

and if a, b, &c., a', &c. have the values before given to them, then the coefficients of the equation are

$$\begin{aligned}
bc'+b'c-2ff' &= \xi\{-\xi^3+4ll'(l+l')(\eta^3+\zeta^3)+(16ll'-2l^2-2l'^2)\xi\eta\zeta, \\
ca'+c'a-2gg' &= \eta\{-\eta^3+4ll'(l+l')(\zeta^3+\xi^3)+(16ll'-2l^2-2l'^2)\xi\eta\zeta, \\
ab'+a'b-2hh' &= \zeta\{-\zeta^3+4ll'(l+l')(\xi^3+\eta^3)+(16ll'-2l^2-2l'^2)\xi\eta\zeta,
\end{aligned}$$

$$\begin{aligned}
gh'+g'h-af'-a'f &= \xi\{(l^2+l'^2)(\xi^3+\eta^3+\zeta^3)+(2l+2l'+8l^2l'^2)\xi\eta\zeta\}+\left(1+4ll'(l+l')\right)\eta^2\zeta^2, \\
hf'+h'f-bg'-b'g &= \eta\{(l^2+l'^2)(\xi^3+\eta^3+\zeta^3)+(2l+2l'+8l^2l'^2)\xi\eta\zeta\}+\left(1+4ll'(l+l')\right)\zeta^2\xi^2, \\
fg'+f'g-ch'-c'h &= \zeta\{(l^2+l'^2)(\xi^3+\eta^3+\zeta^3)+(2l+2l'+8l^2l'^2)\xi\eta\zeta\}+\left(1+4ll'(l+l')\right)\xi^2\eta^2:
\end{aligned}$$

and we thence obtain

$$\begin{aligned}(bc'+b'c-2ff', \ldots, gh'+g'h-af'-a'f, \ldots\between \xi, \eta, \zeta)^2 = & \\ & -(\xi^3+\eta^3+\zeta^3)^2 \\ & +(l^2+l'^2+16ll')(\xi^3+\eta^3+\zeta^3)\xi\eta\zeta \\ & +(6l+6l'+24l^2l'^2)\xi^2\eta^2\zeta^2 \\ & +\left(4+16(l^2l'+ll'^2)\right)(\eta^3\zeta^3+\zeta^3\xi^3+\xi^3\eta^3), =0\end{aligned}$$

as the condition which expresses that a line $\xi x + \eta y + \zeta z = 0$ cuts harmonically its lineo-polar envelopes with respect to the cubic and with respect to a syzygetic cubic.

27. To find the locus of a point such that its second or line polar with respect to the cubic may be a tangent of the Pippian. Let the coordinates of the point be (x, y, z); then if $\xi x + \eta y + \zeta z = 0$ be the equation of the polar, we have

$$\xi : \eta : \zeta = x^2 + 2lyz : y^2 + 2lzx : z^2 + 2lxy,$$

and the line in question being a tangent to the Pippian,

$$-l(\xi^3 + \eta^3 + \zeta^3) + (-1 + 4l^3)\,\xi\eta\zeta = 0.$$

But the preceding values give

$$\begin{aligned} \xi^3 + \eta^3 + \zeta^3 &= (x^3 + y^3 + z^3)^2 + 6l\,(x^3 + y^3 + z^3)\,xyz + 36l^2x^2y^2z^2 + (-2 + 8l^3)\,(y^3z^3 + z^3x^3 + x^2y^3) \\ \xi\eta\zeta &= 4l^2\,(x^3 + y^3 + z^3)\,xyz + (1 + 8l^3)\,x^2y^2z^2 + 2l\,(y^3z^3 + z^3x^3 + x^3y^3); \end{aligned}$$

and we have therefore

$$l\,(x^3 + y^3 + z^3)^2 + (10l^2 - 16l^5)\,(x^3 + y^3 + z^3)\,xyz + (1 + 40l^3 - 32l^6)\,x^2y^2z^2 = 0\,;$$

or introducing U, HU in place of $x^3 + y^3 + z^3$, xyz, the equation becomes

$$-S\,.\,U^2 + (HU)^2 = 0,$$

which is the equation of the locus in question.

28. The locus of a point such that its second or line polar with respect to the cubic is a tangent of the Quippian, is found in like manner by substituting the last-mentioned values of ξ, η, ζ in the equation

$$QU = (1 - 10l^3)\,(\xi^3 + \eta^3 + \zeta^3) - 6l^2\,(5 + 4l^3)\,\xi\eta\zeta.$$

We find as the equation of the locus,

$$\begin{aligned} (1 - 10l^3)\,(x^3 + y^3 + z^3)^2 + 6l\,(1 - 30l^3 - 16l^6)\,(x^3 + y^3 + z^3)\,xyz + 6l^2\,(1 - 104l^3 - 32l^6)\,x^2y^2z^2 \\ - 2\,(1 + 8l^3)^2\,(y^3z^3 + z^3x^3 + x^3y^3) = 0, \end{aligned}$$

where the function on the left-hand side is the octicovariant $\Theta_{\prime\prime}U$ of my Third Memoir, the covariant having been in fact defined so as to satisfy the condition in question. And I have given in the memoir the following expression for $\Theta_{\prime\prime}U$, viz.

$$\begin{aligned} \Theta_{\prime\prime}U = &\,(1 - 16l^3 - 6l^6)\,U^2 \\ &+ (6l\,)\,U\,.\,HU \\ &+ (6l^2\,)\,(HU)^2 \\ &- 2\,(1 + 8l^3)^2\,(y^3z^3 + z^3x^3 + x^3y^3). \end{aligned}$$

Article Nos. 29 to 31.—*Formulæ for the intersection of a cubic curve and a line.*

29. If the line $\xi x + \eta y + \zeta z = 0$ meet the cubic

$$x^3 + y^3 z^3 + 6lxyz = 0$$

in the points

$$(x_1, y_1, z_1), \quad (x_2, y_2, z_2), \quad (x_3, y_3, z_3),$$

then we have

$$x_1 x_2 x_3 : y_1 y_2 y_3 : z_1 z_2 z_3 = \eta^3 - \zeta^3 : \zeta^3 - \xi^3 : \xi^3 - \eta^3.$$

It will be convenient to represent the equation of the cubic by the abbreviated notation $(1, 1, 1, l \between x, y, z)^3 = 0$; we have the two equations

$$\begin{aligned} &(1, 1, 1, l \between x, y, z)^3 = 0, \\ &\xi x + \eta y + \zeta z \qquad\quad = 0; \end{aligned}$$

and if to these we join a linear equation with arbitrary coefficients,

$$\alpha x + \beta y + \gamma z = 0,$$

then the second and third equations give

$$x : y : z = \beta\zeta - \gamma\eta : \gamma\xi - \alpha\zeta : \alpha\eta - \beta\xi;$$

and substituting these values in the first equation, we obtain the resultant of the system. But this resultant will also be obtained by substituting, in the third equation, a system of simultaneous roots of the first and second equations, and equating to zero the product of the functions so obtained[1]. We must have therefore

$$(1, 1, 1, l \between \beta\zeta - \gamma\eta, \gamma\xi - \alpha\zeta, \alpha\eta - \beta\xi)^3 = (\alpha x_1 + \beta y_1 + \gamma z_1)(\alpha x_2 + \beta y_2 + \gamma z_2)(\alpha x_3 + \beta y_3 + \gamma z_3);$$

and equating the coefficients of α^3, β^3, γ^3, we obtain the above-mentioned relations.

30. If a tangent to the cubic

$$x^3 + y^3 + z^3 + 6lxyz = 0$$

at a point (x_1, y_1, z_1) of the cubic meet the cubic in the point (x_3, y_3, z_3), then

$$x_3 : y_3 : z_3 = x_1(y_1^3 - z_1^3) : y_1(z_1^3 - x_1^3) : z_1(x_1^3 - y_1^3).$$

For if the equation of the tangent is $\xi x + \eta y + \zeta z = 0$, then

$$x_1^2 x_3 : y_1^2 y_3 : z_1^2 z_3 = \eta^3 - \zeta^3 : \zeta^3 - \xi^3 : \xi^3 - \eta^3,$$

and

$$\xi : \eta : \zeta = x_1^2 + 2l y_1 z_1 : y_1^2 + 2l z_1 x_1 : z_1^2 + 2l x_1 y_1.$$

[1] This is in fact the general process of elimination given in Schläfli's Memoir, "Ueber die Resultante einer Systemes mehrerer algebraischer Gleichungen," Vienna Trans. 1852. [But the process was employed much earlier, by Poisson.]

These values give

$$\eta^3 - \zeta^3 = (y_1^3 - z_1^3)(y_1^3 + z_1^3 + 6lx_1y_1z_1 - 8l^3x_1^3)$$
$$= (y_1^3 - z_1^3) \times -(1 + 8l^3)\, x_1^3,$$

since (x_1, y_1, z_1) is a point of the cubic; and forming in like manner the values of $\zeta^3 - \xi^3$ and $\xi^3 - \eta^3$, we obtain the theorem.

31. The preceding values of (x_3, y_3, z_3) ought to satisfy

$$(x_1^2 + 2ly_1z_1)\, x_3 + (y_1^2 + 2lz_1x_1)\, y_3 + (z_1^2 + 2lx_1y_1)\, z_3 = 0,$$
$$x_3^2 + y_3^2 + z_3^2 + 6lx_3y_3z_3 = 0\,;$$

in fact the first equation is satisfied identically, and for the second equation we obtain

$$x_3^2 + y_3^2 + z_3^2 = x_1^3(y_1^3 - z_1^3)^3 + y_1^3(z_1^3 - x_1^3)^3 + z_1^3(x_1^3 - y_1^3)^2$$
$$= -x_1^9(y_1^3 - z_1^3) - y_1^9(z_1^3 - x_1^3) - z_1^9(x_1^3 - y_1^3)$$
$$= (x_1^3 + y_1^3 + z_1^3)(y_1^3 - z_1^3)(z_1^3 - x_1^3)(x_1^3 - y_1^3),$$
$$x_3y_3z_3 = x_1y_1z_1(y_1^3 - z_1^3)(z_1^3 - x_1^3)(x_1^3 - y_1^3),$$

and consequently

$$x_3^3 + y_3^3 + z_3^3 + 6lx_3y_3z_3 = (x_1^3 + y_1^3 + z_1^3 + 6lx_1y_1z_1)(y_1^3 - z_1^3)(z_1^3 - x_1^3)(x_1^3 - y_1^3) = 0,$$

which verifies the theorem. It is proper to add (the remark was made to me by Professor Sylvester) that the foregoing values

$$x_3 : y_3 : z_3 = x_1(y_1^3 - z_1^3) : y_1(z_1^3 - x_1^3) : z_1(x_1^3 - y_1^3)$$

satisfy *identically* the relation

$$\frac{x_3^3 + y_3^3 + z_3^3}{x_3y_3z_3} = \frac{x_1^3 + y_1^3 + z_1^2}{x_1y_1z_1}.$$

Article Nos. 32 to 34.—*Formulæ for the Satellite line and point.*

32. The line $\xi x + \eta y + \zeta z = 0$ meets the cubic

$$x^3 + y^3 + z^3 + 6lxyz = 0$$

in three points, and the tangents to the cubic at these points meet the cubic in three points lying in a line, which has been called the Satellite line of the given line.

To find the equation of the satellite line; suppose that (x_1, y_1, z_1), (x_2, y_2, z_2), (x_3, y_3, z_3) are the coordinates of the point in which the given line meets the cubic; then we have, as before,

$$(1, 1, 1, l \between \beta\zeta - \gamma\eta,\ \gamma\xi - \alpha\zeta,\ \alpha\eta - \beta\xi)^3 = (\alpha x_1 + \beta y_1 + \gamma z_1)(\alpha x_2 + \beta y_2 + \gamma z_2)(\alpha x_3 + \beta y_3 + \gamma z_3).$$

The equation of the three tangents is

$$\left.\begin{aligned}\Pi = &[(x_1^2 + 2ly_1z_1)\,x + (y_1^2 + 2lz_1x_1)\,y + (z_1^2 + 2lx_1y_1)\,z]\\ \times &[(x_2^2 + 2ly_2z_2)\,x + (y_2^2 + 2lz_2x_2)\,y + (z_2^2 + 2lx_2y_2)\,z]\\ \times &[(x_3^2 + 2ly_3z_3)\,x + (y_3^2 + 2lz_3x_3)\,y + (z_3^2 + 2lx_3y_3)\,z]\end{aligned}\right\} = 0,$$

and if we put

$$F = (\xi^3 + \eta^3 + \zeta^3)^2 - 24l^2(\xi^3 + \eta^3 + \zeta^3)\,\xi\eta\zeta + (-24l - 48l^4)\,\xi^2\eta^2\zeta^2 + (-4 + 32l^3)(\eta^3\zeta^3 + \zeta^3\xi^3 + \xi^3\eta^3),$$

(F is the reciprocant FU of my Third Memoir), then we have identically

$$F\,.\,U - \ \Pi = (\xi x + \eta y + \zeta z)^2(\xi' x + \eta' y + \zeta' z),$$

and the equation of the satellite line is $\xi' x + \eta' y + \zeta' z = 0$. In fact the geometrical theory shows that we must have

$$F\,.\,U - N\Pi = (\xi x + \eta y + \zeta z)^2(\xi' x + \eta' y + \zeta' z),$$

and it is then clear that N is a mere number. To determine its value in the most simple manner, write $l = 0$, $y = 0$, $x = \zeta$, $z = -\xi$, we have then $F\,.\,U - N\Pi = 0$, where

$$F = \xi^6 + \eta^6 + \zeta^6 - 2\eta^3\zeta^3 - 2\zeta^3\xi^3 - 2\xi^3\eta^3, \quad U = \zeta^3 - \xi^3.$$

The value of Π is $\Pi = F\,.\,U$, and we thus obtain $N = 1$. For, substituting the above values,

$$\begin{aligned}\Pi &= (x_1^2\zeta - z_1^2\xi)(x_2^2\zeta - z_2^2\xi)(x_3^2\zeta - z_3^2\xi)\\ &= \quad \zeta^3 x_1^2x_2^2x_3^2\\ &\quad - \zeta^2\xi\,(x_1^2x_2^2x_3^2 + \&\text{c.})\\ &\quad + \zeta\xi^2\,(x_1z_2^2z_3^2 + \&\text{c.})\\ &\quad - \ \xi^3 z_1^2z_2^2z_3^2,\end{aligned}$$

and we have

$$\begin{aligned}x_1x_2x_3 \qquad &= \eta^3 - \zeta^3,\\ x_1x_2z_3 + \&\text{c.} &= \ 3\zeta^2\xi,\\ x_1z_2z_3 + \&\text{c.} &= -3\zeta\xi^2,\\ z_1z_2z_3 \qquad &= \ \xi^3 - \eta^3,\end{aligned}$$

and thence

$$\begin{aligned}x_1^2x_2^2z_3^2 + \&\text{c.} &= 9\zeta^4\xi^2 + 6\zeta\xi^2(\eta^3 - \zeta^3) = 3\zeta^4\xi^2 + 6\zeta\xi^2\eta^3,\\ x_1^2z_2^2z_3^2 + \&\text{c.} &= 9\zeta^2\xi^4 - 6\zeta^2\xi(\xi^3 - \eta^3) = 3\zeta^2\xi^4 + 6\zeta^2\xi\eta^3,\end{aligned}$$

and consequently

$$\begin{aligned}\Pi = &\quad \zeta^3(\eta^3 - \zeta^3)^2\\ &- \zeta^2\xi\,.\,3\zeta\xi^2(\zeta^3 + \eta^3)\\ &+ \zeta\xi^2\,.\,3\zeta^2\xi(\xi^3 + \eta^3)\\ &- \ \xi^3(\xi^3 - \eta^3)^2\\ = &(\zeta^3 - \xi^3)(\xi^6 + \eta^6 + \zeta^6 - 2\zeta^3\eta^3 - 2\zeta^3\xi^3 - 2\xi^3\eta^3).\end{aligned}$$

Now considering the equation

$$F\,.\,U-\Pi=(\xi x+\eta y+\zeta z)^2(\xi' x+\eta' y+\zeta' z),$$

in order to find ξ', η', ζ' it will be sufficient to find the coefficients of x^3, y^3, z^3 in the function on the left-hand side of the equation. The coefficient of x^3 in Π is

$$\begin{aligned}&(x_1^2+2ly_1z_1)(x_2^2+2lz_2x_2)(x_3^2+2lx_3y_3)\\ =\;& \quad x_1^2x_2^2x_3^2\\ &+2l\,(x_1^2x_2^2y_3z_3+\&\text{c.})\\ &+4l^2\,(x_1^2y_2z_2y_3z_3+\&\text{c.})\\ &+8l^3\quad y_1y_2y_3z_1z_2z_3\,;\end{aligned}$$

and it is easy to see that representing the function

$$(1,\,1,\,1,\,l\between\beta\zeta-\gamma\eta,\;\gamma\xi-\alpha\zeta,\;\alpha\eta-\beta\xi)^3$$

by

$$(\mathrm{a},\,\mathrm{b},\,\mathrm{c},\,\mathrm{f},\,\mathrm{g},\,\mathrm{h},\,\mathrm{i},\,\mathrm{j},\,\mathrm{k},\,\mathrm{l}\between\alpha,\,\beta,\,\gamma)^3,$$

the symmetrical functions can be expressed in terms of the quantities a, b, &c., and that the preceding value of the coefficient of x^3 in Π is

$$\begin{aligned}&\mathrm{a}^2\\ &+2l\;(9\mathrm{hj}-6\mathrm{al})\\ &+4l^2\,(6\mathrm{gk}-3\mathrm{fj}-3\mathrm{hi}+3\mathrm{l}^2)\\ &+8l^3\;\mathrm{bc}\,;\end{aligned}$$

and substituting for a, &c. their values, this becomes

$$\begin{aligned}&(\eta^3-\zeta^3)^2\\ &+2l\;\{-9\,(\xi\eta^2+2l\eta\zeta^2)(\zeta^2\xi+2l\zeta\eta^2)\}\\ &+4l^2\,\{-6\,(\zeta\xi^2+2l\xi\eta^2)(\xi^2\eta+2l\xi\zeta^2)\\ &\qquad+3\,(\eta\zeta^2+2l\zeta\xi^2)(\zeta^2\xi+2l\zeta\eta^2)\\ &\qquad+3\,(\xi\eta^2+2l\eta\zeta^2)(\eta^2\zeta+2l\eta\xi^2)\}\\ &+8l^3\,(\xi^3-\eta^3)(\zeta^3-\xi^3),\end{aligned}$$

and reducing, we obtain for the coefficient of x^3 in Π the following expression,

$$\begin{aligned}&(\eta^3-\zeta^3)^2\\ &-18l\;\;\xi^2\eta^2\zeta^2\\ &-24l^2\,(\xi^3+\eta^3+\zeta^3)\,\xi\eta\zeta\\ &-24l^3\,(\eta^3\zeta^3+\zeta^3\xi^3+\xi^3\eta^3)\\ &+\;8l^3\,(\xi^3-\eta^3)(\zeta^3-\xi^3).\end{aligned}$$

Now the coefficient of x^3 in $F\,.\,U$ is simply F, which is equal to

$$\begin{aligned}&\xi^6+\eta^6+\zeta^6-2\eta^3\zeta^3-2\zeta^3\xi^3-2\xi^3\eta^3\\ &-24l\ \ \xi^2\eta^2\zeta^2\\ &-24l^2(\xi^3+\eta^3+\zeta^3)\,\xi\eta\zeta\\ &-32l^3(\eta^3\zeta^3+\zeta^3\xi^3+\xi^3\eta^3)\\ &-48l^4\ \ \xi^2\eta^2\zeta^2\,;\end{aligned}$$

and subtracting, the coefficient of x^3 in $F\,.\,U-\Pi$ is

$$\begin{aligned}&\xi^6-2\xi^3\eta^3-2\xi^3\zeta^3\\ &-6l\ \ \xi^2\eta^2\zeta^2\\ &-8l^3(\eta^3\zeta^3+\zeta^3\xi^3+\xi^3\eta^3)\\ &-8l^3(\xi^3-\eta^3)(\zeta^3-\xi^3)\\ &-48l^4\xi^2\eta^2\zeta^2,\end{aligned}$$

which is equal to

$$(1+8l^3)\,\xi^2(\xi^4-2\xi\eta^3-2\xi\zeta^3-6l\eta^2\zeta^2).$$

The expression last written down is therefore the value of $\xi^2\xi'$, or dividing by ξ^2 we have ξ', and then the values of η', ζ' are of course known, and we obtain the identical equation

$$F\,.\,U-\Pi=$$

$$(1+8l^3)(\xi x+\eta y+\zeta z)^2\left\{\begin{aligned}&(\xi^4-2\xi\eta^3-2\xi\zeta^3-6l\eta^2\zeta^2)\,x\\ +\,&(\eta^4-2\eta\zeta^3-2\eta\xi^3-6l\zeta^2\xi^2)\,y\\ +\,&(\zeta^4-2\zeta\xi^3-2\zeta\eta^3-6l\xi^2\eta^2)\,z\end{aligned}\right\}$$

and the second factor equated to zero is the equation of the satellite line of $\xi x+\eta y+\zeta z=0$.

33. The point of intersection of the line $\xi x+\eta y+\zeta z=0$ with the satellite line $\xi' x+\eta' y+\zeta' z=0$ is the satellite point of the former line; and the coordinates of the satellite point are at once found to be

$$\begin{aligned}x:y:z=&(\eta^3-\zeta^3)(\eta\zeta+2l\xi^2)\\ :&(\zeta^3-\xi^3)(\zeta\xi+2l\eta^2)\\ :&(\xi^3-\eta^3)(\xi\eta+2l\zeta^2).\end{aligned}$$

34. If the primary line $\xi x+\eta y+\zeta z=0$ is a tangent to the cubic, then (x_1, y_1, z_1) being the coordinates of the point of contact, we have

$$\xi:\eta:\zeta=x_1^2+2ly_1z_1:y_1^2+2lz_1x_1:z_1^2+2lx_1y_1\,;$$

these values give as before

$$\eta^3 - \zeta^3 = -(1+8l^3)\, x_1^3 (y_1^3 - z_1^3);$$

and they give also

$$\eta\zeta + 2l\xi^2 = (1+8l^3)\, y_1^2 z_1^2,$$

and consequently we obtain

$$x : y : z = x_1 (y_1^3 - z_1^3) : y_1 (z_1^3 - x_1^3) : z_1 (x_1^3 - y_1^3),$$

that is, the satellite point of a tangent of the cubic is the point in which this tangent again meets the cubic.

Article Nos. 35 and 36.—*Theorems relating to the satellite point.*

35. If the line $\xi x + \eta y + \zeta z = 0$ be a tangent of the Pippian, then the locus of the satellite point is the Hessian.

Take (x, y, z) as the coordinates of the satellite point, then we have

$$\begin{aligned} x : y : z = {} & (\eta^3 - \zeta^3)(\eta\zeta + 2l\xi^2) \\ : {} & (\zeta^3 - \xi^3)(\zeta\xi + 2l\eta^2) \\ : {} & (\xi^3 - \eta^3)(\xi\eta + 2l\zeta^2); \end{aligned}$$

where the parameters ξ, η, ζ are connected by the equation

$$-l(\xi^3 + \eta^3 + \zeta^3) + (-1 + 4l^3)\,\xi\eta\zeta = 0.$$

We have

$$\begin{aligned} y^3 + z^3 = {} & (\zeta^3 - \xi^3)^3 (\zeta\xi + 2l\eta^2)^3 \\ & + (\xi^3 - \eta^3)^3 (\xi\eta + 2l\zeta^2)^3, \end{aligned}$$

and it is easy to see that the function on the right-hand side must divide by $\eta^3 - \zeta^3$: hence $x^3 + y^3 + z^3$ will also divide by $\eta^3 - \zeta^3$, and consequently by $(\eta^3 - \zeta^3)(\zeta^3 - \xi^3)(\xi^3 - \eta^3)$. We have

$$\begin{aligned} (y^3 + z^3) \div (\eta^3 - \zeta^3) = {} & \xi^3 \left\{ \begin{array}{l} -\zeta^9 - \zeta^6\eta^3 - \zeta^3\eta^6 - \eta^9 \\ + 3\xi^3 (\zeta^6 + \zeta^3\eta^3 + \eta^6) \\ - 3\xi^6 (\zeta^3 + \eta^3) \\ + \ \xi^9 \end{array} \right\} \\ & + 6l\xi^2\eta^2\zeta^2 \{-\zeta^6 - \zeta^3\eta^3 - \eta^6 + 3\xi^3(\zeta^3 + \eta^3) - 3\xi^6\} \\ & + 12l^2\xi\eta\zeta \{-\eta^3\zeta^3(\eta^3 + \zeta^3) + 3\xi^3\eta^3\zeta^3 - \xi^9\} \\ & + 8l^3 \quad \{-\eta^6\zeta^6 + 3\eta^3\zeta^3\xi^6 - (\eta^3 + \zeta^3)\,\xi^9\} \end{aligned}$$

and

$$x^3 \div (\eta^3 - \zeta^3) = (\eta^6 - 2\eta^3\zeta^3 + \zeta^6)(\eta^3\zeta^3 + 6l\xi^2\eta^2\zeta^2 + 12l^4\xi^4\eta\zeta + 8l^3\xi^6).$$

Adding these values and completing the reduction, we find

$$\begin{aligned}(x^3+y^3+z^3)\div(\eta^3-\zeta^3)(\zeta^3-\xi^3)(\xi^3-\eta^3) = &-\xi^6-\eta^6-\zeta^6+2\eta^3\zeta^3+2\zeta^3\xi^3+2\xi^3\eta^3\\ &+18l\quad \xi^2\eta^2\zeta^2\\ &+12l^2(\xi^3+\eta^3+\zeta^3)\,\xi\eta\zeta\\ &+\ 8l^3(\eta^3\zeta^3+\zeta^3\xi^3+\xi^3\eta^3)\,;\end{aligned}$$

and we have also

$$\begin{aligned}xyz\div(\eta^3-\zeta^3)(\zeta^3-\xi^3)(\xi^3-\eta^3) = &\qquad \xi^2\eta^2\zeta^2\\ &+2l\,(\xi^3+\eta^3+\zeta^3)\,\xi\eta\zeta\\ &+4l^2(\eta^3\zeta^3+\zeta^3\xi^3+\xi^3\eta^3)\\ &+8l^3\ \xi^2\eta^2\zeta^2,\end{aligned}$$

and thence

$$\begin{aligned}&\{A\,(x^3+y^3+z^3)+Bxyz\}\div(\eta^3-\zeta^3)(\zeta^3-\xi^3)(\xi^3-\eta^3)\\ &= \quad -A \qquad (\xi^3+\eta^3+\zeta^3)^2\\ &\quad +(12l^2A+4lB)\qquad(\xi^3+\eta^3+\zeta^3)\,\xi\eta\zeta\\ &\quad +\big(18lA+(1+8l^3)\,B\big)\quad \xi^2\eta^2\zeta^2\\ &\quad +\big((4l^2+8l^3)\,A+4l^2B\big)\,(\eta^3\zeta^3+\zeta^3\xi^3+\xi^3\eta^3).\end{aligned}$$

The coefficient of $\eta^3\zeta^3+\zeta^3\xi^3+\xi^3\eta^3$ on the right-hand side will vanish if $(1+2l^3)\,A+l^2B=0$, or, what is the same thing, if $A=l^2$, $B=-(1+2l^3)$; and substituting these values, we obtain

$$\begin{aligned}&\{l^2(x^3+y^3+z^3)-(1+2l^3)\,\xi\eta\zeta\}\div(\eta^3-\zeta^3)(\zeta^3-\xi^3)(\xi^3-\eta^3)\\ &= -l^2\qquad (\xi^3+\eta^3+\zeta^3)\\ &\quad +(-4l+4l^4)\qquad (\xi^3+\eta^3+\zeta^3)\,\xi\eta\zeta\\ &\quad +(-1+8l^3-16l^6)\,\xi^2\eta^2\zeta^2,\end{aligned}$$

or, what is the same thing,

$$\begin{aligned}l^2(x^3+y^3+z^3)-(1+2l^3)\,xyz = &-(\eta^3-\zeta^3)(\zeta^3-\xi^3)(\xi^3-\eta^3)\\ &\times\{-l\,(\xi^3+\eta^3+\zeta^3)+(-1+4l^3)\,\xi\eta\zeta\}^2.\end{aligned}$$

Hence the left-hand side vanishes in virtue of the relation between ξ, η, ζ, or we have

$$l^2(x^3+y^3+z^3)-(1+2l^3)\,xyz=0,$$

which proves the theorem.

36. Suppose that (X, Y, Z) are the coordinates of a point of the Hessian, and let (P, Q, R) be the coordinates of the point in which the tangent to the Hessian at the point (X, Y, Z) again meets the Hessian, or, what is the same thing, the

satellite point *in regard to the Hessian* of the tangent at (X, Y, Z). And consider the conic

$$X(x^2 + 2lyz) + Y(y^2 + 2lzx) + Z(x^2 + 2lxy),$$

which is the first or conic polar of the point (X, Y, Z) in respect of the cubic. The polar (in respect to this conic) of the point (P, Q, R) will be

$$\xi x + \eta y + \zeta z = 0,$$

where

$$\begin{aligned} \xi &= PX + l(RY + QZ), \\ \eta &= QY + l(PZ + RX), \\ \zeta &= RZ + l(QX + PY); \end{aligned}$$

or putting for (P, Q, R) their values,

$$\begin{aligned} \xi &= (Y^3 - Z^3)(X^2 - lYZ), \\ \eta &= (Z^3 - X^3)(Y^2 - lZX), \\ \zeta &= (X^3 - Y^3)(Z^2 - lXY); \end{aligned}$$

and if from these equations and the equation of the Hessian we eliminate (X, Y, Z), we shall obtain the equation in line coordinates of the curve which is the envelope of the line $\xi x + \eta y + \zeta z = 0$. We find, in fact,

$$\xi^3 + \eta^3 + \zeta^3 = (Y^3 - Z^3)(Z^3 - X^3)(X^3 - Y^3)$$

$$\times \begin{cases} \quad l^3\ (X^3 + Y^3 + Z^3)^2 \\ - \quad 3l\ (X^3 + Y^3 + Z^3)XYZ \\ + \quad 9l^2\ X^2Y^2Z^2 \\ + (1 - 4l^3)(Y^3Z^3 + Z^3X^3 + X^3Y^3), \end{cases}$$

$$\xi\eta\zeta \qquad = (Y^3 - Z^3)(Z^3 - X^3)(X^3 - Y^3)$$

$$\times \begin{cases} \quad l^2(X^3 + Y^3 + Z^3)XYZ \\ + (1 - l^3)X^2Y^2Z^2 \\ - \quad l\ (Y^3Z^3 + Z^3X^3 + X^3Y^3); \end{cases}$$

and thence recollecting that

$$HU = l^2(X^3 + Y^3 + Z^3) - (1 + 2l^3)XYZ,$$

we find

$$-l(\xi^3 + \eta^3 + \zeta^3) + (-1 + 4l^3)\xi\eta\zeta = -(Y^3 - Z^3)(Z^3 - X^3)(X^3 - Y^3)(HU)^2,$$

and the equation of the envelope is

$$-l(\xi^3 + \eta^3 + \zeta^3) + (-1 + 4l^3)\xi\eta\zeta = 0,$$

which is therefore the Pippian. We have thus the theorem:

The envelope of the polar of the satellite point *in respect to the Hessian* of the tangent at any point of the Hessian, such polar being in respect of the conic which is the first or conic polar of the point of the Hessian in respect of the cubic, is the Pippian.

Article Nos. 37 to 40.—*Investigations and theorems relating to the first or conic polar of a point of the cubic.*

37. The investigations next following depend on the identical equations

$$\begin{aligned}
&\{\alpha(X^2+2lYZ)+\beta(Y^2+2lZX)+\gamma(Z^2+2lXY)\}\\
&\qquad\times\{-XYZ(x^3+y^3+z^3)+(X^3+Y^3+Z^3)xyz\}\\
&=\{X(x^2+2lyz)+Y(y^2+2lzx)+Z(z^2+2lxy)\}\\
&\qquad\times\{X(Y^3-Z^3)(\gamma y-\beta z)+Y(Z^3-X^3)(\alpha z-\gamma x)+Z(X^3-Y^3)(\beta x-\alpha y)\}\\
&+\{x(X^2+2lYZ)+y(Y^2+2lZX)+z(Z^2+2lXY)\}\\
&\qquad\times\{-(\alpha YZ+\beta ZX+\gamma XY)(Xx^2+Yy^2+Zz^2)+(\alpha X^2+\beta Y^2+\gamma Z^2)(XYz+Yzx+Zxy)\},
\end{aligned}$$

which is easily verified.

I represent the equation in question by

$$K\Upsilon=WL+P\Theta\,;$$

then considering (x, y, z) as current coordinates, and (X, Y, Z) and (α, β, γ) as the coordinates of two given points Σ and Ω, we shall have $U=0$ the equation of the cubic, $W=0$ the equation of the first or conic polar of Σ with respect to the cubic, $P=0$ the equation of the second or line polar of Σ with respect to the cubic. The equation $\Upsilon=0$ is that of a syzygetic cubic passing through the point Σ: the coordinates of the satellite point in respect to this syzygetic cubic of its tangent at Σ are

$$X(Y^3-Z^3) : Y(Z^3-X^3) : Z(X^3-Y^3)\,;$$

and calling the point in question Σ', then $L=0$ is the equation of a line through the points Σ', Ω. The equation $\Theta=0$ is that of a conic, viz. the first or conic polar of Σ with respect to a certain syzygetic cubic

$$-2(\alpha YZ+\beta ZX+\gamma XY)(x^3+y^3+z^3)+(\alpha X^2+\beta Y^2+\gamma Z^2)xyz=0,$$

depending on the points Σ, Ω, or, what is the same thing, the conic $\Theta=0$ is a properly selected conic passing through the points of intersection of the first or conic polars of Σ with respect to any two syzygetic cubics; and lastly, K is a constant coefficient. The equation expresses that the points of intersection of

$$(W=0,\ P=0),\ (W=0,\ \Theta=0),\ (L=0,\ P=0),\ (L=0,\ \Theta=0),$$

lie in the syzygetic cubic $\Upsilon=0$.

The left-hand side of the equation may be written

$$\begin{aligned}&- XYZ\,\{\alpha\,(X^2 + 2lYZ) + \beta\,(Y^2 + 2lZX) + \gamma\,(Z^2 + 2lXY)\}\,(x^3 + y^3 + z^3 + 6lxyz)\\ &+ \quad xyz\,\{\alpha\,(X^2 + 2lYZ) + \beta\,(Y^2 + 2lZX) + \gamma\,(Z^2 + 2lXY)\}\,(X^3 + Y^3 + Z^3 + 6lXYZ);\end{aligned}$$

and it may be remarked also that we have

$$-3XYZ\,\{\alpha\,(X^2 + 2lYZ) + \beta\,(Y^2 + 2lZX) + \gamma\,(Z^2 + 2lXY)\}$$

equal identically to

$$\begin{aligned}&\{X\,(Y^3 - Z^3)(\gamma Y - \beta Z) + Y\,(Z^3 - X^3)(\alpha Z - \gamma X) + Z\,(X^3 - Y^3)(\beta X - \alpha Y)\}\\ &\qquad - (\alpha YZ + \beta ZX + \gamma XY)(X^3 + Y^3 + Z^3 + 6lXYZ).\end{aligned}$$

Hence if we assume

$$X^3 + Y^3 + Z^3 + 6lXYZ = 0,$$

the equation will take the form

$$KU = WL + P\Theta,$$

where the constant coefficient K may be expressed under the two different forms

$$\begin{aligned}K &= -XYZ\,\{\alpha\,(X^2 + 2lYZ) + \beta\,(Y^2 + 2lZX) + \gamma\,(Z^2 + 2lXY)\}\\ &= \tfrac{1}{3}\,\{X\,(Y^3 - Z^3)(\gamma Y - \beta Z) + Y\,(Z^3 - X^3)(\alpha Z - \gamma X) + Z\,(X^3 - Y^3)(\beta X - \alpha Y)\},\end{aligned}$$

and W, L, P, Θ have the same values as before. In the present case the point Σ is a point of the cubic: the equation $W = 0$ represents the first or conic polar of the point in question, and the equation $P = 0$ its second or line polar, which is also the tangent of the cubic. The line $L = 0$ is a line joining the point Ω with the satellite point of the tangent at Σ, or dropping altogether the consideration of the point Ω, is an arbitrary line through the satellite point: the first or conic polar of Σ meets the cubic twice in the point Σ, and therefore also meets it in four other points; the conic $\Theta = 0$ is a conic passing through these four points, and completely determined when the particular position of the line through the satellite point is given. And, as before remarked, $\Theta = 0$ is a conic passing through the points of intersection of the first or conic polars of Σ with respect to any two syzygetic cubics. We have thus the theorem:

The first or conic polar of a point of the cubic touches the cubic at this point, and besides meets it in four other points; the four points in question are the points in which the first or conic polar of the given point in respect of the cubic is intersected by the first or conic polar of the same point in respect to any syzygetic cubic whatever.

38. The analytical result may be thus stated: putting

$$\kappa = \alpha YZ + \beta ZX + \gamma XY, \quad \lambda = \alpha X^2 + \beta Y^2 + \gamma Z^2,$$

or, if we please, considering κ, λ as arbitrary parameters, then the four points lie in the conic

$$(2\kappa X,\ 2\kappa Y,\ 2\kappa Z,\ -\lambda X,\ -\lambda Y,\ -\lambda Z \between x,\ y,\ z)^2 = 0,$$

or, what is the same thing, they are the points of intersection of the two conics

$$Xx^2 + Yy^2 + Zz^2 = 0,$$
$$Xyz + Yzx + Zxy = 0.$$

39. Considering the four points as the angles of a quadrangle, it may be shown that the three centres of the quadrangle lie on the cubic. To effect this, assume that the conic

$$(2\kappa X,\ 2\kappa Y,\ 2\kappa Z,\ -\lambda X,\ -\lambda Y,\ -\lambda Z \between x,\ y,\ z)^2 = 0$$

represents a pair of lines; these lines will intersect in a point, which is one of the three centres in question. And taking x, y, z as the coordinates of this point, we have

$$\begin{aligned} x^2 : y^2 : z^2 : yz : zx : xy = & \ 4\kappa^2\, YZ - \lambda^2 X^2 \\ : & \ 4\kappa^2\, ZX - \lambda^2 Y^2 \\ : & \ 4\kappa^2\, XY - \lambda^2 Z^2 \\ : & \ \lambda^2\, YZ + 2\kappa\lambda X^2 \\ : & \ \lambda^2\, ZX + 2\kappa\lambda Y^2 \\ : & \ \lambda^2\, XY + 2\kappa\lambda Z^2; \end{aligned}$$

and we may, if we please, use these equations to find the relation between κ, λ. Thus in the identical equation $x^2 \,.\, y^2 - (xy)^2 = 0$, substituting for x^2, xy, y^2 their values, and throwing out the factor Z, we find $(4\kappa^3 - \lambda^3)\, XYZ - \kappa\lambda^2 (X^3 + Y^3 + Z^3) = 0$, and thence, in virtue of the equation $X^3 + Y^3 + Z^3 + 6lXYZ = 0$, we obtain

$$4\kappa^3 + 6l\kappa\lambda^2 - \lambda^3 = 0.$$

But the preceding system gives conversely,

$$\begin{aligned} X^2 : Y^2 : Z^2 : YZ : ZX : XY = & \ 4\kappa^2 yz - \lambda^2 x^2 \\ : & \ 4\kappa^2 zx - \lambda^2 y^2 \\ : & \ 4\kappa^2 xy - \lambda^2 z^2 \\ : & \ \lambda^2 yz + 2\kappa\lambda x^2 \\ : & \ \lambda^2 zx + 2\kappa\lambda y^2 \\ : & \ \lambda^2 xy + 2\kappa\lambda z^2. \end{aligned}$$

Hence from the identical relation $X^2 \,.\, Y^2 - (XY)^2 = 0$, substituting for X^2, XY, Y^2 their values, and throwing out the factor z, we find $(4\kappa^3 - \lambda^3)\, xyz - \kappa\lambda^2 (x^3 + y^3 + z^3) = 0$, and thence, in virtue of the equation $4\kappa^3 - \lambda^3 = -6l\kappa\lambda^2$, we obtain

$$x^3 + y^3 + z^3 + 6lxyz = 0,$$

which shows that the point in question lies on the cubic. We have thus the theorem:

The first or conic polar of a point of the cubic touches the cubic at the point, and meets it besides in four points, which are the angles of a quadrangle the centres of which lie on the cubic. In other words, the quadrangle is an inscribed quadrangle.

40. To find the equations of the three *axes* of the quadrangle, that is of the lines through two centres.

We have

$$\begin{aligned}
&(4\kappa^2 YZ - \lambda^2X^2)\,x + (\lambda^2XY + 2\kappa\lambda Z^2)\,y + (\lambda^2ZX + 2\kappa\lambda Y^2)\,z = 0,\\
&(\lambda^2XY + 2\kappa\lambda Z^2)\,x + (4\kappa^2ZX - \lambda^2Y^2)\,y + (\lambda^2YZ + 2\kappa\lambda X^2)\,z = 0,\\
&(\lambda^2ZX + 2\kappa\lambda Y^2)\,x + (\lambda^2YZ + 2\kappa\lambda X^2)\,y + (4\kappa^2XY - \lambda^2Z^2)\,z = 0;
\end{aligned}$$

or arranging these equations in the proper form and eliminating κ^2, $\kappa\lambda$, λ^2, we find

$$\begin{vmatrix}
YZx, & Z^2y + Y^2z, & X(-Xx + Yy + Zz)\\
ZXy, & X^2z + Z^2x, & Y(Xx - Yy + Zz)\\
XYz, & Y^2x + X^2y, & Z(Xx + Yy + Zz)
\end{vmatrix} = 0;$$

or, multiplying out,

$$\begin{aligned}
XYZ\,\{&(Z^3 - Y^3)\,x^3 + (X^3 - Z^3)\,y^3 + (Y^3 - X^3)\,z^3\}\\
&+ x^2yZY^2(-2X^3 + Y^3 + Z^3) + zx^2YZ^2(2X^3 - Y^3 - Z^3)\\
&+ y^2zXZ^2(-2Y^3 + Z^3 + X^3) + xy^2ZX^2(2Y^3 - Z^3 - X^3)\\
&+ z^2xYX^2(-2Z^3 + X^3 + Y^3) + yz^2XY^2(2Z^3 - X^3 - Y^3) = 0.
\end{aligned}$$

We may simplify this result by means of the equation $X^3 + Y^3 + Z^3 + 6lXYZ = 0$, so as to make the left-hand side divide out by XYZ: we thus obtain

$$\begin{aligned}
&(Z^3 - Y^3)\,x^3 + (X^3 - Z^3)\,y^3 + (Y^3 - X^3)\,z^3\\
&+ (-3X^2Y - 6lY^2Z)\,x^2y + (-3Y^2Z - 6lZ^2X)\,y^2z + (-3Z^2X - 6lX^2Y)\,z^2x\\
&+ (3XY^2 + 6lX^2Z)\,xy^2 + (3YZ^2 + 6lY^2X)\,yz^2 + (3ZX^2 + 6lZ^2Y)\,zx^2 = 0;
\end{aligned}$$

or in a different form,

$$\begin{aligned}
&(y^3 - z^3)\,X^3 + (z^3 - x^3)\,Y^3 + (x^3 - y^3)\,Z^3\\
&+ (-3x^2y - 6lz^2x)\,X^2Y + (-3y^2z - 6lx^2y)\,Y^2Z + (-3z^2x - 6ly^2z)\,Z^2X\\
&+ (3xy^2 + 6lyz^2)\,XY^2 + (3yz^2 + 6lzx^2)\,YZ^2 + (3zx^2 + 6lxy^2)\,ZX^2 = 0,
\end{aligned}$$

as the equation of the three axes of the quadrangle.

Article No. 41. *Recapitulation of geometrical definitions of the Pippian.*

In conclusion, I will recapitulate the different modes of generation or geometrical definitions of the Pippian, obtained in the course of the present memoir. The curve in question is:

1. The envelope of the line joining a pair of conjugate poles of the cubic (see Nos. 2 and 13).

2. The envelope of each line of the pair forming the first or conic polar with respect to the cubic of a conjugate pole of the cubic (see Nos. 2 and 14).

3. The envelope of a line which is the polar of a conjugate pole of the cubic, with respect to the conic which is the first or conic polar of the other conjugate pole in respect to any syzygetic cubic (see Nos. 2 and 9).

4. The locus of the harmonic with respect to a pair of conjugate poles of the cubic of the third point of intersection with the Hessian of the line joining the two conjugate poles (see Nos. 2 and 17).

5. The envelope of a line such that its lineo-polar envelope with respect to the cubic breaks up into a pair of lines (see No. 24).

6. The envelope of a line which meets three conics, the first or conic polars of any three points in respect to the cubic, in six points in involution (see No. 22).

7. The envelope of the second or line polar with respect to the cubic, of a point the locus of which is a certain curve of the sixth order in quadratic syzygy with the cubic and Hessian, viz. the curve $-S \,.\, U^2 + (HU)^2 = 0$ (see No. 27).

8. The envelope of a line having for its satellite point a point of the Hessian (see No. 35).

9. The envelope of the polar of the satellite point with respect to the Hessian of the tangent at a point of the Hessian, with respect to the first or conic polar of the point of the Hessian in respect to the cubic (see No. 36).

147.

A MEMOIR ON THE SYMMETRIC FUNCTIONS OF THE ROOTS OF AN EQUATION.

[From the *Philosophical Transactions of the Royal Society of London*, vol. CXLVII. *for the year* 1857, pp. 489—499. Received December 18, 1856,—Read January 8, 1857.]

THERE are contained in a work, which is not, I think, so generally known as it deserves to be, the "Algebra" of Meyer Hirsch [the work referred to is entitled Sammlung von Beispielen Formeln und Aufgaben aus der Buchstabenrechnung und Algebra, 8vo. Berlin, 1804 (8 ed. 1853), English translation by Ross, 8vo. London, 1827] some very useful tables of the symmetric functions up to the tenth degree of the roots of an equation of any order. It seems desirable to join to these a set of tables, giving reciprocally the expressions of the powers and products of the coefficients in terms of the symmetric functions of the roots. The present memoir contains the two sets of tables, viz. the new tables distinguished by the letter (*a*), and the tables of Meyer Hirsch distinguished by the letter (*b*); the memoir contains also some remarks as to the mode of calculation of the new tables, and also as to a peculiar symmetry of the numbers in the tables of each set, a symmetry which, so far as I am aware, has not hitherto been observed, and the existence of which appears to constitute an important theorem in the subject. The theorem in question might, I think, be deduced from a very elegant formula of M. Borchardt (referred to in the sequel), which gives the generating function of any symmetric function of the roots, and contains potentially a method for the calculation of the Tables (*b*), but which, from the example I have given, would not appear to be a very convenient one for actual calculation.

Suppose in general

$$(1,\ b,\ c\ \ldots\between 1,\ x)^{\infty} = (1-\alpha x)(1-\beta x)(1-\gamma x)\ldots,$$

so that

$$-b = \Sigma\alpha,\quad +c = \Sigma\alpha\beta,\quad -d = \Sigma\alpha\beta\gamma,\ \&c.,$$

and if in general

$$(pqr\ldots) = \Sigma\alpha^p\beta^q\gamma^r\ldots,$$

where as usual the summation extends only to the distinct terms, so that e.g. (p^2) contains only half as many terms as (pq), and so in all similar cases, then we have

$$-b=(1),\quad +c=(1^2),\quad -d=(1^3),\ \&\text{c.};$$

and the two problems which arise are, first to express any combination $b^pc^q\ldots$ in terms of the symmetric functions $(l^xm^y\ldots)$, and secondly, or conversely, to express any symmetric function $(l^xm^y\ldots)$ in terms of the combinations $b^pc^q\ldots$.

It will conduce materially to brevity if $1^p2^q\ldots$ be termed the partition belonging to the combination $b^pc^q\ldots$; and in like manner if $l^xm^y\ldots$ be termed the partition belonging to the symmetric function $(l^xm^y\ldots)$, and if the sum of the component numbers of the partition is termed the weight.

Consider now a line of combinations corresponding to a given weight, e.g. the weight 4, this will be

$$\begin{array}{ccccc} e & bd & c^2 & b^2c & b^4 \ \text{(line)} \\ 4 & 13 & 2^2 & 1^22 & 1^4, \end{array}$$

where I have written under each combination the partition which belongs to it, and in like manner a column of symmetric functions of the same weight, viz.

$$\begin{array}{l} (4)\ \text{(column)} \\ (31) \\ (2^2) \\ (21^2) \\ (1^4), \end{array}$$

where, as the partitions are obtained by simply omitting the ( ), I have not separately written down the partitions.

It is at once obvious that the different combinations of the line will be made up of numerical multiples of the symmetric functions of the column; and conversely, that the symmetric functions of the column will be made up of numerical multiples of the combinations of the line; but this requires a further examination. There are certain restrictions as to the symmetric functions which enter into the expression of the combination, and conversely, as to the combinations which enter into the expression of the symmetric function. The nature of the first restriction is most clearly seen by the following Table:

Number of Parts.	Greatest Part.	Combinations with their several Partitions.		Contain Multiples of the Symmetric Functions.	Greatest Part does not exceed	Number of Parts not less than
1	4	e	4	(1^4),	1	4
2	3	bd	13	(1^4), (21^2),	2	3
2	2	c^2	2^2	(1^4), (21^2), (2^2),	2	2
3	2	b^2c	1^22	(1^4), (21^2), (2^2), (31),	3	2
4	1	b^4	1^4	(1^4), (21^2), (2^2), (31), (4)	4	1

Thus, for instance, the combination bd (the partition whereof is 13) contains multiples of the two symmetric functions (1^4), (21^2) only. The number of parts in the partition 13 is 2, and the greatest part is 3. And in the partitions (1^4), (21^2) the greatest part is 2, and the number of parts is not less than 3. The reason is obvious: each term of the developed expression of bd must contain at least as many roots as are contained in each term of d, that is 3 roots, and since the coefficients are linear functions in respect to each root, the combination bd cannot contain a power higher than 2 of any root. The reasoning is immediately applied to any other case, and we obtain

First Restriction.—A combination $b^p c^q \ldots$ contains only those symmetric functions $(l^x m^y \ldots)$, for which the greatest part does not exceed the number of parts in the partition $1^p 2^q \ldots$, and the number of parts is not less than the greatest part in the same partition.

Consider a partition such as $1^2 2$, then replacing each number by a line of units thus,

$$\begin{array}{l}1\\1\\11,\end{array}$$

and summing the columns, we obtain a new partition 31, which may be called the conjugate[1] of $1^2 2$. It is easy to see that the expression for the combination $b^2 c$ (for which the partition is $1^2 2$) contains with the coefficient unity, the symmetric function (31), the partition whereof is the conjugate of $1^2 2$. In fact $b^2 c = (-\Sigma\alpha)^2 (\Sigma\alpha\beta)$, which obviously contains the term $+1\alpha^3\beta$, and therefore the symmetric function with its coefficient $+1\,(31)$; and the reasoning is general, or

THEOREM. A combination $b^q c^q \ldots$ contains the symmetric function (partition conjugate to $1^p 2^q \ldots$) with the coefficient unity, and sign + or − according as the weight is even or odd.

Imagine the partitions arranged as in the preceding column, viz. first the partition into one part, then the partitions into two parts, then the partitions into three parts, and so on; the partitions into the same number of parts being arranged according to the magnitude of the greatest part (the greatest magnitude first), and in case of equality according to the magnitudes of the next greatest part, and so on (for other examples, see the outside column of any one of the Tables). The order being thus completely defined, we may speak of a partition as being prior or posterior to another. We are now able to state a second restriction as follows.

Second Restriction.—The combination $b^p c^q \ldots$ contains only those symmetric functions which are of the form (partition not prior to the conjugate of $1^p 2^q \ldots$).

The terms excluded by the two restrictions are many of them the same, and it might at first sight appear as if the two restrictions were identical; but this is not

[1] The notion of Conjugate Partitions is, I believe, due to Professor Sylvester or Mr Ferrers. [It was due to Mr now Dr Ferrers.]

so: for instance, for the combination bd^2, see Table VII (a), the term (41^3) is excluded by the first restriction, but not by the second; and on the other hand, the term (3^21), which is not excluded by the first restriction, is excluded by the second restriction, as containing a partition 3^21 prior in order to 32^2, which is the partition conjugate to 13^2, the partition of bd^2. It is easy to see why bd^2 does not contain the symmetric function (3^21); in fact, a term of (3^21) is $(\alpha^3\beta^3\gamma)$, which is obviously not a term of $bd^3 = (-\Sigma\alpha)(\Sigma\alpha\beta\gamma)^2$; but I have not investigated the general proof.

I proceed to explain the construction of the Tables (a). The outside column contains the symmetric functions arranged in the order before explained; the outside or top line contains the combinations of the same weight arranged as follows, viz. the partitions taken in order from right to left are respectively conjugate to the partitions in the outside column, taken in order from top to bottom; in other words, each square of the sinister diagonal corresponds to two partitions which are conjugate to each other. It is to be noticed that the combinations taken in order, from left to right, are *not* in the order in which they would be obtained by Arbogast's Method of Derivations from an operand a^x, a being ultimately replaced by unity. The squares above the sinister diagonal are empty (i.e. the coefficients are zero), the greater part of them in virtue of both restrictions, and the remainder in virtue of the second restriction; the empty squares below the sinister diagonal are empty in virtue of the second restriction; but the property was not assumed in the calculation.

The greater part of the numbers in the Tables (a) were calculated, those of each table from the numbers in the next preceding table by the following method, depending on the derivation of the expression for $b^{p+1}c^q\ldots$ from the expression for $b^pc^q\ldots$ Suppose, for example, the column cd of Table V (a) is known, and we wish to calculate the column bcd of Table VI (a). The process is as follows:

Given

2^21	21^3	1^5
1	3	10

we obtain

321	2^3	31^3	2^21^2	21^4	1^6
1	3		2		
		3	6	12	
				10	60
1	3	3	8	22	60

where the numbers in the last line are the numbers in the column bcd of Table VI (a). The partition 2^21, considered as containing a part zero, gives, when the parts are successively increased by 1, the partitions 321, 2^3, 2^21^2, in which the indices of the increased part (i.e. the original part plus unity) are 1, 3, 2; these numbers are taken as multipliers of the coefficient 1 of the partition 2^21, and we thus have the new coefficients 1, 3, 2 of the partitions 321, 2^3, 2^21^2. In like manner the coefficient 3 of the partition 21^3 gives the new coefficients 3, 6, 12 of the partitions 31^3, 2^21^2, 21^4, and the coefficient 10 of the partition 1^5 gives the new coefficients 10, 60 of the partitions 21^4 and 1^6, and finally, the last line is obtained by addition. The process in fact amounts to the multiplication separately of each term of $cd=$

$$1\,(2^21)+3\,(21^3)+10\,(1^5)$$

by $b=(1)$. It would perhaps have been proper to employ an analogous rule for the calculation of the combinations $c^qd^r\ldots$ not containing b, but instead of doing so I availed myself of the existing Tables (b). But the comparison of the last line of each Table (a) (which as corresponding to a combination b^p was always calculated independently of the Tables (b)) with such last line as calculated from the corresponding Table (b), seems to afford a complete verification of both the Tables; and my process has in fact enabled me to detect several numerical errors in the Tables (b), as given in the English translation of the work above referred to. It is not desirable, as regards facility of calculation and independently of the want of verification, to calculate either set of Tables wholly from the other; the rules for the independent calculation of the Tables (b) are fully and clearly explained in the work referred to, and I have nothing to add upon this subject.

The relation of symmetry, alluded to in the introductory paragraph of the present memoir, exists in each Table of either set, and is as follows: viz. the number in the Table corresponding to any two partitions in the outside column and the outside line respectively, is equal to the number corresponding to the same two partitions in the outside line and the outside column respectively. Or, calling the two partitions P, Q, and writing for shortness, combination (P) for the combination represented by the partition P, and for greater clearness, symmetric function (P) (instead of merely (P)) to denote the symmetric function represented by the partition P, we have the following two theorems, viz.

THEOREM. The coefficient in combination (P) of symmetric function (Q) is equal to the coefficient in combination (Q) of symmetric function (P);

and conversely,

THEOREM. The coefficient in symmetric function (P) of combination (Q) is equal to the coefficient in symmetric function (Q) of combination (P).

M. Borchardt's formula, before referred to, is given in the 'Monatsbericht' of the Berlin Academy (March 5, 1885)[1], and may be thus stated; viz. considering the case of n roots, write

$$(1,\ b,\ c,\ \ldots\ k\between 1,\ x)^n=(1-\alpha x)(1-\beta x)\ldots(1-\kappa x)=fx,$$

then

$$\Sigma\left(\frac{1}{1-\alpha x}\;\frac{1}{1-\beta y}\cdots\frac{1}{1-\kappa u}\right)=\frac{1}{k}(-)^n\,\frac{fxfy\ldots fu}{\Pi(x,y,\ldots u)}\,\frac{d}{dx}\,\frac{d}{dy}\cdots\frac{d}{du}\,\frac{\Pi(x,y,\ldots u)}{fxfy\ldots fu},$$

[1] And in *Crelle*, t. liii. p. 195.—Note added 4th Dec. 1857, A. C.

where $\Pi(x, y, \dots u)$ denotes the product of the differences of the quantities $x, y, \dots u$, and on the left-hand side the summation extends to all the different permutations of $\alpha, \beta, \dots \kappa$, or what is the same thing, of $x, y, \dots u$.

Suppose for a moment that there are only two roots, so that

$$(1, b, c \between 1, x)^2 = (1 - \alpha x)(1 - \beta x),$$

then the left-hand side is

$$\frac{1}{(1-\alpha x)(1-\beta y)} + \frac{1}{(1-\alpha y)(1-\beta x)},$$

which is equal to

$$2 + (\alpha+\beta)(x+y) + (\alpha^2+\beta^2)(x^2+y^2) + 2\alpha\beta xy + (\alpha^3+\beta^3)(x^3+y^3) + (\alpha^2\beta+\alpha\beta^2)(x^2y+xy^2) + \&c.,$$

and the right-hand side is

$$\frac{1}{c} \cdot \frac{fxfy}{x-y} \frac{d}{dx} \frac{d}{dy} \frac{x-y}{fxfy},$$

which is equal to

$$\frac{1}{c} \frac{fxfy}{x-y} \left\{ \frac{f'xfy - f'yfx + (x-y)f'xf'y}{(fx)^2 (fy)^2} \right\},$$

and therefore to

$$\frac{1}{c} \cdot \frac{1}{fxfy} \left\{ \frac{f'xfy - f'yfx}{x-y} + f'xf'y \right\};$$

or substituting for fx, fy their values,

$$\frac{f'xfy - f'yfx}{x-y}$$

becomes equal to

$$2c - b^2 - bc(x+y) - 2c^2xy,$$

and $f'xf'y$ is equal to

$$b^2 + 2bc(x+y) + 4c^2xy.$$

The right-hand side is therefore equal to

$$\frac{2 + b(x+y) + 2cxy}{(1+bx+cx^2)(1+by+cy^2)};$$

and comparing with the value of the left-hand side, we see that this expression may be considered as the generating function of the symmetric functions of (α, β), viz. the expression in question is developable in a series of the symmetric functions of (x, y), the coefficients being of course functions of b and c, and these coefficients are (to given numerical factors *près*) the symmetric functions of the roots (α, β).

And in general it is easy to see that the left-hand side of M. Borchardt's formula is equal to

$$[0]+[1](1)(1)'+[2](2)(2)'+[1^2](1^2)(1^2)'+\&c.,$$

where (1), (2), (1^2), &c. are the symmetric functions of the roots $(\alpha, \beta, \ldots \kappa)$, $(1)'$, $(2)'$, $(1^2)'$, &c. are the corresponding symmetric functions of $(x, y, \ldots u)$, and [0], [1], [2], $[1^2]$, &c. are mere numerical coefficients; viz. [0] is equal to $1.2.3\ldots n$, and [1], [2], $[1^2]$, &c. are such that the product of one of these factors into the number of terms in the corresponding symmetric function (1), (2), (1^2), &c. may be equal to $1.2.3\ldots n$. The right-hand side of M. Borchardt's formula is therefore, as in the particular case, the generating function of the symmetric functions of the roots $(\alpha, \beta, \ldots \kappa)$, and if a convenient expression of such right-hand side could be obtained, we might by means of it express all the symmetric functions of the roots in terms of the coefficients.

Tables relating to the Symmetric Functions of the Roots of an Equation.

The outside line of letters contains the combinations (powers and products) of the coefficients, the coefficients being all with the positive sign, and the coefficient of the highest power being unity; thus in the case of a cubic equation the equation is

$$x^3+bx^2+cx+d=(x-\alpha)(x-\beta)(x-\gamma)=0.$$

The outside line of numbers is obtained from that of letters merely by writing 1, 2, 3... for b, c, d..., and may be considered simply as a different notation for the combinations. The outside column contains the different symmetric functions in the notation above explained, viz. (1) denotes $\Sigma\alpha$, (2) denotes $\Sigma\alpha^2$, (1^2) denotes $\Sigma\alpha\beta$, and so on. The Tables (a) are to be read according to the columns; thus Table II (a) means $b^2=1(2)+2(1)^2$, $c=(1^2)$. The Tables (b) are to be read according to the lines; thus Table II (b) means $(2)=-2c+1b^3$, $(1^3)=+1c$.

I (a).

	1
=	b
(1)	− 1

II (a).

	2	1^2
=	c	b^2
(2)		+ 1
(1^2)	+ 1	+ 2

III (a).

	3	12	1^3
=	d	bc	b^3
(3)			− 1
(21)		− 1	− 3
(1^3)	− 1	− 3	− 6

I (b).

	1
=	b
(1)	− 1

II (b).

	2	1^2
=	c	b^2
(2)	− 2	+ 1
(1^2)	+ 1	

III (b).

	3	12	1^3
=	d	bc	b^3
(3)	− 3	+ 3	− 1
(21)	+ 3	− 1	
(1^3)	− 1		

IV (*a*).

=	4 e	13 bd	2^2 c^2	1^22 b^2c	1^4 b^4
(4)					+ 1
(31)				+ 1	+ 4
(2^2)			+ 1	+ 2	+ 6
(21^2)		+ 1	+ 2	+ 5	+ 12
(1^4)	+ 1	+ 4	+ 6	+ 12	+ 24

IV (*b*).

=	4 e	13 bd	2^2 c^2	1^22 b^2c	1^4 b^4
(4)	− 4	+ 4	+ 2	− 4	+ 1
(31)	+ 4	− 1	− 2	+ 1	
(2^2)	+ 2	− 2	+ 1		
(21^2)	− 4	+ 1			
(1^4)	+ 1				

V (*a*).

=	5 f	14 be	23 cd	1^23 b^2d	12^2 bc^2	1^32 b^3c	1^5 b^5
(5)							− 1
(41)						− 1	− 5
(32)					− 1	− 3	− 10
(31^2)				− 1	− 2	− 7	− 20
(2^21)			− 1	− 2	− 5	− 12	− 30
(21^3)		− 1	− 3	− 7	− 12	− 27	− 60
(1^5)	− 1	− 5	− 10	− 20	− 30	− 60	− 120

V (*b*).

=	5 f	14 be	23 cd	1^23 b^2d	12^2 bc^2	1^32 b^3c	1^5 b^5
(5)	− 5	+ 5	+ 5	− 5	− 5	+ 5	− 1
(41)	+ 5	− 1	− 5	+ 1	+ 3	− 1	
(32)	+ 5	− 5	+ 1	+ 2	− 1		
(31^2)	− 5	+ 1	+ 2	− 1			
(2^21)	− 5	+ 3	− 1				
(21^3)	+ 5	− 1					
(1^5)	− 1						

VI (*a*).

=	6 g	15 bf	24 ce	1^24 b^2e	3^2 d^2	123 bcd	1^33 b^3d	2^3 c^3	1^22^2 b^2c^2	1^42 b^4c	1^6 b^6
(6)											+ 1
(51)										+ 1	+ 6
(42)									+ 1	+ 4	+ 15
(3^2)								+ 1	+ 2	+ 6	+ 20
(41^2)							+ 1	...	+ 2	+ 9	+ 30
(321)						+ 1	+ 3	+ 3	+ 8	+ 22	+ 60
(2^3)					+ 1	+ 3	+ 6	+ 6	+ 15	+ 36	+ 90
(31^3)				+ 1	...	+ 3	+ 10	+ 6	+ 18	+ 48	+ 120
(2^21^2)			+ 1	+ 2	+ 2	+ 8	+ 18	+ 15	+ 34	+ 78	+ 180
(21^4)		+ 1	+ 4	+ 9	+ 6	+ 22	+ 48	+ 36	+ 78	+ 168	+ 360
(1^6)	+ 1	+ 6	+ 15	+ 30	+ 20	+ 60	+ 120	+ 90	+ 180	+ 360	+ 720

VI (*b*).

=	6 g	15 bf	24 ce	1^24 b^2e	3^2 d^2	123 bcd	1^33 b^3d	2^3 c^3	1^22^2 b^2c^2	1^42 b^4c	1^6 b^6
(6)	− 6	+ 6	+ 6	− 6	+ 3	− 12	+ 6	− 2	+ 9	− 6	+ 1
(51)	+ 6	− 1	− 6	+ 1	− 3	+ 7	− 1	+ 2	− 4	+ 1	
(42)	+ 6	− 6	+ 2	+ 2	− 3	+ 4	− 2	− 2	+ 1		
(3^2)	+ 3	− 3	− 3	+ 3	+ 3	− 3	...	+ 1			
(41^2)	− 6	+ 1	+ 2	− 1	+ 3	− 3	+ 1				
(321)	− 12	+ 7	+ 4	− 3	− 3	+ 1					
(2^3)	− 2	+ 2	− 2	...	+ 1						
(31^3)	+ 6	− 1	− 2	+ 1							
(2^21^2)	+ 9	− 4	+ 1								
(21^4)	− 6	+ 1									
(1^6)	+ 1										

VII (a).

=	7 h	16 bg	25 cf	$1^2 5$ $b^2 f$	34 de	124 bce	$1^3 4$ $b^3 e$	12^2 bd^2	$2^2 3$ $c^2 d$	$1^2 23$ $b^2 cd$	$1^4 3$ $b^4 d$	12^3 bc^3	$1^3 2^2$ $b^3 c^2$	$1^5 2$ $b^5 c$	1^7 b^7
(7)															− 1
(61)														− 1	− 7
(52)													− 1	− 5	− 21
(43)												− 1	− 3	− 10	− 35
(51^2)											− 1	...	− 2	− 11	− 42
(421)										− 1	− 4	− 3	− 11	− 35	− 105
$(3^2 1)$									− 1	− 2	− 6	− 7	− 18	− 50	− 140
(32^2)								− 1	− 2	− 5	− 12	− 12	− 31	− 80	− 210
(41^3)							− 1	...	...	− 3	− 13	− 6	− 24	− 75	− 210
(321^2)						− 1	− 3	− 2	− 5	− 13	− 34	− 27	− 68	− 170	− 420
$(2^3 1)$					− 1	− 3	− 6	− 7	− 12	− 27	− 60	− 51	− 117	− 270	− 630
(31^4)				− 1	...	− 4	− 13	− 6	− 12	− 34	− 88	− 60	− 150	− 360	− 840
$(2^2 1^3)$			− 1	− 2	− 3	− 11	− 24	− 18	− 31	− 68	− 150	− 117	− 258	− 570	− 1260
(21^5)		− 1	− 5	− 11	− 10	− 35	− 75	− 50	− 80	− 170	− 360	− 270	− 570	− 1200	− 2520
(1^7)	− 1	− 7	− 21	− 42	− 35	− 105	− 210	− 140	− 210	− 420	− 840	− 630	− 1260	− 2520	− 5040

VII (b).

=	7 h	16 bg	25 cf	$1^2 5$ $b^2 f$	34 de	124 bce	$1^3 4$ $b^3 e$	12^2 bd^2	$2^2 3$ $c^2 d$	$1^2 23$ $b^2 cd$	$1^4 3$ $b^4 d$	12^3 bc^3	$1^3 2^2$ $b^3 c^2$	$1^5 2$ $b^5 c$	1^7 b^7
(7)	− 7	+ 7	+ 7	− 7	+ 7	− 14	+ 7	− 7	− 7	+ 21	− 7	+ 7	− 14	+ 7	− 1
(61)	+ 7	− 1	− 7	+ 1	− 7	+ 8	− 1	+ 4	+ 7	− 9	+ 1	− 5	+ 5	− 1	
(52)	+ 7	− 7	+ 3	+ 2	− 7	+ 4	− 2	+ 7	− 3	− 6	+ 2	+ 3	− 1		
(43)	+ 7	− 7	− 7	+ 7	+ 5	+ 2	− 3	− 5	+ 1	+ 3	...	− 1			
(51^2)	− 7	+ 1	+ 2	− 1	+ 7	− 3	+ 1	− 4	− 2	+ 4	− 1				
(421)	− 14	+ 8	+ 4	− 3	+ 2	− 8	+ 3	+ 1	+ 2	− 1					
$(3^2 1)$	− 7	+ 4	+ 7	− 4	− 5	+ 1	...	+ 2	− 1						
(32^2)	− 7	+ 7	− 3	− 2	+ 1	+ 2	...	− 1							
(41^3)	+ 7	− 1	− 2	+ 1	− 3	+ 3	− 1								
(321^2)	+ 21	− 9	− 6	+ 4	+ 3	− 1									
$(2^3 1)$	+ 7	− 5	+ 3	...	− 1										
(31^4)	− 7	+ 1	+ 2	− 1											
$(2^2 1^3)$	− 14	+ 5	− 1												
(21^5)	+ 7	− 1													
(1^7)	− 1														

VIII (*a*). Runs on infrà.

‖	8 i	17 bh	26 cg	1^26 b^2g	35 df	125 bcf	1^35 b^3f	4^2 e^2	134 bde	2^24 c^2e	1^224 b^2ce	1^44 b^4e	23^2 cd^2
(8)													
(71)													
(62)													
(53)													
(4^2)													
(61^2)													
(521)													
(431)													
(42^2)													
(3^22)													+ 1
(51^3)												+ 1	...
(421^2)											+ 1	+ 4	...
(3^21^2)										+ 1	+ 2	+ 6	+ 2
(32^21)									+ 1	+ 2	+ 5	+ 12	+ 5
(2^4)								+ 1	+ 4	+ 6	+ 12	+ 24	+ 12
(41^4)							+ 1	...	...	...	+ 4	+ 17	...
(321^3)						+ 1	+ 3	...	+ 3	+ 7	+ 18	+ 46	+ 12
(2^31^2)					+ 1	+ 3	+ 6	+ 2	+ 11	+ 18	+ 39	+ 84	+ 31
(31^5)				+ 1	...	+ 5	+ 16	...	+ 10	+ 20	+ 55	+ 140	+ 30
(2^21^4)			+ 1	+ 2	+ 4	+ 14	+ 30	+ 6	+ 32	+ 53	+ 114	+ 246	+ 80
(21^6)		+ 1	+ 6	+ 13	+ 15	+ 51	+ 108	+ 20	+ 95	+ 150	+ 315	+ 660	+ 210
(1^8)	+ 1	+ 8	+ 28	+ 56	+ 56	+ 168	+ 336	+ 70	+ 280	+ 420	+ 840	+ 1680	+ 560

‖	1^23^2 b^2d^2	12^23 bc^2d	1^323 b^3cd	1^53 b^5d	2^4 c^4	1^22^3 b^2c^3	1^42^2 b^4c^2	1^62 b^6c	1^8 b^8
(8)									+ 1
(71)								+ 1	+ 8
(62)							+ 1	+ 6	+ 28
(53)						+ 1	+ 4	+ 15	+ 56
(4^2)					+ 1	+ 2	+ 6	+ 20	+ 70
(61^2)				+ 1	...	...	+ 2	+ 13	+ 56
(521)			+ 1	+ 5	...	+ 3	+ 14	+ 51	+ 168
(431)		+ 1	+ 3	+ 10	+ 4	+ 11	+ 32	+ 95	+ 280
(42^2)	+ 1	+ 2	+ 7	+ 20	+ 6	+ 18	+ 53	+ 150	+ 420
(3^22)	+ 2	+ 5	+ 12	+ 30	+ 12	+ 31	+ 80	+ 210	+ 560
(51^3)	...	...	+ 3	+ 16	...	+ 6	+ 30	+ 108	+ 336
(421^2)	+ 2	+ 5	+ 18	+ 55	+ 12	+ 39	+ 114	+ 315	+ 840
(3^21^2)	+ 4	+ 12	+ 30	+ 80	+ 28	+ 68	+ 172	+ 440	+ 1120
(32^21)	+ 12	+ 24	+ 58	+ 140	+ 48	+ 117	+ 284	+ 690	+ 1680
(2^4)	+ 28	+ 48	+ 108	+ 240	+ 90	+ 204	+ 468	+ 1080	+ 2520
(41^4)	+ 6	+ 12	+ 46	+ 140	+ 24	+ 84	+ 246	+ 660	+ 1680
(321^3)	+ 30	+ 58	+ 141	+ 340	+ 108	+ 258	+ 612	+ 1440	+ 3360
(2^31^2)	+ 68	+ 117	+ 258	+ 570	+ 204	+ 453	+ 1008	+ 2250	+ 5040
(31^5)	+ 80	+ 140	+ 340	+ 800	+ 240	+ 570	+ 1320	+ 3000	+ 6720
(2^21^4)	+ 172	+ 284	+ 612	+ 1320	+ 468	+ 1008	+ 2172	+ 4680	+ 10080
(21^6)	+ 440	+ 690	+ 1440	+ 3000	+ 1080	+ 2250	+ 4680	+ 9720	+ 20160
(1^8)	+ 1120	+ 1680	+ 3360	+ 6720	+ 2520	+ 5040	+ 10080	+ 20160	+ 40320

VIII (b). Runs on infrà.

=	8 i	17 bh	26 cg	1^26 b^2g	35 df	125 bcf	1^35 b^3f	4^2 e^2	134 bde	2^24 c^2e	1^224 b^2ce	1^44 b^4e	23^2 cd^2
(8)	− 8	+ 8	+ 8	− 8	+ 8	− 16	+ 8	+ 4	− 16	− 8	+ 24	− 8	− 8
(71)	+ 8	− 1	− 8	+ 1	− 8	+ 9	− 1	− 4	+ 9	+ 8	− 10	+ 1	+ 8
(62)	+ 8	− 8	+ 4	+ 2	− 8	+ 4	− 2	− 4	+ 16	− 4	− 6	+ 2	+ 2
(53)	+ 8	− 8	− 8	+ 8	+ 7	+ 1	− 3	− 4	+ 1	+ 8	− 9	+ 3	− 7
(4^2)	+ 4	− 4	− 4	+ 4	− 4	+ 8	− 4	+ 6	− 8	− 4	+ 4	...	+ 4
(61^2)	− 8	+ 1	+ 2	− 1	+ 8	− 3	+ 1	+ 4	− 9	− 2	+ 4	− 1	− 5
(521)	− 16	+ 9	+ 4	− 3	+ 1	− 8	+ 3	+ 8	− 10	− 4	+ 11	− 3	+ 5
(431)	− 16	+ 9	+ 16	− 9	+ 1	− 10	+ 4	− 8	+ 10	...	− 1	...	− 1
(42^2)	− 8	+ 8	− 4	− 2	+ 8	− 4	+ 2	− 4	...	+ 4	− 2	...	− 2
(3^22)	− 8	+ 8	+ 2	− 5	− 7	+ 5	...	+ 4	− 1	− 2	...	...	+ 1
(51^3)	+ 8	− 1	− 2	+ 1	− 3	+ 3	− 1	− 4	+ 4	+ 2	− 4	+ 1	
(421^2)	+ 24	− 10	− 6	+ 4	− 9	+ 11	− 4	+ 4	− 1	− 2	+ 1		
(3^21^2)	+ 12	− 5	− 9	+ 5	+ 3	− 1	...	+ 2	− 2	+ 1			
(32^21)	+ 24	− 17	...	+ 5	+ 6	− 3	...	− 4	+ 1				
(2^4)	+ 2	− 2	+ 2	...	− 2	...	...	+ 1					
(41^4)	− 8	+ 1	+ 2	− 1	+ 3	− 3	+ 1						
(321^3)	− 32	+ 11	+ 8	− 5	− 3	+ 1							
(2^31^2)	− 16	+ 9	− 4	...	+ 1								
(31^5)	+ 8	− 1	− 2	+ 1									
(2^21^4)	+ 20	− 6	+ 1										
(21^6)	− 8	+ 1											
(1^8)	+ 1												

=	1^23^2 b^2d^2	12^23 bc^2d	1^323 b^3cd	1^53 b^5d	2^4 c^4	1^22^3 b^2c^3	1^42^2 b^4c^2	1^62 b^6c	1^8 b^8
(8)	+ 12	+ 24	− 32	+ 8	+ 2	− 16	+ 20	− 8	+ 1
(71)	− 5	− 17	+ 11	− 1	− 2	+ 9	− 6	+ 1	
(62)	− 9	...	+ 8	− 2	+ 2	− 4	+ 1		
(53)	+ 3	+ 6	− 3	...	− 2	+ 1			
(4^2)	+ 2	− 4	...	...	+ 1				
(61^2)	+ 5	+ 5	− 5	+ 1					
(521)	− 1	− 3	+ 1						
(431)	− 2	+ 1							
(42^2)	+ 1								

IX (*a*). Runs on to p. 430.

	9	18	27	1^27	35	125	1^36	45	135	2^25	1^225	1^45	14^2
‖	j	bi	ch	b^2h	dg	bcg	b^3g	ef	bdf	c^2f	b^2cf	b^4f	be^2
(9)													
(81)													
(72)													
(63)													
(54)													
(71^2)													
(621)													
(531)													
(4^21)													
(52^2)													
(432)													
(3^3)													
(61^3)													
(521^2)													
(431^2)													
(42^21)													
(3^221)													
(32^3)													−1
(51^4)												−1	...
(421^3)											−1	−4	...
(3^21^3)										−1	−2	−6	...
(32^21^2)									−1	−2	−5	−12	−2
(2^41)								−1	−4	−6	−12	−24	−9
(41^5)							−1	...	...	...	−5	−21	...
(321^4)						−1	−3	...	−4	−9	−23	−58	−6
(2^31^3)					−1	−3	−6	−3	−15	−24	−51	−108	−24
(31^6)				−1	...	−6	−19	...	−15	−30	−81	−204	−20
(2^21^5)			−1	−2	−5	−17	−36	−10	−50	−81	−172	−366	−70
(21^7)		−1	−7	−15	−21	−70	−147	−35	−161	−252	−525	−1092	−210
(1^9)	−1	−9	−36	−72	−84	−252	−504	−126	−504	−756	−1512	−3024	−630

=	234 cde	$1^2 34$ b^2de	$12^2 4$ bc^2e	$1^3 24$ b^3ce	$1^5 4$ b^5e	3^3 d^3	123^2 bcd^2	$1^3 3^2$ b^3d^2	$2^3 3$ c^3d	$1^2 2^2 3$ b^2c^2d
(9)										
(81)										
(72)										
(63)										
(54)										
(71^2)										
(621)										
(531)										− 1
$(4^2 1)$									− 1	− 2
(52^2)								− 1	...	− 2
(432)							− 1	− 3	− 3	− 8
(3^3)						− 1	− 3	− 6	− 6	− 15
(61^3)					− 1	...	...	...	...	...
(521^2)				− 1	− 5	...	...	− 2	...	− 5
(431^2)			− 1	− 3	− 10	...	− 2	− 6	− 7	− 19
$(42^2 1)$		− 1	− 2	− 7	− 20	...	− 5	− 17	− 12	− 36
$(3^2 21)$	− 1	− 2	− 5	− 12	− 30	− 3	− 13	− 30	− 27	− 65
(32^3)	− 3	− 7	− 12	− 27	− 60	− 6	− 27	− 64	− 51	− 120
(51^4)	...	...	...	− 4	− 21	...	...	− 6	...	− 12
(421^3)	...	− 3	− 7	− 25	− 75	...	− 12	− 42	− 27	− 85
$(3^2 1^3)$	− 3	− 6	− 17	− 42	− 110	− 6	− 30	− 72	− 64	− 152
$(32^2 1^2)$	− 8	− 19	− 36	− 85	− 200	− 15	− 65	− 152	− 120	− 281
$(2^4 1)$	− 22	− 48	− 78	− 168	− 360	− 36	− 136	− 300	− 234	− 516
(41^5)	...	− 10	− 20	− 75	− 225	...	− 30	− 110	− 60	− 200
(321^4)	− 22	− 54	− 101	− 241	− 570	− 36	− 158	− 372	− 282	− 656
$(2^3 1^3)$	− 60	− 129	− 213	− 459	− 990	− 93	− 333	− 720	− 555	− 1203
(31^6)	− 60	− 155	− 270	− 645	− 1500	− 90	− 390	− 920	− 660	− 1530
$(2^2 1^5)$	− 165	− 350	− 565	− 1200	− 2550	− 240	− 820	− 1740	− 1320	− 2800
(21^7)	− 455	− 945	− 1470	− 3045	− 6300	− 630	− 2030	− 4200	− 3150	− 6510
(1^9)	− 1260	− 2520	− 3780	− 7560	− 15120	− 1680	− 5040	− 10080	− 7560	− 15120

‖	$1^4 23$ b^4cd	$1^6 3$ b^6d	12^4 bc^4	$1^3 2^3$ b^3c^3	$1^5 2^2$ b^5c^2	$1^7 2$ b^7c	1^9 b^9
(9)							− 1
(81)						− 1	− 9
(72)					− 1	− 7	− 36
(63)				− 1	− 5	− 21	− 84
(54)			− 1	− 3	− 10	− 35	− 126
(71^2)		− 1	...	...	− 2	− 15	− 72
(621)	− 1	− 6	...	− 3	− 17	− 70	− 252
(531)	− 4	− 15	− 4	− 15	− 50	− 161	− 504
($4^2 1$)	− 6	− 20	− 9	− 24	− 70	− 210	− 630
(52^2)	− 9	− 30	− 6	− 24	− 81	− 252	− 756
(432)	− 22	− 60	− 22	− 60	− 165	− 455	− 1260
(3^3)	− 36	− 90	− 36	− 93	− 240	− 630	− 1680
(61^3)	− 3	− 19	...	− 6	− 36	− 147	− 504
(521^2)	− 23	− 81	− 12	− 51	− 172	− 525	− 1512
(431^2)	− 54	− 155	− 48	− 129	− 350	− 945	− 2520
($42^2 1$)	− 101	− 270	− 78	− 213	− 565	− 1470	− 3780
($3^2 21$)	− 158	− 390	− 136	− 333	− 820	− 2030	− 5040
(32^3)	− 282	− 660	− 234	− 555	− 1320	− 3150	− 7560
(51^4)	− 58	− 204	− 24	− 108	− 366	− 1092	− 3024
(421^3)	− 241	− 645	− 168	− 459	− 1200	− 3045	− 7560
($3^2 1^3$)	− 372	− 920	− 300	− 720	− 1740	− 4200	− 10080
($32^2 1^2$)	− 656	− 1530	− 516	− 1203	− 2800	− 6510	− 15120
($2^4 1$)	− 1140	− 2520	− 906	− 2016	− 4500	− 10080	− 22680
(41^5)	− 570	− 1500	− 360	− 990	− 2550	− 6300	− 15120
(321^4)	− 1516	− 3480	− 1140	− 2610	− 5940	− 13440	− 30240
($2^3 1^3$)	− 2610	− 5670	− 2016	− 4383	− 9540	− 20790	− 45360
(31^6)	− 3480	− 7800	− 2520	− 5670	− 12600	− 27720	− 60480
($2^2 1^5$)	− 5940	− 12600	− 4500	− 9540	− 20220	− 42840	− 90720
(21^7)	− 13440	− 27720	− 10080	− 20790	− 42840	− 88200	− 181440
(1^9)	− 30240	− 60480	− 22680	− 45360	− 90720	− 181440	− 362880

IX (b). Runs on to p. 432.

=	9 j	18 bi	27 ch	1^27 b^2h	35 dg	125 bcg	1^36 b^3g	45 ef	135 bdf	2^25 c^2f	1^225 b^2cf	1^45 b^4f	14^2 be^2	234 cde	1^234 b^2de
(9)	− 9	+ 9	+ 9	− 9	+ 9	− 18	+ 9	+ 9	− 18	− 9	+ 27	− 9	− 9	− 18	+ 27
(81)	+ 9	− 1	− 9	+ 1	− 9	+ 10	− 1	− 9	+ 10	+ 9	− 11	+ 1	+ 5	+ 18	− 11
(72)	+ 9	− 9	+ 5	+ 2	− 9	+ 4	− 2	− 9	+ 18	− 5	− 6	+ 2	+ 9	+ 4	− 20
(63)	+ 9	− 9	− 9	+ 9	+ 9	...	− 3	− 9	...	+ 9	− 9	+ 3	+ 9	...	− 9
(54)	+ 9	− 9	− 9	+ 9	− 9	+ 18	− 9	+ 11	− 2	− 1	− 7	+ 4	− 11	− 2	+ 13
(71^2)	− 9	+ 1	+ 2	− 1	+ 9	− 3	+ 1	+ 9	− 10	− 2	+ 4	− 1	− 5	− 11	+ 11
(621)	− 18	+ 10	+ 4	− 3	...	− 8	+ 3	+ 18	− 10	− 4	+ 11	− 3	− 14	− 4	+ 13
(531)	− 18	+ 10	+ 18	− 10	...	− 10	+ 4	− 2	− 5	− 8	+ 15	− 4	+ 6	+ 2	− 5
(4^21)	− 9	+ 5	+ 9	− 5	+ 9	− 14	+ 5	− 11	+ 6	+ 1	− 1	...	+ 5	+ 2	− 5
(52^2)	− 9	+ 9	− 5	− 2	+ 9	− 4	+ 2	− 1	− 8	...	+ 6	− 2	+ 1	+ 6	...
(432)	− 18	+ 18	+ 4	− 11	...	− 4	+ 5	− 2	+ 2	+ 6	− 5	...	+ 2	− 8	+ 1
(3^3)	− 3	+ 3	+ 3	− 3	− 6	+ 3	...	+ 3	+ 3	− 3	...	...	− 3	+ 3	...
(61^3)	+ 9	− 1	− 2	+ 1	− 3	+ 3	− 1	− 9	+ 4	+ 2	− 4	+ 1	+ 5	+ 5	− 5
(521^2)	+ 27	− 11	− 6	+ 4	− 9	+ 11	− 4	− 7	+ 15	+ 6	− 15	+ 4	− 1	− 5	+ 1
(431^2)	+ 27	− 11	− 20	+ 11	− 9	+ 13	− 5	+ 13	− 5	...	+ 1	...	− 5	+ 1	+ 2
(42^21)	+ 27	− 19	+ 1	+ 5	− 9	+ 12	− 5	+ 3	− 2	− 6	+ 3	...	+ 1	+ 2	− 1
(3^221)	+ 27	− 19	− 13	+ 12	+ 18	− 7	...	− 7	− 4	+ 3	...	...	+ 3	− 1	
(32^3)	+ 9	− 9	+ 5	+ 2	− 3	− 2	...	+ 1	+ 2	...	...	...	− 1		
(51^4)	− 9	+ 1	+ 2	− 1	+ 3	− 3	+ 1	+ 4	− 4	− 2	+ 4	− 1			
(421^3)	− 36	+ 12	+ 8	− 5	+ 12	− 14	+ 5	− 4	+ 1	+ 2	− 1				
(3^21^3)	− 18	+ 6	+ 11	− 6	− 3	+ 1	...	− 2	+ 2	− 1					
(32^21^2)	− 54	+ 30	+ 5	− 9	− 9	+ 4	...	+ 4	− 1						
(2^41)	− 9	+ 7	− 5	...	+ 3	...	...	− 1							
(41^5)	+ 9	− 1	− 2	+ 1	− 3	+ 3	− 1								
(321^4)	+ 45	− 13	− 10	+ 6	+ 3	− 1									
(2^31^3)	+ 30	− 14	+ 5	...	− 1										
(31^6)	− 9	+ 1	+ 2	− 1											
(2^21^5)	− 27	+ 7	− 1												
(21^7)	+ 9	− 1													
(1^9)	− 1														

=	12^24 bc^2e	1^324 b^3ce	1^54 b^5e	3^3 d^3	123^2 bcd^2	1^33^3 b^3d^2	2^33 c^3d	1^22^23 b^2c^2d	1^423 b^4cd	1^63 b^6d	12^4 bc^4	1^32^3 b^3c^3	1^52^2 b^5c^2	1^72 b^7c	1^9 b^9
(9)	+ 27	− 36	+ 9	− 3	+ 27	− 18	+ 9	− 54	+ 45	− 9	− 9	+ 30	− 27	+ 9	− 1
(81)	− 19	+ 12	− 1	+ 3	− 19	+ 6	− 9	+ 30	− 13	+ 1	+ 7	− 14	+ 7	− 1	
(72)	+ 1	+ 8	− 2	+ 3	− 13	+ 11	+ 5	+ 5	− 10	+ 2	− 5	+ 5	− 1		
(63)	− 9	+ 12	− 3	− 6	+ 18	− 3	− 3	− 9	+ 3	...	+ 3	− 1			
(54)	+ 3	− 4	...	+ 3	− 7	− 2	+ 1	+ 4	...	...	− 1				
(71^2)	+ 5	− 5	+ 1	− 3	+ 12	− 6	+ 2	− 9	+ 6	− 1					
(621)	+ 12	− 14	+ 3	+ 3	− 7	+ 1	− 2	+ 4	− 1						
(531)	− 2	+ 1	...	+ 3	− 4	+ 2	+ 2	− 1							
(4^21)	+ 1	...	...	− 3	+ 3	...	− 1								
(52^2)	− 6	+ 2	...	− 3	+ 3	− 1									
(432)	+ 2	...	...	+ 3	− 1										
(3^3)	...	...	...	− 1											
(61^3)	− 5	+ 5	− 1												
(521^2)	+ 3	− 1													
(431^2)	− 1														

X (*a*). Runs on to p. 436.

‖	10 k	19 bj	28 ci	1^28 b^2i	37 dh	127 bch	1^37 b^3h	46 eg	136 bdg	2^26 c^2g	1^226 b^2cg	1^46 b^4g	5^2 f^2
(10)													
(91)													
(82)													
(73)													
(64)													
(5^2)													
(81^2)													
(721)													
(631)													
(541)													
(62^2)													
(532)													
(4^22)													
(43^2)													
(71^3)													
(621^2)													
(531^2)													
(4^21^2)													
(52^21)													
(4321)													
(3^31)													
(42^3)													
(3^22^2)													
(61^4)													
(521^3)													
(431^3)													
(42^21^2)													
(3^221^2)													
(32^31)													
(2^5)													+ 1
(51^5)												+ 1	...
(421^4)											+ 1	+ 4	...
(3^21^4)										+ 1	+ 2	+ 6	...
(32^21^3)									+ 1	+ 2	+ 5	+ 12	...
(2^41^2)								+ 1	+ 4	+ 6	+ 12	+ 24	+ 2
(41^6)							+ 1	...	...	...	+ 6	+ 25	...
(321^5)						+ 1	+ 3	...	+ 5	+ 11	+ 28	+ 70	...
(2^31^4)					+ 1	+ 3	+ 6	+ 4	+ 19	+ 30	+ 63	+ 132	+ 6
(31^7)				+ 1	...	+ 7	+ 22	...	+ 21	+ 42	+ 112	+ 280	...
(2^21^6)			+ 1	+ 2	+ 6	+ 20	+ 42	+ 15	+ 72	+ 115	+ 242	+ 510	+ 20
(21^8)		+ 1	+ 8	+ 17	+ 28	+ 92	+ 192	+ 56	+ 252	+ 392	+ 812	+ 1680	+ 70
(1^{10})	+ 1	+ 10	+ 45	+ 90	+ 120	+ 360	+ 720	+ 210	+ 840	+ 1260	+ 2520	+ 5040	+ 252

‖	145 bef	235 cdf	$1^{2}35$ $b^{2}df$	$12^{2}5$ $bc^{2}f$	$1^{3}25$ $b^{3}cf$	$1^{5}5$ $b^{5}f$	24^{2} ce^{2}	$1^{2}4^{2}$ $b^{2}e^{2}$	$3^{2}4$ $d^{2}e$	1234 $bcde$
(10)										
(91)										
(82)										
(73)										
(64)										
(5^{2})										
(81^{2})										
(721)										
(631)										
(541)										
(62^{2})										
(532)										
$(4^{2}2)$										
(43^{2})										
(71^{3})										
(621^{2})										
(531^{2})										
$(4^{2}1^{2})$										
$(52^{2}1)$										
(4321)										+ 1
$(3^{3}1)$									+ 1	+ 3
(42^{3})								+ 1	…	+ 3
$(3^{2}2^{2})$							+ 1	+ 2	+ 2	+ 8
(61^{4})						+ 1	…	…	…	…
(521^{3})					+ 1	+ 5	…	…	…	…
(431^{3})				+ 1	+ 3	+ 10	…	…	…	+ 3
$(42^{2}1^{2})$			+ 1	+ 2	+ 7	+ 20	…	+ 2	…	+ 8
$(3^{2}21^{2})$		+ 1	+ 2	+ 5	+ 12	+ 30	+ 2	+ 4	+ 5	+ 21
$(32^{3}1)$	+ 1	+ 3	+ 7	+ 12	+ 27	+ 60	+ 7	+ 16	+ 12	+ 49
(2^{5})	+ 5	+ 10	+ 20	+ 30	+ 60	+ 120	+ 20	+ 45	+ 30	+ 110
(51^{5})	…	…	…	…	+ 5	+ 26	…	…	…	…
(421^{4})	…	…	+ 4	+ 9	+ 32	+ 95	…	+ 6	…	+ 22
$(3^{2}1^{4})$	…	+ 4	+ 8	+ 22	+ 54	+ 140	+ 6	+ 12	+ 12	+ 56
$(32^{2}1^{3})$	+ 3	+ 11	+ 26	+ 48	+ 112	+ 260	+ 18	+ 42	+ 31	+ 128
$(2^{4}1^{2})$	+ 14	+ 32	+ 68	+ 108	+ 228	+ 480	+ 53	+ 114	+ 80	+ 284
(41^{6})	…	…	+ 15	+ 30	+ 111	+ 330	…	+ 20	…	+ 60
(321^{5})	+ 10	+ 35	+ 85	+ 156	+ 368	+ 860	+ 50	+ 120	+ 80	+ 335
$(2^{3}1^{4})$	+ 42	+ 99	+ 210	+ 339	+ 720	+ 1530	+ 144	+ 306	+ 213	+ 735
(31^{7})	+ 35	+ 105	+ 266	+ 462	+ 1092	+ 2520	+ 140	+ 350	+ 210	+ 875
$(2^{2}1^{6})$	+ 130	+ 296	+ 622	+ 990	+ 2082	+ 4380	+ 400	+ 840	+ 570	+ 1900
(21^{8})	+ 406	+ 868	+ 1792	+ 2772	+ 5712	+ 11760	+ 1120	+ 2310	+ 1540	+ 4900
(1^{10})	+ 1260	+ 2520	+ 5040	+ 7560	+ 15120	+ 30240	+ 3150	+ 6300	+ 4200	+ 12600

$1^3 34$ b^3de	$2^3 4$ c^3e	$1^2 2^2 4$ b^2c^2e	$1^4 24$ b^4ce	$1^6 4$ b^6e	13^3 bd^3	$2^2 3^2$ c^2d^2	$1^2 23^2$ b^2cd^2	$1^4 3^2$ b^4d^2	$12^3 3$ bc^3d
									+ 1
								+ 1	...
							+ 1	+ 4	+ 3
						+ 1	+ 2	+ 6	+ 7
					+ 1	+ 2	+ 5	+ 12	+ 12
				+ 1	...	...	...	...	...
			+ 1	+ 6	...	...	...	+ 2	...
		+ 1	+ 4	+ 15	...	...	+ 2	+ 8	+ 7
	+ 1	+ 2	+ 6	+ 20	...	+ 2	+ 4	+ 12	+ 16
+ 1	...	+ 2	+ 9	+ 30	...	...	+ 5	+ 22	+ 12
+ 3	+ 3	+ 8	+ 22	+ 60	+ 3	+ 8	+ 21	+ 56	+ 49
+ 6	+ 6	+ 15	+ 36	+ 90	+ 10	+ 18	+ 42	+ 96	+ 87
+ 10	+ 6	+ 18	+ 48	+ 120	+ 6	+ 15	+ 42	+ 115	+ 87
+ 18	+ 15	+ 34	+ 78	+ 180	+ 18	+ 34	+ 80	+ 188	+ 156
...	...	...	+ 4	+ 25	...	...	...	+ 6	...
+ 3	...	+ 7	+ 32	+ 111	...	...	+ 12	+ 54	+ 27
+ 9	+ 10	+ 27	+ 76	+ 215	+ 6	+ 18	+ 48	+ 132	+ 112
+ 27	+ 18	+ 54	+ 149	+ 390	+ 15	+ 34	+ 99	+ 270	+ 198
+ 48	+ 42	+ 99	+ 236	+ 570	+ 42	+ 80	+ 186	+ 436	+ 358
+ 112	+ 87	+ 198	+ 450	+ 1020	+ 87	+ 156	+ 358	+ 820	+ 645
+ 240	+ 180	+ 390	+ 840	+ 1800	+ 180	+ 310	+ 680	+ 1500	+ 1170
+ 10	...	+ 20	+ 95	+ 330	...	...	+ 30	+ 140	+ 60
+ 76	+ 48	+ 149	+ 416	+ 1095	+ 36	+ 78	+ 236	+ 650	+ 450
+ 132	+ 115	+ 270	+ 650	+ 1580	+ 96	+ 188	+ 436	+ 1032	+ 820
+ 294	+ 228	+ 523	+ 1196	+ 2730	+ 210	+ 370	+ 844	+ 1920	+ 1479
+ 612	+ 468	+ 1008	+ 2172	+ 4680	+ 444	+ 740	+ 1604	+ 3480	+ 2688
+ 215	+ 120	+ 390	+ 1095	+ 2850	+ 90	+ 180	+ 570	+ 1580	+ 1020
+ 775	+ 585	+ 1340	+ 3050	+ 6900	+ 510	+ 880	+ 2000	+ 4520	+ 3390
+ 1566	+ 1194	+ 2547	+ 5436	+ 11610	+ 1092	+ 1776	+ 3792	+ 8100	+ 6180
+ 2030	+ 1470	+ 3360	+ 7560	+ 16800	+ 1260	+ 2100	+ 4760	+ 10640	+ 7770
+ 3990	+ 3015	+ 6330	+ 13290	+ 27900	+ 2700	+ 4280	+ 8980	+ 18840	+ 14220
+ 10080	+ 7560	+ 15540	+ 31920	+ 65520	+ 6720	+ 10360	+ 21280	+ 43680	+ 32760
+ 25200	+ 18900	+ 37800	+ 75600	+ 151200	+ 16800	+ 25200	+ 50400	+ 100800	+ 75600

‖	$1^3 2^2 3$ b^3c^2d	$1^5 23$ b^5cd	$1^7 3$ b^7d	2^5 c^5	$2^2 3^2$ b^2c^4	$1^4 2^3$ b^4c^3	$1^6 2^2$ b^6c^2	$1^8 2$ b^8c	1^{10} b^{10}
(10)									+ 1
(91)								+ 1	+ 10
(82)							+ 1	+ 8	+ 45
(73)						+ 1	+ 6	+ 28	+ 120
(64)					+ 1	+ 4	+ 15	+ 56	+ 210
(5^2)				+ 1	+ 2	+ 6	+ 20	+ 70	+ 252
(81^2)			+ 1	...	...	...	+ 2	+ 17	+ 90
(721)		+ 1	+ 7	...	...	+ 3	+ 20	+ 92	+ 360
(631)	+ 1	+ 5	+ 21	...	+ 4	+ 19	+ 72	+ 252	+ 840
(541)	+ 3	+ 10	+ 35	+ 5	+ 14	+ 42	+ 130	+ 406	+ 1260
(62^2)	+ 2	+ 11	+ 42	...	+ 6	+ 30	+ 115	+ 392	+ 1260
(532)	+ 11	+ 35	+ 105	+ 10	+ 32	+ 99	+ 296	+ 868	+ 2520
$(4^2 2)$	+ 18	+ 50	+ 140	+ 20	+ 53	+ 144	+ 400	+ 1120	+ 3150
(43^2)	+ 31	+ 80	+ 210	+ 30	+ 80	+ 213	+ 570	+ 1540	+ 4200
(71^3)	...	+ 3	+ 22	...	...	+ 6	+ 42	+ 192	+ 720
(621^2)	+ 5	+ 28	+ 112	...	+ 12	+ 63	+ 242	+ 812	+ 2520
(531^2)	+ 26	+ 85	+ 266	+ 20	+ 68	+ 210	+ 622	+ 1792	+ 5040
$(4^2 1^2)$	+ 42	+ 120	+ 350	+ 45	+ 114	+ 306	+ 840	+ 2310	+ 6300
$(52^2 1)$	+ 48	+ 156	+ 462	+ 30	+ 108	+ 339	+ 990	+ 2772	+ 7560
(4321)	+ 128	+ 335	+ 875	+ 110	+ 284	+ 735	+ 1900	+ 4900	+ 12600
$(3^3 1)$	+ 210	+ 510	+ 1260	+ 180	+ 444	+ 1092	+ 2700	+ 6720	+ 16800
(42^3)	+ 228	+ 585	+ 1470	+ 180	+ 468	+ 1194	+ 3015	+ 7560	+ 18900
$(3^2 2^2)$	+ 370	+ 880	+ 2100	+ 310	+ 740	+ 1776	+ 4280	+ 10360	+ 25200
(61^4)	+ 12	+ 70	+ 280	...	+ 24	+ 132	+ 510	+ 1680	+ 5040
(521^3)	+ 112	+ 368	+ 1092	+ 60	+ 228	+ 720	+ 2082	+ 5712	+ 15120
(431^3)	+ 294	+ 775	+ 2030	+ 240	+ 612	+ 1566	+ 3990	+ 10080	+ 25200
$(42^2 1^2)$	+ 523	+ 1340	+ 3360	+ 390	+ 1008	+ 2547	+ 6330	+ 15540	+ 37800
$(3^2 21^2)$	+ 844	+ 2000	+ 4760	+ 680	+ 1604	+ 3792	+ 8980	+ 21280	+ 50400
$(32^3 1)$	+ 1479	+ 3390	+ 7770	+ 1170	+ 2688	+ 6180	+ 14220	+ 32760	+ 75600
(2^5)	+ 2580	+ 5700	+ 12600	+ 2040	+ 4530	+ 10080	+ 22500	+ 50400	+ 113400
(51^5)	+ 260	+ 860	+ 2520	+ 120	+ 480	+ 1530	+ 4380	+ 11760	+ 30240
(421^4)	+ 1196	+ 3050	+ 7560	+ 840	+ 2172	+ 5436	+ 13290	+ 31920	+ 75600
$(3^2 1^4)$	+ 1920	+ 4520	+ 10640	+ 1500	+ 3480	+ 8100	+ 18840	+ 43680	+ 100800
$(32^2 1^3)$	+ 3358	+ 7610	+ 17220	+ 2580	+ 5844	+ 13212	+ 29820	+ 67200	+ 151200
$(2^4 1^2)$	+ 5844	+ 12720	+ 27720	+ 4530	+ 9876	+ 21564	+ 47160	+ 103320	+ 226800
(41^6)	+ 2730	+ 6900	+ 16800	+ 1800	+ 4680	+ 11610	+ 27900	+ 65520	+ 151200
(321^5)	+ 7610	+ 17000	+ 37800	+ 5700	+ 12720	+ 28260	+ 62520	+ 137760	+ 302400
$(2^3 1^4)$	+ 13212	+ 28260	+ 60480	+ 10080	+ 21564	+ 46152	+ 98820	+ 211680	+ 453600
(31^7)	+ 17220	+ 37800	+ 82320	+ 12600	+ 27720	+ 60480	+ 131040	+ 282240	+ 604800
$(2^2 1^6)$	+ 29820	+ 62520	+ 131040	+ 22500	+ 47160	+ 98820	+ 207000	+ 433440	+ 907200
(21^8)	+ 67200	+ 137760	+ 282240	+ 50400	+ 103320	+ 211680	+ 433440	+ 887040	+ 1814400
(1^{10})	+ 151200	+ 302400	+ 604800	+ 113400	+ 226800	+ 453600	+ 907200	+ 1814400	+ 3628800

X (*b*). Runs on to p. 439.

	10	19	28	1^28	37	127	1^37	46	136	2^26	1^226	1^46	5^2	145
=	k	bj	ci	b^2i	dh	bch	b^3h	eg	bdg	c^2g	b^2cg	b^4g	f^2	bef
(10)	− 10	+ 10	+ 10	− 10	+ 10	− 20	+ 10	+ 10	− 20	− 10	+ 30	− 10	+ 5	− 20
(91)	+ 10	− 1	− 10	+ 1	− 10	+ 11	− 1	− 10	+ 11	+ 10	− 12	+ 1	− 5	+ 11
(82)	+ 10	− 10	+ 6	+ 2	− 10	+ 4	− 2	− 10	+ 20	− 6	− 6	+ 2	− 5	+ 20
(73)	+ 10	− 10	− 10	+ 10	+ 11	− 1	− 3	− 10	− 1	+ 10	− 9	+ 3	− 5	+ 20
(64)	+ 10	− 10	− 10	+ 10	− 10	+ 20	− 10	+ 14	− 4	− 2	− 6	+ 4	− 5	− 4
(5^2)	+ 5	− 5	− 5	+ 5	− 5	+ 10	− 5	− 5	+ 10	+ 5	− 15	+ 5	+ 10	− 15
(81^2)	− 10	+ 1	+ 2	− 1	+ 10	− 3	+ 1	+ 10	− 11	− 2	+ 4	− 1	+ 5	− 11
(721)	− 20	+ 11	+ 4	− 3	− 1	− 8	+ 3	+ 20	− 10	− 4	+ 11	− 3	+ 10	− 31
(631)	− 20	+ 11	+ 20	− 11	− 1	− 10	+ 4	− 4	− 4	− 8	+ 15	− 4	+ 10	− 7
(541)	− 20	+ 11	+ 20	− 11	+ 20	− 31	+ 11	− 4	− 7	− 8	+ 18	− 5	− 15	+ 23
(62^2)	− 10	+ 10	− 6	− 2	+ 10	− 4	+ 2	− 2	− 8	...	+ 6	− 2	+ 5	− 8
(532)	− 20	+ 20	+ 4	− 12	− 1	− 3	+ 5	+ 20	− 19	− 4	+ 15	− 5	− 15	+ 10
(4^22)	− 10	+ 10	+ 2	− 6	+ 10	− 12	+ 6	− 14	+ 4	+ 10	− 6	...	+ 5	+ 4
(43^2)	− 10	+ 10	+ 10	− 10	− 11	+ 1	+ 3	− 2	+ 13	− 4	− 3	...	+ 5	− 8
(71^3)	+ 10	− 1	− 2	+ 1	− 3	+ 3	− 1	− 10	+ 4	+ 2	− 4	+ 1	− 5	+ 11
(621^2)	+ 30	− 12	− 6	+ 4	− 9	+ 11	− 4	− 6	+ 15	+ 6	− 15	+ 4	− 15	+ 18
(531^2)	+ 30	− 12	− 22	+ 12	− 9	+ 13	− 5	− 6	+ 15	+ 10	− 19	+ 5	+ 10	− 12
(4^21^2)	+ 15	− 6	− 11	+ 6	− 15	+ 17	− 6	+ 9	− 3	− 1	+ 1	...	+ 5	− 8
(52^21)	+ 30	− 21	+ 2	+ 5	− 9	+ 12	− 5	− 18	+ 18	+ 4	− 17	+ 5	+ 10	− 1
(4321)	+ 60	− 42	− 28	+ 26	+ 3	+ 21	− 12	+ 12	− 15	− 8	+ 7	...	− 5	...
(3^31)	+ 10	− 7	− 10	+ 7	+ 11	− 4	...	+ 2	− 7	+ 4	...	...	− 5	+ 5
(42^3)	+ 10	− 10	+ 6	+ 2	− 10	+ 4	− 2	+ 10	...	− 4	+ 2	...	− 5	...
(3^22^2)	+ 15	− 15	+ 1	+ 7	+ 6	− 7	...	− 9	+ 3	+ 2	...	...	+ 5	− 1
(61^4)	− 10	+ 1	+ 2	− 1	+ 3	− 3	+ 1	+ 4	− 4	− 2	+ 4	− 1	+ 5	− 5
(521^3)	− 40	+ 13	+ 8	− 5	+ 12	− 14	+ 5	+ 16	− 19	− 8	+ 19	− 5	− 5	+ 1
(431^3)	− 40	+ 13	+ 24	− 13	+ 12	− 16	+ 6	− 8	+ 5	...	− 1	...	− 5	+ 5
(42^21^2)	− 60	+ 33	+ 4	− 9	+ 18	− 23	+ 9	− 12	+ 3	+ 8	− 4	...	+ 5	− 1
(3^221^2)	− 60	+ 33	+ 28	− 21	− 24	+ 9	...	...	+ 6	− 4	...	...	+ 5	− 3
(32^31)	− 40	+ 31	− 8	− 7	− 2	+ 5	...	+ 8	− 3	...	...	...	− 5	+ 1
(2^5)	− 2	+ 2	− 2	...	+ 2	...	...	− 2	...	...	...	...	+ 1	
(51^5)	+ 10	− 1	− 2	+ 1	− 3	+ 3	− 1	− 4	+ 4	+ 2	− 4	+ 1		
(421^4)	+ 50	− 14	− 10	+ 6	− 15	+ 17	− 6	+ 4	− 1	− 2	+ 1			
(3^21^4)	+ 25	− 7	− 13	+ 7	+ 3	− 1	...	+ 2	− 2	+ 1				
(32^21^3)	+ 100	− 46	− 12	+ 14	+ 12	− 5	...	− 4	+ 1					
(2^41^2)	+ 25	− 16	+ 9	...	− 4	...	...	+ 1						
(41^6)	− 10	+ 1	+ 2	− 1	+ 3	− 3	+ 1							
(321^5)	− 60	+ 15	+ 12	− 7	− 3	+ 1								
(2^31^4)	− 50	+ 20	− 6	...	+ 1									
(31^7)	+ 10	− 1	− 2	+ 1										
(2^21^6)	+ 35	− 8	+ 1											
(21^8)	− 10	+ 1												
(1^{10})	+ 1													

=	235 cdf	$1^2 35$ b^2df	$12^2 5$ bc^2f	$1^3 25$ b^3cf	$1^5 5$ b^5f	24^2 ce^2	$1^2 4^2$ b^2e^2	$3^2 4$ d^2e	1234 $bcde$	$1^3 34$ b^3de	$2^3 4$ c^3e	$1^2 2^2 4$ b^2c^2e	$1^4 24$ b^4ce	$1^6 4$ b^6e
(10)	− 20	+ 30	+ 30	− 40	+ 10	− 10	+ 15	− 10	+ 60	− 40	+ 10	− 60	+ 50	− 10
(91)	+ 20	− 12	− 21	+ 13	− 1	+ 10	− 6	+ 10	− 42	+ 13	− 10	+ 33	− 14	+ 1
(82)	+ 4	− 22	+ 2	+ 8	− 2	+ 2	− 11	+ 10	− 28	+ 24	+ 6	+ 4	− 10	+ 2
(73)	− 1	− 9	− 9	+ 12	− 3	+ 10	− 15	− 11	+ 3	+ 12	− 10	+ 18	− 15	+ 3
(64)	+ 20	− 6	− 18	+ 16	− 4	− 14	+ 9	− 2	+ 12	− 8	+ 10	− 12	+ 4	...
(5^2)	− 15	+ 10	+ 10	− 5	...	+ 5	+ 5	+ 5	− 5	− 5	− 5	+ 5	...	...
(81^2)	− 12	+ 12	+ 5	− 5	+ 1	− 6	+ 6	− 10	+ 26	− 13	+ 2	− 9	+ 6	− 1
(721)	− 3	+ 13	+ 12	− 14	+ 3	− 12	+ 17	+ 1	+ 21	− 16	+ 4	− 23	+ 17	− 3
(631)	− 19	+ 15	+ 18	− 19	+ 4	+ 4	− 3	+ 13	− 15	+ 5	...	+ 3	− 1	...
(541)	+ 10	− 12	− 1	+ 1	...	+ 4	− 8	− 8	...	+ 5	...	− 1	...	...
(62^2)	− 4	+ 10	+ 4	− 8	+ 2	+ 10	− 1	− 4	− 8	...	− 4	+ 8	− 2	...
(532)	+ 17	− 4	− 13	+ 5	...	− 12	+ 1	+ 1	+ 5	− 1	+ 4	− 2	...	...
($4^2 2$)	− 12	+ 2	+ 2	...	...	+ 2	− 3	+ 2	+ 4	...	− 2	...	...	...
(43^2)	+ 1	− 3	+ 3	...	...	+ 2	+ 3	− 1	− 3	...	...	...	...	...
(71^3)	+ 5	− 5	− 5	+ 5	− 1	+ 6	− 6	+ 3	− 12	+ 6	− 2	+ 9	− 6	+ 1
(621^2)	+ 15	− 19	− 17	+ 19	− 4	− 6	+ 1	− 3	+ 7	− 1	+ 2	− 4	+ 1	
(531^2)	− 4	+ 3	+ 2	− 1	...	+ 2	+ 2	− 3	+ 4	− 2	− 2	+ 1		
($4^2 1^2$)	+ 1	+ 2	− 1	...	...	− 3	+ 3	+ 3	− 3	...	+ 1			
($52^2 1$)	− 13	+ 2	+ 9	− 3	...	+ 2	− 1	+ 3	− 3	+ 1				
(4321)	+ 5	+ 4	− 3	...	...	+ 4	− 3	− 3	+ 1					
($3^3 1$)	− 1	...	...	...	...	− 2	...	+ 1						
(42^3)	+ 4	− 2	...	...	...	− 2	+ 1							
($3^2 2^2$)	− 2	...	...	...	..	+ 1								
(61^4)	− 5	+ 5	+ 5	− 5	+ 1									
(521^3)	+ 5	− 1	− 3	+ 1										
(431^3)	− 1	− 2	+ 1											
($42^2 1^2$)	− 2	+ 1												
($3^2 21^2$)	+ 1													

13^3	2^23^2	1^223^2	1^43^2	12^33	1^32^23	1^523	1^73	2^5	2^23^2	1^42^3	1^62^2	1^82	1^{10}
bd^3	c^2d^2	b^2cd^2	b^4d^2	bc^3d	b^3c^2d	b^5cd	b^7d	c^5	b^2c^4	b^4c^3	b^6c^2	b^8c	b^{10}
+ 10	+ 15	− 60	+ 25	− 40	+ 100	− 60	+ 10	− 2	+ 25	− 50	+ 35	− 10	+ 1
− 7	− 15	+ 33	− 7	+ 31	− 46	+ 15	− 1	+ 2	− 16	+ 20	− 8	+ 1	
− 10	+ 1	+ 28	− 13	− 8	− 12	+ 12	− 2	− 2	+ 9	− 6	+ 1		
+ 11	+ 6	− 24	+ 3	− 2	+ 12	− 3	...	+ 2	− 4	+ 1			
+ 2	− 9	...	+ 2	+ 8	− 4	...	...	− 2	+ 1				
− 5	+ 5	+ 5	...	− 5	...	...	...	+ 1					
+ 7	+ 7	− 21	+ 7	− 7	+ 14	− 7	+ 1						
− 4	− 7	+ 9	− 1	+ 5	− 5	+ 1							
− 7	+ 3	+ 6	− 2	− 3	+ 1								
+ 5	− 1	− 3	...	+ 1									
+ 4	+ 2	− 4	+ 1										
− 1	− 2	+ 1											
− 2	+ 1												
+ 1													

148.

MEMOIR ON THE RESULTANT OF A SYSTEM OF TWO EQUATIONS.

[From the *Philosophical Transactions of the Royal Society of London*, vol. CXLVII. *for the year* 1857, pp. 703—715. Received December 18, 1856,—Read January 8, 1857.]

THE Resultant of two equations such as

$$(a, b, \ldots \between x, y)^m = 0,$$
$$(p, q, \ldots \between x, y)^n = 0,$$

is, it is well known, a function homogeneous in regard to the coefficients of each equation separately, viz. of the degree n in regard to the coefficients $(a, b, \ldots)$ of the first equation, and of the degree m in regard to the coefficients $(p, q, \ldots)$ of the second equation; and it is natural to develope the resultant in the form $kAP + k'A'P' +$ &c., where A, A', &c. are the combinations (powers and products) of the degree n in the coefficients $(a, b, \ldots)$, P, P', &c. are the combinations of the degree m in the coefficients $(p, q, \ldots)$, and k, k', &c. are mere numerical coefficients. The object of the present memoir is to show how this may be conveniently effected, either by the method of symmetric functions, or from the known expression of the Resultant in the form of a determinant, and to exhibit the developed expressions for the resultant of two equations, the degrees of which do not exceed 4. With respect to the first method, the formula in its best form, or nearly so, is given in the *Algebra* of Meyer Hirsch, [for proper title see p. 417], and the application of it is very easy when the necessary tables are calculated: as to this, see my "Memoir on the Symmetric Functions of the Roots of an Equation"(1). But when the expression for the Resultant of two equations is to be calculated without the assistance of such tables, it is I think by far the most simple process to develope the determinant according to the second of the two methods.

1 *Philosophical Transactions*, 1857, pp. 489—497, [147].

Consider first the method of symmetric functions, and to fix the ideas, let the two equations be

$$(a,\ b,\ c,\ d \between x,\ y)^3 = 0,$$
$$(p,\ q,\ r \between x,\ y)^2 = 0.$$

Then writing

$$(a,\ b,\ c,\ d \between 1,\ z)^3 = a\,(1-\alpha z)\,(1-\beta z)\,(1-\gamma z),$$

so that

$$-\frac{b}{a} = \alpha+\beta+\gamma \qquad = (1),$$

$$+\frac{c}{a} = \alpha\beta+\alpha\gamma+\beta\gamma = (1^2),$$

$$-\frac{d}{a} = \alpha\beta\gamma \qquad = (1^3),$$

the Resultant is

$$(p,\ q,\ r \between \alpha,\ 1)^2 \,.\, (p,\ q,\ r \between \beta,\ 1)^2 \,.\, (p,\ q,\ r \between \gamma,\ 1)^2,$$

which is equal to

$$r^3 + qr^2(\alpha+\beta+\gamma) + pr^2(\alpha^2+\beta^2+\gamma^2) + pqr(\alpha^2\beta+\alpha\beta^2+\beta^2\gamma+\beta\gamma^2+\gamma\alpha^2+\gamma^2\alpha) + \&c.;$$

or adopting the notation for symmetric functions used in the memoir above referred to, this is

$$\begin{aligned}
&\{\ r^3\\
&\{+qr^2\ (1)\\
&\begin{cases}+pr^2\ (2)\\ +q^2r\ (1^2)\end{cases}\\
&\begin{cases}+pqr\ (21)\\ +q^3\ (1^3)\end{cases}\\
&\begin{cases}+p^2r\ (2^2)\\ +pq^2\ (21^2)\end{cases}\\
&\{+p^2q\ (2^21)\\
&\{+p^3\ (2^3)\ ,
\end{aligned}$$

the law of which is best seen by dividing by r^3 and then writing

$$\frac{q}{r} = [1], \qquad \frac{p}{r} = [2],$$

and similarly,

$$\frac{q^2}{r^2} = [1^2], \quad \frac{pq}{r^2} = [21], \ \&c\ ;$$

the expression would then become

$$1+[1](1)+[2](2)+[1^2](1^2)+[21](21)+[1^3](1^3)+[2^2](2^2)+[21^2](21^2)+[2^21](2^21)+[2^3](2^3),$$

where the terms within the [ ] and ( ) are simply all the partitions of the numbers 1, 2, 3, 4, 5, 6, the greatest part being 2, and the greatest number of parts being 3. And in like manner in the general case we have all the partitions of the numbers 1, 2, 3,... mn, the greatest part being n, and the greatest number of parts being m.

The symmetric functions (1), (2), (1^2), &c. are given in the Tables (b) of the Memoir on Symmetric Functions, but it is necessary to remark that in the Tables the first coefficient a is put equal to unity, and consequently that there is a power of the coefficient a to be restored as a factor: this is at once effected by the condition of homogeneity. And it is not by any means necessary to write down (as for clearness of explanation has been done) the preceding expression for the Resultant; any portion of it may be taken out directly from one of the Tables (b). For instance, the bracketed portion

$$\begin{aligned}&+pqr\,(21),\\&+q^3\;\;(1^3),\end{aligned}$$

which corresponds to the partitions of the number 3, is to be taken out of the Table III (b). as follows: a portion of this Table (consisting as it happens of consecutive lines and columns, but this is not in general the case) is

=	d	bc
(21)	+3	−1
(1^3)	−1	

;

if in this we omit the sign =, and in the outside line write for homogeneity ad instead of d, and in the outside column, first substituting q, p for 1, 2, then write for homogeneity pqr instead of pq, we have

	ad	bc
pqr	+3	−1
q^3	−1	

viz. $pqr\times(+3ad-1bc)+q^3(-1ad)$, for the value of the portion in question; this is equivalent to

	pqr	q^3
ad	+3	−1
bc	−1	

, or as it may be more conveniently written,

	pqr	q^3
ad	+3	−1
bc	−1	

in which form it constitutes a part of the expression given in the sequel for the Resultant of the two functions in question; and similarly the remainder of the expression is at once derived from the Tables (b) I. to VI.

As a specimen of a mode of verification, it may be remarked that the Resultant quà invariant ought, when operated upon by the sum of the two operations,

$$3a\partial_b + 2b\partial_c + c\partial_d \text{ and } 2p\partial_q + q\partial_r,$$

to give a result zero. The results of the two operations are originally obtained in the forms in the first and second columns, and the first column, and the second column, with all the signs reversed, are respectively equal to the third column, and consequently the sum of the first and second columns vanishes, as it ought to do.

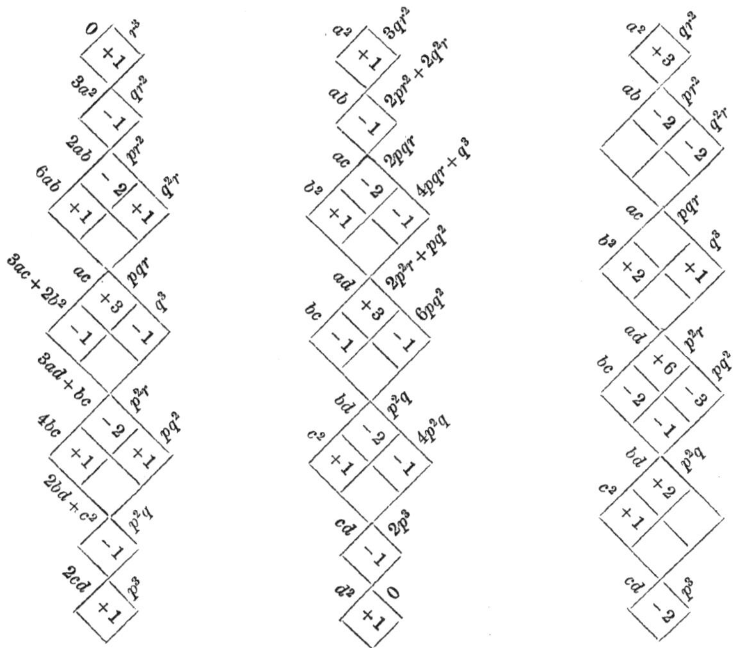

Next to explain the second method, viz. the calculation of the resultant from the expression in the form of a determinant.

Taking the same example as before, the resultant is

$$\begin{vmatrix} & a, & b, & c, & d \\ a, & b, & c, & d, & \\ & & p, & q, & r \\ & p, & q, & r & \\ p, & q, & r & & \end{vmatrix}$$

which may be developed in the form

$$\begin{array}{l} +12\ .\ 345\,\} \\ -13\ .\ 245\,\} \\ \left.\begin{array}{l}+14\ .\ 235 \\ +23\ .\ 145\end{array}\right\} \\ \left.\begin{array}{l}-15\ .\ 234 \\ -24\ .\ 135\end{array}\right\} \\ \left.\begin{array}{l}+25\ .\ 134 \\ +34\ .\ 125\end{array}\right\} \\ -35\ .\ 124\,\} \\ +45\ .\ 123\,\} \end{array}$$

where 12, 13, &c. are the terms of

$$\left(\begin{array}{ccccc} & a, & b, & c, & d \\ a, & b, & c, & d & \end{array}\right)$$

and 123, &c. are the terms of

$$\left(\begin{array}{ccccc} & & p, & q, & r \\ & p, & q, & r & \\ p, & q, & r & & \end{array}\right)$$

viz. 12 is the determinant formed with the first and second columns of the upper matrix, 123 is the determinant formed with the first, second and third columns of the lower matrix, and in like manner for the analogous symbols. These determinants must be first calculated, and the remainder of the calculation may then be arranged as follows:—

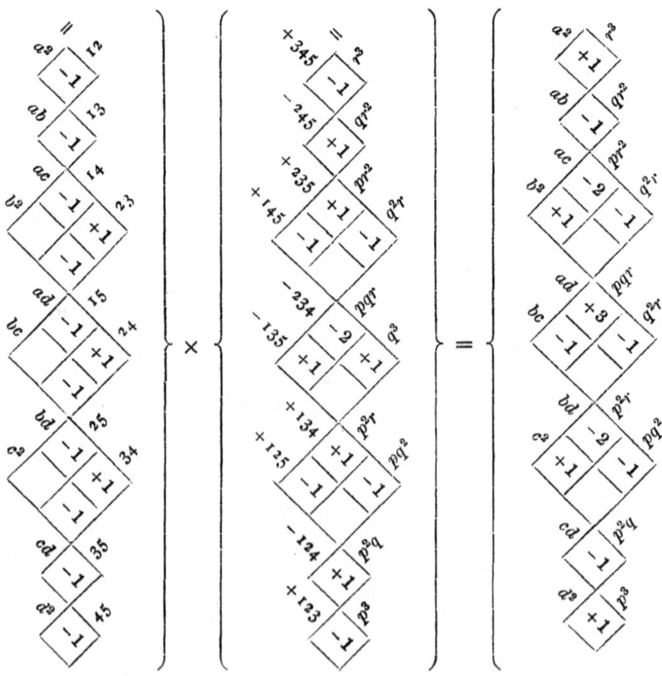

where it is to be observed that the figures in the squares of the third column are obtained from those in the corresponding squares of the first and second columns by the ordinary rule for the multiplication of determinants,—taking care to multiply the dexter lines (i.e. lines in the direction ╲) of the first square by the sinister lines (i.e. lines in the direction ╱) of the second square in order to obtain the sinister lines of the third square. Thus, for instance, the figures in the square

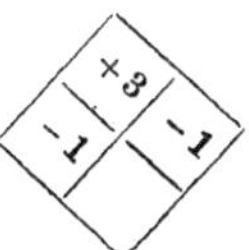

are obtained as follows, viz. the first sinister line $(+3,\ -1)$ by

$$(-1,\ +1)(-2,\ +1) = \quad 2+1=+3,$$

$$(-1,\ +1)(+1,\quad 0) = -1+0=-1,$$

and the second sinister line $(-1,\ 0)$ by

$$(0,\ -1)(-2,\ +1) = 0-1=-1,$$

$$(0,\ -1)(+1,\quad 0) = 0+0= \quad 0.$$

I have calculated the determinants required for the calculation, by the preceding process, of the Resultant of two quartic equations, and have indeed used them for the verification of the expression as found by the method of symmetric functions; as the determinants in question are useful for other purposes, I think the values are worth preserving.

Table of the Determinants of the Matrices,

$$\left(\begin{array}{cccccccc} & & & a, & b, & c, & d, & e \\ & & a, & b, & c, & d, & e & \\ & a, & b, & c, & d, & e & & \\ a, & b, & c, & d, & e, & & & \end{array}\right)$$

and

$$\left(\begin{array}{cccccccc} & & & p, & q, & r, & s, & t \\ & & p, & q, & r, & s, & t & \\ & p, & q, & r, & s, & t & & \\ p, & q, & r, & s, & t & & & \end{array}\right)$$

arranged in the form adapted for the calculation of the Resultant of the two quartic equations $(a, b, c, d, e\between x, y)^4 = 0$, and $(p, q, r, s, t\between x, y)^4 = 0$, viz.

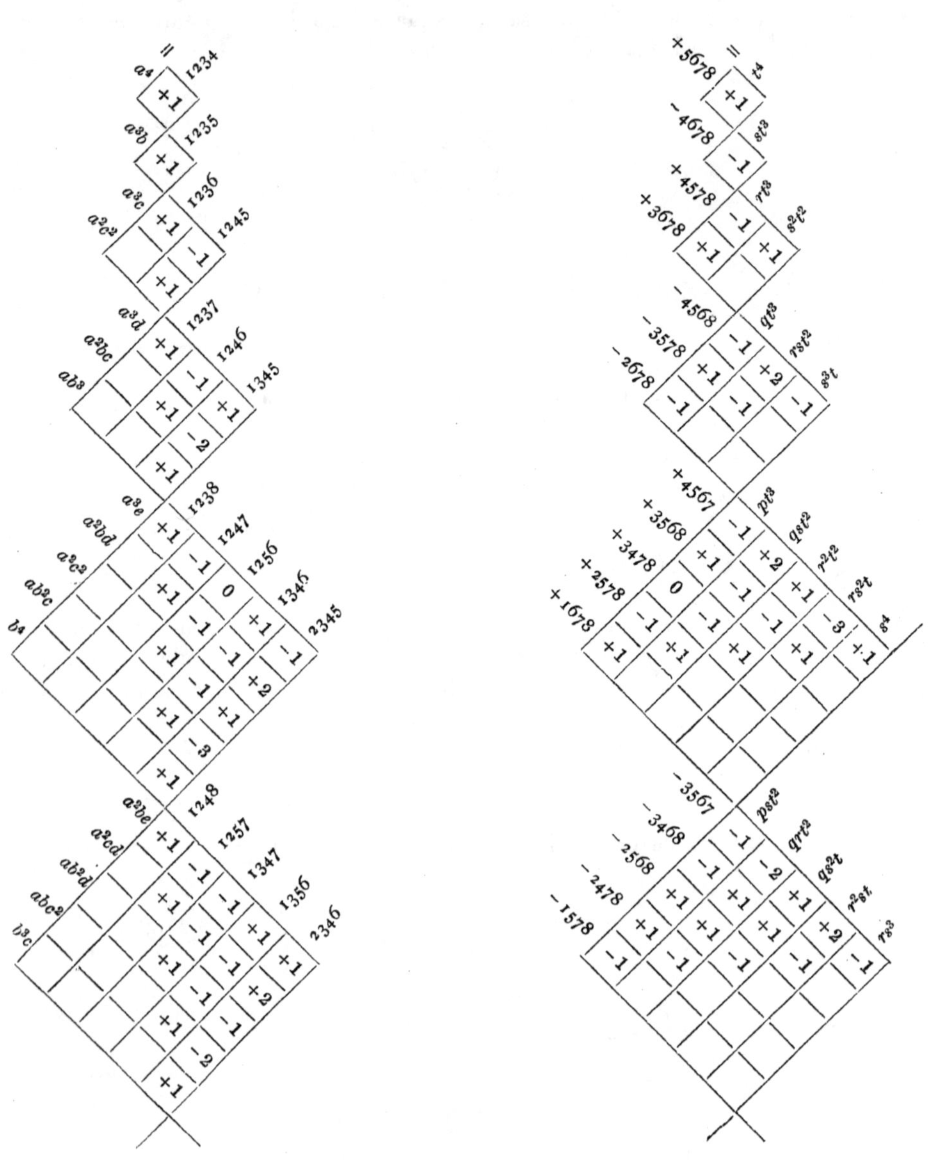

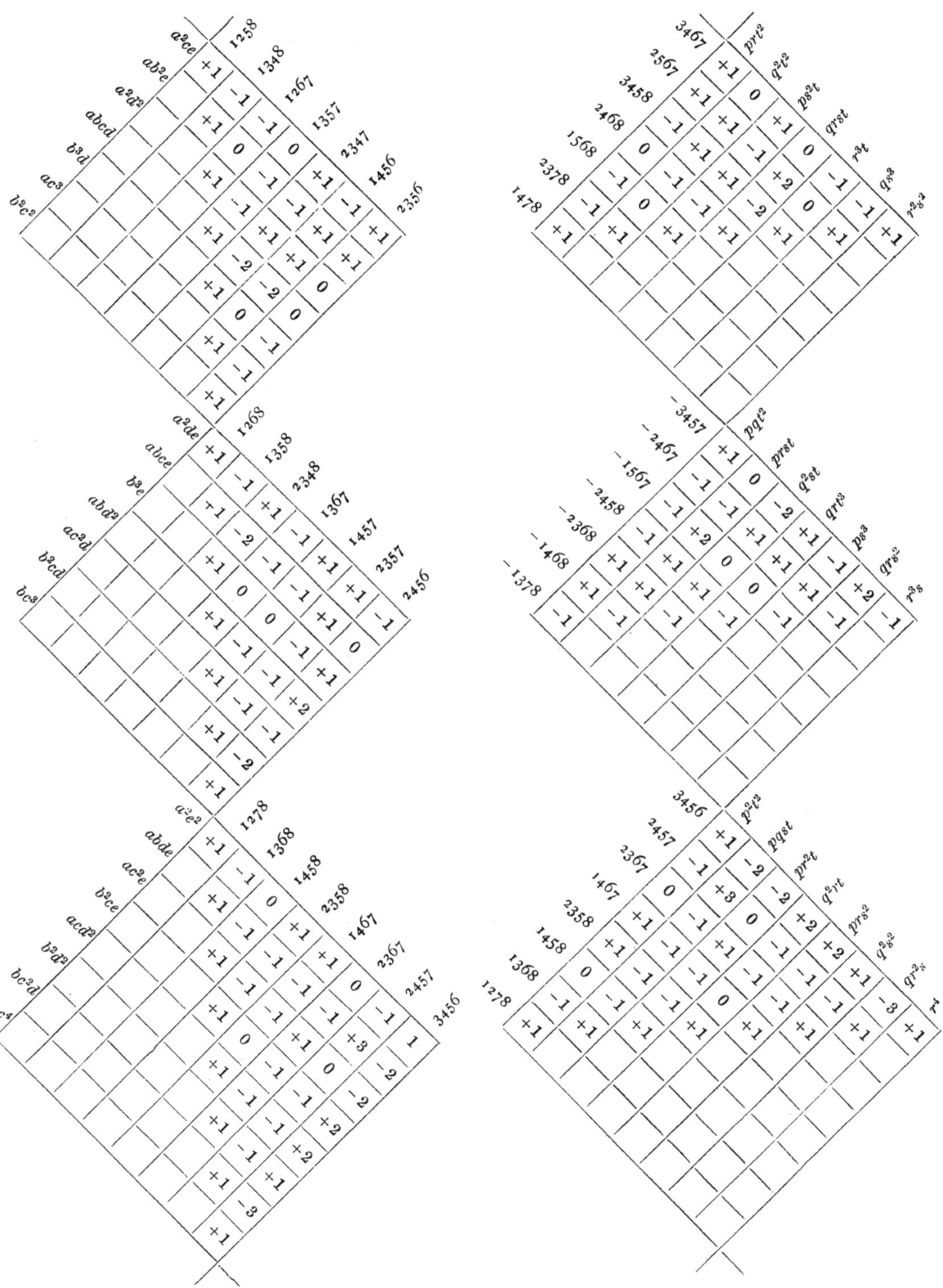

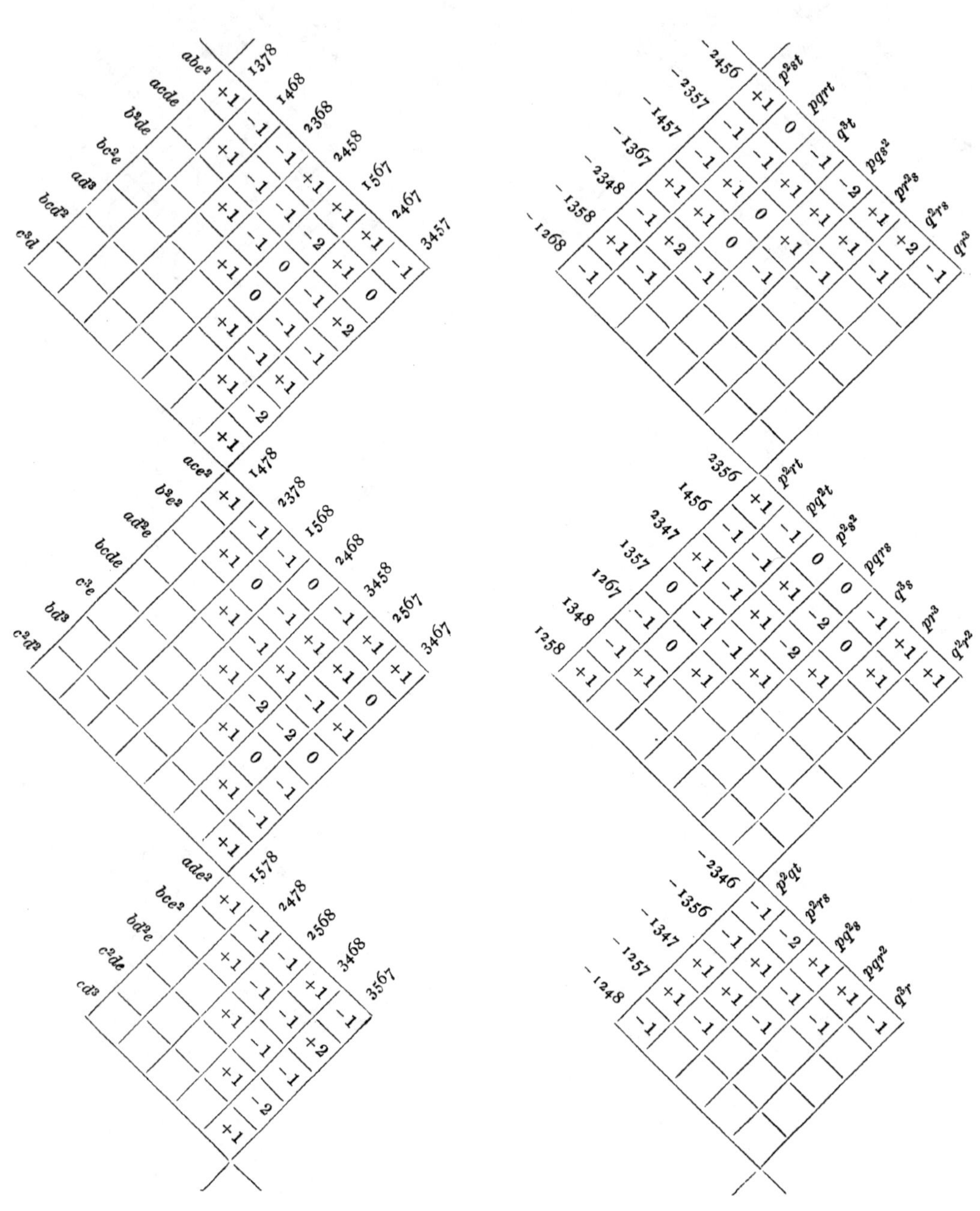

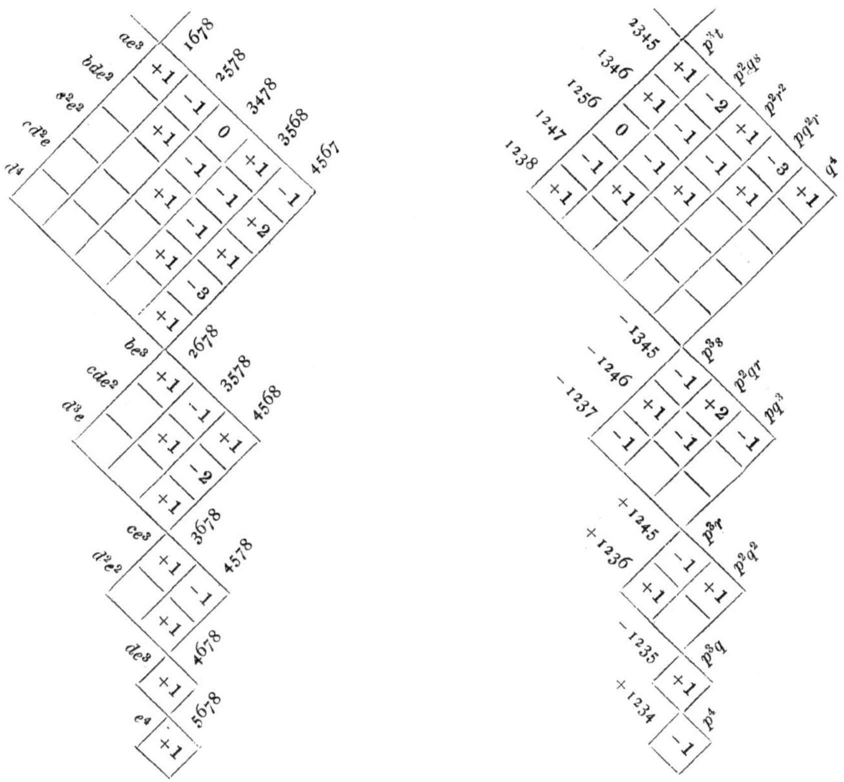

The Tables of the Resultants of two equations which I have calculated are as follows:

Table (2, 2).
Resultant of
$(a, b, c \between x, y)^2$,
$(p, q, r \between x, y)^2$.

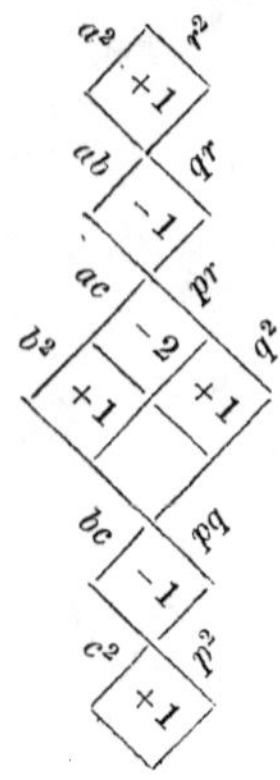

Table (3, 2).
Resultant of
$(a, b, c\ d \between x, y)^2$,
$(p, q, r \between x, y)^2$.

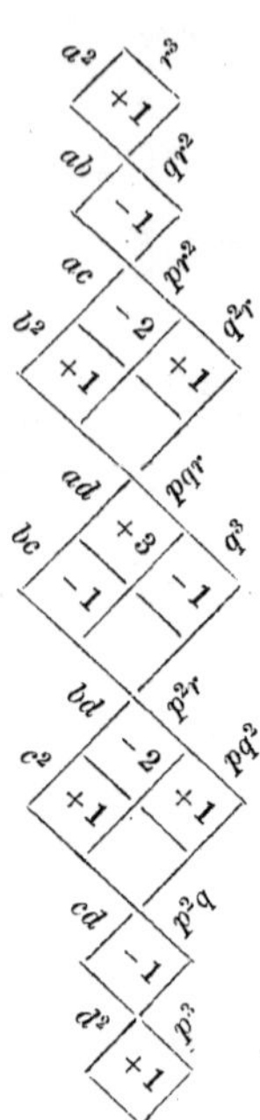

Table (4, 2).
Resultant of
$(a, b, c, d, e \between x, y)^4$,
$(p, q, r \between x, y)^2$.

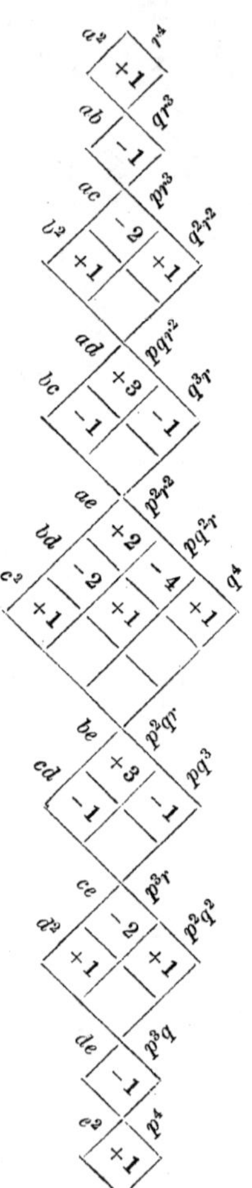

Table (3, 3).
Resultant of
$(a, b, c, d \between x, y)^3$,
$(p, q, r, s \between x, y)^3$.

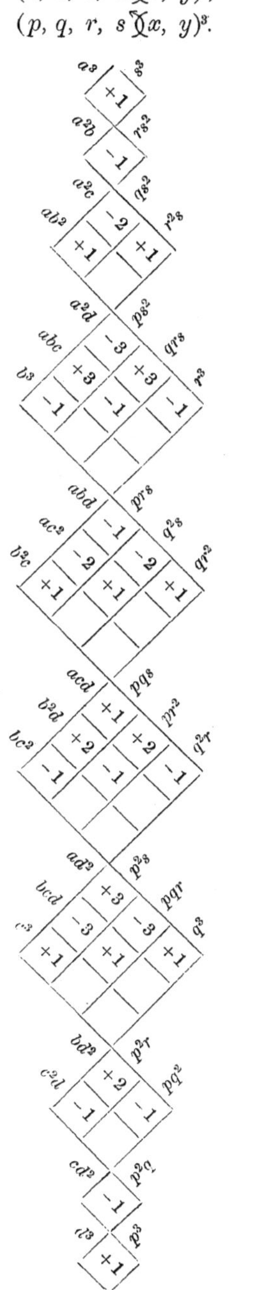

Table (4, 3).
Resultant of
$(a, b, c, d, e \between x, y)^4$,
$(p, q, r, s \between x, y)^3$.

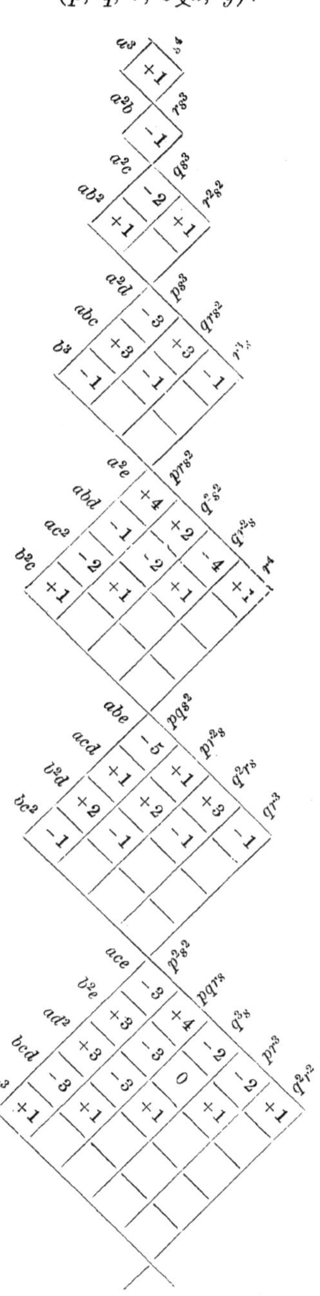

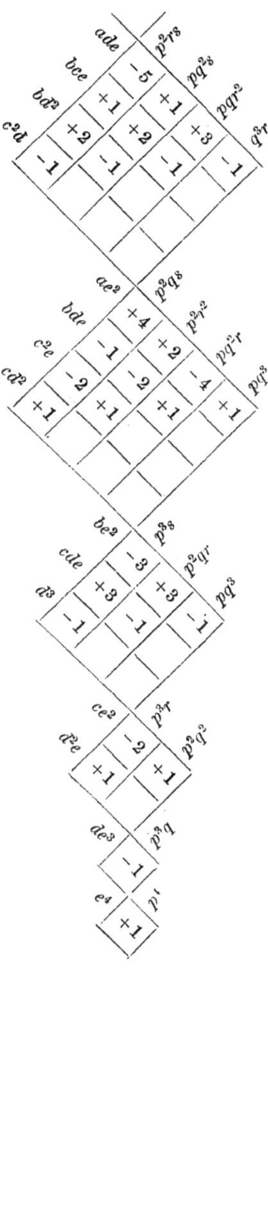

Table (4, 4).

Resultant of

$(a, b, c, d, e \between x, y)^4$,

$(p, q, r, s, t \between x, y)^4$.

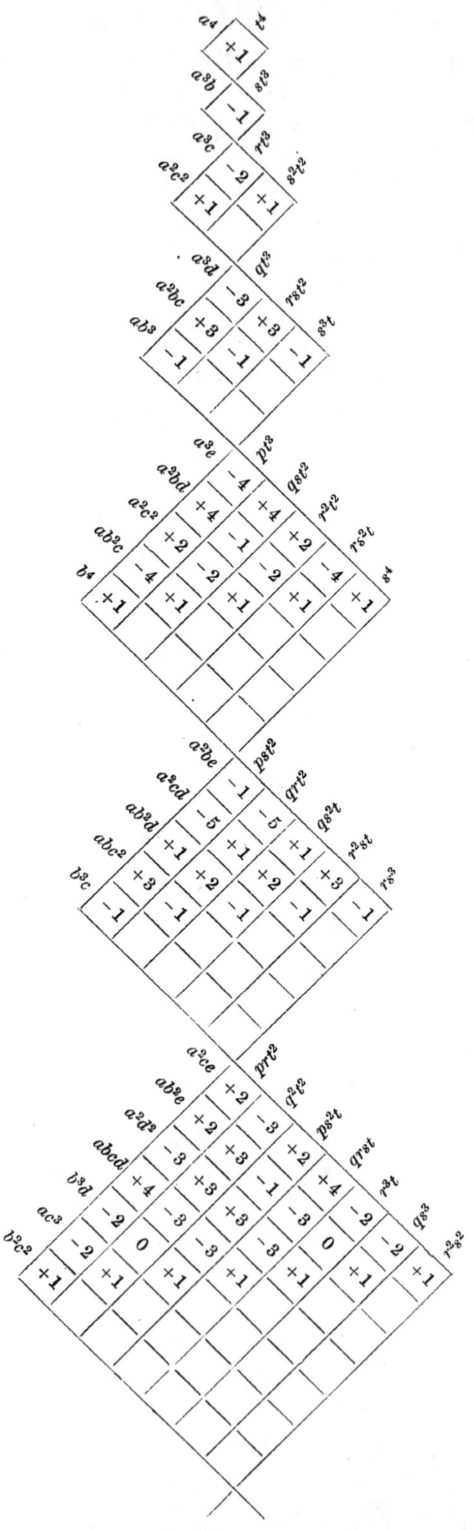

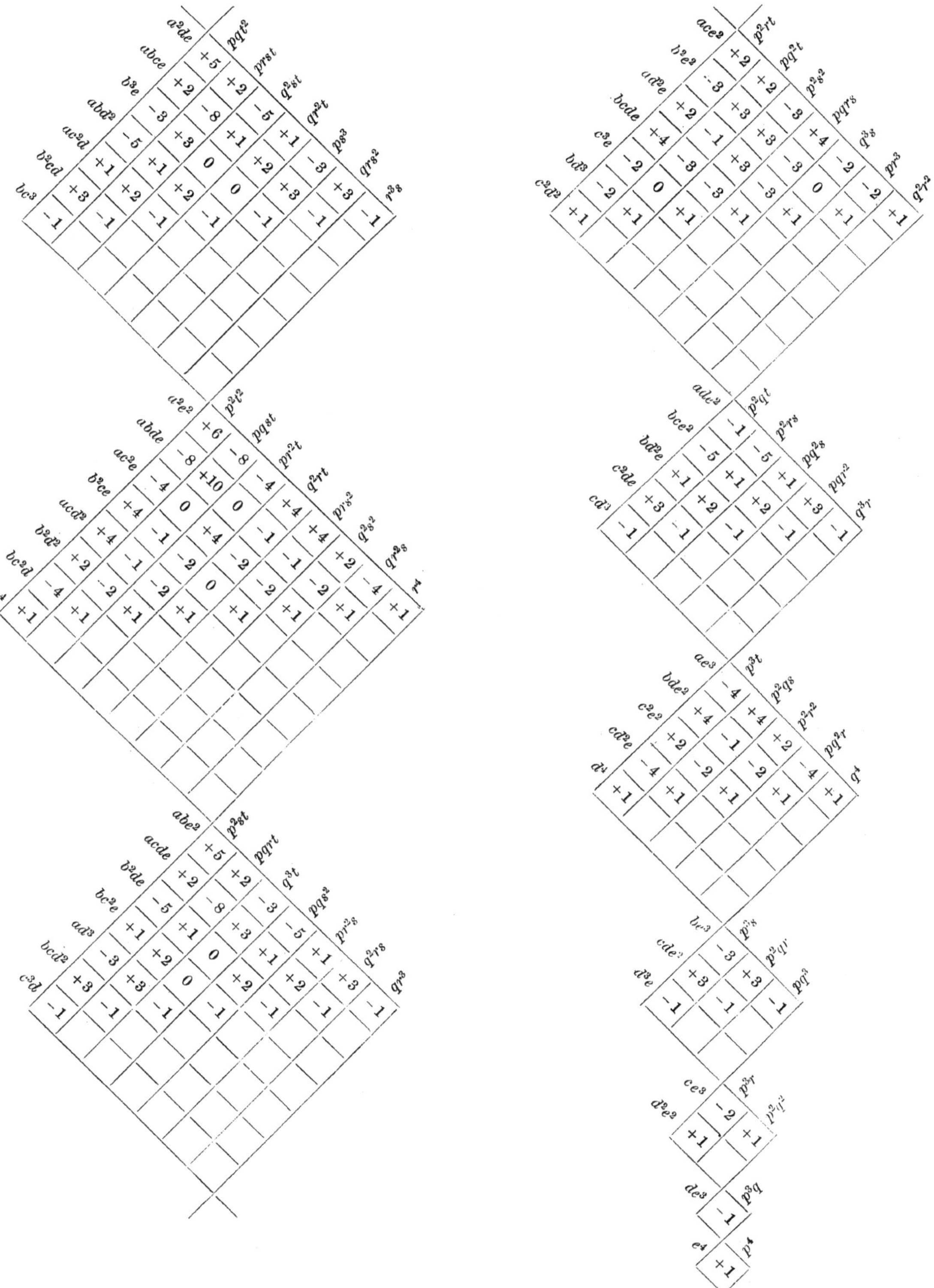

	pqt^2	$prst$	q^2st	qr^2t	ps^3	qrs^2	r^3s
a^2de	+5	+2	−5	+1	−3	+3	−1
$abce$	+2	−8	+1	+2	+3	−1	
b^3e	−3	+3	0	0	−1		
abd^2	−5	+1	+2	−1			
ac^2d	+1	+2	−1				
b^2cd	+3	−1					
bc^3	−1						

	p^2t^2	$pqst$	pr^2t	q^2rt	prs^2	q^2s^2	qr^2s	r^4
a^2e^2	+6	−8	−4	+4	+4	+2	−4	+1
$abde$	−8	+10	0	−1	−1	−2	+1	
ac^2e	−4	0	+4	−2	−2	+1		
b^2ce	+4	−1	−2	0	+1			
acd^2	+4	−1	−2	+1				
b^2d^2	+2	−2	+1					
bc^2d	−4	+1						
c^4	+1							

	p^2st	$pqrt$	q^3t	pqs^2	pr^2s	q^2rs	qr^3
abe^2	+5	+2	−3	−5	+1	+3	−1
$acde$	+2	−8	+3	+1	+2	−1	
b^2de	−5	+1	0	+2	−1		
bc^2e	+1	+2	0	−1			
ad^3	−3	+3	−1				
bcd^2	+3	−1					
c^3d	−1						

	p^2rt	pq^2t	p^2s^2	$pqrs$	q^3s	pr^3	q^2r^2
ace^2	+2	+2	−3	+4	−2	−2	+1
b^2e^2	−3	+3	+3	−3	0	+1	
ad^2e	+2	−1	+3	−3	+1		
$bcde$	+4	−3	−3	+1			
c^3e	−2	0	+1				
bd^3	−2	+1					
c^2d^2	+1						

	p^2qt	p^2rs	pq^2s	pqr^2	q^3r
ade^2	−1	−5	+1	+3	−1
bce^2	−5	+1	+2	−1	
bd^2e	+1	+2	−1		
c^2de	+3	−1			
cd^3	−1				

	p^3t	p^2qs	p^2r^2	pq^2r	q^4
ae^3	−4	+4	+2	−4	+1
bde^2	+4	−1	−2	+1	
c^2e^2	+2	−2	+1		
cd^2e	−4	+1			
d^4	+1				

	p^3s	p^2qr	pq^3
be^3	−3	+3	−1
cde^2	+3	−1	
d^3e	−1		

	p^3r	p^2q^2
ce^3	−2	+1
d^2e^2	+1	

	p^3q
de^3	−1

	p^4
e^4	+1

149.

ON THE SYMMETRIC FUNCTIONS OF THE ROOTS OF CERTAIN SYSTEMS OF TWO EQUATIONS.

[From the *Philosophical Transactions of the Royal Society of London*, vol. CXLVII. *for the year* 1857, pp. 717—726. Received December 18, 1856,—Read January 8, 1857.]

SUPPOSE in general that $\phi=0$, $\psi=0$, &c. denote a system of $(n-1)$ equations between the n variables $(x, y, z, \ldots)$, where the functions ϕ, ψ, &c. are quantics (i.e. rational and integral homogeneous functions) of the variables. Any values $(x_1, y_1, z_1, \ldots)$ satisfying the equations, are said to constitute a set of roots of the system; the roots of the same set are, it is clear, only determinate to a common factor *près*, i.e. only the ratios *inter se* and not the absolute magnitudes of the roots of a set are determinate. The number of sets, or the degree of the system, is equal to the product of the degrees of the component equations. Imagine a function of the roots which remains unaltered when any two sets $(x_1, y_1, z_1, \ldots)$ and $(x_2, y_2, z_2, \ldots)$ are interchanged (that is, when x_1 and x_2, y_1 and y_2, &c. are simultaneously interchanged), and which is besides homogeneous of the same degree as regards each entire set of roots, although not of necessity homogeneous as regards the different roots of the same set; thus, for example, if the sets are (x_1, y_1), (x_2, y_2), then the functions x_1x_2, $x_1y_2+x_2y_1$, y_1y_2 are each of them of the form in question; but the first and third of these functions, although homogeneous of the first degree in regard to each entire set, are not homogeneous as regards the two variables of each set. A function of the above-mentioned form may, for shortness, be termed a symmetric function of the roots; such function (disregarding an arbitrary factor depending on the common factors which enter implicitly into the different sets of roots) will be a rational and integral function of the coefficients of the equations, i.e. any symmetric function of the roots may be considered as a rational and integral function of the coefficients. The general process for the investigation of such expression for a symmetric function of the roots is indicated in Professor Schläfli's Memoir, "Ueber die Resultante eines Systemes mehrerer algebraischer

Gleichungen," *Vienna Transactions*, t. IV. (1852). The process is as follows:—Suppose that we know the resultant of a system of equations, one or more of them being linear; then if $\phi = 0$ be the linear equation or one of the linear equations of the system, the resultant will be of the form $\phi_1 \phi_2 \ldots$, where ϕ_1, ϕ_2, &c. are what the function ϕ becomes upon substituting therein the different sets $(x_1, y_1, z_1 \ldots)$, $(x_2, y_2, z_2 \ldots)$ of the remaining $(n-1)$ equations $\psi = 0$, $\chi = 0$, &c.; comparing such expression with the given value of the resultant, we have expressed in terms of the coefficients of the functions ψ, χ, &c., certain symmetric functions which may be called the fundamental symmetric functions of the roots of the system $\psi = 0$, $\chi = 0$, &c.; these are in fact the symmetric functions of the first degree in respect to each set of roots. By the aid of these fundamental symmetric functions, the other symmetric functions of the roots of the system $\psi = 0$, $\chi = 0$, &c. may be expressed in terms of the coefficients, and then combining with these equations a non-linear equation $\Phi = 0$, the resultant of the system $\Phi = 0$, $\psi = 0$, $\chi = 0$, &c. will be what the function $\Phi_1 \Phi_2 \ldots$ becomes, upon substituting therein for the different symmetric functions of the roots of the system $\psi = 0$, $\chi = 0$, &c. the expressions for these functions in terms of the coefficients. We thus pass from the resultant of a system $\phi = 0$, $\psi = 0$, $\chi = 0$, &c., to that of a system $\Phi = 0$, $\psi = 0$, $\chi = 0$, &c., in which the linear function ϕ is replaced by the non-linear function Φ. By what has preceded, the symmetric functions of the roots of a system of $(n-1)$ equations depend on the resultant of the system obtained by combining the $(n-1)$ equations with an arbitrary linear equation; and moreover, the resultant of any system of n equations depends ultimately upon the resultant of a system of the same number of equations, all except one being linear; but in this case the linear equations determine the ratios of the variables or (disregarding a common factor) the values of the variables, and by substituting these values in the remaining equation we have the resultant of the system. The process leads, therefore, to the expressions for the symmetric functions of the roots of any system of $(n-1)$ equations, and also to the expression for the resultant of any system of n equations. Professor Schläfli discusses in the general case the problem of showing how the expressions for the fundamental symmetric functions lead to those of the other symmetric functions, but it is not necessary to speak further of this portion of his investigations. The object of the present Memoir is to apply the process to two particular cases, viz. I propose to obtain thereby the expressions for the simplest symmetric functions (after the fundamental ones) of the following systems of two ternary equations; that is, first, a linear equation and a quadric equation; and secondly, a linear equation and a cubic equation.

First, consider the two equations

$$(a, b, c, f, g, h \between x, y, z)^2 = 0,$$

$$(\alpha, \beta, \gamma \between x, y, z) = 0,$$

and join to these the arbitrary linear equation

$$(\xi, \eta, \zeta \between x, y, z) = 0,$$

then the two linear equations give

$$x : y : z = \beta\zeta - \gamma\eta : \gamma\xi - \alpha\zeta : \alpha\eta - \beta\xi ;$$

and substituting in the quadratic equation, we have for the resultant of the three equations,

$$(a, b, c, f, g, h \between \beta\zeta - \gamma\eta, \gamma\xi - \alpha\zeta, \alpha\eta - \beta\xi)^2 = 0,$$

which may be represented by

$$(\mathrm{a}, \mathrm{b}, \mathrm{c}, \mathrm{f}, \mathrm{g}, \mathrm{h} \between \xi, \eta, \zeta)^2 = 0,$$

where the coefficients are given by means of the Table.

	a	b	c	f	g	h	
$\mathrm{a} =$		$+\gamma^2$	$+\beta^2$	$-2\beta\gamma$			(ξ^2)
$\mathrm{b} =$	$+\gamma^2$		$+\alpha^2$		$-2\gamma\alpha$		(η^2)
$\mathrm{c} =$	$+\beta^2$	$+\alpha^2$				$-2\alpha\beta$	(ζ^2)
$\mathrm{f} =$	$-\beta\gamma$			$-\alpha^2$	$+\alpha\beta$	$+\gamma\alpha$	$2(\eta\zeta)$
$\mathrm{g} =$		$-\gamma\alpha$		$+\alpha\beta$	$-\beta^2$	$+\beta\gamma$	$2(\zeta\xi)$
$\mathrm{h} =$			$-\alpha\beta$	$+\gamma\alpha$	$+\beta\gamma$	$-\gamma^2$	$2(\xi\eta)$

viz. $\mathrm{a} = b\gamma^2 + c\beta^2 - 2f\beta\gamma$, &c.

But if the roots of the given system are

$$(x_1, y_1, z_1), \quad (x_2, y_2, z_2),$$

then the resultant of the three equations will be

$$(x_1, y_1, z_1 \between \xi, \eta, \zeta) \,.\, (x_2, y_2, z_2 \between \xi, \eta, \zeta) = 0 ;$$

and comparing the two expressions, we have

$$\begin{aligned} \mathrm{a} &= x_1 x_2 , \\ \mathrm{b} &= y_1 y_2 , \\ \mathrm{c} &= z_1 z_2 , \\ 2\mathrm{f} &= y_1 z_2 + y_2 z_1 , \\ 2\mathrm{g} &= z_1 x_2 + z_2 x_1 , \\ 2\mathrm{h} &= x_1 y_2 + x_2 y_1 , \end{aligned}$$

which are the expressions for the six fundamental symmetric functions, or symmetric functions of the first degree in each set, of the roots of the given system.

By forming the powers and products of the second order a^2, ab, &c., we obtain linear relations between the symmetric functions of the second degree in respect to each set of roots. The number of equations is precisely equal to that of the symmetric functions of the form in question, and the solution of the linear equations gives—

$$\begin{aligned}
a^2 &= x_1^2x_2^2,\\
b^2 &= y_1^2y_2^2,\\
c^2 &= z_1^2z_2^2,\\[4pt]
bc &= y_1z_1y_2z_2,\\
ca &= z_1x_1z_2x_2,\\
ab &= x_1y_1x_2y_2,\\[4pt]
4f^2 - 2bc &= y_1^2z_2^2 + y_2^2z_1^2,\\
4g^2 - 2ca &= z_1^2x_2^2 + z_2^2x_1^2,\\
4h^2 - 2ab &= x_1^2y_2^2 + x_2^2y_1^2,\\[4pt]
2af &= x_1y_1z_2x_2 + z_1x_1x_2y_2,\\
2bg &= y_1z_1x_2y_2 + x_1y_1y_2z_2,\\
2ch &= z_1x_1y_2z_2 + y_1z_1z_2x_2,\\[4pt]
4gh - 2af &= x_1^2y_2z_2 + x_2^2y_1z_1,\\
4hf - 2bg &= y_1^2z_2x_2 + y_2^2z_1x_1,\\
4fg - 2ch &= z_1^2x_2y_2 + z_2^2x_1y_1,\\[4pt]
2bf &= y_1^2y_2z_2 + y_2^2y_1z_1,\\
2cg &= z_1^2z_2x_2 + z_2^2z_1x_1,\\
2ah &= x_1^2x_2y_2 + x_2^2x_1y_1,\\[4pt]
2cf &= z_1^2y_2z_2 + z_2^2y_1z_1,\\
2ag &= x_1^2z_2x_2 + x_2^2z_1x_1,\\
2bh &= y_1^2x_2y_2 + y_2^2x_1y_1.
\end{aligned}$$

Proceeding next to the powers and products of the third order a^3, a^2b, &c., the total number of linear relations between the symmetric functions of the third degree in respect to each set of roots exceeds by unity the number of the symmetric functions of the form in question; in fact the expressions for abc, af^2, bg^2, ch^2, fgh, contain, not five, but only four symmetric functions of the roots; for we have

$$\begin{aligned}
abc &= x_1y_1z_1 \,.\, x_2y_2z_2,\\
4af^2 &= (x_1y_1^2x_2z_2^2 + x_2y_2^2x_1z_1^2) + 2x_1y_1z_1x_2y_2z_2,\\
4bg^2 &= (y_1z_1^2y_2x_2^2 + y_2z_2^2y_1x_1^2) + 2x_1y_1z_1x_2y_2z_2,\\
4ch^2 &= (z_1x_1^2z_2y_2^2 + z_2x_2^2z_1y_1^2) + 2x_1y_1z_1x_2y_2z_2,\\
8fgh &= \left.\begin{array}{l}(x_1y_1^2x_2z_2^2 + x_2y_2^2x_1z_1^2)\\ + (y_1z_1^2y_2z_2^2 + y_2^2z_2^2y_1x_1^2)\\ + (z_1x_1^2z_2x_2^2 + z_2x_2^2z_1y_1^2)\end{array}\right\} + 2x_1y_1z_1x_2y_2z_2,
\end{aligned}$$

and consequently the quantities a, b, c, f, g, h, are not independent, but are connected by the equation

$$\text{abc} - \text{af}^2 - \text{bg}^2 - \text{ch}^2 + 2\text{fgh} = 0,$$

an equation, which is in fact verified by the foregoing values of a, &c. in terms of the coefficients of the given system.

The expressions for the symmetric functions of the third degree considered as functions of a, b, c, f, g, h, are consequently not absolutely determinate, but they may be modified by the addition of the term $\lambda(\text{abc} - \text{af}^2 - \text{bg}^2 - \text{ch}^2 + 2\text{fgh})$, where λ is an indeterminate numerical coefficient.

The simplest expressions are those obtained by disregarding the preceding equation for fgh, and the entire system then becomes:

$$\begin{aligned}
\text{a}^3 &= x_1^3 x_2^3,\\
\text{b}^3 &= y_1^3 y_2^3,\\
\text{c}^3 &= z_1^3 z_2^3,\\[4pt]
\text{b}^2\text{c} &= y_1^2 z_1 y_2^2 z_2,\\
\text{c}^2\text{a} &= z_1^2 x_1 z_2^2 x_2,\\
\text{a}^2\text{b} &= x_1^2 y_1 x_2^2 y_2,\\[4pt]
\text{bc}^2 &= y_1 z_1^2 y_2 z_2^2,\\
\text{ca}^2 &= z_1 x_1^2 z_2 x_2^2,\\
\text{ab}^2 &= x_1 y_1^2 x_2 y_2^2,\\[4pt]
\text{abc} &= x_1 y_1 z_1 x_2 y_2 z_2,\\
2\text{a}^2\text{f} &= x_1^2 y_1 z_2 x_2^2 + x_2^2 y_2 z_1 x_1^2,\\
2\text{b}^2\text{g} &= y_1^2 z_1 x_2 y_2^2 + y_2^2 z_2 x_1 y_1^2,\\
2\text{c}^2\text{h} &= z_1^2 x_1 y_2 z_2^2 + z_2^2 x_2 y_1 z_1^2,\\[4pt]
2\text{a}^2\text{g} &= x_1^3 z_2 x_2^2 + x_2^3 z_1 x_1^2,\\
2\text{b}^2\text{h} &= y_1^3 x_2 y_2^2 + y_2^3 x_1 y_1^2,\\
2\text{c}^2\text{f} &= z_1^3 y_2 z_2^2 + z_2^3 y_1 z_1^2,\\[4pt]
2\text{a}^2\text{h} &= x_1^3 z_2^2 x_2 + x_2^3 z_1^2 x_1,\\
2\text{b}^2\text{f} &= y_1^3 x_2^2 y_2 + y_2^3 x_1^2 y_1,\\
2\text{c}^2\text{g} &= z_1^3 y_2^2 z_2 + z_2^3 y_1^2 z_1,\\[4pt]
2\text{bcf} &= y_1^2 z_1 y_2 z_2^2 + y_2^2 z_2 y_1 z_1^2,\\
2\text{cag} &= z_1^2 x_1 z_2 x_2^2 + z_2^2 x_2 z_1 x_1^2,\\
2\text{abh} &= x_1^2 y_1 x_2 y_2^2 + x_2^2 y_2 x_1 y_1^2,\\[4pt]
2\text{bcg} &= y_1 z_1^2 x_2 y_2 z_2 + y_2 z_2^2 x_1 y_1 z_1,\\
2\text{cah} &= z_1 x_1^2 x_2 y_2 z_2 + z_2 x_2^2 x_1 y_1 z_1,\\
2\text{abf} &= x_1 y_1^2 x_2 y_2 z_2 + x_2 y_2^2 x_1 y_1 z_1,
\end{aligned}$$

$$\begin{aligned}
2bch &= y_1^2 z_1 x_2 y_2 z_2 + y_2^2 z_2 x_1 y_1 z_1, \\
2caf &= z_1^2 x_1 x_2 y_2 z_2 + z_2^2 x_2 x_1 y_1 z_1, \\
2abg &= x_1^2 y_1 x_2 y_2 z_2 + x_2^2 y_2 x_1 y_1 z_1,
\end{aligned}$$

$$\begin{aligned}
4af^2 - 2abc &= x_1 y_1^2 z_2^2 x_2 + x_2 y_2^2 z_1^2 x_1, \\
4bg^2 - 2abc &= y_1 z_1^2 x_2^2 y_2 + y^2 z_2^2 x_1^2 y_1, \\
4ch^2 - 2abc &= z_1 x_1^2 y_2^2 z_2 + z_2 x_2^2 y_1^2 z_1,
\end{aligned}$$

$$\begin{aligned}
4bf^2 - 2b^2c &= y_1^3 y_2 z_2^2 + y_2^3 y_1 z_1^2, \\
4cg^2 - 2c^2a &= z_1^3 z_2 x_2^2 + z_2^3 z_1 x_1^2, \\
4ah^2 - 2a^2b &= x_1^3 x_2 y_2^2 + x_2^3 x_1 y_1^2,
\end{aligned}$$

$$\begin{aligned}
4cf^2 - 2bc^2 &= z_1^3 y_2^2 z_2 + z_2^3 y_1^2 z_1, \\
4ag^2 - 2ca^2 &= x_1^3 z_2^2 x_2 + x_2^3 z_1^2 x_1, \\
4bh^2 - 2ab^2 &= y_1^3 x_2^2 y_2 + y_2^3 x_1^2 y_1,
\end{aligned}$$

$$\begin{aligned}
4agh - 2a^2f &= x_1^3 x_2 y_2 z_2 + x_2^3 x_1 y_1 z_1, \\
4bhf - 2b^2g &= y_1^3 x_2 y_2 z_2 + y_2^3 x_1 y_1 z_1, \\
4cfg - 2c^2h &= z_1^3 x_2 y_2 z_2 + z_2^3 x_1 y_1 z_1,
\end{aligned}$$

$$\begin{aligned}
4bgh - 2abf &= y_1^2 z_1 x_2^2 y_2 + y_2^2 z_2 x_1^2 y_1, \\
4chf - 2bcg &= z_1^2 x_1 y_2^2 z_2 + z_2^2 x_2 y_1^2 z_1, \\
4afg - 2cah &= x_1^2 y_1 z_2^2 x_2 + x_2^2 y_2 z_1^2 x_1,
\end{aligned}$$

$$\begin{aligned}
4cgh - 2acf &= y_1 z_1^2 x_2^2 z_2 + y_2 z_2^2 x_1^2 z_1, \\
4ahf - 2bag &= z_1 x_1^2 y_2^2 x_2 + z_2 x_2^2 y_1^2 x_1, \\
4bfg - 2cbh &= x_1 y_1^2 z_2^2 y_2 + x_2 y_2^2 z_1^2 y_1,
\end{aligned}$$

$$\begin{aligned}
8f^2g - 4chf - 2bcg &= z_1^3 x_2 y_2^2 + z_2^3 x_1 y_1^2, \\
8g^2h - 4afg - 2cah &= x_1^3 y_2 z_2^2 + x_2^3 y_1 z_1^2, \\
8h^2f - 4bgh - 2abf &= y_1^3 z_2 x_2^2 + y_2^3 z_1 x_1^2,
\end{aligned}$$

$$\begin{aligned}
8fg^2 - 4chg - 2acf &= z_1^3 x_2^2 y_2 + z_2^3 x_1^2 y_1, \\
8gh^2 - 4afh - 2bag &= x_1^3 y_2^2 z_2 + x_2^3 y_1^2 z_1, \\
8hf^2 - 4bgf - 2cbh &= y_1^3 z_2^2 x_2 + y_2^3 z_1^2 x_1,
\end{aligned}$$

$$\begin{aligned}
8f^3 - 6bcf &= y_1^3 z_2^3 + y_2^3 z_1^3, \\
8g^3 - 6cag &= z_1^3 x_2^3 + z_2^3 x_1^3, \\
8h^3 - 6abh &= x_1^3 y_2^3 + x_2^3 y_1^3.
\end{aligned}$$

Secondly, consider the system of equations

$$\begin{aligned}
(a, b, c, f, g, h, i, j, k, l \between x, y, z)^3 &= 0, \\
(\alpha, \beta, \gamma \between x, y, z) &= 0,
\end{aligned}$$

where the cubic function written at full length is

$$ax^3 + by^3 + cz^3 + 3fy^2z + 3gz^2x + 3hx^2y + 3iyz^2 + 3jzx^2 + 3kxy^2 + 6lxyz.$$

Joining to the system the linear equation

$$(\xi, \eta, \zeta \between x, y, z) = 0,$$

the linear equations give

$$x : y : z = \beta\zeta - \gamma\eta : \gamma\xi - \alpha\zeta : \alpha\eta - \beta\xi,$$

and the resultant is

$$(a, b, c, f, g, h, i, j, k, l \between \beta\zeta - \gamma\eta, \gamma\xi - \alpha\zeta, \alpha\eta - \beta\xi)^3 = 0,$$

which may be represented by

$$(\mathrm{a}, \mathrm{b}, \mathrm{c}, \mathrm{f}, \mathrm{g}, \mathrm{h}, \mathrm{i}, \mathrm{j}, \mathrm{k}, \mathrm{l} \between \xi, \eta, \zeta)^3 = 0,$$

where the coefficients a, b, &c. are given by means of the Table :—

	a	b	c	f	g	h	i	j	k	l	
a =		$+\gamma^3$	$-\beta^3$	$-3\beta\gamma^2$			$+3\beta^2\gamma$				ξ^3
b =	$-\gamma^3$		$+\alpha^3$		$-3\gamma\alpha^2$			$+3\gamma^2\alpha$			η^3
c =	$+\beta^3$	$-\alpha^3$				$-3\alpha\beta^2$			$+3\alpha^2\beta$		ζ^3
f =	$+\beta\gamma^2$				$+\alpha^2\beta$	$-\gamma^2\alpha$	$-\alpha^3$	$-2\alpha\beta\gamma$		$+2\gamma\alpha^2$	$3\eta^2\zeta$
g =		$+\gamma\alpha^2$		$-\alpha^2\beta$		$+\beta^2\gamma$		$-\beta^3$	$-2\alpha\beta\gamma$	$+2\alpha\beta^2$	$3\zeta^2\xi$
h =			$+\alpha\beta^2$	$+\gamma^2\alpha$	$-\beta^2\gamma$		$-2\alpha\beta\gamma$		$-\gamma^3$	$+2\beta\gamma^2$	$3\xi^2\eta$
i =	$-\beta^2\gamma$			$+\alpha^3$		$+2\alpha\beta\gamma$		$+\alpha\beta^2$	$-\gamma\alpha^2$	$-2\alpha^2\beta$	$3\eta\zeta^2$
j =		$-\gamma^2\alpha$		$+2\alpha\beta\gamma$	$+\beta^3$		$-\alpha\beta^2$		$+\beta\gamma^2$	$-2\beta^2\gamma$	$3\zeta\xi^2$
k =			$-\alpha^2b$		$+2\alpha\beta\gamma$	$+\gamma^3$	$+\gamma\alpha^2$	$-\beta\gamma^2$		$-2\gamma^2\alpha$	$3\xi^2\eta$
l =				$-\gamma\alpha^2$	$-\alpha\beta^2$	$-\beta\gamma^2$	$+\alpha^2\beta$	$+\beta^2\gamma$	$+\gamma^2\alpha$		$6\xi\eta\zeta$

viz. $$\mathrm{a} = b\gamma^3 - c\beta^3 - 3f\beta\gamma^2 + 3i\beta^2\gamma, \text{ \&c.}$$

But if the roots of the given system are

$$(x_1, y_1, z_1), \quad (x_2, y_2, z_2), \quad (x_3, y_3, z_3),$$

then the resultant of the three equations may also be represented by

$$(x_1, y_1, z_1 \between \xi, \eta, \zeta) \cdot (x_2, y_2, z_2 \between \xi, \eta, \zeta) \cdot (x_3, y_3, z_3 \between \xi, \eta, \zeta);$$

and comparing with the former expression, we find :

$$\mathrm{a} = x_1x_2x_3,$$

$$\mathrm{b} = y_1y_2y_3,$$

$$\mathrm{c} = z_1z_2z_3,$$

$$3\mathrm{f} = y_1y_2z_3 + y_2y_3z_1 + y_3y_1z_2,$$
$$3\mathrm{g} = z_1z_2x_3 + z_2z_3x_1 + z_3z_1x_2,$$
$$3\mathrm{h} = x_1x_2y_3 + x_2x_3y_1 + x_3x_1y_2,$$

$$3\mathrm{i} = y_1z_2z_3 + y_2z_3z_1 + y_3z_1z_2,$$
$$3\mathrm{j} = z_1x_2x_3 + z_2x_3x_1 + z_3x_1x_2,$$
$$3\mathrm{k} = x_1y_2y_3 + x_2y_3y_1 + x_3y_1y_2,$$

$$6\mathrm{l} = x_1y_2z_3 + x_2y_3z_1 + x_3y_1z_2 + x_1y_3z_2 + x_2y_1z_3 + x_3y_2z_1.$$

But there is in the present case a relation independent of the quantities a, &c., viz. we have $(\alpha, \beta, \gamma \between x_1, y_1, z_1) = 0$, $(\alpha, \beta, \gamma \between x_2, y_2, z_2) = 0$, $(\alpha, \beta, \gamma \between x_3, y_3, z_3) = 0$, and thence eliminating the coefficients (α, β, γ), we find

$$\nabla = x_1y_2z_3 + x_2y_3z_1 + x_3y_1z_2 - x_1y_3z_2 - x_2y_1z_3 - x_3y_2z_1 = 0.$$

By forming the powers and products of the second degree a^2, ab, &c., we obtain 55 equations between the symmetric functions of the second degree in each set of roots. But we have $\nabla^2 = 0 =$ a symmetric function of the roots, and thus the entire number of linear relations is 56, and this is in fact the number of the symmetric functions of the second degree in each set. I use for shortness the sign S to denote the sum of the distinct terms obtained by permuting the different sets of roots, so that the equations for the fundamental symmetric functions are—

$$\mathrm{a} = x_1x_2x_3,$$
$$\mathrm{b} = y_1y_2y_3,$$
$$\mathrm{c} = z_1z_2z_3,$$

$$3\mathrm{f} = S\, y_1y_2z_3,$$
$$3\mathrm{g} = S\, z_1z_2x_3,$$
$$3\mathrm{h} = S\, x_1x_2y_3,$$

$$3\mathrm{i} = S\, y_1z_2z_3,$$
$$3\mathrm{j} = S\, z_1x_2x_3,$$
$$3\mathrm{k} = S\, x_1y_2y_3,$$

$$6\mathrm{l} = S\, x_1y_2z_3;$$

then the complete system of expressions for the symmetric functions of the second order is as follows, viz.

$$\mathrm{a}^2 = x_1^2x_2^2x_3^2,$$
$$\mathrm{b}^2 = y_1^2y_2^2y_3^2,$$
$$\mathrm{c}^2 = z_1^2z_2^2z_3^2,$$

$$\mathrm{bc} = y_1z_1y_2z_2y_3z_3,$$
$$\mathrm{ca} = z_1x_1z_2x_2z_3x_3,$$
$$\mathrm{ab} = x_1y_1x_2y_2x_3y_3,$$

$$3af = S\, x_1y_1x_2y_2z_3x_3,$$
$$3bg = S\, y_1z_1y_2z_2x_3y_3,$$
$$3ch = S\, z_1x_1z_2x_2y_3z_3,$$

$$3bf = S\, y_1^2y_2^2y_3z_3,$$
$$3cg = S\, z_1^2z_2^2z_3x_3,$$
$$3ah = S\, x_1^2x_2^2x_3y_3,$$

$$3cf = S\, y_1z_1y_2z_2z_3^2,$$
$$3ag = S\, z_1x_1z_2x_2x_3^2,$$
$$3bh = S\, x_1y_1x_2y_2y_3^2,$$

$$3ai = S\, x_1y_1z_2x_2z_3x_3,$$
$$3bj = S\, y_1z_1x_2y_2x_3y_3,$$
$$3ck = S\, z_1x_1y_2z_2y_3z_3,$$

$$3bi = S\, y_1^2y_2z_2y_3z_3,$$
$$3cj = S\, z_1^2z_2x_2z_3x_3,$$
$$3ak = S\, x_1^2x_2y_2x_3y_3,$$

$$3ci = S\, y_1z_1z_2^2z_3^2,$$
$$3aj = S\, z_1x_1x_2^2x_3^2,$$
$$3bk = S\, x_1y_1y_2^2y_3^2,$$

$$6al = S\, x_1^2x_2y_2z_3x_3,$$
$$6bl = S\, y_1^2y_2z_2x_3y_3,$$
$$6cl = S\, z_1^2z_2x_2y_3z_3,$$

$$9f^2 - 6bi = S\, y_1^2y_2^2z_3^2,$$
$$9g^2 - 6cj = S\, z_1^2z_2^2x_3^2,$$
$$9h^2 - 6ak = S\, x_1^2x_2^2y_3^2,$$

$$9i^2 - 6cf = S\, y_1^2z_2^2z_3^2,$$
$$9j^2 - 6ag = S\, z_1^2x_2^2x_3^2,$$
$$9k^2 - 6bh = S\, x_1^2y_2^2y_3^2,$$

$$9fg - 3ck = S\, x_1y_1y_2z_2z_3^2,$$
$$9gh - 3ai = S\, y_1z_1z_2x_2x_3^2,$$
$$9hf - 3bj = S\, z_1x_1x_2y_2y_3^2,$$

$$9jk - 3af = S\, x_1^2 x_2 y_2 y_3 z_3,$$
$$9ki - 3bg = S\, y_1^2 y_2 z_2 z_3 x_3,$$
$$9ij - 3ch = S\, z_1^2 z_2 x_2 x_3 y_3,$$

$$9fi - 3bc = S\, y_1^2 y_2 z_2 z_3^2,$$
$$9gj - 3ca = S\, z_1^2 z_2 x_2 x_3^2,$$
$$9hk - 3ab = S\, x_1^2 x_2 y_2 y_3^2,$$

$$3(fj + gk + hi - l^2) = S\, x_1 y_1 z_2 x_2 y_3 z_3,$$

$$3(2fj - gk - hi + l^2) = S\, x_1 y_1 x_2 y_2 z_3^2,$$
$$3(2gk - hi - fj + l^2) = S\, y_1 z_1 y_2 z_2 x_3^2,$$
$$3(2hi - fj - gk + l^2) = S\, z_1 x_1 z_2 x_2 y_3^2,$$

$$3(6fl - 3ki - bg) = S\, x_1 y_1 y_2^2 z_3^2,$$
$$3(6gl - 3ij - ch) = S\, y_1 z_1 z_2^2 x_3^2,$$
$$3(6hl - 3jk - af) = S\, z_1 x_1 x_2^2 y_3^2,$$

$$3(6il - 3fg - ck) = S\, z_1 x_1 y_2^2 z_3^2,$$
$$3(6jl - 3gh - ai) = S\, x_1 y_1 z_2^2 x_3^2,$$
$$3(6kl - 3hf - bj) = S\, y_1 z_1 x_2^2 y_3^2,$$

$$6(-fj - gk - hi + 4l^2) = S\, x_1^2 y_2^2 z_3^2.$$

As an instance of the application of the formulæ, let it be required to eliminate the variables from the three equations,

$$(a,\ b,\ c,\ f,\ g,\ h,\ i,\ j,\ k,\ l \between x,\ y,\ z)^3 = 0,$$
$$(a',\ b',\ c',\ f',\ g',\ h' \between x,\ y,\ z)^2 = 0,$$
$$(\alpha,\ \beta,\ \gamma \between x,\ y,\ z) = 0.$$

This may be done in two different ways; first, representing the roots of the linear equation and the quadric equation by (x_1, y_1, z_1), (x_2, y_2, z_2), the resultant will be

$$(a, \ldots \between x_1,\ y_1,\ z_1)^3 \cdot (a, \ldots \between x_2,\ y_2,\ z_2)^3,$$

which is equal to

$$a^2\, x_1^3 x_2^3 + \&c.,$$

where the symmetric functions $x_1^3 x_2^3$, &c. are given by the formulæ $a'^3 = x_1^3 x_2^3$, &c., in which, since the coefficients of the quadratic equation are (a', b', c', f', g', h'), I have written a′ instead of a. Next, if the roots of the linear equation and the cubic equation are represented by (x_1, y_1, z_1), (x_2, y_2, z_2), (x_3, y_3, z_3), then the resultant will be

$$(a', \ldots \between x_1,\ y_1,\ z_1)^2 \cdot (a', \ldots \between x_2,\ y_2,\ z_2)^2 (a', \ldots \between x_3,\ y_3,\ z_3)^2,$$

which is equal to

$$a'^3\, x_1^2 x_2^2 x_3^2 + \&c.,$$

the symmetric functions $x_1^2 x_2^2 x_3^2$, &c. being given by the formulæ $\mathrm{a}^2 = x_1^2 x_2^2 x_3^2$, &c. The expression for the Resultant is in each case of the right degree, viz. of the degrees 6, 3, 2, in the coefficients of the linear, the quadric, and the cubic equations respectively: the two expressions, therefore, can only differ by a numerical factor, which might be determined without difficulty. The third expression for the resultant, viz.

$$(\alpha, \beta, \gamma \between x_1, y_1, z_1) \cdot (\alpha, \beta, \gamma \between x_2, y_2, z_2) \ldots (\alpha, \beta, \gamma \between x_6, y_6, z_6),$$

(where $(x_1, y_1, z_1), \ldots (x_6, y_6, z_6)$ are the roots of the cubic and quadratic equations) compared with the foregoing value, leads to expressions for the fundamental symmetric functions of the cubic and quadratic equations, and thence to expressions for the other symmetric functions of these two equations; but it would be difficult to obtain the actually developed values even of the fundamental symmetric functions. I hope to return to the subject, and consider in a general point of view the question of the formation of the expressions for the other symmetric functions by means of the expressions for the fundamental symmetric functions.

150.

A MEMOIR ON THE CONDITIONS FOR THE EXISTENCE OF GIVEN SYSTEMS OF EQUALITIES AMONG THE ROOTS OF AN EQUATION.

[From the *Philosophical Transactions of the Royal Society of London*, vol. CXLVII. *for the year* 1857, pp. 727—731. Received December 18, 1856,—Read January 8, 1857.]

It is well known that there is a symmetric function of the roots of an equation, viz. the product of the squares of the differences of the roots, which vanishes when any two roots are put equal to each other, and that consequently such function expressed in terms of the coefficients and equated to zero, gives the condition for the existence of a pair of equal roots. And it was remarked long ago by Professor Sylvester, in some of his earlier papers in the *Philosophical Magazine,* that the like method could be applied to finding the conditions for the existence of other systems of equalities among the roots, viz. that it was possible to form symmetric functions, each of them a sum of terms containing the product of a certain number of the differences of the roots, and such that the entire function might vanish for the particular system of equalities in question; and that such functions expressed in terms of the coefficients and equated to zero would give the required conditions. The object of the present memoir is to extend this theory and render it exhaustive, by showing how to form a series of types of all the different functions which vanish for one or more systems of equalities among the roots; and in particular to obtain by the method distinctive conditions for all the different systems of equalities between the roots of a quartic or a quintic equation, viz. for each system conditions which are satisfied for the particular system, and are not satisfied for any other systems, except, of course, the more special systems included in the particular system. The question of finding the conditions for any particular system of equalities is essentially an indeterminate one, for given any set of functions which vanish, a function syzygetically connected with these will also vanish; the discussion of the nature of the syzygetic relations between the different functions which vanish for any particular system of equalities, and of the order of the system composed of the several conditions for the particular system of equalities, does not enter into the plan of the present memoir. I have referred here to the indeterminateness of the question for the sake of the remark that I have availed myself thereof, to express by means of invariants or covariants the different systems of conditions obtained in the sequel of the memoir; the expressions of the different invariants and covariants referred to are given in my 'Second Memoir upon Quantics,' *Philosophical Transactions*, vol. CXLVI. (1856), [141].

1. Suppose, to fix the ideas, that the equation is one of the fifth order, and call the roots α, β, γ, δ, ϵ. Write $12 = \Sigma\phi(\alpha-\beta)^l$, $12.13 = \Sigma\phi(\alpha-\beta)^l(\alpha-\gamma)^m$, $12.34 = \Sigma\phi(\alpha-\beta)^l(\gamma-\delta)^n$, &c., where ϕ is an arbitrary function and l, m, &c. are positive integers. It is hardly necessary to remark that similar types, such as 12, 13, 45, &c., or as 12.13 and 23.25, &c., denote identically the same sums. Two types, such as 12.13 and 14.15.23.24.25.34.35.45, may be said to be complementary to each other. A particular product $(\alpha-\beta)(\gamma-\delta)$ does or does not enter as a term (or factor of a term) in one of the above-mentioned sums, according as the type 12.34 of the product, or some similar type, does or does not form part of the type of the sum; for instance, the product $(\alpha-\beta)(\gamma-\delta)$ is a term (or factor of a term) of each of the sums 12.34, 13.45.24, &c., but not of the sums 12.13.14.15, &c.

2. If, now, we establish any equalities between the roots, e.g. $\alpha=\beta$, $\gamma=\delta$, the effect will be to reduce certain of the sums to zero, and it is easy to find in what cases this happens. The sum will vanish if each term contains one or both of the factors $\alpha-\beta$, $\gamma-\delta$, i.e. if there is no term the complementary of which contains the product $(\alpha-\beta)(\gamma-\delta)$, or what is the same thing, whenever the complementary type does not contain as part of it, a type such as 12.34. Thus for the sum 14.15.24.25.34.35.45, the complementary type is 12.13.23, which does not contain any type such as 12.34, i.e. the sum 14.15.24.25.34.35.45 vanishes for $\alpha=\beta$, $\gamma=\delta$. It is of course clear that it also vanishes for $\alpha=\beta=\epsilon$, $\gamma=\delta$ or $\alpha=\beta=\gamma=\delta$, &c., which are included in $\alpha=\beta$, $\gamma=\delta$. But the like reasoning shows, and it is important to notice, that the sum in question does not vanish for $\alpha=\beta=\gamma$: and of course it does not vanish for $\alpha=\beta$. Hence the vanishing of the sum 14.15.24.25.34.35.45 is characteristic of the system $\alpha=\beta$, $\gamma=\delta$. A system of roots α, β, γ, δ, ϵ may be denoted by 11111; but if $\alpha=\beta$, then the system may be denoted by 2111, or if $\alpha=\beta$, $\gamma=\delta$, by 221, and so on. We may then say that the sum 14.15.24.25.34.35.45 does not vanish for 2111, vanishes for 221, does not vanish for 311, vanishes for 32, 41, 5.

3. For the purpose of obtaining the entire system of results it is only necessary to form Tables, such as the annexed Tables, the meaning of which is sufficiently explained by what precedes: the mark (×) set against a type denotes that the sum represented by the complementary type vanishes, the mark (○) that the complementary type does not vanish, for the system of roots denoted by the symbol at the top or bottom of the column; the complementary type is given in the same horizontal line with the original type. It will be noticed that the right-hand columns do not extend to the foot of the Table; the reason of this of course is, to avoid a repetition of the same type. Some of

the types at the foot of the Tables are complementary to themselves, but I have, notwithstanding this, given the complementary type in the form under which it naturally presents itself.

4. The Tables are:

Table for the equal Roots of a Quartic.

211	22	31	4			211	22	31	4
×	×	×	×	12 . 13 . 14 . 23 . 24 . 34		o	o	o	o
o	×	×	×	12	13 . 14 . 23 . 24 . 34	o	o	o	×
o	×	×	×	12 . 13	14 . 23 . 24 . 34	o	o	o	×
o	o	×	×	12 . 34	13 . 14 . 23 . 24	o	o	×	×
o	×	×	×	12 . 13 . 14	23 . 24 . 34	211	22	31	4
o	o	×	×	12 . 13 . 24	14 . 23 . 34				
o	×	o	×	12 . 13 . 23	14 . 24 . 34				
211	22	31	4						

Table for the equal Roots of a Quintic.

2111	221	311	32	41	5			2111	221	311	32	41	5
×	×	×	×	×	×	12 . 13 . 14 . 15 . 23 . 24 . 25 . 34 . 35 . 45		o	o	o	o	o	o
o	×	×	×	×	×	12	13 . 14 . 15 . 23 . 24 . 25 . 34 . 35 . 45	o	o	o	o	o	×
o	×	×	×	×	×	12 . 13	14 . 15 . 23 . 24 . 25 . 34 . 35 . 45	o	o	o	o	o	×
o	o	×	×	×	×	12 . 34	13 . 14 . 15 . 23 . 24 . 25 . 35 . 45	o	o	o	o	×	×
o	×	×	×	×	×	12 . 13 . 14	15 . 23 . 24 . 25 . 34 . 35 . 45	o	o	o	o	o	×
o	o	×	×	×	×	12 . 13 . 45	14 . 15 . 23 . 24 . 25 . 34 . 35	o	o	o	o	×	×
o	o	×	×	×	×	12 . 13 . 24	14 . 15 . 23 . 25 . 34 . 35 . 45	o	o	o	o	×	×
o	×	o	×	×	×	12 . 13 . 23	14 . 15 . 24 . 25 . 34 . 35 . 45	o	o	o	×	×	×
o	×	×	×	×	×	12 . 13 . 14 . 15	23 . 24 . 25 . 34 . 35 . 45	o	o	o	×	o	×
o	o	o	×	×	×	12 . 13 . 14 . 23	15 . 24 . 25 . 34 . 35 . 45	o	o	o	×	×	×
o	o	×	×	×	×	12 . 13 . 14 . 25	15 . 23 . 24 . 34 . 35 . 45	o	o	o	×	×	×
o	o	×	×	×	×	12 . 13 . 24 . 34	14 . 15 . 23 . 25 . 35 . 45	o	o	o	o	×	×
o	o	×	×	×	×	12 . 14 . 23 . 35	13 . 15 . 24 . 25 . 34 . 45	o	o	o	o	×	×
o	o	o	o	×	×	12 . 13 . 23 . 45	14 . 15 . 24 . 25 . 34 . 35	o	o	×	×	×	×
o	o	o	×	×	×	12 . 13 . 14 . 15 . 23	24 . 25 . 34 . 35 . 45	2111	221	311	32	41	5
o	o	o	×	×	×	12 . 13 . 14 . 23 . 24	15 . 25 . 34 . 35 . 45						
o	o	o	×	×	×	12 . 13 . 14 . 23 . 25	15 . 24 . 34 . 35 . 45						
o	o	o	×	×	×	12 . 13 . 15 . 24 . 34	14 . 23 . 25 . 35 . 45						
o	o	o	o	×	×	12 . 13 . 14 . 23 . 45	15 . 24 . 25 . 34 . 35						
o	o	×	×	×	×	12 . 15 . 23 . 34 . 45	13 . 14 . 24 . 25 . 35						
2111	221	311	32	41	5								

The two Tables enable the discussion of the theory of the equal roots of a quartic or quintic equation: first for the quartic:

5. In order that a quartic may have a pair of equal roots, or what is the same thing, that the system of roots may be of the form 211, the type to be considered is

$$12.13.14.23.24.34;$$

this of course gives as the function to be equated to zero, the discriminant of the quartic.

6. In order that there may be two pairs of equal roots, or that the system may be of the form 22, the simplest type to be considered is

14.24.34;

this gives the function

$$\Sigma(\alpha-\delta)(\beta-\delta)(\gamma-\delta)\,(x-\alpha y)^2(x-\beta y)^2(x-\gamma y)^2,$$

which being a covariant of the degree 3 in the coefficients and the degree 6 in the variables, can only be the cubicovariant of the quartic.

7. In order that the quartic may have three equal roots, or that the system of roots may be of the form 31, we may consider the type

13.14.23.24,

and we obtain thence the two functions

$$\Sigma(\alpha-\gamma)\,(\alpha-\delta)(\beta-\gamma)(\beta-\delta),$$
$$\Sigma(\alpha-\gamma)^2(\alpha-\delta)(\beta-\gamma)(\beta-\delta)^2,$$

which being respectively invariants of the degrees 2 and 3, are of course the quadrinvariant and the cubinvariant of the quartic. If we had considered the apparently more simple type

12.34,

this gives the function

$$\Sigma(\alpha-\beta)^2(\gamma-\delta)^2,$$

which is the quadrivariant, but the cubinvariant is not included under the type in question.

8. Finally, if the roots are all equal, or the system of roots is of the form 4, then the simplest type is

12;

and this gives the function

$$\Sigma(\alpha-\beta)^2(x-\gamma y)^2(x-\delta y)^2,$$

a covariant of the degree 2 in the coefficients and the degree 4 in the variables; this is of course the Hessian of the quartic.

Considering next the case of the quintic:

9. In order that a quintic may have a pair of equal roots, or what is the same thing, that the system of roots may be of the form 2111, the type to be considered is

12.13.14.15.23.24.25.34.35.45;

this of course gives as the function to be equated to zero, the discriminant of the quintic.

10. In order that the quintic may have two pairs of equal roots, or that the system of roots may be 221, the simplest type to be considered is

$$14.15.24.25.34.35.45;$$

a type which gives the function

$$\Sigma\,(\alpha-\delta)(\alpha-\epsilon)(\beta-\delta)(\beta-\epsilon)(\gamma-\delta)(\gamma-\epsilon)(\delta-\epsilon)^2\,(x-\alpha y)^3\,(x-\beta y)^3\,(x-\gamma y)^3.$$

This is a covariant of the degree 5 in the coefficients and of the degree 9 in the variables; but it appears from the memoir above referred to, that there is not any irreducible covariant of the form in question; such covariant must be a sum of the products (No. 13)(No. 20), (No. 13)(No. 14)², (No. 15)(No. 16) (the numbers refer to the Covariant Tables given in the memoir), each multiplied by a merely numerical coefficient. These numerical coefficients may be determined by the consideration that there being two pairs of equal roots, we may by a linear transformation make these roots 0, 0, ∞, ∞, or what is the same thing, we may write $a=b=e=f=0$, the covariant must then vanish identically. The coefficients are thus found to be 1, −4, 50, and we have for a covariant vanishing in the case of two pairs of equal roots,

$$\begin{aligned} &1\ (\text{No. }13)(\text{No. }20)\\ -\ &4\ (\text{No. }13)(\text{No. }14)^2\\ +\,&50\ (\text{No. }15)(\text{No. }16)\end{aligned}$$

[or in the new notation $AH-4AB^2+50CD$].

In fact, writing $a=b=e=f=0$, and rejecting, where it occurs, a factor x^2y^2, the several covariants become functions of cx, dy; and putting, for shortness, x, y instead of cx, dy, the equation to be verified is

$$\begin{aligned} &1\,.\,10(x+y)(6x^4+8x^3y+28x^2y^2+8xy^3+6y^4)\\ -\ &4\,.\,10(x+y)(3x^2+2xy+3y^2)^2\\ +\ &50(6x^2+8xy+6y^2)(x^3+x^2y+xy^2+y^3)=0;\end{aligned}$$

and dividing out by $(x+y)$ and reducing, the equation is at once seen to be identically true.

11. In order that the quintic may have three equal roots, or that the system of roots may be of the form 311, the simplest type to be considered is

$$12.13.23.45;$$

this gives the function

$$\Sigma\,(\alpha-\beta)^2\,(\beta-\gamma)^2\,(\gamma-\alpha)^2\,(\delta-\epsilon)^4,$$

which being an invariant, and being of the fourth degree in the coefficients, must be the quartinvariant of the quintic [that is No. 19, $=G$]. The same type gives also the function

$$\Sigma\,(\alpha-\beta)^2\,(\beta-\gamma)^2\,(\gamma-\alpha)^2\,(\delta-\epsilon)^2\,(x-\delta y)^2\,(x-\epsilon y)^2,$$

which is a covariant of the degree 4 in the coefficients and the degree 4 in the variables; and it must vanish when $a = b = c = 0$, this can only be the covariant

$$3\,(\text{No. } 20) - 2\,(\text{No. } 14)^2, \ [= 3H - 2B^2],$$

which it is clear vanishes as required.

12. In order that the quintic may have three equal roots and two equal roots, or that the system of roots may be of the form 32, the simplest type to be considered is

$$12.13.14.15,$$

which gives the function

$$\Sigma\,(\alpha-\beta)\,(\alpha-\gamma)\,(\alpha-\delta)\,(\alpha-\epsilon)\,(x-\beta y)^3\,(x-\gamma y)^3\,(x-\delta y)^3\,(x-\epsilon y)^3,$$

a covariant of the degree 4 in the coefficients, and the degree 12 in the variables; and it must vanish when $a = b = c = 0$, $e = f = 0$; this can only be the covariant

$$3\,(\text{No. } 13)^2\,(\text{No. } 14) - 25\,(\text{No. } 15)^2, \ [= 3A^2B - 25C^2],$$

which it is clear vanishes as required.

13. In order that the quintic may have four equal roots, or that the system may be of the form 41, the simplest type to be considered is

$$12.34,$$

which gives the function

$$\Sigma\,(\alpha-\beta)^2\,(\gamma-\delta)^2\,(x-\epsilon y)^2,$$

a covariant of the degree 2 in the coefficients, and of the same degree in the variables; this can only be the covariant (No. 14), [$= B$].

14. Finally, in order that all the roots may be equal, or that the system of roots may be of the form 5, the type to be considered is

$$12;$$

and this gives the function

$$\Sigma\,(\alpha-\beta)^2\,(x-\gamma y)^2\,(x-\delta y)^2\,(x-\epsilon y)^2,$$

a covariant of the degree 2 in the coefficients, and the degree 6 in the variables, and this can only be the Hessian (No. 15), [$= C$].

It will be observed that all the preceding conditions are distinctive; for instance, the covariant which vanishes when the system of roots is of the form 311, does not vanish when the system is of the form 221, or of any other form not included in the form 311.

151.

TABLES OF THE STURMIAN FUNCTIONS FOR EQUATIONS OF THE SECOND, THIRD, FOURTH, AND FIFTH DEGREES.

[From the *Philosophical Transactions of the Royal Society of London,* vol. CXLVII. *for the year* 1857, pp. 733—736. Received December 18, 1856,—Read January 8, 1857.]

THE general expressions for the Sturmian functions in the form of determinants are at once deducible from the researches of Professor Sylvester in his early papers on the subject in the *Philosophical Magazine,* and in giving these expressions in the Memoir 'Nouvelles Recherches sur les Fonctions de M. Sturm,' *Liouville,* t. XIII. p. 269 (1848), [65], I was wrong in claiming for them any novelty. The expressions in the last-mentioned memoir admit of a modification by which their form is rendered somewhat more elegant; I propose on the present occasion merely to give this modified form of the general expression, and to give the developed expressions of the functions in question for equations of the degrees two, three, four, and five.

Consider in general the equation

$$U=(a,\ b,\ \dots\ j,\ k\between x,\ 1)^n,$$

and write

$$P=(a,\ b,\ \dots\ \ j\between x,\ 1)^{n-1},$$

$$Q=(b,\ \ \dots\ j,\ k\between x,\ 1)^{n-1},$$

then supposing as usual that the first coefficient a is positive, and taking for shortness n_1, n_2, &c. to represent the binomial coefficients $\frac{n-1}{1}$, $\frac{n-1\,.\,n-2}{1\,.\,2}$, &c. corresponding to the index $(n-1)$, the Sturmian functions, each with its proper sign, are as follows, viz.

$$U,\ P,\ \begin{vmatrix} P, & Q \\ a, & b \end{vmatrix},\quad -\begin{vmatrix} xP, & P, & xQ, & Q \\ a, & ., & b, & . \\ n_1b, & a, & n_1c, & b \\ n_2c, & n_1b, & n_2d, & n_1c \end{vmatrix},$$

$$+\begin{vmatrix} x^2P, & xP, & P, & x^2Q, & xQ, & Q \\ a, & ., & ., & b, & ., & . \\ n_1b, & a, & ., & n_1c, & b, & . \\ n_2c, & n_1b, & a, & n_2d, & n_1c, & b \\ n_3d, & n_2c, & n_1b, & n_3e, & n_2d, & n_1c \\ n_4e, & n_3d, & n_2c, & n_4f, & n_3e, & n_2d \end{vmatrix}, \text{ \&c.}$$

where the terms containing the powers of x, which exceed the degrees of the several functions respectively, vanish identically (as is in fact obvious from the form of the expressions), but these terms may of course be omitted *ab initio*.

The following are the results which I have obtained; it is well known that the last or constant function is in each case equal to the discriminant, and as the expressions for the discriminant of equations of the fourth and fifth degrees are given, Tables No. 12 and No. 26 [Q', see 143] in my 'Second Memoir upon Quantics'(¹), I have thought it sufficient to refer to these values without repeating them at length.

Table for the degree 2.

The Sturmian functions for the quadric $(a, b, c \between x, 1)^2$ are

$a+1$	$b+2$	$c+1$

$(\ \ldots\ \between x, 1)^2$,

$a+1$	$b+1$

$(\ \ldots\ \between x, 1)$,

$ac-1$
b^2+1

Table for the degree 3.

The Sturmian functions for the cubic $(a, b, c, d \between x, 1)^3$ are

$a+1$	$b+3$	$c+3$	$d+1$

$(\ \ldots\ \between x, 1)^3$,

¹ *Philosophical Transactions*, t. CXLVI. p. 101 (1856), [141].

(| $a+1$ | $b+2$ | $c+1$ | $\between x,\ 1)^2$,

(| $ac-2$
 b^2+2 | $ad-1$
 $bc+1$ | $\between x,\ 1)$,

| a^2d^2+1
 $abcd+6$
 ac^3-4
 bd^3-4
 b^2c^2-3 |

Table for the degree 4.

The Sturmian functions for the quartic $(a,\ b,\ c,\ d,\ e \between x,\ 1)_4$ are

(| $a+1$ | $b+4$ | $c+6$ | $d+4$ | $e+1$ | $\between x,\ 1)^4$,

(| $a+1$ | $b+3$ | $c+3$ | $d+1$ | $\between x,\ 1)^3$,

(| $ac-3$
 b^2+3 | $ad-3$
 $bc+3$ | $ae-1$
 $bd+1$ | $\between x,\ 1)^2$,

3 (| a^2ce-1
 a^2d^2+3
 ab^2e+1
 $abcd-14$
 ac^3+9
 b^3d+8
 b^2c^2-6 | a^2de+1
 $abce-4$
 abd^2-1
 ac^2d+3
 b^3e+3
 b^2cd-2 | $\between x,\ 1)$,

− | a^3e^3+1
 &c.
 Disct. Tab.
 No. 12. |

Table for the degree 5.

The Sturmian functions for the quintic $(a,\ b,\ c,\ d,\ e,\ f \between x,\ 1)^5$ are

(| $a+1$ | $b+5$ | $c+10$ | $d+10$ | $e+5$ | $f+1$ | $\between x,\ 1)^5$,

(	$a+1$	$b+4$	$c+6$	$d+4$	$e+1$	$\between x,\ 1)^4$,

(	$ac-4$ b^2+4	$ad-6$ $bc+6$	$ae-4$ $bd+4$	$af-1$ $be+1$	$\between x,\ 1)^3$,

2 (	$a^2ce\ -\ 8$ $a^2d^2\ +18$ $ab^2e\ +\ 8$ $abcd-76$ $ac^3\ +48$ $b^3d\ +40$ $b^2c^2\ -30$	$a^2cf\ -\ 2$ $a^2de\ +12$ $ab^2f\ +\ 2$ $abce\ -42$ $abd^2\ -12$ $ac^2d\ +32$ $b^3e\ +30$ $b^2cd\ -20$	$a^2df\ +\ 3$ $abcf\ -11$ $abde\ -\ 3$ $ac^2e\ +\ 8$ $b^3f\ +\ 8$ $b^2ce\ -\ 5$	$\between x,\ 1)^2$,

2 ($\between x,\ 1)$,

a^3cf^2	$-$	2	a^3df^2	$+$	3
a^3def	$+$	24	a^3e^2f	$-$	8
a^3e^3	$-$	32	a^2bcf^2	$-$	11
$a^2b^2f^2$	$+$	2	a^2bdef	$+$	58
a^2bde^2	$+$	264	a^2be^3	$+$	8
a^2bcef	$-$	52	a^2c^2ef	$+$	104
a^2bd^2f	$-$	96	a^2cd^2f	$-$	156
a^2c^2df	$+$	64	a^2cde^2	$-$	96
$a^2c^2e^2$	$+$	352	a^2d^3e	$+$	108
a^2cd^2e	$-$	938	ab^3f^2	$+$	8
a^2d^4	$+$	432	ab^2cef	$-$	266
ab^3ef	$+$	28	ab^2d^2f	$-$	8
ab^2ce^2	$-$	970	ab^2de^2	$+$	35
ab^2d^2e	$+$	120	abc^2df	$+$	584
abc^2de	$+$	2480	abc^2e^2	$+$	120
ab^2cdf	$+$	264	$abcd^2e$	$-$	360
$abcd^3$	$-$	1440	ac^4f	$-$	288
abc^3f	$-$	192	ac^3de	$+$	160
ac^4e	$-$	960	b^4ef	$+$	120
ac^3d^2	$+$	640	b^3cdf	$-$	320
b^4df	$-$	160	b^3ce^2	$-$	75
b^4e^2	$+$	450	b^3d^2e	$+$	200
b^3cde	$-$	1400	b^2c^3f	$+$	180
b^3d^3	$+$	800	b^2c^2de	$-$	100
b^3c^2f	$+$	120			
b^2c^3e	$+$	600			
$b^2c^2d^2$	$-$	400			

a^4f^4+1 + &c. Disct. Tab. No. 26, [Q'].

152.

A MEMOIR ON THE THEORY OF MATRICES.

[From the *Philosophical Transactions of the Royal Society of London*, vol. CXLVIII. *for the year*, 1858, pp. 17—37. Received December 10, 1857,—Read January 14, 1858.]

THE term matrix might be used in a more general sense, but in the present memoir I consider only square and rectangular matrices, and the term matrix used without qualification is to be understood as meaning a square matrix; in this restricted sense, a set of quantities arranged in the form of a square, e.g.

$$\begin{array}{|ccc|} (\; a\,, & b\,, & c \;) \\ a'\,, & b'\,, & c' \\ a''\,, & b''\,, & c'' \end{array}$$

is said to be a matrix. The notion of such a matrix arises naturally from an abbreviated notation for a set of linear equations, viz. the equations

$$\begin{aligned} X &= ax \;\; + by \;\; + cz\;, \\ Y &= a'x \; + b'y \; + c'z\,, \\ Z &= a''x + b''y + c''z, \end{aligned}$$

may be more simply represented by

$$(X,\; Y,\; Z) = \begin{array}{|ccc|} (\; a\,, & b\,, & c \; \rangle\!\!\langle \\ a'\,, & b'\,, & c' \\ a''\,, & b''\,, & c'' \end{array} x,\; y,\; z),$$

and the consideration of such a system of equations leads to most of the fundamental notions in the theory of matrices. It will be seen that matrices (attending only to those of the same order) comport themselves as single quantities; they may be added, multiplied or compounded together, &c.: the law of the addition of matrices is precisely similar to that for the addition of ordinary algebraical quantities; as regards their multiplication (or composition), there is the peculiarity that matrices are not in general convertible; it is nevertheless possible to form the powers (positive or negative, integral or fractional) of a matrix, and thence to arrive at the notion of a rational and integral function, or generally of any algebraical function, of a matrix. I obtain the remarkable theorem that any matrix whatever satisfies an algebraical equation of its own order, the coefficient of the highest power being unity, and those of the other powers functions of the terms of the matrix, the last coefficient being in fact the determinant; the rule for the formation of this equation may be stated in the following condensed form, which will be intelligible after a perusal of the memoir, viz. the determinant, formed out of the matrix diminished by the matrix considered as a single quantity involving the matrix unity, will be equal to zero. The theorem shows that every rational and integral function (or indeed every rational function) of a matrix may be considered as a rational and integral function, the degree of which is at most equal to that of the matrix, less unity; it even shows that in a sense, the same is true with respect to any algebraical function whatever of a matrix. One of the applications of the theorem is the finding of the general expression of the matrices which are convertible with a given matrix. The theory of rectangular matrices appears much less important than that of square matrices, and I have not entered into it further than by showing how some of the notions applicable to these may be extended to rectangular matrices.

1. For conciseness, the matrices written down at full length will in general be of the order 3, but it is to be understood that the definitions, reasonings, and conclusions apply to matrices of any degree whatever. And when two or more matrices are spoken of in connexion with each other, it is always implied (unless the contrary is expressed) that the matrices are of the same order.

2. The notation

$$\left(\begin{array}{ccc} a, & b, & c \\ a', & b', & c' \\ a'', & b'', & c'' \end{array}\right)\!\!\!\between\!(x,\ y,\ z)$$

represents the set of linear functions

$$((a,\ b,\ c\between x,\ y,\ z),\ (a',\ b',\ c'\between x,\ y,\ z),\ (a'',\ b'',\ c''\between x,\ y,\ z)),$$

so that calling these $(X,\ Y,\ Z)$, we have

$$(X,\ Y,\ Z)=\left(\begin{array}{ccc} a, & b, & c \\ a', & b', & c' \\ a'', & b'', & c'' \end{array}\right)\!\!\!\between\!(x,\ y,\ z)$$

and, as remarked above, this formula leads to most of the fundamental notions in the theory.

3. The quantities (X, Y, Z) will be identically zero, if all the terms of the matrix are zero, and we may say that

$$\begin{array}{|ccc|} \multicolumn{3}{c}{(\ 0, \quad 0, \quad 0\ )} \\ 0, & 0, & 0 \\ 0, & 0, & 0 \end{array}$$

is the matrix zero.

Again, (X, Y, Z) will be identically equal to (x, y, z), if the matrix is

$$\begin{array}{|ccc|} \multicolumn{3}{c}{(\ 1, \quad 0, \quad 0\ )} \\ 0, & 1, & 0 \\ 0, & 0, & 1 \end{array}$$

and this is said to be the matrix unity. We may of course, when for distinctness it is required, say, the matrix zero, or (as the case may be) the matrix unity *of such an order*. The matrix zero may for the most part be represented simply by 0, and the matrix unity by 1.

4. The equations

$$(X, Y, Z) = \begin{array}{|ccc|} \multicolumn{3}{c}{(\ a\ ,\quad b\ ,\quad c\ \ )\!\!(} \\ a', & b', & c' \\ a'', & b'', & c'' \end{array} x, y, z), \quad (X', Y', Z') = \begin{array}{|ccc|} \multicolumn{3}{c}{(\ \alpha\ ,\quad \beta\ ,\quad \gamma\ \ )\!\!(} \\ \alpha', & \beta', & \gamma' \\ \alpha'', & \beta'', & \gamma'' \end{array} x, y, z)$$

give

$$(X + X', Y + Y', Z + Z') = \begin{array}{|ccc|} \multicolumn{3}{c}{(\ a+\alpha\ ,\quad b+\beta\ ,\quad c+\gamma\ \ )\!\!(} \\ a'+\alpha', & b'+\beta', & c'+\gamma' \\ a''+\alpha'', & b''+\beta'', & c''+\gamma'' \end{array} x, y, z)$$

and this leads to

$$\begin{array}{|ccc|} \multicolumn{3}{c}{(\ a+\alpha\ ,\quad b+\beta\ ,\quad c+\gamma\ )} \\ a'+\alpha', & b'+\beta', & c'+\gamma' \\ a''+\alpha'', & b''+\beta'', & c''+\gamma'' \end{array} = \begin{array}{|ccc|} \multicolumn{3}{c}{(\ a\ ,\quad b\ ,\quad c\ )} \\ a', & b', & c' \\ a'', & b'', & c'' \end{array} + \begin{array}{|ccc|} \multicolumn{3}{c}{(\ \alpha\ ,\quad \beta\ ,\quad \gamma\ )} \\ \alpha', & \beta', & \gamma' \\ \alpha'', & \beta'', & \gamma'' \end{array}$$

as a rule for the addition of matrices; that for their subtraction is of course similar to it.

5. A matrix is not altered by the addition or subtraction of the matrix zero, that is, we have $M \pm 0 = M$.

The equation $L = M$, which expresses that the matrices L, M are equal, may also be written in the form $L - M = 0$, i.e. the difference of two equal matrices is the matrix zero.

6. The equation $L = -M$, written in the form $L + M = 0$, expresses that the sum of the matrices L, M is equal to the matrix zero, the matrices so related are said to be *opposite* to each other; in other words, a matrix the terms of which are equal but opposite in sign to the terms of a given matrix, is said to be opposite to the given matrix.

7. It is clear that we have $L+M=M+L$, that is, the operation of addition is commutative, and moreover that $(L+M)+N=L+(M+N)=L+M+N$, that is, the operation of addition is also associative.

8. The equation

$$(X,\ Y,\ Z)=\left(\begin{array}{ccc} a, & b, & c \\ a', & b', & c' \\ a'', & b'', & c'' \end{array}\right)(mx,\ my,\ mz)$$

written under the forms

$$(X,\ Y,\ Z)=m\left(\begin{array}{ccc} a, & b, & c \\ a', & b', & c' \\ a'', & b'', & c'' \end{array}\right)(x,\ y,\ z)=\left(\begin{array}{ccc} ma, & mb, & mc \\ ma', & mb', & mc' \\ ma'', & mb'', & mc'' \end{array}\right)(x,\ y,\ z)$$

gives

$$m\left(\begin{array}{ccc} a, & b, & c \\ a', & b', & c' \\ a'', & b'', & c'' \end{array}\right)=\left(\begin{array}{ccc} ma, & mb, & mc \\ ma', & mb', & mc' \\ ma'', & mb'', & mc'' \end{array}\right)$$

as the rule for the multiplication of a matrix by a single quantity. The multiplier m may be written either before or after the matrix, and the operation is therefore commutative. We have it is clear $m(L+M)=mL+mM$, or the operation is distributive.

9. The matrices L and mL may be said to be similar to each other; in particular, if $m=1$, they are equal, and if $m=-1$, they are opposite.

10. We have, in particular,

$$m\left(\begin{array}{ccc} 1, & 0, & 0 \\ 0, & 1, & 0 \\ 0, & 0, & 1 \end{array}\right)=\left(\begin{array}{ccc} m, & 0, & 0 \\ 0, & m, & 0 \\ 0, & 0, & m \end{array}\right),$$

or replacing the matrix on the left-hand side by unity, we may write

$$m=\left(\begin{array}{ccc} m, & 0, & 0 \\ 0, & m, & 0 \\ 0, & 0, & m \end{array}\right);$$

the matrix on the right-hand side is said to be the single quantity m considered as *involving the matrix unity.*

11. The equations

$$(X,\ Y,\ Z)=\left(\begin{array}{ccc} a, & b, & c \\ a', & b', & c' \\ a'', & b'', & c'' \end{array}\right)(x,\ y,\ z),\quad (x,\ y,\ z)=\left(\begin{array}{ccc} \alpha, & \beta, & \gamma \\ \alpha', & \beta', & \gamma' \\ \alpha'', & \beta'', & \gamma'' \end{array}\right)(\xi,\ \eta,\ \zeta),$$

give

$$(X,\ Y,\ Z) = \left(\begin{array}{ccc} A\ , & B\ , & C \\ A', & B', & C' \\ A'', & B'', & C'' \end{array}\right)\between(\xi,\ \eta,\ \zeta) = \left(\begin{array}{ccc} a\ , & b\ , & c \\ a', & b', & c' \\ a'', & b'', & c'' \end{array}\right)\between\left(\begin{array}{ccc} \alpha\ , & \beta\ , & \gamma \\ \alpha', & \beta', & \gamma' \\ \alpha'', & \beta'', & \gamma'' \end{array}\right)\between(\xi,\ \eta,\ \zeta),$$

and thence, substituting for the matrix

$$\left(\begin{array}{ccc} A\ , & B\ , & C \\ A', & B', & C' \\ A'', & B'', & C'' \end{array}\right)$$

its value, we obtain

$$\left(\begin{array}{ccc} (a\ ,b\ ,c\ \between\alpha,\alpha',\alpha''), & (a\ ,b\ ,c\ \between\beta,\beta',\beta''), & (a\ ,b\ ,c\ \between\gamma,\gamma',\gamma'') \\ (a',b',c'\between\alpha,\alpha',\alpha''), & (a',b',c'\between\beta,\beta',\beta''), & (a',b',c'\between\gamma,\gamma',\gamma'') \\ (a'',b'',c''\between\alpha,\alpha',\alpha''), & (a'',b'',c''\between\beta,\beta',\beta''), & (a'',b'',c''\between\gamma,\gamma',\gamma'') \end{array}\right) = \left(\begin{array}{ccc} a\ ,& b\ ,& c \\ a', & b', & c' \\ a'', & b'', & c'' \end{array}\right)\between\left(\begin{array}{ccc} \alpha\ , & \beta\ , & \gamma \\ \alpha', & \beta', & \gamma' \\ \alpha'', & \beta'', & \gamma'' \end{array}\right)$$

as the rule for the multiplication or composition of two matrices. It is to be observed, that the operation is not a commutative one; the component matrices may be distinguished as the first or further component matrix, and the second or nearer component matrix, and the rule of composition is as follows, viz. any *line* of the compound matrix is obtained by combining the corresponding *line* of the first or further component matrix successively with the several *columns* of the second or nearer compound matrix.

[We may conveniently write

$$\begin{array}{c|ccc} & (\alpha,\ \alpha',\ \alpha''), & (\beta,\ \beta',\ \beta''), & (\gamma,\ \gamma',\ \gamma'') \\ \hline (a\ ,\ b\ ,\ c\ ) & " & " & " \\ (a',\ b',\ c') & " & " & " \\ (a'',\ b'',\ c'') & " & " & " \end{array}$$

to denote the left-hand side of the last preceding equation.]

12. A matrix compounded, either as first or second component matrix, with the matrix zero, gives the matrix zero. The case where any of the terms of the given matrix are infinite is of course excluded.

13. A matrix is not altered by its composition, either as first or second component matrix, with the matrix unity. It is compounded either as first or second component matrix, with the single quantity m considered as involving the matrix unity, by multiplication of all its terms by the quantity m: this is in fact the before-mentioned rule for the multiplication of a matrix by a single quantity, which rule is thus seen to be a particular case of that for the multiplication of two matrices.

14. We may in like manner multiply or compound together three or more matrices: the order of arrangement of the factors is of course material, and we may

distinguish them as the first or furthest, second, third, &c., and last or nearest component matrices: any two consecutive factors may be compounded together and replaced by a single matrix, and so on until all the matrices are compounded together, the result being independent of the particular mode in which the composition is effected; that is, we have $L.MN = LM.N = LMN$, $LM.NP = L.MN.P$, &c., or the operation of multiplication, although, as already remarked, not commutative, is associative.

15. We thus arrive at the notion of a positive and integer power L^p of a matrix L, and it is to be observed that the different powers of the same matrix are convertible. It is clear also that p and q being positive integers, we have $L^p . L^q = L^{p+q}$, which is the theorem of indices for positive integer powers of a matrix.

16. The last-mentioned equation, $L^p . L^q = L^{p+q}$, assumed to be true for all values whatever of the indices p and q, leads to the notion of the powers of a matrix for any form whatever of the index. In particular, $L^p . L^0 = L^p$ or $L^0 = 1$, that is, the 0th power of a matrix is the matrix unity. And then putting $p = 1$, $q = -1$, or $p = -1$, $q = 1$, we have $L . L^{-1} = L^{-1} . L = 1$; that is, L^{-1}, or as it may be termed the inverse or reciprocal matrix, is a matrix which, compounded either as first or second component matrix with the original matrix, gives the matrix unity.

17. We may arrive at the notion of the inverse or reciprocal matrix, directly from the equation

$$(X, Y, Z) = \left(\begin{array}{ccc} a, & b, & c \\ a', & b', & c' \\ a'', & b'', & c'' \end{array}\right)(x, y, z),$$

in fact this equation gives

$$(x, y, z) = \left(\begin{array}{ccc} A, & A', & A'' \\ B, & B', & B'' \\ C, & C', & C'' \end{array}\right)(X, Y, Z) = \left(\left(\begin{array}{ccc} a, & b, & c \\ a', & b', & c' \\ a'', & b'', & c'' \end{array}\right)^{-1}\right)(X, Y, Z),$$

and we have, for the determination of the coefficients of the inverse or reciprocal matrix, the equations

$$\left(\begin{array}{ccc} A, & A', & A'' \\ B, & B', & B'' \\ C, & C', & C'' \end{array}\right)\left(\begin{array}{ccc} a, & b, & c \\ a', & b', & c' \\ a'', & b'', & c'' \end{array}\right) = \left(\begin{array}{ccc} 1, & 0, & 0 \\ 0, & 1, & 0 \\ 0, & 0, & 1 \end{array}\right),$$

$$\left(\begin{array}{ccc} a, & b, & c \\ a', & b', & c' \\ a'', & b'', & c'' \end{array}\right)\left(\begin{array}{ccc} A, & A', & A'' \\ B, & B', & B'' \\ C, & C', & C'' \end{array}\right) = \left(\begin{array}{ccc} 1, & 0, & 0 \\ 0, & 1, & 0 \\ 0, & 0, & 1 \end{array}\right),$$

which are equivalent to each other, and either of them is by itself sufficient for the complete determination of the inverse or reciprocal matrix. It is well known that if ∇ denote the determinant, that is, if

$$\nabla = \begin{vmatrix} a\ , & b\ , & c \\ a', & b', & c' \\ a'', & b'', & c'' \end{vmatrix}$$

then the terms of the inverse or reciprocal matrix are given by the equations

$$A = \frac{1}{\nabla}\begin{vmatrix} 1, & 0\ , & 0 \\ 0, & b', & c' \\ 0, & b'', & c'' \end{vmatrix}, \qquad B = \frac{1}{\nabla}\begin{vmatrix} 0\ , & 1, & 0 \\ a', & 0, & c' \\ a'', & 0, & c'' \end{vmatrix}, \ \&c.$$

or what is the same thing, the inverse or reciprocal matrix is given by the equation

$$\left(\begin{array}{ccc} a\ , & b\ , & c \\ a', & b', & c' \\ a'', & b'', & c'' \end{array}\right)^{-1} = \frac{1}{\nabla}\left(\begin{array}{ccc} \partial_a\nabla, & \partial_{a'}\nabla, & \partial_{a''}\nabla \\ \partial_b\nabla\ , & \partial_{b'}\nabla, & \partial_{b''}\nabla \\ \partial_c\nabla\ , & \partial_{c'}\nabla\ , & \partial_{c''}\nabla \end{array}\right)$$

where of course the differentiations must in every case be performed as if the terms a, b, &c. were all of them independent arbitrary quantities.

18. The formula shows, what is indeed clear *à priori*, that the notion of the inverse or reciprocal matrix fails altogether when the determinant vanishes: the matrix is in this case said to be indeterminate, and it must be understood that in the absence of express mention, the particular case in question is frequently excluded from consideration. It may be added that the matrix zero is indeterminate; and that the product of two matrices may be zero, without either of the factors being zero, if only the matrices are one or both of them indeterminate.

19. The notion of the inverse or reciprocal matrix once established, the other negative integer powers of the original matrix are positive integer powers of the inverse or reciprocal matrix, and the theory of such negative integer powers may be taken to be known. The theory of the fractional powers of a matrix will be further discussed in the sequel.

20. The positive integer power L^m of the matrix L may of course be multiplied by any matrix of the same degree: such multiplier, however, is not in general convertible with L; and to preserve as far as possible the analogy with ordinary algebraical functions, we may restrict the attention to the case where the multiplier is a single quantity, and such convertibility consequently exists. We have in this manner a matrix cL^m, and by the addition of any number of such terms we obtain a rational and integral function of the matrix L.

21. The general theorem before referred to will be best understood by a complete development of a particular case. Imagine a matrix

$$M = \left(\begin{array}{cc} a, & b \\ c, & d \end{array}\right),$$

and form the determinant

$$\left|\begin{array}{ll} a - M, & b \\ c & , d - M \end{array}\right|,$$

the developed expression of this determinant is

$$M^2 - (a+d)\, M^1 + (ad - bc)\, M^0 \,;$$

the values of M^2, M^1, M^0 are

$$\left(\begin{array}{cc} a^2 + bc\ , & b\,(a+d) \\ c\,(a+d), & d^2 + bc \end{array}\right),\quad \left(\begin{array}{cc} a, & b \\ c, & d \end{array}\right),\quad \left(\begin{array}{cc} 1, & 0 \\ 0, & 1 \end{array}\right),$$

and substituting these values the determinant becomes equal to the matrix zero, viz. we have

$$\left|\begin{array}{ll} a - M, & b \\ c & , d - M \end{array}\right| = \left(\begin{array}{cc} a^2 + bc\ , & b\,(a+d) \\ c\,(a+d), & d^2 + bc \end{array}\right) - (a+d) \left(\begin{array}{cc} a, & b \\ c, & d \end{array}\right) + (ad - bc) \left(\begin{array}{cc} 1, & 0 \\ 0, & 1 \end{array}\right)$$

$$= \left(\begin{array}{ll} (a^2 + bc) - (a+d)\,a + (ad - bc), & b\,(a+d) - (a+d)\,b \\ c\,(a+d) - (a+d)\,c & , d^2 + bc - (a+d)\,d + ad - bc \end{array}\right) = \left(\begin{array}{cc} 0, & 0 \\ 0, & 0 \end{array}\right);$$

that is

$$\left|\begin{array}{ll} a - M, & b \\ c & , d - M \end{array}\right| = 0,$$

where the matrix of the determinant is

$$\left(\begin{array}{cc} a, & b \\ c, & d \end{array}\right) - M \left(\begin{array}{cc} 1, & 0 \\ 0, & 1 \end{array}\right),$$

that is, it is the original matrix, diminished by the same matrix considered as a single quantity involving the matrix unity. And this is the general theorem, viz. the determinant, having for its matrix a given matrix less the same matrix considered as a single quantity involving the matrix unity, is equal to zero.

22. The following symbolical representation of the theorem is, I think, worth noticing: let the matrix M, considered as a single quantity, be represented by $\tilde{M}$, then writing 1 to denote the matrix unity, $\tilde{M}\,.\,1$ will represent the matrix M, considered as a single quantity involving the matrix unity. Upon the like principles of notation, $\tilde{1}\,.\,M$ will represent, or may be considered as representing, simply the matrix M, and the theorem is

$$\text{Det.}\ (\tilde{1}\,.\,M - \tilde{M}\,.\,1) = 0.$$

23. I have verified the theorem, in the next simplest case of a matrix of the order 3, viz. if M be such a matrix, suppose

$$M = \left(\begin{array}{ccc} a, & b, & c \\ d, & e, & f \\ g, & h, & i \end{array}\right),$$

then the derived determinant vanishes, or we have

$$\left|\begin{array}{ccc} a-M, & b\quad, & c \\ d\quad, & e-M, & f \\ g\quad, & h\quad, & i-M \end{array}\right| = 0,$$

or expanding

$$M^3 - (a+e+i)\,M^2 + (ei+ia+ae-fh-cg-bd)\,M - (aei+bfg+cdh-afh-bdi-ceg) = 0\,;$$

but I have not thought it necessary to undertake the labour of a formal proof of the theorem in the general case of a matrix of any degree.

24. If we attend only to the general form of the result, we see that any matrix whatever satisfies an algebraical equation of its own order, which is in many cases the material part of the theorem.

25. It follows at once that every rational and integral function, or indeed every rational function of a matrix, can be expressed as a rational and integral function of an order at most equal to that of the matrix, less unity. But it is important to consider how far or in what sense the like theorem is true with respect to irrational functions of a matrix. If we had only the equation satisfied by the matrix itself, such extension could not be made; but we have besides the equation of the same order satisfied by the irrational function of the matrix, and by means of these two equations, and the equation by which the irrational function of the matrix is determined, we may express the irrational function as a rational and integral function of the matrix, of an order equal at most to that of the matrix, less unity; such expression will however involve *the coefficients of the equation satisfied by the irrational function*, which are functions (in number equal to the order of the matrix) of the terms, assumed to be unknown, of the irrational function itself. The transformation is nevertheless an important one, as reducing the number of unknown quantities from n^2 (if n be the order of the matrix) down to n. To complete the solution, it is necessary to compare the value obtained as above, with the assumed value of the irrational function, which will lead to equations for the determination of the n unknown quantities.

26. As an illustration, consider the given matrix

$$M = \left(\begin{array}{cc} a, & b \\ c, & d \end{array}\right),$$

and let it be required to find the matrix $L = \sqrt{M}$. In this case M satisfies the equation

$$M^2 - (a+d)\,M + ad - bc = 0\,;$$

and in like manner if

$$L = \begin{pmatrix} \alpha, & \beta \\ \gamma, & \delta \end{pmatrix}$$

then L satisfies the equation

$$L^2 - (\alpha + \delta)\,L + \alpha\delta - \beta\gamma = 0\,;$$

and from these two equations, and the rationalized equation $L^2 = M$, it should be possible to express L in the form of a linear function of M: in fact, putting in the last equation for L^2 its value ($= M$), we find at once

$$L = \frac{1}{\alpha+\delta}\,[M + (\alpha\delta - \beta\gamma)],$$

which is the required expression, involving as it should do the coefficients $\alpha+\delta$, $\alpha\delta - \beta\gamma$ of the equation in L. There is no difficulty in completing the solution; write for shortness $\alpha + \delta = X$, $\alpha\delta - \beta\gamma = Y$, then we have

$$L = \begin{pmatrix} \alpha, & \beta \\ \gamma, & \delta \end{pmatrix} = \begin{pmatrix} \dfrac{a+Y}{X}, & \dfrac{b}{X} \\ \dfrac{c}{X}, & \dfrac{d+Y}{X} \end{pmatrix},$$

and consequently forming the values of $\alpha + \delta$ and $\alpha\delta - \beta\gamma$,

$$X = \frac{a+d+2Y}{X},$$

$$Y = \frac{(a+Y)(d+Y) - bc}{X^2},$$

and putting also $a + d = P$, $ad - bc = Q$, we find without difficulty

$$X = \sqrt{P + 2\sqrt{Q}},$$
$$Y = \sqrt{Q},$$

and the values of α, β, γ, δ are consequently known. The sign of $\sqrt{Q}$ is the same in both formulæ, and there are consequently in all four solutions, that is, the radical $\sqrt{M}$ has four values.

27. To illustrate this further, suppose that instead of M we have the matrix

$$M^2 = \begin{pmatrix} a, & b \\ c, & d \end{pmatrix}^2 = \begin{pmatrix} a^2 + bc, & b(a+d) \\ c(a+d), & d^2 + bc \end{pmatrix},$$

so that $L^2 = M^2$, we find

$$P = (a+d)^2 - 2(ad - bc),$$
$$Q = (ad - bc)^2,$$

and thence $\sqrt{Q} = \pm (ad - bc)$. Taking the positive sign, we have

$$Y = \quad ad - bc,$$
$$X = \pm (a + d),$$

and these values give simply

$$L = \pm \begin{pmatrix} a, & b \\ c, & d \end{pmatrix} = \pm M.$$

But taking the negative sign,

$$Y = - ad + bc,$$
$$X = \pm \sqrt{(a-d)^2 + 4bc},$$

and retaining X to denote this radical, we find

$$L = \begin{pmatrix} \dfrac{a^2 - ad + 2bc}{X}, & \dfrac{b(a+d)}{X} \\ \dfrac{c(a+d)}{X}, & \dfrac{d^2 - ad + 2bc}{X} \end{pmatrix},$$

which may also be written

$$L = \frac{a+d}{X} \begin{pmatrix} a, & b \\ c, & d \end{pmatrix} - \frac{2(ad - bc)}{X} \begin{pmatrix} 1, & 0 \\ 0, & 1 \end{pmatrix},$$

or, what is the same thing,

$$L = \frac{a+d}{X} M - \frac{2(ad - bc)}{X};$$

and it is easy to verify *à posteriori* that this value in fact gives $L^2 = M^2$. It may be remarked that if

$$M^2 = \begin{pmatrix} 1, & 0 \\ 0, & 1 \end{pmatrix}^2 = 1,$$

the last-mentioned formula fails, for we have $X = 0$; it will be seen presently that the equation $L^2 = 1$ admits of other solutions besides $L = \pm 1$. The example shows how the values of the fractional powers of a matrix are to be investigated.

28. There is an apparent difficulty connected with the equation satisfied by a matrix, which it is proper to explain. Suppose, as before,

$$M = \begin{pmatrix} a, & b \\ c, & d \end{pmatrix}$$

so that M satisfies the equation

$$\begin{vmatrix} a-M, & b \\ c \quad , & d-M \end{vmatrix} = 0,$$

or

$$M^2 - (a+d)M + ad - bc = 0,$$

and let $X_{,}$, $X_{,,}$ be the single quantities, roots of the equation

$$\begin{vmatrix} a-X, & b \\ c \quad , & d-X \end{vmatrix} = 0$$

or

$$X^2 - (a+d)X + ad - bc = 0.$$

The equation satisfied by the matrix may be written

$$(M - X_{,})(M - X_{,,}) = 0,$$

in which $X_{,}$, $X_{,,}$ are to be considered as respectively involving the matrix unity, and it would at first sight seem that we ought to have one of the simple factors equal to zero; this is obviously not the case, for such equation would signify that the perfectly indeterminate matrix M was equal to a single quantity, considered as involving the matrix unity. The explanation is that each of the simple factors is an indeterminate matrix, in fact $M - X_{,}$ stands for the matrix

$$\left(\begin{array}{ll} a - X_{,}, & b \\ c \quad , & d - X_{,} \end{array} \right),$$

and the determinant of this matrix is equal to zero. The product of the two factors is thus equal to zero without either of the factors being equal to zero.

29. A matrix satisfies, we have seen, an equation of its own order, involving the coefficients of the matrix; assume that the matrix is to be determined to satisfy some other equation, the coefficients of which are given single quantities. It would at first sight appear that we might eliminate the matrix between the two equations, and thus obtain an equation which would be the only condition to be satisfied by the terms of the matrix; this is obviously wrong, for more conditions must be requisite, and we see that if we were then to proceed to complete the solution by finding the value of the matrix common to the two equations, we should find the matrix equal in every case to a single quantity considered as involving the matrix unity, which it is clear ought not to be the case. The explanation is similar to that of the difficulty before adverted to; the equations may contain one, and only one, common factor, and may be both of them satisfied, and yet the common factor may not vanish. The necessary condition seems to be, that the one equation should be a factor of the other; in the case where the assumed equation is of an order equal or superior to the matrix, then if this equation contain as a factor the equation which is always satisfied by the matrix, the assumed equation will be satisfied identically, and the condition is sufficient as well as necessary: in the other case, where the assumed equation is of an order inferior to that of the matrix, the condition is necessary, but it is not sufficient.

30. The equation satisfied by the matrix may be of the form $M^n = 1$; the matrix is in this case said to be periodic of the nth order. The preceding considerations apply to the theory of periodic matrices; thus, for instance, suppose it is required to find a matrix of the order 2, which is periodic of the second order. Writing

$$M = \left(\begin{array}{cc} a, & b \\ c, & d \end{array}\right),$$

we have

$$M^2 - (a+d)M + ad - bc = 0,$$

and the assumed equation is

$$M^2 - 1 = 0.$$

These equations will be identical if

$$a + d = 0, \quad ad - bc = -1,$$

that is, these conditions being satisfied, the equation $M^2 - 1 = 0$ required to be satisfied, will be identical with the equation which is always satisfied, and will therefore itself be satisfied. And in like manner the matrix M of the order 2 will satisfy the condition $M^3 - 1 = 0$, or will be periodic of the third order, if only $M^3 - 1$ contains as a factor

$$M^2 - (a+d)M + ad - bc,$$

and so on.

31. But suppose it is required to find a matrix of the order 3,

$$M = \left(\begin{array}{ccc} a, & b, & c \\ d, & e, & f \\ g, & h, & i \end{array}\right)$$

which shall be periodic of the second order. Writing for shortness

$$\left|\begin{array}{ccc} a-M, & b, & c \\ d, & e-M, & f \\ g, & h, & i-M \end{array}\right| = -(M^3 - AM^2 + BM - C),$$

the matrix here satisfies

$$M^3 - AM^2 + BM - C = 0,$$

and, as before, the assumed equation is $M^2 - 1 = 0$. Here, if we have $1 + B = 0$, $A + C = 0$, the left-hand side will contain the factor $(M^2 - 1)$, and the equation will take the form $(M^2 - 1)(M + C) = 0$, and we should have then $M^2 - 1 = 0$, provided $M + C$ were not an indeterminate matrix. But $M + C$ denotes the matrix

$$\left(\begin{array}{ccc} a+C, & b, & c \\ d, & e+C, & f \\ g, & h, & i+C \end{array}\right)$$

the determinant of which is $C^3 + AC^2 + BC + C$, which is equal to zero in virtue of the equations $1 + B = 0$, $A + C = 0$, and we cannot, therefore, from the equation $(M^2 - 1)(M + C) = 0$, deduce the equation $M^2 - 1 = 0$. This is as it should be, for the two conditions are not sufficient, in fact the equation

$$M^2 = \begin{pmatrix} a^2 + bd + cg, & ab + be + ch, & ac + bf + ci \\ da + ed + fg, & db + e^2 + fh, & dc + ef + fi \\ ga + hd + ig, & gb + he + ih, & gc + hf + i^2 \end{pmatrix} = 1$$

gives nine equations, which are however satisfied by the following values, involving in reality four arbitrary coefficients; viz. the value of the matrix is

$$\begin{pmatrix} \dfrac{\alpha}{\alpha+\beta+\gamma}, & \dfrac{-(\beta+\gamma)\,\nu\mu^{-1}}{\alpha+\beta+\gamma}, & \dfrac{-(\beta+\gamma)\,\nu\mu^{-1}}{\alpha+\beta+\gamma} \\ \dfrac{-(\gamma+\alpha)\,\mu\nu^{-1}}{\alpha+\beta+\gamma}, & \dfrac{\beta}{\alpha+\beta+\gamma}, & \dfrac{-(\gamma+\alpha)\,\lambda\mu^{-1}}{\alpha+\beta+\gamma} \\ \dfrac{-(\alpha+\beta)\,\mu\nu^{-1}}{\alpha+\beta+\gamma}, & \dfrac{-(\alpha+\beta)\,\nu\lambda^{-1}}{\alpha+\beta+\gamma}, & \dfrac{\gamma}{\alpha+\beta+\gamma} \end{pmatrix}$$

so that there are in all five relations (and not only two) between the coefficients of the matrix.

32. Instead of the equation $M^n - 1 = 0$, which belongs to a periodic matrix, it is in many cases more convenient, and it is much the same thing to consider an equation $M^n - k = 0$, where k is a single quantity. The matrix may in this case be said to be periodic to a factor *près*.

33. Two matrices L, M are convertible when $LM = ML$. If the matrix M is given, this equality affords a set of linear equations between the coefficients of L equal in number to these coefficients, but these equations cannot be all independent, for it is clear that if L be any rational and integral function of M (the coefficients being single quantities), then L will be convertible with M; or what is apparently (but only apparently) more general, if L be any algebraical function whatever of M (the coefficients being always single quantities), then L will be convertible with M. But whatever the form of the function is, it may be reduced to a rational and integral function of an order equal to that of M, less unity, and we have thus the general expression for the matrices convertible with a given matrix, viz. any such matrix is a rational and integral function (the coefficients being single quantities) of the given matrix, the order being that of the given matrix, less unity. In particular, the general form of the matrix L convertible with a given matrix M of the order 2, is $L = \alpha M + \beta$, or what is the same thing, the matrices

$$\begin{pmatrix} a, & b \\ c, & d \end{pmatrix}, \quad \begin{pmatrix} a', & b' \\ c', & d' \end{pmatrix}$$

will be convertible if $a' - d' : b' : c' = a - d : b : c$.

34. Two matrices L, M are skew convertible when $LM = -ML$; this is a relation much less important than ordinary convertibility, for it is to be noticed that we cannot in general find a matrix L skew convertible with a given matrix M. In fact, considering M as given, the equality affords a set of linear equations between the coefficients of L equal in number to these coefficients; and in this case the equations are independent, and we may eliminate all the coefficients of L, and we thus arrive at a relation which must be satisfied by the coefficients of the given matrix M. Thus, suppose the matrices

$$\left(\begin{array}{cc} a, & b \\ c, & d \end{array}\right), \quad \left(\begin{array}{cc} a', & b' \\ c', & d' \end{array}\right)$$

are skew convertible, we have

$$\left(\begin{array}{cc} a, & b \\ c, & d \end{array}\right)\left(\begin{array}{cc} a', & b' \\ c', & d' \end{array}\right) = \left(\begin{array}{cc} aa'+bc', & ab'+bd' \\ ca'+dc', & cb'+dd' \end{array}\right),$$

$$\left(\begin{array}{cc} a', & b' \\ c', & d' \end{array}\right)\left(\begin{array}{cc} a, & b \\ c, & d \end{array}\right) = \left(\begin{array}{cc} aa'+b'c, & a'b+b'd \\ c'a+d'c, & c'b+d'd \end{array}\right),$$

and the conditions of skew convertibility are

$$\begin{aligned} 2aa' + bc' + b'c &= 0, \\ b'(a+d) + b(a'+d') &= 0, \\ c'(a+d) + c(a'+d') &= 0, \\ 2dd' + bc' + b'c &= 0. \end{aligned}$$

Eliminating a', b', c', d', the relation between a, b, c, d is

$$\left|\begin{array}{cccc} 2a, & c\ , & b\ , & . \\ b\ , & a+d, & . & b \\ c\ , & . & a+d, & c \\ . & c\ , & b\ , & 2d \end{array}\right| = 0,$$

which is

$$(a+d)^2(ad-bc) = 0.$$

Excluding from consideration the case $ad - bc = 0$, which would imply that the matrix was indeterminate, we have $a + d = 0$. The resulting system of conditions then is

$$a+d=0, \quad a'+d'=0, \quad aa'+bc'+b'c+dd'=0,$$

the first two of which imply that the matrices are respectively periodic of the second order to a factor *près*.

35. It may be noticed that if the compound matrices LM and ML are similar, they are either equal or else opposite; that is, the matrices L, M are either convertible or skew convertible.

36. Two matrices such as

$$\begin{pmatrix} a, & b \\ c, & d \end{pmatrix}, \quad \begin{pmatrix} a, & c \\ b, & d \end{pmatrix},$$

are said to be formed one from the other by transposition, and this may be denoted by the symbol tr.; thus we may write

$$\begin{pmatrix} a, & c \\ b, & d \end{pmatrix} = \text{tr.} \begin{pmatrix} a, & b \\ c, & d \end{pmatrix}.$$

The effect of two successive transpositions is of course to reproduce the original matrix.

37. It is easy to see that if M be any matrix, then

$$(\text{tr. } M)^p = \text{tr. } (M^p),$$

and in particular,

$$(\text{tr. } M)^{-1} = \text{tr. } (M^{-1}).$$

38. If L, M be any two matrices,

$$\text{tr.} (LM) = \text{tr. } M. \text{ tr. } L,$$

and similarly for three or more matrices, L, M, N, &c.,

$$\text{tr.} (LMN) = \text{tr. } N. \text{ tr. } M. \text{ tr. } L, \text{ \&c.}$$

40. A matrix such as

$$\begin{pmatrix} a, & h, & g \\ h, & b, & f \\ g, & f, & c \end{pmatrix}$$

which is not altered by transposition, is said to be symmetrical.

41. A matrix such as

$$\begin{pmatrix} 0, & \nu, & -\mu \\ -\nu, & 0, & \lambda \\ \mu, & -\lambda, & 0 \end{pmatrix}$$

which by transposition is changed into its opposite, is said to be skew symmetrical.

42. It is easy to see that any matrix whatever may be expressed as the sum of a symmetrical matrix, and a skew symmetrical matrix; thus the form

$$\begin{pmatrix} a\ , & h+\nu, & g-\mu \\ h-\nu, & b\ , & f+\lambda \\ g+\mu, & f-\lambda, & c \end{pmatrix}$$

which may obviously represent any matrix whatever of the order 3, is the sum of the two matrices last before mentioned.

43. The following formulæ, although little more than examples of the composition of transposed matrices, may be noticed, viz.

$$\left(\begin{array}{cc} a, & b \\ c, & d \end{array}\right)\left(\begin{array}{cc} a, & c \\ d, & b \end{array}\right) = \left(\begin{array}{cc} a^2+b^2, & ac+bd \\ ac+bd, & c^2+d^2 \end{array}\right)$$

which shows that a matrix compounded with the transposed matrix gives rise to a symmetrical matrix. It does not however follow, nor is it the fact, that the matrix and transposed matrix are convertible. And also

$$\left(\begin{array}{cc} a, & c \\ b, & d \end{array}\right)\left(\begin{array}{cc} a, & b \\ c, & d \end{array}\right)\left(\begin{array}{cc} a, & c \\ b, & d \end{array}\right) = \left(\begin{array}{cc} a^3+bcd+a(b^2+c^2), & c^3+abd+c(a^2+d^2) \\ b^3+acd+b(a^2+d^2), & d^3+abc+d(b^2+c^3) \end{array}\right)$$

which is a remarkably symmetrical form. It is needless to proceed further, since it is clear that

$$\left(\begin{array}{cc} a, & c \\ b, & d \end{array}\right)\left(\begin{array}{cc} a, & b \\ c, & d \end{array}\right)\left(\begin{array}{cc} a, & c \\ b, & d \end{array}\right)\left(\begin{array}{cc} a, & b \\ c, & d \end{array}\right) = \left(\left(\begin{array}{cc} a, & c \\ b, & d \end{array}\right)\left(\begin{array}{cc} a, & b \\ c, & d \end{array}\right)\right)^2.$$

44. In all that precedes, the matrix of the order 2 has frequently been considered, but chiefly by way of illustration of the general theory; but it is worth while to develope more particularly the theory of such matrix. I call to mind the fundamental properties which have been obtained, viz. it was shown that the matrix

$$M = \left(\begin{array}{cc} a, & b \\ c, & d \end{array}\right),$$

satisfies the equation

$$M^2 - (a+d)M + ad - bc = 0,$$

and that the two matrices

$$\left(\begin{array}{cc} a, & b \\ c, & d \end{array}\right), \quad \left(\begin{array}{cc} a', & b' \\ c', & d' \end{array}\right),$$

will be convertible if

$$a'-d' : b' : c' = a-d : b : c,$$

and that they will be skew convertible if

$$a+d=0, \quad a'+d'=0, \quad aa'+bc'+b'c+dd'=0,$$

the first two of these equations being the conditions in order that the two matrices may be respectively periodic of the second order to a factor *près*.

45. It may be noticed in passing, that if L, M are skew convertible matrices of the order 2, and if these matrices are also such that $L^2=-1$, $M^2=-1$, then putting $N=LM=-ML$, we obtain

$$L^2=-1, \qquad M^2=-1, \qquad N^2=-1,$$
$$L=MN=-NM, \quad M=NL=-NL, \quad N=LM=-ML,$$

which is a system of relations precisely similar to that in the theory of quaternions.

46. The integer powers of the matrix

$$M = \begin{pmatrix} a, & b \\ c, & d \end{pmatrix},$$

are obtained with great facility from the quadratic equation; thus we have, attending first to the positive powers,

$$\begin{aligned} M^2 &= (a+d)\,M - (ad-bc), \\ M^3 &= [(a+d)^2 - (ad-bc)]\,M - (a+d)(ad-bc), \\ &\&c., \end{aligned}$$

whence also the conditions in order that the matrix may be to a factor *près* periodic of the orders 2, 3, &c. are

$$\begin{aligned} a+d &= 0, \\ (a+d)^2 - (ad-bc) &= 0, \\ \&c.; \end{aligned}$$

and for the negative powers we have

$$(ad-bc)\,M^{-1} = -M + (a+d),$$

which is equivalent to the ordinary form

$$(ad-bc)\,M^{-1} = \begin{pmatrix} d, & -b \\ -c, & a \end{pmatrix};$$

and the other negative powers of M can then be obtained by successive multiplications with M^{-1}.

47. The expression for the nth power is however most readily obtained by means of a particular algorithm for matrices of the order 2.

Let h, b, c, J, q be any quantities, and write for shortness $R = -h^2 - 4bc$; suppose also that h', b', c', J', q' are any other quantities, such nevertheless that $h' : b' : c' = h : b : c$, and write in like manner $R' = -h'^2 - 4b'c'$. Then observing that $\frac{h}{\sqrt{R}}, \frac{b}{\sqrt{R}}, \frac{c}{\sqrt{R}}$ are respectively equal to $\frac{h'}{\sqrt{R'}}, \frac{b'}{\sqrt{R'}}, \frac{c'}{\sqrt{R'}}$, the matrix

$$\begin{pmatrix} J\left(\cot q - \frac{h}{\sqrt{R}}\right), & \frac{2bJ}{\sqrt{R}} \\ \frac{2cJ}{\sqrt{R}} & , J\left(\cot q + \frac{h}{\sqrt{R}}\right) \end{pmatrix}$$

contains only the quantities J, q, which are not the same in both systems; and we may therefore represent this matrix by (J, q), and the corresponding matrix with

h', b', c', J', q' by (J', q'). The two matrices are at once seen to be convertible (the assumed relations $h' : b' : c' = h : b : c$ correspond in fact to the conditions,

$$a'-d' : b' : c' = a-d : b : c,$$

of convertibility for the ordinary form), and the compound matrix is found to be

$$\left(\frac{\sin(q+q')}{\sin q \sin q'} JJ',\ q+q'\right);$$

and in like manner the several convertible matrices (J, q), (J', q'), (J'', q'') &c. give the compound matrix

$$\left(\frac{\sin(q+q'+q''\ldots)}{\sin q \sin q' \sin q''\ldots} JJ'J''\ldots,\ q+q'+q''\ldots\right).$$

48. The convertible matrices may be given in the first instance in the ordinary form, or we may take these matrices to be

$$\left(\begin{array}{|cc|} a, & b \\ c, & d \end{array}\right),\quad \left(\begin{array}{|cc|} a', & b' \\ c', & d' \end{array}\right),\quad \left(\begin{array}{|cc|} a'', & b'' \\ c'', & d'' \end{array}\right),\ \text{\&c.}$$

where of course $d-a : b : c = d'-a' : b' : c' = d''-a'' : b'' : c'' =$ &c. Here writing $h = d-a$, and consequently $R = -(d-a)^2 - 4bc$, and assuming also $J = \frac{1}{2}\sqrt{R}$ and $\cot q = \dfrac{d+a}{\sqrt{R}}$, and in like manner for the accented letters, the several matrices are respectively

$$(\tfrac{1}{2}\sqrt{R},\ q)\,(\tfrac{1}{2}\sqrt{R'},\ q'),\ (\tfrac{1}{2}\sqrt{R''},\ q''),\ \text{\&c.},$$

and the compound matrix is

$$\left(\frac{\sin(q+q'+q''\ldots)}{\sin q \sin q' \sin q''\ldots}(\tfrac{1}{2}\sqrt{R})(\tfrac{1}{2}\sqrt{R'})(\tfrac{1}{2}\sqrt{R''})\ldots,\ \ q+q'+q''+\ldots\right).$$

49. When the several matrices are each of them equal to

$$\left(\begin{array}{|cc|} a, & b \\ c, & d \end{array}\right),$$

we have of course $q = q' = q''\ldots$, $R = R' = R''\ldots$, and we find

$$\left(\begin{array}{|cc|} a, & b \\ c, & d \end{array}\right)^n = \left(\frac{\sin nq}{\sin^n q}(\tfrac{1}{2}\sqrt{R})^n,\ nq\right);$$

or substituting for the right-hand side, the matrix represented by this notation, and putting for greater simplicity

$$\frac{\sin nq}{\sin^n q}(\tfrac{1}{2}\sqrt{R})^n = (\tfrac{1}{2}\sqrt{R})\,L, \text{ or } L = \frac{\sin nq}{\sin^n q}(\tfrac{1}{2}\sqrt{R})^{n-1},$$

we find

$$\left(\begin{array}{cc} a, & b \\ c, & d \end{array}\right)^n = \left(\begin{array}{cc} \tfrac{1}{2}L\left(\sqrt{R}\cot nq-(d-a)\right), & Lb \\ Lc & , \tfrac{1}{2}L\left(\sqrt{R}\cot nq+(d-a)\right) \end{array}\right)$$

where it will be remembered that

$$R=-(d-a)^2-4bc \text{ and } \cot q=\frac{d+a}{\sqrt{R}},$$

the last of which equations may be replaced by

$$\cos q+\sqrt{-1}\sin q=\frac{d+a+\sqrt{-R}}{2\sqrt{ad-bc}}.$$

The formula in fact extends to negative or fractional values of the index n, and when n is a fraction, we must, as usual, in order to exhibit the formula in its proper generality, write $q+2m\pi$ instead of q. In the particular case $n=\frac{1}{2}$, it would be easy to show the identity of the value of the square root of the matrix with that before obtained by a different process.

50. The matrix will be to a factor *près*, periodic of the nth order if only $\sin nq=0$, that is, if $q=\frac{m\pi}{n}$ (m must be prime to n, for if it were not, the order of periodicity would be not n itself, but a submultiple of n); but $\cos q=\frac{d+a}{2\sqrt{ad-bc}}$, and the condition is therefore

$$(d+a)^2-4(ad-bc)\cos^2\frac{m\pi}{n}=0,$$

or as this may also be written,

$$d^2+a^2-2ad\cos\frac{2m\pi}{n}+4bc\cos^2\frac{m\pi}{n}=0,$$

a result which agrees with those before obtained for the particular values 2 and 3 of the index of periodicity.

51. I may remark that the last preceding investigations are intimately connected with the investigations of Babbage and others in relation to the function $\phi x=\frac{ax+b}{cx+d}$.

I conclude with some remarks upon rectangular matrices.

52. A matrix such as

$$\left(\begin{array}{ccc} a, & b, & c \\ a', & b', & c' \end{array}\right)$$

where the number of columns exceeds the number of lines, is said to be a broad matrix; a matrix such as

$$\left(\begin{array}{cc} a, & b \\ a', & b' \\ a'', & b'' \end{array}\right)$$

where the number of lines exceeds the number of columns, is said to be a deep matrix.

53. The matrix zero subsists in the present theory, but not the matrix unity. Matrices may be added or subtracted when the number of the lines and the number of the columns of the one matrix are respectively equal to the number of the lines and the number of the columns of the other matrix, and under the like condition any number of matrices may be added together. Two matrices may be equal or opposite, the one to the other. A matrix may be multiplied by a single quantity, giving rise to a matrix of the same form; two matrices so related are similar to each other.

54. The notion of composition applies to rectangular matrices, but it is necessary that the number of lines in the second or nearer component matrix should be equal to the number of columns in the first or further component matrix; the compound matrix will then have as many lines as the first or further component matrix, and as many columns as the second or nearer component matrix.

55. As examples of the composition of rectangular matrices, we have

$$\left(\begin{array}{l} a, b, c \\ d, e, f \end{array} \between \begin{array}{l} a', b', c', d' \\ e', f', g', h' \\ i', j', k', l' \end{array}\right) = \left(\begin{array}{l} (a, b, c \between a', e', i'),\ (a, b, c \between b', f', j')\ (a, b, c \between c', g', k'),\ (a, b, c \between d', h'\ l') \\ (d, e, f \between a', e', i'),\ (d, e, f \between b', f', j')\ (d, e, f \between c', g', k'),\ (d, e, f \between d', h'\ l') \end{array}\right),$$

and

$$\left(\begin{array}{l} a, d \\ b, e \\ c, f \end{array} \between \begin{array}{l} a', b', c', d' \\ e', f', g', h' \end{array}\right) = \left(\begin{array}{l} (a, d \between a', e'),\ (a, d \between b', f'),\ (a, d \between c', g'),\ (a, d \between d', h') \\ (b, e \between a', e'),\ (b, e \between b', f'),\ (b, e \between c', g'),\ (b, e \between d', h') \\ (c, f \between a', e'),\ (c, f \between b', f'),\ (c, f \between c', g'),\ (c, f \between d', h') \end{array}\right).$$

56. In the particular case where the lines and columns of the one component matrix are respectively equal in number to the columns and lines of the other component matrix, the compound matrix is square, thus we have

$$\left(\begin{array}{l} a, b, c \\ d, e, f \end{array} \between \begin{array}{l} a', d' \\ b', e' \\ c', f' \end{array}\right) = \left(\begin{array}{l} (a, b, c \between a', b', c'),\ (a, b, c \between d', e', f') \\ (d, e, f \between a', b', c'),\ (d, e, f \between d', e', f') \end{array}\right)$$

and

$$\left(\begin{array}{l} a', d' \\ b', e' \\ c', f' \end{array} \between \begin{array}{l} a, b, c \\ d, e, f \end{array}\right) = \left(\begin{array}{l} (a', d' \between a, d),\ (a', d' \between b, e),\ (a', d' \between c, f) \\ (b', e' \between a, d),\ (b', e' \between b, e),\ (b', e' \between c, f) \\ (c', f' \between a, d),\ (c', f' \between b, e),\ (c', f' \between c, f) \end{array}\right)$$

The two matrices in the case last considered admit of composition in the two different orders of arrangement, but as the resulting square matrices are not of the same order, the notion of the convertibility of two matrices does not apply even to the case in question.

57. Since a rectangular matrix cannot be compounded with itself, the notions of the inverse or reciprocal matrix and of the powers of the matrix and the whole resulting theory of the functions of a matrix, do not apply to rectangular matrices.

58. The notion of transposition and the symbol tr. apply to rectangular matrices, the effect of a transposition being to convert a broad matrix into a deep one and reciprocally. It may be noticed that the symbol tr. may be used for the purpose of expressing the law of composition of square or rectangular matrices. Thus treating (a, b, c) as a rectangular matrix, or representing it by $\begin{pmatrix} a, & b, & c \end{pmatrix}$, we have

$$\text{tr.}\begin{pmatrix} a', & b', & c' \end{pmatrix} = \begin{pmatrix} a' \\ b' \\ c' \end{pmatrix},$$

and thence

$$\begin{pmatrix} a, & b, & c \end{pmatrix} \text{tr.}\begin{pmatrix} a', & b', & c' \end{pmatrix} = \left(a, b, c \between \begin{matrix} a' \\ b' \\ c' \end{matrix}\right) = (a, b, c \between a', b', c'),$$

so that the symbol

$$(a, b, c \between a', b', c')$$

would upon principle be replaced by

$$\begin{pmatrix} a, & b, & c \end{pmatrix} \text{tr.}\begin{pmatrix} a', & b', & c' \end{pmatrix}:$$

it is however more convenient to retain the symbol

$$(a, b, c \between a', b', c').$$

Hence introducing the symbol tr. only on the left-hand sides, we have

$$\begin{pmatrix} a, & b, & c \\ d, & e, & f \end{pmatrix} \text{tr.}\begin{pmatrix} a', & b', & c' \\ d', & e', & f' \end{pmatrix} = \begin{pmatrix} (a, b, c \between a', b', c'), & (a, b, c \between d', e', f') \\ (d, e, f \between a', b', c'), & (d, e, f \between d', e', f') \end{pmatrix},$$

or to take an example involving square matrices,

$$\begin{pmatrix} a, & b \\ d, & e \end{pmatrix} \text{tr.}\begin{pmatrix} a', & b' \\ d', & e' \end{pmatrix} = \begin{pmatrix} (a, b \between a', b'), & (a, b \between d', e') \\ (d, e \between a', b'), & (d, e \between d', e') \end{pmatrix};$$

it thus appears that in the composition of matrices (square or rectangular), when the second or nearer component matrix is expressed as a matrix preceded by the symbol tr., any *line* of the compound matrix is obtained by compounding the corresponding *line* of the first or further component matrix successively with the several *lines* of the matrix which preceded by tr. gives the second or nearer component matrix. It is clear that the terms 'symmetrical' and 'skew symmetrical' do not apply to rectangular matrices.

153.

A MEMOIR ON THE AUTOMORPHIC LINEAR TRANSFORMATION OF A BIPARTITE QUADRIC FUNCTION.

[From the *Philosophical Transactions of the Royal Society of London*, vol. CXLVIII. *for the year* 1858, pp. 39—46. Received December 10, 1857,—Read January 14, 1858.]

THE question of the automorphic linear transformation of the function $x^2+y^2+z^2$, that is the transformation by linear substitutions, of this function into a function $x_{,}^2+y_{,}^2+z_{,}^2$ of the same form, is in effect solved by some formulæ of Euler's for the transformation of coordinates, and it was by these formulæ that I was led to the solution in the case of the sum of n squares, given in my paper "Sur quelques propriétés des déterminants gauches"(1). A solution grounded upon an *à priori* investigation and for the case of any quadric function of n variables, was first obtained by M. Hermite in the memoir "Remarques sur une Mémoire de M. Cayley relatif aux déterminants gauches"(2). This solution is in my Memoir "Sur la transformation d'une function quadratique en elle-même par des substitutions linéaires"(3), presented under a somewhat different form involving the notation of matrices. I have since found that there is a like transformation of a bipartite quadric function, that is a lineo-linear function of two distinct sets, each of the same number of variables, and the development of the transformation is the subject of the present memoir.

1. For convenience, the number of variables is in the analytical formulæ taken to be 3, but it will be at once obvious that the formulæ apply to any number of variables whatever. Consider the bipartite quadric

$$\left(\begin{array}{ccc} a, & b, & c \\ a', & b', & c' \\ a'', & b'', & c'' \end{array}\right)\!\!\big(x,\ y,\ z\big)\!\big(\mathrm{x},\ \mathrm{y},\ \mathrm{z}\big),$$

[1] *Crelle*, t. XXXII. (1846) pp. 119—123, [52].

[2] *Cambridge and Dublin Mathematical Journal*, t. IX. (1854) pp. 63—67.

[3] *Crelle*, t. L. (1855) pp. 288—299, [136].

which stands for

$$\begin{aligned} &(ax \;\;+by \;\;+cz\;\;)\,\mathrm{x} \\ +\,&(a'x \;+b'y \;+c'z\;)\,\mathrm{y} \\ +\,&(a''x+b''y+c''z)\,\mathrm{z}, \end{aligned}$$

and in which $(x,\ y,\ z)$ are said to be the nearer variables, and $(\mathrm{x},\ \mathrm{y},\ \mathrm{z})$ the further variables of the bipartite.

2. It is clear that we have

$$\left(\begin{array}{|ccc|} a, & b, & c \\ a', & b', & c' \\ a'', & b'', & c'' \end{array}\right)\between x,\ y,\ z\between \mathrm{x},\ \mathrm{y},\ \mathrm{z}) = \left(\begin{array}{|ccc|} a, & a', & a'' \\ b, & b', & b'' \\ c, & c', & c'' \end{array}\right)\between \mathrm{x},\ \mathrm{y},\ \mathrm{z}\between x,\ y,\ z)$$

and the new form on the right-hand side of the equation may also be written

$$\left(\mathrm{tr.}\left(\begin{array}{|ccc|} a, & b, & c \\ a', & b', & c' \\ a'', & b'', & c'' \end{array}\right)\between \mathrm{x},\ \mathrm{y},\ \mathrm{z}\between x,\ y,\ z\right),$$

that is, the two sets of variables may be interchanged, provided that the matrix is transposed.

3. Each set of variables may be linearly transformed: suppose that the substitutions are

$$(x,\ y,\ z) = \left(\begin{array}{|ccc|} l, & m, & n \\ l', & m', & n' \\ l'', & m'', & n'' \end{array}\right)\between x_{\prime},\ y_{\prime},\ z_{\prime})$$

and

$$(\mathrm{x},\ \mathrm{y},\ \mathrm{z}) = \left(\begin{array}{|ccc|} \mathrm{l}, & \mathrm{l}', & \mathrm{l}'' \\ \mathrm{m}, & \mathrm{m}', & \mathrm{m}'' \\ \mathrm{n}, & \mathrm{n}', & \mathrm{n}'' \end{array}\right)\between \mathrm{x}_{\prime},\ \mathrm{y}_{\prime},\ \mathrm{z}_{\prime}).$$

Then first substituting for $(x,\ y,\ z)$ their values in terms of $(x_{\prime},\ y_{\prime},\ z_{\prime})$, the bipartite becomes

$$\left(\left(\begin{array}{|ccc|} a, & b, & c \\ a', & b', & c' \\ a'', & b'', & c'' \end{array}\right)\between\left(\begin{array}{|ccc|} l, & m, & n \\ l', & m', & n' \\ l'', & m'', & n'' \end{array}\right)\between x_{\prime},\ y_{\prime},\ z_{\prime}\between \mathrm{x},\ \mathrm{y},\ \mathrm{z}\right);$$

represent for a moment this expression by

$$\left(\begin{array}{|ccc|} A, & B, & C \\ A', & B', & C' \\ A'', & B'', & C'' \end{array}\right)\between x_{\prime},\ y_{\prime},\ z_{\prime}\between \mathrm{x},\ \mathrm{y},\ \mathrm{z}),$$

then substituting for (x, y, z) their values in terms of $(\mathrm{x}_{,}, \mathrm{y}_{,}, \mathrm{z}_{,})$, it is easy to see that the expression becomes

$$\left(\left(\begin{array}{ccc} \mathrm{l}, & \mathrm{m}, & \mathrm{n} \\ \mathrm{l}', & \mathrm{m}', & \mathrm{n}' \\ \mathrm{l}'', & \mathrm{m}'', & \mathrm{n}'' \end{array}\right)\!\!\!\between\!\!\! \left|\begin{array}{ccc} A, & B, & C \\ A', & B', & C' \\ A'', & B'', & C'' \end{array}\right| \right)\!\!\between x_{,}, y_{,}, z_{,} \!\between \mathrm{x}_{,}, \mathrm{y}_{,}, \mathrm{z}_{,}),$$

and re-establishing the value of the auxiliary matrix, we obtain, as the final result of the substitutions,

$$\left(\begin{array}{ccc} a, & b, & c \\ a', & b', & c' \\ a'', & b'', & c'' \end{array}\right)\!\!\between x, y, z \!\between \mathrm{x}, \mathrm{y}, \mathrm{z}) = \left(\left(\begin{array}{ccc} \mathrm{l}, & \mathrm{m}, & \mathrm{n} \\ \mathrm{l}', & \mathrm{m}', & \mathrm{n}' \\ \mathrm{l}'', & \mathrm{m}'', & \mathrm{n}'' \end{array}\right)\!\!\between\!\! \left|\begin{array}{ccc} a, & b, & c \\ a', & b', & c' \\ a'', & b'', & c'' \end{array}\right| \!\!\between\!\! \left|\begin{array}{ccc} l, & m, & n \\ l', & m', & n' \\ l'', & m'', & n'' \end{array}\right| \right)\!\!\between x_{,}, y_{,}, z_{,} \!\between \mathrm{x}_{,}, \mathrm{y}_{,}, \mathrm{z}_{,}),$$

that is, the matrix of the transformed bipartite is obtained by compounding in order, first or furthest the transposed matrix of substitution of the further variables, next the matrix of the bipartite, and last or nearest the matrix of substitution of the nearer variables.

4. Suppose now that it is required to find the automorphic linear transformation of the bipartite

$$\left(\begin{array}{ccc} a, & b, & c \\ a', & b', & c' \\ a'', & b'', & c'' \end{array}\right)\!\!\between x, y, z \!\between \mathrm{x}, \mathrm{y}, \mathrm{z}),$$

or as it will henceforward for shortness be written,

$$(\Omega \between x, y, z \between \mathrm{x}, \mathrm{y}, \mathrm{z});$$

this may be effected by a method precisely similar to that employed by M. Hermite for an ordinary quadric. For this purpose write

$$x + x_{,} = 2\xi, \quad y + y_{,} = 2\eta, \quad z + z_{,} = 2\zeta,$$
$$\mathrm{x} + \mathrm{x}_{,} = 2\Xi, \quad \mathrm{y} + \mathrm{y}_{,} = 2\mathrm{H}, \quad \mathrm{z} + \mathrm{z}_{,} = 2\mathrm{Z},$$

or, as these equations may be represented,

$$(x + x_{,}, \quad y + y_{,}, \quad z + z_{,}) = 2(\xi, \eta, \zeta),$$
$$(\mathrm{x} + \mathrm{x}_{,}, \quad \mathrm{y} + \mathrm{y}_{,}, \quad \mathrm{z} + \mathrm{z}_{,}) = 2(\Xi, \mathrm{H}, \mathrm{Z});$$

then we ought to have

$$(\Omega \between 2\xi - x, 2\eta - y, 2\zeta - z \between 2\Xi - \mathrm{x}, 2\mathrm{H} - \mathrm{y}, 2\mathrm{Z} - \mathrm{z}) = (\Omega \between x, y, z \between \mathrm{x}, \mathrm{y}, \mathrm{z}).$$

5. The left-hand side is

$$4(\Omega \between \xi, \eta, \zeta \between \Xi, \mathrm{H}, \mathrm{Z}) - 2(\Omega \between x, y, z \between \Xi, \mathrm{H}, \mathrm{Z}) - 2(\Omega \between \xi, \eta, \zeta \between \mathrm{x}, \mathrm{y}, \mathrm{z}) + \Omega \between x, y, z \between \mathrm{x}, \mathrm{y}, \mathrm{z}),$$

and the equation becomes

$$2(\Omega \between \xi, \eta, \zeta \between \Xi, \mathrm{H}, \mathrm{Z}) - (\Omega \between x, y, z \between \Xi, \mathrm{H}, \mathrm{Z}) - (\Omega \between \xi, \eta, \zeta \between \mathrm{x}, \mathrm{y}, \mathrm{z}) = 0,$$

or as it may be written,

$$\left.\begin{array}{r}(\Omega\between\xi,\ \eta,\ \zeta\between\Xi,\ \mathrm{H},\ \mathrm{Z})-(\Omega\between x,\ y,\ z\between\Xi,\ \mathrm{H},\ \mathrm{Z})\\ +(\Omega\between\xi,\ \eta,\ \zeta\between\Xi,\ \mathrm{H},\ \mathrm{Z})-(\Omega\between\xi,\ \eta,\ \zeta\between\mathrm{x},\ \mathrm{y},\ \mathrm{z})\end{array}\right\}=0;$$

or again,

$$\left.\begin{array}{r}(\Omega\between\xi-x,\ \eta-y,\ \zeta-z\between\Xi,\ \mathrm{H},\ \mathrm{Z})\\ +(\Omega\between\xi,\ \eta,\ \zeta\between\Xi-\mathrm{x},\ \mathrm{H}-\mathrm{y},\ \mathrm{Z}-\mathrm{z})\end{array}\right\}=0,$$

or what is the same thing,

$$\left.\begin{array}{r}(\Omega\between\xi-x,\ \eta-y,\ \zeta-z\between\Xi,\ \mathrm{H},\ \mathrm{Z})\\ +(\text{tr. }\Omega\between\Xi-\mathrm{x},\ \mathrm{H}-\mathrm{y},\ \mathrm{Z}-\mathrm{z}\between\xi,\ \eta,\ \zeta)\end{array}\right\}=0,$$

and it is easy to see that the equation will be satisfied by writing

$$\begin{aligned}(\ \Omega\between\xi-x,\ \eta-y,\ \zeta-z)&=(\ \Upsilon\between\xi,\ \eta,\ \zeta),\\ (\text{tr. }\Omega\between\Xi-\mathrm{x},\ \mathrm{H}-\mathrm{y},\ \mathrm{Z}-\mathrm{z})&=-(\text{tr. }\Upsilon\between\Xi,\ \mathrm{H},\ \mathrm{Z}),\end{aligned}$$

where Υ is any arbitrary matrix. In fact we have then

$$\begin{aligned}(\ \Omega\between\xi-x,\ \eta-y,\ \zeta-z\between\Xi,\ \mathrm{H},\ \mathrm{Z})&=(\ \Upsilon\between\xi,\ \eta,\ \zeta\between\Xi,\ \mathrm{H},\ \mathrm{Z}),\\ (\text{tr. }\Omega\between\Xi-\mathrm{x},\ \mathrm{H}-\mathrm{y},\ \mathrm{Z}-\mathrm{z}\between\xi,\ \eta,\ \zeta)&=-(\text{tr. }\Upsilon\between\Xi,\ \mathrm{H},\ \mathrm{Z}\between\xi,\ \eta,\ \zeta)\\ &=-(\ \Upsilon\between\xi,\ \eta,\ \zeta\between\Xi,\ \mathrm{H},\ \mathrm{Z}),\end{aligned}$$

and the sum of the two terms consequently vanishes.

6. The equation

$$(\Omega\between\xi-x,\ \eta-y,\ \zeta-z)=(\Upsilon\between\xi,\ \eta,\ \zeta)$$

gives

$$(\Omega-\Upsilon\between\xi,\ \eta,\ \zeta)=(\Omega\between x,\ y,\ z),$$

and we then have

$$(\Omega+\Upsilon\between\xi,\ \eta,\ \zeta)=(\Omega\between x_{,},\ y_{,},\ z_{,}).$$

In fact the two equations give

$$(2\Omega\between\xi,\ \eta,\ \zeta)=(\Omega\between x+x_{,},\ y+y_{,},\ z+z_{,}),$$

or what is the same thing,

$$2(\xi,\ \eta,\ \zeta)=(x+x_{,},\ y+y_{,},\ z+z_{,}),$$

which is the equation assumed as the definition of $(\xi,\ \eta,\ \zeta)$; and conversely, this equation, combined with either of the two equations, gives the other of them.

7. We have consequently

$$\begin{aligned}(x,\ y,\ z)&=(\Omega^{-1}(\Omega-\Upsilon)\between\xi,\ \eta,\ \zeta),\\ (\xi,\ \eta,\ \zeta)&=((\Omega+\Upsilon)^{-1}\,\Omega\between x_{,},\ y_{,},\ z_{,}),\end{aligned}$$

and thence

$$(x,\ y,\ z)=(\Omega^{-1}(\Omega-\Upsilon)(\Omega+\Upsilon)^{-1}\,\Omega\between x_{,},\ y_{,},\ z_{,}).$$

8. But in like manner the equation

$$(\text{tr. }\Omega\between\Xi-\mathrm{x},\ \mathrm{H}-\mathrm{y},\ \mathrm{Z}-\mathrm{z})=-(\text{tr. }\Upsilon\between\Xi,\ \mathrm{H},\ \mathrm{Z})$$

gives

$$(\text{tr. }\overline{\Omega+\Upsilon}\between\Xi,\ \mathrm{H},\ \mathrm{Z})=(\text{tr. }\Omega\between\mathrm{x},\ \mathrm{y},\ \mathrm{z}),$$

and we then obtain

$$(\text{tr. }\overline{\Omega-\Upsilon}\between\Xi,\ \mathrm{H},\ \mathrm{Z})=(\text{tr. }\Omega\between\mathrm{x}_{,},\ \mathrm{y}_{,},\ \mathrm{z}_{,}).$$

9. In fact these equations give

$$(\text{tr. }2\Omega\between\Xi,\ \mathrm{H},\ \mathrm{Z})=(\text{tr. }\Omega\between\mathrm{x}+\mathrm{x}_{,},\ \mathrm{y}+\mathrm{y}_{,},\ \mathrm{z}+\mathrm{z}_{,}),$$

or

$$2(\Xi,\ \mathrm{H},\ \mathrm{Z})=(\mathrm{x}+\mathrm{x}_{,},\ \mathrm{y}+\mathrm{y}_{,},\ \mathrm{z}+\mathrm{z}_{,});$$

and conversely, this equation, combined with either of the two equations, gives the other of them. We have then

$$(\mathrm{x}\ ,\ \mathrm{y}\ ,\ \mathrm{z}\ )=((\text{tr. }\Omega)^{-1}(\text{tr. }\overline{\Omega+\Upsilon}\between\Xi,\ \mathrm{H},\ \mathrm{Z}),$$

$$(\Xi,\ \mathrm{H},\ \mathrm{Z})=((\text{tr. }\overline{\Omega-\Upsilon})^{-1}\text{ tr. }\Omega\between\mathrm{x}_{,},\ \mathrm{y}_{,},\ \mathrm{z}_{,}),$$

and thence

$$(\mathrm{x},\ \mathrm{y},\ \mathrm{z})=((\text{tr. }\Omega)^{-1}(\text{tr. }\overline{\Omega+\Upsilon})(\text{tr. }\overline{\Omega-\Upsilon})^{-1}\text{tr. }\Omega\between\mathrm{x}_{,},\ \mathrm{y}_{,},\ \mathrm{z}_{,}).$$

10. Hence, recapitulating, we have the following theorem for the automorphic linear transformation of the bipartite

$$(\Omega\between x,\ y,\ z\between\mathrm{x},\ \mathrm{y},\ \mathrm{z}),$$

viz. Υ being an arbitrary matrix, if

$$(x,\ y,\ z)=(\Omega^{-1}(\Omega-\Upsilon)(\Omega+\Upsilon)^{-1}\Omega\between x_{,},\ y_{,},\ z_{,}),$$

$$(\mathrm{x},\ \mathrm{y},\ \mathrm{z})=((\text{tr. }\Omega)^{-1}(\text{tr. }\overline{\Omega+\Upsilon})(\text{tr. }\overline{\Omega-\Upsilon})^{-1}\text{ tr. }\Omega\between\mathrm{x}_{,},\ \mathrm{y}_{,},\ \mathrm{z}_{,}),$$

then

$$(\Omega\between x,\ y,\ z\between\mathrm{x},\ \mathrm{y},\ \mathrm{z})=(\Omega\between x_{,},\ y_{,},\ z_{,}\between\mathrm{x}_{,},\ \mathrm{y}_{,},\ \mathrm{z}),$$

which is the theorem in question.

11. I have thought it worth while to preserve the foregoing investigation, but the most simple demonstration is the verification *à posteriori* by the actual substitution of the transformed values of $(x,\ y,\ z)$, $(\mathrm{x},\ \mathrm{y},\ \mathrm{z})$. To effect this, recollecting that in general $\text{tr. }(A^{-1})=(\text{tr. }A)^{-1}$ and $\text{tr. }ABCD=\text{tr. }D.\ \text{tr. }C.\ \text{tr. }B.\ \text{tr. }A$, the transposed matrix of substitution for the further variables is

$$\Omega\,(\Omega-\Upsilon)^{-1}(\Omega+\Upsilon)\,\Omega^{-1};$$

and compounding this with the matrix Ω of the bipartite, and the matrix

$$\Omega^{-1}\,(\Omega-\Upsilon)(\Omega+\Upsilon)^{-1}\,\Omega$$

of substitution for the nearer variables, the theorem will be verified if the result is equal to the matrix Ω of the bipartite; that is, we ought to have

$$\Omega(\Omega-\Upsilon)^{-1}(\Omega+\Upsilon)\Omega^{-1}\Omega\Omega^{-1}(\Omega-\Upsilon)(\Omega+\Upsilon)^{-1}\,\Omega=\Omega,$$

or what is the same thing,

$$\Omega(\Omega-\Upsilon)^{-1}(\Omega+\Upsilon)\,\Omega^{-1}(\Omega-\Upsilon)(\Omega+\Upsilon)^{-1}\Omega=\Omega;$$

this is successively reducible to

$$\begin{aligned}(\Omega+\Upsilon)\Omega^{-1}(\Omega-\Upsilon) &= (\Omega-\Upsilon)\Omega^{-1}(\Omega+\Upsilon),\\ \Omega^{-1}(\Omega+\Upsilon)\,\Omega^{-1}(\Omega-\Upsilon) &= \Omega^{-1}(\Omega-\Upsilon)\Omega^{-1}(\Omega+\Upsilon),\\ (1+\Omega^{-1}\Upsilon)(1-\Omega^{-1}\Upsilon) &= (1-\Omega^{-1}\Upsilon)(1+\Omega^{-1}\Upsilon),\end{aligned}$$

which is a mere identity, and the theorem is thus shown to be true.

12. It is to be observed that, in the general theorem, the transformations or matrices of substitution for the two sets of variables respectively are not identical, but it may be required that this shall be so. Consider first the case where the matrix Ω is symmetrical, the necessary condition is that the matrix Υ shall be skew symmetrical; in fact we have then

$$\text{tr. }\Omega=\Omega,\ \ \text{tr. }\Upsilon=-\Upsilon,$$

and the transformations become

$$\begin{aligned}(x,\ y,\ z) &= (\Omega^{-1}(\Omega-\Upsilon)(\Omega+\Upsilon)^{-1}\Omega \between x_{,},\ y_{,},\ z_{,}),\\ (\mathrm{x},\ \mathrm{y},\ \mathrm{z}) &= (\Omega^{-1}(\Omega-\Upsilon)(\Omega+\Upsilon)^{-1}\Omega \between \mathrm{x}_{,},\ \mathrm{y}_{,},\ \mathrm{z}_{,}),\end{aligned}$$

which are identical. We may in this case suppose that the two sets of variables become equal, and we have then the theorem for the automorphic linear transformation of the ordinary quadric

$$(\Omega \between x,\ y,\ z)^2,$$

viz. Υ being a skew symmetrical matrix, if

$$(x,\ y,\ z)=(\Omega^{-1}(\Omega-\Upsilon)(\Omega+\Upsilon)^{-1}\Omega \between x_{,},\ y_{,},\ z_{,}),$$

then

$$(\Omega \between x,\ y,\ z)^2=(\Omega \between x_{,},\ y_{,},\ z_{,})^2.$$

13. Next, if the matrix Ω be skew symmetrical, the condition is that the matrix Υ shall be symmetrical; we have in this case tr. $\Omega=-\Omega$, tr. $\Upsilon=\Upsilon$, and the four factors in the matrix of substitution for (x, y, z) are respectively $-\Omega^{-1}$, $-(\Omega-\Upsilon)$, $-(\Omega+\Upsilon)^{-1}$ and $-\Omega$, and such matrix of substitution becomes therefore, as before, identical with that for $(x,\ y,\ z)$; we have therefore the following theorem for the automorphic linear transformation of a skew symmetrical bipartite

$$(\Omega \between x,\ y,\ z \between \mathrm{x},\ \mathrm{y},\ \mathrm{z}),$$

when the transformations for the two sets of variables are identical, viz. Υ being any symmetrical matrix, if

$$(x,\ y,\ z) = (\Omega^{-1}(\Omega - \Upsilon)(\Omega + \Upsilon)^{-1}\Omega \between x_{,},\ y_{,},\ z_{,}),$$

$$(\mathrm{x},\ \mathrm{y},\ \mathrm{z}) = (\Omega^{-1}(\Omega - \Upsilon)(\Omega + \Upsilon)^{-1}\Omega \between \mathrm{x}_{,},\ \mathrm{y}_{,},\ \mathrm{z}_{,}),$$

then

$$(\Omega \between x,\ y,\ z \between \mathrm{x},\ \mathrm{y},\ \mathrm{z}) = (\Omega \between x_{,},\ y_{,},\ z_{,} \between \mathrm{x}_{,},\ \mathrm{y}_{,},\ \mathrm{z}_{,}).$$

14. Lastly, in the general case where the matrix Ω is anything whatever, the condition is

$$\Omega^{-1}\Upsilon = -(\text{tr. }\Omega)^{-1}\text{ tr. }\Upsilon$$

for assuming this equation, then first

$$\Omega^{-1}(\Omega - \Upsilon) = (\text{tr. }\Omega)^{-1}(\text{tr. }\overline{\Omega + \Upsilon}),$$

and in like manner

$$\Omega^{-1}(\Omega + \Upsilon) = (\text{tr. }\Omega)^{-1}(\text{tr. }\overline{\Omega - \Upsilon}).$$

But we have

$$1 = (\text{tr. }\Omega)^{-1}(\text{tr. }\overline{\Omega - \Upsilon})(\text{tr. }\overline{\Omega - \Upsilon})^{-1}\text{ tr. }\Omega,$$

and therefore, secondly,

$$(\Omega + \Upsilon)^{-1}\Omega = (\text{tr. }\overline{\Omega - \Upsilon})^{-1}\text{ tr. }\Omega\ ;$$

and thence

$$\Omega^{-1}(\Omega - \Upsilon)(\Omega + \Upsilon)^{-1}\Omega = (\text{tr. }\Omega)^{-1}(\text{tr. }\overline{\Omega + \Upsilon})(\text{tr. }\overline{\Omega - \Upsilon})^{-1}\text{tr. }\Omega,$$

or the two transformations are identical.

15. To further develope this result, let Ω^{-1} be expressed as the sum of a symmetrical matrix Q_0 and a skew symmetrical matrix $Q_{,}$, and let Υ be expressed in like manner as the sum of a symmetrical matrix Υ_0 and a skew symmetrical matrix $\Upsilon_{,}$. We have then

$$\begin{aligned} \Omega^{-1} &= Q_0 + Q_{,}, \\ (\text{tr. }\Omega)^{-1} = \text{tr. }(\Omega^{-1}) &= Q_0 - Q_{,}, \\ \Upsilon &= \Upsilon_0 + \Upsilon_{,}, \\ \text{tr. }\Upsilon &= \Upsilon_0 - \Upsilon_{,}, \end{aligned}$$

and the condition, $\Omega^{-1}\Upsilon = -(\text{tr. }\Omega)^{-1}\text{ tr. }\Upsilon$, becomes

$$(Q_0 + Q_{,})(\Upsilon_0 + \Upsilon_{,}) = -(Q_0 - Q_{,})(\Upsilon_0 - \Upsilon_{,}),$$

that is,

$$Q_0\Upsilon_0 + Q_{,}\Upsilon_{,} = 0,$$

and we have

$$\Upsilon_0 = -Q_0^{-1}Q_{,}\Upsilon_{,},$$

or as we may write it,

$$\Upsilon_0 = -(\tfrac{1}{2}\{\Omega^{-1} + \text{tr.}\,\Omega^{-1}\})^{-1}(\tfrac{1}{2}\{\Omega^{-1} - \text{tr.}\,\Omega^{-1}\})\Upsilon_,,$$

and thence

$$\Upsilon = -(\tfrac{1}{2}\{\Omega^{-1} + \text{tr.}\,\Omega^{-1}\})^{-1}(\tfrac{1}{2}\{\Omega^{-1} - \text{tr.}\,\Omega^{-1}\})\Upsilon_, \;+\Upsilon_,,$$

where $\Upsilon_,$ is an arbitrary skew symmetrical matrix.

16. This includes the before-mentioned special cases; first, if Ω is symmetrical, then we have simply $\Upsilon = \Upsilon_,$, an arbitrary skew symmetrical matrix, which is right. Next, if Ω is skew symmetrical, then $\Upsilon = -0^{-1}\Omega^{-1}\Upsilon_, + \Upsilon_,$, which can only be finite for $\Upsilon_, = 0$, that is, we have $\Upsilon = -0^{-1}\Omega^{-1}0$, and (the first part of Υ being always symmetrical) this represents an arbitrary symmetrical matrix. The mode in which this happens will be best seen by an example. Suppose

$$\Omega^{-1} = \left(\begin{array}{cc} A\;, & H+\nu \\ H-\nu, & B \end{array}\right),\quad \text{tr.}\,\Omega^{-1} = \left(\begin{array}{cc} A\;, & H-\nu \\ H+\nu, & B \end{array}\right),$$

and write

$$\Upsilon_, = \left(\begin{array}{cc} 0, & \theta \\ -\theta, & 0 \end{array}\right),$$

then we have

$$\Upsilon = -\left(\begin{array}{cc} A, & H \\ H, & B \end{array}\right)^{-1}\left(\begin{array}{cc} 0, & \nu \\ -\nu, & 0 \end{array}\right)\left(\begin{array}{cc} 0, & \theta \\ -\theta, & 0 \end{array}\right) + \left(\begin{array}{cc} 0, & \theta \\ -\theta, & 0 \end{array}\right)$$

$$= -\frac{\nu\theta}{AB-H^2}\left(\begin{array}{cc} -B, & H \\ H, & -A \end{array}\right) + \left(\begin{array}{cc} 0, & \theta \\ -\theta, & 0 \end{array}\right)$$

$$= \left(\begin{array}{cc} \dfrac{\nu B\theta}{AB-H^2}\;, & \dfrac{-\nu H\theta}{AB-H^2} + \theta \\ \dfrac{-\nu H\theta}{AB-H^2} - \theta, & \dfrac{\nu A\theta}{AB-H^2} \end{array}\right) + \left(\begin{array}{cc} 0, & \theta \\ -\theta, & 0 \end{array}\right).$$

When Ω is skew symmetrical, A, B, H vanish; but since their ratios remain arbitrary, we may write κA, κB, κH for A, B, H, and assume ultimately $\kappa = 0$. Writing $\kappa\theta$ in the place of θ, and then putting $\kappa = 0$, the matrix becomes

$$\left(\begin{array}{cc} \dfrac{\nu B\theta}{AB-H^2}, & \dfrac{-\nu B\theta}{AB-H^2} \\ \dfrac{-\nu H\theta}{AB-H^2}, & \dfrac{\nu A\theta}{AB-H^2} \end{array}\right)$$

which, inasmuch as $A : \theta$, $B : \theta$, and $C : \theta$ remain arbitrary, represents, as it should do, an arbitrary symmetrical matrix.

17. Hence, finally, we have the following Theorem for the automorphic linear transformation of the bipartite quadric,

$$(\Omega \between x,\ y,\ z \between \mathrm{x},\ \mathrm{y},\ \mathrm{z}),$$

when the two transformations are identical, viz. if $\Upsilon_{,}$ be a skew symmetrical matrix, and if

$$\Upsilon = -\left(\tfrac{1}{2}\{\Omega^{-1} + \text{tr. } \Omega^{-1}\}\right)\left(\tfrac{1}{2}\{\Omega^{-1} - \text{tr. } \Omega^{-1}\}\right)\Upsilon_{,} + \Upsilon_{,}\,;$$

then if

$$(x,\ y,\ z) = (\Omega^{-1}(\Omega - \Upsilon)(\Omega + \Upsilon)^{-1}\,\Omega \between x_{,},\ y_{,},\ z_{,}),$$

$$(\mathrm{x},\ \mathrm{y},\ \mathrm{z}) = (\Omega^{-1}(\Omega - \Upsilon)(\Omega + \Upsilon)^{-1}\,\Omega \between \mathrm{x}_{,},\ \mathrm{y}_{,},\ \mathrm{z}_{,})\,;$$

we have

$$(\Omega \between x,\ y,\ z \between \mathrm{x},\ \mathrm{y},\ \mathrm{z}) = (\Omega \between x_{,},\ y_{,},\ z_{,} \between \mathrm{x}_{,},\ \mathrm{y}_{,},\ \mathrm{z}_{,})\,;$$

and in particular,

If Ω is a symmetrical matrix, then Υ is an arbitrary skew symmetrical matrix;

If Ω is a skew symmetrical matrix, then Υ is an arbitrary symmetrical matrix.

154.

SUPPLEMENTARY RESEARCHES ON THE PARTITION OF NUMBERS.

[From the *Philosophical Transactions of the Royal Society of London*, vol. CXLVIII. *for the year* 1858, pp. 47—52. Received March 19,—Read June 18, 1857.]

THE general formula given at the conclusion of my memoir, "Researches on the Partition of Numbers"(1), is somewhat different from the corresponding formula of Professor Sylvester[2], and leads more directly to the actual expression for the number of partitions, in the form made use of in my memoir; to complete my former researches, I propose to explain the mode of obtaining from the formula the expression for the number of partitions.

The formula referred to is as follows, viz. if $\frac{\phi x}{fx}$ be a rational fraction, the denominator of which is made up of factors (the same or different) of the form $1-x^m$, and if a is a divisor of one or more of the indices m, and k is the number of indices of which it is a divisor, then

$$\left\{\frac{\phi x}{fx}\right\}_{[1-x^a]} = \ldots + \frac{1}{\Pi(s-1)}(x\partial_x)^{s-1} S\frac{\chi\rho}{\rho - x}\ldots$$

$$= \ldots + \frac{1}{\Pi(s-1)}(x\partial_x)^{s-1} S\frac{\theta x}{[1-x^a]}\ldots$$

where

$$\chi\rho = \text{coeff.}\ \frac{1}{t}\ \text{in}\ t^{s-1}\frac{\rho\phi(\rho e^{-t})}{f(\rho e^{-t})},$$

[1] *Philosophical Transactions*, tom. CXLVI. (1856) p. 127, [140].

[2] Professor Sylvester's researches are published in the *Quarterly Mathematical Journal*, tom. I. [1857, pp. 141—152]; there are some numerical errors in his value of $P(1, 2, 3, 4, 5, 6)\,q$.

in which formula $[1-x^a]$ denotes the irreducible factor of $1-x^a$, that is, the factor which equated to zero gives the prime roots, and ρ is a root of the equation $[1-x^a]=0$; the summation of course extends to all the roots of the equation. The index s extends from $s=1$ to $s=k$; and we have then the portion of the fraction depending on the denominator $[1-x^a]$. In the partition of numbers, we have $\phi x=1$, and the formula becomes therefore

$$\left\{\frac{1}{fx}\right\}_{[1-x^a]}=\ldots+\frac{1}{\Pi(s-1)}(x\partial_x)^{s-1}S\frac{\chi\rho}{\rho-x}\ldots$$

$$=\ldots+\frac{1}{\Pi(s-1)}(x\partial_x)^{s-1}\frac{\theta x}{[1-x^a]},$$

where

$$\chi\rho=\text{coeff.}\,\frac{1}{t}\text{ in }t^{s-1}\frac{\rho}{f(\rho e^{-t})}.$$

We may write

$$fx=\Pi(1-x^m),$$

where m has a given series of values the same or different. The indices not divisible by a may be represented by m, the other indices by ap, we have then

$$fx=\Pi(1-x^n)\,\Pi(1-x^{ap}),$$

where the number of indices ap is equal to k. Hence

$$f(\rho e^{-t})=\Pi(1-\rho^n e^{-nt})\,\Pi(1-\rho^{ap}e^{-apt});$$

or since ρ is a root of $[1-x^a]=0$, and therefore $\rho^a=1$, we have

$$f(\rho e^{-t})=\Pi(1-\rho^n e^{-nt})\,\Pi(1-e^{-apt});$$

and it may be remarked that if $n\equiv\nu$ (mod. a), where $\nu<a$, then instead of ρ^n we may write ρ^ν, a change which may be made at once, or at the end of the process of development.

We have consequently to find

$$\chi\rho=\text{coeff.}\,\frac{1}{t}\text{ in }t^{s-1}\frac{\rho}{\Pi(1-\rho^n e^{-nt})\,\Pi(1-e^{-apt})}.$$

The development of a factor $\dfrac{1}{1-\rho^n e^{-nt}}$ is at once deduced from that of $\dfrac{1}{1-ce^{-t}}$, and is a series of positive powers of t. The development of a factor $\dfrac{1}{1-e^{-apt}}$ is deduced from that of $\dfrac{1}{1-e^{-t}}$, and contains a term involving $\dfrac{1}{t}$. Hence we have

$$\frac{1}{\Pi(1-\rho^n e^{-nt})\,\Pi(1-e^{-apt})}=A_{-k}\frac{1}{t^k}+A_{-(k-1)}\frac{1}{t^{k-1}}\ldots+A_{-1}\frac{1}{t}+A_0+\&\text{c.},$$

and thence

$$\chi\rho=\rho A_{-s}.$$

The actual development, when k is small (for instance $k=1$ or $k=2$), is most readily obtained by developing each factor separately and taking the product. To do this we have

$$\frac{1}{1-ce^{-t}}=\frac{1}{1-c}-\frac{c}{(1-c)^2}t+\frac{c+c^2}{(1-c)^3}\tfrac{1}{2}t^2-\frac{c+4c^2+c^3}{(1-c)^4}\tfrac{1}{6}t^3+\&\text{c.},$$

where by a general theorem for the expansion of any function of e^t, the coefficient of t^f is

$$=\frac{(-)^f}{\Pi f}\frac{1}{1-c(1+\Delta)}0^f$$

$$=\frac{(-)^f}{\Pi f}\left(\frac{1}{1-c}+\frac{c}{(1-c)^2}\Delta\ldots+\frac{c^f}{(1-c)^{f+1}}\right)0^f$$

(where as usual $\Delta 0^f=1^f-0^f$, $\Delta^2 0^f=2^f-2\,.\,1^f+0^f$, &c.) and

$$\frac{1}{1-e^{-t}}=\frac{1}{t}+\frac{1}{2}+\frac{1}{12}t-\frac{1}{720}t^3+\frac{1}{30240}t^5-\&\text{c.},$$

where, except the constant term, the series contains odd powers only and the coefficient of t^{2f-1} is $\frac{(-)^{f+1}B_f}{\Pi 2f}$; B_1, B_2, $B_3\ldots$ denoting the series $\frac{1}{6}$, $\frac{1}{30}$, $\frac{1}{42}\ldots$ of Bernoulli's numbers.

But when k is larger, it is convenient to obtain the development of the fraction from that of the logarithm, the logarithm of the fraction being equal to the sum of the logarithms of the simple factors, and these being found by means of the formulæ

$$\log\frac{1}{1-ce^{-t}}=\log\frac{1}{1-c}-\frac{c}{1-c}t+\frac{c}{(1-c)^2}\frac{t^2}{2}-\frac{c+c^2}{(1-c)^3}\frac{t^3}{6}+\frac{c+4c^2+c^3}{(1-c)^4}\frac{t^2}{24}+\&\text{c.}$$

$$\log\frac{1}{1-e^{-t}}=-\log t+\frac{1}{2}t-\frac{1}{24}t^2+\frac{1}{2880}t^4-\frac{1}{181440}t^6+\&\text{c.}$$

The fraction is thus expressed in the form

$$\frac{1}{\Pi(1-\rho^n)\,\Pi(ap)}\frac{1}{t^k}e^{k_1t+k_2t^2+\ldots};$$

and by developing the exponential we obtain, as before, the series commencing with $A_{-k}\frac{1}{t^k}$.

Resuming now the formula

$$\chi\rho=\rho A_{-s},$$

which gives $\chi\rho$ as a function of ρ, we have

$$\frac{\theta x}{[1-x^a]}=S\frac{\chi\rho}{\rho-x};$$

but this equation gives

$$\chi\rho = \theta\rho \left(\frac{\rho - x}{[1 - x^a]}\right)_{x=\rho},$$

and we have

$$[1 - x^a] = (x - \rho)(x - \rho^{a_2}) \dots (x - \rho^{a_\alpha}),$$

if 1, a_2, ... a_α are the integers less than a and prime to it (α is of course the degree of $[1 - x^a]$). Hence

$$\chi\rho = \theta\rho \frac{-1}{\rho^{\alpha-1}\Pi(1 - \rho^{a_i - 1})},$$

and therefore

$$\theta\rho = -\rho^{\alpha-1}\Pi(1 - \rho^{a_i-1})\chi\rho;$$

or putting for $\chi\rho$ its value

$$\theta\rho = -\rho^{\alpha}\Pi(1 - \rho^{a_i-1}) A_{-s},$$

where α is the degree of $[1 - x^a]$ and a_i denotes in succession the integers (exclusive of unity) less than a and prime to it. The function on the right hand, by means of the equation $[1 - \rho^a] = 0$, may be reduced to an integral function of ρ of the degree $\alpha - 1$, and then by simply changing ρ into x we have the required function θx. The fraction $\dfrac{\theta x}{[1 - x^a]}$ can then by multiplication of the terms by the proper factor be reduced to a fraction with the denominator $1 - x^a$, and the coefficients of the numerator of this fraction are the coefficients of the corresponding prime circulator ( ) per a_q.

Thus, let it be required to find the terms depending on the denominator $[1 - x^3]$ in

$$\frac{1}{(1-x)(1-x^2)(1-x^3)(1-x^4)(1-x^5)(1-x^6)};$$

these are

$$S\frac{\chi_1\rho}{\rho - x}, \quad x\partial_x S\frac{\chi_2\rho}{\rho - x},$$

where

$$\chi_1\rho = \text{coeff. } \frac{1}{t} \text{ in } \frac{\rho}{f(\rho e^{-t})},$$

$$\chi_2\rho = \text{coeff. } \frac{1}{t} \text{ in } t\frac{\rho}{f(\rho e^{-t})}$$

and

$$\frac{1}{f(\rho e^{-t})} = \frac{1}{(1-\rho e^{-t})(1-\rho^2 e^{-2t})(1-\rho^4 e^{-4t})(1-\rho^5 e^{-5t})(1-e^{-3t})(1-e^{-6t})}$$

$$= A_{-2}\frac{1}{t^2} + A_{-1}\frac{1}{t} + \&\text{c.},$$

where it is easy to see that

$$A_{-2}=\frac{1}{18}\frac{1}{(1-\rho)(1-\rho^2)(1-\rho^4)(1-\rho^5)},$$

$$A_{-1}=\frac{1}{(1-\rho)(1-\rho^2)(1-\rho^4)(1-\rho^5)}\left\{\frac{1}{4}-\frac{1}{18}\left(\frac{\rho}{1-\rho}+\frac{2\rho^2}{1-\rho^2}+\frac{4\rho^4}{1-\rho^4}+\frac{4\rho^5}{1-\rho^5}\right)\right\},$$

and we have

$$\theta_2\rho=-\rho^2(1-\rho)A_{-2},$$

$$\theta_1\rho=-\rho^2(1-\rho)A_{-1}.$$

But $[1-\rho^3]=1+\rho+\rho^2=0$. Hence $\rho^3=1$, and therefore

$$(1-\rho)(1-\rho^2)(1-\rho^4)(1-\rho^5)=(1-\rho)^2(1-\rho^2)^2=9.$$

Hence

$$\theta_2\rho=-\frac{1}{162}\rho^2(1-\rho)=\frac{1}{162}(1-\rho^2)=\frac{1}{162}(2+\rho),$$

whence

$$\theta_2 x=\frac{1}{162}(2+x),$$

and the partial fraction is

$$\frac{1}{162}\frac{2+x}{1+x+x^2},$$

which is

$$=\frac{1}{162}\frac{2-x-x^2}{1-x^3},$$

and gives rise to the prime circulator $\frac{1}{162}(2,\,-1,\,-1)$ per 3_q.

The reduction $\theta_1\rho$ is somewhat less simple; we have

$$\theta_1\rho=-\frac{1}{9}\rho^2(1-\rho)\left\{\frac{1}{4}-\frac{1}{18}\left(\frac{\rho}{1-\rho}+\frac{2\rho^2}{1-\rho^2}+\frac{4\rho}{1-\rho}+\frac{5\rho^2}{1-\rho^2}\right)\right\}$$

$$=-\frac{1}{9}\rho^2(1-\rho)\left\{\frac{1}{4}-\frac{5}{18}\frac{\rho}{1-\rho}-\frac{7}{18}\frac{\rho^2}{1-\rho^2}\right\}$$

$$=\frac{1}{9}(1-\rho^2)\left\{\frac{1}{4}-\frac{5}{54}\rho(1-\rho^2)-\frac{7}{54}\rho^2(1-\rho)\right\}$$

$$=\frac{1}{972}(1-\rho^2)(51-10\rho-14\rho^2)$$

$$=\frac{1}{972}(61+4\rho-65\rho^2);$$

hence finally

$$\theta_1\rho = \frac{1}{324}(42 + 23\rho), \quad \theta_1 x = \frac{1}{324}(42 + 23x);$$

and the partial fraction is

$$\frac{1}{324} x\partial_x \frac{42 + 23x}{1 + x + x^2},$$

which is

$$= \frac{1}{324} x\partial_x \frac{42 - 19x - 23x^2}{1 - x^3},$$

and gives rise to the prime circulator $\frac{1}{324} q\,(42, \ -19, \ -23)$ per 3_q.

The part depending on the denominator $1 - x$ is

$$\frac{A_{-1}}{1-x} + \frac{1}{1} x\partial_x \frac{A_{-2}}{1-x} + \frac{1}{1\,.\,2}(x\partial_x)^2 \frac{A_{-3}}{1-x} \dots + \frac{1}{1\,.\,2\,.\,3\,.\,4\,.\,5}(x\partial_x)^5 \frac{A_{-6}}{1-x},$$

where

$$\frac{1}{(1-e^{-t})(1-e^{-2t})(1-e^{-3t})(1-e^{-4t})(1-e^{-5t})(1-e^{-6t})}$$

$$= A_{-6}\frac{1}{t^6} + A_{-5}\frac{1}{t^5} \dots + A_{-1}\frac{1}{t} + \text{\&c.}$$

We have here

$$\log \frac{1}{1 - e^{-t}} = -\log t + \frac{1}{2} t - \frac{1}{24} t^2 + \frac{1}{2880} t^4 - \text{\&c.},$$

and thence the fraction is

$$\frac{1}{720\, t^6} e^{\frac{21}{2} t - \frac{91}{24} t^2 + \frac{455}{576} t^4 - \text{\&c.}}$$

which is equal to

$$\frac{720\, t^6}{1}\left(1 + \frac{21}{2} t + \frac{441}{8} t^2 + \frac{3087}{16} t^3 + \frac{64827}{128} t^4 + \frac{1361367}{1280} t^5 + \dots\right.$$

$$\times \left(1 - \frac{91}{24} t^2 + \frac{8281}{1152} t^4 + \dots\right)$$

$$\times \left(1 + \frac{455}{576} t^4 + \dots\right)$$

$$= \frac{1}{720}\frac{1}{t^6} + \frac{7}{480}\frac{1}{t^5} + \frac{77}{1080}\frac{1}{t^4} + \frac{245}{1152}\frac{1}{t^3} + \frac{43981}{103680}\frac{1}{t^2} + \frac{199577}{345600}\frac{1}{t} + \dots$$

and consequently the partial fractions are

$$\frac{1}{86400}(x\partial_x)^5 \frac{1}{1-x} + \frac{7}{11520}(x\partial_x)^4 \frac{1}{1-x} + \frac{77}{6480}(x\partial_x)^3 \frac{1}{1-x} + \frac{245}{2304}(x\partial_x)^2 \frac{1}{1-x}$$

$$+ \frac{43981}{103680}(x\partial_x)\frac{1}{1-x} + \frac{199577}{345600}\frac{1}{1-x},$$

from which the non-circulating part is at once obtained.

The complete expression for the number of partitions is $P(1, 2, 3, 4, 5, 6)\,q =$

$$
\begin{aligned}
&\frac{1}{1036800}(12q^5 + 630q^4 + 1230q^3 + 110250q^2 + 439810q + 598731)\\
+\;&\frac{1}{4608}(6q^2 + 126q + 581)(1,\ 1)\ \text{pcr}\ 2_q\\
+\;&\frac{1}{162}q \ldots\ldots\ldots\ldots (2,\ -1,\ -1)\ \text{pcr}\ 3_q\\
+\;&\frac{1}{324} \ldots\ldots (42,\ -19,\ -23)\ \text{pcr}\ 3_q\\
+\;&\frac{1}{32} \ldots\ldots\ldots (1,\ 1,\ -1,\ -1)\ \text{pcr}\ 4_q\\
+\;&\frac{1}{25} \ldots\ldots (2,\ 1,\ 0,\ -1,\ -2)\ \text{pcr}\ 5_q\\
+\;&\frac{1}{36} \ldots (2,\ 1,\ -1,\ -2,\ -1,\ 1)\ \text{pcr}\ 6_q.
\end{aligned}
$$

155.

A FOURTH MEMOIR UPON QUANTICS.

[From the *Philosophical Transactions of the Royal Society of London*, vol. CXLVIII. *for the year* 1858, pp. 415—427. Received February 11,—Read March 18, 1858.]

THE object of the present memoir is the further development of the theory of binary quantics; it should therefore have preceded so much of my third memoir, t. 147 (1857), p. 627, [144], as relates to ternary quadrics and cubics. The paragraphs are numbered continuously with those of the former memoirs. The first three paragraphs, Nos. 62 to 64, relate to quantics of the general form $(*\between x, y, \ldots)^m$, and they are intended to complete the series of definitions and explanations given in Nos. 54 to 61 of my third memoir; Nos. 68 to 71, although introduced in reference to binary quantics, relate or may be considered as relating to quantics of the like general form. But with these exceptions the memoir relates to binary quantics of any order whatever: viz. Nos. 65 to 80 relate to the covariants and invariants of the degrees 2, 3 and 4; Nos. 81 and 82 (which are introduced somewhat parenthetically) contain the explanation of a process for the calculation of the invariant called the Discriminant; Nos. 83 to 85 contain the definitions of the Catalecticant, the Lambdaic and the Canonisant, which are functions occurring in Professor Sylvester's theory of the reduction of a binary quantic to its canonical form; and Nos. 86 to 91 contain the definitions of certain covariants or other derivatives connected with Bezout's abbreviated method of elimination, due for the most part to Professor Sylvester, and which are called Bezoutiants, Cobezoutiants, &c. I have not in the present memoir in any wise considered the theories to which the catalecticant, &c. and the last-mentioned other covariants and derivatives relate; the design is to point out and precisely define the different covariants or other derivatives which have hitherto presented themselves in theories relating to binary quantics, and so to complete, as far as may be, the explanation of the terminology of this part of the subject.

62. If we consider a quantic

$$(a, b, \ldots \between x, y, \ldots)^m$$

and an adjoint linear form, the operative quantic

$$(a, b, \ldots \between \partial_\xi, \partial_\eta, \ldots)^m,$$

or more generally the operative quantic obtained by replacing in any covariant of the given quantic the facients $(x, y, \ldots)$ by the symbols of differentiation $(\partial_\xi . \partial_\eta, \ldots)$ (which operative quantic is, so to speak, a contravariant operator), may be termed the *Provector;* and the Provector operating upon any contravariant gives rise to a contravariant, which may of course be an invariant. Any such contravariant, or rather such contravariant considered as so generated, may be termed a *Provectant;* and in like manner the operative quantic obtained by replacing in any contravariant of the given quantic the facients $(\xi, \eta, \ldots)$ by the symbols of differentiation $(\partial_x, \partial_y, \ldots)$ (which operative quantic is a covariant operator), is termed the *Contraprovector;* and the contraprovector operating upon any covariant gives rise to a covariant, which may of course be an invariant. Any such covariant, or rather such covariant considered as so generated, may be termed a *Contraprovectant.*

In the case of a binary quantic,

$$(a, b, \ldots \between x, y)^m,$$

the two theorems coalesce together, and we may say that the operative quantic

$$(a, b, \ldots \between \partial_y, -\partial_x)^m,$$

or more generally the operative quantic obtained by replacing in any covariant of the given quantic the facients (x, y) by the symbols of differentiation $(\partial_y, -\partial_x)$ (which is in this case a covariant operator), may be termed the Provector. And the Provector operating on any covariant gives a covariant (which as before may be an invariant), and which considered as so generated may be termed the Provectant.

63. But there is another allied theory. If in the quantic itself or in any covariant we replace the facients $(x, y, \ldots)$ by the first derived functions $(\partial_\xi P, \partial_\eta P, \ldots)$ of any contravariant P of the quantic, we have a new function which will be a contravariant of the quantic. In particular, if in the quantic itself we replace the facients $(x, y, \ldots)$ by the first derived functions $(\partial_\xi P, \partial_\eta P, \ldots)$ of the Reciprocant, then the result will contain as a factor the Reciprocant, and the other factor will be also a contravariant. And similarly, if in any contravariant we replace the facients $(\xi, \eta, \ldots)$ by the first derived functions $(\partial_x W, \partial_y W, \ldots)$ of any covariant W (which may be the quantic itself) of the quantic U, we have a new function which will be a covariant of the quantic. And in particular if in the Reciprocant we replace the facients $(\xi, \eta, \ldots)$ by the first derived functions $(\partial_x U, \partial_y U, \ldots)$ of the quantic, the result will contain U as a factor, and the other factor will be also a covariant. In the case of a binary quantic $(a, b, \ldots \between x, y)^m$ the two theorems coalesce and we have the following theorem, viz. if in the quantic U or in any covariant the facients (x, y) are replaced by the first derived functions $(\partial_y W, -\partial_x W)$ of a covariant W, the result will be a covariant; and if in the quantic

U the facients (x, y) are replaced by the first derived functions $(\partial_y U, -\partial_x U)$ of the quantic, the result will contain U as a factor, and the other factor will be also a covariant.

Without defining more precisely, we may say that the function obtained by replacing as above the facients of a covariant or contravariant by the first derived functions of a contravariant or covariant is a *Transmutant* of the first-mentioned covariant or contravariant.

64. Imagine any two quantics of the same order, for instance, the two quantics

$$U = (a\,,\ b\,,\ldots \between x,\ y \ldots)^m,$$
$$V = (a',\ b',\ldots \between x,\ y \ldots)^m,$$

then any quantic such as $\lambda U + \mu V$ may be termed an *Intermediate* of the two quantics; and a covariant of $\lambda U + \mu V$, if in such covariant we treat λ, μ as facients, will be a quantic of the form

$$(A,\ B,\ \ldots\ B\grave{},\ A\grave{},\between \lambda,\ \mu)^n,$$

where the coefficients $(A,\ B,\ \ldots\ B\grave{},\ A\grave{})$ will be covariants of the quantics U, V, viz. A will be a covariant of the quantic U alone; $A\grave{}$ will be the same covariant of the quantic V alone, and the other coefficients (which in reference to A, $A\grave{}$ may be termed the *Connectives*) will be covariants of the two quantics; and any coefficient may be obtained from the one which precedes it by operating on such preceding coefficient with the combinantive operator

$$a'\partial_a + b'\partial_b + \ldots,$$

or from the one which succeeds it by operating on such succeeding coefficient with the combinantive operator

$$a\partial_{a'} + b\partial_{b'} + \ldots\,,$$

the result being divided by a numerical coefficient which is greater by unity than the index of μ or (as the case may be) λ in the term corresponding to the coefficient operated upon. It may be added, that any invariant in regard to the facients (λ, μ) of the quantic

$$(A,\ B,\ \ldots\ B\grave{},\ A\grave{}\between \lambda,\ \mu)^n$$

is not only a covariant, but it is also a combinant of the two quantics U, V.

As an example, suppose the quantics U, V are the quadrics

$$(a,\ b,\ c\between x,\ y)^2 \text{ and } (a',\ b',\ c'\between x,\ y)^2,$$

then the quadrinvariant of

$$\lambda U + \mu V \text{ is } (\lambda a + \mu a')(\lambda c + \mu c') - (\lambda b + \mu b')^2,$$

which is equal to

$$(ac - b^2,\ ac' - 2bb' + ca',\ a'c' - b'^2 \between \lambda,\ \mu)^2,$$

and $ac' - 2bb' + ca'$ is the connective of the two discriminants $ac - b^2$ and $a'c' - b'^2$.

65. The law of reciprocity for the number of the invariants of a binary quantic[1], leads at once to the theorems in regard to the number of the quadrinvariants, cubinvariants and quartinvariants of a binary quantic of a given degree, first obtained by the method in the second part of my original memoir[2]. Thus a quadric has only a single invariant, which is of the degree 2; hence, by the law of reciprocity, the number of quadrinvariants of a quantic of the order m is equal to the number of ways in which m can be made up with the part 2, which is of course unity or zero, according as m is even or odd. And we conclude that

The quadrinvariant exists only for quantics of an even order, and for each such quantic there is one, and only one, quadrinvariant.

66. Again, a cubic has only one invariant, which is of the degree 4, and the number of cubinvariants of a quantic of the degree m is equal to the number of ways in which m can be made up with the part 4. Hence

A cubinvariant only exists for quantics of an evenly even order, and for each such quantic there is one, and only one, cubinvariant.

67. But a quartic has two invariants, which are of the degrees 2 and 3 respectively, and the number of quartinvariants of a quantic of the degree m is equal to the number of ways in which m can be made up with the parts 2 and 3. When m is even, there is of course a quartinvariant which is the square of the quadrinvariant, and which, if we attend only to the irreducible quartinvariants, must be excluded from consideration. The preceding number must therefore, when m is even, be diminished by unity. The result is easily found to be

Quartinvariants exist for a quantic of any order, even or odd, whatever, the quadric and the quartic alone excepted; and according as the order of the quantic is

$$6g,\quad 6g+1,\quad 6g+2,\quad 6g+3,\quad 6g+4,\quad 6g+5,$$

the number of quartinvariants is

$$g\,,\quad g\,,\quad g\,,\quad g+1,\quad g\,,\quad g+1.$$

In particular, for the orders

$$2,\quad 3,\quad 4,\quad 5;\quad 6,\quad 7,\quad 8,\quad 9,\quad 10,\quad 11;\quad 12,\ \&c.,$$

the numbers are

$$0,\quad 1,\quad 0,\quad 1;\quad 1,\quad 1,\quad 1,\quad 2,\quad 1,\quad 2;\quad 2,\ \&c.$$

Thus the ninthic is the lowest quantic which has more than one quartinvariant.

68. But the whole theory of the invariants or covariants of the degrees 2, 3, 4 is most easily treated by the method above alluded to, contained in the second part of my original memoir; and indeed the method appears to be the appropriate one for the

[1] Introductory Memoir, [139], No. 20.

[2] Ibid. Nos. 10—17.

treatment of the theory of the invariants or covariants of any given degree whatever, although the application of it becomes difficult when the degree exceeds 4. I remark, in regard to this method, that it leads naturally, and in the first instance, to a special class of the covariants of a system of quantics, viz. these covariants are linear functions of the derived functions of any quantic of the system. (It is hardly necessary to remark that the derived functions referred to are the derived functions of any order of the quantic with regard to the facients.) Such covariants may be termed *tantipartite* covariants; but when there are only two quantics, I use in general the term *lineo-linear*. The tantipartite covariants, while the system remains general, are a special class of covariants, but by particularizing the system we obtain all the covariants of the particularized system. The ordinary case is when all the quantics of the system reduce themselves to one and the same quantic, and the method then gives all the covariants of such single quantic. And while the order of the quantic remains indefinite, the method gives covariants (not invariants); but by particularizing the order of the quantic in such manner that the derived functions become simply the coefficients of the quantic, the covariants become invariants: the like applies of course to a system of two or more quantics.

69. To take the simplest example, in seeking for the covariants of a single quantic U, we in fact have to consider two quantics U, V. An expression such as $\overline{12}\, UV$ is a lineo-linear covariant of the two quantics; its developed expression is

$$\partial_x U \,.\, \partial_y V - \partial_y U \,.\, \partial_x V,$$

which is the Jacobian. In the particular case of two linear functions $(a, b \between x, y)$ and $(a', b' \between x, y)$, the lineo-linear covariant becomes the lineo-linear invariant $ab' - a'b$, which is the Jacobian of the two linear functions.

In the example we cannot descend from the two quantics U, V to the single quantic U (for putting $V = U$ the covariant vanishes); but this is merely accidental, as appears by considering a different lineo-linear covariant $\overline{12}^2 UV$, the developed expression of which is

$$\partial_x^2 U \,.\, \partial_y^2 V - 2\partial_x\partial_y U \,.\, \partial_x\partial_y V + \partial_y^2 U \,.\, \partial_x^2 V.$$

In the particular case of two quadrics $(a, b, c \between x, y)^2$, $(a', b', c' \between x, y)^2$, the lineo-linear covariant becomes the lineo-linear invariant

$$ac' - 2bb' + ca'.$$

If we have $V = U$, then the lineo-linear covariant gives the quadricovariant

$$\partial_x^2 U \,.\, \partial_y^2 U - (\partial_x \partial_y U)^2$$

of the single quantic U (such quadricovariant is in fact the Hessian); and if in the last-mentioned formula we put for U the quadric $(a, b, c \between, x, y)^2$, or what is the same thing, if in the expression of the lineo-linear invariant $ac' - 2bb' + ca'$, we put the two quadrics equal to each other, we have the quadrinvariant

$$ac - b^2$$

of the single quadric.

70. The lineo-linear invariant $ab' - a'b$ of two linear functions may be considered as giving the lineo-linear covariant $\partial_x U \,.\, \partial_y V - \partial_y U \,.\, \partial_x V$ of the two quantics U and V, and in like manner the lineo-linear invariant $ac' - 2bb' + ca'$ may be considered as giving the lineo-linear covariant $\partial_x^2 U \,.\, \partial_y^2 V - 2\partial_x\partial_y U \,.\, \partial_x\partial_y V + \partial_y^2 U \,.\, \partial_x^2 V$ of the quantics U, V. And generally, any invariant whatever of a quantic or quantics of a given order or orders leads to a covariant of a quantic or quantics of any higher order or orders: viz. the coefficients of the original quantic or quantics are to be replaced by the derived functions of the quantic or quantics of a higher order or orders.

71. The same thing may be seen by means of the theory of Emanants. In fact, consider any emanants whatever of a quantic or quantics; then, attending only to the facients of emanation, the emanants will constitute a system of quantics the coefficients of which are derived functions of the given quantic or quantics; the invariants of the system of emanants will be functions of the derived functions of the given quantic or quantics, and they will be covariants of such quantic or quantics; and we thus pass from the invariants of a quantic or quantics to the covariants of a quantic or quantics of a higher order or orders.

72. It may be observed also, that in the case where a tantipartite invariant, when the several quantics are put equal to each other, does not become equal to zero, we may pass back from the invariant of the single quantic to the tantipartite invariant of the system; thus the lineo-linear invariant $ac' - 2bb' + ca'$ of two quadrics leads to the quadrinvariant $ac - b^2$ of a single quantic; and *conversely*, from the quadrinvariant $ac - b^2$ of a single quadric, we obtain by an obvious process of derivation the expression $ac' - 2bb' + ca'$ of the lineo-linear invariant of two quadrics This is in fact included in the more general theory explained, No. 64.

73. Reverting now to binary quantics, two quantics of the same order, even or odd, have a lineo-linear invariant. Thus the two quadrics

$$(a,\ b,\ c \between x,\ y)^2,\ (a',\ b',\ c' \between x,\ y)^2$$

have (it has been seen) the lineo-linear invariant

$$ac' - 2bb' + ca';$$

and in like manner the two cubics

$$(a,\ b,\ c,\ d \between x,\ y)^3,\ (a',\ b',\ c',\ d' \between x,\ y)^3$$

have the lineo-linear invariant

$$ad' - 3bc' + 3cb' - da',$$

which examples are sufficient to show the law.

74. The lineo-linear invariant of two quantics of the same odd order is a combinant, but this is not the case with the lineo-linear invariant of two quantics of the same even order. Thus the last-mentioned invariant is reduced to zero by each of the operations

$$a\partial_{a'} + b\partial_{b'} + c\partial_{c'} + d\partial_{d'}$$

and

$$a'\partial_a + b'\partial_b + c'\partial_c + d'\partial_d\,;$$

but the invariant

$$ac' \;\; - 2bb' + ca'$$

is by the operations

$$a\partial_{a'} + b\partial_{b'} + c\partial_{c'}$$

and

$$a'\partial_a + b'\partial_b + c'\partial_c$$

reduced respectively to

$$2(ac - b^2)$$

and

$$2(a'c' - b'^2).$$

75. For two quantics of the same odd order, when the quantics are put equal to each other, the lineo-linear invariant vanishes; but for two quantics of the same even order, when these are put equal to each other, we obtain the quadrinvariant of the single quantic. Thus the quadrinvariant of the quadric $(a, b, c \between x, y)^2$ is

$$ac - b^2\,;$$

and in like manner the quadrinvariant of the quartic $(a, b, c, d, e \between x, y)^4$ is

$$ae - 4bd + 3c^2.$$

76. When the two quantics are the first derived functions of the same quantic of any odd order, the lineo-linear invariant does not vanish, but it is not an invariant of the single quantic. Thus the lineo-linear invariant of

$$(a, b, c \between x, y)^2$$

and

$$(b, c, d \between x, y)^2$$

is

$$(ad - 2bc + cb =)\, ad - bc,$$

which is not an invariant of the cubic

$$(a, b, c, d \between x, y)^3.$$

But for two quantics which are the first derived functions of the same quantic of any even order, the lineo-linear invariant is the quadrinvariant of the single quantic. Thus the lineo-linear invariant of

$$(a, b, c, d \between x, y)^3$$

and

$$(b, c, d, e \between x, y)^3$$

is

$$(ae - 3bd + 3c^2 - db =)\, ae - 4bd + 3c^2,$$

which is the quadrinvariant of the quartic

$$(a, b, c, d, e \between x, y)^4.$$

77. I do not stop to consider the theory of the lineo-linear covariants of two quantics, but I derive the quadricovariants of a single quantic directly from the quadrinvariant. Imagine a quantic of any order even or odd. Its successive even emanants will be in regard to the facients of emanation quantics of an even order, and they will each of them have a quadrinvariant, which will be a quadricovariant of the given quantic. The emanants in question, beginning with the second emanant, are (in regard to the facients of the given quantic assumed to be of the order m) of the orders $m-2$, $m-4,\ldots$ down to 1 or 0, according as m is odd or even, or writing successively $2p+1$ and $2p$ in the place of m, and taking the emanants in a reverse order, the emanants for a quantic of any odd order $2p+1$ are of the orders $1, 3, 5 \ldots 2p-1$, and for a quantic of any even order $2p$, they are of the orders $0, 2, 4 \ldots 2p-2$. The quadricovariants of a quantic of an odd order $2p+1$, are consequently of the orders $2, 6, 10 \ldots 4p-2$, and the quadricovariants of a quantic of an even order $2p$, are of the orders $0, 4, 8 \ldots 4p-4$. We might in each case carry the series one step further, and consider a quadricovariant of the order $4p+2$, or (as the case may be) $4p$, which arises from the 0th emanant of the given quantic; such quadricovariant is, however, only the square of the given quantic.

78. In the case of a quantic of an evenly even order (but in no other case) we have a quadricovariant of the same order with the quantic itself. We may in this case form the lineo-linear invariant of the quantic and the quadricovariant of the same order: such lineo-linear invariant is an invariant of the given quantic, and it is of the degree 3 in the coefficients, that is, it is a cubinvariant. This agrees with the before-mentioned theorem for the number of cubinvariants.

79. In the case of the quartic $(a, b, c, d, e \between x, y)^4$, the cubinvariant is, by the preceding mode of generation, obtained in the form

$$e\,(ac-b^2)-4d\tfrac{2}{4}\,(ad-bc)+6c\tfrac{1}{6}\,(ae-4bd+3c^2)-4b\tfrac{2}{4}\,(be-cd)+a\,(ce-d^2),$$

which is in fact equal to

$$3\,(ace-ad^2-b^2e+2bcd-c^3)\,;$$

and omitting the numerical factor 3, we have the cubinvariant of the quartic.

80. In the case of a quantic of any order even or odd, the quadrinvariants of the quadricovariants are quartinvariants of the quantic. But these quartinvariants are not all of them independent, and there is no obvious method grounded on the preceding mode of generation for obtaining the number of the independent (asyzygetic) quartinvariants, and thence the number of the irreducible quartinvariants of a quantic of a given order.

81. I take the opportunity of giving some additional developments in relation to the discriminant of a quantic

$$(a, b, \ldots b', a' \between x, y)^m.$$

To render the signification perfectly definite, it should be remarked that the discriminant contains the term $a^{m-1}a'^{m-1}$, and that the coefficient of this term may be taken to be

+ 1. It was noticed in the Introductory Memoir, that, by Joachimsthal's theorem, the discriminant, on putting $a = 0$, becomes divisible by b^2, and that throwing out this factor it is to a numerical factor *près* the discriminant of the quantic of the order $(m-1)$ obtained by putting $a=0$ and throwing out the factor x; and it was also remarked, that this theorem, combined with the general property of invariants, afforded a convenient method for the calculation of the discriminant of a quantic when that of the order immediately preceding is known. Thus let it be proposed to find the discriminant of the cubic

$$(a,\ b,\ c,\ d \between x,\ y)^3.$$

Imagine the discriminant expanded in powers of the leading coefficient a in the form

$$Aa^2 + Ba + C,$$

then this function *quâ* invariant must be reduced to zero by the operation $3b\partial_a + 2c\partial_b + d\partial_c$; or putting for shortness $\nabla = 2c\partial_b + d\partial_c$, the operation is $\nabla + 3b\partial_a$, and we have

$$\left.\begin{aligned} a^2\nabla A + a\,\nabla B &+ \nabla C \\ + a\,6bA &+ 3bB \end{aligned}\right\} = 0,$$

and consequently

$$B = -\frac{1}{3b}\nabla C, \quad A = -\frac{1}{6b}\nabla B, \quad \nabla A = 0.$$

But C is equal to b^2 into the discriminant of $(3b,\ 3c,\ d \between x,\ y)^2$, that is, its value is $b^2(12bd - 9c^2)$, or throwing out the factor 3, we may write

$$C = 4b^3d - 3b^2c^2;$$

this gives

$$B = -\frac{1}{3b}(-6b^2cd + 24b^2cd - 12bc^3),$$

or reducing

$$B = -6bcd + 4c^3;$$

and thence

$$A = -\frac{1}{6b}(-6bd^2 + 12c^2d - 12c^2d),$$

or reducing

$$A = d^2,$$

which verifies the equation $\nabla A = 0$, and the discriminant is, as we know,

$$a^2d^2 - 6abcd + 4ac^3 + 4b^3d - 3b^2c^2.$$

82. If we consider the quantic $(a,\ b, \ldots a' \between x,\ 1)^m$ as expressed in terms of the roots in the form $a(x-\alpha y)(x-\beta y)\ldots$, then the discriminant $(= a^{m-1}a'^{m-1} + \&c.$ as above) is to a factor *près* equal to the product of the squares of the differences of the roots, and the factor may be determined as follows: viz. denoting by $\zeta(\alpha,\ \beta,\ \ldots)$ the product of the squares of the differences of the roots, we may write

$$a^{2m-2}\,\zeta(\alpha,\ \beta,\ \ldots) = N\,(a^{m-1}a'^{m-1} + \&c.),$$

where N is a number; and then considering the equation $x^m - 1 = 0$, we have to determine N the equation

$$\zeta(\alpha, \beta, \ldots) = (-)^{m-1} N.$$

But in general

$$\zeta(\alpha, \beta \ldots) = (-)^{\frac{1}{2}m(m-1)} (\alpha-\beta)(\alpha-\gamma)\ldots(\beta-\alpha)(\beta-\gamma)\ldots$$

and if

$$\phi x = (x-\alpha)(x-\beta)\ldots,$$

then

$$(\alpha-\beta)(\alpha-\gamma)\ldots = \phi'\alpha, \text{ \&c.},$$

or

$$\zeta(\alpha, \beta, \ldots) = (-)^{\frac{1}{2}m(m-1)} \phi'\alpha\phi'\beta\ldots;$$

here

$$\phi x = x^m - 1, \quad \phi' x = m x^{m-1},$$

and therefore

$$\phi'\alpha\phi'\beta\ldots = m^m (\alpha\beta\gamma\ldots)^{m-1},$$

but

$$(-)^m \alpha\beta\gamma\ldots = -1,$$

or

$$\alpha\beta\gamma\ldots = (-)^{m+1} 1,$$

and

$$\phi'\alpha\phi'\beta\ldots = (-)^{(m+1)^2} m^m = (-)^{m+1} m^m;$$

whence

$$\zeta(\alpha, \beta\ldots) = (-)^{m-1+\frac{1}{2}m(m-1)} m^m = (-)^{m-1} N,$$

or

$$N = (-)^{\frac{1}{2}m(m-1)} m^m,$$

and consequently

$$a^{m-2}\zeta(\alpha, \beta, \ldots) = (-)^{\frac{1}{2}m(m-1)} m^m (a^{m-1} a'^{m-1} + \text{\&c.}),$$

or what is the same thing, the value of the discriminant $\square\,(= a^{m-1} a'^{m-1} + \text{\&c.})$ is

$$(-)^{\frac{1}{2}m(m-1)} m^{-m} a^{m-2} \zeta(\alpha, \beta, \ldots).$$

It would have been allowable to define the discriminant so as that the leading term should be

$$(-)^{\frac{1}{2}m(m-1)} a^{m-1} a'^{m-1},$$

in which case the discriminant would have constantly the same sign as the product of the squared differences; but I have upon the whole thought it better to make the leading term of the discriminant always positive.

83. A quantic of an even order $2p$ has an invariant of peculiar simplicity, viz. the determinant the terms of which are the coefficients of the pth differential coefficients, or derived functions of the quantic with respect to the facients; such invariant may also be considered as a tantipartite invariant of the pth emanants. Thus the sextic

$$(a, b, c, d, e, f, g \between x, y)^6$$

has for one of its invariants, the determinant

$$\begin{vmatrix} a, & b, & c, & d \\ b, & c, & d, & e \\ c, & d, & e, & f \\ d, & e, & f, & g \end{vmatrix}$$

The invariant in question is termed by Professor Sylvester the *Catalecticant.*

84. Professor Sylvester also remarked, that we may from the catalecticant form a function containing an indeterminate quantity λ, such that the coefficients of the different powers of λ are invariants of the quantic; thus for the sextic, the function in question is

$$\begin{vmatrix} a & , b & , c & , d-\lambda \\ b & , c & , d+\frac{1}{3}\lambda, & e \\ c & , d-\frac{1}{3}\lambda, & e & , f \\ d+\lambda, & e & , f & , g \end{vmatrix}$$

where the law of formation is manifest; the terms in the sinister diagonal are modified by annexing to their numerical submultiples of λ with the signs + and − alternately, and in which the multipliers are the reciprocals of the binomial coefficients. The function so obtained is termed the *Lambdaic.*

85. If we consider a quantic of an odd order, and form the catalecticant of the penultimate emanant, we have the covariant termed the Canonisant. Thus in the case of the quintic

$$(a, b, c, d, e, f \between x, y)^5,$$

the canonisant is

$$\begin{vmatrix} ax+by, & bx+cy, & cx+dy \\ bx+cy, & cx+dy, & dx+ey \\ cx+dy, & dx+ey, & ex+fy \end{vmatrix}$$

which is equivalent to

$$\begin{vmatrix} y^3, & -y^2x, & yx^2, & -x^3 \\ a, & b, & c, & d \\ b, & c, & d, & e \\ c, & d, & e, & f \end{vmatrix}$$

and a like transformation exists with respect to the canonisant of a quantic of any odd order whatever. The canonisant and the lambdaic (which includes of course the catalecticant) form the basis of Professor Sylvester's theory of the *Canonical Forms* of quantics of an odd and an even order respectively.

86. There is another family of covariants which remains to be noticed. Consider any two quantics of the same order,

$$(a, b, \ldots \between x, y)^m,$$
$$(a', b', \ldots \between x, y)^m,$$

and join to these a quantic of the next inferior order,

$$(u, v, \ldots \between y, -x)^{m-1},$$

where the coefficients $(u, v, \ldots)$ are considered as indeterminate, and which may be spoken of as the adjoint quantic.

Take the odd lineo-linear covariants (viz. those which arise from the odd emanants) of the two quantics; the term arising from the $(2i+1)$th emanants is of the form

$$(A, B, \ldots \between x, y)^{2(m-1-2i)},$$

where $(A, B, \ldots)$ are lineo-linear functions of the coefficients of the two quantics.

Take also the quadricovariants of the adjoint quantic; the term arising from the $(2i-m)$th emanant is of the form

$$(U, V, \ldots \between x, y)^{2(m-1-2i)},$$

where $(U, V, \ldots)$ are quadric functions of the indeterminate coefficients $(u, v, \ldots)$. We may then form the quadrinvariant of the two quantics of the order $2(m-1-2i)$: this will be an invariant of the two quantics and the adjoint quantic, lineo-linear in the coefficients of the two quantics and of the degree 2 in regard to the coefficients $(u, v, \ldots)$ of the adjoint quantic; or treating the last-mentioned coefficients as facients, the result is a lineo-linear m-ary quadric of the form

$$(\mathfrak{A}, \mathfrak{B}, \ldots \between u, v, \ldots)^2,$$

viz. in this expression the coefficients $\mathfrak{A}, \mathfrak{B}, \ldots$ are lineo-linear functions of the coefficients of the two quantics. And giving to i the different admissible values, viz. from $i=0$ to $i=\tfrac{1}{2}m-1$ or $\tfrac{1}{2}(m-1)-1$, according as m is even or odd, the number of the functions obtained by the preceding process is $\tfrac{1}{2}m$ or $\tfrac{1}{2}(m-1)$, according as m is even or odd. The functions in question, the theory of which is altogether due to Professor Sylvester, are termed by him *Cobèzoutiants;* we may therefore say that a cobezoutiant is an invariant of two quantics of the same order m, and of an adjoint quantic of the next preceding order $m-1$, viz. treating the coefficients of the adjoint quantic as the facients of the cobezoutiant, the cobezoutiant is an m-ary quadric, the coefficients of which are lineo-linear functions of the coefficients of the two quantics, and the number of the cobezoutiants is $\tfrac{1}{2}m$ or $\tfrac{1}{2}(m-1)$, according as m is even or odd.

87. If the two quantics are the differential coefficients, or first derived functions (with respect to the facients) of a single quantic

$$(a, b, \ldots \between x, y)^m,$$

then we have what are termed the *Cobezoutoids* of the single quantic, viz. the cobezoutoid is an invariant of the single quantic of the order m, and of an adjoint quantic of the order $(m-2)$; and treating the coefficients of the adjoint quantic as facients, the cobezoutoid is an $(m-1)$ary quadric, the coefficients of which are quadric functions of the coefficients of the given quantic. The number of the cobezoutoids is $\frac{1}{2}(m-1)$ or $\frac{1}{2}(m-2)$, according as m is odd or even.

88. Consider any two quantics of the same order,

$$(a, \ldots \between x, y)^m, \quad (a', \ldots \between x, y)^m,$$

and introducing the new facients (X, Y), form the quotient of determinants,

$$\begin{vmatrix} (a, \ldots \between x\,, y\,)^m, & (a\,, \ldots \between x\,, y\,)^m \\ (a, \ldots \between X, Y)^m, & (a', \ldots \between X, Y)^m \end{vmatrix} \div \begin{vmatrix} x\,, y \\ X, Y \end{vmatrix}$$

which is obviously an integral function of the order $(m-1)$ in each set of facients separately, and lineo-linear in the coefficients of the two quantics; for instance, if the two quantics are

$$(a\,, b\,, c\,, d \between x, y)^3,$$
$$(a', b', c', d' \between x, y)^3,$$

the quotient in question may be written

$$\left(\begin{array}{lll} 3\,(ab'-a'b), & 3\,(ac'-a'c) \quad , & ad'-a'd \\ 3\,(ac'-a'c), & ad'-a'd+9\,(bc'-b'c), & 3\,(bd'-b'd) \\ ad'-a'd\,, & 3\,(bd'-b'd) \quad , & 3\,(cd'-c'd) \end{array}\right\between x, y)^2\,(X, Y)^2.$$

The function so obtained may be termed the *Bezoutic Emanant* of the two quantics.

89. The notion of such function was in fact suggested to me by Bezout's abbreviated process of elimination, viz. the two quantics of the order m being put equal to zero, the process leads to $(m-1)$ equations each of the order $(m-1)$: these equations are nothing else than the equations obtained by equating to zero the coefficients of the different terms of the series $(X, Y)^{m-1}$ in the Bezoutic emanant, and the result of the elimination is consequently obtained by equating to zero the determinant formed with the matrix which enters into the expression of the Bezoutic emanant. In other words, this determinant is the Resultant of the two quantics. Thus the resultant of the last-mentioned two cubics is the determinant

$$\begin{vmatrix} 3\,(ab'-a'b), & 3\,(ac'-a'c) \quad , & ad'-a'd \\ 3\,(ac'-a'c), & ad'-a'd+9\,(bc'-b'c), & 3\,(bd'-b'd) \\ ad'-a'd\,, & 3\,bd'-b'd \quad , & 3\,(cd'-c'd) \end{vmatrix}$$

90. If the two quantics are the differential coefficients or first derived functions (with respect to the facients) of a single quantic of the order m, then we have in like manner the *Bezoutoidal Emanant* of the single quantic; this is a function of the order $(m-2)$ in each set of facients, and the coefficients whereof are quadric functions of the coefficients of the single quantic. Thus the Bezoutoidal emanant of the quartic

$$(a, b, c, d, e \between x, y)^4$$

is

$$\left(\begin{array}{lll} 3(ac-b^2), & 3(ad-bc), & ae-bd \\ 3(ad-bc), & ae+8bd-9c^2, & 3(be-cd) \\ ae-bd, & 3(be-cd), & 3(ce-d^2) \end{array}\right| \between x, y)^2 (X, Y)^2$$

and of course the determinant formed with the matrix which enters into the expression of the Bezoutoidal Emanant, is the discriminant of the single quantic.

91. Professor Sylvester forms with the matrix of the Bezoutic emanant and a set of m facients $(u, v, \ldots)$ an m-ary quadric function, which he terms the *Bezoutiant.* Thus the Bezoutiant of the before-mentioned two cubics is

$$\left(\begin{array}{lll} 3(ab'-a'b), & 3(ac'-a'c), & ad'-a'd \\ 3(ac'-a'c), & ad'-a'd+9(bc'-b'c), & 3(bd'-b'd) \\ ad'-a'd, & 3bd'-b'd, & 3(cd'-c'd) \end{array}\right| \between u, v, w)^2;$$

and in like manner with the Bezoutoidal emanant of the single quantic of the order m and a set of $(m-1)$ new facients $(u, v, \ldots)$, an $(m-1)$ary quadric function, which he terms the *Bezoutoid.* Thus the Bezoutoid of the before-mentioned quartic is

$$\left(\begin{array}{lll} 3(ac-b^2), & 3(ad-bc), & ae-bd \\ 3(ad-bc), & ae+8bd-9c^2, & 3(be-cd) \\ ae-bd, & 3(be-cd), & 3(ce-d^2) \end{array}\right| \between u, v, w)^2.$$

To him also is due the important theorem, that the Bezoutiant is an invariant of the two quantics of the order m and of the adjoint quantic $(u, v, \ldots \between y, -x)^{m-1}$, being in fact a linear function with mere numerical coefficients of the invariants called Cobezoutiants, and in like manner that the Bezoutoid is an invariant of the single quantic of the order m and of the adjoint quantic $(u, v, \ldots \between y, -x)^{m-2}$, being a linear function with mere numerical coefficients of the invariants called Cobezoutoids.

The modes of generation of a covariant are infinite in number, and it is to be anticipated that, as new theories arise, there will be frequent occasion to consider new processes of derivation, and to single out and to define and give names to new covariants. But I have now, I think, established the greater part by far of the definitions which are for the present necessary.

156.

A FIFTH MEMOIR UPON QUANTICS.

[From the *Philosophical Transactions of the Royal Society of London*, vol. CXLVIII. *for the year* 1858, pp. 429—460. Received February 11,—Read March 18, 1858.]

THE present memoir was originally intended to contain a development of the theories of the covariants of certain binary quantics, viz. the quadric, the cubic, and the quartic; but as regards the theories of the cubic and the quartic, it was found necessary to consider the case of two or more quadrics, and I have therefore comprised such systems of two or more quadrics, and the resulting theories of the harmonic relation and of involution, in the subject of the memoir; and although the theory of homography or of the anharmonic relation belongs rather to the subject of bipartite binary quadrics, yet from its connexion with the theories just referred to, it is also considered in the memoir. The paragraphs are numbered continuously with those of my former memoirs on the subject: Nos. 92 to 95 relate to a single quadric; Nos. 96 to 114 to two or more quadrics, and the theories above referred to; Nos. 115 to 127 to the cubic, and Nos. 128 to 145 to the quartic. The several quantics are considered as expressed not only in terms of the coefficients, but also in terms of the roots,—and I consider the question of the determination of their linear factors,—a question, in effect, identical with that of the solution of a quadric, cubic, or biquadratic equation. The expression for the linear factor of a quadric is deduced from a well-known formula; those for the linear factors of a cubic and a quartic were first given in my "Note sur les Covariants d'une fonction quadratique, cubique ou biquadratique à deux indéterminées," Crelle, vol. L. (1855), pp. 285—287, [135]. It is remarkable that they are in one point of view more simple than the expression for the linear factor of a quadric.

92. In the case of a quadric the expressions considered are

$$(a,\ b,\ c\between x,\ y)^2, \tag{1}$$

$$ac-b^2\qquad , \tag{2}$$

where (1) is the quadric, and (2) is the discriminant, which is also the quadrinvariant, catalecticant, and Hessian.

And where it is convenient to do so, I write

$$(1) = U,$$
$$(2) = \square.$$

93. We have

$$(\partial_c, -\partial_b, \partial_a \between x, y)^2 \square = U,$$

which expresses that the evectant of the discriminant is equal to the quadric;

$$(a, b, c \between \partial_y, -\partial_x)^2 U = 4\square,$$

which expresses that the provectant of the quadric is equal to the discriminant;

$$(a, b, c \between bx + cy, -ax - by)^2 = \square U,$$

which expresses that a transmutant of the quadric is equal to the product of the quadric and the discriminant.

94. When the quadric is expressed in terms of the roots, we have

$$a^{-1} U = (x - \alpha y)(x - \beta y),$$
$$a^{-2} \square = -\tfrac{1}{4}(\alpha - \beta)^2;$$

and in the case of a pair of equal roots,

$$a^{-1} U = (x - \alpha y)^2,$$
$$\square = 0.$$

95. The problem of the solution of a quadratic equation is that of finding a linear factor of the quadric. To obtain such linear factor in a symmetrical form, it is necessary to introduce arbitrary quantities which do not really enter into the solution, and the form obtained is thus in some sort more complicated than in the like problem for a cubic or a quartic. The solution depends on the linear transformation of the quadric, viz. if we write

$$(a, b, c \between \lambda x + \mu y, \nu x + \rho y)^2 = (a', b', c' \between x, y)^2,$$

so that

$$a' = (a, b, c \between \lambda, \nu)^2,$$
$$b' = (a, b, c \between \lambda, \nu \between \mu, \rho),$$
$$c' = (a, b, c \between \mu, \rho)^2,$$

then

$$a'c' - b'^2 = (ac - b^2)(\lambda\rho - \mu\nu)^2,$$

an equation which in a different notation is

$$(a, b, c \between x, y)^2 . (a, b, c \between X, Y)^2 - \{(a, b, c \between x, y \between X, Y)\}^2 = \square (Yx - Xy)^2,$$

in which form it is a theorem relating to the quadric and its first and second emanants. The equation shows that

$$(a,\ b,\ c \between x,\ y \between X,\ Y) + \sqrt{-\square}\,(Yx - Xy),$$

where $(X,\ Y)$ are treated as supernumerary arbitrary constants, is a linear factor of $(a,\ b,\ c \between x,\ y)^2$, and this is the required solution.

96. In the case of two quadrics, the expressions considered are

$$(a,\ b,\ c \between x,\ y)^2, \tag{1}$$
$$(a',\ b',\ c' \between x,\ y)^2, \tag{2}$$
$$ac - b^2\ , \tag{3}$$
$$ac' - 2bb' + ca', \tag{4}$$
$$a'c' - b'^2\ , \tag{5}$$

, (6)

$$(ab' - a'b,\quad ac' - a'c\ ,\ bc' - b'c \between x,\ y)^2, \tag{7}$$
$$(\lambda a + \mu a',\quad \lambda b + \mu b'\ ,\ \lambda c + \mu c' \between x,\ y)^2, \tag{8}$$
$$(ac - b^2\ ,\ ac' - 2bb' + ca',\ a'c' - b'^2 \between \lambda,\ \mu)^2, \tag{9}$$

(1) and (2) are the quadrics, (3) and (5) are the discriminants, and (4) is the lineo-linear invariant, or connective of the discriminants; (6) is the resultant of the two quadrics, (7) is the Jacobian, (8) is an intermediate, and (9) is the discriminant of the intermediate. And where it is convenient to do so, I write

$$(1) = U,$$
$$(2) = U',$$
$$(3) = \square,$$
$$(4) = Q,$$
$$(5) = \square',$$
$$(6) = R,$$
$$(7) = H,$$
$$(8) = W,$$
$$(9) = \Theta.$$

97. The Jacobian (7) may also be written in the form

$$\begin{vmatrix} y^2, -yx, & x^2 \\ a\,, & b\,, & c \\ a', & b'\,, & c' \end{vmatrix}.$$

The Resultant (6) may be written in the form

$$\begin{vmatrix} & a\,, & 2b\,, & c \\ a\,, & 2b\,, & c\,, & \\ & a', & 2b', & c' \\ a', & 2b', & c', & \end{vmatrix},$$

and also, taken negatively, in the form

$$4(ab'-a'b)(bc'-b'c)-(ac'-a'c)^2,$$

which is the discriminant of the Jacobian; and in the form

$$4(ac-b^2)(a'c'-b'^2)-(ac'-2bb'+ca')^2,$$

which is the discriminant of the Intermediate.

98. We have the following relations:

$$\begin{aligned} (a,\ b,\ c \between b'x+c'y,\ -a'x-b'y)^2 &= -(a'c'-b'^2) & (a\,,\ b\,,\ c \between x,\ y)^2 \\ &\quad +(ac'-2bb'+ca') & (a',\ b',\ c' \between x,\ y)^2, \\ (a',\ b',\ c' \between bx+cy,\ -ax-by)^2 &= +(ac'-2bb'+ca') & (a\,,\ b\,,\ c \between x,\ y)^2 \\ &\quad -(ac-b^2) & (a',\ b',\ c' \between x,\ y)^2, \end{aligned}$$

and moreover

$$\begin{aligned} &(ac-b^2,\ ac'-2bb'+ca',\ a'c'-b'^2 \between U',\ -U)^2 \\ &\qquad = -\{(ab'-a'b,\ ac'-a'c,\ bc'-b'c \between x,\ y)^2\}^2, \end{aligned}$$

an equation, the interpretation of which will be considered in the sequel.

99. The most important relations which may exist between the two quadrics are:

First, when the connective vanishes, or

$$ac'-2bb'+ca'=0,$$

in which case the two quadrics are said to be *harmonically* related: the nature of this relation will be further considered.

Secondly, when $R=0$, the two quadrics have in this case a common root, which is given by any of the equations,

$$\begin{aligned} x^2 : 2xy : y^2 &= \partial_a R : \partial_b R : \partial_c R \\ &= \partial_{a'} R : \partial_{b'} R : \partial_{c'} R \\ &= bc'-b'c : ca'-c'a : ab'-a'b. \end{aligned}$$

The last set of values express that the Jacobian is a perfect square, and that the two roots are each equal to the common root of the two quadrics.

The preceding values of the ratios $x^2 : 2xy : y^2$ are consistent with each other in virtue of the assumed relation $R=0$, hence in general the functions

$$4\partial_a R \,.\, \partial_c R - (\partial_b R)^2,\ \partial_a R \,.\, \partial_{b'} R - \partial_b R \,.\, \partial_{a'} R, \ \&\text{c},$$

all of them contain the Resultant R as a factor.

It is easy to see that the Jacobian is harmonically related to each of the quadrics; in fact we have identically

$$\begin{aligned} a\,(bc'-b'c) + b\,(ca'-c'a) + c\,(ab'-a'b) &= 0, \\ a'\,(bc'-b'c) + b'\,(ca'-c'a) + c'\,(ab'-a'b) &= 0, \end{aligned}$$

which contain the theorem in question.

100. When the quadrics are expressed in terms of the roots, we have

$$\begin{aligned} a^{-1}\,U &= (x-\alpha\, y)(x-\beta\, y), \\ a'^{-1}\,U' &= (x-\alpha' y)(x-\beta' y), \\ 4\,a^{-2}\,\square &= -(\alpha-\beta)^2, \\ 2\,(aa')^{-1} Q &= 2\alpha\beta + 2\alpha'\beta' - (\alpha+\beta)(\alpha'+\beta'), \\ 4\,a'^{-2}\,\square' &= -(\alpha'-\beta')^2, \\ (aa')^{-2}\,R &= (\alpha-\alpha')(\alpha-\beta')(\beta-\alpha')(\beta-\beta'), \\ (aa')^{-1}\,H &= \begin{vmatrix} y^2, & 2yx, & x^2 \\ 1, & \alpha+\beta, & \alpha\beta \\ 1, & \alpha'+\beta', & \alpha'\beta' \end{vmatrix}. \end{aligned}$$

101. The comparison of the last-mentioned value of R with the expression in terms of the roots obtained from the equation

$$-R = 4\square\,\square' - Q^2,$$

gives the identical equation

$$(\alpha-\beta)^2(\alpha'-\beta')^2 - \{2\alpha\beta + 2\alpha'\beta' - (\alpha+\beta)(\alpha'+\beta')\}^2 = -4\,(\alpha-\alpha')(\alpha-\beta')(\beta-\alpha')(\beta-\beta'),$$

which may be easily verified.

102. We have identically

$$\begin{aligned}
&2\alpha\beta+2\alpha'\beta'-(\alpha+\beta)(\alpha'+\beta')\\
&\quad=2(\alpha-\alpha')(\alpha-\beta')-(\alpha-\beta)(2\alpha-\alpha'-\beta')\\
&\quad=2(\beta-\alpha')(\beta-\beta')-(\beta-\alpha)(2\beta-\alpha'-\beta')\\
&\quad=2(\alpha'-\alpha)(\alpha'-\beta)-(\alpha'-\beta')(2\alpha'-\alpha-\beta)\\
&\quad=2(\beta'-\alpha)(\beta'-\beta)-(\beta'-\alpha')(2\beta'-\alpha-\beta);
\end{aligned}$$

and the equation $Q=ac'-2bb'+ca'=0$ may consequently be written in the several forms

$$\frac{2}{\alpha-\beta}=\frac{1}{\alpha-\alpha'}+\frac{1}{\alpha-\beta'},$$

$$\frac{2}{\beta-\alpha}=\frac{1}{\beta-\alpha'}+\frac{1}{\beta'-\beta},$$

$$\frac{2}{\alpha'-\beta'}=\frac{1}{\alpha'-\alpha}+\frac{1}{\alpha'-\beta},$$

$$\frac{2}{\beta'-\alpha'}=\frac{1}{\beta'-\alpha}+\frac{1}{\beta'-\beta},$$

so that the roots (α, β), (α', β') are harmonically related to each other, and hence the notion of the harmonic relation of the two quadrics.

103. In the case where the two quadrics have a common root $\alpha=\alpha'$,

$$\begin{aligned}
a^{-1}\,U &= (x-\alpha y)(x-\beta y),\\
a'^{-1}\,U' &= (x-\alpha y)(x-\beta' y),\\
4\,a^{-2}\,\square &= -(\alpha-\beta)^2,\\
2\,(aa')^{-1}Q &= (\alpha-\beta)(\alpha-\beta'),\\
4a'^{-2}\,\square' &= -(\alpha-\beta')^2,\\
R &= 0,\\
(aa')^{-1}\,H &= (\beta'-\beta)(x-\alpha y)^2.
\end{aligned}$$

104. In the case of three quadrics, of the expressions which are or might be considered, it will be sufficient to mention

$$(a\ ,\ b\ ,\ c\ \between x,\ y)^2, \qquad (1)$$

$$(a',\ b',\ c'\ \between x,\ y)^2, \qquad (2)$$

$$(a'',\ b'',\ c''\ \between x,\ y)^2, \qquad (3)$$

$$\left|\begin{matrix} a\ , & b\ , & c \\ a', & b', & c' \\ a'', & b'', & c'' \end{matrix}\right|, \qquad (4)$$

where (1), (2), (3) are the quadrics themselves, and (4) is an invariant, linear in the coefficients of each quadric. And where it is convenient to do so, I write

$$(1) = U,$$
$$(2) = U',$$
$$(3) = U'',$$
$$(4) = \Omega.$$

105. The equation $\Omega = 0$ is, it is clear, the condition to be satisfied by the coefficients of the three quadrics, in order that there may be a syzygetic relation $\lambda U + \mu U' + \nu U'' = 0$, or what is the same thing, in order that each quadric may be an intermediate of the other two quadrics; or again, in order that the three quadrics may be *in Involution*. Expressed in terms of the roots, the relation is

$$\begin{vmatrix} 1, & \alpha + \beta, & \alpha\beta \\ 1, & \alpha' + \beta', & \alpha'\beta' \\ 1, & \alpha'' + \beta'', & \alpha''\beta'' \end{vmatrix} = 0;$$

and when this equation is satisfied, the three pairs, or as it is usually expressed, the six quantities $\alpha, \beta; \alpha', \beta'; \alpha'', \beta''$, are said to be in involution, or to form an involution. And the two perfectly arbitrary pairs $\alpha, \beta; \alpha', \beta'$ considered as belonging to such a system, may be spoken of as an involution. If the two terms of a pair are equal, e.g. if $\alpha'' = \beta'' = \theta$, then the relation is

$$\begin{vmatrix} 1, & 2\theta, & \theta^2 \\ 1, & \alpha + \beta, & \alpha\beta \\ 1, & \alpha' + \beta', & \alpha'\beta' \end{vmatrix} = 0;$$

and such a system is sometimes spoken of as an involution of five terms. Considering the pairs (α, β), (α', β') as given, there are of course two values of θ which satisfy the preceding equation; and calling these $\theta_{,}$ and $\theta_{,,}$, then $\theta_{,}$ and $\theta_{,,}$ are said to be the sibiconjugates of the involution $\alpha, \beta; \alpha', \beta'$. It is easy to see that $\theta_{,}, \theta_{,,}$ are the roots of the equation $H = 0$, where H is the Jacobian of the two quadrics U and U' whose roots are (α, β), (α', β'). In fact, the quadric whose roots are $\theta_{,}, \theta_{,,}$ is

$$\begin{vmatrix} y^2, & 2yx, & x^2 \\ 1, & \alpha + \beta, & \alpha\beta \\ 1, & \alpha' + \beta', & \alpha'\beta' \end{vmatrix}$$

which has been shown to be the Jacobian in question. But this may be made clearer as follows:—If we imagine that λ, μ are determined in such manner that the intermediate $\lambda U + \mu U'$ may be a perfect square, then we shall have $\lambda U + \mu U' = a''(x - \theta y)^2$, where θ denotes one or other of the sibiconjugates $\theta_{,}, \theta_{,,}$ of the involution. But the condition in order that $\lambda U + \mu U'$ may be a square is

$$(ac - b^2,\ ac' - 2bb' + ca',\ a'c' - b'^2 \between \lambda, \mu)^2;$$

and observing that the equation $\lambda : \mu = U' : -U$ implies $\lambda U + \mu U' = 0 = a''(x-\theta y)^2$, it is obvious that the function

$$(ac - b^2,\ ac' - 2bb' + ca',\ a'c' - b'^2 \between U',\ -U)^2$$

must be to a factor *près* equal to $(x-\theta_{,}y)^2 (x-\theta_{,,}y)^2$. But we have identically

$$(ac-b^2,\ ac'-2bb'+ca',\ a'c'-b'^2 \between U',\ -U)^2 = -\{(ab'-a'b,\ ac'-a'c,\ bc'-b'c \between x,\ y)^2\}^2,$$

and we thus see that $(x-\theta_{,}y)$, $(x-\theta_{,,}y)$ are the factors of the Jacobian.

106. It has been already remarked that the Jacobian is harmonically related to each of the quadrics U, U'; hence we see that the sibiconjugates $\theta_{,}$, $\theta_{,,}$ of the involution α, β, α', β' are a pair harmonically related to the pair α, β, and also harmonically related to the pair α', β', and this properly might be taken as the definition for the sibiconjugates $\theta_{,}$, $\theta_{,,}$ of an involution of four terms. And moreover, α, β; α', β' being given, and $\theta_{,}$, $\theta_{,,}$ being determined as the sibiconjugates of the involution, if α'', β'' be a pair harmonically related to $\theta_{,}$, $\theta_{,,}$, then the three pairs α, β; α', β'; α'', β'' will form an involution; or what is the same thing, any three pairs α, β; α', β'; α'', β'', each of them harmonically related to a pair $\theta_{,}$, $\theta_{,,}$, will be an involution, and $\theta_{,}$, $\theta_{,,}$ will be the sibiconjugates of the involution.

107. In particular, if α, β be harmonically related to $\theta_{,}$, $\theta_{,,}$, then it is easy to see that $\theta_{,}$, $\theta_{,}$ may be considered as harmonically related to $\theta_{,}$, $\theta_{,,}$, and in like manner $\theta_{,,}$, $\theta_{,,}$ will be harmonically related to $\theta_{,}$, $\theta_{,,}$; that is, the pairs $\theta_{,}$, $\theta_{,}$; $\theta_{,,}$, $\theta_{,,}$ and α, β will form an involution. This comes to saying that the equation

$$\begin{vmatrix} 1, & 2\theta_{,}\,, & \theta_{,}^2 \\ 1, & 2\theta_{,,}\,, & \theta_{,,}^2 \\ 1, & \alpha+\beta, & \alpha\beta \end{vmatrix} = 0$$

is equivalent to the harmonic relation of the pairs α, β; $\theta_{,}$, $\theta_{,,}$; and in fact the determinant is

$$(\theta_{,} - \theta_{,,})\,(2\alpha\beta + 2\theta_{,}\theta_{,,} - (\alpha+\beta)\,(\theta_{,} + \theta_{,,})),$$

which proves the theorem in question.

108. Before proceeding further, it is proper to consider the equation

$$\begin{vmatrix} 1, & \alpha, & \alpha', & \alpha\alpha' \\ 1, & \beta, & \beta', & \beta\beta' \\ 1, & \gamma, & \gamma', & \gamma\gamma' \\ 1, & \delta, & \delta', & \delta\delta' \end{vmatrix} = 0,$$

which expresses that the sets $(\alpha, \beta, \gamma, \delta)$ and $(\alpha', \beta', \gamma', \delta')$ are homographic; for although the homographic equation may be considered as belonging to the theory of the bipartite quadric $(x - \alpha y)(\mathrm{x} - \alpha'\mathrm{y})$, yet the theory of involution cannot be completely discussed except in connexion with that of homography. If we write

$$\begin{aligned} A &= (\beta - \gamma)(\alpha - \delta), & B &= (\gamma - \alpha)(\beta - \delta), & C &= (\alpha - \beta)(\gamma - \delta), \\ A' &= (\beta' - \gamma')(\alpha' - \delta'), & B' &= (\gamma' - \alpha')(\beta' - \delta'), & C' &= (\alpha' - \beta')(\gamma' - \delta'), \end{aligned}$$

then we have

$$A + B + C = 0,$$
$$A' + B' + C' = 0,$$

and thence

$$BC' - B'C = CA' - C'A = AB' - A'B;$$

and either of these expressions is in fact equal to the last-mentioned determinant, as may be easily verified. Hence, when the determinant vanishes, we have

$$A : B : C = A' : B' : C'.$$

Any one of the three ratios $A : B : C$, for instance the ratio $B : C$, =

$$\frac{(\gamma - \alpha)(\beta - \delta)}{(\alpha - \beta)(\gamma - \delta)},$$

is said to be the anharmonic ratio of the set $(\alpha, \beta, \gamma, \delta)$, and consequently the two sets $(\alpha, \beta, \gamma, \delta)$ and $(\alpha', \beta', \gamma', \delta')$ will be homographically related when the anharmonic ratios (that is, the corresponding anharmonic ratios) of the two sets are equal.

If any one of the anharmonic ratios be equal to unity, then the four terms of the set taken in a proper manner in pairs, will be harmonics; thus the equation $\frac{B}{C} = 1$ gives

$$\frac{(\gamma - \alpha)(\beta - \delta)}{(\alpha - \beta)(\gamma - \delta)} = 1,$$

which is reducible to

$$2\alpha\delta + 2\beta\gamma - (\alpha + \delta)(\beta + \gamma) = 0,$$

which expresses that the pairs α, δ and β, γ are harmonics.

109. Now returning to the theory of involution (and for greater convenience taking α, α' &c. instead of α, β &c. to represent the terms of the same pair), the pairs α, α'; β, β'; γ, γ'; δ, δ'; &c. will be in involution if each of the determinants formed with any three lines of the matrix

$$\begin{array}{lll} 1, & \alpha + \alpha', & \alpha\alpha', \\ 1, & \beta + \beta', & \beta\beta', \\ 1, & \gamma + \gamma', & \gamma\gamma', \\ 1, & \delta + \delta', & \delta\delta', \\ \&c. & & \end{array}$$

vanishes: but this being so, the determinant

$$\begin{vmatrix} 1, & \alpha, & \alpha', & \alpha\alpha' \\ 1, & \beta, & \beta', & \beta\beta' \\ 1, & \gamma, & \gamma', & \gamma\gamma' \\ 1, & \delta, & \delta', & \delta\delta' \end{vmatrix}$$

which is equal to

$$\begin{vmatrix} \alpha, & 1, & \alpha+\alpha', & \alpha\alpha' \\ \beta, & 1, & \beta+\beta', & \beta\beta' \\ \gamma, & 1, & \gamma+\gamma', & \gamma\gamma' \\ \delta, & 1, & \delta+\delta', & \delta\delta' \end{vmatrix}$$

will vanish, or the two sets $(\alpha, \beta, \gamma, \delta)$ and $(\alpha', \beta', \gamma', \delta')$ will be homographic; that is, if any number of pairs are in involution, then, considering four pairs and selecting in any manner a term out of each pair, these four terms and the other terms of the same four pairs form respectively two sets, and the two sets so obtained will be homographic.

110. In particular, if we have only three pairs α, α'; β, β'; γ, γ', then the sets $\alpha, \beta, \gamma, \alpha'$ and $\alpha', \beta', \gamma', \alpha$ will be homographic; in fact, the condition of homography is

$$\begin{vmatrix} 1, & \alpha, & \alpha', & \alpha\alpha' \\ 1, & \beta, & \beta', & \beta\beta' \\ 1, & \gamma, & \gamma', & \gamma\gamma' \\ 1, & \alpha', & \alpha, & \alpha\alpha' \end{vmatrix} = 0,$$

which may be written

$$\begin{vmatrix} \alpha, & 1, & \alpha+\alpha', & \alpha\alpha' \\ \beta, & 1, & \beta+\beta', & \beta\beta' \\ \gamma, & 1, & \gamma+\gamma', & \gamma\gamma' \\ \alpha', & 1, & \alpha+\alpha', & \alpha\alpha' \end{vmatrix} = 0,$$

or what is the same thing,

$$\begin{vmatrix} \alpha & , 1, & \alpha+\alpha', & \alpha\alpha' \\ \beta & , 1, & \beta+\beta', & \beta\beta' \\ \gamma & , 1, & \gamma+\gamma', & \gamma\gamma' \\ \alpha'-\alpha, & 0, & 0 & , 0 \end{vmatrix} = 0,$$

so that the first-mentioned relation is equivalent to

$$(\alpha'-\alpha)\begin{vmatrix} 1, & \alpha+\alpha', & \alpha\alpha' \\ 1, & \beta+\beta', & \beta\beta' \\ 1, & \gamma+\gamma', & \gamma\gamma' \end{vmatrix} = 0,$$

and the two sets give rise to an involution. The condition of homography as expressed by the equality of the anharmonic ratios may be written

$$\frac{\alpha-\beta\,.\,\gamma-\alpha'}{\alpha-\gamma\,.\,\alpha'-\beta}=\frac{\alpha'-\beta'\,.\,\gamma'-\alpha}{\alpha'-\gamma'\,.\,\alpha-\beta'};$$

or multiplying out,

$$(\alpha-\beta)(\alpha-\beta')(\alpha'-\gamma)(\alpha'-\gamma')-(\alpha'-\beta)(\alpha'-\beta')(\alpha-\gamma)(\alpha'-\gamma')=0,$$

which is a form for the equation of involution of the three pairs. But this and the other transformations of the equation of involution is best obtained by a different method, as will be presently seen.

111. Imagine now any number of pairs α, α'; β, β'; γ, γ'; δ, δ'; &c. in involution, and let x, y, z, w be the fourth harmonics of the same quantity λ with respect to the pairs α, α'; β, β'; γ, γ' and δ, δ' respectively; then the anharmonic ratios of the set (x, y, z, w) will be independent of λ, or what is the same thing, if x', y', z', w' are the fourth harmonics of any other quantity λ' with respect to the same four pairs, the sets (x, y, z, w) and (x', y', z', w') will be homographic, or we shall have

$$\begin{vmatrix} 1, & x, & x', & xx' \\ 1, & y, & y', & yy' \\ 1, & z, & z', & zz' \\ 1, & w, & w', & ww' \end{vmatrix}=0.$$

It will be sufficient to show this in the case where λ is anything whatever, but λ' has a determinate value, say $\lambda'=\infty$; and since if all the terms α, α', &c. are diminished by the same quantity λ the relations of involution and homography will not be affected, we may without loss of generality assume $\lambda=0$, but in this case

$$x=\frac{2\alpha\alpha'}{\alpha+\alpha'},\quad x'=\tfrac{1}{2}(\alpha+\alpha'),$$

and the equation to be proved is

$$\begin{vmatrix} 1, & \dfrac{\alpha\alpha'}{\alpha+\alpha'}, & \alpha+\alpha', & \alpha\alpha' \\ 1, & \dfrac{\beta\beta'}{\beta+\beta'}, & \beta+\beta', & \beta\beta' \\ 1, & \dfrac{\gamma\gamma'}{\gamma+\gamma'}, & \gamma+\gamma', & \gamma\gamma' \\ 1, & \dfrac{\delta\delta'}{\delta+\delta'}, & \delta+\delta', & \delta\delta' \end{vmatrix}=0,$$

which is obviously a consequence of the equations which express the involution of the four pairs.

A set homographic with x, y, z w, which are the fourth harmonics of any quantity whatever λ with respect to the pairs in involution, α, α'; β, β'; γ, γ'; δ, δ', is said to be homographic with the four pairs, and we have thus the notion of a set of single quantities homographic with a set of pairs in involution. This very important theory is due to M. Chasles.

112. Let r; s; t be the anharmonic ratios of a set $\alpha, \beta, \gamma, \delta$, and let $r_,$; $s_,$; $t_,$ be the anharmonic ratios (corresponding or not corresponding) of a set $\alpha_,, \beta_,, \gamma_,, \delta_,$. And suppose that r'; s'; t'; $r_,'$; $s_,'$; $t_,'$; r''; s''; t''; $r_,''$; $s_,''$; $t_,''$; r'''; s'''; t'''; $r_,'''$; $s_,'''$; $t_,'''$, are the analogous quantities for three other pairs of sets; then an equation such as

$$\left|\begin{array}{cccc} 1, & \dfrac{r}{s}, & \dfrac{r_,}{s_,}, & \dfrac{rr_,}{ss_,} \\ \vdots & & & \end{array}\right| = 0,$$

or as it is more conveniently written,

$$\left|\begin{array}{cccc} ss_, \ , & rs_, \ , & r_,s \ , & rr_, \\ s's_,' \ , & r's_,' \ , & r_,'s' \ , & r'r_,' \\ s''s_,'' \ , & r''s_,'' \ , & r_,''s'' \ , & r''r_,'' \\ s'''s_,''', & r'''s_,''', & r_,'''s''', & r'''r_,''' \end{array}\right| = 0$$

is a relation independent of the particular ratios $r : s$ which have been chosen for the anharmonic ratios of the sets; this is easily shown by means of the equations

$$r + s + t = 0, \quad r_, + s_, + t_, = 0,$$

which connect the anharmonic ratios. The equation in fact expresses a certain relation between four sets $(\alpha, \beta, \gamma, \delta)$ and four other sets $(\alpha_,, \beta_,, \gamma_,, \delta_,)$; a relation which may be termed the relation of the homography of the anharmonic ratios of four and four sets: the notion of this relation is also due to M. Chasles.

113. The general relation

$$\left|\begin{array}{ccc} 1, & \alpha + \beta \ , & \alpha\beta \\ 1, & \alpha' + \beta', & \alpha'\beta' \\ 1, & \alpha'' + \beta'', & \alpha''\beta'' \end{array}\right| = 0$$

may be exhibited in a great variety of forms. In fact, if the determinant is denoted by Υ, then multiplying by this determinant the two sides of the identical equation

$$\left|\begin{array}{ccc} u^2, & -u, & 1 \\ v^2, & -v, & 1 \\ w^2, & -w, & 1 \end{array}\right| = (u-v)(v-w)(w-u),$$

we obtain

$$\Upsilon (u-v)(v-w)(w-u) = \left|\begin{array}{ccc} (u-\alpha\ )(u-\beta\ ), & (v-\alpha\ )(v-\beta\ ), & (w-\alpha\ )(w-\beta\ ) \\ (u-\alpha')(u-\beta'), & (v-\alpha')(v-\beta'), & (w-\alpha')(w-\beta') \\ (u-\alpha'')(u-\beta''), & (v-\alpha'')(v-\beta''), & (w-\alpha'')(w-\beta'') \end{array}\right| .$$

If, for example, $u=\alpha$, $v=\beta$, then we have

$$\Upsilon(\alpha-\beta)=-(\alpha-\alpha')(\alpha-\beta')(\beta-\alpha'')(\beta-\beta'')+(\beta-\alpha')(\beta-\beta')(\alpha-\alpha'')(\alpha-\beta'');$$

and again, if $u=\alpha$, $v=\alpha'$, $w=\alpha''$, then we have

$$\Upsilon=-(\alpha-\beta'')(\alpha'-\beta)(\alpha''-\beta')+(\alpha-\beta')(\alpha'-\beta'')(\alpha''-\beta).$$

Putting $\Upsilon=0$, the two equations give respectively

$$\frac{(\alpha-\alpha')(\beta-\alpha'')}{(\alpha-\alpha'')(\alpha'-\beta)}=\frac{(\alpha-\beta'')(\beta-\beta')}{(\alpha-\beta')(\beta''-\beta)};$$

and

$$(\alpha-\beta'')(\alpha'-\beta)(\alpha''-\beta')=(\alpha-\beta')(\alpha'-\beta'')(\alpha''-\beta),$$

which are both of them well-known forms.

114. A corresponding transformation applies to the equation

$$\begin{vmatrix} 1, & \alpha, & \alpha', & \alpha\alpha' \\ 1, & \beta, & \beta', & \beta\beta' \\ 1, & \gamma, & \gamma', & \gamma\gamma' \\ 1, & \delta, & \delta', & \delta\delta' \end{vmatrix}=0,$$

which expresses the homography of two pairs. In fact, calling the determinant Ψ and representing by V the similar determinant

$$\begin{vmatrix} ss', & -s', & -s, & 1 \\ tt', & -t', & -t, & 1 \\ uu', & -u', & -u, & 1 \\ vv', & -v', & -v, & 1 \end{vmatrix},$$

which, equated to zero, would express the homography of the sets (s, t, u, v) and (s', t', u', v'), we have

$$V\Psi=\begin{vmatrix} (s-\alpha)(s'-\alpha'), & (s-\beta)(s'-\beta'), & (s-\gamma)(s'-\gamma'), & (s-\delta)(s'-\delta') \\ (t-\alpha)(t'-\alpha'), & (t-\beta)(t'-\beta'), & (t-\gamma)(t'-\gamma'), & (t-\delta)(t'-\delta') \\ (u-\alpha)(u'-\alpha'), & (u-\beta)(u'-\beta'), & (u-\gamma)(u'-\gamma'), & (u-\delta)(u'-\delta') \\ (v-\alpha)(v'-\alpha'), & (v-\beta)(v'-\beta'), & (v-\gamma)(v'-\gamma'), & (v-\delta)(v'-\delta') \end{vmatrix},$$

which gives various forms of the equation of homography. In particular, if $s=\alpha$, $s'=\beta'$, $t=\beta$, $t'=\alpha'$, $u=\gamma$, $u'=\delta'$, $v=\delta$, $v'=\gamma$, then

$$V\Psi=\begin{vmatrix} . & . & (\alpha-\gamma)(\beta'-\gamma'), & (\alpha-\delta)(\beta'-\delta') \\ . & . & (\beta-\gamma)(\alpha'-\gamma'), & (\beta-\delta)(\alpha'-\delta') \\ (\gamma-\alpha)(\delta'-\alpha'), & (\gamma-\beta)(\delta'-\beta') & . & . \\ (\delta-\alpha)(\gamma'-\alpha'), & (\delta-\beta)(\gamma'-\beta') & . & . \end{vmatrix},$$

and the right-hand side breaks up into factors, which are equal to each other (whence also $V = \Psi$), and the equation $\dot{\Psi} = 0$ takes the form

$$(\alpha-\gamma)(\beta-\delta)(\alpha'-\delta')(\beta'-\gamma')-(\alpha-\delta)(\beta-\gamma)(\alpha'-\gamma')(\beta'-\delta')=0,$$

which is, in fact, one of the equations which express the equality of the anharmonic ratios of $(\alpha, \beta, \gamma, \delta)$ and $(\alpha', \beta', \gamma', \delta')$.

115. In the case of a cubic, the expressions considered are

$$(a, b, c, d \between x, y)^3, \qquad (1)$$

$$(ac-b^2, ad-bc, bd-c^2 \between x, y)^2, \qquad (2)$$

$$\left(\begin{array}{l} -a^2d+3abc-2b^3 \\ -abd+2ac^2-b^2c \\ +acd-2b^2d+bc^2 \\ +ad^2-3bcd+2c^3 \end{array} \between x, y\right)^3, \qquad (3)$$

$$a^2d^2-6abcd+4ac^3+4b^3d-3b^2c^2, \qquad (4)$$

where (1) is the cubic, (2) is the quadricovariant or Hessian, (3) is the cubicovariant, and (4) is the quartinvariant or discriminant.

And where it is convenient to do so, I write

$$(1) = U,$$
$$(2) = H,$$
$$(3) = \Phi,$$
$$(4) = \square,$$

so that we have

$$\Phi^2 - \square U^2 + 4H^3 = 0.$$

116. The Hessian may be written under the form

$$(ax+by)(cx+dy)-(bx+cy)^2,$$

(which, indeed, is the form under which *quâ* Hessian it is originally given), and under the form

$$\begin{vmatrix} y^2, & -yx, & x^2 \\ a, & b, & c \\ b, & c, & d \end{vmatrix}.$$

The cubicovariant may be written under the form

$$\begin{aligned} &\{2(ac-b^2)x+(ad-bc)y\}(bx^2+2cxy+dy^2) \\ -&\{(ad-bc)x+2(bd-c^2)y\}(ax^2+2bxy+cy^2), \end{aligned}$$

that is, as the Jacobian of the cubic and Hessian; and under the form

$$\tfrac{1}{2}(\partial_a,\ \partial_b,\ \partial_c,\ \partial_d \between y,\ -x)^3 \square,$$

that is, as the evectant of the discriminant.

The discriminant, taken negatively, may be written under the form

$$+4(ac-b^2)(bd-c^2)-(ad-bc)^2,$$

that is, as the discriminant of the Hessian.

117. We have

$$(a,\ b,\ c,\ d \between bx^2+2cxy+dy^2,\ -ax^2-2bxy-cy^2)^3 = U\Phi,$$

which expresses that a transmutant of the cubic is the product of the cubic and the cubicovariant. The equation

$$\{(\partial_a,\ \partial_b,\ \partial_c,\ \partial_d \between y,\ -x)^3\}^2 \square = 2U^2$$

expresses that the second evectant of the discriminant is the square of the cubic.

The equation

$$\begin{vmatrix} d^2 & , -3cd & , -3bd+6c^2 & , -3bc+2ad \\ -3cd & , -3c^3+12bd, & -3ad-6bc & , -3ac+6b^2 \\ -3bd+6c^2 & , -3ad-6bc & , -3b^2+12ac, & -3ab \\ -3bc-12ad, & -3ac+6b^2 & , \quad 3ab & , \quad a^2 \end{vmatrix} = 27\square^2$$

expresses that the determinant formed with the second differential coefficients of the discriminant gives the square of the discriminant.

The covariants of the intermediate $\alpha U+\beta\Phi$ are as follows, viz.

118. For the Hessian, we have

$$\begin{aligned}\tilde{H}(\alpha U+\beta\Phi) &= (1,\ 0,\ -\square \between \alpha,\ \beta)^2 H \\ &= (\alpha^2-\beta^2\square)\,H;\end{aligned}$$

for the cubicovariant,

$$\begin{aligned}\tilde{\Phi}(\alpha U+\beta\Phi) &= (0,\ \square,\ 0,\ -\square^2 \between \alpha,\ \beta)^3 U \\ &\quad +(1,\ 0,\ -\square,\ 0 \between \alpha,\ \beta)^3 \Phi \\ &= (\alpha^2-\beta^2\square)(\alpha\Phi+\beta\square U);\end{aligned}$$

and for the discriminant,

$$\begin{aligned}\tilde{\square}(\alpha U+\beta\Phi) &= (1,\ 0,\ -2\square,\ 0,\ \square^2 \between \alpha,\ \beta)^4 \Phi \\ &= (\alpha^2-\beta^2\square)^2\square,\end{aligned}$$

where on the left-hand sides I have, for greater distinctness, written $\tilde{H}$, &c. to denote the functional operation of taking the Hessian, &c. of the operand $\alpha U+\beta\Phi$.

In particular, if $\alpha = 0$, $\beta = 1$,

$$\tilde{H}\Phi = -\square \,.\, H,$$

$$\tilde{\Phi}\Phi = -\square^2 \,.\, U,$$

$$\tilde{\square}\Phi = \square^3 \,.$$

119. Solution of a cubic equation.

The question is to find a linear factor of the cubic

$$(a, b, c, d \between x, y)^3,$$

and this can be at once effected by means of the relation

$$\Phi^2 - \square U^2 = -4H^3$$

between the covariants. The equation in fact shows that each of the expressions

$$\tfrac{1}{2}(\Phi + U\sqrt{\square}),\quad \tfrac{1}{2}(\Phi - U\sqrt{\square})$$

is a perfect cube, and consequently that the cube root of each of these expressions is a linear function of (x, y). The expression

$$\sqrt[3]{\tfrac{1}{2}(\Phi + U\sqrt{\square})} + \sqrt[3]{\tfrac{1}{2}(\Phi - U\sqrt{\square})}$$

is consequently a linear function of x, y, and it vanishes when $U = 0$, that is, the expression is a linear factor of the cubic.

It may be noticed here that the cubic being $a(x - \alpha y)(x - \beta y)(x - \gamma y)$, then we may write

$$\sqrt[3]{\tfrac{1}{2}(\Phi + U\sqrt{\square})} - \sqrt[3]{\tfrac{1}{2}(\Phi - U\sqrt{\square})} = \tfrac{1}{3}a(\omega - \omega^2)(\beta - \gamma)(x - \alpha y),$$

where ω is an imaginary cube root of unity: this will appear from the expressions which will be presently given for the covariants in terms of the roots.

120. Canonical form of the cubic.

The expressions $\tfrac{1}{2}(\Phi + U\sqrt{\square})$, $\tfrac{1}{2}(\Phi - U\sqrt{\square})$ are perfect cubes; and if we write

$$\tfrac{1}{2}(\Phi + U\sqrt{\square}) = \sqrt{\square}\,\mathrm{x}^3,$$

$$\tfrac{1}{2}(\Phi - U\sqrt{\square}) = -\sqrt{\square}\,\mathrm{y}^3,$$

then we have

$$U = \mathrm{x}^3 + \mathrm{y}^3,$$

$$\Phi = \sqrt{\square}\,(\mathrm{x}^3 - \mathrm{y}^3),$$

and thence also

$$H = -\sqrt[3]{\square}\,\mathrm{xy}.$$

121. When the cubic is expressed in terms of the roots, we have

$$a^{-1}U = (x - \alpha y)(x - \beta y)(x - \gamma y);$$

and then putting for shortness

$$A = (\beta - \gamma)(x - \alpha y),\quad B = (\gamma - \alpha)(x - \beta y),\quad C = (\alpha - \beta)(x - \gamma y),$$

so that

$$A + B + C = 0,$$

we have

$$\begin{aligned} a^{-2}H &= -\tfrac{1}{18}(A^2 + B^2 + C^2) = \tfrac{1}{9}(BC + CA + AB),\\ a^{-3}\Phi &= -\tfrac{1}{27}(B - C)(C - A)(A - B),\\ a^{-4}\square &= -\tfrac{1}{27}(\beta - \gamma)^2(\gamma - \alpha)^2(\alpha - \beta)^2. \end{aligned}$$

122. The covariants H, Φ are most simply expressed as above, but it may be proper to add the equations

$$a^{-2}H = -\tfrac{1}{18}\Sigma(\beta - \gamma)^2(x - \alpha y)^2$$

$$= -\tfrac{1}{9}\left\{\begin{matrix} \alpha^2 + \beta^2 + \gamma^2 - \beta\gamma - \gamma\alpha - \alpha\beta, \\ 6\alpha\beta\gamma - \beta\gamma^2 - \gamma\alpha^2 - \alpha\beta^2 - \beta^2\gamma - \gamma^2\alpha - \alpha^2\beta, \\ \beta^2\gamma^2 + \gamma^2\alpha^2 + \alpha^2\beta^2 - \alpha^2\beta\gamma - \beta^2\gamma\alpha - \gamma^2\alpha\beta \end{matrix}\right\}(x, y)^2$$

$$= -\tfrac{1}{9}\{(\alpha + \omega\beta + \omega^2\gamma)x + (\beta\gamma + \omega\gamma\alpha + \omega^2\alpha\beta)y\}\{(\alpha + \omega^2\beta + \omega\gamma)x + (\beta\gamma + \omega^2\gamma\alpha + \omega\alpha\beta)y\}$$

(where ω is an imaginary cube root of unity),

$$a^{-3}\Phi = \tfrac{1}{27}\Sigma(\alpha - \beta)(\alpha - \gamma)^2(x - \beta y)^2(x - \gamma y)$$

$$= \left\{\begin{matrix} 2(\alpha^3 + \beta^3 + \gamma^3) - 3(\beta\gamma^2 + \gamma\alpha^2 + \alpha\beta^2 + \beta^2\gamma + \gamma^2\alpha + \alpha^2\beta) + 12\alpha\beta\gamma, \\ -2(\alpha^2\beta\gamma + \beta^2\gamma\alpha + \gamma^2\alpha\beta) + 4(\beta^2\gamma^2 + \gamma^2\alpha^2 + \alpha^2\beta^2) - (\beta\gamma^3 + \gamma\alpha^3 + \alpha\beta^3 + \beta^3\gamma + \gamma^3\alpha + \alpha^3\beta), \\ -2(\alpha\beta^2\gamma^2 + \beta\gamma^2\alpha^2 + \gamma\alpha^2\beta^2) + 4(\alpha^3\beta\gamma + \beta^3\gamma\alpha + \gamma^3\alpha\beta) - (\beta^2\gamma^3 + \gamma^2\alpha^3 + \alpha^2\beta^3 + \beta^3\gamma^2 + \gamma^3\alpha^2 + \alpha^3\beta^2), \\ +2(\beta^3\gamma^3 + \gamma^3\alpha^3 + \alpha^3\beta^3) - 3(\alpha\beta^2\gamma^3 + \beta\gamma^2\alpha^3 + \gamma\alpha^2\beta^3 + \alpha\beta^3\gamma^2 + \beta\gamma^3\alpha^2 + \gamma\alpha^3\beta^2) + 12\alpha^2\beta^2\gamma^2 \end{matrix}\right\}(x, y)^3$$

$$= \{(2\alpha - \beta - \gamma)x + (2\beta\gamma - \gamma\alpha - \alpha\beta)y\}\{(2\beta - \gamma - \alpha)x + (2\gamma\alpha - \alpha\beta - \beta\gamma)y\}\{(2\gamma - \alpha - \beta)x + (2\alpha\beta - \beta\gamma - \gamma\alpha)y\}.$$

123. It may be observed that we have $a^{-6}\square U^2 = -\tfrac{1}{27}A^2B^2C^2$, which, with the above values of H, Φ in terms of A, B, C and the equation $A + B + C = 0$, verifies the equation $\Phi^2 - \square U^2 + 4H^3 = 0$, which connects the covariants. In fact, we have identically,

$$\begin{aligned} &(B - C)^2(C - A)^2(A - B)^2 = \\ &-4(A + B + C)^3ABC + (A + B + C)^2(BC + CA + AB)^2 + 18(A + B + C)(BC + CA + AB)ABC \\ &\qquad - 4(BC + CA + AB)^3 - 27A^2B^2C^2, \end{aligned}$$

by means of which the verification can be at once effected.

124. If, as before, ω is an imaginary cube root of unity, then we may write

$$27a^{-3}\Phi \quad = -(B-C)(C-A)(A-B),$$

$$27a^{-3}U\sqrt{\square} = \quad 3(\omega-\omega^2)ABC,$$

and these values give

$$27a^{-3}\tfrac{1}{2}(\Phi+U\sqrt{\square}) = \{(\alpha+\omega^2\beta+\omega\gamma)x+(\beta\gamma+\omega^2\gamma\alpha+\omega\alpha\beta)y\}^3,$$

$$27a^{-3}\tfrac{1}{2}(\Phi-U\sqrt{\square}) = \{(\alpha+\omega\beta+\omega^2\gamma)x+(\beta\gamma+\omega\gamma\alpha+\omega^2\alpha\beta)y\}^3,$$

and we thence obtain

$$\sqrt[3]{\tfrac{1}{2}(\Phi+U\sqrt{\square})}-\sqrt[3]{\tfrac{1}{2}(\Phi-U\sqrt{\square})} = -\tfrac{1}{3}a(\omega-\omega^2)(\beta-\gamma)(x-\alpha y),$$

which agrees with a former result.

125. The preceding formulæ show without difficulty, that each factor of the cubicovariant is the harmonic of a factor of the cubic with respect to the other two factors of the cubic; and moreover, that the factors of the cubic and the cubicovariant form together an involution having for sibiconjugates the factors of the Hessian. In fact, the harmonic of $x-\alpha y$ with respect to $(x-\beta y)(x-\gamma y)$ is $(2\alpha-\beta-\gamma)x+(2\beta\gamma-\gamma\alpha-\alpha\beta)y$, which is a factor of the cubicovariant; the product of the pair of harmonic factors is

$$(2\alpha-\beta-\gamma)x^2+2(\beta\gamma-\alpha^2)xy+(-2\alpha\beta\gamma+\alpha^2\beta+\alpha^2\gamma)y^2;$$

and multiplying this by $\beta-\gamma$, and taking the sum of the analogous expressions, this sum vanishes, or the three pairs form an involution. That the Hessian gives the sibiconjugates of the involution is most readily shown as follows:—the last-mentioned quadric may be written

$$\big(-(\alpha+\beta+\gamma)+3\alpha\big)x^2+2\big(\alpha\beta+\alpha\gamma+\beta\gamma-\alpha(\alpha+\beta+\gamma)\big)xy+\big(-3\alpha\beta\gamma+\alpha(\alpha\beta+\alpha\gamma+\beta\gamma)\big)y^2,$$

which is equal to

$$\left(3\frac{b}{a}+3\alpha\right)x^2+2\left(3\frac{c}{a}-3\frac{b}{a}\alpha\right)xy+\left(3\frac{d}{a}+3\frac{c}{a}\alpha\right)y^2,$$

or, throwing out the factor $3a^{-1}$, to

$$(b+a\alpha,\ \ 2c-2b\alpha,\ \ d+c\alpha\between x,\ y)^2,$$

which is harmonically related to the Hessian

$$(ac-b^2,\ \ ad-bc,\ \ bd-c^2\between x,\ y)^2;$$

and in like manner the other two pairs of factors will be also harmonically related to the Hessian.

126. In the case of a pair of equal roots, we have

$$\begin{aligned} a^{-1}U &= (x-\alpha y)^2(x-\gamma y), \\ a^{-2}H &= -\tfrac{1}{9}(\alpha-\gamma)^2(x-\alpha y)^2, \\ a^{-3}\Phi &= -\tfrac{2}{27}(\alpha-\gamma)^3(x-\alpha y)^3, \\ \square &= 0. \end{aligned}$$

And in the case of all the roots equal, we have

$$a^{-1}U=(x-\alpha y)^3,$$

$$H=0,\quad \Phi=0,\quad \square=0.$$

127. In the solution of a biquadratic equation we have to consider the cubic equation $\varpi^3-M(\varpi-1)=0$. The cubic here is $(1,\ 0,\ -M,\ M)(\varpi,\ 1)^3$, or what is the same thing,

$$(1,\ 0,\ -\tfrac{1}{3}M,\ M)(\varpi,\ 1)^3;$$

the Hessian is

$$M(-\tfrac{1}{3},\ 1,\ -\tfrac{1}{9}M)(\varpi,\ 1)^2;$$

the cubicovariant is

$$M(-1,\ \tfrac{2}{9}M,\ -\tfrac{1}{3}M,\ M+\tfrac{2}{27}M^2)(\varpi,\ 1)^3;$$

and the discriminant is

$$M^2(1-\tfrac{4}{27}M).$$

128. In the case of a quartic, the expressions considered are

$$(a,\ b,\ c,\ d,\ e)(x,\ y)^4, \tag{1}$$

$$ae-4bd+3c^2, \tag{2}$$

$$(ac-b^2,\ 2(ad-bc),\ ae+2bd-3c^2,\ 2(be-cd),\ ce-d^2)(x,\ y)^4, \tag{3}$$

$$ace+2bcd-ad^2-b^2e-c^3, \tag{4}$$

$$\left.\begin{Bmatrix} - & a^2d + 3\,abc - 2\,b^3, \\ - & a^2e - 2\,abd + 9\,ac^2 - 6\,b^2c, \\ - & 5\,abe + 15\,acd - 10\,b^2d, \\ + & 10\,ad^2 - 10\,b^2e, \\ + & 5\,ade + 10\,bd^2 - 15\,bce, \\ + & ae^2 + 2\,bde - 9\,c^2e + 6\,cd^2, \\ + & be^2 - 3\,cde + 2\,d^3 \end{Bmatrix}\right)(x,\ y)^6, \tag{5}$$

where (1) is the quartic, (2) is the quadrinvariant, (3) is the quadricovariant or Hessian, (4) is the cubinvariant, and (5) is the cubicovariant.

And where it is convenient to do so, I write

$$\begin{aligned}(1) &= U,\\ (2) &= I,\\ (3) &= H,\\ (4) &= J,\\ (5) &= \Phi.\end{aligned}$$

The preceding covariants are connected by the equation

$$JU^3 - IU^2H + 4H^3 = -\Phi^2.$$

The discriminant is not an irreducible invariant, its value is

$$\square = I^3 - 27J^2 = a^3e^3 + \&c.,$$

for which see Table No. 12, [p. 272].

129. It is for some purposes convenient to arrange the expanded expression of the discriminant in powers of the middle coefficient c. We thus have

$$\begin{aligned}\square = {} & a^3e^3 - 12\,a^2bde^2 - 27\,a^2d^4 - 6\,ab^2d^2e - 27\,b^4e^2 - 64\,b^3d^3\\ & + c\,(\,54\,a^2d^2e + 54\,ab^2e^2 + 108\,abd^3 + 108\,b^3de)\\ & + c^2\,(-18\,a^2e^2 - 180\,abde + 36\,b^2d^2)\\ & + c^3\,(-54\,ad^2 - 54\,b^2e)\\ & + c^4\,(81\,ae).\end{aligned}$$

130. Solution of a biquadratic equation.

We have to find a linear factor of the quartic

$$(a,\ b,\ c,\ d,\ e \between x,\ y)^4.$$

The equation $JU^3 - IU^2H + 4H^3 = -\Phi^2$, putting for shortness

$$M = \frac{I^3}{4J^2},$$

may be written

$$(1,\ 0,\ -M,\ M \between IH,\ JU)^3 = -\tfrac{1}{4}I^3\Phi^2.$$

Hence, if ϖ_1, ϖ_2, ϖ_3 are the roots of

$$(1,\ 0,\ -M,\ M \between \varpi,\ 1)^3 = 0,$$

the expressions $IH - \varpi_1 JU$, $IH - \varpi_2 JU$, $IH - \varpi_3 JU$ are each of them squares; write

$$\begin{aligned}(\varpi_2 - \varpi_3)(IH - \varpi_1 JU) &= X^2,\\ (\varpi_3 - \varpi_1)(IH - \varpi_2 JU) &= Y^2,\\ (\varpi_1 - \varpi_2)(IH - \varpi_3 JU) &= Z^2,\end{aligned}$$

so that, identically,

$$X^2+Y^2+Z^2=0;$$

and consequently $X+\iota Y$, $X-\iota Y$ are each of them squares. The expression

$$\alpha X+\beta Y+\gamma Z$$

will be a square if only

$$\alpha^2+\beta^2+\gamma^2=0,$$

as may be seen by writing it under the form

$$\tfrac{1}{2}(\alpha+\iota\beta)(X-\iota Y)+\tfrac{1}{2}(\alpha-\iota\beta)(X+\iota Y)-\gamma i\sqrt{X^2+Y^2};$$

and in particular, writing $\sqrt{\varpi_2-\varpi_3}$, $\sqrt{\varpi_3-\varpi_1}$, $\sqrt{\varpi_1-\varpi_2}$ for α, β, γ, the expression

$$(\varpi_2-\varpi_3)\sqrt{IH-\varpi_1 JU}+(\varpi_3-\varpi_1)\sqrt{IH-\varpi_2 JU}+(\varpi_1-\varpi_2)\sqrt{IH-\varpi_3 JU}$$

is a square; and since the product of the different values is a multiple of U^2 (this is most readily perceived by observing that the expression vanishes for $U=0$), the expression is the square of a linear factor of the quartic.

131. To complete the solution: ϖ_1, ϖ_2, ϖ_3 are the roots of the cubic equation

$$(1,\ 0,\ -\tfrac{1}{3}M,\ M)(\varpi,\ 1)^3=0;$$

and hence, putting for shortness,

$$P^3=\tfrac{1}{2}M\{(-1,\ \tfrac{2}{9}M,\ -\tfrac{1}{3}M,\ M+\tfrac{2}{27}M^2)(IH,\ JU)^3+\sqrt{1-\tfrac{4}{27}M}\,(1,\ 0,\ -\tfrac{1}{3}M,\ M)(IH,\ JU)^3,$$

$$Q^3=\tfrac{1}{2}M\{(-1,\ \tfrac{2}{9}M,\ -\tfrac{1}{3}M,\ M+\tfrac{2}{27}M^2)(IH,\ JU)^3-\sqrt{1-\tfrac{4}{27}M}\,(1,\ 0,\ -\tfrac{1}{3}M,\ M)(IH,\ JU)^3,$$

we have (ω being an imaginary cube root of unity)

$$\tfrac{1}{3}(\omega-\omega^2)(\varpi_2-\varpi_3)(IH-\varpi_1 JU)=P-Q;$$

and if

$$P_0^3=\tfrac{1}{2}M\{-1+\sqrt{1-\tfrac{4}{27}M}\},$$

$$Q_0^3=\tfrac{1}{2}M\{-1-\sqrt{1-\tfrac{4}{27}M}\},$$

then

$$\tfrac{1}{3}(\omega-\omega^2)(\varpi_2-\varpi_3)=P_0-Q_0.$$

Hence, multiplying and observing that $(\omega-\omega^2)^2=-3$, we find

$$-\frac{1}{(\omega-\omega^2)^2}(\varpi_2-\varpi_3)^2(IH-\varpi_1 JU)=(P-Q)(P_0-Q_0),$$

and consequently

$$(\varpi_2-\varpi_3)\sqrt{IH-\varpi_1 JU}=(\omega-\omega^2)\sqrt{-(P-Q)(P_0-Q_0)}.$$

We have, in like manner,

$$\tfrac{1}{3}(\omega-\omega^2)(\varpi_2-\varpi_3)(IH-\varpi_1 JU)=\ \ P-\ \ Q,$$
$$\tfrac{1}{3}(\omega-\omega^2)(\varpi_3-\varpi_1)(IH-\varpi_2 JU)=\omega P-\omega^2 Q,$$
$$\tfrac{1}{3}(\omega-\omega^2)(\varpi_1-\varpi_2)(IH-\varpi_3 JU)=\omega^2 P-\omega Q,$$

and
$$\tfrac{1}{3}(\omega-\omega^2)(\varpi_2-\varpi_3)=\ \ P_0-\ \ Q_0,$$
$$\tfrac{1}{3}(\omega-\omega^2)(\varpi_3-\varpi_1)=\omega\ P_0-\omega^2 Q_0,$$
$$\tfrac{1}{3}(\omega-\omega^2)(\varpi_1-\varpi_2)=\omega^2 P_0-\omega\ Q_0,$$
and therefore
$$(\varpi_2-\varpi_3)\sqrt{IH-\varpi_1 JU}=(\omega-\omega^2)\sqrt{-\ \ (P-Q)\ \ \ \ (P_0-Q_0)},$$
$$(\varpi_3-\varpi_1)\sqrt{IH-\varpi_2 JU}=(\omega-\omega^2)\sqrt{-(\omega P-\omega^2 Q)(\omega P_0-\omega^2 Q)},$$
$$(\varpi_3-\varpi_1)\sqrt{IH-\varpi_3 JU}=(\omega-\omega^2)\sqrt{-(\omega^2 P-\omega Q)(\omega^2 P_0-\omega Q_0)}\,;$$
and hence disregarding the common factor $\omega-\omega^2$, the square of the linear factor of the quartic is
$$\sqrt{-(P-Q)(P_0-Q_0)}+\sqrt{-(\omega P-\omega^2 Q)(\omega P_0-\omega^2 Q_0)}+\sqrt{-(\omega^2 P-\omega Q)(\omega^2 P_0-\omega Q_0)},$$
which is the required solution.

It may be proper to add that
$$-\varpi_1=\ \ P_0+\ \ Q_0\,,$$
$$-\varpi_2=\omega\ P_0+\omega^2 Q_0,$$
$$-\varpi_3=\omega^2 P_0+\omega\ Q_0.$$

132. The solution gives at once the canonical form of the quartic; in fact, writing
$$X+\iota Y=2\sqrt{(\varpi_2-\varpi_3)(\varpi_3-\varpi_1)}\sqrt{J}\,\mathrm{x}^2,$$
$$X-\iota Y=2\sqrt{(\varpi_2-\varpi_3)(\varpi_3-\varpi_1)}\sqrt{J}\,\mathrm{y}^2,$$
where X, Y have their former significations, we find, by a simple reduction,
$$IH-\varpi_1 JU=\ \ (\varpi_3-\varpi_1)\,J\,(\mathrm{x}^2+\mathrm{y}^2)^2,$$
$$IH-\varpi_2 JU=-(\varpi_2-\varpi_3)\,J\,(\mathrm{x}^2-\mathrm{y}^2)^2,$$
$$IH-\varpi_3 JU=-\frac{(\varpi_2-\varpi_3)(\varpi_3-\varpi_1)}{\varpi_1-\varpi_2}J\,.\,4\mathrm{x}^2\mathrm{y}^2,$$
and thence putting
$$\theta=-\frac{\varpi_2}{\varpi_1-\varpi_2}=\frac{\tfrac{1}{3}(\omega-\omega^2)(\omega^2 P_0+\omega Q_0)}{(\omega^2 P_0-\omega Q_0)},$$
we have
$$U=\mathrm{x}^4+\mathrm{y}^4+6\theta\mathrm{x}^2\mathrm{y}^2,$$
which is the form required.

133. The Hessian may be written under the form
$$(\partial_e,\ -\partial_d,\ \partial_c,\ -\partial_b,\ \partial_a \between x,\ y)^4 J,$$
that is, as the evectant of the cubinvariant.

The cubicovariant may be obtained by writing the quartic under the form

$$(ax+by,\ bx+cy,\ cx+dy,\ dx+ey\between x,\ y)^3,$$

and then, treating the linear functions as coefficients, or considering this as a cubic, the cubicovariant of the cubic gives the cubicovariant of the quartic.

If we represent the cubicovariant by

$$\Phi=(\mathrm{a},\ \mathrm{b},\ \mathrm{c},\ \mathrm{d},\ \mathrm{e},\ \mathrm{f},\ \mathrm{g}\between x,\ y)^6,$$

then we have identically,

$$\mathrm{ag}-9\mathrm{ce}+8\mathrm{d}^2=0\,;$$

and moreover forming the quadrinavariant of the sextic, we find

$$\mathrm{ag}-6\mathrm{bf}+15\mathrm{ce}-10\mathrm{d}^2=\tfrac{1}{6}\square,$$

where $\square$ is the discriminant of the quartic. From these two equations we find

$$\mathrm{bf}-4\mathrm{ce}+3\mathrm{d}^2=-\tfrac{1}{36}\square,$$

which is an expression given by Mr Salmon: it is the more remarkable as the left-hand side is the quadrinvariant of $(\mathrm{b},\ \mathrm{c},\ \mathrm{d},\ \mathrm{e},\ \mathrm{f}\between x,\ y)^4$, which is *not* a covariant of the quartic. It may be noticed also that we have

$$\mathrm{af}-3\mathrm{be}+2\mathrm{cd}=0,$$

$$\mathrm{bg}-3\mathrm{cf}+2\mathrm{de}=0.$$

134. The covariants of the intermediate

$$\alpha U+6\beta H$$

of the quartic and Hessian are as follows, viz.

The quadrinvariant is

$$\tilde{I}\ (\alpha U+6\beta H)=\ (I,\ 18J,\ 3I^2\between\alpha,\ \beta)^2\,;$$

the cubinvariant is

$$\tilde{J}\ (\alpha U+6\beta H)=\ (J,\ I^2,\ 9IJ,\ -I^3+54J^2\between\alpha,\ \beta)^3;$$

the Hessian is

$$\tilde{H}\,(\alpha U+6\beta H)=\ (1,\ 0,\ -3I\between\alpha,\ \beta)^2 H$$
$$+(0,\ I,\ \ 9J\between\alpha,\ \beta)^2 U\,;$$

and the cubicovariant is

$$\tilde{\Phi}\ (\alpha U+6\beta H)=\ (1,\ 0,\ -9I,\ -54J\between\alpha,\ \beta)^3\,\Phi\,;$$

to which may be added the discriminant, which is

$$\tilde{\square}\,(\alpha U+6\beta H)=(1,\ 0,\ -18\,I,\ 108\,J,\ 81\,I^2,\ 972\,IJ,\ -2916\,J^2\between\alpha,\ \beta)^6\square.$$

135. The expression for the lambdaic is

$$\begin{vmatrix} a, & b, & c-2\lambda \\ b, & c+\lambda, & d \\ c-2\lambda, & d, & e \end{vmatrix} = J + \lambda I - 4\lambda^3.$$

If the determinant is represented by Λ, that is if

$$\Lambda = -4\lambda^3 + \lambda I + J,$$

then if λ_1, λ_2, λ_3 are the roots of the equation $\Lambda = 0$, and if the values of $\partial_a \Lambda$, &c. obtained by writing λ_1 in the place of λ are represented by $\partial_a \Lambda_1$, &c., then if x, y satisfy the equation

$$(a, b, c, d, e \between x, y)^4 = 0,$$

we have identically (X, Y being arbitrary),

$$\frac{(a, b, c, d, e \between X, Y \overset{3}{\between} x, y)}{Xy - Yx}$$

$$= \sqrt{-(\partial_e, -\partial_d, \partial_c, -\partial_b, \partial_a \between X, Y)^4 \Lambda_1}$$

$$+ \sqrt{-(\partial_e, -\partial_d, \partial_c, -\partial_b, \partial_a \between X, Y)^4 \Lambda_2}$$

$$+ \sqrt{-(\partial_e, -\partial_d, \partial_c, -\partial_b, \partial_a \between X, Y)^4 \Lambda_3}$$

a theorem due to Aronhold. I have quoted this theorem in its original form as an application of the lambdaic, but I remark that

$$-(\partial_e, -\partial_d, \partial_c, -\partial_b, \partial_a \between X, Y)^4 \Lambda = -\lambda (a, \ldots \between X, Y)^4 - (ac - b^2, \ldots \between X, Y)^4 = -\lambda U' - H'$$

if U', H' are what U, H become, substituting for (x, y) the new facients (X, Y). Moreover, we have

$$\lambda = -\frac{J\varpi}{I};$$

for substituting this value in the equation $\Lambda = 0$, we obtain the before-mentioned equation $\varpi^3 - M(\varpi - 1) = 0$. We have, therefore,

$$-(\partial_e, -\partial_d, \partial_c, -\partial_b, \partial_a \between X, Y)^4 \Lambda = \frac{J\varpi}{I} U' - H' = -\frac{1}{I}(IH' - J\varpi U'),$$

and the equation becomes

$$\frac{(a, b, c, d, e \between X, Y \overset{3}{\between} x, y)}{Xy - Yx} \sqrt{-I} = \sqrt{IH' - J\varpi_1 U'} + \sqrt{IH' - J\varpi_2 U'} + \sqrt{IH' - J\varpi_3 U'}.$$

Moreover, if $(x-\alpha y)$ be a factor of the quartic, then replacing in the formula y by the value αx, (x, y) will disappear altogether; and then changing (X, Y) into (x, y) where x, y are now arbitrary, we have

$$\frac{(a, b, c, d, e\between x, y)^3(\alpha, 1)}{x-\alpha y}\sqrt{-I}=\sqrt{IH-\varpi_1 JU}+\sqrt{IH-\varpi_2 JU}+\sqrt{IH-\varpi_3 JU},$$

which is a form connected with the results in Nos. 130 and 131.

136. We have

$$\begin{vmatrix} y^4, & -4xy^3, & 6x^2y^2, & -4x^3y, & x^4 \\ & a\ , & 3b\ , & 3c\ , & d \\ a\ , & 3b\ , & 3c\ , & d\ , & \\ & b\ , & 3c\ , & 3d\ , & e \\ b\ , & 3c\ , & 3d\ , & e\ , & \end{vmatrix}=6IH-9JU;$$

it will appear from the formulæ relating to the roots of the quartic, that the expression $6IH-9JU$ vanishes identically when there are two pairs of equal roots, or what is the same thing, when the quartic is a perfect square. The conditions in order that the expression may vanish are obviously

$$\begin{aligned} &6(ac-b^2) : 3(ad-bc) : ae+2bd-3c^2 : 3(be-cd) : 6(ce-d^2) : 9J \\ =\ &\quad a \quad : \quad b \quad : \quad c \quad : \quad d \quad : \quad e \quad : I, \end{aligned}$$

conditions which imply that the several determinants

$$\left\|\begin{matrix} 6(ac-b^2), & 3(ad-bc), & ae+2bd-3c^2, & 3(be-cd), & 6(ce-d^2) \\ a\ , & b\ , & c\ , & d\ , & e \end{matrix}\right\|,$$

all of them vanish. If for a moment we write $6H=(a', b', c', d', e'\between x, y)^4$, then the determinants are

$$\left\|\begin{matrix} a', & b', & c', & d', & e' \\ a\ , & b\ , & c\ , & d\ , & e \end{matrix}\right\|;$$

we have identically

$$\begin{aligned} ad'-a'd &= 3(bc'-b'c), \\ eb'-e'b &= 3(dc'-d'c), \\ ae'-a'e &= 3(bd'-b'd), \end{aligned}$$

and the ten determinants thus reduce themselves to seven determinants only, these in fact being, to mere numerical factors *près*, the coefficients of the cubicovariant; this perfectly agrees with a subsequent result, viz. that the cubicovariant vanishes identically when the quartic is a perfect square.

137. It may be remarked that the equation $6IH - 9JU = 0$ will be satisfied identically if

$$a = \frac{b^2}{c - \phi}, \quad e = \frac{d^2}{c - \phi}, \quad bd = (c - \phi)(c + 2\phi),$$

where ϕ is arbitrary; the quartic is in this case the square of

$$\left(\frac{b}{\sqrt{c - \phi}}, \ \sqrt{c - \phi}, \ \frac{d}{\sqrt{c - \phi}} \between x, \ y\right)^2.$$

If with the conditions in question we combine the equation $I = 0$ (which in this case implies also $J = 0$), we obtain $\phi = 0$, and consequently

$$\frac{a}{b} = \frac{b}{c} = \frac{c}{d} = \frac{d}{e},$$

or the quartic will be a complete fourth power.

It is easy to express in terms of the coefficients a', b', c', d', e' of $6H$ the different determinants

$$\left\| \begin{matrix} a, & b, & c, & d \\ b, & c, & d, & e \end{matrix} \right\|,$$

we have in fact

$$\left\{ \begin{aligned} ae - bd &= \tfrac{1}{2}\left(c' + \frac{1}{\sqrt{3}} \sqrt{a'e' + 4b'd' - 3c'^2}\right), \\ 3(bd - c^2) &= \tfrac{1}{2}\left(c' - \frac{1}{\sqrt{3}} \sqrt{a'e' + 4b'd' - 3c'^2}\right), \\ ac - b^2 &= \tfrac{1}{6} a', \\ ad - bc &= \tfrac{1}{3} b', \\ be - cd &= \tfrac{1}{3} c', \\ ce - d^2 &= \tfrac{1}{6} e', \end{aligned} \right.$$

whence all the above-mentioned determinants will vanish, or the quartic will be a perfect fourth power if only the Hessian vanishes identically.

138. Considering the quartic as expressed in terms of the roots, we have

$$a^{-1}U = (x - \alpha y)(x - \beta y)(x - \gamma y)(x - \delta y);$$

and if we write for shortness

$$\begin{aligned} A &= (\beta - \gamma)(\alpha - \delta), \\ B &= (\gamma - \alpha)(\beta - \delta), \\ C &= (\alpha - \beta)(\gamma - \delta), \end{aligned}$$

which are connected by

$$A + B + C = 0,$$

then we have

$$a^{-2} I = \tfrac{1}{24}(A^2 + B^2 + C^2) = -\tfrac{1}{12}(BC + CA + AB),$$
$$a^{-3} J = \tfrac{1}{432}(B - C)(C - A)(A - B);$$

and for the discriminant we have

$$a^{-6}\Box = \tfrac{1}{256}(\alpha - \beta)^2 (\alpha - \gamma)^2 (\alpha - \delta)^2 (\beta - \gamma)^2 (\beta - \delta)^2 (\gamma - \delta)^2,$$
$$= \tfrac{1}{256} A^2 B^2 C^2,$$

and it is easy by means of a preceding formula to verify the equation $\Box = I^3 - 27J^2$.

139. The formulæ show a very remarkable analogy between the covariants of a cubic and the invariants of a quartic. In fact

For the cubic.	For the quartic.
$A = (\beta - \gamma)(x - \alpha y)$,	$A = (\beta - \gamma)(\alpha - \delta)$,
$B = (\gamma - \alpha)(x - \beta y)$,	$B = (\gamma - \alpha)(\beta - \delta)$,
$C = (\alpha - \beta)(x - \gamma y)$,	$C = (\alpha - \beta)(\gamma - \delta)$;

and then we have corresponding to each other:

For the cubic.	For the quartic.
The Hessian,	The quadrinvariant,
The cubicovariant,	The cubinvariant,
The cubic into the square root of the discriminant.	The discriminant.

140. For the two covariants, we have

$$a^{-2}H = -\tfrac{1}{48}\Sigma(\alpha - \beta)^2 (x - \gamma y)^2 (x - \delta y)^2,$$

and

$$a^{-3}\Phi = -\tfrac{1}{32}\mathfrak{ABC},$$

if for shortness,

$$\mathfrak{A} = (\delta + \alpha - \beta - \gamma,\ -\delta\alpha + \beta\gamma,\ \delta\alpha(\beta + \gamma) - \beta\gamma(\delta + \alpha)\between x, y)^2,$$
$$\mathfrak{B} = (\delta + \beta - \gamma - \alpha,\ -\delta\beta + \gamma\alpha,\ \delta\beta(\gamma + \alpha) - \gamma\alpha(\delta + \beta)\between x, y)^2,$$
$$\mathfrak{C} = (\delta + \gamma - \alpha - \beta,\ -\delta\gamma + \alpha\beta,\ \delta\gamma(\alpha + \beta) - \alpha\beta(\delta + \gamma)\between x, y)^2.$$

141. We have

$$M = \tfrac{27}{8}\frac{(A^2 + B^2 + C^2)^3}{(B - C)^2 (C - A)^2 (A - B)^2};$$

or putting for shortness

$$\Lambda = \tfrac{3}{2}\frac{A^2 + B^2 + C^2}{(B - C)(C - A)(A - B)},$$

we have

$$M = \tfrac{3}{2}(A^2 + B^2 + C^2)\Lambda^2;$$

and it is then easy to deduce

$$\varpi_1 = \Lambda(B - C),$$
$$\varpi_2 = \Lambda(C - A),$$
$$\varpi_3 = \Lambda(A - B);$$

in fact, these values give

$$\varpi_1 + \varpi_2 + \varpi_3 = 0,$$
$$\varpi_1\varpi_2 + \varpi_1\varpi_3 + \varpi_2\varpi_3 = -M,$$
$$\varpi_1\varpi_2\varpi_3 = M,$$

and they are consequently the roots of the equation $\varpi^3 - M(\varpi - 1) = 0$.

142. The leading coefficient of $IH - \varpi_1 JU$ is then equal to a^4 into the following expression, viz.

$$\tfrac{1}{24}(A^2 + B^2 + C^2)\, a^{-2}(ac - b^2) - \tfrac{1}{288}(A^2 + B^2 + C^2)(B - C),$$

which is equal to

$$\tfrac{1}{1152}(A^2 + B^2 + C^2)\{48a^{-2}(ac - b^2) - 4(B - C)\},$$

and the term in { } is

$$8(\alpha\beta + \alpha\gamma + \alpha\delta + \beta\gamma + \beta\delta + \gamma\delta) - 3(\alpha + \beta + \gamma + \delta)^2 - 4(\gamma - \alpha)(\beta - \delta) + 4(\alpha - \beta)(\gamma - \delta),$$

which is equal to

$$-3(\delta + \alpha - \beta - \gamma)^2.$$

But $IH - \varpi_1 JU$ is a square, and it is easy to complete the expression, and we have

$$a^{-4}(IH - \varpi_1 JU) = -\tfrac{1}{384}(A^2 + B^2 + C^2)\{(\delta + \alpha - \beta - \gamma,\ -\delta\alpha + \beta\gamma,\ \delta\alpha(\beta + \gamma) - \beta\gamma(\delta + \alpha)\between x, y)^2\}^2$$
$$a^{-4}(IH - \varpi_2 JU) = -\tfrac{1}{384}(A^2 + B^2 + C^2)\{(\delta + \beta - \gamma - \alpha,\ -\delta\beta + \gamma\alpha,\ \delta\beta(\gamma + \alpha) - \gamma\alpha(\delta + \beta)\between x, y)^2\}^2,$$
$$a^{-4}(IH - \varpi_3 JU) = -\tfrac{1}{384}(A^2 + B^2 + C^2)\{(\delta + \gamma - \alpha - \beta,\ -\delta\gamma + \alpha\beta,\ \delta\gamma(\alpha + \beta) - \alpha\beta(\gamma + \delta)\between x, y)^2\}^2.$$

We have, moveover,

$$\varpi_2 - \varpi_3 = -3\Lambda A,$$
$$\varpi_3 - \varpi_1 = -3\Lambda B,$$
$$\varpi_1 - \varpi_2 = -3\Lambda C,$$

and thence

$$a^{-2}(\varpi_2 - \varpi_3)\sqrt{IH - \varpi_1 JU} = \tfrac{1}{8}(\omega - \omega^2)\frac{A^2 + B^2 + C^2}{(B - C)(C - A)(A - B)}(\beta - \gamma)(\alpha - \delta)$$
$$\times(\delta + \alpha - \beta - \gamma,\quad -\delta\alpha + \beta\gamma,\quad \delta\alpha(\beta + \gamma) - \beta\gamma(\delta + \alpha)\between x, y)^2;$$

and taking the sum of the analogous expressions, we find

$$a^{-2}\{(\varpi_2 - \varpi_3)\sqrt{IH - \varpi_1 JU} + (\varpi_3 - \varpi_1)\sqrt{IH - \varpi_2 JU} + (\varpi_1 - \varpi_2)\sqrt{IH - \varpi_3 JU}\}$$
$$= -\tfrac{1}{4}(\omega - \omega^2)\frac{A^2 + B^2 + C^2}{(B - C)(C - A)(A - B)}(\alpha - \beta)(\beta - \gamma)(\gamma - \alpha)(x - \delta y)^2,$$

which agrees with a former result.

143. The equation $I=0$ gives

$$A : B : C = 1 : \omega : \omega^2,$$

where ω is an imaginary cube root of unity; the factors of the quartic may be said in this case to be *Symmetric Harmonics.*

The equation $J=0$ gives one of the three equations,

$$A=B,\quad B=C,\quad C=A\,;$$

in this case a pair of factors of the quartic are harmonics with respect to the other pair of factors. If we have simultaneously $I=0$, $J=0$, then

$$A=B=C=0,$$

and in this case three of the factors of the quartic are equal.

144. If any two of the linear factors of the quartic are considered as forming, with the other two linear factors, an involution, the sibiconjugates of the involution make up a quadratic factor of the cubicovariant; and considering the three pairs of sibiconjugates, or what is the same thing, the six linear factors of the cubicovariant, the factors of a pair are the sibiconjugates of the involution formed by the other two pairs of factors.

In fact, the sibiconjugates of the involution formed by the equations

$$(x-\alpha y)(x-\delta y)=0,\quad (x-\beta y)(x-\gamma y)=0$$

are found by means of the Jacobian of these two functions, viz. of the quadrics

$$(2,\ -\delta-\alpha,\ \ 2\delta\alpha \between x,\ y)^2,$$
$$(2,\ -\beta-\gamma,\ \ 2\beta\gamma \between x,\ y)^2,$$

which is

$$(\delta+\alpha-\beta-\gamma,\ \ -\delta\alpha+\beta\gamma,\ \ \delta\alpha(\beta+\gamma)-\beta\gamma(\delta+\alpha) \between x,\ y)^2,$$

viz. a quadratic factor of the cubicovariant; and forming the other two factors, there is no difficulty in seeing that any one of these is the Jacobian of the other two.

145. In the case of a pair of equal roots, we have

$$\begin{aligned}
a^{-1}U &= (x-\alpha y)^2(x-\gamma y)(x-\delta y),\\
a^{-2}I &= \tfrac{1}{12}(\alpha-\gamma)^2(\alpha-\delta)^2,\\
a^{-3}J &= -\tfrac{1}{216}(\alpha-\gamma)^3(\alpha-\delta)^3,\\
\square &= 0,\\
a^{-2}H &= -\tfrac{1}{48}\{2(\alpha-\gamma)^2(x-\delta y)^2+2(\alpha-\delta)^2(x-\gamma y)^2+(\gamma-\delta)^2(x-\alpha y)^2\}(x-\alpha y)^2,\\
a^{-3}\Phi &= \tfrac{1}{32}(\gamma-\delta)^2(2\alpha-\gamma-\delta,\ \gamma\delta-\alpha^2,\ \gamma\alpha^2+\delta\alpha^2-2\gamma\alpha\delta \between x,\ y)^2(x-\alpha y)^4.
\end{aligned}$$

In the case of two pairs of equal roots, we have

$$\begin{aligned} a^{-1}U &= (x-\alpha y)^2(x-\gamma y)^2, \\ a^{-2}I &= \tfrac{1}{12}(\alpha-\gamma)^4, \\ a^{-3}J &= -\tfrac{1}{216}(\alpha-\gamma)^6, \\ \square &= 0, \\ a^{-1}H &= -\tfrac{1}{12}(\alpha-\gamma)^2(x-\alpha y)^2(x-\gamma y)^2, \\ \Phi &= 0; \end{aligned}$$

these values give also

$$6IH-9JU=0.$$

146. In the case of three equal roots, we have

$$\begin{aligned} a^{-1}U &= (x-\alpha y)^3(x-\delta y), \\ I &= 0, \quad J=0, \quad \square=0, \\ a^{-2}H &= -\tfrac{1}{48}(\alpha-\delta)^2\{2(x-\delta y)^2+(x-\alpha y)^2\}(x-\alpha y)^2, \\ a^{-3}\Phi &= \tfrac{1}{32}(\alpha-\delta)^3(x-\alpha y)^6; \end{aligned}$$

and in the case of four equal roots, we have

$$\begin{aligned} a^{-1}U &= (x-\alpha y)^4, \\ I &= 0, \quad J=0, \quad \square=0, \\ H &= 0, \quad \Phi=0. \end{aligned}$$

The preceding formulæ, for the case of equal roots, agree with the results obtained in my memoir on the conditions for the existence of given systems of equalities between the roots of an equation.

Addition, 7th October, 1858.

Covariant and other Tables (binary quadrics Nos. 25 bis, 29 A, 49 A, and 50 bis).

Mr Salmon has pointed out to me, that in the Table No. 25 of the simplest octinvariant of a binary quintic[1], the coefficients -210, -17, $+18$ and $+38$ are erroneous, and has communicated to me the corrected values, which I have since verified: the terms, with the corrected values of the coefficients, are [shown in the Table]

No. 25 bis.

[The terms with the erroneous coefficients were abc^2d^2ef, ac^5f^2, $b^4d^2f^2$, bc^3d^3e; the correct values -220, -27, $+22$, and $+74$ of the coefficients are given in the Table Q, No. 25, p. 288.]

[1] Second Memoir, *Philosophical Transactions*, t. CXLVI. (1856) p. 125.

Mr Salmon has also performed the laborious calculation of Hermites' 18-thic invariant of a binary quintic, and has kindly permitted me to publish the result, which is given in the following Table:

No. 29 A.

[This is the Table W No. 29 A given pp. 299—303, the form being slightly altered as appears p. 282.]

Mr Salmon has also remarked to me, that in the Table No. 50 of the cubinvariant of a binary dodecadic[1], the coefficients are altogether erroneous. There was, in fact, a fundamental error in the original calculation; instead of repeating it, I have, with a view to the deduction therefrom of the cubinvariant (see Fourth Memoir, No. 78), first calculated the dodecadic quadricovariant, the value of which is given in the following Table:

No. 49 A.

[For this Table see p. 319.]

It is now very easy to obtain the cubinvariant, which is

No. 50 bis.

[This is the Table No. 50, p. 319, the original No. 50 with coefficients which were altogether erroneous having been omitted.]

[1] Third Memoir, *Philosophical Transactions*, t. CXLVI. (1856) p. 635.

157.

ON THE TANGENTIAL OF A CUBIC.

[From the *Philosophical Transactions of the Royal Society of London*, vol. XLVIII. *for the year* 1858, pp. 461—463. Received February 11,—Read March 18, 1858.]

IN my "Memoir on Curves of the Third Order" ([1]), I had occasion to consider a derivative which may be termed the "tangential" of a cubic, viz. the tangent at the point (x, y, z) of the cubic curve $(*\between x, y, z)^3=0$ meets the curve in a point (ξ, η, ζ), which is the tangential of the first-mentioned point; and I showed that when the cubic is represented in the canonical form $x^3+y^3+z^3+6lxyz=0$, the coordinates of the tangential may be taken to be $x(y^3-z^3) : y(z^3-x^3) : z(x^3-y^3)$. The method given for obtaining the tangential may be applied to the general form $(a, b, c, f, g, h, i, j, k, l\between x, y, z)^3$: it seems desirable, in reference to the theory of cubic forms, to give the expression of the tangential for the general form[2]; and this is what I propose to do, merely indicating the steps of the calculation, which was performed for me by Mr Creedy.

The cubic form is

$$(a, b, c, f, g, h, i, j, k, l\between x, y, z)^3,$$

which means

$$ax^3+by^3+cz^3+3fy^2z+3gz^2x+3hx^2y+3iyz^2+3jzx^2+3kxy^2+6lxyz;$$

and the expression for ξ is obtained from the equation

$$x^2\xi=(b, f, i, c\between (j, f, c, i, g, l\between x, y, z)^2, \quad -(h, b, i, f, l, k\between x, y, z)^2)^3$$
$$-(a, b, c, f, g, h, i, j, k, l\between x, y, z)^3(\mathfrak{C}x+\mathfrak{D}),$$

[1] *Philosophical Transactions*, vol. CXLVII. (1857), [146].

[2] At the time when the present paper was written, I was not aware of Mr Salmon's theorem (Higher Plane Curves, p. 156), that the tangential of a point of the cubic is the intersection of the tangent of the cubic with the first or line polar of the point with respect to the Hessian; a theorem, which at the same time that it affords the easiest mode of calculation, renders the actual calculation of the coordinates of the tangential less important. Added 7th October, 1858.—A. C.

where the second line is in fact equal to zero, on account of the first factor, which vanishes. And $\mathfrak{C}$, $\mathfrak{D}$ denote respectively quadric and cubic functions of (y, z), which are to be determined so as to make the right-hand side divisible by x^2; the resulting value of ξ may be modified by the adjunction of the evanescent term

$$(\mathrm{a}x + \mathrm{h}y + \mathrm{j}z)\,(a, b, c, f, g, h, i, j, k, l \between x, y, z)^3,$$

where a, h, j are arbitrary coefficients; but as it is not obvious how these coefficients should be determined in order to present the result in the most simple form, I have given the result in the form in which it was obtained without the adjunction of any such term.

Write for shortness,

$$\begin{aligned}
P &= (k, l \between y, z),\\
Q &= (b, f, i \between y, z)^2,\\
R &= (l, g, \between y, z),\\
S &= (f, i, c \between y, z)^2,\\
B &= (h, j \between y, z),\\
C &= (k, l, g \between y, z)^2,\\
D &= (b, f, i, c \between y, z)^3,
\end{aligned}$$

so that

$$\begin{aligned}
(h, b, i, f, l, k &\between x, y, z)^2 = (h, P, Q \between x, 1)^2,\\
(j, f, c, i, g, l &\between x, y, z)^2 = (j, R, S \between x, 1)^2,\\
(a, b, c, f, g, h, i, j, k, l &\between x, y, z)^3 = (a, B, C, D \between x, 1)^3.\\
\mathfrak{C}x + \mathfrak{D} &= (\mathfrak{C}, \mathfrak{D} \between x, 1),
\end{aligned}$$

and then for greater convenience writing $(h, 2P, Q \between x, 1)^2$, &c. for $(h, P, Q \between x, 1)^2$, &c., and omitting the $(x, 1)^2$, &c. and the arrow-heads, or representing the functions simply by $(h, 2P, Q)$, &c., we have

$$\begin{aligned}
x^2\xi = {} & b\,(j, 2R, S)^3\\
& - 3f(j, 2R, S)^2 . (h, 2P, Q)\\
& + 3i\,(j, 2R, S) . (h, 2P, Q)^2\\
& - c \quad . (h, 2P, Q)^3\\
& - (a, 3B, 3C, D) . (\mathfrak{C}, \mathfrak{D}),
\end{aligned}$$

which can be developed in terms of the quantities which enter into it. The conditions, in order that the coefficients of x, x^0 may vanish, are thus seen to be

$$D\mathfrak{D} = bS^3 - 3fS^2Q + 3iSQ^2 - cQ^3,$$

$$D\mathfrak{C} - 3C\mathfrak{D} = b\,(6RS^2) - 3f\,(2S^2P + 4RSQ) + 3i\,(2RQ^2 + 4SPQ) - c\,(6PQ^2),$$

and from these we obtain

$\mathfrak{C} =$ (

$bck\ -3$	$big\ +6$	$bcg\ +3$
$bil\ +6$	$cfk\ -6$	$cfl\ -6$
$fik\ +3$	$f^2g\ -6$	$fgi\ -3$
$f^2l\ -6$	$i^2k\ +6$	$i^2l\ +6$

$\between y, z)^2$

$$\mathfrak{D} = \left(\begin{array}{c|c|c|c} b^2c \ -1 & bcf \ -3 & bci \ +3 & bc^2 \ +1 \\ bfi \ +3 & bi^2 \ +6 & cf^2 \ -6 & cfi \ -3 \\ f^3 \ -2 & f^2i \ -3 & fi^2 \ +3 & i^3 \ +2 \end{array}\right)(y, z)^3$$

and substituting these values, the right-hand side of the equation divides by x^2, and throwing out this factor we have the value of ξ; and the values of η, ζ may be thence deduced by a mere interchange of letters. The value for ξ is

x^4	x^3y	x^3z	x^2y^2	x^2yz	x^2z^2	xy^3	xy^2z
bj^3 +1	bj^2l + 6	bgj^2 + 6	$abck$ + 3	$abgi$ − 6	$abcg$ − 3	ab^2c + 1	$abcf$ + 3
ch^3 −1	ch^2k − 6	ch^2l − 6	$abil$ − 6	$acfk$ + 6	$acfl$ + 6	$abfi$ − 3	abi^2 − 6
fhj^2 −3	$fhjl$ −12	$fghj$ −12	af^2l + 6	af^2g + 6	$afgi$ + 3	af^3 + 2	af^2i + 3
h^2ij +3	fj^2k − 6	fj^2l − 6	$afik$ − 3	ai^2k − 6	ai^2l − 6	$bchk$ − 3	$bchl$ −12
	h^2il + 6	gh^2i + 6	bch^2 − 3	$bgjl$ +24	bcj^2 + 3	$bhil$ − 6	$bcjk$ + 9
	$hijk$ +12	$hijl$ +12	$bhij$ + 6	bij^2 + 6	bg^2j +12	$bijk$ +12	$bghi$ − 6
			bjl^2 +12	cfh^2 − 6	$cfhj$ − 6	bl^3 + 8	bgl^2 +24
			chk^2 −12	$chkl$ −24	fjl^2 −24	ck^3 − 8	$bijl$ +18
			f^2hj − 6	f^2j^2 − 6	fg^2h −12	f^2hl + 6	$cfhk$ − 6
			fh^2i − 3	$fghl$ −24	$fgjl$ −24	f^2jk −12	ck^2l −24
			fhl^2 −12	$fgjk$ −24	fij^2 − 3	$fhik$ + 3	f^2gh + 6
			$fjkl$ −24	fjl^2 −24	$ghil$ +24	fkl^2 −24	f^2jl −18
			$hikl$ +24	$ghik$ +24	hi^2j + 6	ik^2l +24	$fghl$ −48
			ijk^2 +12	h^2i^2 + 6	ijl^2 +12		$fhil$ +12
				hil^2 +24			$fijk$ − 9
				$ijkl$ +24			fl^3 −24
							gik^2 +24
							h^2k + 6
							ikl^2 +48

xyz^2	xz^3	y^4	y^3z	y^2z^2	yz^3	z^4
$abci$ − 3	abc^2 − 1	b^2ij + 3	b^2cj + 3	$bcfj$ + 9	bc^2h − 3	bcg^2 + 3 ;
acf^2 + 6	$acfi$ + 3	bck^2 − 3	$bcfh$ − 3	$bchi$ − 9	$bcgl$ + 6	c^2fh − 3
afi^2 − 3	ai^3 − 2	bf^2j − 3	$bckl$ − 3	$bgil$ +18	$bcij$ + 3	$cfgl$ − 6
$bcgh$ − 9	$bcgj$ + 3	$bfhi$ − 3	$bfij$ + 3	$cfkl$ −18	bg^2i + 6	$cfij$ + 3
$bcjl$ +12	bg^3 + 8	$bikl$ + 6	$bgik$ + 3	f^2gl −18	cf^2j + 6	chi^2 + 3
bg^2l +24	$cfgh$ −12	f^3h + 3	bhi^2 − 6	f^2ij − 9	$cfgk$ − 6	fg^2i − 3
$bgij$ + 6	$cfjl$ + 3	f^2kl − 6	bil^2 +12	fhi^2 + 9	$cfhi$ − 3	gi^2l + 6
$cfhl$ −18	cl^3 − 8	fik^2 + 3	cfk^2 − 6	i^2kl +18	cfl^2 −12	i^3j − 3
$cfjk$ + 6	fg^2l −24		f^3j − 6		f^2g^2 − 6	
ckl^2 −24	$fgij$ − 3		f^2gk − 6		$fgil$ − 6	
f^2gj − 6	ghi^2 +12		f^2hi + 9		fi^2j − 9	
fg^2k −24	gil^2 +24		f^2l^2 −12		gi^2k + 6	
$fghi$ + 9	i^2jl − 6		$fikl$ + 6		hi^3 + 6	
fgl^2 −48			i^2k^2 + 6		i^2l^2 +12	
$fijl$ −12						
$gikl$ +48						
hi^2l +18						
i^2jk − 6						
il^3 +24						

and it is not necessary to write down the corresponding values for η, ζ.

158.

A SIXTH MEMOIR UPON QUANTICS.

[From the *Philosophical Transactions of the Royal Society of London,* vol. CXLIX. *for the year* 1859, pp. 61—90. Received November 18, 1858,—Read January 6, 1859.]

I PROPOSE in the present memoir to consider the geometrical theory: I have alluded to this part of the subject in the articles Nos. 3 and 4 of the Introductory Memoir, [139]. The present memoir relates to the geometry of one dimension and the geometry of two dimensions, corresponding respectively to the analytical theories of binary and ternary quantics. But the theory of binary quantics is considered for its own sake; the geometry of one dimension is so immediate an interpretation of the theory of binary quantics, that for its own sake there is no necessity to consider it at all; it is considered with a view to the geometry of two dimensions. A chief object of the present memoir is the establishment, upon purely descriptive principles, of the notion of distance. I had intended in this introductory paragraph to give an outline of the theory, but I find that in order to be intelligible it would be necessary for me to repeat a great part of the contents of the memoir in relation to this subject, and I therefore abstain from entering upon it. The paragraphs of the memoir are numbered consecutively with those of my former Memoirs on Quantics.

147. It will be seen that in the present memoir, the geometry of one dimension is treated of as a geometry of points in a line, and the geometry of two dimensions as a geometry of points and lines in a plane. It is, however, to be throughout borne in mind, that, in accordance with the remarks No. 4 of the Introductory Memoir, the terms employed are not (unless this is done expressly or by the context) restricted to their ordinary significations. In using the geometry of one dimension in reference to geometry of two dimensions considered as a geometry of points and lines in a plane, it is necessary to consider,—1°, that the word point may mean *point* and the word line mean *line*; 2°, that the word point may mean *line* and the

word line mean *point*. It is, I say, necessary to do this, for in such geometry of two dimensions we have systems of points in a line and of lines through a point, and each of these systems is in fact a system belonging to, and which can by such extended signification of the terms be included in, the geometry of one dimension. And precisely because we can by such extension comprise the correlative theorems under a common enunciation, it is not in the geometry of one dimension necessary to enunciate them separately; it may be and very frequently is necessary and proper in the geometry of two dimensions, where we are concerned with systems of each kind, to enunciate such correlative theorems separately. It may, by way of further illustration, be remarked, that in using the geometry of one dimension in reference to geometry of three dimensions considered as a geometry of points, lines, and planes in space, it would be necessary to consider,—1°, that the words point and line may mean respectively *point* and *line*; 2°, that the word line may mean *point in a plane*[1], and the word point mean *line*, viz. the expression points in a line mean *lines through a point and in a plane*; 3rd, that the word line may mean *line* and the word point mean *plane*, viz. the expression points in a line mean *planes through a line*. And so in using the geometry of two dimensions in reference to geometry of three dimensions considered as a geometry of points, lines, and planes in space, it would be necessary to consider,—1°, that the words point, line, and plane may mean respectively *point*, *line*, and *plane*; 2°, that the words point, line, and plane may mean respectively *plane*, *line*, and *point*. But I am not in the present memoir concerned with geometry of three dimensions. The thing to be attended to is, that in virtue of the extension of the signification of the terms, in treating the geometry of one dimension as a geometry of points in a line, and the geometry of two dimensions as a geometry of points and lines in a plane, we do in reality treat these geometries respectively in an absolutely general manner. In particular—and I notice the case because I shall have occasion again to refer to it—we do in the geometry of two dimensions include spherical geometry; the words plane, point, and line, meaning for this purpose, spherical surface, arc (of a great circle) and point (that is, pair of opposite points) of the spherical surface. And in like manner the geometry of one dimension includes the cases of points on an arc, and of arcs through a point.

148. I repeat also a remark which is in effect made in the same No. 4; the coordinates x, y of the geometry of one dimension, and the coordinates x, y, z and ξ, η, ζ of the geometry of two dimensions are only determinate to a common factor *près* (that is, it is the ratios only of the coordinates, and not their absolute magnitudes, which are determinate); hence in saying that the coordinates x, y are equal to a, b, or in writing $x, y = a, b$, we mean only that $x : y = a : b$, and we never as a result obtain $x, y = a, b$, but only $x : y = a : b$. And the like with respect to the coordinates x, y, z and ξ, η, ζ. (In the geometry of two dimensions, $x, y = a, b$, is for this reason considered and spoken of as a single equation.) But when this is once understood, there is no objection to treating the coordinates as if they were completely determinate.

[1] It would be more accurate to say that the word line may mean *point-in-and-with-a plane*, viz. the *locus in quo* of lines through the point and in the plane. Added, June 16, 1859.—A. C.

On Geometry of One Dimension, Nos. 149 to 168.

149. In geometry of one dimension we have the line as a space or *locus in quo*, which is considered as made up of points. The several points of the line are determined by the coordinates (x, y), viz. attributing to these any specific values, or writing $x, y = a, b$, we have a particular point of the line. And we may say also that the line is the *locus in quo* of the coordinates (x, y).

150. A linear equation,

$$(*\between x, y)^1 = 0,$$

is obviously equivalent to an equation of the before-mentioned form $x, y = a, b$, and represents therefore a point. An equation such as

$$(*\between x, y)^m = 0$$

breaks up into m linear equations, and represents therefore a system of m points, or point-system of the order m. The component points of the system, or the linear factors, or the values thereby given for the coordinates, are termed roots. When $m = 1$ we have of course a single point, when $m = 2$ we have a quadric or point-pair, when $m = 3$ a cubic or point-triplet, and so on. The point-system is the only figure or locus occurring in the geometry of one dimension. The quantic $(*\between x, y)^m$, when it is convenient to do so, may be represented by a single letter U, and we then have $U = 0$ for the equation of the point-system.

151. The equation

$$(*\between x, y)^m = 0$$

may have two or more of its roots equal to each other, or generally there may exist any systems of equalities between the roots of the equation, or what is the same thing, the system may comprise two or more coincident points, or any systems of coincident points. In particular, when the discriminant vanishes the equation will have a pair of equal roots, or the system will comprise a pair of coincident points; in the case of the quadric $(a, b, c\between x, y)^2 = 0$, the condition is $ac - b^2 = 0$, or as it may be written, $a, b = b, c$; in the case of the cubic

$$(a, b, c, d\between x, y)^3 = 0,$$

the condition is

$$a^2d^2 - 6abcd + 4ac^3 + 4b^3d - 3b^2c^2 = 0.$$

The preceding is the only special case for a quadric: for a cubic we have besides the special case where the three roots are equal, or the cubic reduces itself to three coincident points; the conditions for this are

$$ac - b^2 = 0, \quad ad - bc = 0, \quad bd - c^2 = 0,$$

equivalent to the two conditions

$$a : b = b : c = c : d.$$

For equations of a higher order the analytical question is considered, and as regards the quartic and the quintic respectively completely solved, in my "Memoir on the Conditions for the Existence of given Systems of Equalities between the Roots of an Equation" (1).

152. Any covariant of the equation

$$(* \between x, y)^m = 0,$$

equated to zero, gives rise to a point-system connected in a definite manner with the original point-system. And as regards the invariants, the evanescence of any invariant implies a certain relation between the points of the system; the identical evanescence of any covariant implies relations between the points of the system, such that the derived point-system obtained by equating the covariant to zero is absolutely indeterminate. The like remarks apply to the covariants or invariants of two or more equations, and the point-systems represented thereby.

153. In particular, for the two point-pairs represented by the quadric equations

$$(a, b, c \between x, y)^2 = 0,$$
$$(a', b', c' \between x, y)^2 = 0,$$

if the lineo-linear invariant vanishes, that is, if

$$ac' - 2bb' + ca' = 0,$$

we have the harmonic relation,—the two point-pairs are said to be harmonically related to each other, or the two points of the one pair are said to be harmonics with respect to the two points of the other pair. The analytical theory is fully developed in the "Fifth Memoir upon Quantics" (2). The chief results, stated under a geometrical form, are as follows:

1°. If either of the pairs and one point of the other pair are given, the remaining point of such other pair can be found.

2°. A point-pair can be found harmonically related to any two given point-pairs.

154. The last of the two theorems gives rise to the theory of involution. The two given point-pairs, viewed in relation to the harmonic pair, are said to be an involution of four points; and the points of the harmonic pair are said to be the (double or) sibiconjugate points of the involution. A system of three or more pairs, such that the third and every subsequent pair are each of them harmonically related to the sibiconjugate points of the first and second pairs, is said to be a system in involution. In particular, for three pairs we have what is termed an involution of six points; and it is clear that when two pairs and a point of the third pair are given, the remaining point of the third pair can be determined. And in like manner

1 *Philosophical Transactions*, vol. CXLVII. (1857), pp. 727—731, [150].

2 *Philosophical Transactions*, vol. CXLVIII. (1858), pp. 429—462, [156].

for a greater number of pairs, when two pairs and a point of each of the other pairs are given, the remaining point of each of the other pairs can be determined. Two points of the same pair are said to be conjugate to each other; or if we consider two pairs as given, then the points of the third or any subsequent pair are said to be conjugate to each other in respect to the given pairs. This explains the expression sibiconjugate points; in fact, the two pairs being given, either sibiconjugate point is, as the name imports, conjugate to itself. In other words, any two pairs and one of the sibiconjugate points considered as a pair of coincident points, form a system in involution, or involution of five points.

155. The three point-pairs, $U=0$, $U'=0$, $U''=0$, will be in involution when the quadrics U, U', U'' are connected by the linear relation or syzygy $\lambda U+\lambda' U'+\lambda'' U''=0$. This property, or the relation

$$\begin{vmatrix} a, & b, & c \\ a', & b', & c' \\ a'', & b'', & c'' \end{vmatrix} = 0$$

to which it gives rise, might have been very properly adopted as the definition of the relation of involution, but I have on the whole preferred to deduce the theory of involution from the harmonic relation. The notion, however, of the linear relation or syzygy of three or more point-systems gives rise to a much more general theory of involution, but this is a subject that I do not now enter upon; it may, however, be noticed, that if $U=0$, $U'=0$ be any two point-systems of the same order, then we may find a point-system $U''=0$ of the same order, in involution with the given point-systems (that is, satisfying the condition $\lambda U+\lambda' U'+\lambda'' U''=0$), and such that the point-system $U''=0$ comprises a pair of coincident points; this is obviously an extension of the notion of the sibiconjugate points of an ordinary involution.

156. It was remarked in the Fifth Memoir, that the theories of the anharmonic ratio and of homography belong analytically to the subject of bipartite (lineo-linear) binary quantics; this may be further illustrated geometrically as follows: we may imagine two distinct spaces of one dimension, or lines, one of them the *locus in quo* of the coordinates (x, y), and the other the *locus in quo* of the coordinates (x, y), which are absolutely independent of, and are not in anywise related to, the coordinates of the first-mentioned system. There is no difficulty in the conception of this; for we may in a plane or in space of three dimensions imagine any two lines, and study the relations of analogy between the points of the one line *inter se*, and the points of the other line *inter se*, without in anywise adverting to the space of two or three dimensions which happens to be the common *locus in quo* of the two lines. It is proper to remark, that in speaking of the spaces of one dimension, which are the *loci in quibus* of the coordinates (x, y) and (x, y) respectively, as being each of them a line, we imply a restriction which is altogether unnecessary; the words line and point may, in regard to the two figures respectively, be used in different significations; for instance, one of the spaces may be a *line* and the points in it *points*; while the other of the spaces may be a *point* and the points in it *lines*, or it may be a *line* and the points in it *planes*.

157. A lineo-linear equation

$$(x - ay)(\mathrm{x} - \alpha \mathrm{y}) = 0$$

denotes then the two points (x, $y = a$, 1) and (x, $\mathrm{y} = \alpha$, 1) existing irrespectively of each other in distinct spaces, and only by the equation itself brought into an ideal connexion; and any invariantive relation between the coefficients of any such bipartite function denotes geometrically a relation between a point-system in the space which is the *locus in quo* of the coordinates (x, y), and a point-system in the space which is the *locus in quo* of the coordinates (x, y); for instance, the equation

$$\begin{vmatrix} 1, & a, & \alpha, & a\alpha \\ 1, & b, & \beta, & b\beta \\ 1, & c, & \gamma, & c\gamma \\ 1, & d, & \delta, & d\delta \end{vmatrix} = 0$$

is the relation of homography between the four points (a, 1), (b, 1), (c, 1), (d, 1) in the first line, and the four points (α, 1), (β, 1), (γ, 1), (δ, 1) in the second line. The analytical theory is discussed in the Fifth Memoir; and, in particular, it is there shown, that writing

$$\begin{aligned} A &= (d-a)(b-c), & \mathfrak{A} &= (\delta-\alpha)(\beta-\gamma), \\ B &= (d-b)(c-a), & \mathfrak{B} &= (\delta-\beta)(\gamma-\alpha), \\ C &= (d-c)(a-b), & \mathfrak{C} &= (\delta-\gamma)(\alpha-\beta), \end{aligned}$$

then the condition may be expressed under any one of the forms

$$A : B : C = \mathfrak{A} : \mathfrak{B} : \mathfrak{C},$$

equations which denote the equality of the anharmonic ratios of the two point-systems.

158. The number of points in each system may be four, or any greater number; the homographic relation is then conveniently expressed under the form

$$\begin{Vmatrix} 1\ , & 1\ , & 1\ , & 1\ , & 1, \ldots \\ a\ , & b\ , & c\ , & d\ , & e\ , \\ \alpha\ , & \beta\ , & \gamma\ , & \delta\ , & \epsilon\ , \\ a\alpha, & b\beta, & c\gamma, & d\delta, & e\epsilon, \end{Vmatrix} = 0.$$

The relation is such that given three points of the one system and the corresponding three points of the other system, then to any fourth point whatever of the first system there can be found a corresponding fourth point of the second system. It is to be observed, however, that two systems of four points homographically related to each other, always correspond together in four different ways, viz. the two systems being (a, b, c, d) and (α, β, γ, δ); then if the four points of the first system are (a, b, c, d), the corresponding four points of the second system may be taken in the four several orders, (α, β, γ, δ), (β, α, δ, γ), (γ, δ, α, β), (δ, γ, β, α).

159. What precedes is not to be understood as precluding the existence of a relation between the spaces which are the *loci in quibus* of the coordinates (x, y) and (x, y) respectively: not only may these be spaces of the same kind, but they may be one and the same space or line; and the points of the two systems may then be points of the same kind; and further, the coordinates (x, y) and (x, y) may belong to the same system of coordinates, that is, the equations $(x, y = a, 1)$ and (x, y $= a$, 1) may denote one and the same point.

160. If the two point-systems are systems of the same kind, and are in one and the same line, then there are in general two points of the first system which coincide each of them with the corresponding point of the second system; such two points may be said to be the sibiconjugate points of the homography. In particular, the two sibiconjugate points of the homography may coincide together.

161. A system in involution affords an example of two homographic systems in the same line; in fact, taking arbitrarily a point out of each pair, the points so obtained form a system which is homographic with the system formed with the other points of the several pairs; and in this case the sibiconjugate points of the involution are also the sibiconjugate points of the homography. Thus if A and A', B and B', C and C', D and D' are pairs of the system in involution, then (A, B, C, D) and (A', B', C', D') will be homographic point-systems; and, as a particular case, (A, B, C, C') and (A', B', C', C) will be homographic point-systems. It is proper to notice that if F is a sibiconjugate point of the involution, then (A, B, F, F) and (A', B', F, F) are not (what at first sight they appear to be) homographic point-systems.

162. Imagine an involution of points; take on the line which is the *locus in quo* of the point-system a point O, and consider the point-system formed by the harmonics of O in respect to the several pairs of the involution; and in like manner take on the line any other point O', and consider the point-system formed by the harmonics of O' in respect to the several pairs of the involution; these two point-systems are homographically related to each other.—See Fifth Memoir, No. 111.

163. Two involutions may be homographically related to each other; in fact, take on the line which is the *locus in quo* of the first involution a point O, and consider the point-system formed by the harmonics of O in relation to the several pairs of the involution; take in like manner on the line which is the *locus in quo* of the second involution a point Q, and consider the point-system formed by the harmonics of Q with respect to the several pairs of the involution; then if the two point-systems are homographically related, the two involutions are said to be themselves homographically related: the last preceding article shows that the nature of the relation does not in anywise depend on the choice of the points O and Q. And it is not necessary that, as regards the two involutions respectively, the words line and point should have the same significations.—See Fifth Memoir, No. 111.

164. Four or more tetrads of points in a line may be homographically related to the same number of tetrads in another line. This is the case when the an-

harmonic ratios of the tetrads of the first system are homographically related to the anharmonic ratios of the tetrads of the second system. And it is not material which of the three anharmonic ratios of a tetrad of either system is selected, provided that the same selection is made for each of the other tetrads of the same system. The order of the points of a tetrad must be attended to, but there are in all four admissible permutations of the points of a tetrad, viz. if A, B, C, D are the points of a tetrad, then (A, B, C, D), (B, A, D, C), (C, D, A, B), (D, C, B, A) may be considered as one and the same tetrad. Any three tetrads whatever in the second system may correspond to any three tetrads of the first system; and then given a fourth tetrad of the first system, and three out of the four points of the corresponding tetrad of the second system, the remaining point of the tetrad may be determined. The words line and point need not, as regards the two systems of tetrads respectively, be understood in the same significations.—See Fifth Memoir, No. 112.

165. The foregoing theory of the harmonic relation shows that if we have a point-pair

$$(a, b, c \between x, y)^2 = 0,$$

the equation of any other point-pair whatever can be expressed, and that in two different ways, in the form

$$(a, b, c \between x, y)^2 + (lx + my)^2 = 0\,;$$

the points $(lx + my = 0)$ corresponding to the two admissible values of the linear function being in fact the harmonics of the point-pair in respect to the given point-pair $(a, b, c \between x, y)^2 = 0$, or what is the same thing, the sibiconjugate points of the involution formed by the two point-pairs (see Fifth Memoir, No. 105). The point-pair represented by the equation in question does not in itself stand in any peculiar relation to the given point-pair $(a, b, c \between x, y)^2 = 0$; but when thus represented it is said to be inscribed in the given point-pair, and the point $lx + my = 0$ is said to be the axis of inscription. And the harmonic of this point with respect to the given point-pair (that is, the other sibiconjugate point of the involution of the two point-pairs) is said to be the centre of inscription[1].

166. We may, if we please, (x', y') and θ being constants, exhibit the equation of the inscribed point-pair in the form

$$(a, b, c \between x, y)^2 (a, b, c \between x', y')^2 \sin^2\theta - (ac - b^2)(xy' - x'y)^2 = 0,$$

where we have for the axis of inscription and centre of inscription respectively, the equations

$$\begin{aligned} &xy' - x'y &&= 0, \\ &(a, b, c \between x, y \between x', y') &&= 0\,; \end{aligned}$$

[1] The words inscribed, inscription, are used not in opposition to, but as identical with, the words circumscribed, circumscription; and in like manner (*post*, Nos. 203 et seq.) as regards conics.

or in the equivalent form,

$$(a,\ b,\ c\between x,\ y)^2\,(a,\ b,\ c\between x',\ y')^2\cos^2\theta - \{(a,\ b,\ c\between x,\ y\between x',\ y')\}^2 = 0,$$

where we have for the axis of inscription and the centre of inscription respectively, the equations

$$\begin{aligned}(a,\ b,\ c\between x,\ y\between x',\ y') &= 0,\\ xy' - x'y &= 0.\end{aligned}$$

167. The equivalence of the two forms depends on the identical equation

$$(a,\ b,\ c\between x,\ y)^2\,(a,\ b,\ c\between x',\ y')^2 - \{(a,\ b,\ c\between x,\ y\between x',\ y')\}^2 = (ac-b^2)\,(xy'-x'y)^2,$$

which is in fact the equation mentioned, Fifth Memoir, No. 95. If, for shortness, we write

$$\begin{aligned}&(a,\ b,\ c\between x,\ y)^2 &&= 00,\\ &(a,\ b,\ c\between x,\ y\between x',\ y') &&= 01 = 10,\\ &\&c.,\end{aligned}$$

then the equation may be represented in the form

$$\begin{vmatrix} 00, & 01 \\ 10, & 11 \end{vmatrix} = (ac-b^2)\begin{vmatrix} x, & y \\ x, & y' \end{vmatrix}^2.$$

168. There is a like equation for the three sets $(x,\ y)$, $(x',\ y')$, $(x'',\ y'')$; the right-hand side here vanishes, for there are not columns enough to form therewith a determinant, and the equation is

$$\begin{vmatrix} 00, & 01, & 02 \\ 10, & 11, & 12 \\ 20, & 21, & 22 \end{vmatrix} = 0,$$

an equation which may also be written in the form

$$\cos^{-1}\frac{01}{\sqrt{00}\,\sqrt{11}} + \cos^{-1}\frac{12}{\sqrt{11}\,\sqrt{22}} = \cos^{-1}\frac{02}{\sqrt{00}\,\sqrt{22}},$$

as it is easy to verify by reducing this equation to an algebraical form. The various formulæ have been given in relation to the establishment of the notion of distance in the geometry of one dimension, but it will be convenient to defer the consideration of this theory so as to discuss it in connexion with geometry of two dimensions.

On Geometry of Two Dimensions, Nos. 169 to 208.

169. In geometry of two dimensions we have the plane as a space or *locus in quo*, which is considered under two distinct aspects, viz. as made up of points, and as made up of lines. The several points of the plane are determined by means of

the point-coordinates (x, y, z), viz. attributing to these any specific values, or writing $x, y, z = a, b, c$, we have a particular point of the plane; and in like manner the several lines of the plane are determined by the line-coordinates (ξ, η, ζ), viz. attributing to these any specific values, or writing $\xi, \eta, \zeta = \alpha, \beta, \gamma$, we have a particular line of the plane. And we may say that the plane is the *locus in quo* of the point-coordinates (x, y, z), and of the line-coordinates (ξ, η, ζ). It is not necessary to consider separately the analytical theories of point-coordinates and of line-coordinates; for the theory of the former in relation to points and lines respectively is identical with the theory of the latter in relation to lines and points respectively; but it is necessary to show how either system of coordinates, say the system of point-coordinates, is applicable to both points and lines, or in fact all loci whatever, and to explain the mutual relation of the two systems of coordinates.

170. Considering then point-coordinates, the equations

$$x, y, z = a, b, c,$$

determine, as already mentioned, a point.

A linear equation

$$(* \between x, y, z)^1 = 0$$

determines a line, viz. the line which is the locus of all the points, the coordinates of which satisfy this equation. And in like manner an equation

$$(* \between x, y, z)^m = 0$$

determines a curve of the mth order, viz. the curve which is the locus of all the points, the coordinates of which satisfy this equation. In particular, an equation of the second degree

$$(* \between x, y, z)^2 = 0$$

determines a conic.

171. If the quantic breaks up into rational factors, then the equation of the curve is satisfied by equating to zero any one of these factors, or the curve breaks up into curves of a lower order, and the order of the entire curve is equal to the sum of the orders of the component curves. In particular, a curve of any order may break up into a system of lines, the number of lines being of course equal to the order of the curve, and any two or more of these lines may coincide with each other. A curve not thus breaking up into curves of a lower order is said to be a proper curve.

172. Returning to the linear equation and expressing the coefficients, the equation is

$$(\xi, \eta, \zeta \between x, y, z) = 0,$$

or, what is the same thing,

$$\xi x + \eta y + \zeta z = 0;$$

and we say as a definition, that the coordinates (line-coordinates) of this line are (ξ, η, ζ).

173. But the same equation, considering (x, y, z) as constant coefficients, and (ξ, η, ζ) as line-coordinates, is the equation of a point, viz. the point which is the locus (envelope) of all those points the coordinates of which satisfy the equation in question; and such point is precisely the point, the coordinates (point-coordinates) of which are (x, y, z). In fact, if (ξ, η, ζ) are considered as variable parameters connected by the equation $\xi x + \eta y + \zeta z = 0$, then taking (X, Y, Z) as current point-coordinates, the equation $\xi X + \eta Y + \zeta Z = 0$ is satisfied by writing (x, y, z) for (X, Y, Z); or the several lines the coordinates whereof are (ξ, η, ζ), all pass through the point (x, y, z).

174. Hence recapitulating, the equation

$$(\xi, \eta, \zeta \between x, y, z) = 0,$$

or

$$\xi x + \eta y + \zeta z \quad = 0,$$

considering (x, y, z) as current point-coordinates, and (ξ, η, ζ) as constant coefficients, is the equation of a line the coordinates (line-coordinates) of which are (ξ, η, ζ); and the same equation, considering (ξ, η, ζ) as current line-coordinates, and (x, y, z) as constant coefficients, is the equation of a point the coordinates (point-coordinates) of which are (x, y, z).

175. The expression, the point (a, b, c), means the point whose point-coordinates are (a, b, c); and in like manner the expression, the line (α, β, γ), means the line whose line-coordinates are (α, β, γ). The last-mentioned expression may, without any impropriety or risk of ambiguity, be employed when we are dealing with point-coordinates; but it is of course always allowable, and it is frequently better, to substitute for the definition the thing signified, and say the line having for its equation $\alpha x + \beta y + \gamma z = 0$, or more briefly, the line $\alpha x + \beta y + \gamma z = 0$. It will be proper to do this in the following articles, Nos. 176 to 184, which contain some formulæ in point-coordinates relating to the theory of the point and the line.

176. The condition that the point (a, b, c) may lie in the line

$$\alpha x + \beta y + \gamma z = 0,$$

is of course

$$\alpha a + \beta b + \gamma c = 0.$$

177. The equation of the line passing through the points (a, b, c), (a', b', c'), is

$$\begin{vmatrix} x, & y, & z \\ a, & b, & c \\ a', & b', & c' \end{vmatrix} = 0;$$

and if in this equation (a', b', c') are considered as indeterminate, we have the equation of a line subjected to the single condition of passing through the point (a, b, c). The equation contains apparently two arbitrary parameters, but these in fact reduce themselves to a single one.

178. The coordinates of the point of intersection of the lines

$$\alpha x + \beta y + \gamma z = 0,$$
$$\alpha' x + \beta' y + \gamma' z = 0,$$

are given by the equations

$$x, y, z = \beta\gamma' - \beta'\gamma, \quad \gamma\alpha' - \gamma'\alpha, \quad \alpha\beta' - \alpha'\beta;$$

and if in these equations we consider α', β', γ' as indeterminate, we have the coordinates of a point subjected to the single condition of lying in the line $\alpha x + \beta y + \gamma z = 0$; the result, as in the last case, contains in appearance two arbitrary parameters, but these really reduce themselves to a single one.

179. The condition in order that the points (a, b, c), (a', b', c'), (a'', b'', c'') may lie in a line is

$$\begin{vmatrix} a\ , & b\ , & c \\ a', & b', & c' \\ a'', & b'', & c'' \end{vmatrix} = 0,$$

which may also be expressed by the equations

$$a'', b'', c'' = \lambda a + \mu a', \quad \lambda b + \mu b', \quad \lambda c + \mu c',$$

where λ, μ are arbitrary multipliers; these equations give therefore the coordinates of an indeterminate point in the line joining the points (a, b, c) and (a', b', c').

180. The condition that the lines

$$\alpha x + \beta y + \gamma z = 0,$$
$$\alpha' x + \beta' y + \gamma' z = 0,$$
$$\alpha'' x + \beta'' y + \gamma'' z = 0$$

may meet in a point is

$$\begin{vmatrix} \alpha\ , & \beta\ , & \gamma \\ \alpha', & \beta', & \gamma' \\ \alpha'', & \beta'', & \gamma'' \end{vmatrix} = 0,$$

a relation which may also be expressed by the equations

$$\alpha'', \beta'', \gamma'' = l\alpha + m\alpha', \quad l\beta + m\beta', \quad l\gamma + m\gamma';$$

where l, m are arbitrary multipliers; and substituting these values in the equation $\alpha'' x + \beta'' y + \gamma'' z = 0$, we have for the equation of a line subjected to the single condition of passing through the point of intersection of the lines $\alpha x + \beta y + \gamma z = 0$, $\alpha' x + \beta' y + \gamma' z = 0$, the equation

$$l(\alpha x + \beta y + \gamma z) + m(\alpha' x + \beta' y + \gamma' z) = 0,$$

which is, in fact, at once obtained by the consideration that the values of (x, y, z) which satisfy simultaneously the equations $\alpha x+\beta y+\gamma z=0$ and $\alpha' x+\beta' y+\gamma' z=0$, satisfy also the equation in question.

181. The equation of the line passing through the point of intersection of the lines $\alpha x+\beta y+\gamma z=0$ and $\alpha' x+\beta' y+\gamma' z=0$, and also through the point (a, b, c), is obviously

$$\begin{vmatrix} \alpha x+\beta y+\gamma z, & \alpha' x+\beta' y+\gamma' z \\ \alpha a+\beta b+\gamma c, & \alpha' a+\beta' b+\gamma' c \end{vmatrix}=0,$$

which, or the equivalent form

$$\frac{\alpha x+\beta y+\gamma z}{\alpha a+\beta b+\gamma c}=\frac{\alpha' x+\beta' y+\gamma' z}{\alpha' a+\beta' b+\gamma' c},$$

is usually the most convenient one; but it is to be observed that the equation can also be written in the forms

$$\begin{vmatrix} x\ , & y\ , & z \\ a\ , & b\ , & c \\ \beta\gamma'-\beta'\gamma, & \gamma\alpha'-\gamma'\alpha, & \alpha\beta'-\alpha'\beta \end{vmatrix}=0,$$

and

$$\begin{vmatrix} bz-cy, & cx-az, & ay-bx \\ \alpha\ , & \beta\ , & \gamma \\ \alpha'\ , & \beta'\ , & \gamma' \end{vmatrix}=0,$$

or in the form

$$(\beta\gamma'-\gamma\beta')(bz-cy)+(\gamma\alpha'-\gamma'a)(cx-az)+(\alpha\beta'-\beta\alpha')(ay-bx)=0,$$

which might also be represented by

$$\begin{vmatrix} \alpha, & \beta, & \gamma \\ \alpha', & \beta', & \gamma' \end{vmatrix}\begin{vmatrix} x, & y, & z \\ a, & b, & c \end{vmatrix}=0.$$

182. To find the coordinates of the point of intersection of the line joining the points (a, b, c), (a', b', c'), with the line $\alpha x+\beta y+\gamma z=0$, we have

$$x,\ y,\ z=\lambda a+\mu a',\quad \lambda b+\mu b',\quad \lambda c+\mu c',$$

where λ, μ are given by

$$\lambda(\alpha a+\beta b+\gamma c)+\mu(\alpha a'+\beta b'+\gamma c')=0.$$

The preceding are elementary formulæ of almost constant occurrence; it may be proper to add to them the formulæ which follow.

183. To find the equation of the line passing through the point of intersection of the lines

$$\alpha_1 x+\beta_1 y+\gamma_1 z=0,\quad \alpha_2 x+\beta_2 y+\gamma_2 z=0,$$

and the point of intersection of the lines

$$\alpha_3 x + \beta_3 y + \gamma_3 z = 0, \quad \alpha_4 x + \beta_4 y + \gamma_4 z = 0.$$

Write for shortness $u_1 = \alpha_1 x + \beta_1 y + \gamma_1 z$, &c.; then we have identically

$$\begin{vmatrix} u_1, & u_2, & u_3, & u_4 \\ \alpha_1, & \alpha_2, & \alpha_3, & \alpha_4 \\ \beta_1, & \beta_2, & \beta_3, & \beta_4 \\ \gamma_1, & \gamma_2, & \gamma_3, & \gamma_4 \end{vmatrix} = 0,$$

and the two equations

$$\begin{vmatrix} u_1, & u_2, & ., & . \\ \alpha_1, & \alpha_2, & \alpha_3, & \alpha_4 \\ \beta_1, & \beta_2, & \beta_3, & \beta_4 \\ \gamma_1, & \gamma_2, & \gamma_3, & \gamma_4 \end{vmatrix} = 0, \qquad \begin{vmatrix} ., & ., & u_3, & u_4 \\ \alpha_1, & \alpha_2, & \alpha_3, & \alpha_4 \\ \beta_1, & \beta_2, & \beta_3, & \beta_4 \\ \gamma_1, & \gamma_2, & \gamma_3, & \gamma_4 \end{vmatrix} = 0$$

are consequently equivalent to each other, and each of them represents the required line. It is easy to deduce the form

$$x \begin{vmatrix} \alpha_1, & \alpha_2, & ., & . \\ ., & ., & \alpha_3, & \alpha_4 \\ \beta_1, & \beta_2, & \beta_3, & \beta_4 \\ \gamma_1, & \gamma_2, & \gamma_3, & \gamma_4 \end{vmatrix} + y \begin{vmatrix} \beta_1, & \beta_2, & ., & . \\ \alpha_1, & \alpha_2, & \alpha_3, & \alpha_4 \\ ., & ., & \beta_3, & \beta_4 \\ \gamma_1, & \gamma_2, & \gamma_3, & \gamma_4 \end{vmatrix} + z \begin{vmatrix} \gamma_1, & \gamma_2, & ., & . \\ \alpha_1, & \alpha_2, & \alpha_3, & \alpha_4 \\ \beta_1, & \beta_2, & \beta_3, & \beta_4 \\ ., & ., & \gamma_3, & \gamma_4 \end{vmatrix} = 0.$$

184. The condition in order that the points of intersection of the lines $u_1 = 0$, $u_2 = 0$, of the lines $u_3 = 0$, $u_4 = 0$, and of the lines $u_5 = 0$, $u_6 = 0$ (where, as before, u_1 denotes $\alpha_1 x + \beta_1 y + \gamma_1 z$, &c.) may lie in the same line, is

$$\begin{vmatrix} \alpha_1, & \alpha_2, & \alpha_3, & \alpha_4, & ., & . \\ \beta_1, & \beta_2, & \beta_3, & \beta_4, & ., & . \\ \gamma_1, & \gamma_2, & \gamma_3, & \gamma_4, & ., & . \\ ., & ., & \alpha_3, & \alpha_4, & \alpha_5, & \alpha_6 \\ ., & ., & \beta_3, & \beta_4, & \beta_5, & \beta_6 \\ ., & ., & \gamma_3, & \gamma_4, & \gamma_5, & \gamma_6 \end{vmatrix} = 0,$$

which is of course really symmetrical with respect to the six sets. The last formula was given by me, *Cambridge Mathematical Journal*, t. IV. (1845), p. 18, [9].

185. Instead of the term point of a curve, it will be convenient to use the term 'ineunt' of the curve.

The line through two consecutive ineunts of the curve is the tangent at the ineunt. The point of intersection of two consecutive tangents is the ineunt on the tangent.

The equation of a curve in point-coordinates, or as it may be termed the point-equation of the curve, is the relation which exists between the point-coordinates of any ineunt of the curve.

The equation of a curve in line-coordinates, or line-equation of the curve, is the relation which exists between the line-coordinates of any tangent of the curve.

186. It has been mentioned, that the order of a curve is the degree of its point-equation: in like manner the class of a curve is the degree of its line-equation; and in the same way that a curve, as represented by a point-equation, may break up into curves having the order of the entire curve for the sum of their orders, so a curve as represented by a line-equation may break up into curves having the class of the entire curve for the sum of their classes. And, in particular, a curve may break up into a system of points, the number of points being equal to the class of the curve, and two or more of these points may coincide together.

187. A line is a curve of the order one and class zero; a point is a curve of the order zero and class one. A proper conic is a curve of the order two and class two; but when the conic breaks up into a pair of lines, the class sinks to zero; and when the conic breaks up into a pair of points, the order sinks to zero. It is to be noticed that a point, or system of points, cannot be represented by an equation in point-coordinates, nor a line or system of lines by an equation in line-coordinates. We may say, in general, that a curve is both a point-curve and a line-curve, but a point or system of points is a line-curve only, and a line or system of lines is a point-curve only.

188. The points of intersection (common ineunts) of two curves are the points the coordinates of which satisfy simultaneously the point-equations of the two curves. Hence the number of common ineunts is equal to the product of the orders of the two curves; and, in particular if one of the curves be a line, the number of points of intersection (common ineunts) is equal to the order of the curve. In like manner the common tangents of the two curves are the lines the coordinates of which satisfy simultaneously the line-equations of the two curves. Hence the number of common tangents is equal to the product of the classes of the two curves; and, in particular, if one of the curves be a point, the number of common tangents (tangents to the curve through the point) is equal to the class of the curve. Since the tangent is the line through two consecutive ineunts, it besides meets the curve, assumed to be of the order m, in $(m-2)$ points; and in like manner we may from any ineunt of a curve of the class n draw $(n-2)$ tangents to the curve.

189. The point-equation of a line passing through the points (x', y', z') and (x'', y'', z'') is, as already noticed,

$$\begin{vmatrix} x\,, & y\,, & z \\ x', & y', & z' \\ x'', & y'', & z'' \end{vmatrix} = 0.$$

Suppose that (x, y, z) are the coordinates of a point (ineunt) of the curve $U=0$ the coordinates of the consecutive ineunt will be $(x+dx, y+dy, z+dz)$, and the line joining these two points will be the tangent to the curve at the point (x, y, z). Take (X, Y, Z) as current point-coordinates, the equation of the tangent is

$$\begin{vmatrix} X & , Y & , Z \\ x & , y & , z \\ x+dx, & y+dy, & z+dz \end{vmatrix} = 0,$$

or, what is the same thing,

$$X(ydz - zdy) + Y(zdx - xdz) + Z(xdy - ydx) = 0.$$

But since U is a homogeneous function of (x, y, z), we have

$$x\partial_x U + y\partial_y U + z\partial_z U = mU = 0;$$

and since $(x+dx, y+dy, z+dz)$ is a point of the curve, we have

$$dx\partial_x U + dy\partial_y U + dz\partial_z U = 0;$$

and from these two equations

$$ydz - zdy : zdx - xdz : xdy - ydx = \partial_x U : \partial_y U : \partial_z U,$$

and the equation of the tangent consequently is

$$X\partial_x U + Y\partial_y U + Z\partial_z U = 0.$$

190. Take (ξ, η, ζ) as the line-coordinates of the tangent, then the equation of the tangent is

$$\xi X + \eta Y + \zeta Z = 0;$$

and comparing the two forms, we have

$$\xi : \eta : \zeta = \partial_x U : \partial_y U : \partial_z U;$$

and if from these equations and the equation $U=0$ (the point-equation of the curve) we eliminate (x, y, z), we obtain an equation between (ξ, η, ζ), which is the line-equation of the curve. We may, if we please, present the system under the form

$$\begin{aligned} \partial_x U + \lambda\xi &= 0, \\ \partial_y U + \lambda\eta &= 0, \\ \partial_z U + \lambda\zeta &= 0, \\ U &= 0, \end{aligned}$$

or, what is more simple, under the form

$$\begin{aligned} \partial_x U + \lambda\xi &= 0, \\ \partial_y U + \lambda\eta &= 0, \\ \partial_z U + \lambda\zeta &= 0, \\ \xi x + \eta y + \zeta z &= 0, \end{aligned}$$

and from either system eliminate x, y, z and λ.

191. If the point-equation of a conic be

$$(a,\ b,\ c,\ f,\ g,\ h \between x,\ y,\ z)^2 = 0,$$

then its line-equation is

$$-\begin{vmatrix} & \xi, & \eta, & \zeta \\ \xi & a, & h, & g \\ \eta & h, & b, & f \\ \zeta & g, & f, & c \end{vmatrix} = 0,$$

or writing

$$\begin{aligned} \mathfrak{A} &= bc - f^2, \\ \mathfrak{B} &= ca - g^2, \\ \mathfrak{C} &= ab - h^2, \\ \mathfrak{F} &= gh - af, \\ \mathfrak{G} &= hf - bg, \\ \mathfrak{H} &= fg - ch, \end{aligned}$$

and, to complete the system,

$$K = abc - af^2 - bg^2 - ch^2 + 2fgh,$$

then the line-equation of the conic is

$$(\mathfrak{A},\ \mathfrak{B},\ \mathfrak{C},\ \mathfrak{F},\ \mathfrak{G},\ \mathfrak{H} \between \xi,\ \eta,\ \zeta)^2 = 0.$$

192. The quantities $\mathfrak{A}$, &c. satisfy the relations

$$K^2 = \mathfrak{A}\mathfrak{B}\mathfrak{C} - \mathfrak{A}\mathfrak{F}^2 - \mathfrak{B}\mathfrak{G}^2 - \mathfrak{C}\mathfrak{H}^2 + 2\mathfrak{F}\mathfrak{G}\mathfrak{H},$$

$$\begin{aligned} &\mathfrak{A}a + \mathfrak{H}h + \mathfrak{G}g = K, &&\mathfrak{H}a + \mathfrak{B}h + \mathfrak{F}g = 0\ , &&\mathfrak{G}a + \mathfrak{F}h + \mathfrak{C}g = 0\ , \\ &\mathfrak{A}h + \mathfrak{H}b + \mathfrak{G}f = 0\ , &&\mathfrak{H}h + \mathfrak{B}b + \mathfrak{F}f = K, &&\mathfrak{G}h + \mathfrak{F}b + \mathfrak{C}f = 0\ , \\ &\mathfrak{A}g + \mathfrak{H}f + \mathfrak{G}c = 0\ , &&\mathfrak{H}g + \mathfrak{B}f + \mathfrak{F}c = 0\ , &&\mathfrak{G}g + \mathfrak{F}f + \mathfrak{C}c = K, \end{aligned}$$

and moreover

$$\begin{aligned} &Ka = \mathfrak{B}\mathfrak{C} - \mathfrak{F}^2, &&Kf = \mathfrak{G}\mathfrak{H} - \mathfrak{A}\mathfrak{F}, \\ &Kb = \mathfrak{C}\mathfrak{A} - \mathfrak{G}^2, &&Kg = \mathfrak{H}\mathfrak{F} - \mathfrak{B}\mathfrak{G}, \\ &Kc = \mathfrak{A}\mathfrak{B} - \mathfrak{H}^2, &&Kh = \mathfrak{F}\mathfrak{G} - \mathfrak{C}\mathfrak{H}. \end{aligned}$$

193. A system of points in a line is said to be a range, and a system of lines through a point is said to be a pencil. The theories of ranges and pencils, considered irrespectively of each other, are in fact a single theory, constituting the geometry of one dimension. It has been seen how in geometry of one dimension a range of points and a pencil of lines, although considered (as they must be considered) as existing in distinct spaces, may nevertheless stand in certain relations to each other. In geometry of two dimensions, the range and pencil may of course coexist in one and the same plane as their common *locus in quo*; and such coexistence occurs in fact very frequently: thus if we have a line and a point, and if lines are drawn

joining the point with the several points of the line, these lines constitute a pencil, and the points of the line constitute a range, and such pencil and range are homographically related.

194. The theory of homography in geometry of two dimensions may be made to depend upon the corresponding theory in geometry of one dimension, or what is the same thing, upon the homography of ranges or pencils. For consider two figures existing in distinct planes or spaces of two dimensions, any four points (not in a line) in the second figure may correspond to any four points (not in a line) in the first figure; and when this is so, we may, by the process about to be explained, given any other point of the first figure, construct the corresponding point of the second figure; and the two figures are then, by definition, homographically related. Suppose that the points A', B', C', D' of the second figure correspond respectively to the points A, B, C, D of the first figure, and let E be any other point of the first figure; suppose that E' is the corresponding point of the second figure; the pencils AB, AC, AD, AE and $A'B'$, $A'C'$, $A'D'$, $A'E'$ should be homographic to each other, that is, E' must lie on a given line through A'; and in like manner the pencils BA, BC, BD, BE and $B'A'$, $B'C'$, $B'D'$, $B'E'$ should be homographic to each other, that is, E' must lie on a given line through B'. And then, as a theorem, CA, CB, CD, CE and $C'A'$, $C'B'$, $C'D'$, $C'E'$, or DA, DB, DC, DE and $D'A'$, $D'B'$, $D'C'$, $D'E'$ will be homographic pencils, that is, the construction will be a determinate one whichever two of the four points are selected for the points A and B. The foregoing construction leads to an analytical relation, which I think constitutes a better foundation of the theory. Consider the first plane as the *locus in quo* of the coordinates (x, y, z), and the second plane as the *locus in quo* of the coordinates (X, Y, Z), these two coordinate systems being absolutely independent of each other. Consider any point of the first plane and a corresponding point of the second plane such that its coordinates (X, Y, Z) are given linear functions of the coordinates (x, y, z) of the point in the first plane. Any figure whatever in the first plane gives rise to a corresponding figure in the second plane, and the two figures are said to be homographic to each other. To a point of the first figure there corresponds in the second figure a point, to a line a line, to a range of points or pencil of lines, a homographic range of points or pencil of lines; the line or point which is the *locus in quo* of the range or pencil in the one figure corresponding with the line or point which is the *locus in quo* of the range or pencil in the other figure. And generally, to any curve of any order and class in the first figure, and to its ineunts and tangents, there correspond in the second figure a curve of the same order and class, and the ineunts and tangents of such curve.

195. It is to be remarked, that it is not by any means necessary that the word plane, or the words point and line, or consequently the words order and class, should have the same significations as regards the two figures respectively. The theory of homography, as above explained, in fact comprises what is commonly termed the theory of homography and also the theory of reciprocity.

196. Let the word plane have the ordinary signification as regards the two figures respectively; and Suppose, first, that the words point and line, and therefore order

and class, have also the ordinary significations as regards the two figures respectively: we have here the ordinary theory of homography, in which, to any range of points or pencil of lines in the first figure, there corresponds a homographic range of points or pencil of lines in the second figure, and to a curve of any order and class in the first figure there corresponds a curve of the same order and class in the second figure.

197. We may, as a specialization giving rise to further developments, assume that the two figures exist in one and the same plane. There is here in general a triangle, each of whose angles or sides, considered as a point or line in the first figure, corresponds to itself, considered as a point or line in the second figure: such triangle may be called the sibiconjugate triangle. Any one point of the plane, considered as belonging to the first figure, may correspond to any other point of the plane, considered as belonging to the second figure, and the second figure can be completely constructed by means of the sibiconjugate triangle and such pair of corresponding points. In certain special cases the sibiconjugate triangle becomes wholly or in part indeterminate; thus if the two figures are identical, each point of the plane, considered as belonging to the first figure, coincides with itself, considered as belonging to the second figure. But I reserve the further discussion of the theory of homography for another occasion.

198. Suppose, secondly, that in the foregoing general theory, as regards the first figure, the words point and line, and therefore order and class, signify *point* and *line*, *order* and *class*; while as regards the second figure, the words point and line signify *line* and *point* respectively, and therefore the words order and class, *class* and *order* respectively. We have in the present case the ordinary theory of reciprocity, viz. using all the words in the same significations as regards the two figures respectively; to a point in the first figure there corresponds in the second figure a line; to a line, a point; to a range of points or pencil of lines, a pencil of lines or range of points; to a curve of any order and class, and its ineunts and tangents, a curve of the same class and order, and the tangents and ineunts of such curve.

199. As a specialization giving rise to further developments, we may assume that the two figures exist in one and the same plane. In this case, the points which, considered indifferently as belonging to the first or the second figure, lie upon the corresponding lines in the second or first figure, generate a conic which may be termed the pole-conic; and the lines which, considered indifferently as belonging to the first or the second figure, pass through the corresponding points in the second or first figure, envelope a conic which may be termed the polar-conic, and these two conics have double contact with another. The further consideration of this subject is reserved for another occasion; but I remark that in the particular case where the two conics coincide, we have the ordinary theory of poles and polars in regard to a conic; a theory, which, in a different point of view, may be considered as arising out of the harmonic relation, and which must here be noticed.

200. Consider a conic and a point; any line through the point meets the conic in two points (ineunts of the conic), and the harmonic in relation to these two points of the given point has for its locus a line which is the polar of the given point.

The polar passes through the points of contact of the conic with the tangents through the given point.

In like manner considering the conic and a line; from any point of the line we may draw two tangents to the conic, and the harmonic of the given line with respect to the two tangents envelopes a point which is the pole of the given line. The pole is the point of intersection of the tangents of the conic at the points of intersection with the given line.

The polars of the several points of a line envelope a point which is the pole of the line; and the poles of the several lines through a point generate a line which is the polar of the point; and this proposition shows how the theory of poles and polars gives rise to a theory of reciprocity.

201. If the point-equation of a conic be

$$(a,\ b,\ c,\ f,\ g,\ h\between x,\ y,\ z)^2 = 0,$$

the point-equation of the polar with respect to this conic of the point $(x',\ y',\ z')$ is

$$(a,\ b,\ c,\ f,\ g,\ h\between x,\ y,\ z\between x',\ y',\ z') = 0.$$

But it has been seen that the line-equation of the same conic is

$$(\mathfrak{A},\ \mathfrak{B},\ \mathfrak{C},\ \mathfrak{F},\ \mathfrak{G},\ \mathfrak{H}\between \xi,\ \eta,\ \zeta)^2 = 0,$$

and the line-equation of the pole with respect to this conic of the line $(\xi',\ \eta',\ \zeta')$ (that is, the line whose point-equation is $\xi'x + \eta'y + \zeta'z = 0$) is

$$(\mathfrak{A},\ \mathfrak{B},\ \mathfrak{C},\ \mathfrak{F},\ \mathfrak{G},\ \mathfrak{H}\between \xi,\ \eta,\ \zeta\between \xi',\ \eta',\ \zeta') = 0,$$

in other words, the point-coordinates of the pole are

$$\mathfrak{A}\xi' + \mathfrak{H}\eta' + \mathfrak{G}\zeta',\quad \mathfrak{H}\xi' + \mathfrak{B}\eta' + \mathfrak{F}\zeta',\quad \mathfrak{G}\xi' + \mathfrak{F}\eta' + \mathfrak{C}\zeta'.$$

202. If $U = 0$, $V = 0$ be the point-equations of any two curves of the same order, then λ, μ being arbitrary coefficients,

$$\lambda U + \mu V = 0$$

is the equation of a curve of the same order passing through the points of intersection (common ineunts) of the two curves; such curve is said to be in involution with the given curves. The discussion of the general theory of involution is reserved for another occasion.

203. In particular, if $U = 0$ be the equation of a conic, and $P = 0$, $Q = 0$ the equations of two lines, then

$$U + \lambda PQ = 0$$

is the equation of a conic passing through the points of intersection of the conic with the two lines; and if the two lines coincide, then

$$U + \lambda P^2 = 0$$

is the equation of a conic having double contact with the conic $U=0$ at its points of intersection with the line $P=0$. Such conic is said to be inscribed in the conic $U=0$; the line $P=0$ is the axis of inscription; this line has the same pole with respect to each of the two conics, and the pole is termed the centre of inscription: the relation of the two conics is completely expressed by saying that the four common ineunts coincide in pairs upon the axis of inscription, and that the four common tangents coincide in pairs through the centre of inscription; it is consequently a similar relation in regard to ineunts and tangents respectively; and it is to be inferred *à priori*, that if $\Upsilon=0$ be the line-equation of the conic $U=0$, and $\Pi=0$ the line-equation of the centre of inscription, then the line-equation of the inscribed cone is $\Upsilon+\mu\Pi^2=0$.

204. To verify this, I remark that if the equation of the axis of inscription be

$$\xi'x+\eta'y+\zeta'z=0,$$

then (*ante*, No. 201) we have for the line-equation of the centre of inscription

$$\Pi=(\mathfrak{A},\ldots\between\xi,\ \eta,\ \zeta\between\xi',\ \eta',\ \zeta')=0.$$

The line-equation of the inscribed conic is in the first instance obtained in the form

$$(\mathfrak{A},\ldots\between\xi,\ \eta,\ \zeta)^2+\lambda\,(a,\ldots\between\eta\zeta'-\eta'\zeta,\ \zeta\xi'-\zeta'\xi,\ \xi\eta'-\xi'\eta)^2=0\,;$$

but we have identically,

$$(\mathfrak{A},\ldots\between\xi,\ \eta,\ \zeta)^2(\mathfrak{A},\ldots\between\xi',\ \eta',\ \zeta')^2-\{(\mathfrak{A},\ldots\between\xi,\ \eta,\ \zeta\between\xi',\ \eta',\ \zeta')\}^2$$
$$=K\,(a,\ldots\between\eta\zeta'-\eta'\zeta,\ \zeta\xi'-\zeta'\xi,\ \xi\eta'-\xi'\eta)^2,$$

and the equation thus becomes

$$[K+\lambda\,(\mathfrak{A},\ldots\between\xi',\ \eta',\ \zeta')^2]\,(\mathfrak{A},\ldots\between\xi,\ \eta,\ \zeta)^2-\lambda\,\{(\mathfrak{A},\ldots\between\xi',\ \eta',\ \zeta'\between\xi,\ \eta,\ \zeta)\}^2=0,$$

which is of the form in question.

205. Take $(x',\ y',\ z')$ as the point-coordinates of the centre of inscription, the equation of the axis of inscription is

$$(a,\ b,\ c,\ f,\ g,\ h\between x,\ y,\ z\between x',\ y',\ z')=0\,;$$

and we may, if we please, exhibit the equation of the inscribed conic in the form

$$(a,\ldots\between x,\ y,\ z)^2\,(a,\ldots\between x',\ y',\ z')^2\cos^2\theta-\{(a,\ldots\between x,\ y,\ z\between x',\ y',\ z')\}^2=0,$$

where θ is a constant. This equation may also be written

$$(a,\ldots\between x,\ y,\ z)^2\,(a,\ldots\between x',\ y',\ z')^2\sin^2\theta-(\mathfrak{A},\ldots\between yz'-y'z,\ zx'-z'x,\ xy'-x'y)^2=0,$$

the two forms being equivalent in virtue of the identity,

$$(a,\ldots\between x,\ y,\ z)^2\,(a,\ldots\between x',\ y',\ z')^2-\{(a,\ldots\between x',\ y',\ z'\between x,\ y,\ z)\}^2$$
$$=(\mathfrak{A},\ldots\between yz'-y'z,\ zx'-z'x,\ xy'-x'y)^2.$$

206. The line-coordinates (ξ', η', ζ') of the axis of inscription are

$$ax' + hy' + gz', \quad hx' + by' + fz', \quad gx' + fy' + cz',$$

and we thence deduce the relation

$$(\mathfrak{A}, \ldots \between \xi', \eta', \zeta')^2 = K(a, \ldots \between x', y', z')^2.$$

In order that the form

$$(a, \ldots \between x, y, z)^2 (a, \ldots \between x', y', z')^2 \cos^2\theta - \{(a, \ldots \between x', y', z' \between x, y, z)\}^2 = 0$$

may agree with the originally assumed form

$$(a, \ldots \between x, y, z)^2 + \lambda(\xi' x + \eta' y + \zeta' z)^2,$$

or what is the same thing,

$$(a, \ldots \between x, y, z)^2 + \lambda\{(a, \ldots \between x, y, z \between x', y', z')\}^2 = 0,$$

we must have

$$\lambda = \frac{-1}{(a, \ldots \between x', y', z')^2 \cos^2\theta},$$

which may also be written

$$\lambda = \frac{-K}{(\mathfrak{A}, \ldots \between \xi', \eta', \zeta')^2 \cos^2\theta},$$

or what is the same thing,

$$K + \lambda(\mathfrak{A}, \ldots \between \xi', \eta', \zeta')^2 - \lambda(\mathfrak{A}, \ldots \between \xi', \eta', \zeta')^2 \sin^2\theta = 0;$$

and we thence, by a preceding formula, obtain the line-equation of the inscribed conic, viz.

207. The point-equation being

$$(a, \ldots \between x, y, z)^2 (a, \ldots \between x', y', z')^2 \cos^2\theta - \{(a, \ldots \between x, y, z \between x', y', z')\}^2 = 0,$$

or

$$(a, \ldots \between x, y, z)^2 (a, \ldots \between x', y', z')^2 \sin^2\theta - (\mathfrak{A}, \ldots \between yz' - y'z, zx' - z'x, xy' - x'y)^2 = 0,$$

equivalent in virtue of

$$(a, \ldots \between x, y, z)^2 (a, \ldots \between x', y', z')^2 - \{(a, \ldots \between x, y, z \between x', y', z')\}^2$$
$$= (\mathfrak{A}, \ldots \between yz' - y'z, zx' - z'x, xy' - x'y)^2;$$

then the corresponding forms of the line-equation are

$$(\mathfrak{A}, \ldots \between \xi, \eta, \zeta)^2 (\mathfrak{A}, \ldots \between \xi', \eta', \zeta')^2 \sin^2\theta - \{(\mathfrak{A}, \ldots \between \xi, \eta, \zeta \between \xi', \eta', \zeta')\}^2 = 0,$$

and

$$(\mathfrak{A}, \ldots \between \xi, \eta, \zeta)^2 (\mathfrak{A}, \ldots \between \xi', \eta', \zeta')^2 \cos^2\theta - K(a, \ldots \between \eta\zeta' - \eta'\zeta, \zeta\xi' - \zeta'\xi, \xi\eta' - \xi'\eta)^2 = 0,$$

equivalent to each other in virtue of the before mentioned identity

$$(\mathfrak{A}, \ldots \between \xi, \eta, \zeta)^2 (\mathfrak{A}, \ldots \between \xi', \eta', \zeta')^2 - \{(\mathfrak{A}, \ldots \between \xi, \eta, \zeta \between \xi', \eta', \zeta')\}^2$$
$$= K(a, \ldots \between \eta\zeta' - \eta'\zeta, \zeta\xi' - \zeta'\xi, \xi\eta' - \xi'\eta)^2.$$

208. Write for shortness

$$\begin{aligned}(a, \ldots \between x, y, z)^2 &= 00, \\ (a, \ldots \between x, y, z \between x', y', z) &= 01 = 10, \\ \&c.,\end{aligned}$$

then we have identically,

$$\left|\begin{matrix} 00, & 01, & 02 \\ 10, & 11, & 12 \\ 20, & 21, & 22 \end{matrix}\right| = K \left|\begin{matrix} x\,, & y\,, & z \\ x', & y', & z' \\ x'', & y'', & z'' \end{matrix}\right|^2$$

and if the determinant on the right hand vanishes, that is if (x, y, z), (x', y', z'), (x'', y'', z'') are points in a line, then we have

$$\left|\begin{matrix} 00, & 01, & 02 \\ 10, & 11, & 12 \\ 20, & 21, & 22 \end{matrix}\right| = 0,$$

an equation, which, as already remarked, is equivalent to

$$\cos^{-1}\frac{01}{\sqrt{00}\sqrt{11}} + \cos^{-1}\frac{12}{\sqrt{11}\sqrt{22}} = \cos^{-1}\frac{02}{\sqrt{00}\sqrt{22}}.$$

The foregoing investigations in relation to the inscribed conic are given for the sake of the application thereof to the theory of distance, and it has been necessary to make use of analytical formulæ of some complexity which are introduced out of their natural place.

On the Theory of Distance, Nos. 209 to 229.

209. I return to the geometry of one dimension. Imagine in the line or *locus in quo* of the range of points, a point-pair, which I term the Absolute. Any point-pair whatever may be considered as inscribed in the Absolute, the centre and axis of inscription being the sibiconjugate points of the involution formed by the points of the given point-pair and the points of the Absolute; the centre and axis of inscription *quà* sibiconjugate points are harmonics with respect to the Absolute. A point-pair considered as thus inscribed in the Absolute is said to be a *point-pair circle*, or simply a *circle*; the centre of inscription and the axis of inscription are termed the centre and the axis. Either of the two sibiconjugate points may be considered as the centre, but the selection when made must be adhered to. It is proper to notice that, given the centre and one point of the circle, the other point of the circle is determined in a unique manner. In fact the axis is the harmonic of the centre in respect to the Absolute, and then the other point is the harmonic of the given point in respect to the centre and axis.

210. As a definition, we say that the two points of a circle are equidistant from the centre. Now imagine two points P, P'; and take the point P'' such that P, P'' are a circle having P' for its centre; take in like manner the point P''' such that

P', P''' are a circle having P'' for its centre; and so on: and again in the opposite direction, a point $P^{\backprime}$ such that P', $P^{\backprime}$ are a circle having P for its centre; a point $P^{\backprime\backprime}$ such that P, $P^{\backprime\backprime}$ are a circle having $P^{\backprime}$ for its centre, and so on. We have a series of points... $P^{\backprime\backprime}$, $P^{\backprime}$, P, P', P'', ... at equal intervals of distance: and if we take the points P, P' indefinitely near to each other, then the entire line will be divided into a series of equal infinitesimal elements; the number of these elements included between any two points measures the distance of the two points. It is clear that, according to the definition, if P, P', P'' be any three points taken in order, then

$$\text{Dist.}\,(P,\ P') + \text{Dist.}\,(P',\ P'') = \text{Dist.}\,(P,\ P''),$$

which agrees with the ordinary notion of distance.

211. To show how the foregoing definition leads to an analytical expression for the distance of two points in terms of their coordinates, take

$$(a,\ b,\ c \between x,\ y)^2 = 0$$

for the equation of the Absolute. The equation of a circle having the point $(x',\ y')$ for its centre is

$$(a,\ b,\ c \between x,\ y)^2\,(a,\ b,\ c \between x',\ y')^2 \cos^2\theta - \{(a,\ b,\ c \between x,\ y \between x',\ y')\}^2 = 0;$$

and consequently if $(x,\ y)$, $(x'',\ y'')$ are the two points of the circle, then

$$\frac{(a,\ b,\ c \between x,\ y \between x',\ y')}{\sqrt{(a,\ b,\ c \between x,\ y)^2}\,\sqrt{(a,\ b,\ c \between x',\ y')^2}} = \frac{(a,\ b,\ c \between x',\ y' \between x'',\ y'')}{\sqrt{(a,\ b,\ c \between x',\ y')^2}\,\sqrt{(a,\ b,\ c \between x'',\ y'')^2}},$$

an equation which expresses that the points $(x'',\ y'')$ and $(x,\ y)$ are equidistant from the point $(x',\ y')$. It is clear that the distance of the points $(x,\ y)$ and $(x',\ y')$ must be a function of

$$\frac{(a,\ b,\ c \between x,\ y \between x',\ y')}{\sqrt{(a,\ b,\ c \between x,\ y)^2}\,\sqrt{(a,\ b,\ c \between x',\ y')^2}},$$

and the form of the function is determined from the before-mentioned property, viz. if P, P', P'' be any three points taken in order, then

$$\text{Dist.}\,(P,\ P') + \text{Dist.}\,(P',\ P'') = \text{Dist.}\,(P,\ P'').$$

This leads to the conclusion that the distance of the points $(x,\ y)$, $(x',\ y')$ is equal to a multiple of the arc having for its cosine the last-mentioned expression (see *ante*, No. 168); and we may in general assume that the distance is equal to the arc in question, viz. that the distance is

$$\cos^{-1}\frac{(a,\ b,\ c \between x,\ y \between x',\ y')}{\sqrt{(a,\ b,\ c \between x,\ y)^2}\,\sqrt{(a,\ b,\ c \between x',\ y')^2}},$$

or, what is the same thing,

$$\sin^{-1}\frac{(ac - b^2)\,(xy' - x'y)}{\sqrt{(a,\ b,\ c \between x,\ y)^2}\,\sqrt{(a,\ b,\ c \between x',\ y')^2}}.$$

It follows that the two forms

$$(a,\ b,\ c\between x,\ y)^2\,(a,\ b,\ c\between x',\ y')^2\cos^2\theta-\{(a,\ b,\ c\between x,\ y\between x',\ y')\}^2=0,$$
$$(a,\ b,\ c\between x,\ y)^2\,(a,\ b,\ c\between x',\ y')^2\sin^2\theta-(ac-b^2)\,(xy'-x'y)\qquad=0,$$

of the equation of a circle, each of them express that the distances of the two points from the centre are respectively equal to the arc θ; or, if we please, that θ is the radius of the circle.

212. When $\theta=0$, we have

$$xy'-x'y=0,$$

an equation which expresses that $(x,\ y)$ and $(x',\ y')$ are one and the same point. When $\theta=\tfrac{1}{2}\pi$, we have

$$(a,\ b,\ c\between x,\ y\between x',\ y')=0,$$

an equation which expresses that the points $(x,\ y)$ and $(x',\ y')$ are harmonics with respect to the Absolute. The distance between any two points harmonics with respect to the Absolute is consequently a quadrant, and such points may be said to be quadrantal to each other. The quadrant is the unit of distance.

213. The foregoing is the general case, but it is necessary to consider the particular case where the Absolute is a pair of coincident points. The harmonic of any point whatever in respect to the Absolute is here a point coincident with the Absolute itself: the definition of a circle is consequently simplified; viz. any point-pair whatever may be considered as a circle having for its centre the harmonic of the Absolute with respect to the point-pair; we may, as before, divide the line into a series of equal infinitesimal elements, and the number of elements included between any two points measures the distance between the two points. As regards the analytical expression, in the case in question $ac-b^2$ vanishes, or the distance is given as the arc to an evanescent sine. Reducing the arc to its sine and omitting the evanescent factor, we have a finite expression for the distance. Suppose that the equation of the Absolute is

$$(qx-py)^2=0,$$

or what is the same thing, let the Absolute (treated as a single point) be the point $(p,\ q)$, then we find for the distance of the points $(x,\ y)$ and $(x',\ y')$ the expression

$$\frac{xy'-x'y}{(qx-py)\,(qx'-py')};$$

or, introducing an arbitrary multiplier,

$$\frac{(q\alpha-p\beta)\,(xy'-x'y)}{(qx-py)\,(qx'-py')},$$

which is equal to

$$\frac{\beta x-\alpha y}{qx-py}-\frac{\beta x'-\alpha y'}{qx'-py'}.$$

It is hardly necessary to remark, that in the present case the notion of the quadrantal relation of two points has altogether disappeared, and that the unit of distance is arbitrary.

214. Passing now to geometry of two dimensions, we have here to consider a certain conic, which I call the Absolute. Any line whatever determines with the Absolute (cuts it in) two points which are the Absolute in regard to such line considered as a space of one dimension, or *locus in quo* of a range of points, and in like manner any point whatever determines with the Absolute (has for tangents of the Absolute through the point) two lines which are the Absolute in regard to such point considered as a space of one dimension, or *locus in quo* of a pencil of lines. The foregoing theory for geometry of one dimension establishes the notion of distance as regards each of these ranges and pencils considered apart by itself; in order to bring the different ranges and pencils into relation with each other, it is necessary to assume that the quadrant which is the unit of distance for these several systems respectively, is one and the same distance for each system (of course, when, as in the analytical theory, we actually represent the quadrant by the ordinary symbol $\frac{1}{2}\pi$, the above assumption is tacitly made; but substituting the thing signified for the definition, and looking at the quadrant merely as the distance between two points, or as the case may be, lines, harmonically related to the point-pair, or as the case may be, line-pair, constituting the Absolute, the assumption is at once seen to be an assumption, and it needs to be made explicitly). But the assumption being made, the foregoing theory of distance in geometry of one dimension enables the comparison not only of the distances of points upon different lines, or of lines through different points, but of the distances of points on a line and of lines through a point. The pole of any line in relation to the Absolute may be termed simply the pole, and in like manner the polar of any line in relation to the Absolute may be termed simply the polar, and we have the theorem that the distance of two points or lines is equal to the distance of their polars or poles, or what is the same thing, that the distance of two poles and the distance of the two corresponding polars are equal. And we may, as a definition, establish the notion of the distance of a point from a line, viz. it is the complement of the distance of the polar of the point from the line, or what is the same thing, the complement of the distance of the point from the pole of the line. The distance of a pole and polar is therefore the complement of zero, that is, it is the quadrant.

215. It has, by means of the preceding assumption as to the quadrant, been possible to establish the notion of distance, without the assistance of the circle, but this figure must now be considered. A conic inscribed in the Absolute is termed a circle; the centre of inscription (or point of intersection of the common tangents) and the axis of inscription (or line of junction of the common ineunts) are the centre and axis of the circle. All the points of a circle are equidistant from the centre; all the tangents are equidistant from the axis, and this distance is the complement of the former distance.

216. These properties of the circle lead immediately to the analytical expressions for the distances of points or lines in terms of the coordinates. In fact, take

$$(a,\ b,\ c,\ f,\ g,\ h\between x,\ y,\ z)^2=0$$

for the point-equation of the Absolute; its line-equation will be

$$(\mathfrak{A},\ \mathfrak{B},\ \mathfrak{C},\ \mathfrak{F},\ \mathfrak{G},\ \mathfrak{H}\between \xi,\ \eta,\ \zeta)^2=0.$$

The point-equation of the circle having the point $(x',\ y',\ z')$ for its centre, is

$$(a,\dots\between x,\ y,\ z)^2\,(a,\dots\between x',\ y',\ z')^2\cos^2\theta-\{(a,\dots\between x,\ y,\ z\between x',\ y',\ z')\}^2=0,$$

or

$$(a,\dots\between x,\ y,\ z)^2\,(a,\dots\between x',\ y',\ z')^2\sin^2\theta-(\mathfrak{A},\dots\between yz'-y'z,\ zx'-z'x,\ xy'-x'y)^2=0,$$

from which (by the same reasoning as for the case of geometry of one dimension) it follows that the distance of the points $(x,\ y,\ z)$, $(x',\ y',\ z')$ is

$$\cos^{-1}\frac{(a,\dots\between x,\ y,\ z\between x',\ y',\ z')}{\sqrt{(a,\dots\between x,\ y,\ z)^2}\,\sqrt{(a,\dots\between x',\ y',\ z')^2}},$$

or what is the same thing,

$$\sin^{-1}\frac{\sqrt{(\mathfrak{A},\dots\between yz'-y'z,\ zx'-z'x,\ xy'-x'y)^2}}{\sqrt{(a,\dots\between x,\ y,\ z)^2}\,\sqrt{(a,\dots\between x',\ y',\ z')^2}};$$

and it appears from the cosine formula (see *ante*, No. 208), that if P, P', P'' be points on the same line, then we have, as we ought to have,

$$\text{Dist.}\ (P,\ P')+\text{Dist.}\ (P',\ P'')=\text{Dist.}\ (P,\ P'').$$

217. In like manner, the line-equation of the same circle, the line-coordinates of the axis being $(\xi',\ \eta',\ \zeta')$, is

$$(\mathfrak{A},\dots\between \xi,\ \eta,\ \zeta)^2\,(\mathfrak{A},\dots\between \xi',\ \eta',\ \zeta')^2\sin^2\theta-\{(\mathfrak{A},\dots\between \xi,\ \eta,\ \zeta\between \xi',\ \eta',\ \zeta')\}^2=0,$$

or

$$(\mathfrak{A},\dots\between \xi,\ \eta,\ \zeta)^2\,(\mathfrak{A},\dots\between \xi',\ \eta',\ \zeta')^2\cos^2\theta-K\,(a,\dots\between \eta\zeta'-\eta'\zeta,\ \zeta\xi'-\zeta'\xi,\ \xi\eta'-\xi'\eta)^2=0,$$

from which it follows that the distance of the lines $(\xi,\ \eta,\ \zeta)$ and $(\xi',\ \eta',\ \zeta')$ is

$$\cos^{-1}\frac{(\mathfrak{A},\dots\between \xi,\ \eta,\ \zeta\between \xi',\ \eta',\ \zeta')}{\sqrt{(\mathfrak{A},\dots\between \xi,\ \eta,\ \zeta)^2}\,\sqrt{(\mathfrak{A},\dots\between \xi',\ \eta',\ \zeta')^2}},$$

or what is the same thing

$$\sin^{-1}\frac{\sqrt{K\,(a,\dots\between \eta\zeta'-\eta'\zeta,\ \zeta\xi'-\zeta'\xi,\ \xi\eta'-\xi'\eta)^2}}{\sqrt{(\mathfrak{A},\dots\between \xi,\ \eta,\ \zeta)^2}\,\sqrt{(\mathfrak{A},\dots\between \xi,\ \eta',\ \zeta')^2}}.$$

218. And we may from the first formula of either set, deduce for the distance of the point (x, y, z) and the line (ξ', η', ζ'), the expression

$$\sin^{-1}\frac{\sqrt{K}\,(\xi' x + \eta' y + \zeta' z)}{\sqrt{(a, \ldots\between x, y, z)^2}\,\sqrt{(\mathfrak{A}, \ldots\between \xi', \eta', \zeta')^2}},$$

as may be easily seen by writing $\mathfrak{A}\xi' + \mathfrak{H}\eta' + \mathfrak{G}\zeta', \ldots$ for x', y', z', or $ax + hy + gz, \ldots$ for ξ, η, ζ, and putting $\sin^{-1}$ for $\cos^{-1}$.

219. It may be noticed that there are certain lines, viz. the tangents of the Absolute, in regard to which, considered as a space of one dimension, the Absolute is a pair of coincident points; and in like manner certain points, viz. the ineunts of the Absolute, in regard to which, considered as a space of one dimension, the Absolute is a pair of coincident lines.

220. We may, in particular, suppose that the Absolute, instead of being a proper conic, is a pair of points. The line through the two points may be called the Absolute line; such line is to be considered as a pair of coincident lines. Any point whatever determines with the Absolute, two lines, viz. the lines joining the point with the two points of the Absolute; this line-pair is the Absolute for the point considered as a space of one dimension or *locus in quo* of a pencil of lines, and the theory of the distances of lines through a point is therefore precisely the same as in the general case. But any line whatever determines with the Absolute (meets the Absolute line in) a pair of coincident points, which pair of coincident points is the Absolute in regard to such line considered as a space of one dimension or *locus in quo* of a range of points, and the theory of the distance of points on a line is therefore the theory before explained for this special case. But we cannot, in the same way as before, compare the distances of points upon different lines, since we have not in the present case the quadrant as a unit of distance. The comparison must be made by means of the circle, viz. in the present case any conic passing through the two points of the Absolute is termed a circle, and the point of intersection of the tangents to the circle at the two points of the Absolute (or what is the same thing, the pole of the Absolute line in respect to the circle) is the centre of the circle. The Absolute line itself may, if it is necessary to do so, be considered as the axis of the circle. It is assumed that the points of the circle are all of them equidistant from the centre, and by this assumption we are enabled to compare distances upon different lines. In fact we may, by a construction precisely similar to that of *Euclid*, Book I. Prop. II., from a given point A draw a finite line equal to a given finite line BC, and thence also upon a given line through A, determine the finite line AD equal to the given finite line BC. Since the unit of distance for points on a line is arbitrary, we cannot of course compare the distances of points with the distances of lines. The distance of a point from a line does, however, admit of comparison with the distance of two points; we have only to assume as a definition that the distance of a point from a line is the distance of the point from the point of intersection of the line with the quadrantal line through the point.

221. As regards the analytical theory, suppose that the point-coordinates of the two points of the Absolute are (p, q, r), (p_0, q_0, r_0), then the line-equation of the Absolute is

$$2(p\xi + q\eta + r\zeta)(p_0\xi + q_0\eta + r_0\zeta) = 0;$$

so that we have $\mathfrak{A} = 2pp_0$, $\mathfrak{B} = 2qq_0$, $\mathfrak{C} = 2rr_0$, $\mathfrak{F} = qr_0 + rq_0$, $\mathfrak{G} = rp_0 + pr_0$, $\mathfrak{H} = pq_0 + qp_0$, and thence $K = 0$; but

$$K(a, b, c, f, g, h \between x, y, z)^2 = \begin{vmatrix} x, & y, & z \\ p, & q, & r \\ p_0, & q_0, & r_0 \end{vmatrix}^2$$

where obviously

$$\begin{vmatrix} x, & y, & z \\ p, & q, & r \\ p_0, & q_0, & r_0 \end{vmatrix} = 0$$

is the equation of the Absolute line.

222. The expression for the distance of the two points (x, y, z), (x', y', z') is given as the arc to an evanescent sine; but reducing the arc to its sine, and omitting the evanescent factor, the resulting expression is

$$\sqrt{2 \begin{vmatrix} x, & y, & z \\ x', & y', & z' \\ p, & q, & r \end{vmatrix} \begin{vmatrix} x, & y, & z \\ x', & y', & z' \\ p_0, & q_0, & r_0 \end{vmatrix}} \div \begin{vmatrix} x, & y, & z \\ p, & q, & r \\ p_0, & q_0, & r_0 \end{vmatrix} \begin{vmatrix} x', & y', & z' \\ p, & q, & r \\ p_0, & q_0, & r_0 \end{vmatrix}$$

and the expression for the distance of the two lines (ξ, η, ζ), (ξ', η', ζ') is

$$\cos^{-1} \frac{(p\xi + q\eta + r\zeta)(p_0\xi' + q_0\eta' + r_0\zeta') + (p\xi' + q\eta' + r\zeta')(p_0\xi + q_0\eta + r_0\zeta)}{\sqrt{2(p\xi + q\eta + r\zeta)(p_0\xi + q_0\eta + r_0\zeta)}\ \sqrt{2(p\xi' + q\eta' + r\zeta')(p_0\xi' + q_0\eta' + r_0\zeta')}},$$

or, what is the same thing,

$$\sin^{-1} \frac{(qr_0 - rq_0)(\eta\zeta' - \eta'\zeta) + (rp_0 - pr_0)(\zeta\eta' - \zeta'\eta) + (pq_0 - qp_0)(\xi\eta' - \xi'\eta)}{\sqrt{2(p\xi + q\eta + r\zeta)(p_0\xi + q_0\eta + r_0\zeta)}\ \sqrt{2(p\xi' + q\eta' + r\zeta')(p_0\xi' + q_0\eta' + r_0\zeta')}};$$

and finally, the expression for the distance of the point x, y, z from the line (ξ', η', ζ'), reducing the arc to its sine and omitting the evanescent factor, is

$$(\xi' x + \eta' y + \zeta' z) \div \begin{vmatrix} x, & y, & z \\ p, & q, & r \\ p_0, & q_0, & r_0 \end{vmatrix} \sqrt{2(p\xi' + q\eta' + r\zeta')(p_0\xi' + q_0\eta' + r_0\zeta')}.$$

223. If in the above formula we put $(p, q, r) = (1, i, 0)$, $(p_0, q_0, r_0) = (1, -i, 0)$, where as usual $i = \sqrt{-1}$, then the line-equation of the Absolute is $\xi^2 + \eta^2 = 0$, or what is the same thing, the Absolute consists of the two points in which the line $z = 0$ intersects the line-pair $x^2 + y^2 = 0$; the last-mentioned line-pair, as passing through the Absolute, is by definition a circle; it is in fact the circle radius zero, or an evanescent circle. If we put also the coordinate z equal to unity, then the preceding assumption as to the coordinates of the points of the Absolute must be understood to mean only $x : y : 1 = 1 : i : 0$, or $1 : -i : 0$; that is, we must have x and y infinite, and, as before, $x^2 + y^2 = 0$, or in other words, the Absolute will consist of the points of intersection of the line infinity by the evanescent circle $x^2 + y^2 = 0$. With the values in question,

224. The expression for the distance of the points (x, y) and (x', y') is

$$\sqrt{(x - x')^2 + (y - y')^2};$$

that for the distance of the lines (ξ, η, ζ) and (ξ', η', ζ') is

$$\cos^{-1} \frac{\xi\xi' + \eta\eta'}{\sqrt{\xi^2 + \eta^2}\sqrt{\xi'^2 + \eta'^2}}$$

$$= \sin^{-1} \frac{\xi\eta' - \xi'\eta}{\sqrt{\xi^2 + \eta^2}\sqrt{\xi'^2 + \eta'^2}},$$

which may also be written

$$= \tan^{-1} \frac{\xi}{\eta} - \tan^{-1} \frac{\xi'}{\eta'};$$

and the expression for the distance of the point (x, y) from the line (ξ', η', ζ') is

$$\frac{\xi' x + \eta' y + \zeta'}{\sqrt{\xi'^2 + \eta'^2}},$$

which are obviously the formulæ of ordinary plane geometry, (x, y) being ordinary rectangular coordinates.

225. The general formulæ suffer no *essential* modification, but they are greatly simplified in form by taking for the point-equation of the Absolute

$$x^2 + y^2 + z^2 = 0,$$

or, what is the same, for the line-equation

$$\xi^2 + \eta^2 + \zeta^2 = 0.$$

In fact, we then have for the expression of the distance of the points (x, y, z), $(x', y' z')$,

$$\cos^{-1} \frac{xx' + yy' + zz'}{\sqrt{x^2 + y^2 + z^2}\sqrt{x'^2 + y'^2 + z'^2}};$$

for that of the lines (ξ, η, ζ), (ξ', η', ζ'),

$$\cos^{-1} \frac{\xi\xi' + \eta\eta' + \zeta\zeta'}{\sqrt{\xi^2 + \eta^2 + \zeta^2}\sqrt{\xi'^2 + \eta'^2 + \zeta'^2}};$$

and for that of the point (x, y, z) and the line (ξ', η', ζ'),

$$\sin^{-1} \frac{\xi' x + \eta' y + \zeta' z}{\sqrt{x^2 + y^2 + z^2}\sqrt{\xi'^2 + \eta'^2 + \zeta'^2}}.$$

226. Suppose (x, y, z) are ordinary rectangular coordinates in space satisfying the condition

$$x^2 + y^2 + z^2 = 1,$$

the point having (x, y, z) for its coordinates will be a point on the surface of the sphere, and (the last-mentioned equation always subsisting) the equation $\xi x + \eta y + \zeta z = 0$ will be a great circle of the sphere; and since we are only concerned with the ratios of ξ, η, ζ, we may also assume $\xi^2 + \eta^2 + \zeta^2 = 1$. We may of course retain in the formulæ the expressions $x^2 + y^2 + z^2$ and $\xi^2 + \eta^2 + \zeta^2$, without substituting for these the values unity, and it is in fact convenient thus to preserve all the formulæ in their original forms. We have thus a system of spherical geometry; and it appears that the Absolute in such system is the (spherical) conic, which is the intersection of the sphere with the concentric cone or evanescent sphere $x^2 + y^2 + z^2 = 0$. The circumstance that the Absolute is a proper conic, and not a mere point-pair, is the real ground of the distinction between spherical geometry and ordinary plane geometry, and the cause of the complete duality of the theorems of spherical geometry.

227. I have, in all that has preceded, given the analytical theory of distance along with the geometrical theory, as well for the purpose of illustration, as because it is important to have the analytical expression of a distance in terms of the coordinates; but I consider the geometrical theory as perfectly complete in itself: the general result is as follows, viz. assuming in the plane (or space of geometry of two dimensions) a conic termed the Absolute, we may by means of this conic, by descriptive constructions, divide any line or range of points whatever, and any point or pencil of lines whatever, into an infinite series of infinitesimal elements, which are (as a definition of distance) assumed to be equal; the number of elements between two points of the range or two lines of the pencil, measures the distance between the two points or lines; and by means of the quadrant, as a distance which exists as well with respect to lines as points, we are enabled to compare the distance of two lines with that of two points; and the distance of a point and a line may be represented indifferently as the distance of two points, or as the distance of two lines.

228. In ordinary spherical geometry, the general theory undergoes no modification whatever; the Absolute is an actual conic, the intersection of the sphere with the concentric evanescent sphere.

229. In ordinary plane geometry, the Absolute degenerates into a pair of points, viz. the points of intersection of the line infinity with any evanescent circle, or what is the same thing, the Absolute is the two circular points at infinity. The general theory is consequently modified, viz. there is not, as regards points, a distance such as the quadrant, and the distance of two lines cannot be in any way compared with the distance of two points; the distance of a point from a line can be only represented as a distance of two points.

230. I remark in conclusion, that, *in my own point of view*, the more systematic course in the present introductory memoir on the geometrical part of the subject of quantics, would have been to ignore altogether the notions of distance and metrical geometry; for the theory in effect is, that the metrical properties of a figure are not the properties of the figure considered *per se* apart from everything else, but its properties when considered in connexion with another figure, viz. the conic termed the Absolute. The original figure might comprise a conic; for instance, we might consider the properties of the figure formed by two or more conics, and we are then in the region of pure descriptive geometry: we pass out of it into metrical geometry by fixing upon a conic of the figure as a standard of reference and calling it the Absolute. Metrical geometry is thus a part of descriptive geometry, and descriptive geometry is *all* geometry, and reciprocally; and if this be admitted, there is no ground for the consideration, in an introductory memoir, of the special subject of metrical geometry; but as the notions of distance and of metrical geometry could not, without explanation, be thus ignored, it was necessary to refer to them in order to show that they are thus included in descriptive geometry.

NOTES AND REFERENCES.

101. No. V. of this paper gives a correction of a formula (18) in the paper 8, On Lagrange's Theorem.

102. I refer to this paper in my "Note on Riemann's paper 'Versuch einer allgemeinen Auffassung der Integration und Differentiation,' *Werke*, pp. 331—344." *Math. Ann.* t. XVI. (1880), pp. 81—82, for the sake of pointing out the connexion which it has with this paper of Riemann's (contained, as the Editors remark, in a MS. of his student time dated 14 Jan. 1847, and probably never intended for publication): the idea is in fact the same, Riemann considered a function of $x+h$ expanded in a doubly infinite, necessarily divergent, series of integer or fractional powers of h, according to an assigned law: and he thence deduces a theory of fractional differentiation.

114. This Memoir on Steiner's extension of Malfatti's problem is referred to by Clebsch in the paper "Anwendung der elliptischen Functionen auf ein Problem der Geometrie des Raumes," *Crelle*, t. LIII. (1857), pp. 292—308: it is there shown that my fundamental equations, p. 67, are the algebraical integrals of a system of equations

$$\frac{dy}{\sqrt{Y}}+\frac{dz}{\sqrt{Z}}=0, \quad \frac{dz}{\sqrt{Z'}}+\frac{dx}{\sqrt{X'}}=0, \quad \frac{dx}{\sqrt{X''}}+\frac{dy}{\sqrt{Y''}}=0,$$

the integrals of which become comparable when the quartic functions under the square roots differ only by constant factors; and expressing that this is so, he obtains the relations which I assumed to exist between the coefficients α, β, γ, δ, &c., under which the equations admit of solution by quadratics only. And he is thereby led to reduce the problem, not to the foregoing system of fundamental equations, but to other equations connecting themselves with the *usual* form of the Addition-theorem; and with a view thereto to develope a new solution of the Problem.

115, 116. The theory is further developed in my Memoir "On the Porism of the in-and-circumscribed Polygon," *Phil. Trans.* t. CLI., for 1861.

119. I attach some value to the process here explained: the most simple application is that referred to at the end of the paper, for the factorial binomial theorem; to multiply $m+n$ by $m+n-1$, we multiply the m by $(m-1)+n$, and the n by $m+(n-1)$, thus obtaining the result in the form $m(m-1)+2mn+n(n-1)$, and so in other cases.

121. The papers and works relating to the Question are

1. Boole. Proposed Question in the Theory of Probabilities, *Camb. and Dubl. Math. Jour.* t. VI. (1851), p. 286.

2. Cayley. 121, Note on a Question in the Theory of Probabilities, *Phil. Mag.* t. VI. (1853), p. 259.

3. Boole. Solution of a Question in the Theory of Probabilities, *Phil. Mag.* t. VII. (1854), pp. 29—32.

4. Boole. An Investigation of the Laws of Thought, on which are founded the Mathematical Theories of Logic and Probabilities, 8vo. London and Cambridge, 1854 (see in particular pp. 321—326).

5. Wilbraham. On the Theory of Chances developed in Prof. Boole's Laws of Thought, *Phil. Mag.* t. VII. (1854), pp. 465—476.

6. Dedekind. Bemerkungen zu einer Aufgabe der Wahrscheinlichkeitsrechnung, *Crelle*, t. L. (1855), pp. 268—271;

viz. Boole proposed the question in 1, I gave my solution in 2, Boole objected to it in 3, and gave without explanation or demonstration his solution, referring to his then forthcoming work 4, which contains (pp. 321—326) his investigation. Wilbraham in 5 defended my solution, and criticised Boole's: and finally Dedekind in 6 (which does not refer to 4 or 5) completed my solution, by determining the sign of a radical, and establishing between the data, as conditions of a possible experience, the relations $p-\beta q$ and $q-\alpha p$ neither of them negative.

I remark that although Boole in 1, 3, and 4 speaks throughout of "causes," yet it would seem that he rather means "concomitant events": I think that in his point of view the more accurate enunciation of the question would be—The probabilities of two events A and B are α and β respectively; the probability that if the event A present itself the event E will accompany it is p, and the probability that if the event B present itself the event E will accompany it is q; moreover it is assumed that the event E cannot appear in the absence of both the events A and B: required the probability of the event E.

He makes no assumption as to the independence *inter se* of A, and B: and moreover, in thus regarding A and B as events instead of causes, there is no room for regarding E as a consequence of one or the other of A and B, or of both of them.

In my solution I regard A and B as causes: I assume that they are independent causes; and further that either or both of them may act efficiently so as to produce the event E, but that the event E cannot happen unless at least one of them act efficiently, viz. it cannot happen in consequence of the conjoint separately inefficient action of the two causes. On these assumptions it appears to me that my solution, as completed by Dedekind, is correct. This would not preclude the correctness of Boole's solution, if according to what precedes we consider it as the solution of a different question: but I am unable to understand it.

I resume my own solution, completing it according to Dedekind. I write with him u instead of ρ for the required probability of the event E; the equations of the text thus are

$$p=\lambda+(1-\lambda)\mu\beta,\quad q=\mu+(1-\mu)\lambda\alpha;\quad u=\lambda\alpha+\mu\beta-\lambda\mu\alpha\beta,$$

and we thence deduce

$$u-\beta q=(1-\beta)\lambda\alpha,\quad u-\alpha p=(1-\alpha)\mu\beta;$$

and then eliminating λ, μ, we find

$$u=\frac{u-\beta q}{1-\beta}+\frac{u-\alpha p}{1-\alpha}-\frac{(u-\beta q)(u-\alpha p)}{(1-\beta)(1-\alpha)},$$

or as this equation may be written

$$u^2-u(1-\alpha\beta+\alpha p+\beta q)+(1-\beta)\alpha p+(1-\alpha)\beta q+\alpha\beta pq=0;$$

say we have

$$u=\tfrac{1}{2}(1-\alpha\beta+\alpha p+\beta q-\rho),$$

where

$$\begin{aligned}\rho^2&=(1-\alpha\beta+\alpha p+\beta q)^2-4(1-\beta)\alpha p-4(1-\alpha)\beta q-4\alpha\beta pq,\\&=(1-2\alpha+\alpha\beta+\alpha p-\beta q)^2+4\alpha(1-\alpha)(1-\beta)(1-p),\\&=(1-2\beta+\alpha\beta-\alpha p+\beta q)^2+4\beta(1-\beta)(1-\alpha)(1-q),\\&=(1-\alpha\beta+\alpha p-\beta q)^2-4\alpha(1-\beta)(p-\beta q),\\&=(1-\alpha\beta-\alpha p+\beta q)^2-4\beta(1-\alpha)(q-\alpha p),\end{aligned}$$

and hence also

$$\lambda=\frac{\tfrac{1}{2}(1-\alpha\beta+\alpha p-\beta q-\rho)}{(1-\beta)\alpha},$$

$$\mu=\frac{\tfrac{1}{2}(1-\alpha\beta-\alpha p+\beta q-\rho)}{(1-\alpha)\beta}.$$

Here p, q, α, β, as probabilities, are none of them negative or greater than 1; p is the probability that, A acting, E will happen; and βq is the probability that B will act and E happen. But if A act, then even if B does not act, E may happen, or B may act and E happen, that is p is greater than or at least equal to βq, say $p-\beta q$ is not negative. And similarly $q-\alpha p$ is not negative. We thus have as conditions of a possible experience, $p-\beta q$ and $q-\alpha p$ neither of them negative.

The formulæ show that ρ^2 is real; and then further, taking for ρ its positive value, it at once appears that we have u, λ, μ no one of them negative or greater than 1, viz. the values are such as these quantities, as probabilities, ought each of them to have: and we have thus a real solution.

Boole in 1 after remarking that the quadratic equation in u may be written in the form

$$\frac{(1-\alpha p'-u)(1-\beta q'-u)}{1-u}=\alpha'\beta' \quad (p'=1-p, \text{ \&c.})$$

says that this is certainly erroneous; for in the particular case $p=1$, $q=0$ it gives $u=1$ or $u=\alpha(1-\beta)$, whereas the value should be $u=\alpha$. But observe that $p=1$, $q=0$, give $q-\alpha p, =-\alpha$, a negative value, so that the solution does not apply. If we further examine the meaning, A is a cause such that if it act then $(p=1)$ the event is sure to happen; and B is a cause(?) such that if it act then $(q=0)$ the event is sure not to happen; this is self-contradictory unless we make the new assumption that the causes A and B cannot both act. It is remarkable that even in this case my solution gives the plausible result $u=\alpha(1-\beta)$, viz. the probability of the event is the product of the probabilities of A acting, and B not acting.

In further illustration, and at the same time to examine Boole's solution, I write as follows:

	Wilbraham.	Boole.	Cayley.
ABE	ξ	$x\,y\,s\,t$	$\alpha\beta\,(1-\lambda'\mu')$
ABE'	ξ'	$x\,y\,s'\,t'$	$\alpha\,\beta\,\lambda'\mu'$
$A'BE$	η	$x'y\,s'\,t$	$\alpha'\beta\mu$
$A'BE'$	η'	$x'y\,s'\,t'$	$\alpha'\beta\mu'$
$AB'E$	ζ	$x\,y'\,s\,t'$	$\alpha'\beta'\lambda$
$AB'E'$	ζ'	$x\,y'\,s'\,t'$	$\alpha\,\beta'\lambda'$
$A'B'E$	0	0	0
$A'B'E'$	σ'	$x'y'\,s'\,t'$	$\alpha'\beta'$

where in the first column the accent denotes negation: ABE means that the events A, B, E all happen, ABE' that A and B each happen, E' does not happen, and so for the other symbols. And in like manner in the third and fourth columns, where the unaccented letters denote probabilities, an accented letter is the probability of the contrary event, $x'=1-x$, &c.

By hypothesis E cannot happen unless either A or B happen, that is Prob. $A'B'E=0$, or writing $A'B'E$ for the probability (and so in other cases) say $A'B'E=0$. And I then (with Wilbraham) denote the probabilities of the other seven combinations of events by ξ, ξ', η, η', ζ, ζ' and σ'; and (as before) the required probability of the event E by u.

The data of the Problem are $1=1$, $A=\alpha$, $B=\beta$, $AE=\alpha p$, $BE=\beta q$, and we have thence to find $E=u$, where on the left-hand side of the first equation 1 means $ABE+ABE'+\text{\&c.}=\xi+\xi'+\eta+\eta'+\zeta+\zeta'+\sigma'$, and similarly A means

$$ABE+ABE'+AB'E+AB'E', =\xi+\xi'+\zeta+\zeta', \text{ \&c.};$$

we thus have

$$\begin{aligned}
\xi + \xi' + \eta + \eta' + \zeta + \zeta' + \sigma' &= 1\ ,\\
\xi + \xi' + \zeta + \zeta' &= \alpha\ ,\\
\xi + \xi' + \eta + \eta' &= \beta\ ,\\
\xi + \zeta &= \alpha p,\\
\xi + \eta &= \beta q,\\
\xi + \eta + \zeta &= u,
\end{aligned}$$

six equations for the determination of the eight quantities ξ, ξ', η, η', ζ, ζ', σ', and u.

For the determination of u, it is therefore necessary to find or assume two more equations: in my solution this is in effect done by giving to ξ, ξ', η, η', ζ, ζ', σ' the values in the fourth column, values which satisfy the six equations, and establish the two additional relations

$$\frac{\xi'}{\eta'} = \frac{\zeta'}{\sigma'}, \quad \frac{\xi + \xi'}{\eta + \eta'} = \frac{\zeta + \zeta'}{\sigma'},$$

or, as these may be written,

$$\frac{A\,BE'}{A'BE'} = \frac{A\,B'E'}{A'B'E'}, \quad \frac{A\,B}{A'B} = \frac{A\,B'}{A'B'};$$

these then are assumptions implicitly made in my solution; they amount to this, that the events A, B are treated as independent, *first* in the case where E does *not* happen; secondly in the case where it is not observed whether E does or does not happen.

Boole in his solution introduces what he calls logical probabilities (but what these mean, I cannot make out): viz. these are Prob. $A = x$, or say simply $A = x$; and similarly, $B = y$, $AE = s$, $BE = t$; then in the case ABE we have A, B, AE, BE, and the logical probability is taken to be $xyst$; and we obtain in like manner the other terms of the third column. And then taking ξ, ξ', η, η', ζ, ζ', σ' to be proportional to the terms of the third column, say $V\xi = xyst$, &c. and substituting in the six equations, we have six equations for the determination of x, y, s, t, V, u, and we thus arrive at the value of the required probability u.

But the assumed values of ξ, ξ', &c. give further

$$\frac{\xi}{\eta} = \frac{\zeta}{\sigma'}, \quad \frac{\xi'}{\eta'} = \frac{\zeta'}{\sigma'}, \quad \text{that is} \quad \frac{A\,BE}{A'BE} = \frac{A\,B'E}{A'B'E} \quad \text{and} \quad \frac{A\,BE'}{A'BE'} = \frac{A\,B'E'}{A'B'E'},$$

which are assumptions made in Boole's solution. Wilbraham remarks that the second of these assumed equations, though perfectly arbitrary, is perhaps not unreasonable: it asserts that in those cases where E does not happen, the relation of independence exists between A and B, that is, provided E does not happen, A is as likely to happen whether B happens or does not happen. But that the first of these equations appears to him not only arbitrary, but eminently anomalous: no one (he thinks) can contend that it is either deduced from the data of the problem, or that the mind by the operation of any law of thought recognises it as a necessary or even a reasonable assumption.

To complete Boole's solution: the equations easily give

$$\frac{s'tx'y}{u-\alpha p}=\frac{st'xy'}{u-\beta q}=\frac{s't'}{1-u}=V;$$

and

$$\frac{s't'x'}{1-\alpha p'-u}=\frac{s't'y'}{1-\beta q'-u}=\frac{stxy}{\alpha p+\beta q-u}=V;$$

and multiplying together the first three values, and also the second three values, we have in each case the same numerator $ss'^2tt'^2xx'yy'$, and we thus obtain the equation

$$(u-\alpha p)(u-\beta q)(1-u)-(1-\alpha p'-u)(1-\beta p'-u)(\alpha p+\beta q-u)=0,$$

which, the term in u^2 disappearing, is a quadric equation; it is in fact

$$u^2(-1+\alpha p'+\beta q')+u\{1+\alpha(p-p')+\beta(q-q')-\alpha^2pp'-\beta^2qq'+\alpha\beta(-1+2p'q')\}$$
$$+\{-\alpha p-\beta q+\alpha^2pp'+\beta^2qq'+\alpha\beta(1-p'q')-(\alpha p+\beta q)\alpha\beta p'q'\}=0;$$

or, what is more simple, if we write with Boole $\alpha p=a$, $\beta q=b$, $1-\alpha p'=a'$, $1-\beta q'=b'$, $\alpha p+\beta y=c'$, then the equation is $(u-a)(u-b)(1-u)-(a'-u)(b'-u)(c'-u)=0$, that is

$$(1-a'-b')u^2-\{ab-a'b'+(1-a'-b')c'\}u+(ab-a'b'c')=0,$$

giving

$$u=\frac{ab-a'b'+(1-a'-b')c'+Q}{2(1-a'-b')},$$

where

$$Q^2=\{ab-a'b'+(1-a'-b')c'\}^2-4(1-a'-b')(ab-a'b'c').$$

We have as conditions which must be satisfied by the data, that each of the quantities a', b', c' is greater than each of the quantities a, b; or say, each of the quantities $1-\alpha p'$, $1-\beta q'$, $\alpha p+\beta q$ greater than each of the quantities αp, βq: Q^2 is then real, and taking Q positive, we have u equal to or greater than each of the three quantities and greater than each of the two quantities. The difficulties which I find in regard to this solution have been already referred to.

139. See volume I. Notes and References 13, 14, 15, 16 and 100. I have in the last of these noticed that the terms covariant and invariant were due to Sylvester: and I have referred to papers by Boole, Eisenstein, Hesse, Schläfli and Sylvester. Anterior to the present memoir 139 we have other papers by Boole and Sylvester, one by Hermite (with other papers not directly affecting the theory), a paper by Salmon, and a very important memoir by Aronhold: it will be convenient to give a list as follows:

Boole.

1. Researches on the theory of analytical transformations with a special application to the reduction of the general equation of the second order, *Camb. Math. Jour.* t. II. 1841, pp. 64—73.

2. Exposition of a general theory of linear transformations, Part I. *Camb. Math. Jour.* t. III. 1843, pp. 1—20.

Exposition of a general theory of linear transformations, Part II. *Camb. Math. Jour.* t. III. 1843, pp. 106—119.

3. Notes on linear transformations, *Camb. and Dubl. Math. Jour.* t. IV. 1845, pp. 166—171.

4. On the theory of linear transformations, *Camb. and Dubl. Math. Jour.* t. VI. 1851, pp. 87—106.

5. On the reduction of the general equation of the *n*th degree, *Camb. and Dubl. Math. Jour.* t. VI. 1851, pp. 106—113.

6. Letter to the Editor (reply to Prof. Sylvester), *Camb. and Dubl. Math. Jour.* t. VI. pp. 284, 285.

Sylvester.

1. On the intersections, contacts and other relations of two conics expressed by indeterminate coordinates, *Camb. and Dubl. Math. Jour.* t. V. 1850, pp. 262—282.

2. On a new class of theorems in elimination between quadratic functions, *Phil. Mag.* t. XXXVII. 1850, pp. 213—218.

3. On certain general properties of homogeneous functions, *Camb. and Dubl. Math. Jour.* t. VI. 1851, pp. 1—17.

4. On the intersections of two conics, *Camb. and Dubl. Math. Jour.* t. VI. 1851, pp. 18—20.

5. Reply to Prof. Boole's Observations contained in the November Number of the Journal, *Camb. and Dubl. Math. Jour.* t. VI. 1851, pp. 171—174.

6. Sketch of a memoir on Elimination, Transformation and Canonical forms, *Camb. and Dubl. Math. Jour.* t. VI. 1851, pp. 186—200.

7. On the general theory of Associated Algebraical forms, *Camb. and Dubl. Math. Jour.* t. VI. 1851, pp. 18—20.

8. On Canonical forms, 8vo. London, Bell, 1851.

9. On a remarkable discovery in the theory of Canonical forms and of hyperdeterminants, *Phil. Mag.* t. II. 1851, pp. 391—410.

10. On the Principles of the Calculus of Forms. Part I. Generation of Forms. Sect 1. On Simple Concomitance. 2. On Complex Concomitance. 3. On Commutants. Notes in Appendix (1), (2), (3), (4), (5), (6), (7), (8), *Camb. and Dubl. Math. Jour.* t. VI. 1852, pp. 52—97.

11. On the Principles of the Calculus of Forms. Sect. 4. Reciprocity, also Properties and Analogies of Certain Invariants &c. 5. Applications and Extension of the theory of the Plexus. 6. On the partial differential equations to Concomitants, Orthogonal and Plagional Invariants, &c. Notes in Appendix (9), (10), (11). Postscript, *Camb. and Dubl. Math. Jour.* t. VI. 1852, pp. 179—217.

12. Note on the Calculus of Forms, *Camb. and Dubl. Math. Jour.* t. VIII. 1853 pp. 62—64.

13. On the Calculus of Forms otherwise the theory of Invariants. Sect. 7. On Combinants, *Camb. and Dubl. Math. Jour.* t. VIII. 1853, pp. 256—269.

14. On the Calculus of Forms otherwise the theory of Invariants. Sect. 7. Continued. 8. On the reduction of a sextic function of two variables to its canonical form, *Camb. and Dubl. Math. Jour.* t. IX. 1854, pp. 85, 103.

Salmon. Exercises in the Hyperdeterminant Calculus, *Camb. and Dubl. Math. Jour.* t. IX. 1854, pp. 19—33.

Hermite. Sur la théorie des fonctions homogènes à deux indéterminées, *Camb. and Dubl. Math. Jour.* t. IX. 1854, pp. 172—217.

Aronhold. Zur Theorie der homogenen Functionen von drei Variabeln, *Crelle* t. XXXIX. 1850, pp. 140—159.

In the present Memoir 139, dropping altogether the consideration of linear transformations, I start from the notion of certain operations upon the constants and facients of a quantic, viz. if to fix the ideas we consider the case of a binary quantic $(a, b, \ldots b', a' \between x, y)^m$, then there is an operation $\{y\partial_x\}, = a\partial_b + 2b\partial_c \ldots + mb'\partial_{a'}$ which performed upon the quantic is tantamount to the operation $y\partial_x$: and similarly an operation $\{x\partial_y\}, = mb\partial_a + (m-1)c\partial_b \ldots + a'\partial_{b'}$ which performed upon the quantic is tantamount to the operation $x\partial_y$. Or, what is the same thing, there are two operations $\{y\partial_x\} - y\partial_x$, and $\{x\partial_y\} - x\partial_y$ each of which performed upon the quantic reduces it to zero: to use an expression subsequently introduced, say each of these is an *annihilator* of the quantic. The assumed definition is that any function of the coefficients and variables which is reduced to zero by each of these operators, is a Covariant: and in particular if the function contain the coefficients only (in which case obviously the operators may be reduced to $\{y\partial_x\}$ and $\{x\partial_y\}$ respectively) the function is an Invariant.

I believe I actually arrived at the notion by the simple remark, say that $a\partial_b + 2b\partial_c$ operating upon $ac - b^2$ reduced it to zero, and that the same operation performed upon $ax^2 + 2bxy + cy^2$ reduced it to $2axy + 2by^2$ which is $= y\partial_x \{ax^2 + 2bxy + cy^2\}$. But the earliest published mention of the notion is in the year 1852 in Note 7 of Sylvester's paper on the Principles of the Calculus of Forms (Sylvester 10). Here, connecting it with the theory of linear transformations, he writes "There is one principle of *paramount* importance which has not been touched upon in the preceding pages,... The principle now in question consists in introducing the idea of *continuous* or *infinitesimal* variation into the theory. To fix the ideas suppose C to be a function of the coefficients of $\phi(x, y, z)$ such that it remains unaltered when x, y, z become respectively fx, gy, hz, where $fgh = 1$. Next suppose that C does not alter when x becomes $x + ey + \epsilon z$, where e, ϵ are indefinitely small; it is easily and obviously demonstrable that if this be true for e, ϵ indefinitely small, it must be true for all values of e, ϵ. Again suppose that C alters neither when x receives such infinitesimal increment, y and z remaining constant, nor when y and z separately receive corresponding increments z, x and x, y in the respective cases remaining constant. ...C will remain constant for any concurrent linear transformations of x, y, z when the modulus is unity. This all-important principle...also *instantaneously* gives the necessary and sufficient conditions to which an invariant of any given order of any homogeneous function whatever is subject, and thereby reduces the problem of discovering invariants to a definite form." And in section 6 of the same paper (Sylvester 11) referring to the Note, he writes "This method may also be extended to concomitants generally. M. Aronhold as I collect from private information was the first to think of the application of this method to the subject: but it was Mr Cayley who communicated to me the equations which define the invariants of functions of two variables. The method by which I obtain these equations and prove their sufficiency is my own, but I believe has been adopted by Mr Cayley in a Memoir about to be published in *Crelle's Journal* [? 100]. I have also recently been informed of a paper about to appear in *Liouville's Journal* from the pen of M. Eisenstein, where it appears that the same idea and mode of treatment have been made use of. Mr Cayley's communication to me was made in the early part of December last [1851] and my method (the result of a remark made long before) of obtaining these and the more general equations and of demonstrating their sufficiency imparted a few weeks subsequently—I believe between January and February of the present year [1852]," and then applying the principle to the binary quadric, he proceeds to consider the theory of the operator $a\dfrac{d}{db}+2b\dfrac{d}{dc}+3c\dfrac{d}{dd}+\ldots$, and the other operator with the coefficients in the reverse order, as applied to an invariant ϕ of the quantic. The theory of these operators was thus familiar to Sylvester in 1852, but it was in nowise made the foundation of the structure.

I notice as contained in the paper Boole (4), what is probably the first statement of the "provectant" process of forming an invariant; for example, from the quartic function $(a, b, c, d, e\between x, y)^4$ he derives

$$\tfrac{1}{48}(a, b, c, d, e\between \partial_y, -\partial_x)^4 . (a, b, c, d, e\between x, y)^4 = ae - 4bd + 3c^2, \text{ the quadrinvariant;}$$

and similarly from the Hessian $(ac-b^2,\ 2(ad-bc),\ ae+2bd-3c^2,\ 2(be-cd),\ ce-d^2\between x, y)^4$ is derived the cubinvariant $ace - ad^2 - b^2e + 2bcd - 3c^3$. Mention is also made of the function $A(\beta\delta-\gamma^2)+B(\beta\gamma-\alpha\delta)+C(\alpha\gamma-\beta^2)$, ($A$, B, C given quadric functions, α, β, γ, δ given cubic functions of (a, b, c, d, e, f)), which is the octinvariant Q of the binary quintic.

The papers of Sylvester contain a great number of important results which will some of them be referred to in connexion with the later Memoirs on Quantics.

Hermite's discovery of the invariant of the degree 18 of the quintic, and the demonstration of his law of reciprocity are both given in the Memoir by him which is above referred to.

147. Upon looking at any one of the Tables, for instance VIII (a), it will be noticed (1) that the partition symbols in the outside top line and left-hand column respectively are differently arranged, (2) that the numbers of each pair of equal numbers (see the Memoir) are not symmetrically situate, and (3) that the table is what may be called a half-square; viz. the squares above (or, in the case of a (b) table, those below) the sinister diagonal are all vacant; the squares in the sinister diagonal itself are all occupied by units ($+1$ or -1 as the case may be). It is possible (and that in many ways) to give the same arrangement to the partition-symbols in the outside line and column respectively, and at the same time to retain the half-square form of the table: or (what is far more important) we may with Faà di Bruno, give the same arrangement to the partition-symbols, and at the same time make the table symmetrical, viz. cause the two numbers of each pair of equal numbers to be symmetrically situate in regard to the dexter diagonal of the square—but we cannot at the same time retain accurately the half-square form of the table. The general principle is that in the outside column (or line) the partition-symbols which are conjugate to each other have symmetrical positions, while the self-conjugate symbols are collected at the middle of the column (or line); there is then in regard to these self-conjugate symbols a sort of dislocation of the sinister diagonal, the units which belong to them being transferred to the dexter diagonal, and in the sinister diagonal replaced by zeros, for instance at the crossing of the two diagonals we may have

1	0
0	1

instead of

	1
1	

. A Table thus arranged may be called Symmetric.

Again as remarked by Fiedler, the two corresponding tables (a) and (b) may be united into a single table; the sinister diagonal is the same for each of them, and if we then insert into the (b) table below the sinister diagonal the numbers of the (a) table, we have a table which is to be read according to the lines for the numbers above and in the sinister diagonal; and according to the columns for the numbers in and below the same diagonal. This may be called a United table: it may be unsymmetric, or be rearranged so as to be made symmetric.

The tables have been rearranged as above, and extended to the order 14: I give the following references.

Fiedler. Elemente der Neueren Geometrie &c. (1862), pp. 73 et seq. (II. to X, (a) and (b) united, unsymmetric).

Faà di Bruno. Sur les Fonctions Symétriques, *Comptes Rendus*, t. 76 (1873), pp. 163—168 (II to VIII, (b), symmetric, there is some error in VIII, inasmuch as it is presented without the dislocation of the sinister diagonal).

——— Théorie des Fonctions Binaires, 8vo. Turin &c. 1876. II to XI (b) symmetric.

Durfee. Tables of the Symmetric Functions of the Twelfthic, *Amer. Math. Jour.* t. v. (1882), pp. 45—60. XII (a) and (b) unsymmetric.

Rehorovsky. Tafeln der symmetrischen Functionen der Wurzeln und der Coefficienten-Combinationen vom Gewichte eilf und zwölf. Wien, Denks. t. 26 (1883), pp.

53—60. XI (*a*) and (*b*), XII (*a*) and (*b*): unsymmetric, united. Is referred to in the next mentioned paper.

Durfee. The Tabulation of Symmetric Functions, *Amer. Math. Jour.* t. V. (1882), pp. 348, 349. XII (*a*) and (*b*); symmetric, united.

MacMahon. Symmetric Functions of the 13[ic], *Amer. Math. Jour.* t. VI. (1884), pp. 289—300. XIII (*b*); symmetric.

Cayley. Symmetric Functions of the roots for the degree 10 for the Form $1 + bx + \frac{cx^2}{1 \,.\, 2} + \ldots = (1 - \alpha x)(1 - \beta x)(1 - \gamma x) \ldots$ *Amer. Math. Jour.* t. VII. (1885), pp. 47—56. II to X (*b*), unsymmetric. The calculation of the tables for this new form (MacMahon's) of the coefficients afforded a complete verification of the (*b*) tables, showing that there was not a single error in these tables as published in the *Philosophical Transactions.*

Durfee. Symmetric Functions of the 14[ic], *Amer. Math. Jour.* t. IX. (1887), pp. 278—292. XIV (*b*) symmetric, the arrangement is different from and seemingly better than that in the tables XII (*b*) and XIII (*b*).

MacMahon. Properties of a Complete Table of Symmetric Functions, *Amer. Math. Jour.* t. X. (1888), pp. 42—46.

——— Memoir on a New Theory of Symmetric Functions, *Amer. Math. Jour.* t. XI. (1889), pp. 1—36. (*a*) and (*b*) Tables for the weights 1 to 6 and their several partitions. To explain this, observe that the general idea is to ignore the coefficients altogether, regarding them as merely particular symmetric functions of the roots: thus the (*b*) table for the weight 4 (partition 1^4) is in fact the table IV (*b*) giving the symmetric functions (4), (31), (2^2), (21^2), (1^4) in terms of (1^4), $(1^3)(1)$, $(1^2)^2$, $(1^2)(1)^2$, $(1)^4$, that is in terms of the combinations e, bd, c^2, b^2c, b^4 of the coefficients, but that the other tables weight 4 to a different partition, give the values of symmetric functions (combinations of the foregoing) which are expressible in terms of other symmetric functions of the roots: for instance weight 4 (partition 21^2) gives (4), (31), (2^2), and (21^2) in terms of (21^2), $(21)(1)$, $(2)(1^2)$ and $(2)(1)^2$. A leading idea in this valuable memoir is that of the "Separations" of a Partition.

150. The theory is developed in an incomplete form. If to fix the ideas we consider a quintic equation $(a, b, c, d, e, f \between x, 1)^5 = 0$, then a single equality $\alpha = \beta$ between the roots implies a onefold relation between the coefficients (a, b, c, d, e, f): this is completely and precisely expressed by means of a single equation ($\nabla = 0$, where ∇ is the discriminant, $= a^4f^4 + \&c.$). Similarly a system of two equalities $\alpha = \beta = \gamma$, or $\alpha = \beta$, $\gamma = \delta$ as the case may be, implies a twofold relation between the coefficients (a, b, c, d, e, f) and the question arises, to determine the order of this twofold relation, and to find how it can be completely and precisely expressed, whether by two equations $A = 0$, $B = 0$, or if need be by a larger number of equations $A = 0$, $B = 0$, $C = 0$, &c. between the coefficients; this is not done in the memoir,

but what is done is only to find two or more equations satisfied in virtue of the system of the two equalities between the roots. And similarly in the case of a system of more than two equalities. See my paper 77, where this notion of the order of a system of equations was established.

152. The next later memoir on the theory of Matrices, so far as I am aware is that by Laguerre, "Sur le Cacul des Systèmes Lineaires," *Jour. Ec. Polyt.* t. XXV. (1867), pp. 215—264. A "système lineaire" is what I called a matrix, and the mode of treatment is throughout very similar to that of my memoir; in particular we have in it my theorem of the equation satisfied by a matrix of any order. The memoir contains a theorem relating to the integral functions of *two* matrices A, B of the same order, viz. this is expressible in the form $m+pA+qB+rAB$. For later developments see the papers by Sylvester in the *American Mathematical Journal.*

158. The notion of the "Absolute" was I believe first introduced in the present memoir. In reference to the theory of distance founded upon it and here developed, I refer to the papers

Klein, Ueber die sogenannte Nicht-Euklidische Geometrie, *Math. Ann.* t. IV. (1871), pp. 573—625.

Cayley, On the Non-Euclidian Geometry, *Math. Ann.* t. V. (1872), pp. 630—634.

Klein, Ueber die sogenannte Nicht-Euklidische Geometrie, *Math. Ann.* t. VI. (1873), pp. 112—145.

In his first paper Klein substitutes, for my $\cos^{-1}$ expression for the distance between two points, a logarithmic one; viz. in linear geometry if the two fixed points are A, B then the assumed definition for the distance of any two points P, Q is

$$\text{dist.}\,(PQ)=c\log\frac{AP\,.\,BQ}{AQ\,.\,BP};$$

this is a great improvement, for we at once see that the fundamental relation, $\text{dist.}\,(PQ)+\text{dist.}\,(QR)=\text{dist.}\,(PR)$, is satisfied: in fact we have

$$\text{dist.}\,(QR)=c\log\frac{AQ\,.\,BR}{AR\,.\,BQ},$$

and thence

$$\text{dist.}\,(PQ)+\text{dist.}\,(QR)=c\log\frac{AP\,.\,BR}{AR\,.\,BQ},=\text{dist.}\,PR.$$

But in my Sixth Memoir, the question arises, what is meant by "coordinates": if in linear geometry (x, y) are the coordinates of a point P, does this mean that $x:y$ is the ratio of the distances in the ordinary sense of the word of the point P from two fixed points A, B: and if so, does the notion of distance in the new sense ultimately depend on that of distance in the ordinary sense? And similarly in Klein's definition, do AP, BQ, AQ, BP denote distances in the ordinary sense of the word, and if so does the notion of distance in the new sense ultimately depend on that of distance in the ordinary sense?

As to my memoir, the point of view was that I regarded "coordinates" not as distances or ratios of distances, but as an assumed fundamental notion, not requiring or admitting of explanation. It recently occurred to me that they might be regarded as mere numerical values, attached arbitrarily to the point, in such wise that for any given point the ratio $x : y$ has a determinate numerical value, and that to any given numerical value of $x : y$ there corresponds a single point. And I was led to interpret Klein's formulæ in like manner; viz. considering A, B, P, Q as points arbitrarily connected with determinate numerical values a, b, p, q, then the logarithm of the formula would be that of $(a-p)(b-q) \div (a-q)(b-p)$. But Prof. Klein called my attention to a reference (p. 132 of his second paper) to the theory developed in Staudt's Geometrie der Lage, 1847 (more fully in the Beiträge zur Geometrie der Lage, Zweites Heft, 1857). The logarithm of the formula is $\log(A, B, P, Q)$, and, according to Staudt's theory (A, B, P, Q), the anharmonic ratio of any four points, has independently of any notion of distance the fundamental properties of a numerical magnitude, viz. any two such ratios have a sum and also a product, such sum and product being each of them a like ratio of four points determinable by purely descriptive constructions. The proof is easiest for the product: say the ratios are (A, B, P, Q) and (A', B', P', Q'): then considering these as given points we can construct R, such that $(A', B', P', Q') = (A, B, Q, R)$: the two ratios are thus (A, B, P, Q) and (A, B, Q, R), and we say that their product is (A, B, P, R) {observe as to this that introducing the notion of distance, the two factors are $\frac{AP \,.\, BQ}{AQ \,.\, BP}$ and $\frac{AQ \,.\, BR}{AR \,.\, BQ}$ and thus their product $= \frac{AP \,.\, BR}{AR \,.\, BP}$, which is (A, B, P, R), which is the foundation of the definition}. Next for the sum, we construct $Q_{,}$ such that $(A', B', P', Q') = (A, B, P, Q_{,})$; the sum then is $(A, B, P, Q) + (A, B, P, Q_{,})$; and if we then construct S such that (A, A), $(Q, Q_{,})$, (B, S) are an involution, we say that $(A, B, P, Q) + (A, B, P, Q_{,}) = (A, B, P, S)$. {Observe as to this that again introducing the notion of distance the last mentioned equation is $\frac{AP \,.\, BQ}{AQ \,.\, BP} + \frac{AP \,.\, BQ_{,}}{AQ_1 \,.\, BP} = \frac{AP \,.\, BS}{AS \,.\, BP}$, that is $\frac{BQ}{AQ} + \frac{BQ_{,}}{AQ_{,}} = \frac{BS}{AS}$, which expresses that S is determined as above; in fact the equation $\frac{b-q}{a-q} + \frac{b-q_{,}}{a-q_{,}} = \frac{b-s}{a-s}$ is readily seen to be equivalent to

$$\begin{vmatrix} 1, & b+s\ , & bs \\ 1, & 2a\ , & a^2 \\ 1, & q+q_1, & qq_1 \end{vmatrix} = 0\}.$$

It must however be admitted that, in applying this theory of Staudt's to the theory of distance, there is at least the appearance of arguing in a circle, since the construction for the product of the two ratios, is in effect the assumption of the relation,

$$\text{dist. } PQ + \text{dist. } QR = \text{dist. } PR.$$

I may refer also to the Memoir, Sir R. S. Ball "On the theory of the Content," *Trans. R. Irish Acad.* vol. XXIX. (1889), pp. 123—182, where the same difficulty is discussed. The opening sentences are—"In that theory [Non-Euclidian Geometry] it seems as if we try to replace our ordinary notion of distance between two points by the logarithm of a certain anharmonic ratio. But this ratio itself involves the notion of distance measured in the ordinary way. How then can we supersede the old notion of distance by the Non-Euclidian notion, inasmuch as the very definition of the latter involves the former?"

An extensive list of papers is given, Halsted, Bibliography of Hyper-Space and of Non-Euclidean Geometry, *Amer. Math. Jour.* t. I. (1878), pp. 261—276 and 384—385, also t. II. (1879), pp. 65—70.

END OF VOL. II.

CAMBRIDGE:

PRINTED BY C. J. CLAY, M.A. AND SONS,

AT THE UNIVERSITY PRESS.